Language, Music, and the Brain

A Mysterious Relationship

Strüngmann Forum Reports

Julia Lupp, series editor

The Ernst Strüngmann Forum is made possible through the generous support of the Ernst Strüngmann Foundation, inaugurated by Dr. Andreas and Dr. Thomas Strüngmann.

This Forum was supported by funds from the Deutsche Forschungsgemeinschaft (German Science Foundation) and the Stiftung Polytechnische Gesellschaft

Language, Music, and the Brain
A Mysterious Relationship

Edited by
Michael A. Arbib

Program Advisory Committee:
Michael A. Arbib, W. Tecumseh Fitch, Peter Hagoort,
Lawrence M. Parsons, Uwe Seifert, and Paul F. M. J. Verschure

The MIT Press
Cambridge, Massachusetts
London, England

© 2013 Massachusetts Institute of Technology and
the Frankfurt Institute for Advanced Studies

Series Editor: J. Lupp
Assistant Editor: M. Turner
Photographs: U. Dettmar
Lektorat: BerlinScienceWorks

Supplemental material available online: www.esforum.de/sfr10/

All rights reserved. No part of this book may be reproduced in any form by electronic or mechanical means (including photocopying, recording, or information storage and retrieval) without permission in writing from the publisher.

MIT Press books may be purchased at special quantity discounts for business or sales promotional use. For information, please email special_sales@mitpress.mit.edu or write to Special Sales Department, The MIT Press, 55 Hayward Street, Cambridge, MA 02142.

The book was set in TimesNewRoman and Arial.
Printed and bound in the United States of America.

Library of Congress Cataloging-in-Publication Data

Language, music, and the brain : a mysterious relationship / edited by Michael A. Arbib.
  pages cm. — (Strüngmann Forum reports)
Includes bibliographical references and index.
ISBN 978-0-262-01810-4 (hardcover : alk. paper) 1. Brain. 2. Language and culture. 3. Music—Psychological aspects. 4. Cognitive neuroscience. I. Arbib, Michael A., editor of compilation.
QP376.L32 2013
612.8'2—dc23

2013002294

10 9 8 7 6 5 4 3 2 1

Contents

The Ernst Strüngmann Forum		vii
List of Contributors		ix
Preface		xi

Part 1: An Expanded Perspective

1	Five Terms in Search of a Synthesis Michael A. Arbib	3
2	A Cross-Cultural Perspective on the Significance of Music and Dance to Culture and Society: Insight from BaYaka Pygmies Jerome Lewis	45
3	Cross-Cultural Universals and Communication Structures Stephen C. Levinson	67

Part 2: Action, Emotion, and the Semantics

4	Shared Meaning, Mirroring, and Joint Action Leonardo Fogassi	83
5	Emotion in Action, Interaction, Music, and Speech Klaus R. Scherer	107
6	Neural Correlates of Music Perception Stefan Koelsch	141
7	Film Music and the Unfolding Narrative Annabel J. Cohen	173
8	Semantics of Internal and External Worlds Uwe Seifert, Paul F. M. J. Verschure, Michael A. Arbib, Annabel J. Cohen, Leonardo Fogassi, Thomas Fritz, Gina Kuperberg, Jônatas Manzolli, and Nikki Rickard	203

Part 3: Structure

9	The Infrastructure of the Language-Ready Brain Peter Hagoort and David Poeppel	233
10	Musical Syntax and Its Relation to Linguistic Syntax Fred Lerdahl	257
11	An Integrated View of Phonetics, Phonology, and Prosody D. Robert Ladd	273
12	Multiple Levels of Structure in Language and Music Sharon Thompson-Schill, Peter Hagoort, Peter Ford Dominey, Henkjan Honing, Stefan Koelsch, D. Robert Ladd, Fred Lerdahl, Stephen C. Levinson, and Mark Steedman	289

Part 4: Integration

13 **Neural Mechanisms of Music, Singing, and Dancing** 307
    Petr Janata and Lawrence M. Parsons

14 **Sharing and Nonsharing of Brain Resources for Language and Music** 329
    Aniruddh D. Patel

15 **Action, Language, and Music: Events in Time and Models of the Brain** 357
    Michael A. Arbib, Paul F. M. J. Verschure, and Uwe Seifert

16 **Computational Modeling of Mind and Music** 393
    Paul F. M. J. Verschure and Jônatas Manzolli

17 **The Neurobiology of Language, Speech, and Music** 417
    Jonathan Fritz, David Poeppel, Laurel Trainor, Gottfried Schlaug, Aniruddh D. Patel, Isabelle Peretz, Josef P. Rauschecker, John Halle, Francesca Stregapede, and Lawrence M. Parsons

Part 5: Development, Evolution, and Culture

18 **Communication, Music, and Language in Infancy** 463
    Sandra E. Trehub

19 **Evolving the Language- and Music-Ready Brain** 481
    Michael A. Arbib and Atsushi Iriki

20 **Birdsong and Other Animal Models for Human Speech, Song, and Vocal Learning** 499
    W. Tecumseh Fitch and Erich D. Jarvis

21 **Culture and Evolution** 541
    Ian Cross, W. Tecumseh Fitch, Francisco Aboitiz, Atsushi Iriki, Erich D. Jarvis, Jerome Lewis, Katja Liebal, Bjorn Merker, Dietrich Stout, and Sandra E. Trehub

Bibliography 563

Subject Index 651

The Ernst Strüngmann Forum

Founded on the tenets of scientific independence and the inquisitive nature of the human mind, the Ernst Strüngmann Forum is dedicated to the continual expansion of knowledge. Through its innovative communication process, the Ernst Strüngmann Forum provides a creative environment within which experts scrutinize high-priority issues from multiple vantage points.

This process begins with the identification of a theme. By nature, a theme constitutes a problem area that transcends classic disciplinary boundaries. It is of high-priority interest requiring concentrated, multidisciplinary input to address the issues involved. Proposals are received from leading scientists active in their field and are selected by an independent Scientific Advisory Board. Once approved, a steering committee is convened to refine the scientific parameters of the proposal, establish the framework for the Forum, and select the participants. Approximately one year later, the central meeting, or Forum, is held to which circa forty experts are invited.

The activities and discourse that define a Forum begin well before the week in Frankfurt and conclude when the resulting ideas are turned over to the scientific community in published form. Throughout each stage, focused dialog from multiple vantage points is the means by which issues are approached. Often, this requires relinquishing long-established perspectives or overcoming disciplinary idiosyncrasies, which can otherwise inhibit joint examination. When this is accomplished, however, synergy results and new insights emerge to inform future research.

A Forum can best be imagined as a dynamic think tank. Although a framework is provided to address the central theme, each individual participant contributes to how the issues are prioritized and approached. There are no lectures or presentations; instead, information on key thematic areas is made available in advance to provide the starting point for discussion.

The theme for this Ernst Strüngmann Forum, "Language, Music and the Brain: A Mysterious Relationship," began its development in 2009. Initiated by Michael A. Arbib, the proposal was approved and a steering committee was subsequently convened from April 11–13, 2010. Members of this committee, W. Tecumseh Fitch, Peter Hagoort, Lawrence M. Parsons, Uwe Seifert, and Paul F. M. J. Verschure, joined Arbib to refine the proposal, construct the framework, and select the participants to the Forum. From May 8–13, 2011, the Forum was held in Frankfurt am Main with the goal of exploring the mechanisms (neuronal, psychological, and anthropological), principles, and processes that underlie language and music.

This volume conveys the multifaceted discourse that took place between diverse experts, each of whom assumed an active role. It contains two types of contributions. The first provides background information to various aspects of the overall theme. These chapters, originally written before the Forum, have since been extensively reviewed and revised to provide current understanding of the topics. The second (Chapters 8, 12, 17, and 21) are intended to be summaries of the group discussions that transpired at the Forum. These chapters are not consensus documents nor are they proceedings. They are meant to transfer the essence of the working group discussions and to expose open questions, points of disagreement, and areas in need of future enquiry.

An endeavor of this kind creates its own unique group dynamics and puts demands on everyone who participates. Each invitee contributed not only their time and congenial personality, but a willingness to probe beyond that which is evident. For this, I extend my gratitude to each person.

I also wish to thank the steering committee, the authors of the background papers, the reviewers of the papers, and the moderators of the individual working groups: Larry Parsons, Peter Hagoort, Paul Verschure, and Tecumseh Fitch. To draft a report during the week of the Forum and bring it to its final form in the months thereafter is never a simple matter. For their efforts and tenacity, I am especially grateful to Jonathan Fritz, David Poeppel, Sharon Thompson-Schill, Uwe Seiffert, Paul Verschure, and Ian Cross. Most importantly, I extend my sincere appreciation to Michael Arbib for his commitment to this project.

A communication process of this nature relies on institutional stability and an environment that guarantees intellectual freedom. The generous support of the Ernst Strüngmann Foundation, established by Dr. Andreas and Dr. Thomas Strüngmann in honor of their father, enables the Ernst Strüngmann Forum to conduct its work in the service of science. The Scientific Advisory Board ensures the scientific independence of the Forum, and the Frankfurt Institute of Advanced Studies shares its vibrant intellectual setting with the Forum. Supplementary financial support for this theme was received from the German Science Foundation.

Long-held views are never easy to put aside. Yet, when attempts are made in this direction, when the edges of the unknown begin to appear and knowledge gaps are able to be defined, the act of formulating strategies to fill these gaps becomes a most invigorating exercise. On behalf of everyone involved, I hope that this volume will convey a sense of this lively exercise.

Julia Lupp, Program Director
Ernst Strüngmann Forum
Frankfurt Institute for Advanced Studies (FIAS)
Ruth-Moufang-Str. 1, 60438 Frankfurt am Main, Germany
http://esforum.de

List of Contributors

Francisco Aboitiz   Centro de Investigaciones Médicas Escuela de Medicina, Universidad Católica de Chile, Santiago, Chile
Michael A. Arbib   University of Southern California, USC Brain Project, Los Angeles, CA 90089–2520, U.S.A.
Annabel J. Cohen   Department of Psychology, University of Prince Edward Island, Charlottetown, Prince Edward Island, C1A 4P3, Canada
Ian Cross   Centre for Music and Science, Faculty of Music University of Cambridge, Cambridge, CB3 9DP, U.K.
Peter Ford Dominey   INSERM U846, 69675 Bron, Cedex France
W. Tecumseh Fitch   Department of Cognitive Biology, University of Vienna, 1090 Vienna, Austria
Leonardo Fogassi   Department of Neurosciences, University of Parma, 43100 Parma, Italy
Jonathan Fritz   Institute for Systems Research, University of Maryland, College Park, MD 20742, U.S.A.
Thomas Fritz   Max Planck Institute for Human Cognitive and Brain Sciences, 04103 Leipzig, Germany
Peter Hagoort   Max Planck Institute for Psycholinguistics, 6500 AH Nijmegen, The Netherlands
John Halle   Bard Conservatory of Music, Annandale-on-Hudson, NY 12504-5000, U.S.A.
Henkjan Honing   ILLC Institute for Logic, Language and Computation, University of Amsterdam, 1090 GE Amsterdam, The Netherlands
Atsushi Iriki   Laboratory for Symbolic Cognitive Development, RIKEN Brain Science Institute, Wako-shi Saitama 351-0198, Japan
Petr Janata   Center for Mind and Brain, Davis, CA 95618, U.S.A.
Erich D. Jarvis   Department of Neurobiology, Duke University Medical Center, NC 27710, U.S.A.
Stefan Koelsch   Department of Educational Sciences and Psychology, Freie Universität Berlin, 14195 Berlin, Germany
Gina Kuperberg   Department of Psychology, Tufts University, MA 02155, U.S.A.
D. Robert Ladd   School of Philosophy, Psychology, and Language Sciences, University of Edinburgh, Edinburgh EH8 9AD, U.K.
Fred Lerdahl   Department of Music, Columbia University, New York, NY 10027, U.S.A.
Stephen C. Levinson   Max Planck Institute for Psycholinguistics, NL-6525 AH Nijmegen, The Netherlands
Jerome Lewis   Department of Anthropology, University College London, London WC1H 0BW, U.K.

Katja Liebal Department of Evolutionary Psychology, Freie Universität Berlin, 14195 Berlin, Germany
Jônatas Manzolli Institute of Arts, Department of Música, NIC, University of Campinas, 13087-500 Campinas SP, Brazil
Bjorn Merker Fjälkestadsvägen 410-82, 29491, Sweden
Lawrence M. Parsons Department of Psychology, University of Sheffield, Sheffield S10 2TP, U.K.
Aniruddh D. Patel Department of Psychology, Tufts University, Medford, MA 02115, U.S.A.
Isabelle Peretz Laboratory for Brain, Music and Sound Research, Université de Montréal Brams-Pavillon, Montreal, Quebec H2V 4P3, Canada
David Poeppel Department of Psychology, New York University, New York, NY 10003, U.S.A.
Josef P. Rauschecker Institute for Cognitive and Computational Sciences, Georgetown University Medical Center, Washington, D.C. 20057, U.S.A.
Nikki Rickard School of Psychology and Psychiatry, Monash University, Victoria, 3145, Australia
Klaus R. Scherer Swiss Centre for Affective Sciences, Université de Genève, 1205 Geneva, Switzerland
Gottfried Schlaug Department of Neurology, Harvard Medical School, Boston, MA 02215, U.S.A.
Uwe Seifert Musikwissenschaftliches Institut, Universität Köln, 50923 Köln, Germany
Mark Steedman Informatics, Forum 4.15, University of Edinburgh, Edinburgh EH8 9AB, U.K.
Dietrich Stout Department of Anthropology, Emory University, Atlanta, GA 30322, U.S.A.
Francesca Stregapede Individual Differences in Language Processing Group, Max Planck Institute for Psycholinguistics, 6500 AH Nijmegen, The Netherlands
Sharon Thompson-Schill Department of Psychology, University of Pennsylvania, Philadelphia, PA 19104–6241, U.S.A.
Laurel Trainor Department of Psychology, Neuroscience and Behaviour, McMaster University, Hamilton, Ontario L8S 4K1, Canada
Sandra E. Trehub Department of Psychology, University of Toronto, BRAMS, North Mississauga, Ontario L5L 1C6, Canada
Paul F. M. J. Verschure SPECS, Center of Autonomous Systems and Neurorobotics, Universitat Pompeu Fabra and Catalan Institute of Advanced Studies (ICREA), 08018 Barcelona, Spain

Preface

Language, music, and the brain. Very few people are specialists in all three fields but most have an interest in them all. This book should appeal to those who, whether or not they are specialists, want to gain a fresh perspective on these three themes as we explore the relationships between them.

Just about every reader of this book is an expert in speaking at least one language, and many will be fluent in another as well—able to engage, that is, in the give and take of conversation as well as to listen to talks and follow the dialog in movies. All of us can also read and write, and so when we think of language we may equally think of patterns of sound or of words on the page or screen. Moreover, we gesture as we talk, and some may use signed language to "talk" without using the voice.

When it comes to music, most of us are expert listeners, even if relatively few of us are expert singers or instrumentalists, but most of us maybe unable to read or write a musical score. Thus, when we think of music, we are more likely to think of patterns of sound than of the notational systems that represent them. Nonetheless, we have likely engaged in group singing or, at least, singing in the shower, have danced to music, or tapped out a rhythm abstracted from a complex musical performance.

Whatever our level of expertise, we understand both language and music as auditory patterns to be made and generated, but also as embedded in a variety of multimodal behaviors (not just vocal) that play a key role in a range of social behaviors.

What of "the brain"? Each of us has one, but most readers may know little about the detailed structure or function of a typical human brain or of the lessons to be learned by comparing the brains of humans with those of other species. Indeed, many may use the terms "brain" and "mind" interchangeably. In this book, the study of how the brain underpins the mind is a continual theme.

This volume is the result of the tenth Ernst Strüngmann Forum—a think tank dedicated to the continual expansion of knowledge. For me, this extended process of discourse began in 2009, when Wolf Singer encouraged me to submit a proposal. I wanted to define a topic that related to my expertise and yet would enable me to move beyond it into areas where I could learn much more. My own expertise lies in the use of computational modeling to understand the brain mechanisms that link vision and action, and in exploring their relation to the evolution of the ability of the human brain to support language. Inspired by the 2008 meeting, "The Mirror System Hypothesis: On Being Moved," organized by the musicologist Uwe Seifert at the University of Cologne, I decided that music could add that new dimension to the study of brain and language. The resultant proposal was shaped around four themes: diversity of song, syntax, semantics of internal and external worlds, and evolution.

After its approval, under the direction of Julia Lupp and with the other members of the steering committee—Tecumseh Fitch, Peter Hagoort, Larry Parsons, Uwe Seifert, and Paul Verschure—the proposal was refined and a framework constructed to guide the discussions at the Forum. This work also entailed identifying the thematic areas for the background papers (which provided information in advance of the Forum) and selecting the participants. Here is a brief overview of the four themes that constituted the framework to this Forum:

- *Song and Dance as a Bridge between Music and Language*: Language and music are not just "patterns of sound." They are part of our embodied interaction with the physical and social world. Not only do we speak and listen, we move our hands and bodies in ways that enrich our conversation. Not only do we listen to a musical performance, we may dance or, at least, move our bodies to the music. When we engage in singing, elements of music, language, action, perception, emotion, and social interaction may all come together. This richness of embodied experience frames our exploration of how music and language are related to the brain.
- *Multiple Levels of Structure from Brains to Behavior to Culture*: Language has its own distinct sets of basic sounds (consonant and vowels, phonemes, and syllables), as well as its own ways of combining these sounds into words and of putting words together to form phrases, sentences, and other utterances. Languages also vary to a great extent. What can be said about the varied structures of phonology, morphology, and grammar and the brain mechanisms that support them? Turning to music, to what extent do its hierarchical structures—and the associated brain mechanisms—parallel those of language, and to what extent are they specific to music?
- *Semantics of Internal and External Worlds*: Language can be used to share descriptions of states of the external world (i.e., what is happening and who did what and to whom) as well as speech acts that involve past memories, plans for the future, and adventures in worlds of the imagination. Music has limited power of this kind and yet can touch the internal world of our emotions in diverse ways. The challenge here is to go beyond this simplistic matching of language and music to external and internal meanings, respectively, and to seek to characterize the diversity of semantics expressed in music and language, the overlap and differences between them, the ways they can enrich each other, and the brain mechanisms which support them.
- *Culture and Evolution*: To expand our all-too-limited view of language and music, we need to understand how they vary from culture to culture and to employ the frameworks of evolution and comparative neurobiology to do so. Delineating the communication systems of other primates

may help us formulate hypotheses about the communication abilities of our common ancestors, the changes in brain and body that led to modern human brains (biological evolution), and the rich social systems they make these possible (cultural evolution). How can the study of brain mechanisms that support vocal learning and that which has been called "song" in more distant species inform our understanding? What is the relation between song, music, and language in our own evolution?

These themes are explored in greater depth in Chapter 1, where an overview is provided of their development throughout the book. As will become clear, they provided a way of structuring the Forum; however, the flow of discussion cut across them in various ways, as one would expect.

My thanks to the authors of the background papers, which laid the foundations for the week's discussion in Frankfurt, and to all—but especially the moderators and rapporteurs—who traveled to Frankfurt from near and far to help carry that discussion forward, and then invested much effort to bring this volume to its final form. My thanks to the Forum team for their efforts in making the Forum such a success and in preparing this volume. Last but by no means least, let me acknowledge that the Ernst Strüngmann Forum was made possible through the generous support of the Ernst Strüngmann Foundation, inaugurated by Andreas and Thomas Strüngmann. It is thus to the Strüngmanns, above all, that we, the authors of this volume, owe a debt of gratitude for this opportunity to spend a memorable week of intellectual exchange together, forging new connections between our varied pools of expertise to remove at least some of that mystery that challenges us to probe the relationship between language, music, and the brain.

—Michael A. Arbib, La Jolla, December 2012

Part 1: An Expanded Perspective

1

Five Terms in Search of a Synthesis

Michael A. Arbib

Abstract

One may think of music and language as patterns of sound, and this view anchors much of the research on these topics. However, one may also view the making of music and the use of language within the broader perspective of our interaction with the physical and social world and their impact on the inner world of the emotions. This chapter sets forth a framework to explore five terms—language, music, and brain as well as the action–perception cycle and emotion—so as to bridge our informal understanding of these terms and the challenge of defining precise scientific analyses within an integrative framework. It then provides a chapter-by-chapter overview of the architecture of the book to show how themes emerge in different chapters, and how the chapters work together to provide a more-or-less coherent narrative. A discussion of Tan Dun's *Water Passion after St. Matthew* frames several high-level questions designed to encourage readers to fuse the reality of their personal experience with the insights of well-focused studies to gain an ever fuller understanding of the relationships between language, music, and the brain.

Introduction

In this chapter, I will introduce the various themes that informed the design of the tenth *Ernst Strüngmann Forum on Language, Music and the Brain: A Mysterious Relationship*, and will outline the results of that Forum, of which this book is the record. Our strategy is to view music and language not only as patterns of sound but also in terms of action and perception; we further want to understand how these functions are achieved within human brains and through group interaction. We will argue for considering action and perception within an integrated cycle—our emotions and motivation, goals, plans, and actions affect what we perceive, and what we perceive affects our goals, plans, and actions as well as our emotions and motivation. However, in asserting, for example, that "Our strategy is to view music and language…in terms of action and perception," I must add the caveat that this does not mean that every chapter will view music and language in these terms. My hope is that even when reading chapters which focus, for example, on the structure of music and language *in abstracto*, readers will be able to enrich their understanding by assessing whether an explicit linkage to a more action-oriented approach could benefit, and benefit from, extension of such studies. It is in the very nature of science that it succeeds by focusing on parts of the whole. The challenge is to determine which the "right" parts are, and how lessons gained from the study of separated parts may provide a firm basis for study of the larger system formed when the parts are combined. More specifically:

1. Having noted what music and language share, and having noted the dazzling heterogeneity of languages and musical genres, we must nonetheless characterize in what ways music and language differ.
2. We must then discover which of those shared and distinctive properties can now be linked to neuroscience.
3. We must also develop a framework within which the neuroscience-ready pieces of the puzzle can be integrated to express their relationships to each other and to the human reality of the experience of music and language in action, perception, and social interaction.

Language and music are immensely complex, while neuroscience as a science must deal with circumscribed problems. So the challenge, again, is to delimit aspects of music and language (some distinctive, some shared) for which one can define specific aspects for study in cognitive science and neuroscience. The brain is a very complex system, and its complexities slowly unfold as we delve further into its diverse structures and functions.

The chapter approaches our subject in three phases: First, we seek to capture key notions concerning our five terms. We discuss the action–perception cycle, emotion, language, and music in a way that links our personal experience to more focused concepts that may be candidates for scientific analysis. We then introduce basic concepts from neuroscience that shape our understanding of structures and functions of the brain.

Second, we tour the architecture of the book, examining the contents of the "background" chapters to lay the foundations which present a view of the deliberations of the four discussion groups at the Forum. The themes of the discussion groups were: *Semantics of Internal and External Worlds* (Seifert et al., Chapter 7), *Multiple Levels of Structure in Language and Music* (Thompson-Schill et al., Chapter 12), *Song as a Bridge between Music and Language* (Fritz et al., Chapter 17), and *Culture and Evolution* (Cross et al., Chapter 21).

Finally, a discussion of Tan Dun's *Water Passion after St. Matthew* frames several high-level questions designed to encourage readers to fuse the reality of personal experience with the insights of well-focused studies to gain an ever fuller understanding of the relationships between language, music, and the brain.

The Five Terms

In preparation for the focal meeting of the Forum, which was held in Frankfurt am Main from May 8 through 13, 2011, I asked the participants to send me their thoughts on five terms of great relevance to the Forum—not only *language*, *music*, and *brain* but also *action* and *emotion*. Here, I distill my own thoughts in combination with responses received from Francisco Aboitiz, Franklin Chang, Annabel Cohen, Peter Dominey, Leo Fogassi, John Halle, Henkjan Honing, Petr Janata, Stefan Koelsch, Gina Kuperberg, Bob Ladd, Marc Leman, Fred Lerdahl, Stephen Levinson, Katja Liebal, Jônatas Manzolli, Bjorn Merker, Aniruddh Patel, David Poeppel, Josef Rauschecker, Nikki Rickard, Klaus Scherer, Ricarda Schubotz, Uwe Seifert, Mark Steedman, Dietrich Stout, Francesca Stregapede, Sharon Thompson-Schill, Laurel Trainor, Sandra Trehub and Paul Verschure—to all of whom my thanks.

I will address, in turn, the action–perception cycle and emotion, while noting that motivation and emotion themselves have a strong impact on what we perceive and how we act. Thereafter, I will consider music and language in their relation to action, perception, and emotion, briefly examining the question: "Is there a language–music continuum?" Finally, for this section, the brain is considered as a physical, dynamic, adaptive biological entity as distinct from "another word for mind." The challenge here is to see how we can gain direct results about human brains and how we can augment these insights with findings on the brains of monkeys, songbirds, and other species—even though these creatures have neither music nor language.

The Action–Perception Cycle

An action combines a movement with a goal, where actions are distinguished from "mere" movement by assuming that the former is always goal-directed. Putting one's hand down to touch the table and having it fall unattended under gravity may involve a similar movement, but only the former is goal-directed. Moreover, goals may be conscious or unconscious, with the latter more likely when the action is just a part of some overall but familiar behavior.

Further, *praxis* is different from *communication*, distinguishing action on objects in the environment directed to practical ends from actions designed to affect the "mental state" of other agents.

Rather than separating perception and action in terms of stimulus and response, we find it valuable to emphasize the *action–perception cycle* (Neisser 1976; Arbib 1989; for auditory perception in relation to music, song and dance, see Janata and Parsons, Chapter 13, this volume; Fuster 2004). In general, actions will be directed by nested, hierarchical planning structures based on motivation and emotion, intentions, goals, desires, and beliefs, but our plans may change as the action unfolds. Here, indeed, is an interesting contrast between language and music. A natural model for language is conversation (see

Levinson, Chapter 3), where the plan (implicit in the different plans of the participants) is exceptionally dynamic, as the response of A to what B has just said may drastically change what B says next. By contrast, music often emerges from a group activity in which some people play instruments and others sing or dance, but in which, despite differing roles and a measure of improvisation, the overall structure is preplanned by tradition (see Lewis, Chapter 2) or an explicit musical score.

In spoken language, the major impact of action on the observer is auditory, but co-speech gestures provide a visual counterpoint. For sign language, vision alone carries the message. But the sensory impact of language and music extends beyond the obvious. Evelyn Glennie has been profoundly deaf since the age of 12, yet is a world-renowned percussionist who plays barefoot in order to "feel" the music. Glennie (1993) says that she has taught herself to "hear" with parts of her body other than her ears. We all have access to these sensory data, but we pay less attention to them and, indeed, Glennie's brain may well have adapted to make fuller use of these cues.

Actions are compositional in the sense that they are assembled from a number of well-rehearsed movements of the skeletal motor system but may be adapted to better meld with the actions before and after them (cf. coarticulation in speech) and are dynamically modulated to match changing circumstances. Perception finds its primary role in guiding action, without denying that humans can contemplate the past, plan for the future, develop abstractions, and think just for the fun of it in ways that make *action-oriented perception* the core of our behavior—both developmentally and in evolutionary terms—rather than its be-all and end-all.

The granularity of action can vary from small (e.g., a single saccadic eye movement, saying a single vowel, or singing a single note) to extended (e.g., catching a ball, speaking a sentence, or singing a melody) and so on up many levels of complexity to the preparation of a four-course dinner, an hour's conversation, a novel, or a symphony. (Consider the discussion of Tan Dun's *Water Passion after St. Matthew* in the later section, "A Passion for an Integrative Perspective.") Expectation plays a crucial role in performance; at any given level, the action and its sensory consequences should be congruent with one's goals and the sensory information in the environment. This may operate at all levels, from adjusting a dance step to the feel of the floor against our feet, the timing of the music and the expression on our partner's face to the way in which failure to meet a high-level goal (just missing a train, for example) may radically affect our emotional state, the material that now demands our attention, and the planning of new subgoals if we are still to attain our original higher-level goal (e.g., getting to a particular destination in time for dinner). Such shaping of attention does not preclude (though it does limit) our awareness of unexpected events. Hence our perceptions may serve simply to check and modulate our current behavior or set us off on new courses of behavior.

A key issue, then, is the extent to which the structures used for embodied action and perception are also used for language and music. Some would argue that language can best be understood in relation to physical gesture and that the essence of music cannot be separated from the dance; others insist that the term "embodiment" has little explanatory power. One could, for example, say that the essence of Wittgenstein's *Tractatus Logico-Philosophicus* or a Beethoven symphony is "orthogonal" to embodiment. Yes, the child acquires its first words through gesture as it learns to interact with the physical and social world, and, yes, the basic musical experience is linked to bodily movement, but these are but the first rungs of the ladder; the essence of music and language are to be found not in these basic properties but rather in the way in which the human brain can produce and perceive complex hierarchical structures which interweave form and meaning, appreciating each as enriched by the other. Nonetheless, looking ahead to our discussion of "brain," it may be useful to compare the brain to an orchestra. It is not the case that only one instrument will play at a time (the solo is the exception), nor is it the case that a particular instrument will only be used in a particular type of performance. Instead, different tasks (orchestral pieces) will invoke different instruments playing in different styles in different combinations at different times. Similarly, music and language may "play" the same brain regions as each other, though perhaps with some regions invoked very little in one as compared to another. Assessing this issue of shared resources was a continuing theme of the Forum (see especially Patel, Chapter 14). One word of caution, however: Unlike most orchestras, a brain does not have a "conductor." There is no single region of the brain that coordinates what all the other brain regions are doing. Here, perhaps, the analogy with a jazz group may be more apt, where the improvisations of each player guide, and are guided by, those of the others. The point of this digression is that even if music and language required the evolution of brain regions that are not required for praxis, the appreciation and performance of language and music will most certainly invoke "ancient" brain mechanisms as well.

Perception invokes *perceptual schemas* and pattern recognition: recognizing a leopard when we see (even a partially obscured picture of) one, despite receiving a different retinal image every time we see a leopard. In evolutionary terms rooted in the relation of prey and predator, however, we need not only to recognize the leopard but also where it is and how it is behaving, and all this in relation to other aspects of the scene. A narrow focus on what is perceived may ground reactive behavior (e.g., running away from leopards, grasping food wherever it is, and putting it in one's mouth), in which case perceptual schemas not only activate appropriate *motor schemas* but also provide appropriate parameters that allow motor control to be adjusted to current circumstances. Autonomic emotions such as fear and greed might bias such reactive behavior. However, our concern with action also embraces cases in which the goal is in some sense consciously held so that deliberative planning is possible in which a sequence of actions directed toward subgoals leading to a desirable state can be composed, after a search through a limited range of possible futures. Such deliberative search for plans demands the flexible deployment of multiple schemas that are less directly motor or perceptual, evokes and updates varied forms of memory (Fuster 2009), and is computationally demanding. The evolution of this capability has driven, and perhaps has also been driven by, the evolution of brain mechanisms that support music and language, not solely as auditory forms but also as basic forms of human social interaction. Again, we can ask to what extent our human ability for language and music exploits those mechanisms we share with other animals, though perhaps with greater memory capacity, and to what extent the genetic underpinnings of genuinely new capabilities had to evolve. Contrast the approach to evolution of language taken by Arbib and Iriki (Chapter 19) with that of Fitch and Jarvis (Chapter 20): the former links language to the manual skills shared with other primates whereas the latter emphasizes vocal learning that humans share with songbirds.

Here it may be worth a short digression to consider my approach to schema theory (building on the foundation of Arbib 1975) from three different perspectives:

- "Basic" schema theory: In general, schema theory offers a form of cooperative computation in which perceptual, motor, and other schemas cooperate and compete in mediating our embodied interaction with the world, determining both what we perceive and how we act. The VISIONS model (Hanson and Riseman 1978) demonstrated how interacting instances of perceptual schemas compete and cooperate to yield the overall interpretation of a visual scene. Schemas that are inconsistent with each other compete, reducing their confidence levels, whereas schemas that support each other may cooperate to become better established (e.g., seeing a window in a location under a region that might be a roof increases one's confidence that the region is indeed a roof, while also activating the hypothesis that a larger region containing both of them is a house). A network of schemas in long-term memory provides the active knowledge base for the establishment, with increasing confidence, of an interpretation in visual working memory of the scene being seen. Similar concepts were used in HEARSAY (Lesser et al. 1975), a classic model of the processes involved in speech understanding. A complementary effort (Jeannerod and Biguer 1982; Arbib 1981) examined how perceptual schemas may serve not only to recognize an object but also to pass parameters of the object to motor schemas. In this case, recognizing the location, size, and orientation of an object provides the necessary data for motor schemas to guide the reach-to-grasp action for that object.

- "Neural" schema theory: The model of dexterity, though not that of scene understanding, has been elaborated to the level of interacting neural networks registered against neurophysiological recordings from single cells of the macaque brain (Fagg and Arbib 1998). The distinction is crucial. We may look at a certain class of animal and human behaviors and characterize it by a single schema. In this context and with this initial state, input patterns will relate to observed behavior in a succinctly summarized way. Going further, schema theory offers us a framework in which to decompose schemas into smaller schemas and assess the patterns of competition and cooperation which enable the assemblage not only to capture the behavior specified in the original schema but also to make predictions about behaviors that may be observed in novel situations. The process may be repeated and refined as novel predictions are tested and the results of these tests used to update the model. However, such a process model might be developed without any reference to data from neuroscience. A schema theory model becomes a *neural* schema theory model when hypotheses are made on localization of schemas in the brain (testable by lesion studies and/or brain imaging) or when the dynamics of a schema is elaborated in terms of networks of biologically plausible neurons (testable by data from, e.g., animal neurophysiology; see Arbib, Verschure, and Seifert, Chapter 15).

- "Social" schema theory: Both basic and neural schema theory chart processes within the head of a single bird, monkey, or human. We need, however, to distinguish the processes that support the individual's use of, for example, English from the patterns of English shared by a community. To bridge this divide, Mary Hesse and I (Arbib and Hesse 1986) introduced the notion of social schemas to characterize patterns of social reality, collective representations such as those exhibited by people who share a common language or religion or realm of social practice (see Lewis, Chapter 2). The patterns of behavior exhibited by members of a community then may cohere (more or less) to provide social schemas which define an external social reality. These social schemas, in turn, shape the development of new "internal" schemas in the brain of a child or other newcomer; they become a member of the community to the extent that their internal schemas yield behavior compatible with the community's social schemas. Conversely, to the extent that changes in the internal schemas of individuals "catch on" with other members of the community, the social schemas of that community change in turn.

Let us return to "internal schemas," for which it is worth stressing the important role of *automatization*. With sufficient practice, a compound action directed at a certain goal, the tuning and assemblage of those actions, and the ability to modify them to circumstances that vary within a certain range are automatically evoked without conscious attention. When learning a language, finding an appropriate word and pronouncing it correctly are demanding tasks.

With increasing fluency, neither of these requires conscious effort, and we are able to focus on the right way to express what we intend to say as we engage in animated conversation. Even an accomplished pianist must practice long and hard to keep the schemas for producing and sequencing chords at a certain level of perfection, but this frees the musician to concentrate on higher levels of phrasing and emotional expression during a public performance.

An ongoing debate concerns whether perception–action couplings (as in the "parameter passing" linking perceptual and motor schemas) are present from birth or rest on subsequent development on the basis of experience. Southgate et al. (2009) report comparable sensorimotor alpha-band activity attenuation in electroencephalographs for nine-month-olds and adults while observing hand–object interactions, whereas Oztop et al. (2004) offer a computational model of how the mirror system may become attuned to specific types of grasp during the first nine months postpartum. Drilling down a level below Piaget's theory of the development of schemas (Piaget 1971), it seems most plausible that neonatal imitation (Meltzoff and Moore 1977) and the brain mechanisms which support it are quantitatively different from those engaged when a caregiver assists a child master a novel task (Zukow-Goldring 2012).

When we investigate other species, an interesting question surfaces: How can we determine which actions are directed at changing the behavior of another individual (MacKay 1972)? For example, if individual A starts to run, individual B might observe this and start to run as well, resulting in a playful chase between the two. How could we tell whether A started to run with or without the intention to invite B to play? To differentiate one from the other on a behavioral level, a human observer might monitor A's gaze direction and persistence in case the addressed individual does not react, but this may not be sufficient to differentiate "mere" behavior from the communication of intended actions. Although action implies self-caused body movement, we may include emotional responses as actions in the social context, as they can have a strong influence on the behavior of others. Tomasello et al. (1997) have suggested *ontogenetic ritualization* as one possible mechanism for the transition from a praxic to a communicative action. For example, A might carry out action X to get B to carry out action Y, but if B comes to initiate Y before X is completed, a truncated form X' of the *praxic* action X may become established as a *communicative* action, a request to do Y (for further discussion, see Arbib et al. 2008).

Emotion

Most goal-directed behaviors, including music and language, require hierarchical structuring of action plans that span multiple timescales. Fridja and Scherer (2009) argue that emotions have a strong motivational force and produce states of action readiness and prepare action tendencies which help the organism adapt to or address important events in their lives. Emotions "color" or "label" every experience with an affective value, although it can be imperceptibly mild. Furthermore, the distinction between emotions and feelings can be helpful to distinguish between a biologically conditioned basic set of emotions and the feelings one can report verbally (see, e.g., LeDoux 2000). Moreover, the fundamentals of emotion are action-oriented, driving approach (desire, anger) or withdrawal (sadness, disgust). Although emotion prepares action readiness and different, possibly conflicting, action tendencies, execution of even highly emotional impulsive behaviors (e.g., aggression or flight) may be more or less sensitive to situational and normative context and strategic choice.

Emotion can be analyzed under two headings (Arbib and Fellous 2004), of which I have so far emphasized the former:

- *Internal* aspects of emotions: As just discussed, these frequently contribute to the organization of behavior (prioritization, action selection, attention, social coordination, and learning). For example, the actions one is likely to perform will differ depending on whether one is angry or sad.
- *External* aspects of emotions: Emotional expression for communication and social coordination may include bodily posture, facial expression, and tone of voice. If we see that someone is angry, we will interact with them more cautiously than we would otherwise, or not at all.

These two aspects have coevolved (Darwin 1872/1998). Animals need to survive and perform efficiently within their ecological niche, and in each case the patterns of coordination will greatly influence the suite of relevant emotions and the means whereby they are communicated. The emotional state sets the framework in which the choice (whether conscious or unconscious) of actions will unfold. However, emotions are also embedded in the action–perception cycle, so that one's emotions may change as the consequence of one's actions become apparent, and our perception of these consequences may well rest on our perception of the emotional response of others to our behavior.

Karl Pribram (1960) quipped that the brain's so-called limbic system is responsible for the four Fs: feeding, fighting, fleeing, and reproduction. It is interesting that three of the four have a strong social component. The animal comes with a set of basic "drives" for hunger, thirst, sex, self-preservation, etc., and these provide the basic "motor," motivation, for behavior. Motivated behavior not only includes bodily behavior (as in feeding and fleeing, orofacial responses, and defensive and mating activities), but also autonomic output (e.g., heart rate and blood pressure), and visceroendocrine output (e.g., adrenaline, release of sex hormones). Thus emotions not only provide well-springs of action and inaction but are also closely associated with bodily states (Damasio 1999; James 1884/1968) and motivation. Our autonomic and visceromotor states lie at the heart [sic] of our emotional repertoire. However, the emotions that we talk about and perceive in others are both more restricted than this (how many people perceive another's cortisol level?) yet also more subtle, intertwining these basic motivations with our complex cognitions of social role and interactions, as in the cases of jealousy and pride. We may feel an emotion or simply recognize it. Contrast *I feel happy* with *I observe that you are smiling and infer that you are happy*. In the former, the emotion is felt, whereas in the latter the emotion is identified. Some would suggest that the former case involves mirror neurons (see Fogassi, Chapter 4), whereas the latter does not. Facial expressions (Ekman 1992) reliably represent basic emotions (whether there are basic emotions and, if so, what they are remains controversial; one list might include love, happiness, anger, sadness, fear and disgust; for further discussion, see Ekman 1999) as well as more complex emotions (e.g., guilt, shame, gratitude). Surprisingly, voluntarily forming a facial expression can generate emotion-specific autonomic nervous system activity (Levenson et al. 1990; Hennenlotter et al. 2009).

Till now, this discussion has emphasized the role of emotions in framing detailed planning of behavior, and emotional expression as a means for better coordinating the behavior of a group of conspecifics. In Chapter 5, Scherer contrasts utilitarian emotions with aesthetic emotions. On his account, *utilitarian emotions* help a person adapt to relevant events, especially through speech in social interaction, whereas *aesthetic emotions* are generally desired and sought through engagement with cultural practices, such as music, art, or literature. However, I wonder whether it is an *emotion* that is to be assessed as utilitarian or aesthetic, or whether it is an *instance of experienced emotion* that might be so classified. For example, when we consider the ritual use of music (Lewis, Chapter 2), the emotions aroused in members of a group may be strictly utilitarian in that the participants are emotionally preparing themselves for, e.g., the rigors of the hunt, whereas speech may arouse a similar emotion whether through a utilitarian warning of danger or as an aesthetic frisson of danger in a work of literature.

Perhaps the distinction can be addressed in evolutionary terms. Emotions evolved as potent modulators of the action–perception cycle, both within an individual and across a group. However, perhaps in part through the vagaries of biological evolution but increasingly through the ratcheting of cultural evolution (Tomasello 1999b), our ancestors discovered that being in a certain emotional state was enjoyable in and of itself, whether or not it met the current demands of everyday survival. Again, the need of the caregiver to induce happiness in a child, or of a warrior to strike terror into an enemy—both eminently utilitarian—evolved to make the "manipulation" of the emotions of others for nonutilitarian ends yield an increased palette of emotional experiences that were aesthetic, even if they engaged the same emotions per se as were involved in the types of behavior that provided adaptive pressures for the emergence of the modern human brain. An analogy: One can understand the adaptive advantages of a sense of smell that lets one judge the proximity of a predator, or tell whether fruit is or is not yet ripe, or estimate the receptivity of a potential mate. However, the system that evolved to support these abilities is not limited to detection of such specific odors but in fact has the potential to learn to recognize any smell within a vast range of combinations not only of these primal odors but of the chemical signatures which themselves combine to characterize this primal set. Similarly, basic emotions may be related to activation of brain areas (relating insula to disgust, amygdala to fear, and so on, but only as a first approximation), but perhaps the very same evolutionary processes that increased the scope of language and music also allowed the elicitation and experience of totally new emotional experiences, strengthened by their integration into increasingly complex cognitive contexts.

Emotions serve to set goals for further behaviors, though emotional states with long time constants tend to be referred to as moods. These processes again operate across multiple timescales and therefore levels of action hierarchies. While English usage often treats emotions and feelings as contents of consciousness, progress in linking emotions to neural and visceromotor processes suggests that consciously expressible emotions are the "tip of the iceberg." Thus we need to embed our analysis of "conscious feelings of emotion" within a much wider context of a great variety of motivational systems that include consummatory (e.g., thirst and hunger), "informational" (e.g., fear, curiosity, and pain), and social signals (e.g., anger, jealousy, and sexual arousal). Such analysis must also include subtle social emotions (e.g., *Schadenfreude*) and, with new developments in art making possible new aesthetic experiences, the experience of a familiar emotion in a novel context or a novel blend to be savored only by the aesthetic connoisseur.

As noted earlier, animals have expressive actions (Darwin 1872/1998) which play a major role in social communication and cooperation. These carry over into expression through language and music in humans because physiological processes such as respiration and muscle tone push expression into different directions. However, in humans, external factors such as social norms or listener expectations may pull the expression in more culturally normed directions. Our feelings may be inchoate, but we hear others label our emotions on the basis of observing our expression and those of others, and so we come to associate emotions (labeled according to our culture) with certain words and certain classes of facial and bodily expression. As a result, we come to recognize other's emotions through our own, so it is harder to imagine emotions in bats or computers. Most relevantly for this Forum, emotions can be invoked by the combination of words in language and the pattern of sounds in music. However, these two means of emotional expression may differ greatly in their cultural dependence and the neural pathways involved. While emotions in their "autonomic" form appear to be nonsymbolic, nonintensional, and nonpropositional, emotion in humans, at least, frequently has an "intensional" component of propositional attitude: anger is anger about some fact; fear and desire are fear and desire of possibilities. In intact humans, autonomic and intensional emotions inevitably interact, since intensional emotions reflect intensional conflict which may, in turn, give rise to autonomic responses. Such diversity makes it clear that the term "emotion" needs careful qualification, or replacement by a set of more restricted terms, if scientific analysis is to be fruitful.

A psychiatrist may see interesting dissociations in the features of emotion spared and impaired across different disorders (schizophrenia, major depressive disorders, autism spectrum disorders, anxiety disorders), and how they interact with various cognitive systems. Kuperberg et al. (2011) explore interactions between emotion and cognition by examining its impact on online language processing in the brain, seeing effects of emotional context on the semantic processing of other emotional words and on neutral words. The findings suggest that schizophrenia is associated with (among other factors) a specific neural deficit related to the online evaluation of emotionally valent, socially relevant information.

To study emotion in species other than humans is often a highly unsatisfying adventure, since a similar looking expression in a nonhuman species need not serve the same function as it does in humans (Waller and Dunbar 2005). Still, there is some interesting research going on in this field. Parr et al. (2008) use the matching-to-sample task to investigate whether chimpanzees are able to assign certain facial expressions to "emotional" situations with either a positive or negative valence. Vick et al. (2007) applied the *facial action coding system* (FACS) to measure and describe facial movements in chimpanzees, and FACS is now available for macaques (Parr et al. 2010) and gibbons (Waller et al. 2012). To measure the emotional state of a nonhuman species is, however, difficult, particularly with the use of noninvasive methods. It gets even more complicated when we are interested in so-called social emotions. For example, to see whether a chimpanzee empathizes with another who is being emotionally harmed, Liebal measures whether the observing individual will help more after witnessing the harming compared to situations when no harm occurred. Still, for this particular example it is not possible to tell whether the observing individual feels *with* someone or *for* someone; in other words, the differentiation between empathy and sympathy.

Language

Language is not equivalent to speech. Language can exist as spoken language or signed language and, as such, can exploit voice, hands, and face in varied combinations. In each case (e.g., voice/audition, hand/vision), there is duality of patterning: basic but meaningless patterns (e.g., phonemes, syllables, handshapes) can be combined to form basic units (words or morphemes) which either have meaning in themselves or combine with other units to extend their meaning. These units constitute the lexicon. Constructions then specify how such units can be combined into larger structures (e.g., phrases), and these or other constructions can be employed repeatedly to create larger structures. Each human language thus supports a *compositional semantics* whereby from the meanings of words and phrases and the *constructions* that combine them, we can infer the meaning (more or less) of the whole phrase or sentence even when it has never before been experienced. Recursion plays a crucial role in that structures formed by applying certain *constructions* may contribute to units to which those same constructions can again be applied. As a result, language can express explicitly a vast diversity of *propositions*, providing an explicit means for description and thereby for social sharing of goal states and action hierarchies that describe interactions between objects (both animate and inanimate). Of course, language can support other so-called speech acts as well as diverse social functions. Language, then, is a shared system of communication distinguished from other forms of communication by, at least, an open lexicon (a store of meaningful words/symbols) and a syntax with compositional semantics at the core. Other features include phonology—building meaningful symbols from a stock of meaningless symbols (duality of patterning)—and prosody (see Ladd, Chapter 11). Language use obeys rules of pragmatics, and languages have a number of design features (e.g., interchangeability, discreteness, displacement) as identified by Hockett (1960b).

Language builds on a basic grounding in sensorimotor experience. Over time (evolutionary and developmental), this capability became extended to increasingly abstract notions, thus providing the basis for reasoning, mathematics, and analogical reasoning in general. This can be considered as the linguistic bootstrapping of cognition.

Recent work on sign language—especially studies of Al-Sayyid Bedouin Sign Language (Sandler et al. 2005; Aronoff et al. 2008) which provide evidence that phonological structure emerges gradually—can inform our ideas of language structure and the relation between language and speech. This, in turn, will help provide insights about the possible origins of language (is it a meaningful question to ask whether language or speech came first?) and hence about the evolutionary links between language and music (see Arbib and Iriki, Chapter 19, as well as Fitch and Jarvis, Chapter 20).

Language is a learned capacity. It is still a matter of debate whether we are innately programmed to learn language per se, or whether an ability to acquire a more limited range of praxic and communicative actions through imitation was what, through cultural evolution, came to serve the learning of language by humans in the last 100,000 years, long after the core genome of *Homo sapiens* was in place. For Darwin, language learning consists of an instinctive tendency to acquire an art, a design that is not peculiar to humans but seen in other species such as song-learning birds. Humans have evolved circuitry, termed the "phonological loop" (Baddeley 2003), which allows one to generate, maintain activated, and learn complex articulatory patterns that convey increasingly elaborate messages, though one may debate whether this is "speech first" or "manual gesture" first, or some combination thereof (Aboitiz 2012).

While some approaches to language give primacy to the speaker's production and hearer's comprehension of grammatically well-formed sentences, it is nonetheless the fact that normal language use is replete with utterances which are at least in part ungrammatical. Yet grammaticality can serve for language a role analogous to that which frictionless planes, massless springs, rigid bodies, etc., play within physics. We appeal to them, despite our awareness that they are at variance with how the real world appears to us. We do so based on the recognition that such idealizations, as Chomsky (2000) observes, are at the heart of the procedure we follow in attempting to discover reality, the real principles of nature. Insofar as they allow us to come up with real answers and ask coherent questions with respect to the objects in question, it is reasonable to accept their utility. Nonetheless, for non-Chomskians, the emphasis on syntax and the downgrading of meaning may be the wrong simplification, especially if our concern is with evolution, acquisition, or neural processes. An alternative approach, *construction grammar* (Croft 2001; Goldberg 2003) regards each construction of the grammar as essentially combining form and meaning: combine elements in the specified way and you will derive a new meaning from the constituent meanings in a fashion specified by the construction. It must be stressed, however, that a grammar (whether of these or other formal classes) is a symbolic abstraction, so the question of which grammatical framework (perhaps yet to be defined) is best suited to frame the search for neural correlates of language remains open.

Some scholars observe that humans possess (nontrivial) recursive grammars but that animals do not, and they conclude that recursion is distinctively human and is seen as the locus of the evolutionary advance in primate evolution that gave rise to language (Hauser et al. 2002). Others reject this view, noting that hierarchical actions seen in nonhumans exhibit recursive properties, so that it is the mapping from praxic structure to communicative structure that was essential to the emergence of language (Arbib 2006b). Consider the discrete, multilevel, and productive (Lashley 1951) nature of many manual actions, as exemplified, for example, in tool use and tool making (Stout 2011), in which consistent products are generated from inherently variable raw materials and action outcomes. This can be seen as a precursor for language recursion if we regard different instances of detaching a flake, which do vary in many details, as instances of the same "thing" in the way that "noun phrases" form a category even if they comprise different words or meanings. However, unlike language where the capacity of embedding of constructions (as in recursion) is limited by working memory capacity, in praxis the external world provides its own "working memory," so that the depth of iteration may thus be very large indeed.

Language production requires action, and this also leads to theories of the importance of the motor system in decoding speech, or other manifestations of language (consider the parallels between movements of the limb and movements of the articulators exploited in articulatory phonology; Goldstein et al. 2006). Parallels can be suggested between the types of mechanisms that must be engaged to comprehend and execute action sequences and those that are required to understand and produce language. There may be similar neural costs when one views an action that violates expectations and roles (e.g., ironing bread) and when one violates expectations of roles around verbs during language comprehension. Nonetheless, having a brain capable of recursive planning of action does not a priori imply that one has a brain capable of employing a language-like ability for the open-ended description of such plans and actions. In addition, there is a crucial recursive concept that is rather different from those just described: the concept of knowledge of other minds (de Villiers and Pyers 2002), in particular the kind of knowledge of other minds that is needed to formulate plans involving arbitrary numbers of cooperating agents (Tomasello 2009).

It is clear that language is unique to humans. What is less clear is when children become language users. The first word is often considered a sign of language onset, but similar vocal or gestural signals would receive different interpretations in the case of nonhuman species. For some scholars, word combinations mark the true onset of language, but language, if we are to call it that, is extremely limited at the two-word stage. There is increasing research aimed at identifying so-called precursors of language. At one time, skills demonstrable in the newborn period were considered innate, but there is abundant evidence that newborns' responsiveness to sound patterns, including specific languages, is influenced by prenatal exposure. That is not to argue against innate skills, only to illustrate the difficulty of obtaining definitive evidence of innateness. There is every reason to believe that innate abilities and dispositions underlie language acquisition. These abilities and dispositions may well include perceptual and learning biases, but social factors will probably emerge as the critical foundation. Thus the notion of a universal grammar providing innate knowledge of the general patterns that can be adopted by the syntax of any human language seems increasingly untenable.

Music

Humans engage in music for fun, for enjoyment, for religious ceremonies, for cure and therapy, for entertainment, for identification with social and cultural groups, for money, and for art, to mention just a few possibilities. However, it is very hard (if not impossible) to provide an exhaustive definition of music, especially if one aims at a very broad definition that is not limited to, for example, the Western tradition of tonal music. It is thus useful to distinguish "musicality" as a natural, inborn quality that is an outcome of our biology, from various "musics," each of which is a culturally determined phenomenon based on that very biology (Honing 2010, 2011b). In other words, this is a distinction between a general "musical" cognitive capacity and particular cultural manifestations of music.

Music is a continuously updated modern form of artistic expression yet provides (at least in the view of Darwin) a prehistoric record of the original bases of human communication through rhythmic, chanted, repetitive yet varied vocalization. Music is a temporally structured acoustic artifact that can within seconds touch the human emotion system at its deepest levels, but it is not only acoustic. How does playing a musical instrument relate to other forms of embodied action? How does practical action relate to or differ from musical action? We may think of musical action in terms of gesture. Musical gesture is a composite phenomenon made up of melodic contour, rhythmic shape, dynamics, tempo, and tone color. The varying contributions of these largely independent factors make musical gesture difficult to formalize. Perhaps a theory of dance could be developed employing tools from music analysis (e.g., the interaction between grouping and meter, the headed hierarchy of time-span reduction, and the rise and fall of tension in prolongational structure). Blaesing et al. (2010) provides a number of essays which put dance in a neurocognitive perspective.

The production of music requires the action of singing or playing a musical instrument, and music listening may entail decoding these actions in more or less detail. Palmer (1997) assesses music performance in terms of structural and emotional factors that contribute to performers' conceptual interpretations and the hierarchical and associative retrieval influences, style-specific "syntactic" influences, and constraints on the range of planning as well as the fine motor control evidenced in music performance. He concludes that music performance is not unique in its underlying cognitive mechanisms. Repp (1992) assessed the diversity and commonality in different performances of the same piece of music, based in this case on analysis of timing microstructure in Schumann's *Träumerei*. He reported that the grouping structure, which prescribes the location of major tempo changes, and timing function, which represents a natural manner of executing such changes, seem to provide the two major constraints under which pianists operate. He showed that within these constraints, there is room for much individual variation, and there are always exceptions to the rules.

The capacity to be emotionally aroused by rhythm is a basic property of music (though some genres may downplay it). Music also excites us at a higher level, challenging learned expectations, drawing on cultural traditions, and demanding rapid drawing and redrawing of melodic and temporal schemas. When harmony, form, improvisation, and lyrics are added, we may even experience awe. The reliance of the emotional response at this level of higher cortical processes makes music unique to humans. The profundity of music shows us how unique the human brain is. Music provides one of the most powerful ways we have to investigate the mysteries of the human brain though, alas, most investigations are still limited to the lower reaches of musical effect and affect.

A further distinction worth making is between the narrow musical capacity (features that are unique to music) and the broad musical capacity (features that music shares with other cognitive capacities). Not all musics project meter, though most do; thus musical meter does not appear to be culturally universal, though responsiveness to rhythm is seen in infants and may indeed be part of musicality. Infants are naturally equipped to engage with the music of any culture (Trehub 2003, 2000). However, the possibility of meter, the forms that metrical grids can take, and most of the cognitive principles by which meter arises all belong to musicality (and, to a lesser extent, to language readiness). Aside from a few birds (notably, the cockatoo "Snowball," Patel et al. 2009; for earlier evidence, see the closing sequence of the 1988 movie "The Young Einstein"), nonhuman animals do not manifest metricality. Some musical genres have nothing that could be called harmony, drumming often makes no use of melody, and for at least some musical forms it is difficult to identify properties we might want to call rhythmic.

In modern Western society, and many others, much of music production has become a specialized activity performed by people with specific training for such a task, as is necessary for highly sophisticated musical art. In this sense, the general audience has often taken a more passive position, as appreciators of art. The materialization of devices for recording and playing music now permits a solitary appreciation of music, but it might be argued that social interaction between the (composer and) interpreter and listener is still there, but temporally dislocated. Sloboda (2000) stresses the levels of expertise required to perform Western art music; expert musical performance is not just a matter of technical motor skill, it also involves the ability to generate expressively different performances of the same piece of music according to the nature of intended structural and emotional communication. Nonetheless, we sing our favorite songs, we go dancing, we may drum rhythmically on a table top, perhaps we sing hymns in church, so that even if we are not at a professional level, we are nonetheless experienced (but unpolished) performers of music. We are also expert listeners, able to recognize hundreds of pieces on the basis of just a few initial chords: the "plink" instead of the "blink" (Krumhansl 2010). In many non-Western cultures, however, music is generally an activity performed by the whole group and is not related to our conception of art (see Lewis, Chapter 2). Within those cultures, music is a matter of social (and ritual) activity rather than individual enjoyment. The collectiveness of music could be related to interpersonal coordination (including dance) and this coherent collective behavior to the maintenance of group identity (Cross 2006). In any case, we may contrast the coordination of near simultaneous performance of people engaged together in music with the alternation of overt action of two people engaged in that basic form of language, the conversation (Levinson, Chapter 3).

Two interesting observations from Morocco (Katherine E. Hoffman, pers. comm.) add further perspective. First, the chant of a muezzin is not considered by Muslims to be music, and attempts to meld such chant with explicitly musical forms are considered deeply offensive by some, but a welcome innovation by others. Here we may contrast Gregorian chant, which is certainly considered a musical form. Second, in the past, at Berber celebrations such as weddings, a group of women from the village would sing traditional songs in which a simple verse might be repeated over and over, but with variations as fun is made of the idiosyncrasies of individual people of the village. Increasingly, however, professional musicians are now hired for such occasions. This may yield more polished performances, but they no longer engage the particular activities of the villagers, and women lose one of their few outlets for social expression outside the home.

Is There a Language–Music Continuum?

The above discussion of musical structure and varied musics makes it clear that we must avoid defining music too narrowly in terms of a *necessary conjunction* of specific properties. Rather, we can assess how different musics coordinate their individual subset of the properties inherent in musicality, but then we discover that many of these properties are relevant to analysis of languages. This motivates a question discussed at great length at the Forum: "Is there a language–music continuum?" Perhaps there is. Nonetheless, it seems that the key to being a language is the coupling of lexicon and grammar to provide a compositional semantics which (among other properties) supports the communication of an unbounded set of propositions. By contrast, music is not an assemblage of words, for words are abstract symbols which convey factual meaning. As Menuhin and Davis claim: (1979:1) "Music touches our feelings more deeply than words and makes us respond with our whole being." Well, this may not be true of all musics, but it does perhaps serve to anchor the "music end" of the continuum. In any case, voice and emotion have played a central role in the coevolution of music and language and still represent vital and powerful elements in present-day speech communication and music making.

Merker (2002) observes that both music and language are communication systems with the capacity to generate unbounded pattern diversity by combining a relatively small set of discrete and nonblending elements into larger individuated pattern entities and that each is a diagnostic trait of *H. sapiens*. No such communication system exists in our closest relatives among the apes. It can be debated whether birdsong is such a system and, if so, whether it is intrinsically similar to human music in some strict sense. Merker speculates that music (song) may be a necessary but not sufficient antecedent of language; others hold to the contrary or suggest that early stages of each supported the evolution of the other. In any case, returning to the action–perception cycle, we reiterate that praxis, too, has the capacity to generate unbounded pattern diversity by combining familiar elements into larger individuated patterns. However, any communication here is a side effect, and the set of "motor schemas" being combined is neither small nor nonblending. Another interesting distinction is the way in which praxic action may depend strongly on the currently observable state of the physical environment relative to the actor, whereas such a linkage in language or music is located more in the realm of social perception of other's mental states (to simplify the relevant diversity for the needs of the present discussion).

Emotional responses to language are generally in relation to the semantic content rather than to the specific syntactic structuring or mechanistic details of delivery. Nonetheless, syntax and sensorimotor coordination indisputably contribute to the aesthetics of language and comprehension. Moreover, prosody allows the mode of delivery to change the meaning of a message dramatically. The tone of voice in which one says, "that was a great job," may determine whether it is heard as a compliment or an insult. Linguistic content modulates what listeners infer about a speaker's emotional state; knowledge of emotional state affects the interpretation of an utterance. However, that does not exclude the possibility that certain nonlinguistic vocal features do have fairly direct emotional correlates, even if those emotional correlates can be affected by the linguistic context. Here the links to music seem particularly relevant, as in studies indicating that certain intervals (e.g., the minor third) or timbres (e.g., that of the xylophone) have an intrinsic emotional valence. Low, loud, dissonant sounds evoke fear; rapid, higher, consonant sounds evoke friendliness or joy. But why? Mothers around the world talk and sing to infants using a cooing tone of voice and higher pitch than when interacting with adults. Infants prefer these higher-pitched vocalizations and mothers sing in different styles to help prelinguistic infants regulate their emotional state. Across cultures, songs sung while playing with babies are fast, high, and contain exaggerated rhythmic accents; lullabies are lower, slower, and softer. Talking to people of all ages, we use falling pitches to express comfort; relatively flat, high pitches to express fear; and large bell-shaped pitch contours to express joy and surprise. All this suggests important bridges between music and mechanisms for prosody that may be shared with language. Hearing music with an unfamiliar structure, listeners base their emotional reactions largely on such sound features. Alternation between consonance and dissonance is a powerful device: consonance can be very beautiful and resolution of dissonance especially poignant.

Music and language are viewed as one in several cultures. At least in some ways, "language" and "music" are rather different aspects of the same domain, or two poles of a continuum. We speak with rhythm, timbre, and melody. Diane Deutsch's musical illusions (Deutsch 1995)—e.g., when you play the phrase "sometimes behave so strangely" on a repeated loop, it eventually sounds as if sung[1]—show that spoken language can also be perceived as song (although we often do not hear it in our everyday lives, presumably because we are not used to listen to speech musically, whereas babies do). Once an individual puts emphasis into his/her utterances, the speech becomes more song-like (Martin Luther King's speeches are a nice example of how it is often difficult to discern whether someone is singing or speaking). The F_0 contour of spoken information is important in tone languages; even in nontonal languages, phrase boundaries, for example, are marked by musical features such as ritardando and rise in F_0 frequency. The acoustical features that characterize spoken phonemes are

---

[1] This loop can be heard on Radiolab, http://www.podtrac.com/pts/redirect.mp3/audio.wnyc.org/radiolab/radiolab042106a.mp3

identical to those that characterize musical timbres, and affective information in speech is coded with such musical features.

Many art forms, such as rap music or recitatives, are both song and speech. In many musical styles, music is structured according to a syntactic system (involving phrase-structure grammar/recursion; every first movement of a sonata, for example, fulfills these criteria). Moreover, music has not only emotional meaning: it has extramusical meaning (i.e., meaning due to reference to an object of the extramusical world) with all the sign qualities that we know from language (in Charles S. Peirce's terms "iconic," "indexical," and "symbolic") as well as intramusical meaning (i.e., meaning due to reference of one musical element, or group of musical elements, to another musical element, or group of musical elements).

In language, especially in speech and sign, a great deal of conveyed meaning is not merely denotational for possible and actual physical aspects of the world, but deals strongly with its interpersonal aspects. In English, these aspects of meaning are to a considerable extent conveyed by prosodic structure, from which various signs are conveyed by pitch contour, accent/segment alignment, lengthening, and pausing marking various distinctions of contrast, topic/comment status, speaker/hearer origin, and so on. (In other languages, particularly tone languages, discourse particles do similar work, often in interaction with prosody.)

While the functions of music are broad, language is clearly representational, and its sound does not have to be "appreciated" or "felt." Language provides the conduit through which the phenomenology of emotion can be expressed. Verbal self-report nevertheless provides a limited account of emotion; language offers us glimpses of another's stream of consciousness, but its timescale is broad and may not capture dynamic fluctuations, and other methods are required in conjunction with language to detect changes of which we may not be aware.

Lerdahl (cf. Jackendoff and Lerdahl 2006) emphasizes (a) hierarchical rhythmic structure, (b) hierarchical melodic structure, and (c) affective meaning as being defining elements of human music. He considers this particular combination to be effectively unique to humans as well as common to most activities we might want to call "music." A broader definition of music or "song" as a complex, learned vocalization (Fitch 2006a) may also be useful pragmatically in comparative studies (though we do learn to recite, e.g., poems or memorize lines of a play). Unlike music, language does not typically involve regular rhythm structures; the most important difference may be that linguistic semantics is symbolic (in the sense of C. S. Peirce as elaborated by Deacon 2003) whereas the meaning of music is only secondarily symbolic (it assumes symbolic meaning only in the context of linguistic culture). Indeed, the difference in rhythm probably also relates to this semantic difference: rhythmic constraints would interfere with linguistic communication whereas generation of expectation by predictable rhythms is a mechanism for communication of musical (especially affective) meaning. Thus many scholars posit a shared evolutionary origin for language and music in nonsymbolic vocal signaling (in the context of many neural "preadaptations" associated with other complex, intentional behaviors) serving some important social function, followed by subsequent divergence. Others focus more on the emergence of gestural communication from manual praxis (for further discussion, see Cross et al., Chapter 21).

Emotions are related to music both intrinsically (minor keys have been associated to sadder moods compared to major keys) and extrinsically (a song played in the background at a crucial point in our life is capable of eliciting the feelings experiences at that moment even years later). Recently, it has been observed that the tone with which sadness is expressed in speech is similar to the minor third, which indicates that music might mimic this natural speech pitch to convey sadness, thus pointing to the existence of a threefold association among language, emotions, and music. Even though music can bear an objective or a subjective relationship to the emotions experienced, the detection of musical tones is independent of cultural cues (Balkwill and Thompson 1999).

Poetic language has a number of similarities to music. They both may be (though this is not always so) characterized by a predefined structure. In poetry, meter dictates the pattern of stressed and unstressed syllables, whereas in music it determines the duration of a beat. They both display a mixture of repetition and change or progression: in poetry, specific sounds are repeated either at the end of words (rhyme), at the beginning of words (alliteration), along a poetic line (assonance), or entire lines (parallelism); in music, individual sounds (notes) or sequences of sounds are repeated either partially or entirely.

Linguistic human communication includes an articulatory (phonological) component as well as a prosodic component. Prosody evidences emotion and relies on factors like intensity and pitch. Songbirds have evolved what some consider a musically based communication, precisely by manipulating pitch and rhythm. But does this constitute "music" any more than ape calls are language? Well, it goes beyond chicken calls, but it can be argued that the song of the nightingale is not music until heard by a musician who reinterprets it. Some have argued that songbirds (in this case, starlings) are the only animals that have been shown to be able to learn recursive grammars (Gentner et al. 2006), but the tests were quite inadequate and a simple finite-state process could explain the limited experimental data adduced to date (Corballis 2007). In any case, this structure has no links to a compositional semantics, though placing different timbres (types of sound) and pitches on a temporal scaffold in humans immensely extends the cognitive–affective space in which music operates. The combinatorial possibilities are sculpted to yield probabilistic structures that guide our understanding, appreciation, and emotional responsiveness to individual sounds and their combinations. Timbres vary in their semantic and affective connotations. Pitches can be sequenced so as to mimic vocal and linguistic contours with affective connotations. Simultaneous combinations of aspects of neural architecture are intrinsic to properties of mind, and which aspects are mere kludges, contingent on the undirected nature of that search. At any moment, complex patterns of excitation and inhibition link huge populations of neurons in different regions. While "initial" anatomy (the wiring up of the immature brain that occurs over the course of embryonic development) provides the framework for these computations, it is learning from experience—mediated in part by plasticity of synapses (connection points between neurons)—which determines the details of the wiring that makes these interactions possible and constrains them. Perhaps surprisingly, the "initial" anatomy can be strongly influenced by the experience of the embryo within the womb.

Thereafter, changes in cortical structure and function during infancy cloud the interpretation of studies aimed at uncovering the neural underpinnings of language, music, emotion, and action. Unquestionably, young brains are more flexible than older brains, but there remain many open questions about the trade-off between experiential and maturational factors. Recordings of event-related potentials (ERPs) have illuminated remarkable abilities on the part of sleeping newborns, including beat detection, the perception of pitch invariance across timbre, and the perception of interval invariance across absolute pitch level. Moreover, research with near-infrared spectroscopy (NIRS) has revealed newborn sensitivity to regularities in syllable sequences. What can we make of this apparent "knowledge" in the absence of conscious awareness? How can we interpret such preattentive processing in immature organisms who cannot deploy their attention voluntarily?

Returning to our discussion of emotion in relation to physiology and the brain: in evolutionary terms, emotion emerged from the processing of internal and external information in the crudest sense as an early-warning system. Humoral control (noradrenalin in fast-acting systems; serotonin in slow-acting mood regulation over longer timescales) plays an important role (Kelley 2005). These neuromodulators are broadcast to varied targets in the brain as part of a fast reaction system centered on the hypothalamus and the autonomic nervous system (but invoking other systems in appetitive and consummatory behavior) for the "primordial emotions" or drives. These include Pribram's 4 Fs of feeding, fighting, fleeing, and reproduction (Pribram 1960); and on this basis we may view emotion as derived from the binding of survival-related perceptual systems (pain, hunger, satiety, comfort, danger) to mid- and high-level cognitive systems. Thus, emotion expresses internal states at the levels of *both perception and action* (which, of course, are integrated in the action–perception cycle), which covers a large range from primitive to abstract levels. As we seek to expand on this basis in studying the role of, for example, the amygdala and orbitofrontal cortex in mammals, we find it increasingly hard to separate emotional processes from the cognitive systems that shape and are shaped upon them, modulating compositions from the palette of bodily correlates of the primordial emotions. (For a diagram of some of the regions of the human brain of most interest to us, see Figure 8.2.) A crucial task for *social neuroscience* is to understand what networks of neural mechanisms serve to integrate the internal and external aspects of emotion. Proceeding from the microlevel of synapses, we move to the macrolevel as we develop the theme of social cognitive neuroscience in relation to social schemas: the mix of genetic and social inheritance.

Different parts of the brain have different functions, but a given schema (perhaps based on conceptual decomposition of some externally defined behavioral or psychological function) may involve the interaction of many regions of the brain and, conversely, any brain region may contribute to the implementation of a range of schemas. Thus, when an area involved in a function is damaged, cooperation between other areas may yield (partial) restoration of function. From an evolutionary point of view, we find that many "old" regions of the brain survive from those of our very distant ancestors, but that newer centers add new functions in themselves and, through the new connections, allow new functions to exploit activity in the older regions.

When we study certain nonhuman creatures, such as monkeys or songbirds, we may use single cell recordings, which employ microelectrodes to see how the firing of single neurons correlates across several tasks with various sensory, motor, and other features. Cumulative studies can then form detailed hypotheses about how neurons work together in specific circuits. Such data can even exhibit the short-term dynamics of working memory and the long-term dynamics of learning and memory. By contrast, human brain imaging methods smear activity of millions of neurons across several seconds to yield the hemodynamic response. Positron emission tomography (PET) and functional magnetic resonance imaging (fMRI) have been interpreted mainly via "boxology," localization of cognitive and other functions to one or a few brain regions. Basically, these techniques are based on the notion that when neurons are more actively engaged in a task, more blood will flow to support their metabolism. PET and fMRI are often used to compare blood flow across the brain when a human subject performs task A and task B. Morphing different brains to some standard and averaging across subjects can yield a three-dimensional brain map of statistical significance of the hypothesis that a particular brain region X is, in general, more active in task A than task B. The problem is that many papers interpret such data as if they imply that if the level of significance exceeds some threshold, then "X is engaged in task A but not task B." However, the truth is more likely to be that X is a crucial part of the cooperative computation involved in *both* A and B, but that the relative level of activity (and, indeed, the engagement of different circuits within X) will vary between tasks. Where fMRI smears the time course of neural activity but gains some precision of spatial location (but still very crude relative to the localization of individual neurons), ERPs yield insights into the millisecond-by-millisecond time course of neurocognitive processing, recording the average potentials of cortical neurons from various leads attached to the scalp. However, each lead is giving its own weighted average of the whole brain's activity, and thus localization is very limited. The combination of fMRI and ERP may in some cases be used to suggest that certain components of the ERP signal correspond to, and can thus be localized by and provide the timing of, certain peaks of fMRI activation.

Neuroscience is a vast and rich domain of study, so vast that specialists may work in areas that remain mutually incomprehensible unless and until some new study shows the necessity of integrating data and hypotheses from distinct subdomains. For example, people studying "brain mechanisms" of music and language may focus on lesion and imaging studies in a search for correlates between tasks and brain regions. Brodmann (1909) used differences in cell types and layering to distinguish 52 areas of human cerebral cortex, with each area occurring in both the left and right hemisphere. Broca's area, long seen as a classic component of the brain's language system, is usually taken to comprise Brodmann areas 44 and 45 of the *left* hemisphere, though some may append other areas in their definition (Grodzinsky and Amunts 2006). Such studies may support notions of localization that are even coarser than this. It is rare, then, for such studies to address "how a brain region works" by actually addressing data on how cells of different types are connected within a region, let alone the neurochemistry of the synapses which mediate the dynamic interaction of those cells. Indeed, in most cases, data on such details are unavailable for studies of the human brain. Instead, neuroscientists use animals of very different species to explore fine details of neural activity which can be expected to hold (perhaps in modified form) in humans as well. For example, our basic insight into how one neuron signals to other neurons by propagating spikes (transient changes of membrane potential) along its axon ("output line") comes from studies of the giant axon of the squid (Hodgkin and Huxley 1952).

Eight of the chapters in this volume take the study of the brain as their primary focus. A quick review of what data and methods each chapter uses will help the reader assess the scope of neuroscience in relation to the study of language and music. A principal conclusion is that much, much more needs to be done to apply the results from animal studies if we are to extract the full implications of high-level correlational analysis of the human brain. Let's look first at those chapters that focus on animal studies:

- Chapter 4: *Shared Meaning, Mirroring, and Joint Action* by Leonardo Fogassi details neurophysiology of macaque neurons, with a special emphasis on mirror neurons, both those involved in manual and orofacial actions and those related to emotions; comparable fMRI studies of humans.
- Chapter 15: *From Action to Language and Music in Models of the Brain* by Michael A. Arbib, Paul F. M. J. Verschure, and Uwe Seifert presents computational models addressing neurophysiological data from various mammalian species (such as rat, monkey) to show how circuitry within and across brain regions (examples include hippocampus, cerebellum, and cerebral cortex) achieves various functions.

Five Terms in Search of a Synthesis 29

- Chapter 19: *Evolving the Language- and Music-Ready Brain* by Michael A. Arbib and Atsushi Iriki presents further neurophysiology of macaque neurons plus data on behavior and communication of nonhuman primates, with some attention to how new behaviors might emerge as connections are modified by gene expression and circuit reorganization to yield new neural systems without changing the underlying genome.
- Chapter 20: *Birdsong and Other Animal Models for Human Speech, Song, and Vocal Learning* by W. Tecumseh Fitch and Erich D. Jarvis discusses neurophysiology and genetic correlates of vocal learning in songbirds, with comparative data on other vertebrate groups. (Note that the techniques of genetic analysis are proving of increasing importance in probing disorders of the human brain.)

The other four focus on studies of human brains:

- Chapter 6: *Neural Correlates of Music Perception* by Stefan Koelsch uses electroencephalography (EEG), magnetoencephalography (MEG), galvanic skin reponse, PET and fMRI to study the time course of processing musical "syntax" and "semantics" as well as certain emotional correlates.
- Chapter 9: *The Infrastructure of the Language-Ready Brain* by Peter Hagoort and David Poeppel provides an architecture for language processing based on ERP and fMRI data on humans but is enriched by discussion of data on the two auditory processing streams of the macaque and by comparative neuroanatomy of the arcuate fasciculus in human, chimpanzee, and macaque. Particular attention is paid to the possible relevance of neural rhythms in the differential processing of music and language in the left and right hemispheres.
- Chapter 13: *Neural Mechanisms of Music, Singing, and Dancing* by Petr Janata and Lawrence M. Parsons explores the relation of brain mechanisms serving music, song, and dance using EEG, ERPs, fMRI, and PET. They also present melodic intonation therapy for rehabilitation of speech following stroke, which is based on the idea that word representations might be accessed more easily via song. Moreover, fMRI studies of duet percussion performances mark a start at extending brain measures from the individual to pairs and ensembles.
- Chapter 14: *Sharing and Nonsharing of Brain Resources for Language and Music* by Aniruddh D. Patel compares language and music processing based primarily on ERPs and fMRI data, but appeal is made to animal studies of the neurophysiology of auditory processing.

Of course, these themes are revisited in various combinations in the four chapters based on the group discussions.

The Architecture of the Book

With an exposition of our five themes firmly in place, we now embark on a chapter-by-chapter tour which shows how these ideas structured the Forum and this book.

Part 1: An Expanded Perspective

Part 1 is designed to escape the view that language is just a sound pattern made up of words, and that music is just a different family of sound patterns that taps more directly into our emotions (though understanding and questioning that difference plays an important role in what follows).

Chapter 1: Five Terms in Search of a Synthesis (Michael A. Arbib)

Here the focus is on five key terms: the action–perception cycle, emotion, language, music and brain. Its basic premise is that communicating with language and making music are forms of action with their complementary forms of perception. Thus our aim of understanding the ways in which music and language each relate to the brain, and to each other, requires us to assess ways in which the human brain's capacity to support music and language relate to more general capabilities we share with the brains of other creatures—not only action and perception but also motivation and emotion, and various forms of learning. In this framework, we may hope to understand music and language in a broad action-oriented perspective: The sound patterns of vocal music integrate with the actions of playing music and their embodiment in dance. Similarly, when we speak, our vocal production is accompanied by facial and manual gesture, while deaf communities have full human languages that exploit face and hands with no role for the voice.

Chapter 2: A Cross-Cultural Perspective on the Significance of Music and Dance to Culture and Society (Jerome Lewis)

Much of this book is rooted in studies of music and language shaped primarily by the study of the English language and the Western tradition of tonal music. This chapter serves as a primary corrective to this view. There is no single conception of language or music that truly cuts across all cultures. All societies have music and dance, but some have no general, separate terms for them. Some have specific names for different performances that involve music and dance; others use the same word for music making, singing, dancing and, often, for ritual as well. This chapter explores, in cross-cultural perspective, the crucial role of music (and its integration of language through song) and dance, however named, in the self-definition of human social groups. Lewis argues that participation in music and dance activity grounds "foundational cultural schemas" affecting multiple cultural domains: from cosmology to architectural style, or from hunting and gathering techniques to political organization.

Chapter 3: Cross-Cultural Universals and Communication Structures (Stephen C. Levinson)

Some linguists have argued that humans are endowed with an innate "universal grammar" which in some sense inscribes the essentials of language in the infant's brain quite apart from childhood experience. Levinson argues that the diversity of languages makes such a hypothesis highly unlikely. Instead, language is rooted in the capacity for vocal learning hooked up to a distinctive ethology for communicative interaction (though we might suggest a more general modality of learning since language extends far beyond speech). Turning from a linguistic emphasis on isolated sentences, the chapter focuses on face-to-face communication; here, multimodality of conversational interaction is emphasized with the varied involvement of body, hand, and mouth. This suggests (and is certainly consistent with Chapter 2) that the origins of music should also be sought in joint action, but with the emphasis on affective quality and simultaneous expression rather than on turn-taking. The deep connection of language to music can best be seen in song.

Part 2: Action, Emotion, and Semantics

As song demonstrates, language and music can interact in varied ways, and thus a continuing debate at the Forum was the extent to which language and music are truly distinct (though sharing a limited set of resources) or whether they are variations on a single human theme. Could music and language just be different terms for ranges along a single continuum? We shall see much further discussion in this book, but one agreement is that language can express propositions in a way that music-without-words cannot, and that music may connect with emotions in a very different way from that elicited by words alone. Chapters 2 and 3 have emphasized that music and language often take the form of joint action, in which the performance of one person will be recognized and acted upon in the performance of another.

Chapter 4: Shared Meaning, Mirroring, and Joint Action (Leonardo Fogassi)

This chapter takes up the study of the brain and shows how discoveries made by studying the neurophysiology of the neurons of macaque monkeys led to new insights into the human brain, as studied by imaging activity of different brain regions. The key idea relates mirror neurons (observed in monkeys) and mirror mechanisms (observed in humans). These are engaged not only in the performance of some behavior but also during observation of this behavior when performed by others. Here, Fogassi shows how different types of mirror systems may be engaged in social cognitive functions such as understanding of goal-directed motor acts, intention, and emotion by providing an immediate, automatic component of that understanding based on matching biological stimuli with internal somatomotor or visceromotor representations. Mirror systems can cooperate with other cortical circuits which are involved when understanding of another's behavior requires inferential processes. The chapter shows how this mechanism of mirroring, initially evolved for action understanding in nonhuman primates, could have been exploited for other functions involving interindividual interactions, including language and music. Later chapters, starting with Chapter 6, will look at brain correlates of a single human engaged in perception of stimuli or production of actions in the domains of language or music, but the ideas from this chapter alert us to the search for neural correlates of the social interactions charted in Chapters 2 and 3, when people join together in music and dance, or engage in conversation, so that the state of each brain is continually tuned to the behavior of the other's.

*Chapter 5: Emotion in Action, Interaction, Music,
and Speech (Klaus R. Scherer)*

Scherer advocates a componential appraisal model of emotion and demonstrates its role in understanding the relationships between action, interaction, music, and speech, with particular emphasis on the motor expression component of emotion. Brief nonverbal displays of emotion (affect bursts) are seen as providing an important element in the evolution of human communication based on *speech and gesture* and, probably in parallel, for *singing and music*. A dynamic model of the auditory communication of emotions distinguishes the function of expression as symptom of actor state, symbol of a message, and appeal to the listener. Evidence is then marshaled for the similarity of the expressive cues used to convey specific emotions in both speech and music. *Utilitarian emotions*, helping to adapt to relevant events that happen to the person, especially through speech in social interaction, are distinguished from *aesthetic emotions* which are generally desired and sought out through engagement with cultural practices such as music, art, or literature.

Chapter 6: Neural Correlates of Music Perception (Stefan Koelsch)

Focusing on studies of simple aspects of music processing in the human brain, Koelsch provides an overview of neural correlates of music-syntactic and music-semantic processing, as well as of music-evoked emotions. However, the notion of syntax studied here is restricted to harmonic progression; a somewhat larger-scale notion of musical syntax is offered in Chapter 10. These three aspects of music processing are often intertwined. For example, a musical event which is music-syntactically irregular not only evokes "syntactic analysis" in the perceiver, but may also evoke processing of meaning an emotional response, or decoding of the producer's intentions, etc. Since the neural correlates of these processes overlap with those engaged during the perception of language, Koelsch argues that "music" and "language" are two poles of a music–language continuum. Experience can greatly affect the brain's responses. One study showed different patterns of relative activation of working memory resources for certain language–music comparisons in musicians as compared to nonmusicians. The results suggest that musical expertise leads to a network comprising more structures (or new patterns of coordination/coupling between structures) that underlie tonal working memory, which shows a considerable overlap with the functional network subserving verbal working memory, but also substantial differences between both systems. There may be a parallel in the increased activation of the mirror system exhibited when dancers observe performances in the genre in which they are expert as distinct from a genre with which they are unfamiliar (Calvo-Merino et al. 2005).

Chapter 7: Film Music and the Unfolding Narrative (Annabel J. Cohen)

In this chapter the focus shifts from the small-scale structure in music to a consideration of the large-scale structure of narrative, assessing the role of music in narrative film. Unlike most other sensory information in a film (the visual scenes, sound effects, dialog, and text), music is typically directed to the audience and not to the characters in the film. Particular attention is given to the interplay between the emotional experience of the audience (referred to as internal semantics) and the external "reality" of the film (referred to as external semantics) as a basis for assessing where music, and film music in particular, lives with respect to these two domains. The concept of the *working narrative* is introduced as the audience's solution to the task of integrating and making sense out of the two sources of information provided in the film situation: the sensory information (including the acoustic information of music) and the information based on experience including a story grammar. Cohen's *congruence-association model with the working narrative* accommodates the multimodal context of film while giving music its place.

Chapter 8: Semantics of Internal and External Worlds
(Uwe Seifert, Paul F. M. J. Verschure, Michael A. Arbib, Annabel J. Cohen, Leonardo Fogassi, Thomas Fritz, Gina Kuperberg, Jônatas Manzolli, and Nikki Rickard)

This group report analyzes the similarities and differences of meaning in language and music, with a special focus on the neural underpinning of meaning. In particular, factors such as emotion that are internal to an agent are differentiated from factors that arise from the interaction with the external environment and other agents (such as sociality and discourse). This world axis (from internal to external worlds) is complemented by three other axes: the affective-propositional axis; the sensorimotor-symbolic axis; and the structure axis (from small- to large-scale structure). Common structure–function relationships in music and language and their neuronal substrate are addressed, with emphasis on expectation and prediction. Special emphasis is placed on how discourse and narrative relate to emotion and appraisal. Neurocinematics is studied for its focus on large-scale structure where music and language strongly interact.

Part 3: Structure

It is common in linguistics to distinguish *phonology* (the sound patterns of a language, or their equivalent for a signed language), *syntax* (e.g., the hierarchical structure that characterizes how words of various categories are assembled into sentences), and *semantics* (patterns of meaning which, in many cases, permit the meaning of phrases and sentences to be built up from the meanings of words plus the constructions whereby words and larger assemblages are combined). It remains controversial as to how these three areas can best be characterized, and to what extent the description of one can be separated from that of the others. What does seem uncontroversial, however, is that there is no easy correspondence between the structures of language and the structures of music. Thus, great care should be used when using terms like "musical syntax" or "musical semantics" to avoid reading into them properties of language that they do not share.

Chapter 9: The Infrastructure of the Language-Ready Brain (Peter Hagoort and David Poeppel)

This chapter sketches the cognitive architecture of both language comprehension and production, as well as the neurobiological infrastructure that makes the human brain language-ready. The focus is on spoken language (although intriguing data exist on the neural correlates of using sign language), since that compares most directly to processing the sound patterns of music (the relation to song and dance will occupy us in Chapter 13). With this focus, language processing consists of a complex and nested set of processes to get from sound to meaning (in comprehension) or meaning to sound (in production). Hagoort and Poeppel briefly present a selection of the major constituent operations, from fractionating the input into manageable units to combining and unifying information in the construction of meaning. It then offers a partial delineation of "brain networks" for speech–sound processing, syntactic processing, and the construction of meaning, leaving aside for now the overlap and shared mechanisms in the various processes within the neural architecture for language processing. Finally, they highlight some possible relations between language and music that arise from this architecture.

*Chapter 10: Musical Syntax and Its Relation to
Linguistic Syntax (Fred Lerdahl)*

Music is meaningful, but there is no musical counterpart to the lexicon or semantics of language, nor are there analogs of parts of speech or syntactic phrases. Here, Lerdahl seeks to establish a notion of musical syntax at a more fundamental level, starting from the view that syntax can be broadly defined as the hierarchical organization of discrete sequential objects generating a potentially infinite set of combinations from a relatively small number of elements and principles (thereby extending not only to linguistic syntax in the usual sense but also to a syntax of phonology). Sequences of musical events receive three types of structure: groupings, grids, and trees. Lerdahl describes the formation of successive structural levels, using a Beatles song as illustration, and discusses issues of sequential ordering, the status of global structural levels, contour, and the question of psychological musical universals. Intriguingly, broad correspondences are made between musical syntax and linguistic phonology, not musical syntax and linguistic syntax.

*Chapter 11: An Integrated View of Phonetics,
Phonology, and Prosody (D. Robert Ladd)*

To inform us about the language end of the parallel between musical syntax and linguistic phonology offered in Chapter 10, Ladd argues that "phonetics, phonology and prosody" do not constitute three separate subsystems of language. Rather, linguistic sound systems have both phonetic and phonological aspects, and there is little justification for defining prosody as some kind of separate channel that accompanies segmental sounds. For example, one is mistaken to think of lexical tone (as manifested in, e.g., Chinese) as prosodic. Instead, Ladd argues, the essence of prosody is the structuring of the stream of speech and that, rather than looking for specific local acoustic cues that *mark* a boundary or a stressed syllable, we should be looking for cues that *lead the perceiver to infer* structures in which a boundary or a given stressed syllable are present. Analogs in music abound (e.g., harmonic cues to meter mean that a note can be structurally prominent without being louder or longer or otherwise acoustically salient), and Ladd suggests that our understanding of music may well inform research on linguistic prosody rather than the reverse. Controversially, Ladd questions the view that duality of patterning—the construction of meaningful units (e.g., words) out of meaningless ones (e.g., phonemes)—is a central design feature of language, suggesting that the fact that words are composed of phonemes is arguably just a special case of the pervasive abstract hierarchical structure of language. If this view can be upheld, the issue of whether music exhibits duality of patterning can be seen as the wrong question, along with the question of whether birdsong is more like phonology or more like syntax. Instead, it suggests that, evolutionarily, music and language are both built on the ability to assemble elements of sound into complex patterns, and that what is unique about human language is that this elaborate combinatoric system incorporates compositional referential semantics.

Chapter 12: Multiple Levels of Structure in Language and Music (Sharon Thompson-Schill, Peter Hagoort, Peter Ford Dominey, Henkjan Honing, Stefan Koelsch, D. Robert Ladd, Fred Lerdahl, Stephen C. Levinson, and Mark Steedman)

This group report explores the notion that a central principle common to all human musical systems and all languages, but one that is not characteristic of (most) other domains, is their foundation in hierarchical structure. In the end, this type of structural framework is extended to include action. Written English and scores of Western tonal music provide key examples, but the variation of structures between languages and genres is noted. Nonetheless, it is argued (controversially) that the meaning representations which languages express are the same for all languages. Long-range dependencies show a similarity in the way in which sentences and music are mapped onto meaning, though sentence-internal semantics are very different from "intramusical meaning." This leads to a discussion of ambiguity which in the case of music may relate to assigning a note to a key, grouping of elements, and ascription of rhythmic structure. A future challenge is to understand how extrasyntactic sources of information (e.g., context, world knowledge) play their role in disambiguation. In addition, auditory processing is assessed in relation to language and music. Because the functions of language and music are different, it is hopeless to impose one system's labels (e.g., semantics, syntax, and phonology) on the other; there is overlap at the level of analyzing what kinds of information can be extracted from the acoustic signal, what functions each type of information supports (within language and music), and what are the similarities and differences between the types of information used—and the means by which they are integrated—in these two domains. Apart from a brief mention of the possible relevance of different neural rhythms to this process (see Chapter 9), the brain is not discussed in this chapter.

Part 4: Integration

Although the relation of music and language is a theme that recurs throughout the book, it is in Part 4 that we build on what we have learned in Parts 2 and 3 to offer a more explicitly integrated view of how language and music relate to action and emotion and the brain mechanisms that support them.

Chapter 13: Neural Mechanisms of Music, Singing, and Dancing
(Petr Janata and Lawrence M. Parsons)

Chapter 6 introduced us to the neural correlates of human processing of simple patterns of musical notes and their relation to emotion, whereas Chapter 9 reviewed key brain mechanisms related to the production and perception of language. Extending our survey of human brain mechanisms, in this chapter Janata and Parsons consider song (bringing together music and language) and dance (bringing together music and bodily action). As emphasized in Chapter 2, song and dance have been important components of human culture for millennia. The chapter enumerates processes and functions of song that span multiple timescales, ranging from articulatory processes at the scale of tens and hundreds of milliseconds to narrative and cultural processes that span minutes to years. A meta-analysis of pertinent functional neuroimaging studies identifies brain areas of interest for future studies and assesses them in terms of perception–action cycles (what I have called, in this chapter, action–perception cycles, but since it is a cycle, the order does not matter). To date, most research on song has focused on the integration of linguistic and musical elements, whether in the binding together of pitch and syllables or, at a more temporally extended level, in the binding of melodies and lyrics, often with an eye toward the question of whether melodic information facilitates retention of linguistic information, particularly in individuals who have suffered a neurological insult. The evidence supports a view that merging novel linguistic and melodic information is an effortful process, one that is dependent on the context in which the information is associated, with social context aiding retention. Turning from song (music and language) to dance (music and bodily action), Janata and Parsons chart the interacting network of brain areas active during spatially patterned, bipedal, rhythmic movements that are integrated in dance. The way in which BA 44 is activated in the right hemisphere (left BA 44 is part of Broca's area, classically associated with speech production) across contrasts of various conditions supports a role for this region in both elementary motor sequencing and in dance, during both perception and production. Such findings may support the hypothesis that Broca's area and its right homologue may support supralinguistic sequencing and syntax operations (though this does not preclude differential patterns of involvement of subregions as the task varies).

Chapter 14: Sharing and Nonsharing of Brain Resources
for Language and Music (Aniruddh D. Patel)

This chapter offers a framework for the study of language–music relations in the brain which can address the fact that there are striking dissociations between brain mechanisms supporting language and music as well as evidence for similar processing mechanisms. Patel proposes three distinct ways in which language and music can be dissociated by neurological abnormalities, yet have closely related cortical processing mechanisms. He proposes that this relationship can occur when the two domains use a related functional computation, and this computation relies on either: (a) the same brain network, but one domain is much more robust to impairments in this network than the other; (b) the interaction of shared brain networks with distinct, domain-specific brain networks; or (c) separate but anatomically homologous brain networks in opposite cerebral hemispheres. These proposals are used to explore relations between language and music in the processing of relative pitch, syntactic structure (but recall the earlier caveat on the need to distinguish language syntax and music "syntax"), and word articulation in speech versus song, respectively.

Chapter 15: Action, Language, and Music: Events in Time and Models of the Brain (Michael A. Arbib, Paul F. M. J. Verschure, and Uwe Seifert)

Many accounts linking music and language to the brain represent the brain simply as a network of "boxes," roughly corresponding to brain regions, each of which has an active role in providing a specific resource. Once we move away from the auditory periphery, there are even fewer models that offer finer-grain explanations of the underlying circuitry that supports these resources and their interaction. This chapter offers a bridge to future research by presenting a tutorial on a number of models which link brain regions to the underlying networks of neurons in the brain, paying special attention to processes which support the organization of events in time, though emphasizing more the sequential ordering of events than the organization of sequential order within a hierarchical framework. Its tour of models of the individual brain is complemented by a brief discussion of the role of brains in social interactions. Models are offered for the integration of cerebral activity with that in other brain regions such as cerebellum, hippocampus, and basal ganglia. Throughout, implications for future studies linking music and language to the brain are discussed. Particular emphasis is given to the fact that the brain is a learning machine continually reshaped by experience.

Chapter 16: Computational Modeling of Mind and Music (Paul F. M. J. Verschure and Jônatas Manzolli)

This chapter assesses how the broadening of concepts of music in recent years might assist us in understanding the brain, and vice versa. Particular attention is paid to non-Western musical traditions of group music-making and to computer-mediated music. Computational models implemented in robots provide new ways of studying embodied cognition, including a role in grounding both language and music. In particular, models of perception, cognition, and action have been linked to music composition and perception in the context of *new media art* where an *interactive environment* functions as a kind of "laboratory" for computational models of cognitive processing and interactive behavior. The distributed adaptive control architecture is then introduced as an integrated framework for modeling the brain's neural computations, providing the integrated real-time, real-world means to model perception, cognition, and action. Verschure and Manzolli conclude with an overview of RoBoser, Ada, and XIM, which provide alternatives to the formalistic paradigm of music composition by having the musical output of a computational system sonify the dynamics of complex autonomous real-world interactions.

Chapter 17: The Neurobiology of Language, Speech, and Music (Jonathan Fritz, David Poeppel, Laurel Trainor, Gottfried Schlaug, Aniruddh D. Patel, Isabelle Peretz, Josef P. Rauschecker, John Halle, Francesca Stregapede, and Lawrence M. Parsons)

To clarify the domain-specific representations in language and music and the common domain-general operations or computations, it is essential to understand the neural foundations of language and music, including the neurobiological and computational "primitives" which form the basis for both at perceptual, cognitive, and production levels. This group report summarizes the current state of knowledge, exploring results from recent studies of input codes, learning and development, brain injury, and plasticity as well as the interactions between perception, action, and prediction. These differing perspectives offer insights into language and music as auditory structures and point to underlying common and distinct mechanisms as well as future research challenges. The discussions summarized in this report centered on two issues: First, how can we be precise and explicit about how to relate and compare the study of language and music in a neurobiological context? Second, what are the neural foundations that form the basis for these two dimensions of human experience, and how does the neural infrastructure constrain ideas about the complex relation between language and music? For example, to what extent is neural circuitry shared between music and language and to what extent are different circuits involved? The section on neurobiological constraints and mechanisms focuses explicitly on some of the neural mechanisms that are either reasonably well understood or under consideration, distinguishing between data pointing to domain-specific neural correlates versus domain-general neural correlates.

Part 5: Development, Evolution, and Culture

As stressed in Part 1, languages and musics vary from culture to culture. How, then, does a child become attuned to the music and language of a particular culture (the developmental question)? How did brains evolve so that the human brain is language- and music-ready (the question of biological evolution)? In particular, what "ingredients" of language and music are in some sense prespecified in the genome and which reflect historical patterns of sociocultural change (the question of cultural evolution)? This last section of the book addresses these questions and offers partial answers. Many of the open challenges concerning the evolutionary processes that underwrite the relationships between language, music and brain are highlighted for future enquiry.

Chapter 18: Communication, Music, and Language in Infancy
(Sandra E. Trehub)

Music is considered here as a mode of communication that has particular resonance for preverbal infants. Infants detect melodic, rhythmic, and expressive nuances in music and in the intonation patterns of speech. They have ample opportunity to use those skills because mothers shower them with melodious sounds, both sung and spoken. Infants are sensitive to distributional information in such input, proceeding readily from culture-general to culture-specific skills. Mothers' goals in child-rearing are well known, but their intuitive didactic agenda is often ignored. Regardless of the amiable and expert tutoring that most infants receive, Trehub observes that their progress from avid consumers of music and speech to zealous producers is remarkable.

Chapter 19: Evolving the Language- and Music-Ready Brain
(Michael A. Arbib and Atsushi Iriki)

This chapter returns to the theme initiated by Chapter 4, asking what can be learned about the human brain by comparing it to the monkey brain and placing human and monkey within an evolutionary framework. It focuses on the evolution of the language-ready brain, offering triadic niche construction as the framework in which to see the interaction between the environmental niche, the cognitive niche, and the neural potential latent in the genome at any stage of evolution. This framework enriches the presentation of the mirror system hypothesis, which traces an evolutionary path from mirror neurons for the recognition of manual actions (introduced in Chapter 4), via systems supporting increasingly complex forms of imitation, to the emergence of pantomime, protosign, and protospeech. This hypothesis is briefly contrasted with the Darwinian musical protolanguage hypothesis, which roots the evolution of language ability in a birdsong-like ability coupled to increasing cognitive complexity. Arbib and Iriki stress the linkage of both language and music to outward bodily expression and social interaction, and conclude with an all-too-brief discussion of the evolution of the music-ready brain.

Chapter 20: Birdsong and Other Animal Models for Human Speech, Song, and Vocal Learning (W. Tecumseh Fitch and Erich D. Jarvis)

Where Chapter 19 addresses the evolution of the language-ready brain by pondering how human and monkey might have evolved from a common ancestor, this chapter focuses on *vocal learning* as a key ingredient of both music and language, and thus asks what we can learn about the human brain by looking at other species which (unlike nonhuman primates) exhibit vocal learning. It focuses on the comparative biology of birdsong and human speech from behavioral, biological, phylogenetic, and mechanistic perspectives. Fitch and Jarvis claim that song-learning birds and humans have evolved similar, although not identical, vocal communication behaviors due to shared "deep homologies" in nonvocal brain pathways and associated genes from which the vocal pathways are derived. The convergent behaviors include complex vocal learning, vocal-learning critical periods, dependence on auditory feedback to develop and maintain learned vocalizations, and rudimentary features for vocal syntax and phonology. They argue that to develop and maintain function of the novel vocal-learning pathways, there were convergent molecular changes on some of the same genes in the evolution of songbird and human, including FoxP2 and axon guidance molecules. The unique parts of the brain pathways that control spoken language in humans and "song" in distantly related song-learning birds are seen to have evolved as specializations of a preexisting system that controls movement and complex motor learning inherited from their (very distant) common ancestor.

Chapter 21: Culture and Evolution (Ian Cross, W. Tecumseh Fitch, Francisco Aboitiz, Atsushi Iriki, Erich D. Jarvis, Jerome Lewis, Katja Liebal, Bjorn Merker, Dietrich Stout, and Sandra E. Trehub)

This final group report explores the relationships between language and music in evolutionary and cultural context. Language and music are characterized pragmatically in terms of features that appear to distinguish them (such as language's compositional propositionality as opposed to music's foregrounding of isochronicity), and those that they evidently share. The chapter considers those factors that constitute proximate motivations for humans to communicate through language and music, ranging from language's practical value in the organization of collective behavior to music's significant role in eliciting and managing prosocial attitudes. It then reviews possible distal motivations for music and language, in terms of the potentially adaptive functions of human communication systems. The chapter assesses what advantages might accrue to flexible communicators in the light of ethological and archaeological evidence concerning the landscape of selection. Subsequently, the possible evolutionary relationships between music and language are explored, evaluating six possible models of their emergence. The roles of culture and of biology in the evolution of communication systems are discussed within the framework of triadic niche construction, and the chapter concludes by surveying comparative and phylogenetic issues that might inform further research.

A Passion for an Integrative Perspective

> [You] may be very surprised to hear that the symphony is not only music, but that it always tells a story. Which has a beginning, a middle, and an end. Except for the unfinished symphony, which has a beginning. [Long pause] As I was saying…(Danny Kaye, *The Little Fiddle*)

Like a symphony, this book tells a story. This story has a beginning (the chapters which survey different domains of current research bridging between two or three of our themes of music, language and the brain), a middle (the chapters which each report on a group discussion building on that background). [Long pause.] The discussions offered new bridges between our themes and opened up novel directions for further research (though, alas, some discussants were reluctant to see their livelier ideas committed to paper). These new directions have no obvious end in sight but will continue to challenge and enlighten us for many years to come. It is the task of this chapter to encourage each reader not just to read individual chapters in isolation, but to look actively for connections between chapters—whether to discover that a question raised by one chapter is indeed answered by another, or to formulate a new question posed by the juxtaposition of ideas from these and other chapters. To this end, I recount an outstanding musical experience.

On August 4, 2012, I attended a performance of Tan Dun's *Water Passion after St. Matthew* presented for the La Jolla Music Society.[2] It builds on both the Great Schema of the Bible—as Mary Hesse and I (Arbib and Hesse 1986) called it in homage to Northrop Frye's (1982) charting of the Bible as the Great Code for understanding Western literature—and the Western musical tradition, even though both were foreign to Tan Dun. He was born in 1957, was cut off from Western music and ideas (other than a form of Marxism) by the Cultural Revolution, and gained his spiritual foundation from his grandmother's Buddhism. He first heard Bach's music when he was twenty and was captivated by Bach's organ music and the chorales from the St. Matthew Passion. His own *Water Passion* reflects not only these two Western traditions (Bible and Bach) but also his experience with Peking Opera and other aspects of Chinese music and with Buddhism.

So we are already challenged here by the confluence of four (at least) social schemas: Christianity, Buddhism, music as related to that of Bach, and music as related to the Peking Opera. How does the narrative (as sung by a bass taking the part of John, Jesus, Judas, and Peter; a soprano taking the part of the Devil, Judas, and Peter [again]; and the chorus singing for the crowd, Pilate, and more) emerge from the Gospel according to St. Matthew as reflected through a (partially) Buddhist sensibility? Tan Dun asserted that to develop his

---

[2] What follows is based on an interview of Tan Dun by Cho-Liang Lin prior to the performance, the performance itself (magical), and the program notes written by Are Guzelimian, Senior Director and Artistic Advisor at Carnegie Hall.

narrative he viewed 43 movies based on the story of Jesus. He also read a Bible printed in four languages, including German (his first European language), English, and Chinese and said that reading the translations gave him a more cross-cultural understanding of the story. So how do these multiple influences coalesce into the final libretto? And how does the libretto shape the music, and how does the music shape the libretto?

Tan used the sound of water as an integral element for his *Passion*, inspired by the symbolism of birth, creation, and re-creation, with the water cycle (precipitation and evaporation) as a symbol of resurrection. Besides the interaction of the narrative and the music (extended to include the sound of water), there is a powerful visual image. The sloping stage is defined by a cross that is comprised of seventeen transparent water bowls. To the left and above the cross bar is the female chorus; to the right (as seen from the audience) is the male chorus. At left front is the bass singer and a lone violinist; to the right is the soprano and a cellist. And at each point of the cross (save the one closest to the audience, near where the conductor stands), there is a percussionist. The first words of this Passion are "a sound is heard in water" and here, barely audible, the percussionists scoop water and drop it back into the nearest vessel. As the piece proceeds, each plays a range of percussion instruments, but also plays with the water, and with percussive sounds through the water. Much of the singing and string playing is recognizably "Western," yet the invocation of the sound of water, and the occasional use of Chinese and Tibetan singing styles and of Chinese instruments, and of the chorus as a nonvocal instrument (clicking stones together at one time; shaking thunder sheets at another) adds richness to a composition for just two string players and three percussionists.

At the end, the Resurrection is followed by the classic lines "a time to love, a time of peace, a time to dance, a time of silence" from Ecclesiastes. Then, after the last instrumental and vocal note is sounded, the stage goes dark. The only sound heard is that of water. A performer splashes in each of the seventeen vessels, his or her face illuminated by light shining up from below the water.

What will it take to understand what went on in the mind of the composer that finally coalesced in this *Water Passion*?

What goes on in the minds of the performers as they coordinate their behavior with each other and with the movements of the conductor, following the libretto and a score which allows for, even encourages, some measure of improvisation?

And what goes on in the minds of us, the audience, as our eyes dart back and forth between stage and libretto, and as our attention shifts from performer to performer, and as we assimilate the words of the text, the many sounds of a music that is Western and Oriental, vocal and instrumental, extended by the sound of water and a strong visual structure, each of us with our varying knowledge of the traditions that informed the composition of the piece?

Will our answers to any of these questions be any more satisfying if our talk of "mind" is reinforced by hard data about the "brain"? To return to my guiding metaphor, how can our understanding of diverse processes be orchestrated to enrich our understanding of each whole?

The grand answers to these questions are beyond the reach of current linguistics, musicology, and neuroscience. Science must break large-scale problems down to smaller ones amenable to precise experiments and/or clear theoretical expression or modeling. Too often, we find ourselves focused so intently upon a sub-sub-problem defined by the tradition of a particular research community that we forget to look up from the workbench and assess how our work can be linked with the work of others to begin to chart the overall territory. But being overly focused is no worse than succumbing to a holistic approach which can only address grand themes at the price of losing any chance of understanding the diverse processes whose competition and cooperation make the dynamic adaptability of the whole possible.

My hope is that—like the members of Tan Dun's audience, whose eyes flicked back and forth between page and stage while assimilating the unfolding soundscape—readers of the book will find here the tools to flick back and forth between the reality of personal experience and the insights of well-focused studies to gain an ever fuller understanding of the relationships between language, music and the brain.

2

A Cross-Cultural Perspective on the Significance of Music and Dance to Culture and Society

Insight from BaYaka Pygmies

Jerome Lewis

Abstract

The concepts associated with what English speakers recognize as music and dance are not shared cross-culturally. In some societies there are no general terms for music and dance; instead, specific names describe different performances that involve music and dance. In other societies the same word is used to refer to music-making, singing, dancing, and often to ceremony or ritual as well. Despite such differences, every social group has its music, and this music is somehow emblematic of a group's identity. This chapter explores how this observation can be explained from a cross-cultural perspective: What do music and dance do for human social groups? Why are music and dance so universally central to a group's self-definition?

It is suggested that participation in music and dance activities provides experiences of aesthetic principles which in turn may influence "foundational cultural schemas" affecting multiple cultural domains: from cosmology to architectural style, from hunting and gathering techniques to political organization. Such dance and musical participation inculcates culture not as a text or set of rules, but as a profound aesthetic orientation. Foundational cultural schemas may thus be better understood as aesthetic orientations that influence our everyday decisions and behavior by seducing us to conform to them using our aesthetic sense, enjoyment of harmony, desire to cooperate, curiosity, and pleasure-seeking propensities. Musical foundational schemas may have extraordinary resilience, and this resilience is likely due to their special aesthetic, incorporative, adaptive, and stylistic qualities that ensure continuity with change.

A Cross-Cultural Perspective

> We may say that each social group has its music.—B. Nettl (2000:465)

In isolated places where people do not have widespread contact with other cultures, music may be fairly homogeneous across a particular society, even though it is composed of a range of styles for different occasions. As the result of frequent cultural contact and musical exchange between members of different societies, many social groups today have a range of different types of music which they have come to call their own. The Blackfoot people of North America, for example, say that they have Indian and white music, which Nettl refers to as "bimusicality" (Nettl 2000). His teachers in Persian classical music claimed proficiency in many musical traditions, just as they could speak several foreign languages competently while still regarding them as foreign; this Nettl refers to as "multimusicality" (Nettl 2000). In large multicultural nation states such as the United States of America, different ethnic groups (e.g., Native Americans, Polish Americans, Hispanic Americans, or Italian Americans) use music and dance performances as a key marker of ethnicity. Cutting across such ethnic identities may be other groupings, for example, peer groups, which express themselves through musical affiliations to "rap," "reggae," "hard rock," "country," or "punk" and their associated dance and aesthetic styles. Between generations in the same social group there may also be different types of music: while my parents mostly enjoy "classical," I tend toward "world," but my son enjoys "jazz." We all have our own favorite type of music, which is somehow indicative of the sort of person we are and where we are "at." In this chapter I attempt to unravel how music and dance do this.

There are many kinds of music and dance, and many ways of conceptualizing of them. The concepts associated with what English speakers recognize as music and dance are not shared cross-culturally. In some societies there are no general terms for music and dance, but rather specific names for different performances that involve music and dance. When Japanese researchers first began to analyze dance apart from the specific repertoire to which it belonged, they had to invent a word for "dance" (Ohtani 1991). Seeger (1994) describes how the Suyá of the Amazon forest do not distinguish movement from sound since both are required for a correct performance. A single word *ngere* means to dance and to sing because, as the Suyá say, "They are one." In Papua New Guinea, anthropologists have struggled to talk with their informants about dance independently of music. As in most of the local languages, the lingua franca, *Tok Pisin*, has one term *singsing* that is used interchangeably to refer to singing or dancing, or both. The Blackfoot term *saapup* rolls music, dance, and ceremony into one (Nettl 2000:466).

While this conflation by people in other cultures could be interpreted as lacking sophistication, it actually offers a profound insight into the nature of music. To appreciate why requires us to consider our ethnocentric biases. It helps to begin with language. Those of us who have experienced the training required for literacy tend to prioritize words as the containers of meaning in an act of communication. By contrast, many people with no education in literacy tend to think of words as just one part of the exchange between people engaged in communication. To understand what is being communicated, they, like us when we are in conversation, pay a great deal of attention to gesture, pantomime, and body language as well as to the context of the conversation, the social relationships between speakers as well as their personal and cultural histories. As a result of schooling in reading and writing, we tend to ignore these less easily documented aspects of the "message" and give priority to the words exchanged in our representations of an act of communication.

Some researchers avoid this. Kendon, for example, uses audiovisual recordings of conversations to make microanalyses of the relationship between speech and body movement (Kendon 1972, 1980). He, like others (e.g., McNeill 1985, 1992; Schegloff 1984), have found that speech and gesture are produced together and should therefore be considered as two aspects of a single process: "Speakers combine, as if in a single plan of action, both spoken and gestural expression" (Kendon 1997:111). Kendon further notes that gestures may have a morphology and show, to a limited extent, at least some compositionality (Kendon 1997:123). Furthermore, there is a tendency for the gesture phrase to begin before the spoken phrase to which the gesture is contributing semantic information, thereby indicating their mutual co-construction in the mind of the speaker. While there is historical and cultural variation in the extent to which gesture is cultivated or restrained in conversation, gesture in speech is a human universal.

As sign languages illustrate, the language faculty is multimodal. Speech is but one mode. Language's "ecological niche" (see Levinson, this volume) includes speech and gesture in a face-to-face interaction between two people raising and lowering their voices, anticipating each other's utterances, reading subtle facial expressions and body language, attributing intentions to each other, timing their interjections, and turn-taking. These two people have a history between them that contains cultural and ideological elements central to interpreting the meaning that emerges from their communicative interaction. Our focus on speech abstracted from language's ecological niche is an artifice of writing. Has our ability to write music or to record and listen again to the sounds of a musical performance independently of its production blinkered us to the full context of musical production, and has this led us to focus on the sounds of "music" to the exclusion of other aspects?

The relationship between music and dance parallels that between speech and gesture. Just as speech is composed of linguistic and gestural components, music necessarily includes a gestural component—a rhythmical movement of the body we call "dance," or "percussion," or the "playing" an instrument. Music, like language, is multimodal. Many deaf people, for instance, enjoy dancing by feeling the rhythm in their bodies. Just as there can be language without speech, there can be music without sound. Musical behavior can be expressed through voice or other body movements that range from simple swaying to dancing, or from percussive tapping, stamping, or clapping to the skillful manipulation of purpose-built objects such as drums, flutes, violins, or pianos. Evidence from neuroimaging shows that attentive listening to musical sounds engages, to a certain extent, aspects of the action system in the brain (Brown and Martinez 2007; Grahn and Brett 2007; Janata et al. 2002b). In effect, whenever we attend to music, our bodies prepare to dance. Kubik (1979:228) put it succinctly: "Music is a pattern of sound as well as a pattern of body movement, both in creating this sound and in responding to it in dance."

From an anthropological perspective, musical, like linguistic, meaning emerges from its total context—one that includes the sounds, body movements, and symbols as well as the "who," "where," "why," "when," and "how" of its performance. To understand and appreciate a musical moment, much may be involved: the social relations of the musicians and other participants, the staging of their performance, the choice of venue and songs, the music's tempo and structural characteristics, the atmosphere of the occasion, the emotional entrainment that occurs between participants, the smells, the colors of costumes or decorations, the moves of the dancers, the resonance of symbolic connections made to myth, religious ideology, environment or ordinary life, and so on. This wealth of information is nonetheless absent in the musical notation that represents the music being performed and is only partially represented in audio or film recordings.

In most parts of the world, and for most of human history, music exists only because of the social relations that enable its performance. Recorded and written music, in conjunction with increased musical specialization in our own society, has made the idea of musical appreciation being separate from its performance seem normal to European or American scientists. From a cross-cultural and historical perspective, this is an anomaly. Extracting "music" from the social context of performance is to miss the point of music. As Levinson (this volume) observes, "the motivation for and structural complexity of music may have its origins in joint action rather than in abstract representations or solitary mentation."

Meaning and Function in Music

The most common Western folk theory to account for musical meaning and function emphasizes its role in expressing sentiment and nonverbal ideals. This view continues to underpin many of the theoretical approaches taken in studies of music cross-culturally. Often referred to as "expressionism," it informs both social and cognitive accounts of the relationship between music and language and is based on the presumed distinction between musically encoded feeling and linguistically encoded thought.

Reality is more nuanced. Like language, music is a universal human behavior that combines gestural and sonic elements. Both are multimodal and, as Levinson points out (this volume), expectancy, prosody, and paralanguage in speech and song are bridges between language and music. These connections are exploited in certain communicative styles that mix language and music to capitalize on the range of expressive possibilities offered. For example, formalized political oratory, such as the Maori *haka*, combines speech, chant, gesture, and dance to reinforce the statement; traditional forms of lamentation in many societies mix distinctive gestures, dance, song, and speech in formulaic ways (Feld 1982; Feld and Fox 1994:39–43); storytelling, such as that of BaYaka Pygmies, combine linguistic, mimetic dance, and musical forms into a single communicative event.

Consider the fable *Sumbu a we* (chimpanzee you will die), which is found online in the supplemental information to this volume, Example 1 (http://www.esforum.de/sfr10/lewis.html). To "tell" this story, some voices narrate; others mimic the chimpanzee's part, while still others sing the initiation songs. For an additional example, watch as Mongemba describes an elephant hunt by taking full advantage of a range of expressive modes, which illustrate the importance of both speech and gesture for eliciting meaning and are suggestive of the connection of gesture to dance and speech to song (Example 2, http://www.esforum.de/sfr10/lewis.html).

There are, however, differences. Music tends to *formulaicness* (Richman 2000:304), since preexisting formulae—rhythms, riffs, themes or motifs—are cyclically repeated, often with slight variation or embellishment. Music thus tends to repeat the same utterances over and over, filled more with redundancies than explicit messages. By contrast, language continually produces novel utterances through the recomposition of words and gestures to create new meanings. Where language is based on units with fairly restricted shared meanings, music is constructed from units with multilayered, fluctuating, or no meaning. While both combine implicit embodied meanings (dance and gesture) and explicit sung or spoken meanings, music tends to prioritize the implicit and nonverbal, whereas language prioritizes the explicit and the verbal.

In musical contexts, extracting meaning from what is predominantly nonverbal presents a methodological problem. Using words to discuss a sequence of mostly nonverbal sounds and actions is challenging since the meanings contained within are performed nonverbally precisely because they are most effectively transmitted in this way. Music and dance "generate certain kinds of social experience that can be had in no other way….Perhaps, like Levi-Strauss's 'mythical thought,' they can be regarded as primary modeling systems for the organization of social life…" (Blacking 1985:65).

Feld and Fox (1994:35) typify some of these social organizational functions provided by music "as an emblem of social identity…, as a medium for socialization…, as a site of material and ideological production…, as a model for social understandings and evocations of place and history…, as a modality for the construction and critique of gender and class relations..., and as an idiom for metaphysical experience." Other functions could be added such as group communication, individual and group display, sexual selection, keeping dangerous wild animals away, infant and child socialization and learning, a framing for ritual, or a means to mark episodes or changes of status in ceremonies, or the suspension of normal social behaviors as in carnival or spirit possession. Music can also transmit meaning propositionally. For example, it can greet or mark arrivals or departures, deaths, births, and other events, or it can signify social status or announce changes such as a new king or a newly married couple.

The structures, practices, and meanings of music are culturally determined and thus the meaning, function, or significance of particular music can only be understood in relation to its structural properties and specific cultural context. While an exposition of these properties and context may permit the inference of meanings and functions to music and dance, as with any discussion of nonverbal communication, it can rarely be complete, definitive, or certain. The descriptions and discussion that follow must therefore be understood as approximations.

The Stick Dance of Bhaktapur

Early functionalist understanding of music in anthropology was dominated by Radcliffe-Brown's theory, developed in his ethnography on the Andaman Islanders. He argued that an orderly social existence requires the transmission and maintenance of culturally desirable sentiments. Each generation is inculcated with these sentiments, which are revitalized in adults through participation in music and dance (Radcliffe-Brown 1922:233–234). Radcliffe-Brown's emphasis on the importance of "sentiment" echoed earlier views put forward by Spencer in his article on the "Origins and Function of Dance" (Spencer 1857). Spencer suggested that, in addition to the verbal understandings and representations of the ideals of a society, the highest ideals of a society are nonverbal and that their expression is the basis of the nonverbal arts. If this is so, then how does the repeated experience of music and dance lead people to experience desirable sentiments and "the highest ideals" of a society?

Based on work by Maurice Bloch, the musicologist Richard Widdess (2012) perceptively connected insights from cognitive anthropology to ethnomusicology. Bloch (1998) argued that culture is composed of bodies of expert knowledge and associated skills structured similarly to other knowledge-skill complexes, such as driving a car. Such expertise is acquired, stored, and recovered in mainly nonlinguistic ways to be used efficiently. Drivers, for instance, can chat to their passengers while remaining in full control of the car. Here, "nonlinguistic" means that such expert knowledge is not formulated in natural language and not governed by rules of linear succession characteristic of linguistic grammars (Bloch 1998:10–11). Instead, the underlying nature of cultural knowledge is more akin to the notion of "schema" or "model" in cognitive psychology and should thus be termed "cultural models." Bloch explained that the Merina of Madagascar are constantly evaluating (as he now also evaluates) whether an area of forest will make a good swidden field for agriculture. This complex series of appraisals concerning vegetation types, hydrology, slope, landscape, soil, and so on takes just a few seconds. This is an example of expertly applying a cultural "ideal model."

Bloch questions the reliance of anthropologists on the verbal explanations of their informants, because such statements are a transformation of nonlinguistic cultural models into linguistic form. Since the underlying structures of culture may be beyond the capacity of its bearers to formulate linguistically, Bloch suggests that what researchers are actually doing is selecting statements that correspond to aspects of these cultural models in approximate ways. They are able to do this because they have internalized those models not so much by asking questions as by participant observation. Anthropology's characteristically long field research (18–24 months) provides the investigator with the opportunity of becoming an expert in the aspects of the new culture they are studying.

Widdess recognized a similar process that occurs during ethnomusicological research. Through learning to sing or play music—a complex skill normally transmitted by observation and practice, not language, and by taking part in musical performance—the ethnomusicologist becomes aware of the range of meanings that music elicits for people in the society concerned: "These can be located in relation to culture-specific concepts, functions, social and political dynamics and historical trajectories of music, as well as in embodied experience and metaphorical accounts of it" (Widdess 2012:88–89).

To illustrate this, Widdess uses his analysis of the meanings contained within the stick dance performed annually in the Nepalese town of Bhaktapur (Widdess 2012). He demonstrates, in particular, how the very structure of the music is related to culturally contextual meanings. The music during the procession is composed of two sections: A and B, which are cyclically repeated, plus a short invocation at the start, during section B, or at the end of the procession. Section A is typified by a slow 8-beat meter, whereas B has a fast 6-beat meter. Widdess maps these structural features of the music onto the following: The drum rhythm at the beginning of A echoes a seasonal song whose words express the affective meanings of the procession. The slow beat organizes a walking dance through the narrow streets with dancers clashing sticks every 7th beat. Once the procession of dancers reaches a square, crossroad, or open space, the music changes from A to the fast 6-beat of B, giving the dancers the opportunity to exhibit their energy and skill to onlookers. The invocation piece is played at the start and end of the procession, and in front of every temple passed.

The circularity of the music (from A to B to A to B…interspersed with invocations) mirrors the circularity of the processional route around the town.

This, in turn, is suggestive of temple worship practices being applied to the town itself and of Hindu-Buddhist cosmology of reincarnation and rebirth. Other aspects of the music's structure, particularly its elements of structural compression, can be seen as the sonic equivalent of local architectural temple and fountain styles, reflecting a particular concept of space as mandala-like concentric rings, where the most powerful divinities are compressed into the smallest central regions. The music and dance exploit this aspect of divine power by employing similar compression and intensification in the music's movement between the slower, but moving Section A, and the fast, but stationary Section B. In the chanted invocations, the dance is explicitly dedicated to the invocation of divine power.

The isomorphism between musical, material, visual, and conceptual patterns of meanings in the music described by Widdess were never verbally expressed to him. But, according to cognitive anthropological theory, it would be surprising if they were. Widdess argues that such flexibility in meaning in music is characteristic of what cognitive anthropologists, such as Bloch (1998) or Shore (1996), call "foundational cultural schemas"—cultural models that cross the boundaries of cultural and sensory domains, rather than focus specifically on one thing—such as what makes a good swidden.

Musical performances involve a huge range of potential meanings and functions: from the sound and structure of the music itself, to the social and political relationships it establishes among performers, to the way it refracts culture-specific concepts, history, or identity. As such, musical styles are promising candidates to illuminate cultural analyses since "the highly specialized, schematic structures of music, and their realization through performance in context...offer fertile ground for the discovery of cross-domain, nonlinguistic cultural models and cultural meanings" (Widdess 2012:94). The politically egalitarian BaYaka hunter-gatherers, whom I have studied in Northern Congo, offer an example of how such foundational cultural schemas can be uncovered through an analysis of musical activity, further demonstrating how music extends well beyond the realm of sound.

A Central African BaYaka Pygmy Hunter-Gatherer Perspective

When the BaYaka[1] discuss the extent to which other Pygmy groups are "real" forest people, they often focus on the extent of their skill in performing ritual. For example, in 2006 when I played some 50-year-old recordings of Mbuti music, made by Colin Turnbull in the 1950s on the eastern border of the dense

---

[1] They are also referred to as Mbendjele and number some 15,000–20,000 individuals occupying around five million hectares of remote forest in Northern Congo and the border area of Central African Republic. Since it is easier for English speakers to pronounce and remember "BaYaka," I use this more encompassing term. Mbendjele use it to refer to all Central African hunter-gatherers in a similar way to the academic term "Pygmy."

forest covering the Congo Basin, to BaYaka friends over a thousand miles to the west, they immediately exclaimed: "They must be BaYaka since they sing just like us!" To grasp this requires a better understanding of what music means to BaYaka people and their culture, as well as what they do with it.

To the BaYaka, music is potent and productive; it has power. When the BaYaka set out to net-hunt, for example, women alternate a sung vowel with a blow on a single note flute to enchant the forest. They explain that this makes the animals feel *kwaana*—soft, relaxed, and tired—so that they may be more easily caught in the nets. Before a planned elephant hunt, women sing *Yele* late into the night. Extended mesmeric singing and dance styles are combined with a secret drink to facilitate certain women to enter a trance. While in *Yele* trance, these women say that their spirits travel over the forest to locate elephants, and that they "tie the elephants' spirits down" so that they can be later killed by the men. In the morning, the women tell the men where to go to find the elephants that they have tied up. The general principle implied is that music and dance enchant sentient beings, making them relaxed, happy, and open. In the case of animals, this makes them easier to kill; in the case of people, music makes them more willing to give up things when asked. During large group ritual performances, this principle is used to acquire things from other people within the group as well as from outsiders, such as local farmers.

Such rituals are a regular feature of camp life and are called *mokondi massana*, literally "spirit play." During a *mokondi massana*, people, and then spirits, dance to complex interweaving vocal melodies interlocked into a dense yodeled and hocketed[2] polyphony that overlaps with a percussive polyrhythm made by clapping and drumming. To attract forest spirits (*mokondi*) out of the forest to play and dance with the human group, this music must be beautifully performed. Although there are many other contexts in which people make music, spirit plays are the most appreciated and valued musical event of the BaYaka. Their neighbors share this appreciation and consider the BaYaka to be the most accomplished musicians in the region. In fact, the BaYaka perform the major life-cycle rituals for their neighbors in return for copious alcohol, "smoke," and food.

The BaYaka have an egalitarian social organization of the type described by Woodburn as "immediate return" (Woodburn 1982). In a society where it is rude to ask questions (not easy for a researcher), rude to tell someone else what to do (men cannot order their wives, parents cannot order their children), and there are no social statuses that carry authority, it is often difficult to understand how anything gets done. Yet somehow, day after day, the camp spontaneously organizes itself to find sufficient food without an elder or leader directing people to act. People organize themselves sensitively in relation to what others announce they are doing, so that their actions are complementary. This

---

[2] Yodeling is a singing style that alternates between a chest and a head voice. Hocket is a technique in which singers sing alternately to complete a single melody.

apparent spontaneity forced me to look obliquely for clues and be sensitive to ways in which BaYaka organize themselves and transmit knowledge, without giving special status or authority to any particular individual. Observing the inability of egalitarian societies to judge new innovations, Brunton (1989) provocatively suggested that such societies are inherently unstable, their practices haphazard or accidental assemblages, and their continued existence fortuitous. Yet given the long survival of such societies in the ethnographic record, something else is clearly going on.

Elsewhere (Lewis 2008), I have examined how a taboo complex called *ekila*, based on the separation of different kinds of blood (menstrual blood and the blood of killing animals), serves to inculcate specific gendered roles and ideological orientations without reference to authority figures. My analysis of BaYaka music builds on this work and demonstrates another key way that the BaYaka learn the organizational principles of their society. These principles are rarely made explicit, yet consistently—across communities in widely different areas and even speaking different languages (Baka, BaAka, BaYaka, Mikaya, Mbuti, Efe)—I have observed that many of the same organizational practices are based on *ekila*-like taboos (concerning different types of blood) as well as the common cultural institution of spirit plays (the primary site for the interlocked polyphonic singing style).

BaYaka Musical Socialization

To understand why a musical education can be a cultural one requires ethnography. This musical-*cum*-cultural inculcation begins before birth. As of 24 weeks, a normally developing fetus hears the world around its mother. Just as the pregnant mother regularly sings as she goes about her daily activities or when she immerses herself in the group of women singing these intertwining melodies late into the night, so too does her unborn child (Montermurro 1996). If the endorphins that this experience produces in the mother are shared with her fetus, as Verney and Weintraub (2002:63, 159) claim, powerful associations between the sounds heard and pleasure are established *in utero*. This prenatal acoustic and emotional reinforcement would be very effective at inculcating both the desire to participate in singing and the development of a knowledge base for later use.

Regular immersion in the rhythm and melodies of BaYaka polyphony continues after birth as the baby is sung lullabies, or dances along on the mother's back, or sits in her lap when the women sing together in a tight group of intertwined bodies as the forest spirits are enticed into camp. During performances, mothers often "dance" small babies by exploiting their standing reflex long before they can walk. The baby's motor development for dancing is encouraged together with its rhythmic and vocal development. Any infant or small child that makes an attempt at musical performance is immediately praised and encouraged to continue regardless of the quality of their performance. Women's daily activities are often musically coordinated, from harmonizing water baling when dam fishing, to the distinctive *Yele* yodels that are sung as women move around the forest looking for food (Example 3, http://www.esforum.de/sfr10/lewis.html). As such, these activities provide frequent opportunities for children to engage in musical play.

Whenever babies or infants cry excessively, their caregiver begins yodeling louder than the baby and often firmly pats a percussive rhythm on their back. This action is surprisingly effective in quieting even the most distraught baby and reinforces the association of the melodies with comfort and homeliness. The frequency with which I have observed babies and infants experiencing this intense musical involvement—literally having these melodies and rhythms drummed into the prelinguistic body—suggests that it might be an important element of musical development. It seems to institute a process that ensures the development of fine musical skills and a keen sense of rhythm necessary for later participation in this sophisticated singing style.

Such implicit learning is tested as soon as infants begin to walk and participate more independently in music making. Sitting next to its mother or further away with other children, an infant begins to fine-tune its listening skills as it mimics what it hears. In this manner, children progressively acquire the repertoire of formulas that must be used to participate appropriately in the polyphony in the absence of explicit instruction. This imitation is actively encouraged with praise and so the infant is further stimulated to participate. Explicit intergenerational teaching is rare, though it does happen. Instead, peer group imitation is the major avenue for the transmission of key skills.

While there is no general word for music in BaYaka, *massana* encompasses what we would recognize as musical activities, but it also refers to any type of cooperative, playful activity. Ritual song and dance styles are generically referred to as *eboka*, each with a specific name, and BaYaka differentiate between the verbs to sing (*bo.yemba*), to dance (*bo.bina*), and to play/do ritual (*bo.sane*). *Massana* includes any activity that involves groups of children cooperating to have fun and can range from casual play to structured role-play games, to spirit-play (*mokondi massana*) ritual performances. During *Massana*, the children (or accompanying adults) summon mysterious forest spirits into camp to bless them with joy, laughter, food, and health (for further information, see Lewis 2002:124–195). *Massana* extends the social nexus of music and dance to one that encompasses cooperation, play, mime, speech, and ritual.

One of the most important venues for BaYaka children to learn ritual and musical interaction is during the performance of the children's spirit play called *Bolu* (Lewis 2002:132–136). Bolu leads directly into adult spirit play. It is like a prototype, containing all the basic elements of adult spirit plays, including its own forest spirit (Bolu) and secret area (*njaŋga*) to which the spirit is called from the forest by the initiates; in this case, boys between the ages of three to eight years old. *Bolu*'s secret area creates a space for sharing secrets, which cultivates the same-sex solidarity so central to BaYaka culture and social organization. Meanwhile, similarly aged girls dance up and down the camp singing *Bolu* songs.

A successful performance requires boys and girls, as separate groups, to cooperate and coordinate in doing different but complementary tasks. The singing and dancing is built up until the leafy, cloth-covered spirit, called *Bolu*, is attracted into camp. The dancing and singing boys must then ensure that the girls do not dance too close to the *Bolu* spirit. Keeping *Bolu* in camp makes people happy, and this keeps the forest open and generous so that food will come.

Although not explicitly stated, the basic structure of spirit plays involving both sexes (a minority are gender exclusive) mirrors the gendered division of labor, thus reinforcing the principle that a life of plenty is best achieved through the successful combination of gendered differences and gendered production. Men call the spirit out of the forest to the secret *njaŋga* area and prepare it to dance. Women entice it out of the secret area and into the human space by their beautiful singing and seductive dancing, thus enabling all to enjoy the euphoria that the spirit brings. This gendered pattern of interaction resonates with gendered productive activities in diverse domains: from making children to eating dinner (for a more detailed account, see Lewis 2008). Men say they must repeatedly deposit semen in a woman's womb for her to make it into a beautiful baby, which she then returns after birth to the man and his clan, who give it a name. Men take raw meat from dangerous forest animals and it is cooked by women in order for it to be tasty and safely consumed to sustain the camp. The principle seems to be that men bring things from the outside to the inside; once inside, women transform the thing by making it beautiful and safe for all.

Acquiring competence in the BaYaka musical style simultaneously provides the small children the context for developing competency in a particular style of gendered coordination. As Blacking (1985:64–65) astutely observed: "Movement, dance, music and ritual can usefully be treated as modes of communication on a continuum from the non-verbal to the verbal. All four modes can express ideas that belong to other spheres of human activity: social, political, economic, religious and so on." Spirit plays are perhaps the most important cultural institution of the BaYaka, since their performance leads to familiarity and competence in so many other domains of activity.

Mokondi Massana: Spirit Plays

The performance of spirit plays forms BaYaka persons in very particular ways, most explicitly during the initiation ceremonies into the secret society responsible for each of the spirit plays.³ Each has its sacred path, secret lore, and defined group of initiates responsible for preparing the spirit play and calling the spirit out of the forest. In these secret societies, hidden knowledge is shared: among women, this involves catching the spirits of game animals so men can kill them, using "sexiness" to control and manage men, and maintaining fertility, childbirth, and healthy child-rearing; for men, this concerns hunting, honey collecting, traveling in the forest (night-walking, high-speed displacement, invisibility, etc), and making themselves "awesome" (impressive, handsome, and fearsome). Only in a musical context will different groups communicate their qualities, claims, and issues explicitly. All these point to important ways in which participation in different spirit plays forms BaYaka persons (Lewis 2002:124–195 provides more detail). To support my analysis of music as a foundational cultural schema, let us examine some of the underlying principles.

BaYaka are explicit about the importance of performing spirit plays and will encourage their performance if a few days have passed without one. After announcing to the camp that such-and-such spirit play should be danced, people are called by the initiates to assemble together in the middle of camp and "mix themselves together" (*bosanganye njo*) both physically, by laying legs and arms over each other, and acoustically, by interlocking their different sung vowel-sound melodies. Arom (1978:24) refers to this as "pure" music since the songs rarely have words. To get an idea of this style, a video of two young women singing *Maolbe* is provided (Example 4, http://www.esforum.de/sfr10/lewis.html). Sometimes a phrase will be called out by whoever starts the song, but then the singing proceeds without words. Sometimes several different spirit plays are performed on the same day and, if there are enough young people in camp, they may be performed every evening.

From time to time during the dense polyphony of spirit play, some participants (male or female) stand up to clown and dance. Often BaYaka will criticize singers who are not singing energetically enough or those who sit apart from others or who are chatting or sleeping. When things are going just right, they might shout "Great joy of joys!" (*bisengo!*), "Just like that!" (*to bona!*), "Again! Again!" (*bodi! bodi!*), "Take it away!" (*tomba!*), or "Sing! Dance!" (*pia massana!*).

Established spirit plays have special, mostly secret, vocabularies for congratulating moments of fine performance. There is much creativity and variation in the details of each spirit play, concerning who is eligible to join, the secret lore, the appearance and dance of the forest spirit, the songs, rhythms, and dance steps of participants. Structurally, however, spirit plays resemble one another: membership is through initiation (*bo.gwie*) to a sacred path (*njanga*)

---

³ In my research area there are over 20 different spirit plays. Tsuru (1998) counted more than 50 different spirit plays (called *me*) among Baka Pygmies along a 200 km stretch of road in Cameroon.

where a forest spirit (*mokondi*) is called and its blessing and secret knowledge shared in exchange for polyphonic interlocked hocketed singing and dancing (*massana*).

The characteristics of this ritual system are shared across a range of Pygmy groups that speak different languages and are dispersed over Western Central Africa: the Baka and BaGyeli in Gabon and Cameroon; the BaAka in Central African Republic and Northern Congo; and the BaYaka, Luma, and Mikaya in Northern Congo. These groups form an international network of certain spirit plays across the region.

During these rituals different groups form to animate, organize, and perform the spirit plays. These groups can be comprised of the children from the camp, who sometimes use their songs to claim things (mostly desirable foods) from adults in the camp, or the men as they express their solidarity to the women, or the women to the men. Other performances involve establishing communication between the camp and game animals, or the camp and the forest as a sentient being. Like other animists, BaYaka society includes the forest and animals around them.

Spirit plays structure the wider society by ensuring that small camps dispersed throughout the forest come together to form larger communities from time to time. This aggregation and dispersal of people is organized and motivated by the social opportunities afforded by performing spirit plays. From the smallest social unit, spirit plays regularly bring camp members together. Once in a while they draw neighboring camps together for a special event, such as to celebrate an elephant kill. In the dry season, commemoration ceremonies (*eboka*) bring people together in greater numbers than any other event. These *eboka* are the most important social events of the year: marriages are arranged, news from across the forest is exchanged, old friends meet, and so do old enemies. How, and in which spirit plays, you participate defines your age and gender, as well as the specialist skills you may have, such as animal spirit catcher or elephant hunter. Only during spirit plays (and particularly in their sacred areas) do BaYaka publicly offer each other advice or elaborate on the particular qualities and strengths of the group brought together by the forest spirit.

BaYaka songs often begin with a phrase or sentence to indicate which repertoire of melodies can be used, but then proceed entirely based on hocketed vowel sounds. There is an initial message followed by an embodied message. During the women-only spirit play of *Ngoku*, the united body of the singing women dances arm-in-arm up and down the central area of camp. As they begin a new song, whoever stopped the last song sings out a line—such as, "you are all our children!," "let's fuck!," "we like young men!" or "the vagina always wins, the penis is already tired!"—to tell the other women which melodies to sing. Asserting themselves to their husbands individually in this way could be misunderstood, but as a united group of beautiful, sexy, but unavailable women they speak as "Woman" to the men (Finnegan 2009 expands on this theme). These rude songs do embarrass men and are a key way in which women demonstrate and impose their power in relation to men. Men, on the other hand, speak as "Man" to the women during spirit plays, such as *Sho* or *Ejengi*, by emphasizing brawn—male dances are strong, mysterious, and awesome. As they stamp up and down the camp, bound together as one, they frighten but also attract, making themselves desired but respected.[4] This process of assertion and counter-assertion is central in maintaining egalitarian relations between the gender groups.

In these spirit plays, different groups in society are able to define and express themselves as a group to the rest of society. Individuals passing through these institutions explore what these identities mean as they move through life. By singing as one, no individual can be held responsible for what is sung. A large group of people trying to speak as one tends to produce "a speaker"; otherwise what they say is difficult to hear. By singing as one, the corporate body speaks and is understood. BaYaka take full advantage of the possibilities for group communication that musical performance affords.

BaYaka explicitly use spirit plays to enchant those who witness them. They say that the beauty of it makes an onlooker "go soft." Sharing sound with the forest establishes a relationship of care and concern between the human group and the forest. Since persons who care for each other share on demand, sharing song with the forest legitimates any demands people make, so that the forest can be expected to share its bounty (e.g., pigs or elephants) with people. Such singing is not considered as spiritual but as instrumental—like a hunting technique. Similarly, music is used by the Mbendjele to enchant and make their Bilo farmer neighbors generous. All the Bilo's key ceremonies are conducted for them by the Mbendjele, who extract huge amounts of goods for doing so. A full description would be lengthy. The key point is that BaYaka use music to establish communication between groups across ethnic and species boundaries, as well as within their own society.

The implicit principle is that when many people speak at once, their message is incoherent and the language may not even be understood. If, however, many sing together, their message is reinforced. In speech, one body communicates; in music, many bodies can do so. Spirit plays happen often, but the experience is quite different to, for example, the listening of music on your stereo player at home after work. Spirit plays involve energetic, intense, full-bodied participation (Figure 2.1) that requires you to contribute as best you can, and in distinctive ways that relate to both the spirit play being performed and the music's structure.

---

[4] Watching football fans chanting in unison, or soldiers singing as they march, activates a similar principle. I know this is well appreciated in conflict situations as reflected in popular stereotypes of "war dances." The key point is that forms of group dance are very much about communicating as groups not individuals.

Figure 2.1 After fourteen hours of performance these young women are still going strong. They use their hands to clap out polyrhythms that accompany their hocketed and yodeled polyphony to attract forest spirits into camp. Pembe, Congo-Brazzaville 2010. Photo by Jerome Lewis.

The Role of Musical Structure in Inculcating Culture

Ethnomusicologist Simha Arom (1978, 1985) analyzed this distinctive and complexly organized style to show that its structure is based on repeated interlocked "melodic modules." When listening to the wealth of sound and melody this style produces, it is easy to think that each voice sings randomly, but a sophisticated underlying musical organization constrains and directs innovation and creativity. Each participant's life-long musical apprenticeship has ensured that this musical deep structure is so effectively inculcated that each singer knows how variations can be executed and when to integrate them into the song.

More recently, Kisliuk (2001) built on this to emphasize how creative BaYaka music is despite this rigorous organization. She describes how BaAka Pygmies in the Lobaye forest (Central African Republic) use musical performance as a way to explore modernity, by adopting missionary songs and other music. Over time, Kisliuk notes (2001:188) that new songs, such as hymns, are transformed by "elaborating on a theme until eventually it is engulfed in a flurry of kaleidoscopic improvisations, countermelodies, and elaborations," effectively becoming increasingly BaAka in style. This constant embellishment, variation, and recombination of the "melodic modules" occurs within their own music as well, creating huge potential for variation each time a song is performed and leading to the creation of new musical repertoires and the extension of existing ones. Kisliuk refers to this underlying pattern as a distinctive BaAka "socio-aesthetic" that encourages people to engage with new environmental stimuli in a dialogic way. Through the performance process, Pygmies "colonize" the new, first exploring it in its own terms, then successively incorporating it or discarding it.

What is fascinating is that the music's deep structure enables, even encourages, great variation and creativity in its surface manifestations—the performed spirit play or song being sung—while respecting a coherent deep pattern that remains mostly below the surface. In this sense it manages to be conservative, yet hugely creative and innovative. This freedom within constraint enables each individual to interpret the deep structure according to their current predispositions, experience, and needs. It is not a rigid or dogmatic imposition but an aesthetic orientation that drives sound into increasing complexity in a uniquely Pygmy way. I am not attributing causality to either the individual or the musical deep structure, but rather to the interaction of the two in particular life circumstances. Music does not dictate cultural orientations, but rather familiarizes participants with these culturally specific ways of organizing themselves, shows them to be effective, and then leaves it up to the individual and group to make them relevant to the current moment, or not. With these caveats in mind I will illustrate how singing an interlocked, hocketed polyphony has certain phenomenological consequences on people who do so.[5]

Participating appropriately in a song composed of different parts sung by different people simultaneously involves musical, political, psychological, and economic training. Anyone can start or stop a song, though there are particular conventions to follow. There is no hierarchy among singers, no authority organizing participation; all must be present and give of their best. All must share whatever they have. Each singer must harmonize with others but avoid singing the same melody; if too many sing the same part, the polyphony dissolves. Thus each singer has to hold their own and resist being entrained into the melodies being sung around them. Learning to do this when singing cultivates a particular sense of personal autonomy: one that is not selfish or self-obsessed, but is keenly aware of what others are doing and seeks to complement this by doing something different.

---

[5] Interestingly, Blacking provides evidence from the Venda that the *Tshikona* polyphonic "national song," which is sung by all, played on 20 pipes, and accompanied by four drummers, "is valuable and beautiful to the Venda, not only because of the quantity of people and tones involved, but because of the quality of the relationships that must be established between people and tones whenever it is performed...[*Tshikona* creates] a situation that generates the highest degree of individuality in the largest possible community of individuals. *Tshikona* provides the best of all possible worlds, and the Venda are fully aware of its value...of all shared experiences in Venda society, a performance of *Tshikona* is said to be the most highly valued" (Blacking 1973/2000:51).

Musical skill could be understood as priming participants to culturally appropriate ways of interacting with others, so that the choices each makes do not need explicit justification, since they are instinctive, based on an aesthetic feeling of what one ought to do. This aesthetic sense wills a person to do it, even though there is no force obliging them to do so. This is a key aspect of the unspoken grammar of interaction, which is a central dynamic organizing daily camp life in a society where no one, not even parents to their children, can oblige others to do their will.

Recognizing melodic modules in the music, then deciding where to fit your particular module into the interlocked rhythm is an aesthetic decision that has similarities with the types of decisions people make when hunting and gathering. I have observed how people, as they walk down a forest path, take great pleasure in discussing what they see and what it means. In particular, people remarked on regularly occurring conjunctions of features that indicated a resource to extract from the forest. No one ever said this to me, but being successful in identifying these conjunctions utilizes decision-making skills that are similar to those used when successfully applying a melodic module at the right time in a particular song.

A criss-crossing of narrow animal trails in leafy but relatively open undergrowth, for example, indicates the presence of small, tasty antelope-like animals called duikers. In such an "environmental melody," the melodic module to choose is to squat down and mimic a duiker's call, so that the duikers come out of the undergrowth to within reach of your spear! Such cross-domain similarities between the application of musical knowledge and the application of subsistence knowledge are suggestive. Though the apprenticeship required for each activity is different and leads to the acquisition of different areas of knowledge (musical melodies or hunting strategies), the manner in which this knowledge is deployed in daily decision making has a striking structural resemblance.

These resemblances go further. The musically acquired aesthetic predisposition to sing a melodic line different from your neighbor (if too many sing the same melody the polyphony is lost) makes for efficient hunting and gathering when transformed into an economic aesthetic: do something different from others. If everyone goes hunting in the same area of forest, there is a risk that there would be nothing to eat.

Modes of musical participation are so intimately integrated into everyday life in these Pygmy communities that each person's physical and social development has been profoundly influenced by music. In such an egalitarian cultural context, where explicit teaching is rare, these modes of music and dance participation are one of the major avenues for learning the cultural grammars of interaction. By learning how to join in the song appropriately, each person is also learning how to behave appropriately in a range of other contexts. By regularly repeating this same process during performances over a lifetime, a particular BaYaka way of doing things is repeatedly inculcated, almost subliminally, to each generation without recourse to authority figures.

In summary, musical participation requires the cultivation of special skills that are useful in a range of other domains of cultural activity, such as politics and economics. Indeed, the organizational similarities between activities in these different domains confirm music as a truly central foundational cultural schema, since it is the primary source for propagating this particular Pygmy cultural aesthetic.

This explains why music and ritual are so preoccupying for the BaYaka and other hunter-gatherers, and important when they want to know how like themselves other hunter-gatherers are. In the BaYaka case, they implicitly seem to recognize that performing these rituals and their accompanying musical repertoires has pedagogic, political, economic, social, and cosmological ramifications that serve to reproduce key cultural orientations they consider central to BaYaka personhood and cultural identity. Music and ritual involve an interactive, creative process. The deep structure interacts with the natural/social environment and people's characters/experiences to produce an aesthetic negotiation that manifests as a unique sound and corresponding series of body movements, as well as a particular cultural approach to ritual, politics, and economy. This is why the BaYaka, listening to Mbuti singing their songs over 1000 miles away and in a different language, could immediately hear the structural similarity, and explains what led them to exclaim that the Mbuti must also be BaYaka.

Conclusion

Dance and musical performance offer a privileged window into the structure of foundational cultural schemas[6] and their influence on people's everyday decisions and behavior. They do so by seducing us to conform to them using our aesthetic sense, enjoyment of harmony, desire to cooperate, curiosity, and pleasure-seeking propensities.

Such foundational cultural schemas have the potential to resonate with multiple meanings. This, in turn, enables them to continue to be applicable and useful when things change. Flexibility is crucial for foundational cultural schemas to be relevant over long periods of time, adapting to changing circumstances and new situations, providing guidance but not direction, continuity despite variation, and a means of ordering and making sense out of novelty.

Perhaps the combination of constancy in structure and style with creativity in output offers a partial account of why musically organized foundational

---

[6] The extent to which Pygmies represent a special case of this is unclear. Pygmies are clearly very sophisticated musicians so it is not surprising that culture is so musically influenced. In less musical societies, culture may be less influenced by music. Further research is clearly required.

cultural schemas can be so resilient. If they are to be meaningful for each generation, they must be able to adapt flexibly to new contexts and resonate across different domains. They must be able to frame the way people act and think rather than determine what they do or say; otherwise they will not cope with change and may be abandoned because irrelevant. A distinctive musical style does this very effectively, by being able to adapt to new circumstances without losing relevance or continuity. Meanings can be held within music propositionally (e.g., the ringing of church bells to announce a newly married couple, or their silence indicating something went wrong) and implicationally or structurally (e.g., expressing the joy and happiness of the event, or the quiet shame of public rejection). The key is that musical meaning is diverse, interactive, situated, multilayered, and wonderfully stretchy.

Music's role in the cultural transmission of enduring aesthetic, economic, social, and political orientations is remarkable. The dense interlocked hocketing of the BaYaka's and other Pygmy's vocal polyphony is probably many thousands of years old. Upon hearing Mbuti music, the BaYaka immediately recognized that the Mbuti were "real forest people" like themselves, even though genetic studies suggest that they last lived together around 18–20 thousand years ago (Bahuchet 1996). Victor Grauer takes this even further: He suggests that this unique and distinctive style, shared only by San Bushman groups and Central African Pygmies,[7] extends back to the time when they were both the same people. According to the genetic studies he quotes (Chen et al. 2000), this was between 75–100 thousand years ago (Grauer 2007).

These studies imply that musical foundational schemas may have extraordinary resilience. I argue that this resilience is due to their special aesthetic, incorporative, adaptive, and stylistic qualities which ensure continuity despite change. As the recognition of Mbuti by BaYaka attests, the schemas survive, even when language, technology, and geographical location all change.

If, as Patel as well as Fitch and Jarvis (this volume) suggest, many of the same brain resources are used for language, music performance, and perception, then the claim that language and music are at either end of a human communicative continuum seems plausible. There are, however, also important differences; most notably, the way that language makes greater use of the left hemisphere of the brain and music the right. From the material presented here, one might speculate that music, because of its aesthetic qualities, may have special significance for understanding the way culture is held in human brains and transmitted down the generations. Language, by contrast, is more concerned with the immediate contingencies of current human interaction; music is adapted to long-term orientations that determine the aesthetics or culturally appropriate forms that this interaction can take. Though both are based on similar brain resources, music and language have adapted to provide human beings with different cognitive advantages: one set is biased toward long-term

---

[7] This connection is disputed by Olivier and Furniss (1999).

interaction and cohesion of social groups, the other to the specifics of individual interactions.

Acknowledgments

I would like to thank the participants of this Ernst Strüngmann Forum for their stimulating comments, particularly Michael Arbib and two anonymous reviewers for their critical engagement with the material presented here.

3

Cross-Cultural Universals and Communication Structures

Stephen C. Levinson

Abstract

Given the diversity of languages, it is unlikely that the human capacity for language resides in rich universal syntactic machinery. More likely, it resides centrally in the capacity for vocal learning combined with a distinctive ethology for communicative interaction, which together (no doubt with other capacities) make diverse languages learnable. This chapter focuses on face-to-face communication, which is characterized by the mapping of sounds and multimodal signals onto speech acts and which can be deeply recursively embedded in interaction structure, suggesting an interactive origin for complex syntax. These actions are recognized through Gricean intention recognition, which is a kind of "mirroring" or simulation distinct from the classic mirror neuron system. The multimodality of conversational interaction makes evident the involvement of body, hand, and mouth, where the burden on these can be shifted, as in the use of speech and gesture, or hands and face in sign languages. Such shifts having taken place during the course of human evolution. All this suggests a slightly different approach to the mystery of music, whose origins should also be sought in joint action, albeit with a shift from turn-taking to simultaneous expression, and with an affective quality that may tap ancient sources residual in primate vocalization. The deep connection of language to music can best be seen in the only universal form of music, namely song.

Introduction

To approach the issues surrounding the relationship between language and music tangentially, I argue that the language sciences have largely misconstrued the nature of their object of study. When language is correctly repositioned as a quite elaborate cultural superstructure resting on two biological columns, as it were, the relationship to music looks rather different.

This chapter puts forth the following controversial position: Languages vary too much for the idea of "universal grammar" to offer any solid explanation of our exclusive language capacity. Instead we need to look directly for our biological endowment *for* language, communication, and culture. Part of this may involve the neural circuitry that is activated in language use (see Hagoort and Poeppel, this volume), although the innate nature of this is still unresolved, since it apparently develops in part parallel to the learning of language (Brauer et al. 2011b). Two systems, however, clearly contribute to our native language-ready capacities: (a) an evolved set of interactive abilities, which makes it possible to learn the cultural traditions we call languages, and (b) a specialized vocal-learning system (an auditory-vocal loop). These two systems have distinct neurocognitive bases and different phylogenetic histories. Judging from traces of parallel material culture, system (a) is well over 1.5 million years old—a time period when system (b) was not yet in place. Here I concentrate on system (a), our interactive abilities, because its contribution to linguistic capacity has not been properly appreciated. I begin with a brief description of this story and then explore its implications for language, music, and their interrelation.

Language Diversity and Its Implications

Let us begin with the observation that human communication systems are unique in the animal world in varying across social groups on every level of form and meaning. There are some 7000 languages, each differing in sound systems, syntax, word formation, and meaning distinctions. New information about the range of language diversity and its historical origins has undercut the view that diversity is tightly constrained by "universal grammar" or a language-specialized faculty or mental module (Evans and Levinson 2009). Common misconceptions, enshrined in the generative approach to language universals, are that all languages use syntactic phrase structure as the essential foundation for expressing, for example, grammatical relations, or that all languages use CV syllables (a Consonant followed by a Vowel), or have the same basic set of word classes (e.g., noun, verb, adjective). Instead, some languages make little or no use of surface phrase structure or immediate constituents, not all languages use CV syllables, and many languages have word classes (like ideophones, classifiers) that are not found in European languages. The entire apparatus of generative grammar fails to have purchase in languages that lack phrase structure (e.g., the so-called "binding conditions" that control the distribution of reflexives and reciprocals). Nearly all language universals posited by generative grammarians have exceptions in one set of languages or another.

The other main approach to language universals, due to Greenberg (1966), escapes this dilemma by aiming for strong statistical tendencies rather than exceptionless structural constraints. The claim would then be that if most languages follow a tendency for specific structures, this reflects important biases in human cognition. Greenberg suggested, for example, that languages tend to have "harmonic" word orders, so that if a language has the verb at the end of the clause, it will have postpositions that follow the noun phrase (rather than prepositions that precede the noun phrase). Dryer (2008) has recently tested a great range of such predictions, with apparently good support. These generalizations rely on sampling many languages, both related and unrelated; one can hardly avoid related languages because a few large language families account for most of the languages of the world. One problem that then arises is that related languages may be similar just because they have inherited a pattern from a common ancestor. One recent solution has been to control for relatedness by looking at, for example, word order wholly within large language families. It turns out that the Greenbergian generalizations about harmonic word order do not hold: language change within language families often does not respect the postulated strong biases, and language families show distinctly different tendencies of their own (Dunn et al. 2011).

The upshot is that although there are clear tendencies for languages to have certain structural configurations, much of this patterning may be due entirely to cultural evolution (i.e., to inheritance and elaboration during the processes of historical language change and diversification). All the languages of the world outside Africa ultimately derive (judging from genetic bottlenecks) from a very small number that left Africa at the time of the diaspora of modern humans not later than ca. 70,000 years ago, with the possible proviso that interbreeding with Neanderthals and Denisovans (now known to have occurred) could have amplified the original diversity (Dediu and Levinson 2013).[1]

There are three important implications. First, we have underestimated the power of cultural transmission: using modern bioinformatic techniques we can now show that languages can retain strong signals of cultural phylogeny for 10,000 years or more (Dunn et al. 2005; Pagel 2009). Consequently, language variation may tell us more about historical process than about innate constraints on the language capacity. Those seeking parallels between music and language be warned: in neither case do we have a clear overview of the full range of diverse cultural traditions, universal tendencies within each domain, and intrinsic connections across those tendencies. Over the last five years, linguists have made significant progress in compiling databases reflecting (as yet still in a patchy way) perhaps a third of the linguistic diversity in the world, but no corresponding database of ethnomusicological variation is even in progress.[2]

Second, the observed diversity is inconsistent with an innate "language capacity" or universal grammar, which specifies the structure of human language in anything like the detail imagined, for example, by the "government and binding" or "principles and parameters" frameworks in linguistics (Chomsky

---

[1] The recently discovered Denisovans were a sister clade to Neanderthals, present in eastern Eurasia ca. 50 kya; they contributed genes to present-day Papuans, just like Neanderthals did to western Eurasians.

[2] Useful leads will be found in Patel (2008:17ff), who points out that cultural variability in scale structure (numbers from 2–7 tones, differently spaced) and rhythm (2008:97ff) makes strong universals impossible. See also Nettl (2000), who quotes approvingly Herzog's title "Musical Dialects: A Non-Universal Language."

1981; Baker 2001). Even the scaled-back Minimalist program makes claims about phrase structure that are ill-fitted to the diversity of languages. There is no doubt a general "language readiness" special to the species, but this does not seem well captured by the major existing linguistic frameworks. We seem to be left with general architectural properties of languages (e.g., the mapping of phonology to syntax, and syntax to semantics), abstract Hockettian "design features," and perhaps with stronger universals at the sound and meaning ends (i.e., phonology and semantics, the latter pretty unexplored) than in morphosyntax.

Third, since the diversity rules out most proposed linguistic universals, we need to look elsewhere than "universal grammar" for the specific biological endowment that makes language possible for humans and not, apparently, for any other species. Apart from our general cognitive capacities, the most obvious feature is the anatomy and neurocognition of the vocal apparatus, and our vocal-learning abilities, rare or even unique among the primates.[3] These input/output specializations may drive the corresponding neurocognition, the loop between motor areas and the temporal lobe. They may even, during human development, help build the arcuate fasciculus (the fiber bundle that links the frontal lobes with the temporal lobes, i.e., very approximately, Broca's and Wernicke's areas; Brauer et al. 2011b).[4] The neural circuitry involved in language processing may thus have been "recycled," rather than evolved, for the function (Dehaene and Cohen 2007).

Only slightly less obvious is a set of abilities and propensities that are the essential precondition for language: advanced theory of mind (ToM) and cooperative motivations and abilities for coordinated interaction, which together form the background to social learning and makes possible both culture in general and the learning of specific languages. These aspects of cognition and, in particular, the grasp of Gricean communication (meaning$_{nn}$) seem to have their own neural circuitry, distinct from vocal circuitry and mirror neuron circuitry (Noordzij et al. 2010).[5] These interactional abilities are much more central to language than previously thought; together with vocal learning, they provide the essential platform both for cultural elaboration of language and for infants to bootstrap themselves into the local linguistic system. Correspondingly, they may play some

---

[3] See, however, Masataka and Fujita (1989) for monkey parallels.

[4] The crucial experiment—checking on the development of these structures in deaf children and home-signers—has not to my knowledge been done.

[5] Gricean signaling (producing a noninstrumental action whose sole purpose is to have its intention recognized) is a kind of second-order mirror system: perception (decoding) depends on (a) seeing the noninstrumental character of the signal, and (b) simulating what effect the signaler intended to cause in the recipient just by recognition of that intention. Consider my signaling to you at breakfast that you have egg on your chin just by energetically rubbing my chin: a first-order mirror interpretation is that I have, say, egg on my chin; a second-order one is that I'm telling you that it is on your chin.

parallel role in musical learning and performance. They have preceded modern language in both ontogeny and phylogeny, which we turn to next.

The Timescale of the Evolution of Speech and Language

In a metastudy drawing on the most recent discoveries, we have argued that the origins of these vocal abilities can be traced back over half a million years to *Homo heidelbergensis*, who exhibited a modern human vocal tract, modern breathing control, modern audiograms, and the FOXP2 variants inherited in common by his descendants: Neanderthals, Denisovans, and modern humans (Dediu and Levinson 2013). At 1.6 mya, *H. erectus* lacked these vocal specializations but exhibited control of fire and complex tool traditions (the Acheulian or Mode 2 type), arguing for a communication system able to support advanced cultural learning. Such a system presupposes the cooperative interaction style of humans, which in turn relies on advanced ToM capacities. Therefore, *H. erectus* (or *H. ergaster* as some prefer to call the African variant) had some quite advanced form of language that was less vocally specialized than that used by *H. heidelbergensis*. *H. erectus*, in turn, is the presumed descendant of *H. habilis*, who already used a varied stone tool kit at 2.5 mya, with the first stone tools in use as early as 3.4 mya. Thus social learning and cooperative communication have deep phylogenetic roots.

Speech and language, as we know them, evolved in the million years time between 1.5 mya–0.5 mya (Dediu and Levinson 2013). Modern language is thus of a much greater antiquity than usually assumed (Klein 1999; Chomsky 2007, 2010 presume the last 50–100 kya). Nevertheless, from a geological or genetic time perspective, a million years pales in comparison to the 50 million years existence of birdsong or bat echolocation: language has been able to develop so fast because much of its complexity was outsourced to cultural evolution. There is, however, a deep biological infrastructure for human language: the vocal-auditory system, on the one hand, and the cooperative communicative instincts, on the other, on which cultural elaboration is based.

Most importantly, a fully cooperative communication system and interaction style evolved gradually in our line over the three million years of tool use leading up to *H. heidelbergensis*. Judging from the development of material culture, ToM capacities and advanced cooperative abilities are very ancient, tied to increasing encephalization and group size. They are the crib for language in both phylogeny and ontogeny.

Two controversial issues should be raised here. First, the Darwinian view that speech evolved not for language but for musical use, with songbirds as the animal model, has recently been revived, for example, by Mithen (2005) and Fitch (2009b). This view is, to my mind, a nonstarter. As just explained, our speech system evolved after at least a million years of functional communication geared to handing on cultural learning and tool traditions. The preconditions for culture involved prolonged infancy, intensive cooperative social interaction, and the large social groups that motivated increased encephalization. It is not plausible that all this developed without some kind of protolanguage. Thus language (perhaps in gestural form) preceded speech, not the other way around as Darwin had imagined. Thus, songbirds probably do not provide the right analogy; vocal learners among more social species (e.g., sea mammals) may provide a better animal model (for a contrary view, see Fitch and Jarvis, this volume).

The second controversial issue is indeed the possible gestural origin of language (cf. Arbib 2005a, b). If language was carried by a medium other than fully articulate and voluntarily controlled speech for a million years or more, gesture is a prime candidate. Call and Tomasello (2007) make a good case for gesture being the voluntary, flexible medium of communication for apes, with vocal calls being more reflex. On phylogenetic grounds, then, one might indeed argue for gestural precursors to language. However, first, there is no specialization of the hand for communication that parallels the evolution of the vocal system, which one would expect if it played such a crucial role. Second, the human communication system is properly thought of as based on hand + mouth, allowing greater loading of the hand (as in sign languages) or of the mouth (as in spoken languages), but always involving both. It is therefore likely that this joint system has great antiquity. What seems plausible is that during the million years preceding *H. heidelbergensis* (by which time speech was fully formed), the burden of communication was shifted relatively from hand to mouth. More generally, human communication is intrinsically multimodal, as reflected, for example, in the general purpose nature of Broca's area (Hagoort 2005).

The Interactive Niche

Every language is learned in face-to-face interaction in a special context that is unique to the species. Most animals avert gaze except in aggression; in contrast, within restrictions, mutual gaze is tolerated or even required in many kinds of human interaction (for a cross-cultural study, see Rossano et al. 2009). This is a token of the *presumption of cooperation* which operates (again with limitations) in intragroup human interaction—a persistent puzzle from an evolutionary point of view (Boyd and Richerson 2005). Under this cooperative envelope, interaction consists of a sequenced exchange of actions, following specific turn-taking rules geared to the structure of minimal contributions (e.g., clauses in spoken communication), and which permit one action at a time (see Sacks et al. 1974). The expectation is that each such action unit is tied to the prior one by a "logic" of action: a request is met with a compliance or denial; a greeting by a greeting; a question with an answer or evasion; a pointing by a gaze following; and so forth. The structure of action sequences can be complex, arguably as or more complex than anything seen in natural language syntax, as we shall see. Yet an elementary system of this kind is visible in the earliest prelinguistic mother–infant interaction ("proto-conversation," Bruner 1975; for resonances in the musical domain, see Malloch and Trevarthen 2009).

This interactional envelope is the context in which the great bulk of language use occurs; monologue is the exception, and in some societies hardly occurs at all. Narrative, likewise, plays a small role, statistically speaking, in the use of language. The basic niche for language is the tit-for-tat of informal conversation: action–response, action–response, and so on. It is intriguing to wonder what the equivalent natural ecological niche for music might be; perhaps Western music, with its division of performers and audience, is entirely misleading, like comparing lecturing to the natural niche for language use.

Two aspects of this interactional envelope are much more complex and intricate than meets the eye. The first is turn-taking. The fact that turns at talking alternate seems at first quite trivial, but consider this: the gap between turns is on average 200 ms across a wide variety of languages (and the mode offset between turns is 0 ms, without any gap at all, in all languages tested; Stivers et al. 2009). Since it takes at least 600 ms to crank up the speech production system, speakers must be anticipating the last words of their interlocutors' turns; they must also predict the content in order to respond appropriately (direct EEG measurement suggests actual launch of production is quite a bit earlier than this on average). The whole system is built on predicting what the other will say part way through the saying of it, and since what is said has all the open-endedness that syntax delivers, this is no mean feat. This system exerts tremendous cognitive demands: comprehension and production must run in parallel, at least part of the time (see Figure 3.1). It is perhaps not fanciful to imagine that this universal pace or fixed metabolism of the turn-taking system was set up early in the phylogeny of protolanguage, so that we inherit a system ill-adapted to the complexity of the structures we now thrust into these short turns.

The second complexity is action sequencing: questions expect answers, requests compliance, offers acceptances, etc. This requires recognizing a turn as a question, early enough in its production to allow time to formulate the response. This recognition might be imagined to be based on syntax or lexical cues, but corpus work shows, for example, that most questions in English are yes-no questions, and most of these are in declarative form with falling intonation. So, most questions are recognized by means other than direct linguistic

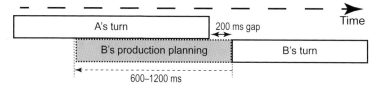

Figure 3.1 Overlap of comprehension and production processes in conversation.

cues, for example by noting that B is making a statement about a subject for which he knows I have more knowledge (e.g., "You've had breakfast"). The same holds for most kinds of speech acts: they don't come wrapped in some canonical flag. This problem of "indirect speech acts" has been neglected since the 1970s, but it is *the* fundamental comprehension problem: the speech act is what the hearer needs to extract to respond in the tight temporal frame required by the turn-taking system. Likewise, the whole function of language is often misconstrued: the job of language is not to deliver abstract propositions but to deliver speech acts.

Since the job of language is to deliver actions explains, of course, why speech comes interleaved with nonverbal actions in any ordinary interchange: I say "Hi"; you smile and ring up my purchases saying, "You know you can get two of these for the price of one"; I explain that one will do and hand you a bill; you say "Have a good day." Words and deeds are the same kind of interactional currency, which is why language understanding is just a special kind of action understanding. In cooperative interaction, responses are aimed at the underlying action goals or plans. Consider the telephone call in Example 3.1 (Schegloff 2007:30):

1.	Caller:	Hi.	
2.	Responder:	Hi.	
3.	Caller:	Whatcha doin'?	
4.	Responder:	Not much.	(3.1)
5.	Caller:	Y'wanna drink?	
6.	Responder:	Yeah.	
7.	Caller:	Okay.	

Line 3 might look like an idle query, but it is not treated as one: the response "Not much" clearly foresees the upcoming invitation in line 5 and makes clear that there is not much impediment to such an invitation, which is then naturally forthcoming. Conversation analysts call turns like line 3 a "pre-invitation," and show with recurrent examples that such turns have the character of conditional offers. Just as the caller can hardly say at line 5 "Oh just asking," so the responder will find it hard to refuse the invitation she has encouraged by giving the "go-ahead" at line 4 (although a counterproposal might be in order at line 6). The underlying structure of such a simple interchange, translated into hierarchical action goals, might look something like the sketch in Figure 3.2, where *Whatcha doin'* acquires its pre-invitation character from a projection of what it might be leading to.

Action attribution thus plays a key role in the use of language and is based quite largely on unobservables, like the adumbrated next action if I respond to this one in such a way. The process is clearly based on advanced ToM capacities, and beyond that on the presumption that my interlocutor has designed his turn precisely to be transparent in this regard. This, of course, is Grice's insight, his theory of meaning$_{nn}$: human communicative signals work

Cross-Cultural Universals and Communication Structures 75

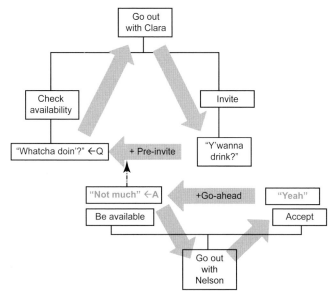

Figure 3.2 Action assignment based on plan recognition (see Example 3.1). Arrows indicate direction of inference from what is said to what is likely to come up next, which then "colors" the interpretation of the present turn.

by presenting an action designed to have its intention recognized, where that recognition exhausts the intention. In recent work, we have tried to isolate the neural circuitry involved in just this process and find it distinct from either the language circuitry or the mirror neuron circuits (Noordzij et al. 2010): we find overlapping areas of activation in the posterior superior temporal sulcus, in mirror-like fashion, in both signaler and receiver, interpreted as signaler's simulation of recipient's inferencing.[6]

The inferential character of action ascription makes it a complex process. However, an unexpected further order of complexity is that it has a quasi

---

[6] As a way to generalize over these observations and the classic mirror neuron system, it may be helpful to think (in a slightly different way than Arbib 2005b) of a *hierarchy of action–perception mirror loops*, as follows:

degree 0 (intra-organism): *Action–perception feedback*, as in proprioception or auditory feedback of one's own production, allows cybernetic feedback. Highly evolved systems include echolocation in bats and cetaceans.

degree 1 (cross-organism): *Classic mirror neuron system:* other's action recognition and self-action use overlapping neural resources. This can be further distinguished into degree 1.1 instinctive systems and degree 1.2 learned systems. Mouth mirror neurons might offer a route to vocal learning (Arbib 2005b:118).

degree 2 (cross-organism): *Gricean simulation systems:* applies to actions that self-advertise that they are signals (noninstrumental actions), so *discounting* mirror neuron systems of degree 1. Works by the recipient simulating what the signaler calculated the recipient would think/feel (that being the noninstrumental intention).

syntax (Levinson 1981). Consider the following simple exchange in Example 3.2 (Merritt 1976):

$$
\begin{array}{lll}
A: & Q_1 & \text{"May I have a bottle of Mich?"} \\
B: & Q_2 & \text{"Are you twenty one?"} \\
A: & A_2 & \text{"No."} \\
B: & A_1 & \text{"No."}
\end{array}
\quad (3.2)
$$

This has a pushdown stack character: Q_1 is paired with A_1, but Q_2–A_2 intervenes. Many further levels of embedding are possible, and they can be characterized, of course, by the phrase-structure-grammar in Example 3.3:

$$
\begin{array}{l}
Q\&A \rightarrow Q\ (Q\&A)\ A \\
Q\&A \rightarrow Q\ A
\end{array}
\quad (3.3)
$$

What is interesting is that this kind of center embedding has been thought to be one of the pinnacles of human language syntax. An exhaustive search of all available large language corpora has yielded, however, the following finding: the greatest number of recursive center embeddings in spoken languages is precisely 2, whereas in written languages the number is maximally 3 (Karlsson [2007] has found exactly 13 cases in the whole of Western literature).[7]

In contrast, it is trivial to find examples of center embeddings of 3 or greater depth in interaction structure. Example 3.4 (abbreviated from Levinson 1983:305) shows one enquiry embedded within another, and a "hold-OK" sequence (labeled 3) within that:

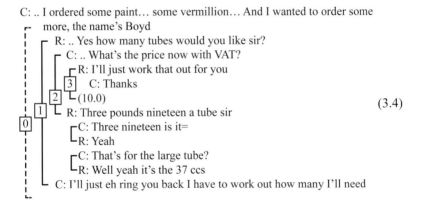

(3.4)

Examples 6 deep or more can arguably be found in conversation. When one finds a domain in which a cognitive facility is most enhanced, it is reasonable to assume that this is the home in which it originally developed. The implication is that core recursion—as expressed in center embedding—has its origin

---

[7] More precisely, Karlsson (2007) calls one center embedding "degree 1," a center embedding within a center embedding "degree 2," and shows that degree 2 is the maximal attested depth for spoken languages, degree 3 for written texts.

in interaction systems, not in natural language syntax. Exactly parallel arguments can, I believe, be made for so-called cross-serial dependencies, vanishingly rare in syntax but exhibited recurrently in conversational structure.[8] Why exactly it is so much easier to keep track of discontinuous dependencies in joint action than in solitary performance remains unclear; the mental registers required would seem to be the same, but the distributed production clearly helps cognition in some way.

More generally, the implication is that we have minds engineered for extraordinary coordination in joint action (see Fogassi, this volume). It may be interesting to reconsider music in this light: not as, in origin, a solitary enterprise or a performance to a passive audience, but as an interchange between actors, where guessing the next phrase is crucial to coming in on time, where one performer "answers" another, where the basic units are seen as "actions" rather than formal objects, and where extremes of coordination carry a deep satisfaction of their own (see Janata and Parsons, this volume). This suggests improvisational jazz as the model, not the sonata or the lullaby.

Language and Music

Sixty years ago the great anthropologist Levi-Strauss pointed out that music is the central mystery of anthropology.[9] Nothing has changed since. Contrary to the Chomskyan idea that language is a late evolutionary freak, a spandrel from some other evolutionary development, the fossil and archaeological record actually shows a steady, slow accumulation of culture which was only made possible by some increasingly sophisticated mode of communication, already essentially modern and primarily in the vocal channel by 0.5 million years ago (pretty much as Pinker and Bloom 1990 imagined on more slender evidence). But what is the story for music?

To what extent could music be parasitic on language, or more broadly on our communicative repertoire? First, some basic points. In small-scale societies with simple technology, music often equals song: that is to say, music only occurs with language. It is often imagined that music always involves instruments, but again small-scale societies often have no instruments, in some cases also avoiding any form of ancillary percussion (as in the elaborate, but purely vocal, range of song styles of Rossel Island, Papua New Guinea).[10] Phoneticians often distinguish language, prosody, and paralanguage, where the latter two are suprasegmental properties of speech (roughly tonal and wide

---

[8] Cross-serial dependencies have the form A1–B1–A2–B2 where the linkages cross over. Example 3.1 contains such a pattern, but I leave that as an exercise for the reader.

[9] Compare Darwin (1871:333) "As neither the enjoyment nor the capacity of producing musical notes are faculties of the least direct use to man in reference to his ordinary habits of life, they must be ranked amongst the most mysterious with which he is endowed."

[10] The Rossel Island observations come from my own ethnographic work.

timbre qualities, respectively), only partially bound into the linguistic system in rule-governed ways (see Ladd, this volume). Song is in a sense just language in a special, marked suprasegmental register or style or genre. Rossel Islanders, for example, do not have a category of "music" that would place each of their named types of song style (e.g., *tpile we*, "operetta"; *ntamê*, "sacred chants"; *yaa*, "laments") in opposition to speech of other types (e.g., *wii*, "fast-declaimed poetry"). It is thought-provoking to realize that "music" seems to be an ethnocentric category (Nettl 2000:466).

We are hampered, as mentioned, by having no ethnomusical databases that cover the world, but it is likely that song is in the unmarked case not a solo performance, but a joint activity involving a chorus (Nettl 2000).[11] Most of the song styles on Rossel Island, for example, are joint performances sung in unison, with the exception of laments (*yaa*) which are composed and sung by individuals, typically at funerals. This contrasts with normal conversation, which is composed from individual, short turns with rapid alternation. Song is thus a marked genre, in being predominantly jointly performed in unison (which is a rare, but observable occurrence in conversation, as in greetings or joint laughter). In some circumstances, but not all, song is like speech-giving, a performance by a set of performers with a designated audience. Linguistic systems make a lot of distinctions between speakers, addressees, auditors, and the like, originally explored by Goffman (1981, Chapter 3). For example, when I say, "The next candidate is to come in now," the syntax projects a second speech event, indicating that I am instructing you to go and ask the candidate on my behalf (see Levinson 1988). The same distinctions are relevant for song: both a song and a speech may be authored by one individual on behalf of another (the principal) and performed by a third (as in the praise songs of West Africa; see Charry 2000). In Rossel Island laments, author, principal, and mouthpiece are identical; sacred hymns (*ntamê*), however, are composed by the gods and sung by elder males to a precise formula, to a male-only audience.

Song, surely the original form of music,[12] makes clear the possibly parasitic nature of music on language: the tonal and rhythmic structure must to some extent be fitted to the structure of the language. The language of the lyrics determines both aspects of the fine-grained structure, the affectual quality matched to the words, and the overall structure, for example, the timing of subunits and nature of the ending (e.g., the number of verses).

The perspective adopted here, emphasizing the role of language in its primordial conversational niche, also suggests a possible take on the cultural (and possibly biological) evolution of music. The motivation for and structural complexity of music may have its origins in joint action rather than in abstract representations or solitary mentation. It may also rely on Gricean reflexive

---

[11] Patel (2008:371), however, reports one Papuan society where song is largely private and covert.

[12] The assumption makes the prediction that no cultural system of musical genres will be found without song genres (see Nettl 2000).

mirroring or simulation to achieve the empathy that seems to drive it,[13] together with the apparently magical coordination through prediction which is one source of the pleasure it gives. The rhythmic properties may owe at least something to the rapidity of turn-taking, the underlying mental metabolism, and the interactional rhythms that are set up by turn-taking. The multimodality of human communication allows the natural recruitment of additional channels, whether multiple voices or instrumental accompaniments, and of course dance (Janata and Parsons, this volume).

Still, few will be satisfied with the notion that music is, even in origin, just a special kind of speech (see Wallin et al. 2000; Morley 2011). They will point to the existence of (largely) independent cultural traditions of instrumental music, to the special periodic rhythms of music, and to its hotline to our emotions (Patel 2008). One speculation might be based on the Call and Tomasello (2007:222) argument that in the Hominidae, with the sole exception of humans, vocal calls are instinctive, reflex, and affectual ("Vocalizations are typically hardwired and used with very little flexibility, broadcast loudly to much of the social group at once—who are then infected with the emotion;" see also Scherer, this volume), in contrast to the gestural system which is more intentionally communicative and socially manipulative. If language began in the gestural channel and slowly, between 1.5–0.5 million years ago, moved more prominently into the vocal channel, it is possible that the vocal channel retains an ancient, involuntary, affective substrate. Note, for example, how laughter and crying exhibit periodic rhythms of a kind not found in language and are exempt from the regular turn-taking of speech. It could be this substrate to which music appeals, using all the artifice that culture has devised to titillate this system. Drugs—culturally developed chemicals—work by stimulating some preexisting reward centers. The ancient affective call system could be the addiction music feeds.[14]

Conclusion

The theory of language, properly reconstructed, yields much of the complexity of linguistic structure over to cultural evolution, seeking biological roots primarily in the auditory-vocal system and the species-special form of

---

[13] Gricean reflexive intentions may play a larger role in music than is obvious: Performers "work" an audience, intending to induce a feeling partly by affective evocation, but partly by getting the audience to realize that is what they are trying to do. This accounts for the difference between a recording and a live performance—the performer tries to persuade the particular audience to adopt the affective state intended. Thus all three types of action–perception loop mentioned in footnote 6 may apply equally well to music.

[14] A recent study of musical "chills" shows striking similarity with cocaine highs, providing "neurochemical evidence that intense emotional responses to music involve ancient reward circuitry" (Salimpoor et al. 2011).

communicational abilities in cooperative interaction. What is peripheral in current linguistic theory (speech and pragmatics) should be central; what is central in much theory (syntax) may be more peripheral. Syntaxes are, I have suggested, language-specific cultural elaborations with partial origins in the interactional system, within bounds set by aspects of general cognition (Christiansen and Chater 2008). Viewed in this light, the relation of language to music shifts. The vocal origins of music may ultimately be tied to the instinctual affective vocal system found in apes, while the joint action and performance aspects may be connected to the interactional base for language. Just as syntaxes are artifacts honed over generations of cultural evolution, so are the great musical traditions.

Acknowledgment

Special thanks are due to Michael Arbib for helpful comments on both a long abstract and the draft paper, as well as to Penelope Brown and Peter Hagoort for helpful early comments. I also owe a diffuse debt to Ani Patel's stimulating 2010 Nijmegen Lectures and to discussions at the Strüngmann Forum meeting, where this paper was discussed.

Part 2: Action, Emotion, and the Semantics

4

Shared Meaning, Mirroring, and Joint Action

Leonardo Fogassi

Abstract

Mirroring the behavior of others implies the existence of an underlying neural system capable of resonating motorically while this behavior is observed. This chapter aims to show that many instances of "resonance" behavior are based on a mirror mechanism that gives origin to several types of mirror systems, such as those for action and intention, understanding, and empathy. This mechanism provides an immediate, automatic kind of understanding that matches biological stimuli with internal somatomotor or visceromotor representations. It can be associated with other cortical circuits when an understanding of others' behavior implies inferential processes.

Properties of the mirror system are described in monkeys, where it was originally discovered, as well as in humans. Discussion follows on how the system can be involved in social cognitive functions such as understanding of goal-directed motor acts, intention, and emotions. The possible involvement of this system/mechanism in mirroring language and music is then discussed, and it is suggested that it initially evolved for action understanding in nonhuman primates and could have been exploited for other functions involving interindividual interactions.

Introduction

The capacity to mirror others applies to several types of behaviors. Overtly, it can simply be manifested as a kind of automatic resonance, such as in contagious behavior, mimicry, or in the chameleon effect. Perhaps the most diffuse and adaptively relevant mirroring behavior is imitation. This term, if used in a broad sense, includes several types of different processes, ranging from true imitation (i.e., to copy the form of an action; see Whiten et al. 2009) to action facilitation (e.g., an increase in frequency of an observed behavior that belongs to the observer's repertoire) to emulation, which consists in reproducing the goal of an observed behavior independent of the means used to achieve it. Depending on the situation or context, one or more of these imitative processes may be involved. For example, to extract an object from a closed container that can be opened in different, equally efficient ways, it is not necessary to adopt the exact same strategy used by a demonstrator (Whiten 1998). However, to learn a new guitar chord, the exact finger posture and sequence shown by the expert must be reproduced. While imitative processes such as action facilitation or neonatal imitation have been observed in nonhuman primates, instances of true imitation in nonhuman primates have been described mainly in apes. Monkeys show a very limited imitative capacity (Visalberghi and Fragaszy 2001).

These examples suggest that mirroring others involves very different behaviors, although the underlying neural capacity of "resonating" motorically in them all can be recognized (Rizzolatti et al. 2002). This motor resonance could consist of at least two different neural mechanisms: one related to simple movements or meaningless gestures, the other to meaningful and transitive motor acts. The neural mechanisms underlying these different types of motor resonance have been examined over the last twenty years at the single neuron and population levels. At the single neuron level, many studies in monkeys have demonstrated that the capacity to resonate during the observation of others' actions is provided by a peculiar category of visuomotor neurons (mirror neurons), which play an important role in enabling monkeys to understand the actions of others. A similar system has been demonstrated in humans, mainly at the population level. Here, action understanding is not only enabled, it appears to be (possibly as a result of new evolutionary pressures) involved in manifesting the result of the neural resonance mechanism, as in imitation.

I begin with a description of the basic features of the mirror system in monkeys and humans. I specifically address the role of the system in intention and emotion understanding and discuss the mirroring function in language and music, including its possible neural substrate.

The Mirror System in Monkeys

For a long time, it was assumed that the motor cortex (i.e., the posterior part of the frontal lobe) could be divided into two main subdivisions: Brodmann's area 4 and area 6 (Brodmann 1909). From a functional perspective, these areas correspond to the primary motor and premotor cortex, respectively. However, the premotor cortex itself can be subdivided into a mosaic of areas, as shown by parcellation studies (Figure 4.1; see also Matelli et al. 1985, 1991). Accordingly, the ventral part of monkey premotor cortex (PMV) is composed of two areas: F4, caudally, and F5, rostrally (Figure 4.1a).

Area F4 controls goal-directed axial and proximal movements (Fogassi et al. 1996), whereas area F5 controls both hand and mouth goal-directed movements, such as grasping, manipulation, biting, etc. (Rizzolatti et al. 1988). In addition to the purely motor neurons in area F5, there are two classes of

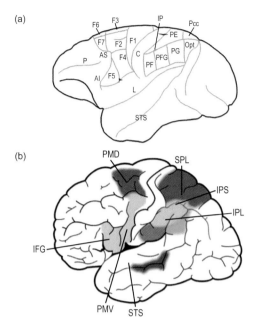

Figure 4.1 (a) Lateral view of the monkey brain showing the parcellation of the agranular frontal and posterior parietal cortices. Agranular frontal areas are defined according to Matelli et al. (1985, 1991). F1 corresponds to Brodmann's area 4 (primary motor cortex), F3 and F6 to the mesial part of Brodmann's area 6 (supplementary and pre-supplementary motor cortex), F2 and F7 to the dorsal part of Brodmann's area 6 (dorsal premotor cortex, PMD), F4 and F5 to the ventral part of Brodmann's area 6 (ventral premotor cortex, PMV). All posterior parietal areas (labeled "P" followed by a letter) are defined according to Pandya and Seltzer (1982) and Gregoriou et al. (2006). Areas PF, PFG, PG, and Opt constitute the inferior parietal lobule (IPL), PE and Pec the superior parietal lobule. AI: inferior arcuate sulcus; AS: superior arcuate sulcus; C: central sulcus; IP: intraparietal sulcus; L: lateral fissure; P: principal sulcus; STS: superior temporal sulcus. (b) Schematic of cortical areas that belong to the parietofrontal mirror system. Yellow indicates transitive distal movements; purple depicts reaching movements; orange denotes tool use; green indicates intransitive movements; blue denotes a portion of the STS that responds to observation of upper-limb movements. IFG: inferior frontal gyrus; IPS: intraparietal sulcus; SPL: superior parietal lobule. Modified from Cattaneo and Rizzolatti (2009).

visuomotor neurons. One class is made up of canonical neurons that respond to object presentation and grasping of the same object (Murata et al. 1997; Raos et al. 2006). The other consists of mirror neurons that activate both when the monkey executes a goal-directed motor act (e.g., grasping) and when it observes another individual (either a human being or a conspecific) performing the same act (Figure 4.2; see also Gallese et al. 1996; Rizzolatti et al. 1996).

The response of these neurons is very specific; they do not respond to the presentation of a simple object or to an observed act that is mimicked without the target. Visual response is also independent of several details of the observed

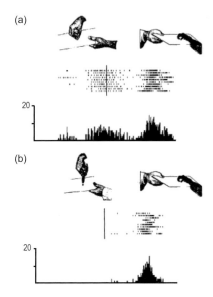

Figure 4.2 Response of a mirror neuron during observation and execution of a hand-grasping motor act. (a) The neuron shows a visual response when an experimenter grasps a piece of food in front of the monkey and when the monkey grasps the same piece of food from the experimenter's hand. The silence between the visual and motor response corresponds to the time in which the experimenter approaches the monkey with the plate of food before the monkey grasps it. (b) The experimenter extends a piece of food using pliers to the monkey who then grasps it. Here there is no visual response, because the observed act is performed with a nonbiological effector. In both (a) and (b) the rasters and histograms are aligned to the moment in which the experimenter's hand or the pliers touches the food, respectively. Abscissae: time; ordinates: spikes per bin; bin width: 20 ms (after Rizzolatti et al. 1996).

motor act: the type of object that is the target of the observed motor act, the location in space and direction of this act, the hand used by the observed agent (see, however, Caggiano et al. 2009). Mirror neurons also discharge when the final part of the observed act is not visible to the observing monkey, but the animal knows in advance that there is an object behind the obstacle which prevents it from seeing the consequences of the observed motor act (Umiltà et al. 2001). This latter finding is considered one of the crucial demonstrations that mirror neurons do not simply code the visual features of a biological movement, but are instead relevant for the comprehension of the goal-relatedness of the motor act. As an additional demonstration of this concept, it has been shown that a particular class of mirror neurons responds not only during observation of a motor act but also during listening to the sound produced by this act (Kohler et al. 2002). Some mirror neurons, for example, activate both when the monkey observes the act of breaking a peanut and when it only hears the noise produced by the act. As expected, these "audiovisual" mirror neurons activate during execution of the same, noisy, motor act.

From these examples it is clear that the most important property of mirror neurons is to match the observation or hearing of a motor act belonging to the monkey motor repertoire with its execution. It has been proposed that when the input related to the observed or listened motor act has access (in the observer) to the internal motor representation of a similar motor act, this input achieves its meaning. In other words, the *internal motor knowledge* of an individual constitutes the basic framework for understanding the motor acts of others. I use the term "internal motor knowledge" because this knowledge is not built through the hierarchical elaboration of sensory input but rather through the motor system. Of course, we can "recognize" motor acts by simply using the visual or auditory representations of the biological objects that are observed or heard. However, we do not "internally" understand the meaning of these stimuli.

As is typical of other types of functions that require visuomotor integration, mirror neurons are the result of a process based on a specific anatomical circuit. The main areas belonging to this circuit are the anterior part of the superior temporal sulcus (STSa), the rostral part of inferior parietal cortex, and the PMV. STSa contains neurons that discharge during the observation of biological actions and, more specifically, of hand motor acts, but they are not endowed with motor properties (Perrett et al. 1989). This region is anatomically linked with area PFG (Figure 4.1) of the inferior parietal lobule (IPL), which in turn is connected to area F5 (Bonini et al. 2010; Rozzi et al. 2006). These two links allow the matching, observed in mirror neurons, between the visual representation of a hand motor act and its motor representation. Remarkably, mirror neurons have also been found in area PFG (Fogassi et al. 2005; Gallese et al. 2002; Rozzi et al. 2008) and have properties very similar to those of premotor mirror neurons. This similarity is not surprising for two reasons: First, because of the reciprocal connections between the two areas, neurons belonging to them share several properties. Second, evidence that neurons of IPL also have strong motor properties (see Rozzi et al. 2008) suggests that the matching between visual representations (of observed acts) and motor representations (of the same acts) may, in principle, also occur in this area.

All of the evidence reviewed above shows that mirror neurons activate in relation to goal-directed motor acts. These acts can also be defined as transitive movements because they are directed toward a target, in contrast to intransitive ones which lack a target. Is there any evidence for neural mechanisms underlying other types of mirroring that involve intransitive movements? As discussed, mirror neurons in the monkey do not respond to mimicked motor acts. Even when mimicking evokes their activation, this is normally weaker than when it is obtained during observation of the same act interacting with a target (Kraskov et al. 2009). There is, however, an exception: among mirror neurons that respond to the observation of mouth motor acts, there is a class that is selectively activated by the observation and execution of monkey communicative gestures (Ferrari et al. 2003; see also below). Thus, from the evidence collected in monkeys, it can be concluded that mirror neurons respond to the observation of meaningful movements, where the word "meaningful" includes both transitive movements (goal-directed motor acts) and gestures endowed with meaning. The fact that mirror neurons for intransitive movements concern only the orofacial motor representation could be related to their role in communication.

In summary, the presence of mirror neurons responding—even when the motor act is hidden or the sound of it is heard—is a strong indication for suggesting that monkey mirror neurons underpin the understanding of goal-related motor acts. However, whether their discharge also has a role in triggering social interactions is not known. Recent data suggest, albeit indirectly, that the modulation of the response of mirror neurons to visual cues could be related to the behavioral reaction of the observing individual. Caggiano et al. (2009) studied mirror neuron response while monkeys observed an experimenter performing a grasping act that took place within the monkey's reaching space (peripersonal space) or outside of it (extrapersonal space). Their results show that 50% of the mirror neurons recorded with this paradigm were sensitive to the distance at which the agent performed the observed motor act; half of these responded more strongly when the observed motor act was performed close to the monkey (peripersonal neurons), the other half when it was performed far from the monkey (extrapersonal neurons). This modulation has been proposed to be related to the possible subsequent interaction of the observer with other individuals whose behavior takes place inside or outside the observer's reaching space.

Mirroring and Shared Attention

Shared attention involves the interaction between two individuals and can be instrumental in achieving joint activity between individuals. One mechanism that may contribute to the capacity for shared attention is gaze following (i.e., orienting one's eyes and head in the same direction in which another individual is looking). Gaze-following behavior has been described in both apes and monkeys (Ferrari et al. 2000; Tomasello et al. 1998). What could be the neural underpinning of it?

Similar to the visual recognition of effectors, such as the forelimb or the body, the observation of eye position is known to activate neurons in the STS (Perrett et al. 1992). Recently, however, the presence of mirror neurons for eye movements in the lateral intraparietal area (LIP) has been demonstrated. This area, located inside the intraparietal sulcus, is part of a circuit involving the frontal eye field; it plays a crucial role in organizing intended eye movements (Andersen and Buneo 2002). Most LIP neurons have a visual receptive field and discharge when the monkey gazes in the direction of the receptive field. A subset has also been found to discharge when a monkey observes the picture of another monkey looking at the neuron-preferred direction (Shepherd et al. 2009), although their discharge is weaker when the observed monkey gazes in the opposite direction. Interestingly, in this study, the picture is presented frontally, outside the neuron-preferred spatial direction. This means that the signal elicited by gazing at the other monkey is processed by high-order visual areas and fed to LIP, which in principle could use it to orient the eyes toward the same direction that the observed monkey is looking. This finding suggests that a system involved in the control of eye movements toward spatial targets is endowed with a mirror mechanism. Just as the mirror system for hand and mouth motor acts, this system is goal-related; in this case, the goal is a given spatial location. Shepherd et al. (2009), however, did not investigate whether, as in the hand and mouth mirror system, the coded goal can be broad (e.g., left or right) or more specific (e.g., in relation to a given gaze amplitude).

Thus, shared attention may be partly subserved by a cortical mechanism that is strongly involved in the control of voluntary eye movements. Although the gaze-following reaction allows an individual to share the same target with another, it can only partly explain joint attention. Other factors are required, in particular the capacity of the two individuals to share intentional communication.

The Mirror System in Humans and a Comparison between Monkeys and Humans

Electrophysiological and neuroimaging studies have demonstrated the existence of a mirror system in humans: when the action of others are observed, a cortical network becomes active when the same motor acts are executed (Rizzolatti and Craighero 2004). This network is made up of two main nodes: the frontal node includes the premotor cortex and the posterior part of the inferior frontal gyrus (IFG); the caudal node is formed by the inferior parietal cortex and, in part, the superior parietal cortex (Figure 4.1b). On the basis of anatomical and other functional considerations, the ventral premotor/IFG and inferior parietal areas belonging to this system can be considered homologous to area F5 and area PFG of the monkey, respectively. Finally, in agreement with the findings on STS in the monkey, a sector of STS is activated when motor acts are observed, but not during the execution of the same motor acts. When comparing monkey and human data, one must be cautious because it is difficult to record from single neurons in humans. This is only possible in neurosurgical patients, and only for a very limited time (for a single neuron study on mirror properties, see Mukamel et al. 2010). Furthermore, it is important to note that while the most commonly employed neuroimaging technique, fMRI, basically reveals a presynaptic activity, single neuron data express neuronal output (i.e., the outcome of presynaptic integration). Therefore, although the resolution of electrophysiological and neuroimaging techniques is progressively improving, these methods primarily support inferences at the population level.

Mirroring Intransitive Movements in Monkeys and Humans

Information about mirror responses to intransitive movements in monkeys is very restricted. Just as in the response to transitive motor acts, mirror neurons that respond to the observation of mimed motor acts have only been recorded in area F5 (Kraskov et al. 2009). Note, however, that the monkeys used for this study were previously trained to use a tool to catch food that was out of reach. Umiltà et al. (2008) have demonstrated that F5 neurons code grasping that is performed with tools, both during observation and execution, once the monkeys are trained to use those tools; this confirms the capacity of motor and mirror neurons to code the goal at an abstract level. Thus, it is possible that the mirror responses to mimed motor acts are due to a high abstraction capacity of the premotor cortex. To date, during observation of mimed motor acts, strong mirror responses have not been reported in the parietal area PFG, but it is possible that the sample did not include this type of neuron. This may be because mirror neurons are distributed more sparsely in PFG than in F5.

Although the proportion of neurons responding to intransitive motor acts or orofacial gestures in F5 is small (Ferrari et al. 2003 and previous section), one can hypothesize that from this area, neurons capable of coding goal-directed acts evolved to assign goal-directedness to pantomimes as well as to "ritualize" such acts, thus transforming the original meaning of their discharge (see Arbib 2005a).

In contrast to monkey studies, human data have revealed that cortical activation can be elicited through the observation of meaningful as well as meaningless movements. For example, transcranial magnetic stimulation (TMS) studies[1] show that the observation of meaningless movements elicits a resonance in the motor cortex, and that this activation corresponds somatotopically to that of the effector performing the observed movements (Fadiga et al. 1995). In neuroimaging studies, observation of meaningless movements appears to activate a dorsal premotor-parietal circuit that is different from the circuit activated by observing goal-directed motor acts (Grèzes et al. 1998). In contrast,

---

[1] TMS is carried out by giving magnetic pulses through a coil located on the subject's scalp, which produces an electrical field in the cortex underlying the coil, thus modifying the excitability of the neuronal population of this cortical sector. When applied to the motor cortex, pulse delivery at a given intensity elicits overt movements (motor evoked potentials, MEPs), so that it is possible to map, on the scalp, the motor representation of the activated effector. When used in research, the intensity of magnetic pulses is often kept at threshold level to study the enhancement produced by a specific task performed by the subject. For example, if a subject is asked to imagine raising his index finger, the contemporaneous TMS stimulation, at threshold, of the motor field involved in this movement will induce a MEP enhancement, whereas stimulation of the field involved in the opposite movement will determine a MEP decrease.

observation of pantomimed motor acts activates the same premotor and IFG regions as those activated by the same act when directed to a target (Buccino et al. 2001; Grèzes et al. 1998; for more on the importance of pantomime in human evolution, see Arbib and Iriki, this volume, and Arbib 2005a). As far as the inferior parietal cortex is concerned, miming of functional motor acts activates the same sectors of the parietal cortex that are active during the observation of goal-related motor acts. In particular, the observation of symbolic gestures appears to activate both the ventral premotor and the inferior parietal cortex; however, the latter involves more posterior sectors than those activated by the observation of goal-directed motor acts (Lui et al. 2008). These reports indicate that the observation of intransitive movements may partly activate the same regions that belong to the classical "mirror system"; however, different regions of frontal and parietal cortices may be involved, depending on whether these movements are meaningless or meaningful.

For intransitive gestures, we need to consider those belonging to sign language. Neuroimaging studies have shown that the basic linguistic circuit is activated by the production and comprehension of sign language (see MacSweeney et al. 2008). In two single-case neuropsychological studies in deaf signers (Corina et al. 1992; Marshall et al. 2004), a dissociation was found between a clear impairment in the use of linguistic signs and a preserved use of pantomime or nonlinguistic signs. Although in normally hearing nonsigners there is overlap between some of the cortical sectors activated by observation of pantomime and those involved in language, it may well be possible that, in deaf signers, a cortical reorganization could have created clearly separate circuits for signing and pantomiming, due to the need to use forelimb gestures for linguistic communication.

Mirroring Implies the Retrieval of "First-Person" Knowledge

A fundamental aspect of the mirror mechanism is that the resonance of the motor system, during observation of motor acts, normally occurs when the observed actions are already present in the observer's motor repertoire. That is, the observer has "first-person" knowledge of observed acts. Motor acts which do not belong to the observer's motor repertoire should thus not activate the motor system. Precisely this has been observed in an fMRI study by Buccino et al. (2004a). Here, participants were presented with video clips of mouth gestures performed by a man, a monkey, and a dog. Two types of gestures were shown: the act of biting a piece of food and oral silent communicative gestures (e.g., speech reading, lip smacking, barking). Results showed that ingestive gestures performed by an individual of another species (e.g., a dog or a monkey) activate the human mirror system. For communicative gestures, those performed by a human activated the mirror system (particularly the IFG); those performed by non-conspecifics only weakly activated it (monkey gesture) or did not activate it at all (silent barking). In the case of silent barking, only higher-order visual areas showed signal increase. These findings indicate that while all observed acts activate higher-order visual areas, only those that are known motorically by an individual—either because the acts were learned or already part of the innate motor repertoire—enter their motor network. Visual areas, such as STS, only provide the observer with a visual description of the observed act.

In summary, Buccino et al.'s study suggests that what we call "understanding" can be related to different mechanisms. The kind of automatic understanding of actions that I have described until now is based on a pragmatic, first-person knowledge. Other types of understanding, including that which results as an outcome of inferential reasoning (discussed below), are based on mechanisms that may allow discrimination between different behaviors, but which are not related to motor experience.

Intention Coding and Emotion Understanding

Actions are organized as sequences of motor acts and are aimed at an ultimate goal (e.g., eating). Motor acts that make up an action sequence, however, have subgoals (e.g., grasping an object). The ultimate goal of an action is thus achieved when fluently linked motor acts are executed (see Jeannerod 1988). The action's final goal corresponds to the motor intention of the acting agent. Although there has been rich evidence for the existence of neurons coding the goal of motor acts, the neural organization underlying action goals has, until recently, been poorly investigated.

It has now been shown that monkey premotor and parietal motor neurons play an important role in coding the intention of others (Bonini et al. 2010; Fogassi et al. 2005). During the execution of grasping for different purposes, neuronal discharge varies depending on the final goal of the action in which the grasping is embedded (e.g., grasping a piece of food for eating purposes vs. grasping an object on which to place food). Interestingly, during the observation of a grasping motor act that is embedded in different actions, the visual discharge of mirror neurons is also modulated according to the action goal. These data suggest that when the context in which another's actions unfold is unambiguous, mirror neurons may play an important role in decoding the intentions of others. A long-standing debate exists, however, as to the level and type of neural mechanisms involved in coding the intentions of others (see Gallese and Goldman 1998; Jacob and Jeannerod 2005; Saxe and Wexler 2005). The main theoretical positions are:

1. The so-called *theory theory*: ordinary people understand others' intentions through an inferential process that allows the observer to link internal states with behavior, thus enabling the observer to forecast the mental state of the observed individual.

2. The *simulation theory*: individuals are able to decode intentions of others because, by observing their behavior, they can reproduce it internally, as if they were simulating that specific behavior.

Only the second theory implies a role of the motor system in this mental function; the first implies the involvement of other cognitive circuits. While many have argued for the exclusive truth of one theory or another, it is possible that both processes are involved under certain situations. For example, when observing someone struggling with a door, we may "automatically" assume that the immediate intention is to open it (simulation theory). If, however, the door is the front door of our house and we do not know this person, we might infer that he is planning to burglarize our house (theory theory), since burglary is not part of our own action repertoire (Arbib 2012). The data described here on mirror neurons provide evidence for neural mechanisms and support the simulation theory part of this scenario; that is, showing how actions of others may be internally reproduced by the activity of the observer's motor system which codes his own intentions. In other words, the parieto-premotor mirror neuron circuit may involve a primitive form of intention understanding that occurs automatically without any type of inference process. This function requires both the activation of mirror neurons specific to a given motor act as well as contextual and mnemonic information. Note, however, that before action execution, contextual and mnemonic information per se is not enough to elicit the differential discharge of mirror neurons. In fact, this discharge is present only when the observed agent executes a *specific motor act* capable of eliciting a mirror neuron response.

A similar basic process of intention understanding has been shown in human subjects as they observe motor acts performed within different contexts (Iacoboni et al. 2005). In this study, participants were presented with different video clips showing:

- Two different contexts: one shows an array of objects arranged as if a person is just about to have tea; the other shows the same objects arranged as if a person has just had tea.
- Two ways of grasping a cup (motor act) without context (empty background).
- Two ways of grasping a cup but this time within one of the two contexts (intention condition).

Iacoboni et al.'s hypothesis was that if the mirror system is modulated by the observed motor act as well as by the global action in which this act is embedded, the presentation of the same act within a cueing context should produce a *higher* activation than when the context is absent and, possibly, a *different* activation when viewing the grasping actions in the two contexts. Independent of whether participants were simply asked to observe the video clips or to observe them to figure out the intention underlying the grasping act within the context, results showed that the "intention condition" selectively activated the right IFG, when compared with the other two conditions. This suggests that decoding motor intention in humans can occur automatically, without the need for inferential reasoning. This automatic form of intention understanding occurs frequently in daily life; however, as mentioned above, there are situations in which a reasoning process is necessary to comprehend the final goal of the observed behavior of another individual. This is typical of ambiguous situations. For example, in an fMRI study by Brass et al. (2007), participants had to observe unusual actions performed in plausible and implausible situations. Here, activation, which resulted from the subtraction between implausible and plausible situations, occurred in the STS and, less reliably, in the anterior frontomedian cortex. These two regions, together with the cortex of the temporoparietal junction, are considered to belong to a "mentalizing" system involved in inferential processes, based on pure visual elaboration of the stimuli. Thus, one can hypothesize that during observation of intentional actions, when the task specifically requires inferences to understand the agent's intentions, an additional network of cortical regions, besides the mirror system, may activate (Figure 4.3).

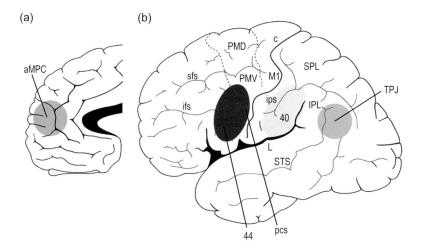

Figure 4.3 Anterior mesial (a) and lateral view (b) of the human brain. The colored regions indicate the approximate cortical sectors that belong to the mirror system (red and yellow ovals) as well as to the so-called "mentalyzing" circuit (green and light blue circles). aMPC: anterior part of the medial prefrontal cortex; C: central sulcus; ifs: inferior frontal sulcus; IPL: inferior parietal lobule; ips: inferior parietal sulcus; M1: primary motor cortex; pcs: precentral sulcus; PMV: ventral premotor cortex; PMD: dorsal premotor cortex; sfs: superior frontal sulcus; SPL: superior parietal lobule; STS: superior temporal sulcus; TPJ: temporoparietal junction. Area 40 corresponds to the supramarginal gyrus, that is the rostral half of the human inferior parietal lobule, area 44 constitutes the posterior part of Broca's area.

Representation of Other and Self by the Motor System

The discharge of mirror neurons appears to represent others' actions and intentions through a matching process that occurs at the single neuron level. The reason why an observer can recognize another's action is because the same neural substrate is activated, regardless of whether the observer thinks of performing (or actually do perform) an action or whether he sees another person perform this same action. Activation of the mirror system alone does not allow the observer to discriminate between his and the other's action. One explanation may be that a difference in the intensity or timing exists between the visual and motor response of mirror neurons, thus providing the observer with information on sense of agency. To date, however, this has not been demonstrated. There are sensory cues, however, that tell the observer who is acting. For example, proprioceptive and tactile signals are only at work when the observer is acting, whereas non-egocentric visual information reaches the observer only during observation. In joint actions, the observer's motor representations are activated by his own as well as the other's actions. Thus, one can predict that the visual discharge of mirror neurons of the observer could come first, activating the corresponding motor representation; thereafter, a motor activation related to his behavioral reaction should follow, accompanied by proprioceptive, tactile, and egocentric visual feedback. This process is not as simple as it appears. Usually, humans internally anticipate the consequences of motor acts so that the two neuronal populations—mirror neurons and purely motor neurons involved in the motor reactions—could be active almost at the same time. Furthermore, when considering the capacity of mirror neurons to predict the intentions of others, we need to account for the comparison that takes place between the predicted consequences of another's behavior and the actual performance. These processes have not been thoroughly investigated at the single neuron level. Interestingly, a recent study, which used a multidimensional recording technique, found that both premotor and parietal cortex neuronal populations can show some distinction of self action from that of the other (Fujii et al. 2008).

In humans, studies that used different paradigms (e.g., reciprocal imitation, taking leadership in action, or evaluation of observed actions performed by themselves or others) came to the conclusion that, beyond a shared neural region (mirror system), other cortical regions are selectively activated for the sense of agency. Among these, the inferior parietal cortex, the temporoparietal junction, and the prefrontal cortex seem to be crucial structures for distinguishing between own versus others' action representations (Decety and Chaminade 2003; Decety and Grèzes 2006).

Mirror System and Emotions

The mirror-matching mechanism appears to be most suitable in explaining the basic human capacity for understanding the emotions of others. In his fascinating book, *The Expression of the Emotions in Man and Animals*, Darwin (1872) described, in a vivid and detailed way, the primary emotional reactions and observed how similar they are among different species as well as, in the human species, among very different cultures.

Emotions are crucial for our behavioral responses. They are controlled by specific brain structures belonging to the limbic system and involving many cortical and subcortical sectors, such as cingulate and prefrontal cortex, amygdala, hypothalamus, medial thalamic nucleus, and orbitofrontal cortex (Figure 4.4a; see also LeDoux 2000; Rolls 2005b; Koelsch this volume). Although very much linked to the autonomous nervous system, the skeletomotor system is also involved in the expression of emotions. A good example, emphasized by Darwin, is the association between facial expressions (and also some other body gestures) and specific emotions. This link is so strong in humans that we recognize immediately the type of emotion felt by another individual, just by viewing facial expressions. This highlights the importance of signals about others' emotions for our own behavior. These signals are advantageous because they allow us to avoid danger, to achieve benefits, and to create interindividual bonds. Therefore, the mechanisms that underlie the understanding of others' emotions—the core of empathy—constitute an important issue for research.

There are different theories about how we understand the feelings of others, and most are based on the decoding of facial expressions. One theory maintains that the understanding of others' emotions occurs through inferential elaboration, based on emotion-related sensory information: a certain observed facial expression means happiness, another sadness, and so on. Another, very different theory holds that we can understand emotions because emotion-related sensory information is directly mapped onto neural structures that, when active, determine a similar emotional reaction in the observer. This theory implies a "mirroring" of the affective state of the other individual and involves a partial recruitment of the visceromotor output associated with specific facial expressions. In fact, the neural structures related to affective states are also responsible for visceromotor outputs (i.e., motor commands directed to visceral organs).

Results from several studies suggest that a neural mechanism similar to that used by the mirror system for action understanding is also involved in the understanding of emotions. For example, in subjects instructed to observe and imitate several types of emotional facial expressions, Carr et al. (2003) demonstrated that the IFG and the insular cortex were activated. Frontal activation would be expected on the basis of the motor resonance (discussed above). Regarding the insular cortex (see Figure 4.4b), this region is the target of fibers that convey information about an individual's internal body state (Craig 2002), in addition to olfactory, taste, somatosensory, and visual inputs. With respect to the motor side, according to older data in humans (Penfield and Faulk 1955; Showers and Lauer 1961; for a recent demonstration, see Krolak-Salmon et al. 2003), it has been reported (Caruana et al. 2011) that

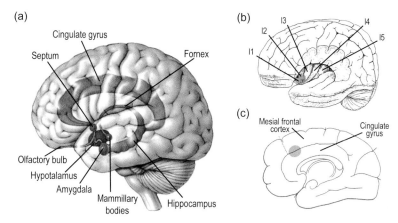

Figure 4.4 (a) Location of the main regions forming the limbic circuit (dark gray) overimposed on a transparent lateral view of the left human hemisphere. Most of these regions lay close to the hemispheric midline. (b) Lateral view of a human hemisphere showing the insular cortex after removal of the ventral part of frontal and parietal cortex, which normally covers it. Note that the insula is subdivided in several lobules (I1 to I5). The green oval indicates the approximate location of the anterior part of the insula, where fMRI shows overlap of activation during observation and feeling of disgust and pain. (c) Mesial view of a human hemisphere. The yellow oval indicates the approximate location of the cortical sector where fMRI shows overlap of activation during observation and feeling of disgust and pain. This sector encompasses the cingulate and paracingulate gyrus. Modified after Pinel (2006) and Umiltà (1995).

microstimulation of the monkey insula can produce both somatomotor and visceral responses (e.g., change in heart and respiration frequency, effects on the gastroenteric system).

The relevance of the insular cortex for the understanding of emotions has been further elucidated in various fMRI studies. One of these (Wicker et al. 2003) investigated areas activated by disgust. Here, participants had, in one condition, to smell pleasant or disgusting odorants and, in another, to observe video clips of actors smelling disgusting, pleasant, and neutral odorants and expressing the corresponding emotions. The most important result was that the same sectors of the anterior insula and (to a lesser degree) anterior cingulate cortex (see Figure 4.4c) were activated when a participant was exposed to disgusting odorants and when disgust was observed. No such overlap was found, however, in the insula for pleasant stimuli. Interestingly, clinical studies show that insular lesions produce deficits in recognizing disgust expressed by others (Adolphs et al. 2003; Calder et al. 2000). As far as the cingulate cortex is concerned, this phylogenetically old region is subdivided (a) along its rostro-caudal axis into a posterior, granular and an anterior, agranular sectors and (b) along its dorso-ventral axis into four sectors, from that adjacent to the corpus callosum to the paracingulate gyrus. Whereas the ventral part (area 24) is more endowed with motor functions, the anterior dorsal part is activated by painful stimuli; it contributes to the processing of aversive olfactory and gustatory stimuli. During a neurosurgical operation, Hutchison et al. (1999) report that a single neuron in this cortical sector responded both when the patient received a painful stimulus to his finger as well as when he saw the surgeon receive the same stimulus.

An fMRI study by Singer et al. (2004) investigated empathy for pain: Participants were couples. Female partners were scanned while their male partners stood just outside the scanner, with only their hands visible to the female partner. Two different cues informed the female partner whether she was going to receive a light painful stimulus ("self" condition) or whether her partner was going to receive it ("other" condition). Among the areas activated in the two conditions, the anterior insula and anterior cingulate cortex (Figure 4.4b and 4.4c) showed overlapping activations. The empathic scales constructed with the subjects' responses to specific questionnaires revealed that there was a significant positive correlation between the degree of empathy and intensity of activation in these cortical regions. In a similar, subsequent study (Singer et al. 2006), both male and female participants were scanned while they observed another individual receiving a painful stimuli, similar to the experiment described above. The observed individual was one of two actors who, before scanning, had played a game with subjects (one acted as a fair player, the other as unfair). For female participants, the results of this study replicated the previous study. For male participants, however, insula activation was present only when a fair player was observed to feel pain. When an unfair player was observed to experience pain, the nucleus accumbens, a reward-related area, was activated. This activation correlates with a desire for retaliation, as assessed in a post-scanning interview.

Altogether, these findings suggest that humans understand disgust (and most likely other emotions) through a direct mapping mechanism that recruits a first-person neural representation of the same feelings. These feelings are normally associated with precise visceromotor reactions. Intensity of activation can vary, however, depending on how the observed emotion is embodied. It can also be modulated by other cognitive factors, thus involving the activation of other areas not directly related to the observed emotion. Thus, it can be concluded that a mirror system, different from the one activated by actions but working in a similar way, may come into play during the understanding of others' emotions. This kind of mechanism is very likely a necessary prerequisite for establishing empathic relations with others, but per se is not sufficient. To share an emotional state with another does not always elicit the same reactions. If, for instance, we see a person expressing pain, this does not mean that we automatically feel compassion. Compassion may occur with higher probability if the person in pain is a friend or a relative. However, it is much less likely to happen if the other person is not known to us or is an enemy, as demonstrated by Singer et al. (2006).

Mirroring and Language

Many anatomical and functional data indicate a possible homology between human Broca's area (a crucial component of the language system), or part of it, and the ventral premotor area F5, although there is debate over which sector of F5 is the true homologue of Broca's area (Petrides et al. 2005; Rizzolatti and Arbib 1998). Coupled with these findings, the property of the motor system to code goals and the presence of mirror neurons in it prompted the idea that these basic functions could be good candidates for explaining how dyadic communication, and then language, evolved. However, whether this evolution was originally grounded in gesture (Rizzolatti and Arbib 1998) or vocalization (Fogassi and Ferrari 2004, 2007) is a matter of debate (for a discussion on gesture, see Arbib and Iriki, this volume).

In terms of vocalization, consider the following: As reported above, in monkey area F5, there is a category of mirror neurons that is specifically activated when motor acts performed with the mouth—some of which respond to communicative monkey gestures—are observed or executed (Ferrari et al. 2003). Second, a subset of F5 motor neurons has been recently reported that activate during the production of trained calls (Coudé et al. 2011). For many years, call production was considered an attribute of emotionally related medial cortical areas. However, the data of Coudé et al. suggest that PMV (already endowed with a neural machinery for the control of voluntary hand and mouth actions) could also be involved in the voluntary control of the combination of laryngeal and articulatory movements to produce vocalizations that are not simply spontaneously driven by a stimulus. These findings, together with the known capacity of PMV to control laryngeal movements (Hast et al. 1974; Simonyan and Jürgens 2003), suggest that this cortical sector could constitute a prototype for primate voluntary vocal communication. It has still to be established whether this neuronal activity related to call production is paralleled by the presence of mirror activity for the perception of these same calls. While mirror neurons related to orofacial communicative gestures are already present at the phylogenetic monkey level, neurons endowed with the property to produce and perceive vocalizations may have appeared later in evolution.

Interestingly, as reported in chimpanzees, vocalizations can be combined with meaningful gestures (Leavens et al. 2004a). This suggests the possibility that the "gestural" and "vocal" theories of language evolution could, at some level, converge. In fact, in the hominin lineage, the frequency of combined vocalizations and gestures could have increased and, as the orofacial articulatory apparatus became more sophisticated, this apparatus may have achieved a leading role in communication. Note, however, that the link between spoken language and gestures is still present in our species and that a reciprocal influence between these modalities has been clearly demonstrated (McNeill 1992; Gentilucci et al. 2004b; Gentilucci and Corballis 2006).

Unlike the language frontal area, there is presently no clear evidence for a possible homologue of Wernicke's area (for some possibilities, see Arbib and Bota 2003). In addition to the anatomical location of Broca's area (which is similar to that of F5 in the monkey precentral cortex), fMRI experiments show that this cortical sector becomes active when subjects perform complex finger movements and during imitation of hand motor acts (Binkofski et al. 1999; Buccino et al. 2004b; Iacoboni et al. 1999). This indicates that this area, beyond controlling mouth motor acts, contains a hand motor representation as well. Other fMRI studies show that Broca's area activates when hand and mouth motor acts are observed (Binkofski et al. 1999; Buccino et al. 2004b; Iacoboni et al. 1999). This indicates that, beyond controlling mouth motor acts, this area also contains a hand motor representation. Other fMRI studies show that Broca's area activates during the observation of hand and mouth motor acts (Buccino et al. 2001) or listening to the noise produced by some of these acts (Gazzola et al. 2006). This is reminiscent of the presence of audiovisual mirror neurons in macaque F5, which are activated by sound produced by the observed motor act (Kohler et al. 2002). Because of all these activations, Broca's area has been included in the frontal node of the mirror system. In addition, Broca's area is known to activate while listening to words (Price et al. 1996).

This evidence leads to the hypothesis that if a mirror mechanism is involved in speech perception, the motor system should "resonate" during listening to verbal material. This is exactly the finding of a TMS study (Fadiga et al. 2002) in which motor evoked potentials (MEPs) were recorded from the tongue muscles of normal volunteers who were instructed to listen to acoustically presented words, pseudo-words, and bitonal sounds. In the middle of words and pseudo-words there was either a double "*f*" or a double "*r*": "*f*" is a labiodental fricative consonant that, when pronounced, requires virtually no tongue movements; "*r*" is a linguo-palatal fricative consonant that, in contrast, requires the involvement of marked tongue muscles to be pronounced. TMS pulses were given to the left motor cortex of the participants during stimulus presentation, exactly at the time in which the double consonant was produced by the speaker. The results show that listening to words and pseudo-words containing the double "*r*" determined a significant increase in the amplitude of MEPs recorded from the tongue muscles with respect to listening to words and pseudo-words containing the double "*f*" and bitonal sounds. Furthermore, activation during word listening was higher than during listening to pseudo-words. This strongly suggests that phonology and, perhaps, (partly) semantics are processed by the motor system. These findings appear to be in line with Liberman's motor theory of speech perception (Liberman and Mattingly 1985), which maintains that our capacity to perceive speech sounds is based on shared representations of speech motor invariants between the sender and the receiver. While it is still being debated whether semantic attributes of words are understood directly through this mechanism or whether this mirror activity is not necessary for this function, the possibility that a mirror-matching mechanism is fundamental for phonological perception is quite compelling. Other theoretical approaches contrast with this view (see, e.g., Lotto et al. 2009). Among the arguments, there are data in Broca's aphasics which show that these patients can be as good as normal in word comprehension or, although impaired in speech discrimination, can be good in speech recognition. Furthermore, there are lesions involving the left frontoparietal system that leave speech recognition intact.

Interestingly, a strict link between language and the motor system has been provided by studies in which subjects had to listen to sentences containing action verbs—*I grasp a glass* or *I kick a ball*—contrasted with abstract sentences—*I love justice* (Tettamanti et al. 2005). The subtraction between action-related sentences and abstract sentences produced a somatotopic activation of premotor areas, in sectors corresponding to the activation of the effectors involved in the listened sentence. Thus, verbal material related to action verbs activates not only Broca's area but also the corresponding motor representations in the whole premotor cortex. It is worth noting that words related to nonhuman actions do not elicit activation of the motor cortex (Pulvermüller and Fadiga 2009).

In addition to phonology and semantics, the other fundamental property of human language is syntax (Hagoort and Poeppel, this volume). Lesion studies demonstrate that Broca's area and the left perisylvian cortex play an important role in grammar processing (Caplan et al. 1996). Patients with lesions to the inferior frontal cortex also have difficulty in ordering pictures into well-known sequences of actions (Fazio et al. 2009). Neuroimaging investigations show that Broca's region (BA 44 or pars opercularis and BA 45 or pars triangularis) in the inferior frontal cortex and Wernicke's region in the superior temporal cortex are more strongly active in response to complex sentences than to simple control sentences. Furthermore, these areas are active when listening to hierarchically nested sequences (Bahlmann et al. 2008). Sequential organization is typical of actions, and various aspects of it are coded by several cortical regions (Tanji 2001; Tanji and Hoshi 2008; Bonini et al. 2010; Bonini et al. 2011; also discussed below). It is not clear whether and how the structure of sequential motor organization could have been exploited for linguistic construction. One hypothesis holds that during evolution, the more the motor system became capable of flexibly combining motor acts to generate a greater number of actions, the more it approximated a linguistic-like syntactic system. Such a capability could have extended to the combination of larynx and mouth movements in phono-articulatory gestures for communicative purposes (Fogassi and Ferrari 2007).

Mirroring in Dance and Music

For the mirror-matching mechanism to occur, motor representations (either of acts or sounds) must be part of the observer's internal motor repertoire. This repertoire allows us to understand many goal-directed actions and meaningful gestures. Dance gestures constitute one type that is easily recognized. Although many of the observed motor synergies performed by an expert dancer can be represented in the observer's motor system, the steps specific to a given type of dance are unknown to naïve observers. Thus, motor resonance should be higher when an observer is also an expert dancer. In fact, neuroimaging studies have demonstrated that during the observation of dance steps, activation of the observer's motor system is higher when the observed steps belong to the observer's own motor experience (Calvo-Merino et al. 2005). Furthermore, the observation and imagination of rehearsed dance steps produce higher motor activation than steps that are not rehearsed (Cross et al. 2006). Activation of the mirror system is thus greater when the observing subject has more motor experience in the observed motor skills. This result is important because it demonstrates a mirror activation not only for actions, but also for meaningful gestures (e.g., dance steps), which are probably derived from goal-directed actions through a process of ritualization (Arbib 2005a).

The issue of the relation between dance recognition and the structure of the motor system underlying this recognition involves not only the capacity to "resonate" during observation of single gestures but also the capacity to recognize sequences of gestures. This issue has been partly addressed above for the understanding of intentional actions and language. How, though, does the motor system build and, as a consequence, recognize this sequence? A similar question is also valid for dance. It is very likely that we recognize the single gestures that form dance steps using our mirror system; however, a process is then needed to allow the appropriate sequence to be internally reconstructed. Studies on imitation learning of playing instruments may further our understanding of the neural circuitry involved in this process. The basic neural organization for encoding and programming motor acts or meaningful gestures is represented by the parietofrontal circuits, and this organization comes into play both during execution and observation. However, when individuals are required, for instance, to imitate a guitar or a piano chord, neuroimaging studies indicate that there is also a strong involvement of prefrontal cortex (Buccino et al. 2004b). Prefrontal cortex has been classically considered to have a major role in action planning. When it is necessary to build sequences of motor acts, however, its role becomes crucial (Figure 4.5).

Music is another function that may imply mirroring. Although music cannot be strictly considered as a kind of dyadic communication, it can involve a sender and a receiver of a message. Music can, of course, be experienced alone. However, music is often shared with other people. When we attend a concert, for example, both our auditory and visual systems become deeply engaged. The sensory inputs elaborated inside our cortex activate several higher-order neural structures: high-order sensory and motor areas as well as the emotional system. Because of the presence of mirror neurons, the motor system comes into play as we observe the motor acts of musicians as they perform on

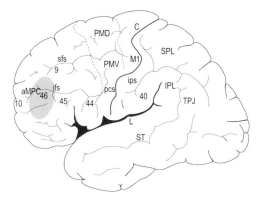

Figure 4.5 Lateral view of a human hemisphere. The gray oval indicates the approximate prefrontal sector involved in action planning and organization. This part of the cortex appears to be important in organizing the sequence of motor acts composing an action. Abbreviations as in Figure 4.3. Area 44 and 45 constitute Broca's area.

their instruments while concurrently listening to the auditory outcome of these gestures. It is quite obvious that this observation-listening process may activate hand or mouth mirror neurons as well as audiovisual mirror neurons; thus we resonate motorically even while remaining still. In addition, we often tap our fingers or toes to the rhythm of the music, which in turn is the result of repeated motor acts. As discussed for dance, sequential behavior in music is fundamental. The same circuitry that is involved in sequential actions and gestures may also be involved in music production and perception.

A specific case of production or perception in music is denoted by musicians themselves, who must synchronize many elements to produce a perfect musical piece. As discussed for joint actions, to play music with others we require:

- a neural mechanism that permits the same motor representations (mirror mechanism) to be shared with co-players,
- a mechanism(s) to distinguish our own actions from those of the others, and
- a mechanism that allows coordination (see also Levinson, this volume).

Several studies have demonstrated a retrieval of the motor system while listening to sounds. For example, participants listening to both familiar and unfamiliar sounds show (when activation from the latter is subtracted from the former) an activation of the superior temporal cortex as well as the supplementary motor area and IFG (Peretz et al. 2009). Listening to and reproducing isochronous or complex rhythms activate several areas of the premotor cortex (pre-SMA, SMA proper, premotor cortex) and subcortical structures (basal ganglia, cerebellum) (Chen et al. 2009). The role of PMD seems to be more related to the use of the metric structure of sound for movement organization, whereas PMV may transform known melodies into movement. This has also been described in an area between PMD and PMV, recruited during passive listening.

It is common knowledge that music can evoke several different emotions (for a review, see Juslin and Vastfjall 2008b; see also Scherer and Koelsch, both this volume). Sound intensity, fluctuation of timing, and timbre, for example, may elicit emotions such as happiness, sadness, and fear. The issue here is whether music can evoke emotional mirroring.

Gestures of players may elicit a kind of mirroring. One would expect this effect to be stronger in accomplished players. This mirroring, in turn, may evoke emotional feeling in a listener that draws upon the listener's own past experiences of playing. Postural gestures of the player during performance emphasize musical message and may elicit a contagion in the observer/listener. Molnar-Szakacs and Overy (2006) have proposed that while listening to expressive music, the mirror system for actions will be activated and relayed to the limbic system through the insular cortex, which, as described above, is a crucial structure for the representation of subjective states. Other authors make a parallel between acoustic features of music and emotional speech, suggesting that in the listener there is a mechanism that mimics the perceived emotion internally, in response to specific stimulus feature (Juslin 2001). In line with this theory, Koelsch et al. (2006) found that listening to music activates brain areas related to a circuitry that serves the formation of premotor representations for vocal sound production. They conclude that this could reflect a mirror function mechanism, by which listeners may mimic the emotional expression of the music internally. All of these considerations suggest that listening to music and observing performers produce both an activation of the motor system and an emotional mirroring, involving the insula and areas of the limbic system, such as cingulate cortex and amygdala.

Another important aspect of music that is strongly related to the motor system is represented by song. In animals studies, song production has been investigated in birds (see Fitch and Jarvis, this volume), although birdsong is different from a human song with words. The presence of mirror neurons activated by both the production and listening of the species-specific song was recently demonstrated in a telencephalic nucleus of a swamp sparrow (Prather et al. 2008). These mirror neurons, however, do not appear to code the goal of a motor act, as in phonological resonance; instead, they map sequences of heard notes onto the brain motor invariants used to produce them. Interestingly, disrupting auditory feedback does not alter this singing-related neuronal activity, which clearly indicates its motor nature. The fact that these neurons innervate striatal structures that are important for song learning suggests their possible role in vocal acquisition.

The presence of mirror neurons in song suggests that resonance mechanisms are probably parsimonious solutions which have evolved in the vertebrate brain several times to process complex biological information. Indeed,

Fitch and Jarvis (this volume) view birdsong and human vocal learning as convergently evolved systems.

Another interesting aspect of birdsong is that it, like speech, is sequentially organized. Since sequential organization is typical of actions, speech, and music, it would be interesting to elucidate better the neural mechanisms that underlie both production and recognition of motifs in singing birds. In particular, it would be helpful to understand whether the structure of birdsong is more comparable to the syntax or phonology of language (Yip 2006).

Action Sequence Organization and Language Structure: A Comparison

Sequential organization is a typical feature of motor actions. Motor acts endowed with a meaning (the motor goal) form the basis for action (Jeannerod 1988; Bernstein 1996; Rosenbaum et al. 2007). A similarity of this organization with that typical of language can be suggested at two different, but not mutually exclusive, levels. At the first level, motor acts seem to play the role of the words within a phrase. As the motor acts, words are the first minimal element of language endowed with a meaning. The meaning of a sentence is given by the sequential organization of words. By changing the position of the words, the meaning of the sentence changes or is missing. Similarly, if the order of the motor acts in an action sequence is changed, the action goal may also change.

At a second, more motoric, level, syllable production is the result of the execution of orofacial motor acts in combination with the appropriate contraction of larynx muscles. The higher-level neural control, underlying this combination, intervenes to organize a fluent link between syllables, exactly as occurs in the neural control of forelimb or mouth actions. The apparent difference between syllables and motor acts is that the former are devoid of meaning. However, one could always argue that the achievement of the specific configuration necessary to pronounce a syllable represents a goal per se.

While we do not know the neural mechanism that underlies the organization of syllables into words, and of words into phrases, the issue of sequential organization in the monkey motor system at the single neuron level has been addressed in two main series of studies. The first assessed the responses of neurons in mesial (SMA/F3, pre-SMA/F6) and prefrontal cortex, while monkeys executed sequences of movements (Tanji 2001; Tanji and Hoshi 2008), such as turning, pulling, pushing, or traversing trajectories in a maze. These studies show that the neurons of these cortical sectors were activated for different aspects of the task. Some categories of neurons code the type of sequence; others denote the order of a movement within a sequence; still others code the final location of the movement series. The second series of studies (Fogassi et al. 2005; Bonini et al. 2010; Bonini et al. 2011) assessed the responses of parietal and premotor neurons during execution and observation of natural action sequences. Results show that motor and mirror neurons from these areas can be differently activated depending on the specific action sequence in which the motor act they code is embedded.

Together, these two series of studies suggest that sequential actions are organized under the control of the premotor-parietal motor system and prefrontal cortex. Part of this neural substrate has a major role in linking motor acts for the achievement of an action goal, while part is involved in coding the order in which various motor elements can appear in a sequence. The mechanisms used by these cortical circuits could have been exploited during the evolution of the organization for syntactic structure. Note that the above described activation of motor structures during language and music production and perception is a good, although indirect, confirmation of this hypothesis.

Relation between Aphasia, Amusia, and Apraxia

Many human neuroimaging studies report an activation of Broca's area and its right homologue during the processing of syntax in both language and music (Chen et al. 2008a; Maess et al. 2001; Patel 2003). Patients with damage to Broca's area can, however, show both aphasia and amusia (Alajouanine 1948; Patel 2005). Interestingly, many Broca's patients can also show limb apraxia, although aphasia and apraxia are not necessarily associated (De Renzi 1989). A recent study showed that such patients, unlike another group of apraxic patients with parietal damage, present a deficit in gesture recognition (Pazzaglia et al. 2008). The presence of three possible syndromes with a lesion to the same region raises the question of whether there is a shared common mechanism or structure (see Fadiga et al. 2009). A possible sharing of similar circuits between language and music comes also from some rehabilitation studies. It has been reported that dyslexic children and nonfluent aphasic patients can benefit from music therapy, as demonstrated not only by an improvement in behavioral scores but also by a change in the white matter of corticocortical connections between superior temporal cortex and premotor cortex/inferior frontal gyrus, in particular the arcuate fasciculus (Schlaug et al. 2009).

In conclusion, it is possible that language and music may partially share neural circuits, due to a possible common motor substrate and organization. The fact that cortical regions included in the mirror system for actions are also activated during the processing of language and music-related gestures lends further support to this hypothesis.

Acknowledgments

I thank D. Mallamo for his initial help in preparing the illustrations. This work has been supported by the Italian PRIN (2008) the Italian Institute of Technology (RTM), and the ESF Poject CogSys.

5

Emotion in Action, Interaction, Music, and Speech

Klaus R. Scherer

Abstract

This chapter highlights the central role of emotion in understanding the relationships between action, interaction, music, and speech. It is suggested that brief nonverbal displays of emotion (affect bursts) may have played an important part in the evolution of human communication based on *speech and gesture*, and, probably in parallel, *singing and music*. After a brief account of the evolutionary development of emotion, the nature and architecture of the emotion system is discussed, advocating a componential appraisal model. Particular emphasis is given to the component of motor expression. A dynamic model of emotion communication distinguishes the function of expression as symptom (of actor state), symbol (of a message), and appeal (to the listener), highlights the role of distal and proximal cues in the process, and describes the differential types of coding (biological push vs. sociocultural pull) of the expressive signs. A brief overview of research on vocal emotion expression in speech and music provides evidence for the similarity of the expressive cues used to convey specific emotions. A distinction is proposed between *utilitarian emotions*, which help adaptation to relevant events that happen to a person, especially through speech in social interaction, and *aesthetic emotions*, which are generally desired and sought out through engagement with cultural practices such as music, art, or literature. In conclusion, some evidence supporting the proposal that affect bursts might have been the starting point for the joint evolution of language and music is reviewed.

Introduction

> Music is nothing else but wild sounds civilized into time and tune. Such is the extensiveness thereof, that it stoopeth as low as brute beasts, yet mounteth as high as angels. (Thomas Fuller, 1640)

There are many theories on the origin of language and music, postulating a large variety of possible mechanisms. As these different accounts are not mutually exclusive, the safest position might be that many factors contributed to the development of what we now know as speech and music. In many of these accounts, the *expression of emotion* is seen as a precursor to these pinnacles of human evolution. This is not surprising as emotion is an evolutionarily old system, shared with many animal species, that plays a major role in behavioral adaptation and social interaction. Many of the pioneering scholars who advocated this view—Rousseau and Herder in the 18th century (see Moran and Gode 1986) and later Helmholtz (1863/1954), Darwin (1872/1998; see also Menninghaus 2011), and Jespersen (1922)—have also assumed strong links between the development of music and language through protospeech and protosinging. Thus, it seems fruitful to examine the role of emotion in the evolution of these communication systems, and of the brain that controls them, to understand their mysterious relationship. Music encompasses an enormous variety of sound combinations that carry extremely subtle emotional meanings. As for language, although it has evolved abstract syntactic and semantic rule systems to represent meaning, much of the communicative power of speech is linked to its capacity to express moment-to-moment changes of extremely subtle affective states. Most likely, the evolution of the complex human brain accounts for both the wide range and subtlety of human emotionality and the sophisticated control systems that allow the social communication of emotion through an intrinsic connection of the systems of language and vocal music to their carrier: the human voice.

This chapter aims at providing a foundation for further exploration of these relationships. First, the nature and function of emotion are discussed, followed by the description of a probable architecture of the human emotion system in the form of a component process model. Special emphasis is placed on emotional expression, and the pertinent interdisciplinary research on the communication of emotion via speech prosody and music is reviewed. Finally, it is proposed that multimodal affect bursts might be reasonably considered as a starting point for the joint evolution of speech and language, and the plausibility of this hypothesis is discussed on the basis of selected evidence. While there are occasional references to the neural underpinnings of the phenomena discussed, I cannot do justice in this chapter to these complex issues, and the reader is thus referred to more appropriate sources: recent neuroscience approaches to speech and music are addressed by Hagoort and Poeppel, Parsons and Janata, Koelsch, and Patel, all this volume; details on work concerning the brain circuits of emotion can be found in Davidson et al. (2003:Part 1; see also LeDoux 2000).

Function and Evolution of Emotion

Many early theorists have emphasized that emotion can disrupt or interrupt coordinated behavior sequences leading to a redirection of behavior. This is particularly true for strong emotions which can engender sudden motivational shifts. However, milder emotions, such as pleasure or anxiety, can be superimposed on plans and behavior sequences without necessarily interrupting them. This point is important in discussing music-induced emotions, which only rarely produce violent emotions or interrupt and redirect the listener's behavior. Consequently, we need to consider the functions of a wide variety of emotional processes.

Emotion can be seen as an interface between the organism and its environment, mediating between constantly changing situations and events and the individual's behavioral responses (Scherer 1984). The major aspects of this process are threefold:

1. Evaluation of the relevance of environmental stimuli, events, or constraints for the organism's needs, plans, or preferences in specific situations.
2. Preparation, both physiological and psychological, of actions that are appropriate for dealing with these stimuli.
3. Communication of reactions, states, and intentions by the organisms to the social surroundings.

It is the last of these aspects—communication—that is most relevant to language and music.

In evolutionary history, emotions have replaced rigid reflex-like stimulus response patterns with a greater flexibility of behavioral inventories, supporting cognitive evaluation processes to assess stimuli and events. This decoupling of stimulus and response also makes it possible to constantly reevaluate complex stimuli and/or situations without much time delay, since preparation of the behavioral reaction is part of the emotion process (Scherer 1984, 2001).

In socially organized species, emotions are often visible in rudimentary patterns of motor expression to communicate both the reaction and the behavioral intention of the individual to the social surroundings. This allows other organisms to predict the most likely behavior of the person expressing the emotion and to plan their own behavior accordingly. In turn, this provides feedback about the likely reaction of others to the intentional movement or expression, allowing appropriate changes in the behavioral plans of the individual expressing emotion.

Darwin (1872/1998) emphasized that particular types of functional expression of emotion in the form of intention movements seem to have been selectively developed in the course of evolution for purposes of communication, and may have laid the building blocks for the development of speech and music. This coevolutionary process was made possible by the continuously increasing complexity of the human central nervous system and the accompanying changes in the peripheral nervous system underlying emotional expression (see Hebb's [1949] suggestion of a positive correlation between cognitive sophistication and emotional differentiation across species).

The preceding account focuses mainly on the behavioral flexibility afforded by the emotion system and the important communicative functions in social interaction which guide the action tendencies of interaction partners and other group members. Of course there are many more factors involved in the evolution of the emotion mechanism (see Arbib and Fellous 2004; Nesse and Ellsworth 2009; Tooby and Cosmides 1990).

What, then, is an emotion? Frijda and Scherer (2009) propose a functional approach to answer this question, pointing out that most scholars seem to agree with the following constituents:

1. Emotions are elicited when something relevant happens to the organism that has a direct bearing on its needs, goals, values, and general well-being. Relevance is determined by the appraisal of events on a number of criteria; in particular their novelty or unexpectedness, their intrinsic pleasantness or unpleasantness, and their consistency with current motivations.
2. Emotions prepare the organism to respond to important events in its life and thus have a strong motivational force, producing states of so-called action readiness.
3. Emotions engage the entire organism by preparatory tuning of the somatovisceral and motor systems and thus involve several components and subsystems of the organism. These tend to cohere to a certain degree in emotion episodes, sometimes to the point of becoming highly synchronized.
4. Emotions bestow what is called control precedence. This means that the states of action readiness which emotions have prepared can claim (not always successfully) priority in the control of behavior and experience.

Thus, "an emotion" is defined here as *a bounded episode characterized by an emergent pattern of synchronization between different components that prepare adaptive responses to relevant events as defined by the four criteria outlined above* (cf. Scherer 2005). In this generality, the definition covers a wide variety of very different forms of emotion, including proto-emotions shared by many animals, thus constituting a very early achievement in evolution. Many ethologists and biologists do not like to talk about emotions in animals, as they often assume that emotions imply conscious feelings. This is not true for the minimal definition given above. Reducing the mechanisms involved to the essentials (e.g., in analyzing the emotions of an "organismus simplicissimus"; see Scherer 1984:299–301) illustrates the evolutionary continuity of the emotion system.

The phylogenesis of a highly sophisticated emotion system building on the functional characteristics described above, but adding the ability to self-reflect and report on experiential feelings, has required major evolutionary changes in brain organization from protohominids to *Homo sapiens*, particularly in the

Table 5.1 Design features of emotion and required evolutionary changes for the development of the complex human emotion system.

Design features	Necessary evolutionary changes
Rapid detection of relevant events and evaluation of expected consequences (often in the absence of sufficient information)	More powerful analysis and evaluation mechanisms (with respect to memory, learning, association, inference, and prediction), particularly the development of more fine-grained appraisal checks (e.g., attribution of causality, compatibility with social norms and values).
Preparation of highly synchronized response organization in the service of adaptive action	More flexible motor control systems for exploration and manipulation responses.
Connected multimodal signaling of reactions and intentions to the social environment	Complex circuitry for sending and receiving socio-emotional messages, especially a high degree of voluntary control of vocal and facial expression for the purpose of strategic communication.
Constant recursive monitoring through reappraisal, response adaptation, and emotion regulation	Integrative brain representation of appraisal results and somatosensory feedback of bodily reactions, accessible as conscious feeling, for regulation purposes. Development of the capacity of categorization and labeling of feelings for social sharing.

context of complex social organization and interaction. Table 5.1 illustrates some of the necessary evolutionary changes required for the human emotion system.

The Architecture of the Emotion System

Despite a growing convergence toward a functional view of emotion, as outlined above, there are appreciable differences in currently proposed emotion theories. The following discussion provides a brief overview.

Basic or *discrete emotion theories* have their roots in Darwin's (1872/1998) description of the appearance of characteristic expressions for eight families of emotions and the functional principles underlying their production (serviceable associated habits). Thus, Tomkins (1962) and his disciples Ekman (1972) and Izard (1977) developed theories which suggest that a specific type of event triggers a specific "affect program" corresponding to one of the universal basic emotions and produces characteristic patterns of expressions and physiological responses.

Dimensional/constructivist emotion theories often invoke the proposal by James (1884/1968:19) that "the bodily changes follow directly the perception of the exciting fact, and that our feeling of the same changes as they occur *is* the emotion." Current proponents of this position generally interpret this in the sense that no specific determinants for emotion differentiation are required, so an individual feeling bodily arousal is free to construe situational meaning. This position is frequently adopted by dimensional emotion theorists. Thus, Russell (2003) proposes that "core affect," defined as a position in a valence x arousal affect space, can be construed as a felt emotion on the basis of motivational or contextual factors. Constructivist accounts have also been proposed by scholars concerned with the social role of emotion and the influence of culture.

Appraisal theories of emotion date back to Aristotle, Descartes, Spinoza, and Hume, who assumed that the major emotions (as indexed by the respective words in the language) are differentiated by the type of evaluation or judgment a person makes of the eliciting event. However, the term *appraisal* was first used in this specific sense in the 1960s and only saw active theoretical development in the early 1980s (see the historical review by Schorr 2001). Appraisal theories assume that emotions are elicited and differentiated by people's subjective evaluation of the significance of events for their well-being and goal achievement. While critics have misrepresented the appraisal process as requiring effortful, conscious, and conceptual analysis, theorists in this tradition have consistently emphasized that appraisal processes often occur in an automatic, unconscious, and effortless fashion. Leventhal and Scherer (1987) have suggested that the type of processing with respect to content (types of appraisal) and level of processing (sensorimotor, schematic, and conceptual) is determined by the need to arrive at a conclusive evaluation result (yielding a promising action tendency). If automatic, effortless, unconscious processes do not produce a satisfactory result, more controlled, effortful, and possibly conscious mechanisms are brought into play. For example, "biologically prepared" stimuli such as snakes (Öhman and Mineka 2001) or baby faces (Brosch et al. 2007) can be processed in a rapid, effortless fashion by subcortical circuitry. In contrast, more complex, novel stimuli are likely to be processed at a higher conceptual level, involving propositional knowledge and underlying cultural meaning systems that most probably involve the prefrontal cortex. This architecture, reminiscent of the superposition of different levels of brain structure in evolution, allows us to apply the model to many different species as well as different developmental stages of organisms.

Appraisal theories highlight the dynamic character of the emotion process (an episode rather than a state), producing a rich variety of changing experiences which can ultimately be bounded, categorized and labeled, especially when the emotional experience is shared via verbal communication (Scherer 2004a, 2009).

Despite differences with respect to detail, there is now wide agreement among most emotion researchers on the central role of appraisal in eliciting and differentiating emotion episodes and on multicomponential response structures including subjective experience (Frijda 2007; Moors et al. 2013). The *component process model* (CPM) is one of the earliest and most elaborated models proposed in this tradition (Scherer 1984, 2001, 2009) and will serve here as an illustration of the general approach.

Figure 5.1 shows the architecture of the CPM, which generates dynamic emotion processes following events that are highly pertinent to the needs, goals, and values of an individual. The model is dynamic in that all of its components are subject to constant change given the various feedforward and feedback inputs. This is also true for the triggering "event" (a term used as a placeholder for objects encountered, consequences of one's own behavior or that of others, as well as naturally occurring phenomena), since events generally unfold over time and may thus require frequent reappraisal. Even without changes in the event itself, appraisals will constantly change given the sequential processing of different criteria on different levels of brain organization, leading to frequent reappraisal. Most importantly, the processes described by the model are recursive; that is, the results of later processing stages are fed back to earlier stages where they may produce modifications of prior outcomes. For example, if I appraise the behavior of a stranger as insulting, resulting in a tendency toward aggression, the feedback of the respective motor preparatory changes to the appraisal module may lead to reappraisal with respect to the appropriateness of the response to the nature and strength of the provocation. These recursive processes will continue until the system has reached a new equilibrium and no more major changes are occurring, ending the particular emotion episode.

The timescale of these recursive processes is highly variable depending on the nature and temporal organization of the event structure and the internal appraisal and response processes. Generally, however, emotion episodes, by definition, are expected to be relatively brief, a matter of minutes rather than hours. In particular dramatic circumstances, different emotion episodes that

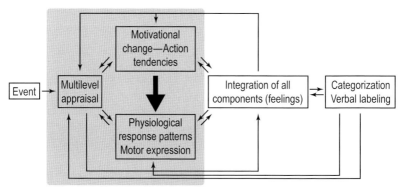

Figure 5.1 Schematic view of the component process model (after Scherer 2009).

have different objects, may enchain an event to the emotion episode, as for example in the course of a competitive football match.

The event and its consequences are appraised using a set of criteria with multiple levels of processing (the appraisal component). The result will generally have a motivational effect (often changing or modifying the motivational state that existed before the event occurred) as well as effects in the autonomic nervous system (e.g., in the form of cardiovascular and respiratory changes) and in the somatic nervous system (in the form of motor expression in face, voice, and body). The neural correlates of all these component changes are constantly fused in a multimodal integration between the participating brain areas (with continuous updating as events and appraisals change), thus giving rise to distributed representations. Parts of these representations are likely to become conscious and subject to assignment to fuzzy emotion categories, which may then lead to labeling with emotion words, expressions, or metaphors (Scherer 2009:1322–1324).

The central module of the model represents the appraisal of a set of fundamental criteria (stimulus evaluation checks) on different levels of brain organization in a parallel processing fashion. Figure 5.2 shows the processes within the gray-shaded panel in Figure 5.1 in detail. In particular, the horizontal panel labeled "Appraisal processes" shows the different groups of appraisal criteria (with the stimulus evaluation checks within respective groups), organized in the theoretically expected sequence (see Scherer 2001, 2009), together with the respective cognitive structures that are recruited in these appraisal processes (downward arrows represent the input of the different cognitive structures into the appraisal process, e.g., retrieval of past experiences of a similar kind from memory; upward arrows represent modification of the current content of these structures by the appraisal results, e.g., attention being redirected by a relevance appraisal). The horizontal panels below the appraisal level show three response components and the final integration of all changes in the feeling component. The bold downward arrows illustrate the central assumption of the model in each phase of the process: the appraisal results sequentially and cumulatively affect all response domains. The dotted upward arrows represent the changes induced in this fashion (which are fed back to the appraisal module and the cognitive structures subserving and regulating the appraisal process).

Based on phylogenetic, ontogenetic, and microgenetic considerations (see Scherer 1984:313–314; Scherer et al. 2004), the CPM predicts that appraisal occurs in sequence, following a fixed order (a claim that has been supported by strong experimental evidence based on brain activity, peripheral measures, and expression patterns; for an overview, see Scherer 2009). Thus, Grandjean and Scherer (2008) showed that both the nature and the timing of the emotion-constitutive appraisal processes can be experimentally examined by objective measures of electrical activity in the brain. The method of choice is the use of the encephalographic (EEG) recordings of brain activity using topographical map and wavelet analysis techniques for optimal temporal resolution (which

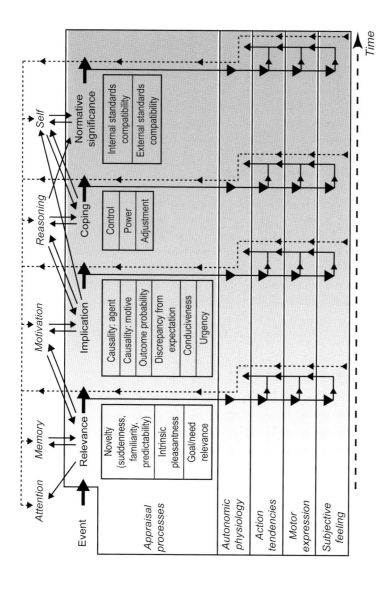

Figure 5.2 Process view of the CPM predictions on the cognitive mechanisms involved in appraisal and its efferent effects on the response patterning in different components, highlighting the proposed recursive, sequential, and cumulative process (reprinted from Sander et al. 2005 with permission from Elsevier).

would be impossible to attain with standard fMRI techniques). This approach allows theory-guided investigation of the nature of appraisal-generated emotion elicitation, both for automatic, unconscious processing as well as for higher, controlled levels of processing, in much greater detail than has previously been the case. The results of two separate experiments on the timing of the different, experimentally manipulated checks (shown in Figure 5.3) show that the different appraisal checks (a) have specific brain state correlates, (b) that occur rapidly in a brief time window after stimulation, and (c) that occur in sequential rather than parallel fashion. Furthermore, the results shown in Figure 5.3 suggest that the duration of processing of different checks varies, as they do not achieve preliminary closure at the same time, strongly supporting model predictions. The results also imply that early appraisal checks, including novelty and intrinsic pleasantness detection, can occur in an automatic, unconscious mode of processing, whereas later checks, specifically goal conduciveness, may rely on more extensive, effortful, and controlled processing. As shown in Figure 5.2, the appraisal mechanism requires interaction between many cognitive functions to compare the features of stimulus events to representations in memory, the self-concept, as well as expectations and motivational urges (see Grandjean and Scherer 2008 for details). Given the complexity of these interactions and the lack of focused neuroscience research on appraisal processes, the underlying neural circuits cannot yet be specified.

The motivational change produced by the appraisal of an event mediates the changes in the other components. If, for example, you are peacefully strolling through a forest, the sudden appearance of a bear will immediately produce a host of changes in physiological symptoms (to prepare for rapid motor action)

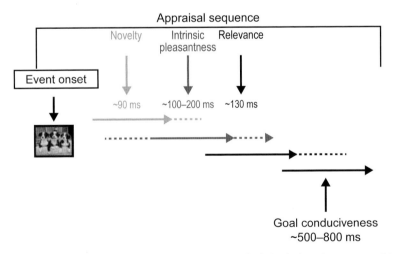

Figure 5.3 Onset and duration of major early appraisal checks based on topographical map and wavelet analysis of brain activity (high-resolution EEG data) from two studies in which appraisal checks were experimentally manipulated (after Scherer 2009).

and motor expression, which will be experienced as fear. (Note that this corresponds to the order proposed by James [1884/1968], if one considers that he meant only the feeling component when proposing to define "emotion"; see Scherer 2005:696). However, action tendencies do not necessarily imply gross motor behavior. Many appraisal results, such as novelty, may generate information search as a new action tendency, whereas experiencing pleasure from engaging in an intrinsically agreeable activity, such as basking in the sun on the beach, yields a desire to continue this pleasant activity.

Contrary to basic emotion theories (see review in Scherer and Ellgring 2007), the CPM does not assume the existence of a limited set of discrete emotions, but considers the possibility of an infinite number of different types of emotion episodes. This is important to understand the many subtle differentiations between members of an emotion family (e.g., the generic term anger may stand for any one of the following: annoyance, exasperation, fury, gall, indignation, infuriation, irritation, outrage, petulance, rage, resentment, vexation, etc.), as well as the fact that the very same event can make one person happy and another sad, depending on goals and values (e.g., winners and losers in a tennis match). However, certain so-called *modal* emotions (Scherer 1994b), such as anger, do occur more frequently and engender more or less stereotypical responses to a frequently occurring type of event or stimulus (the detailed CPM predictions for the some modal emotions—that largely correspond to basic emotions—are shown in Scherer 2009:Table 1).

Here it is important to reiterate that the term "feeling" should be used to denote a component of emotion but it should *not* be used as a synonym for the term "emotion" (the latter having led to major confusion, as shown by the century-old debate concerning the James-Lange theory; Scherer 2005). The feeling component serves to monitor and regulate the component process in emotion and enables the individual to communicate their emotional experience to others (as mentioned above, this is particularly developed in the human emotion system and may have no, or only a rudimentary, equivalent in nonhuman species). This could be instrumental in language and music since it allows the organism to regulate and adjust its communication depending on its own emotional experience, which in turn can reflect that of others. If it is to serve a monitoring function, subjective experience needs to be based on an integrated brain representation of the continuous patterns of change in the components and their coherence or synchronization. As schematically shown in Figure 5.4, the CPM conceptualizes this notion (and the role of consciousness, categorization, and verbalization) with the help of a Venn diagram. Figure 5.4 is not a process chart with input and output functions but rather serves to illustrate the fact that only parts of the component changes in an emotion situation become conscious, and only parts of these can be verbalized. Circle A illustrates the integrated brain representation of changes in all components of an emotion episode in some form of monitoring structure. Circle B represents the part of this integrated representation that becomes conscious and corresponds to

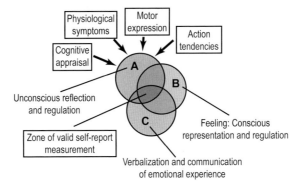

Figure 5.4 Schematic illustration of the brain representation of synchronized mental and bodily changes during an emotion episode and the relationships to emotion regulation, consciousness, and verbalization (from Scherer 2004a, reprinted with permission from Cambridge University Press).

nonverbal "feelings" or *qualia*. Circle C denotes the categorization and verbalization (labeling) of feelings (e.g., in social sharing of emotional experiences). As shown in Figure 5.4, verbal labels may often cover only a small part of the underlying emotion process in a particular case, as we may not find appropriate words to cover all aspects of our experience (the nonoverlapping part of the circle represents surplus meaning). Also, the choice of a label or category may be influenced by nonconscious aspects of the integrated representation of component changes (which explains why C intersects with A even in its nonconscious parts). In addition, other factors (e.g., social constraints or strategic considerations) may determine the choice of a label. This is of particular importance for labeling music-induced emotions, which often proves difficult due to the dearth of highly differentiated sociocultural labeling conventions in this domain.

Regulation is part and parcel of the emotion process. Any change in one of the components of the process is immediately fed back into the appraisal module and reevaluated, resulting in a regulation of the process. This can occur at any of the levels of processing discussed above (see also Figure 5.4), and the criteria for regulation can be based on simple homeostatic processes at very low levels of automatic processing or in voluntary efforts to change an emotion or its expression on the level of conscious value judgments.

The Motor Expression of the Emotion Process

Because production of speech and music generally involves a high degree of affective expressivity, let us turn our attention to the component of motor expression. Much of the human fascination with emotion has been fueled by the ubiquity and dramatic nature of emotional expression in face, voice, gesture, and posture. Unless it is suppressed, controlled, or strategically manipulated

(Scherer 2000), emotional expression externalizes and signals internal emotion processes, and this is of paramount importance in social interaction.

As discussed above, adaptive behavioral responses require coordination at the physiological and motor level, and emotional processes are believed to achieve such coordination (Scherer 2009). Coordination at the motor level involves the synchronization of different muscle systems resulting in specific behavioral patterns aimed at modifying physical environmental conditions (like running away from danger) or stimulating another's perceptual system (social signaling). Coordination of the different muscles can have effects on multiple expressive modalities (e.g., voice, face, and gestures), producing so-called multimodal signals. Multimodality has evolved to make signals more efficient by adapting to the constraints imposed by transmission in variable physical environments and by the receiver's psychobiological makeup (Guilford and Dawkins 1991; Rowe 1999). If a conspecific is threatening me and I am willing to fight, the credibility of this expressed intention may lead the antagonist to withdraw. Much of animal communication is based on this principle of prevention and "behavioral negotiation," for example in dominance fights (Hauser 1996).

While many of the constituent components of the emotion process remain invisible to the outside observer, as soon as there are such external manifestations, these become *sign vehicles* (Morris 1971) that can be interpreted as symptoms of the actor's reactions, feelings, and behavioral intentions. This, in turn, motivates the expressor to manipulate or fake these signals for strategic purposes. The complex relationship between true and false signals has been hotly debated in ethology (Hauser 1996; Scherer 1985).

Indeed, expression of any kind is only one half of communication, with perception being the other. Here, in extension of the CPM, a new synthetic model of emotion communication is proposed: the *dynamic tripartite emotion expression and perception* (TEEP) model, based on a modified Brunswikian lens model (Brunswik 1956; Scherer 2003). The model, shown in Figure 5.5, illustrates how the sender continuously expresses ongoing emotion processes through a multitude of distal cues to the observer, who perceives these as proximal cues and probabilistically attributes the emotion processes unfolding in the sender. The degree to which the proximal cues capture the information content of the distal cues depends on the quality of the transmission channel and the response characteristics of sensory organs. The model is dynamic as it reflects the process nature of the underlying emotion episodes (a fundamental architectural property of the CPM, as described above). In contrast to the general assumption in the literature that a stable emotional "state" is expressed and recognized, the model assumes that the event, the appraisals, and the consequent response patterns continuously change (as do, in consequence, the observer attributions).

The model is "tripartite" as it calls attention to the fact that any sign has three functions (Bühler 1934/1988; Scherer 1988):

1. It provides *symptoms* of an ongoing emotional process in the sender.

2. It signals the emotion and thus *appeals* to the observer.
3. It *symbolizes* or represents meaning in the respective species or group (due to the ritualization of the link between symptom and appeal, or the existence of a shared code for encoding and decoding).

It should be noted that the respective elliptic shapes in Figure 5.5 are not active elements of the dynamic model but serve to highlight these three functions. This is particularly important in the case of the symbolic function, which is obvious for language signs for which Bühler developed his model, but is also essential to understand the communication process in the case of nonverbal emotion expressions (reminding us of the importance of the sociocultural context and the existence of shared codes, even in music or speech prosody).

The TEEP model highlights the fact that the production of the distal expressive cues and their proximal interpretation are determined both by psychobiological mechanisms and the rules or expectations generated by the sociocultural context. In consequence, the model distinguishes between push and pull effects on the production side and schematic recognition and inference rules on the perception side (Scherer 1988; Scherer and Kappas 1988). Push effects are motor response patterns due to physiological changes within the individual and to the preparation of motor actions as a consequence of appraisal. They usually

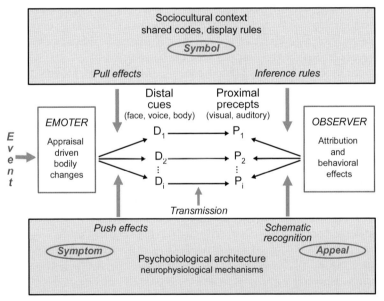

Figure 5.5 The *tripartite emotion expression and perception* (TEEP) model showing the communication of emotion through nonverbal cues in face, voice, body, or musical instruments. TEEP provides a framework to empirically assess cue validity and observer perception capacity. The shaded panel in the lower part represents neurophysiological mechanisms reflecting biological adaptation; the upper panel depicts sociocultural prescription for cue production (see text).

have a rapid onset and are direct and uncontrolled externalizations of internal processes. They are "pushed" into existence by internal changes. Examples of expressions exclusively due to push effects are affect bursts (i.e., brief, discrete, sudden expressions as a consequence of emotionally charged events; Scherer 1994a) or infant grunts. Push effects are supposed to occur universally, but their concrete appearances are relatively idiosyncratic, and thus subject to individual differences. On the perception side, it can be assumed that organisms have, in the course of evolution, developed schematic recognition mechanisms (the extreme form being the innate releasing mechanisms of Lorenz; see Hauser 1996) for the quasi-automatic detection of meaningful patterns in the behavior of others.

By comparison, pull effects are expressive configurations that are part of a socially shared communication code serving as socio-communicative signals used to inform or influence other group members. The distal cue characteristics are "pulled" into the shape required by the social context. Individuals learn through socialization to employ specific patterns of responses for communicating effectively, or deceptively, their internal processes and behavioral intentions to other people. In this sense, pull effects mainly reflect cultural and linguistic rules and expectations. Examples of pure pull effects are conventionalized emotion expressions and "affect emblems" (expressions having a shared cultural meaning), such as "yuk" and "ouch" (similar to visual emblems; Ekman and Friesen 1969). As a consequence of their highly conventionalized form, pull effects show little interindividual variation but a large degree of intercultural variability. These responses can be decoded effectively only if they respect social rules and adhere to the fixed socially shared code. In consequence, on the perception side we expect mechanisms built on sociocultural inference rules.

These aspects of the TEEP model reflect the specific evolution of speech and music in humans, as compared to the more basic psychobiological processes, which are likely to be at least partially shared with nonhuman species. This is particularly obvious in the case of the social embedding of the communication processes (e.g., different people in different groups interiorizing different registers, which may be invoked in different contexts, or different musical genres and rule systems having developed in different cultures). So far, due to lack of space, the view proposed here has centered on the individual and his/her internal representation of social factors, but as many other chapters in this book demonstrate, a more integrated view of the different systems needs to be adopted, given their interaction in conversation (language) or shared musical performance (e.g., singing hymns, dancing together, playing in an orchestra), including the interesting phenomenon of interpersonal entrainment of emotion processes (see chapters by Levinson, Fogassi, Cross et al., all this volume).

The TEEP model allows us to compare systematically the role that speech and music play in the communication of emotion by the two expression systems. Table 5.2 illustrates some of the hypotheses that can be entertained with

Table 5.2 Comparison of the emotion communication functions in speech and music.

Situation/ Orientation	Symptom orientation	Symbol orientation	Appeal orientation
Speech: Social interaction (generally spontaneous speech)	*Speaker*: real emotional arousal (push) expressed in voice or strategic faking (pull)	*Speaker*: may use established pragmatic codes for emotional messages (pull)	*Speaker*: may want to produce specific effect in listener (e.g., rhetoric)
	Listener: tries to infer authentic state of speaker	*Listener*: tries to decode intended message	*Listener*: emotional contagion or empathy (or adverse reactions), action tendencies
Music: Performance (generally scripted reproduction of a musical score)	*Performer*: real emotional arousal due to identification with character or composer (push) expressed in performance or following scripted emotional expression (pull)	*Performer*: uses prototypically scripted emotional expressions to render composer's intentions (pull)	*Performer*: attempts to produce intended affective reactions in listeners
	Listener: is affected by evidence of symptoms for authentic affect	*Listener*: attempts to decode the intended emotional meaning on the basis of established musical conventions	*Listener*: generally seeks to experience the emotional mood of specific pieces or procure enjoyment feelings

respect to the similarities and differences. Although the production of distal cues (and the perception of the respective proximal cues) may share many similarities, due to the assumed coevolution of music and speech (see below), both the situational functions and the interests of senders and receivers appear to be somewhat different. Thus, brief overviews on the expression and perception of emotion in voice and speech, on the one hand, and emotional effects of music, on the other, will now be provided.

The Vocal Communication of Emotion in Speech

Expression of emotion in the face has been an extremely popular subject for research ever since Darwin's pioneering volume (Darwin 1872/1998), and there is an extraordinary amount of evidence showing that emotions are differentially expressed by specific patterns of facial muscle movements. Furthermore, there is overwhelming evidence that observers can recognize these expressions with an accuracy that is far better than chance, even across cultural boundaries (although there seems to be a slight in-group advantage; see Scherer et al.

2011). Furthermore, bodily movements, gestures, or posture are also major modes of emotion communication (Dael et al. 2012).

While there has been less work on vocal expression, there is ample evidence that the major emotions are also differentially expressed and recognized in voice and speech. Many studies have measured the acoustic voice cues that characterize the expressions of different emotions (for reviews, see Scherer 2003; Scherer et al. 2003; Juslin and Laukka 2003; Juslin and Scherer 2005). Recently, we have extended this research to more subtle emotions than the small set basic emotions, as well as identifying the acoustic parameters which index the affective dimensions of arousal, valence, and potency/control (Goudbeek and Scherer 2010).

Mainstream linguists and phoneticians have tended to avoid the study of what has been called "non-, para- or extralinguistic communication," referring to those parts of communication that do not involve structural elements of the language code, often with the justification that such "emotional" or "attitudinal" aspects of speech (including phenomena such as voice quality, intonation, rhythm, or tempo) did not fall into the research domain of speech scientists (some notable exceptions notwithstanding, e.g., Bolinger 1964; Crystal 1974; Crystal and Quirk 1964; Wennerstrom 2001). Much of the recent work on prosody and intonation (see Ladd, this volume) has focused on genuinely linguistic or phonetic issues such as tone contrast or the syntactic function of intonation (Gussenhoven 2004; Ladd 2008). There have been some promising early studies exploring the links between phonology, linguistics, and pragmatics, on the one hand, and psychological studies on emotion expression, on the other (Ladd et al. 1985; Frick 1985), and it would seem most useful to continue in this vein. Bänziger and Scherer (2005) have reviewed the literature on the influence of emotions on intonation patterns (more specifically fundamental frequency [F_0] or pitch contours). A number of authors claim that specific intonation patterns reflect specific emotions, whereas others have found little evidence to support this claim and thus argue that F_0/pitch and other vocal aspects are continuously, rather than categorically, affected by emotions and/or emotional arousal. Using a new, quantitative coding system for the assessment of F_0 contours, the authors found that in actor-portrayed emotional expressions, mean level and range of F_0 in the contours vary strongly as a function of the degree of activation of the portrayed emotions. In contrast, there was comparatively little evidence for qualitatively different contour shapes for different emotions.

How well are the vocal cues identified in this research recognized by observers, allowing accurate inference of the target emotions? A recent review (Scherer et al. 2011) shows that recognition accuracy is much higher than chance, although somewhat lower than for facial expression (except in the case of anger, which is better recognized in the voice).

There is renewed interest in nonlinguistic vocalizations. Schröder (2003) showed that affect bursts (as defined by Scherer 1994a; see below), presented without context, can convey a clearly identifiable emotional meaning. In addition, Hawk et al. (2009) found that accuracy scores for nonlinguistic affective vocalizations and facial expressions were almost equivalent across nine emotions and that both were generally higher than the accuracy for speech-embedded prosody. Simon-Thomas et al. (2009) found higher accuracy scores when using only iconic vocal expressions or vocal bursts. Sauter et al. (2010) examined the recognition of nonverbal emotional vocalizations, such as screams and laughs, across two widely different cultural groups (Western participants and participants from isolated Namibian villages). Vocalizations that communicated the so-called basic emotions (anger, disgust, fear, joy, sadness, and surprise) were bidirectionally recognized. Patel et al. (2011) analyzed short affect bursts (sustained /a/ vowels produced by ten professional actors for five emotions) and found three components that explain acoustic variations due to emotion: "tension," "perturbation," and "voicing frequency." Focusing on voice production mechanisms, Sundberg et al. (2011), showed that each of the emotions studied in this corpus appears to possess a specific combination of acoustic parameters reflecting a specific mixture of physiological voice control parameters.

Communication of Emotion through Music

A similar analysis can be made for music; the major difference is that distal cues have been produced by a composer, although the issue to what extent performers need to develop some of the appropriate emotional feeling to perform a piece convincingly is very complex and hotly debated (Roach 1985). Although the scope of this chapter does not allow a review of this literature, the reader is referred to Ball (2010) for an overview. Evidence for the claim that emotion expression in speech and music is rather similar was reported in a meta-analysis of emotion–expression studies on voice and music reproduced by Juslin and Laukka (2003). Table 5.3 shows the distal cue patterns that were found in their analysis, showing a remarkably large overlap in the results between voice and music studies. The likelihood that a common code underlies these similarities is reinforced by the fact that these patterns correspond rather well to the predictions made by Scherer (1986) for vocal emotion expression on the basis of the CPM (see Juslin and Laukka 2003:801).

This large scale meta-analysis of the results in the pertinent literature on emotional expression in speech and music supports the claim that human listeners tend to make very stable emotion attributions to a number of basic acoustic parameters, even in meaningless sound sequences. Scherer and Oshinsky (1977) electronically synthesized tone sequences with systematic manipulation of amplitude and pitch variation, pitch level and contour, tempo, envelope, and filtration rated on emotional expressiveness; tone sequences were then presented either as filtered speech or musical melodies. The results

Table 5.3 Summary of cross-modal patterns of acoustic cues for discrete emotions (after Juslin and Laukka 2003, with permission from the *Psychological Bulletin*).

Emotion	Acoustic cues (vocal expression/music performance)
Anger	Fast speech or rate tempo
	High voice intensity or sound level
	Much variability in voice intensity or sound level
	Much high-frequency energy
	High F_0/pitch level
	Much F_0/pitch variability
	Rising F_0/pitch contour
	Fast voice onsets or tone attacks
	Microstructural irregularity
Fear	Fast speech or rate tempo
	Low voice intensity or sound level (except in panic fear)
	Much variability in voice intensity or sound level
	Little high-frequency energy
	High F_0/pitch level
	Little F_0/pitch variability
	Rising F_0/pitch contour
	A lot of microstructural irregularity
Happiness	Fast speech rate or tempo
	Medium high voice intensity or sound level
	Medium high-frequency energy
	High F_0/pitch level
	Much F_0/pitch variability
	Rising F_0/pitch contour
	Fast voice onsets or tone attacks
	Very little microstructural regularity
Sadness	Slow speech rate or tempo
	Low voice intensity or sound level
	Little variability in voice intensity or sound level
	Little high-frequency energy
	Low F_0/pitch level
	Little F_0/pitch variability
	Falling F_0/pitch contour
	Slow voice onsets or tone attacks
	Microstructural irregularity
Tenderness	Slow speech rate or tempo
	Low voice intensity or sound level
	Little variability in voice intensity or sound level
	Little high-frequency energy
	Low F_0/pitch level
	Little F_0/pitch variability
	Falling F_0/pitch contours
	Slow voice onsets or tone attacks
	Microstructural regularity

showed that two-thirds to three-quarters of the variance in the emotion attributions can be explained by the manipulation of the acoustic cues. Table 5.4 shows the attributions that were significantly associated with the respective levels of the manipulated parameters (which generally correspond to the patterns shown in Table 5.3). The assumption is that the interpretation of acoustic cues in simple tone sequences and in composed music correspond to the way in which different emotions are expressed in human and animal behavior—through iconic coding based on sound and dynamic movement (see also Bowling et al. 2010).

Whether the same explanation holds for the emotional effects of more music-specific features such as tonality is not immediately apparent. Thus, Bowling et al. (2010:491) cite many references showing that "other things being equal (e.g., intensity, tempo, and rhythm), music using the intervals of the major scale tends to be perceived as relatively excited, happy, bright, or martial, whereas music using minor scale intervals tends to be perceived as more subdued, sad, dark, or wistful." Scherer and Oshinsky (1977) also manipulated major and minor mode, in addition to the acoustic parameters shown in Table 5.4, and found that the major mode was associated with pleasantness and happiness, the minor mode with disgust and anger. Bowling et al. (2010) examined

Table 5.4 Emotion attributions significantly associated with acoustic parameters (after Scherer and Oshinsky 1977:Table V, with permission from Springer Science + Business Media).

Acoustic parameters of tone sequences	Direction of effect	Emotion rating scales listed in decreasing order of associative strength
Amplitude variation	Small	Happiness, pleasantness, activity
	Large	Fear
Pitch variation	Small	Disgust, anger, fear, boredom
	Large	Happiness, pleasantness, activity, surprise
Pitch contour	Down	Boredom, pleasantness, sadness
	Up	Fear, surprise, anger, potency
Pitch level	Low	Boredom, pleasantness, sadness
	High	Surprise, potency, anger, fear, activity
Tempo	Slow	Sadness, boredom, disgust
	Fast	Activity, surprise, happiness, pleasantness, potency, fear, anger
Envelope	Round	Disgust, sadness, fear, boredom, potency
	Sharp	Pleasantness, happiness, surprise, activity
Filtration cutoff	Intermediate (few harmonics)	Pleasantness, boredom, happiness, sadness
	High (many harmonics)	Potency, anger, disgust, fear, activity, surprise

the hypothesis that major and minor tone stimuli elicit different affective reactions because their spectra are similar to the spectra of voiced speech uttered in different emotions. Comparing the spectra of voiced segments in excited and subdued speech (using fundamental frequency and frequency ratios as measures), they found that the spectra of major intervals are indeed more similar to spectra found in excited speech, whereas the spectra of particular minor intervals are more similar to the spectra of subdued speech. They conclude that the characteristic affective impact of major and minor tone collections arises from associations routinely made between particular musical intervals and voiced speech.

The similarity between the acoustic structure of speech vocalizations and musical sounds is further underlined by recent results reported by Gill and Purves (2009). These authors asked why humans employ only a few of the enormous number of possible tone combinations to create music. They report evidence that the component intervals of the most widely used scales throughout history and across cultures are those with the greatest overall spectral similarity to a harmonic series, suggesting that humans prefer tone combinations that reflect the spectral characteristics of human vocalizations.

While the emphasis in this chapter has thus far been on the parallelism between emotion expression in speech and music, there are of course differences (as shown in Table 5.2). Although vocal expressions generally occur during the actual unfolding of an emotion process in the individual (with the emotion preceding and often preparing an appropriate action tendency, often in a social interaction), emotional expressiveness is constructed in music by a composer and realized by a performer with the explicit purpose of achieving a certain effect in the listener. A further difference (due to the function of emotion perception) is that in normal social interactions, an observer of a particular expression will attempt to infer the underlying emotion to adapt his/her own behavior in the interaction accordingly, generally without feeling the same emotion as the expresser. In some cases, an observer may feel an emotion that is diagonally opposed to the one perceived (e.g., a menacing anger expression may produce a feeling of fear). In contrast, while a music listener may attempt to infer what emotion being represented by music, this is rarely the explicit aim of listening to music, rather it is to feel an emotional effect (using emotion in a very wide sense of affective arousal). These differences in function might lead to different types of emotions being communicated by speech and music. Thus, certain emotions might be more frequently and more powerfully expressed (and evoked) by music whereas others might be better and more frequently expressed in speech. If this is indeed the case, there might be special mechanisms whereby certain types of emotions are elicited in the listener by music. These two issues will be briefly discussed below. Note that the focus here is on passive listening to recorded or live music rather than on active performances of music in social settings and for social occasions. These forms of music making and the emotional consequences are described elsewhere (see Lewis, and Cross et al., this volume).

What Types of Emotion Does Music Evoke?

In trying to disentangle the role of emotion in the evolutionary process, with special emphasis on music, it is important to consider the potential differences between different kinds of emotions. One option is to distinguish between *utilitarian* and *aesthetic* emotions (Scherer 2004b). The former are those usually studied in emotion research (e.g., such as anger, fear, joy, disgust, sadness, shame, or guilt). These are utilitarian in the sense of having major functions in the adaptation and adjustment of individuals to events that have important consequences for their well-being by preparing action tendencies (fight, flight), recovery and reorientation (grief work), enhancement of motivation (joy, pride), and social obligations (reparation), etc. Generally, such utilitarian emotions are relatively high-intensity reactions, often involving a synchronization of many subsystems, including changes in the endocrine, hormonal, and autonomous nervous systems as well as in the somatic nervous system, which are driven by the appraisals in the central nervous system. Because of this strong involvement of different bodily systems, it has become fashionable to consider emotions as "embodied states."

Goal relevance and coping potentially play a much less important role in aesthetic emotions. An aesthetic experience is one where the appreciation of the intrinsic qualities of a piece of visual art or a piece of music is of paramount importance. This corresponds in many ways to Kant's (1790/2001) well-known definition of aesthetic experience as *interesseloses Wohlgefallen* (disinterested pleasure), or William James (1884/1968) attempt to grapple with the distinction between "coarse" and "subtle" or "refined" emotions. Frijda and Sundararajan (2007) suggest that refined emotions tend to occur in situations in which goals directly relevant to survival or fundamental well-being are not at the center of the individual's attention, proposing that such emotions are more felt than acted upon, do not show strong physiological arousal, and cannot be appropriately described by basic emotion labels. Unfortunately, these brief remarks cannot do justice to the complexity of the issues involved in the discussion of the aesthetics of music (see Scruton 2009).

Nonetheless, music (and other forms of art) can produce physiological and behavioral changes (for reviews, see Hodges 2010; Västfjäll 2010). These changes seem to serve the behavioral readiness or preparation of specific, adaptive action tendencies (Frijda 1986), mostly with respect to attention, sharpening of sensory perception, and arousal regulation (stimulation vs. relaxation) rather than fight/flight reactions. They are not proactive but rather diffusely reactive. For example, the most commonly reported bodily symptoms for intense aesthetic experiences are goose bumps, shivers, tingling of the spine, or moist eyes. The greater intensity of the subjective feeling and the potential embodiment distinguish aesthetic emotions from simple preferences (see Scherer 2005), which consist of brief judgments about achieving behavioral goals. In other words, compared to preferences, aesthetic emotions are based on more comprehensive appraisal, including outward-facing reactions.

Music-induced emotions are special because (a) these emotions are desired and actively sought out rather than endured, (b) the appraisal criteria are weighted differently, (c) there are different types of goals, including aesthetic goals (see Hargreaves and North 2010), and (d) there is less system synchronization in the sense described by the CPM (see above) and a predominance of diffusely reactive behavioral and physiological responses. Finally, listening to music is likely to generate, more frequently and more consistently, the type of emotions glossed as aesthetic in comparison to that of basic emotions (see Zentner and Eerola 2010b; Sloboda 2010). However, emotions relating to nostalgia, love, wonder, and transcendence are experienced equally as often in nonmusical everyday life contexts as in musical contexts (Zentner et al. 2008:515).

Ball (2010:262) cites multiple sources which suggest that the emotions elicited by music cannot ever be put into words because they have a quality of their own for which we have no words. The CPM model of emotion proposed here fits nicely with this ideographic notion of a multitude of ineffable feelings, because it allows for an infinity of different emotions as a result of the integration of appraisal results and response patterns, only very few of which can be expressed by words (Scherer 1984). This would seem to make a taxonomy of musically induced emotions impossible. Yet, as shown in Figure 5.4, we often use existing labels from the vocabulary to refer to our emotions (especially to share them), even if only part of the underlying processes are reflected. Both this empirical fact and the need for scientific description require attempts to study the words most suitable to describe music-induced feelings.

What words do people use to describe what they actually feel when listening to music (as compared to what they think the music expresses)? Based on extensive empirical work (identifying which labels listeners to different kinds of music prefer to use to refer to their emotional experiences), we have developed and validated the *Geneva Emotional Music Scale* (GEMS), which consists of nine scales: wonder, transcendence, tenderness, nostalgia, peacefulness, power, joyful activation, tension, and sadness. In a study in which listeners evaluated their emotional reactions to a set of standard musical stimuli in the laboratory as well as in a free-listening task in their homes, using basic emotion scales, dimensional ratings, and the GEMS, we were able to show that the listeners preferred to report their emotional reaction on the GEMS scale, that they agreed more with their judgments on the GEMS scale, and that the GEMS ratings more successfully discriminated the musical excerpts (Zentner et al. 2008).

Recent experience using short forms of GEMS (see Zentner and Eerola 2010b:206) in different contexts (e.g., at a festival for contemporary music) suggests that the range of eligible emotions may depend on the type of music, even though Zentner et al. (2008) showed validity of GEMS for classical jazz, rock, and pop music. Responses of habitual listeners to contemporary music suggest that we need to include more epistemic (knowledge and insight related) emotions in the tools used to study the emotional effect of music.

In conclusion, it seems plausible that voice quality and prosody in speech may communicate utilitarian emotions more frequently or more efficiently (given that speech is often the means of pursuing goals in social interaction), whereas (apart from the special cases of poetry and theater) aesthetic emotions might be more readily elicited by vocal and instrumental music. A further difference might be that the major aim in the vocal communication of emotion is the recognition of the emotion encoded by the speaker, whereas in the case of vocal and instrumental music, the target might be the elicitation of a specific set of emotions intended to be felt by the listener, irrespective of the emotions expressed in the music. Thus, the furor in a Handel aria does not necessarily require the identification of the exact emotion of the character (let alone the production of this emotion in the listener); its intent is to produce enjoyment and fascination in the listener. In addition to utilitarian and aesthetic emotions, one might wish to distinguish another class of social or relational emotions such as love and tenderness. These emotions might be well served by both speech and music (it may be no accident that Juslin and Laukka's [2003] comparison included tenderness in addition to the four major basic emotions).

How Does Music Generate Emotion?

Utilitarian emotions generally occur because something important happens to the person. When we experience an emotion while listening to music, there is also something that happens to us, but the underlying elicitation mechanisms tend to be somewhat different. Scherer and Zentner (2001) have attempted to identify the production rules that underlie this process, arguing that, generally, a multitude of factors interact to produce the emotional effect. They distinguish structural (e.g., tones, intervals, chords, melodies, fugues, rhythm), performance (e.g., physical appearance, expression, technical and interpretative skills, current affective and motivational state), listener (e.g., musical expertise, stable dispositions, current motivational and mood state), and context (e.g., location and event) features.

These features all concern aspects of a composer's score that the performer needs to make audible. They can be subdivided into segmental and suprasegmental types. Segmental features consist of the acoustic characteristics that are the building blocks of musical structure, such as individual sounds or tones produced by the singing voice or specific musical instruments (for a review of musical acoustics, see Benade 1990). Suprasegmental features consist of systematic changes in sound sequences over time, such as intonation and amplitude contours in speech. In music, comparable features are melody, tempo, rhythm, harmony, and other aspects of musical structure and form. While iconic coding (e.g., tempo, rising/falling contours) plays an important role, suprasegmental features seem to transmit emotional information primarily through symbolic coding, which is based on a process of historically evolved, sociocultural conventionalization (see Kappas et al. 1991; Scherer 1988; Sloboda 1992).

Music can evoke emotion like any other event that serves or interferes with an individual's goal. One gets angry when neighbors play Mozart on their powerful sound system at 2 a.m. Yet it is not the music that produces the emotion, but rather the inconsiderate behavior of the neighbors. What we are interested in is the way in which the nature and quality of music as such elicits specific emotions; that is, as argued above, aesthetic emotions. In addition to the production rules referred to above, Scherer and Zentner (2001) identified a number of *routes* whereby music can elicit such *musical emotions*:

1. Specific types of appraisal (such as novelty, unexpectedness, pleasantness).
2. Music-related memory associations.
3. Contagion and empathy.
4. Entrainment and proprioceptive feedback.
5. Facilitation of preexisting emotions.

For a similar proposal, see Juslin and Västfjäll (2008a). Scherer and Coutinho (2013) provide detailed examples of how the different appraisal checks in the CPM model can apply to emotion generation in listening to music and extend the original proposal by Scherer and Zentner.

Coevolution of Speech, Singing, and Music Based on Affect Expression

After this overview of the architecture of the emotion system and the nature of emotional expression and communication through speech and music, we can now turn to the issue that was briefly sketched out in the introduction and which has discussed by many pioneering scholars: the assumption that there is a close link between the evolution of music and language. This hypothesis seems very plausible as there are, as shown above, many vital and powerfully shared elements between these expressive systems in present-day speech communication and music making (see also Brown 2000; Mithen 2005). Music has developed sophisticated structures and tools (such as instruments) for the production of a virtually limitless variety of sound combinations, together with the invention of elaborate systems for composing and annotating music. As for language, complex syntactic and semantic rule systems have evolved for the representation of meaning, and various nonvocal means as has the production and transmission of language. During the past decades, most scientific analyses of language and music have focused on these formal systems, but interest in the voice as a carrier of meaning in speech and music (singing) is growing, particularly with respect to the voice as a medium of emotional expression.

As proposed earlier (Scherer 1991), the vocal expression of emotion may play a major role in this coevolution. Elaborating on the original proposal, it is suggested here that *affect bursts* may have been a major precursor in the parallel evolution of speech and music. Affect bursts are defined as very brief, discrete, nonverbal expressions of emotion, coordinated between body, face, and voice, triggered by clearly identifiable events (Scherer 1994a). This evolutionary relic, shared with many animal species, is still with us. Consider, for example, a facial/vocal expression of disgust upon seeing a hairy black worm emerge from an oyster shell that one is about to bring close to one's mouth. While reactions may differ, for most people there will be a brief burst of facial and vocal activity that is directly triggered by the visual information and the evaluation of the significance of the worm's appearance.

Affect bursts may involve facial and bodily movement as well as vocalization; the following discussion will, however, focus on the vocal channel. It is instructive to consider animal vocalizations, the large majority of which are affective or motivational in nature. Ever since the pioneering work by Darwin (1872/1998), students of animal communication have demonstrated the important role of vocalization in the expression of affect (Marler and Tenaza 1977; Morton 1977; Scherer 1985; Tembrock 1975; Hauser 1996). While the structure of vocal call systems in animal species studied to date have also evolved into much more sophisticated systems involving complex patterning, syntax-like sequential structures, and representational reference (e.g., reliably indicating certain types of predator; see Seyfarth et al. 1980a; Scherer 1988; Jarvis and Fitch, this volume), it is not impossible for the origin of animal vocalization to also be traced to affect burst vocalizations. (This does not exclude the potential involvement of other factors; see Fogassi, and Arbib and Iriki, both this volume).

Empirical research has found a high degree of evolutionary continuity in vocal affect expression. Morton (1977) has suggested a system of *motivational-structural* rules that seem to hold across many species of mammals. According to these rules, the fundamental frequency (F_0) of vocalizations increases with diminishing size and power of the animal, and tonality of the sounds increases with increasing fear and tendency toward flight. Translating these rules to emotional aspects of human utterances, we would expect F_0 level to depend on submission/fear and spectral noise on aggression/anger. Morton does not make predictions for intonation contours except for rise-fall or fall contours in the case of moderate anger. He emphasizes the need to distinguish different degrees of anger and fear because the effect on vocalization may change dramatically. This is very much in line with the insistence of appraisal theorists to allow for many different varieties of emotion beyond the basic emotions, including important the distinctions between different family members—hot and cold anger, anxiety and panic fear, or sadness and despair—each of which have very different vocal signatures (Banse and Scherer 1996).

In human speech we still find the rudiments of nonlinguistic human affect vocalizations, often referred to as "interjections," such as "aua," "ai," "oh," and "ii," which are quite reminiscent of animal vocalizations. These may have been more or less "domesticated" by a specific phonological system (see Scherer 1994a; Wundt 1900); however, as shown in the oyster worm example, they can occur as purely push-based affect bursts. Affect bursts constitute the extreme push pole of the continuum of Figure 5.5; they are exclusively determined by the effects of physiological arousal and are thus the closest to a purely psychobiological expression of emotion.

In the case of the oyster worm, "raw" disgust sounds are "pushed out" during the initial appearance of the worm. In public settings, this will probably be quickly superseded by a predominance of pull effects, especially in a communal setting of oyster eating (particularly in "good" company). The term vocal *affect emblem* refers to brief vocal expressions, representing the extreme pull pole of the push-pull continuum, that are almost exclusively determined by sociocultural norms (for an explanation of push and pull effects, see the description of the TEEP model above). One would expect a large number of intermediate cases that show at least some degree of direct physiological effect, but which at the same time are subject to sociocultural shaping. Between these two extremes there are a large number of intermediate cases (i.e., nonverbal vocal expressions) that are spontaneously triggered by a particular affect-arousing event and which show at least some degree of componential synchronization, but are at the same time subject to shaping by control and regulation pull effects (Scherer 1994a, 2000). Affect emblems constitute an especially fertile ground for examining how behavior that originated as spontaneous affect bursts acquired shared, symbolic character. Ekman and Friesen (1969) have postulated the following requirements for the status of an emblem:

1. Existence of a verbal "translation."
2. Social agreement on its meaning.
3. Intentional use in interaction.
4. Mutual understanding of the meaning.
5. Sender assumes responsibility in emblem production.

Thus, the evolutionary path from affect bursts to vocal emblems may provide a rudimentary account of the establishment of referential meaning starting with pure emotion expressions. The details of these evolutionary processes will need to be worked out, including the adaptive pressures that drove this evolutionary process (for suggestions in this direction, see Smith et al. 2010). But clearly, the advantages of referential communication through stable, well-defined shared sign systems for social interaction and collaboration, largely independent of individual differences and situational contexts, are such that it is easy to imagine the advantages that the adoption of ever more sophisticated code systems engenders, fueling the evolutionary pressure for further development. It should be noted that, as suggested above, most affect bursts are multimodal and thus have facial, vocal, gestural, and postural components. This multicomponentiality obviously reinforces the potential for the development of referential meaning (e.g., deictic functions of eye movements or gestures). As always, other factors are likely to have contributed, such as bipedalism and opposable thumbs facilitating praxic action and thereby deictic gesturing (see Arbib and Iriki, this volume).

Affect bursts generally consist of single sounds or repeated sounds. In protohuman and human species, these have evolved into more complex sound sequence patterns, showing the rudiments of syntax, and melody-like intonation patterns (with singing possibly predating speech and a subsequent parallelism in development, at least, with respect to the pragmatic functions such as emotion signaling). Of particular interest in this respect is the evolution of rhythm and rhythm perception to ever more complex forms (Bispham 2006; Fitch 2011). Mithen (2005) has hypothesized that Neanderthals possibly used a form of protomusical language (the "Hmmmmm communication system") that was (a) holistic, because it relied on whole phrases rather than words, rather like music; (b) manipulative, because it focused on manipulating behavior of others rather than the transmission of information; (c) multimodal, because it used the body as well as the voice; (d) musical, because it used the variations in pitch, rhythm, and timbre for emotion expression, care of infants, sexual display, and group bonding; and (e) mimetic, because it involved a high degree of mime and mimicry of the natural world. Similarly, Brown (2000) has argued for what he calls a "musilanguage" model of music evolution. Nonetheless, all this is subject to debate given that the origin of language and of music is still shrouded in mystery (see Cross et al., this volume, and reviews by Johansson 2005; Patel 2008).

As mentioned at the outset, there is a strong likelihood that several mechanisms have contributed to the evolutionary development of speech and music (see Fitch and Jarvis, Arbib and Iriki, Lewis, and Cross et al., this volume). Considering the way that humans use even today affect vocalizations as a means of communication (i.e., in cases of lack of speech ability or language differences), and that certain songs resemble conventionalized affect vocalizations (e.g., wailing patterns in mourning rituals), one of these mechanisms might well have been the use of affect vocalizations as building blocks for more sophisticated communication systems that extend referentially beyond the iconic signaling of the sender's emotion. As mentioned, ethologists have shown that expression and impression are closely linked (see Andrew 1972). Thus, Leyhausen (1967) has shown that evolutionary pressure for "impression" (e.g., signal clarity) can affect expression patterns, and Hauser (1996) reviews extensive evidence showing that animals often manipulate their expressive behavior to produce a certain impression (pull effects). In the process of conventionalization and ritualization, expressive signals may be shaped by the constraints of transmission characteristics, limitations of sensory organs, fostering the evolution of more abstract language and music systems. Just as newer neocortical structures with highly cognitive modes of functioning have been superimposed on older "emotional" structures such as the limbic system and modified them in the process, the evolution of human speech and singing has made use of the more primitive, analog vocal affect signaling system. A central requirement for this process is the ability to control finely the production of vocalization and permit voluntary elicitation (Scherer 2000). Jürgens (2009) shows that the production of innate vocal patterns, such as the nonverbal emotional vocal utterances of humans and most nonhuman mammalian vocalizations (which are generally linked to motivation and emotion), seems to be generated by the reticular formation of pons and medulla oblongata. In contrast, the control structures for learned vocalization patterns, probably developed later, involve the motor cortex with its feedback loops. The fact that these dual-pathway control structures are shared across mammalian species suggests an evolutionarily old biphasic development with primitive vocal affect burst mechanisms being complemented by more sophisticated control structures for the production of complex sounds and sounds sequences.

The control of vocal sequences seems to have also developed rather early in mammalian evolution. Ouattaraa et al. (2009) have demonstrated the remarkable capacity of Campbell's monkeys to concatenate vocalizations into context-specific call sequences, providing an extremely complex example of "protosyntax" in animal communication. Finally, as to the control of articulation, Kay et al. (1998), studying the mammalian hypoglossal canal which transmits the nerve that supplies the muscles of the tongue, present evidence which suggests that the vocal capabilities of Neanderthals may have been the same as those of humans today—at least 400,000 years ago. This implies that human vocal abilities may have appeared much earlier in time than the first archaeological evidence for symbolic behavior.

To date, we have little insight into the factors that might have fostered the acquisition of complex control structures for vocalization. One interesting approach would be to examine the possibility that this process has occurred in coevolution with the control and regulation of emotion itself. Thus, Porges (1997, 2001) has argued that a more advanced form of the brainstem regulatory centers of the vagus and other related cranial nerves is directly linked to an expanded ability to express emotions, which in turn may determine proximity, social contact, and the quality of communication. These in turn, together with the ability to regulate emotion in general, required for smooth social interaction and collaboration, are part and parcel of the development of a social *engagement system*. Porges maintains that the somatomotor components of the vagal system contribute to the regulation of behaviors involved in exploration of the social environment (e.g., looking, listening, ingesting) as well as behaviors involved in acknowledging social contact (e. g., facial and head gestures, vocalizing).

In making use of vocalization, which continued to serve as a medium for emotion expression as the production system for the highly formalized systems of language and music, the functions of the two systems became by necessity strongly intermeshed. Thus, in speech, changes in fundamental frequency (F_0) contours, formant structure, or characteristics of the glottal source spectrum can, depending on the language and the context, serve to communicate phonological contrasts, syntactic choices, pragmatic meaning, or emotional expression. The strongest pull factor for vocal paralinguistics is, of course, language itself. The intonation contours proscribed by language serve as pull factors for any kind of speech.

In addition, the existence of marking in an arbitrary signal system like language adds the possibility of using violations for communicative purposes. In the evolution of protospeech, coding rules may have changed or new rules added. Thus, Scherer et al. (1984) suggested two general principles underlying the coding of emotional information in speech: covariation and configuration. The *covariation* principle assumes a continuous, but not necessarily linear, relationship between some aspect of the emotional response and a particular acoustic variable. For example, Ladd et al. (1985) suggested that the F_0 range shows an evolutionarily old covariance relationship (based on push effects) with attitudinal and affective information such that a larger range communicates more intense emotional meaning. They used vocal resynthesis to show that a narrow F_0 range was heard as a sign of sadness or of absence of specific speaker attitudes whereas wide F_0 range was consistently judged as expressing high arousal, producing attributions of strong negative emotions such as annoyance or anger (see also Scherer and Oshinsky 1977). In contrast, almost all linguistic descriptions assume that intonation involves a number of categorical distinctions, analogous to contrasts between segmental phonemes or between grammatical categories. In consequence, the *configuration* principle implies that the specific meaning conveyed by an utterance is actively inferred by the listener from the total prosodic configuration, such as "falling intonation contour," and the linguistic choices in the context. Arguing that the pragmatic use of intonation follows configurational rules, which occur later in evolution, Scherer et al. (1984) empirically showed that rising and falling contours take on different pragmatic meaning in Y/N and WH questions.

While a number of authors have claimed that specific intonation patterns reflect specific emotions (e.g., Fonagy 1983), others have found little evidence supporting this claim and argue that F_0/pitch and other vocal aspects are continuously, rather than categorically, affected by emotions and/or emotional arousal (Pakosz 1983). The results reported in the study by Bänziger and Scherer (2005) mentioned above (mean level and range of F_0 vary as a function of arousal—possibly a continuously coded push effect—but little evidence for qualitatively different contour shapes) suggest that intonation cues are configurational and that they may at least partly be determined by pull effects.

Similarly, in music, melody, harmonic structure, or timing may reflect sophisticated configurational principles used by composers, depending on specific traditions of music, to communicate specific emotional meaning (Huron 2006; Meyer 1956; Seashore 1938/1967). Yet, at the same time, in all traditions of music, we find copious evidence for direct covariance relationships with emotional expression (Scherer 1995). Informal observation suggests that some early forms of music (with some rudiments apparently surviving in the folk music of certain cultures) contain elements of nature sounds and vocal interjections that sound very much like affect burst representations. Most importantly, the meta-analysis performed by Juslin and Laukka (2003) shows that many of the patterns of emotional impact on acoustic parameters also hold for the expression of emotion in music. Similarly, Ilie and Thompson (2006; 2011), focusing on the valence, energy, and tension dimensions of emotion, provided evidence that experiential and cognitive consequences of acoustic stimulus manipulations in speech and music segments are overlapping (but not completely identical), suggesting that music and speech draw on a common emotional code. It can be argued that this evidence is more plausibly explained by the assumption of a common precursor than by the idea of a convergence of independently evolved systems, given that spontaneously spoken language and music are generally used in different contexts.

Much of the emotion mechanism and its important expression component have evolved in the context of *utilitarian* functions of adaptive preparation of responses, in particular in social contexts that require face-to-face interaction and communication. It should be noted that motor expression is only one component of emotion and that the different behavioral consequences of different emotions may well have produced adaptive pressures. Thus, being afraid may yield an immense variety of behaviors that are not only different emotions but also different ways of acting out that emotion, whether freeze, or flight, or fight; the latter two involve an infinite variety when taking into account the nature of the current environment. In the course of biological and cultural coevolution, other functions of emotional expression and impression appeared, particularly in the domain of social empathy and bonding, magic/religious ritual, play and pleasure, as well as the pursuit of beauty and knowledge, which may be responsible for the elicitation of aesthetic and epistemic (knowledge-related) emotions. The work reported in the literature (e.g., Wallin et al. 2000; and Brown 2000) and other chapters in this volume (see Lewis, Fitch and Jarvis, Cross et al., Arbib and Iriki) reveal many different factors that may have served as potential starting points for the development of language and music from affect bursts. For example, gesture and dance, the rule structure of birdsong, the need for social interaction across distances, communication with herding animals, social festivities and rituals, magic effects and incipient religion, and the need for affect control—including of motor expression—for social regulation and strategic signaling are all important factors. These explanations are not necessarily mutually exclusive; many of the factors mentioned may have influenced the coevolution of language, speech, and music at different points in evolutionary history.

Conclusion and Outlook

This chapter has seconded early proposals that music and language share a common evolutionary predecessor in the form of the primitive nonverbal expressions of emotion called affect bursts. Based on a componential appraisal theory of emotion, it has been proposed that emotions have a highly adaptive purpose in terms of preparing the individual for appropriate behavioral responses to an event or stimulus. When considering the evolution of music, the most important of these components is motor expression—a multimodal process that is highly synchronized between the different channels, both in speech and music performance.

Starting from these theoretical assumptions, it is suggested that speech, singing, music, and dancing share a similar functional architecture for the expression of emotion. A distinction is made between utilitarian and aesthetic emotions, where the former constitute the "garden variety" of emotions (e.g., anger, fear, or disgust) and the latter are those elicited by art, music, films or literature. Aesthetic emotions differ in that they are generally desired and actively pursued; the responses they elicit are not motivated by central goals for safety, achievement, or dominance, nor are they dependent on the coping potential of the individual. Although they may generate bodily responses like utilitarian emotions, this is not so much to prepare the individual for action readiness but a diffusive reaction linked to sharpening the sensorium, allocate attention, or regulate arousal. Aesthetic emotions are reactive rather than proactive.

Although quite a bit of research has been done to explore the expression of emotion through music and the emotional response to music, there are many areas in need of theory and empirical research, some of which have been briefly identified in this chapter. Closer examination of the role of affect burst in music is needed which could include the study of its historical use in music as well as comparisons of affect bursts in different genres of music and cultures. There is also an overall need to compare the expression of emotion in speech and music to determine similarities and differences between underlying production mechanisms, acoustic characteristics, and their impact on the listener. A new model of emotion communication, the TEEP model, is suggested as a promising theoretical framework for this purpose. Such efforts will provide further insight into the role of affect bursts in the evolution of language/speech and music and possibly help identify at what point they became distinct from each other. Research on the expression of emotion has tended to focus on the face, with much less study on expression via body movements, voice gestures, or posture. To understand the emotional expression of music, there is a need for better techniques to explore multimodal expression. These should have a strong theoretical grounding and specify the hypothesized mechanisms of production they are seeking to investigate.

In the past, most scientific analyses of language and music have focused on formal systems, stressing competence rather than performance. Greater research effort is needed on the voice as the primary human instrument for language and music production and as a medium of emotional expression. This should include the study of previously neglected elements of vocal expression such as prosody and segmentation. Finally, there is a need to explore the relative importance of the different structural features of music as originally proposed by Scherer and Zentner (2001) and the possible interaction between these features in the emotional impact of music (for an interesting example of an experimental approach, see Costa et al. 2004).

The consideration of the evolution and current architecture of emotion, and its function in action and social interaction, provides an important contribution to the exploration of the mysterious relationship between language, music, and the brain. The central argument is that the link between language and music is essentially the pragmatic aspect of expressing emotions, moods, attitudes, and rhetoric intentions through the means of segmental features like timbre and suprasegmental features like timing, rhythm, and melody. The origin of this relationship in evolutionary history can be sought in the appearance of an externalization of the synchronization of organismic subsystems at a stage in which the priority of prepared response mechanisms was replaced by an emphasis on system regulation to allow for social engagement and interaction. Affect bursts have been identified as "living fossils" of such early expressions encouraging the idea that speech, singing, music, and dancing evolved in parallel, and probably in close interaction with the evolution of appropriate brain structures, from this early precursor. Although other factors described in this volume have most certainly contributed to these evolutionary processes, the evidence briefly outlined above suggests granting a special role to affective expression for the following reasons: As an innate mechanism for spontaneous expression in face, voice, and body, affective expression may have been the earliest communicative mechanism in place, rapidly followed by control structures for learned behaviors. Both affect bursts and controlled vocalizations are widely shared across many species. Finally, the production mechanisms, at least for spontaneous expressions, are located in the subcortical regions of the mammalian brain.

6

Neural Correlates of Music Perception

Stefan Koelsch

Abstract

This chapter provides an overview of neural correlates of music-syntactic and music-semantic processing, as well as of music-evoked emotions. These three aspects of music processing are often intertwined. For example, a music-syntactically irregular musical event does not only evoke processes of syntactic analysis in the perceiver, but might also evoke processing of meaning, an emotional response, decoding of the producer's intentions, etc. In addition, it becomes clear that the neural correlates of these processes show a strong overlap with the processes engaged during the perception of language. These overlaps indicate that "music" and "language" are different aspects, or two poles, of a single continuous domain: the music–language continuum.

Musical Syntax

The regularity-based arrangement of musical elements into sequences is referred to here as *musical syntax* (see also Riemann 1877; Patel 2003; Koelsch 2005; Koelsch and Siebel 2005). It is not useful, however, to conceptualize musical syntax as a unitary concept because there are different categories of syntactic organization. Such syntactic organization can emerge from regularities based on local dependencies, from regularities involving long-distance dependencies, from regularities established on a short-term basis that do not require long-term knowledge, and from regularities that can only be represented in a long-term memory format, etc. Therefore, different cognitive processes have to be considered when thinking about (different categories of) musical syntax. In this section, I begin with a discussion of the cognitive processes that can be involved in the processing of musical syntax and then describe neural correlates of some of these processes.

Cognitive Processes

What are the cognitive (sub)processes involved in processing different categories of musical syntax? Below I briefly enumerate such processes, mainly referring to tonal music (other kinds of music do not necessarily involve all of these features; see Fritz et al. 2009). The ordering of the enumerated processes does not reflect a temporal order of music-syntactic processing; the processes may partly happen in parallel:

1. *Element extraction*: Elements such as tones and chords (or phonemes and words in language) are extracted from the continuous stream of auditory information. In homophonic and polyphonic music, representation of a current melodic and harmonic event is established (with the harmonic event coloring the melodic event). With regard to the temporal structure, a tactus (or "beat") is extracted. (The tactus is represented by the most salient time periodicity with which musical elements occur, corresponding to the rate at which one might clap, or tap to the music.)

2. *Knowledge-free structuring*: Representation of structural organization is established online (on a moment-to-moment basis) without obligatory application of long-term knowledge. For example, in auditory oddball paradigms, with sequences such as .....–... –......–....–.. etc., it is possible to establish a representation where "." is a high-probability standard and "–" is a low-probability deviant *without any prior knowledge* of any regularity underlying the construction of the sequence. Likewise, listening to a musical passage in one single key, an individual can establish a representation of the tones of a key and detect out-of-key tones (thus also enabling the determination of key membership) based on information stored in the auditory sensory memory: In-key tones become standard stimuli, such that any out-of-key tone (e.g., any black piano key producing a tone within a sequence of C major) represents a deviant stimulus ("auditory oddball"). These auditory oddballs elicit a brain-electric response referred to as the mismatch negativity (MMN). The processes that underlie the establishment of models representing such regularities include grouping and Gestalt principles. With regard to the melodic structure of a piece, grouping is required to assemble single tones to a melodic contour. In terms of temporal structure, grouping serves to extract the meter of a piece, as well as of rhythmic patterns. For a discussion on the establishment of a tonal "hierarchy of stability" of tones and chords based on Gestalt principles, see Koelsch (2012).

3. *Musical expectancy formation.* The online models described in Pt. 2 are based on a moment-to-moment basis without long-term knowledge of rules or regularities. By contrast, music-syntactic processing may also involve representations of regularities that are stored in a

long-term memory format (e.g., probabilities for the transition of musical elements such as chord functions). Such representations can be modeled computationally as fragment models: n-gram model, Markov model, chunking, or PARSER models (for details, see Rohrmeier and Koelsch 2012).

The important difference between *knowledge-free structuring* and *musical expectancy* is that the former is based on psychoacoustic principles and information stored in the auditory sensory memory, whereas the latter is based on long-term memory (this does not exclude that during the listening to a piece, experience based on knowledge-free structuring is immediately memorized and used throughout the musical piece). With regard to tonal music, Rohrmeier (2005) found, in a statistical analysis of the frequencies of diatonic chord progressions occurring in Bach chorales, that the supertonic was five times more likely to follow the subdominant than to precede it (Rohrmeier and Cross 2008). Such statistical properties of the probabilities for the transitions of chord functions are learned implicitly during the listening experience (Tillmann 2005; Jonaitis and Saffran 2009) and stored in a long-term memory format.

Importantly, with regard to major-minor tonal music, the interval structure of a chord function (e.g., whether a chord is presented in root position or as a sixth [first inversion] or six-four [second inversion] chord) determines the statistical probabilities of chord transitions. For example, six-four chords often have dominant character: a dominant that does not occur in root position is unlikely to indicate the arrival of a cadence; a tonic presented as a sixth chord is unlikely to be the final chord of a chord sequence; the same holds for a tonic in root position with the third in the top voice (for details, see Caplin 2004). In this sense, a chord function parallels a lexeme, a chord in root position parallels a lemma, and the different inversions of a chord parallel word inflections.

On the metrical and tonal grid that is established due to knowledge-free structuring, musical expectancies for subsequent structural elements are formed on the basis of implicit knowledge. Note that such musical expectancies are different from the expectancies (or predictions) formed as a result of knowledge-free structuring, because the latter are formed on the basis of acoustic similarity, acoustic regularity, and Gestalt principles; long-term memory representations of statistical probabilities are not required. Importantly, making predictions based on the processes of *knowledge-free structuring* and *musical expectancy* can only represent local dependencies—not long-distance dependencies, which is discussed next.

4. *Structure building*: Tonal music is hierarchically organized (for an example of music that does not have a hierarchical structure, see Fritz

et al. 2009). Such hierarchical organization gives rise to building and representing structures that involve long-distance dependencies on a phrase-structure level (i.e., structures based on context-free grammar). Such hierarchical structures may involve recursion, and they can best be represented graphically as tree structures. The processing and representation of such structures requires (auditory) working memory. Two approaches have so far developed systematic theoretical accounts on hierarchical structures of music: Lerdahl's combination of the *generative theory of tonal music* and his *tonal pitch space theory* (TPS; see Lerdahl, this volume), and Rohrmeier's *generative syntax model* (GTM; Rohrmeier 2011). To date, no neurophysiological investigation has tested whether individuals perceive music cognitively according to tree structures. Similarly, behavioral studies on this topic are extremely scarce (Cook 1987a, b; Bigand et al. 1996; Lerdahl and Krumhansl 2007). Whether tree structures have a psychological reality in the cognition of listeners of (major-minor tonal) music remains an open question.

5. *Structural reanalysis and revision*: During the syntactic processing of a sequence of elements, perceivers often tend to structure elements in the most likely way, for example, with regard to language, based on thematic role assignment, minimal attachment, and late closure (for details, see Sturt et al. 1999). However, a listener may also recognize that an established hierarchical model needs be revised; that is, with regard to the representation of a hierarchical organization using a tree structure, the headedness of branches, the assignment of elements to branches, etc., may have to be modified. To give an example with regard to language comprehension, the beginning of "garden-path" sentences (i.e., sentences which have a different syntactic structure than initially expected) suggests a certain hierarchical structure, which turns out to be wrong: "He painted the wall with cracks" (for further examples, see Townsend and Bever 2001). Note that while building a hierarchical structure, there is always ambiguity as to which branch a new element might belong: whether a new element belongs to a left- or a right-branching part of a tree, whether the functional assignment of a node to which the new element belongs is correct, etc. However, once a branch has been identified with reasonable certainty, it is represented as a branch, not as a single element. If this branch (or at least one node of this branch), however, subsequently turns out not to fit into the overall structure (e.g., because previously assumed dependencies break), the structure of the phrase, or sentence, must be reanalyzed and revised.

6. *Syntactic integration*: As mentioned above, several syntactic features constitute the structure of a sequence. In tonal music, these include melody, harmony, and meter. These features must be integrated by a listener to establish a coherent representation of the structure, and thus

to understand the structure. For example, a sequence of chord functions is only "syntactically correct" when played on a certain metric grid; when played with a different meter, or with a different rhythm, the same sequence might sound less correct (or even incorrect), because, for example, the final tonic no longer occurs on a heavy beat.

In many listeners, the simultaneous operation of melody, meter, rhythm, harmony, intensity, instrumentation, and texture evokes feelings of pleasure. After the closure of a cadence, and particularly when the closure resolves a previous breach of expectancy and/or previous dissonances, the integrated representation of the (simultaneous) operation of all syntactic features is perceived as particularly pleasurable and relaxing.

7. *Large-scale structuring*: The cognitive processes described above are concerned with the processing of phrase structure (i.e., processing of phrases that close with a cadence). Musical pieces, however, usually consist of numerous phrases, and thus have large-scale structures: verse and chorus in a song, the A–B–A(′) form of a Minuet, the parts of a sonata form, etc. When listening to music from a familiar musical style with such organization, these structures can be recognized, often with the help of typical forms and transitions. Such recognition is the basis for establishing a representation of the large-scale structuring of a piece.

Neural Correlates of Music-Syntactic Processing

To date, neurophysiological studies on music-syntactic processing have utilized the classical theory of harmony, according to which chord functions are arranged within harmonic sequences according to certain regularities. Here I describe neural correlates of music-syntactic processing that have been obtained through such studies. These studies show a remarkable similarity between neural correlates of music- and language-syntactic processing.

In major-minor tonal music, chord functions (Figure 6.1a) are arranged within harmonic sequences according to certain regularities. Chords built on the tones of a scale fulfill different functions. A chord built on the first–scale tone is called the tonic (I), the chord on the fifth-scale tone is the dominant (V). Normally, a chord constructed on the second scale tone of a major scale is a minor chord; however, when this chord is changed to be major, it can be interpreted as the dominant of the dominant (V/V or secondary dominant) (see square brackets in Figure 6.1a). One example for a regularity-based arrangement of chord functions is that the dominant is followed by the tonic (V–I), particularly at a possible end of a chord sequence; a progression from the dominant to the dominant of the dominant (V–V/V) is less regular (and seen as unacceptable as a marker of the end of a harmonic sequence). The left sequence in Figure 6.1b ends on a regular dominant-tonic (V–I) progression; in the right sequence

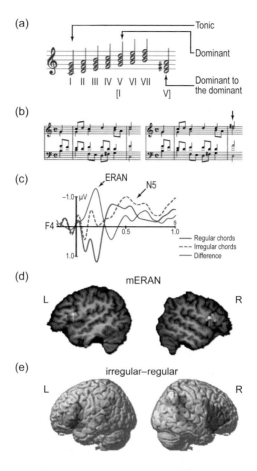

Figure 6.1 Neural correlates of music-syntactic processing. (a) Chord functions are created from the chords built on the tones of a scale. (b) The left sequence ends on a regular dominant-tonic (V–I) progression. The final chord in the right-hand sequence (see arrow) is a dominant of the dominant; this chord function is irregular, especially at the end of a harmonic progression (sound examples are available at www.stefan-koelsch.de/TC_DD). (c) Electric brain potentials (in μV) elicited by the final chords of the two sequence types presented in (b) (recorded from a right frontal electrode site, F4, from twelve subjects). Both sequence types were presented in pseudorandom order equiprobably in all twelve major keys. Brain responses to irregular chords clearly differ from those to regular chords. The first difference between the two black waveforms is maximal at about 0.2 s after the onset of the chord (this is best seen in the red difference wave, which represents regular, subtracted from irregular chords) and has a right frontal preponderance. This early right anterior negativity (ERAN) is usually followed by a later negativity, the N5. (d) With MEG, the magnetic equivalent of the ERAN was localized in the inferior frontolateral cortex. (e) fMRI data obtained from twenty subjects using a similar chord sequence paradigm. The statistical parametric maps show areas that are more strongly activated during the processing of irregular than during the processing of regular chords. Reprinted with permission from Koelsch (2005).

of Figure 6.1b, the final chord is the dominant to the dominant (arrow). (Sound examples of the sequences can be downloaded from www.stefan-koelsch.de.)

Figure 6.1c shows electric brain potentials elicited by the final chords of the two sequence types presented in Figure 6.1b recorded from a right frontal electrode site (F4) from twelve subjects (for details on how to obtain such potentials, see Koelsch 2012). Both sequence types were presented in pseudorandom order equiprobably in all twelve major keys. Brain responses to irregular chords clearly differ from those to regular chords. The first difference between the two black waveforms is maximal at about 0.2 s after the onset of the chord (this is best seen in the red difference wave, which represents regular, subtracted from irregular chords) and has a right frontal preponderance. This early right anterior negativity (ERAN) is usually followed by a later negativity, the N5 potential (short arrow; for details about the polarity of evoked potentials, see Koelsch 2012).

With magnetoencephalography (MEG), the magnetic equivalent of the ERAN was localized in the inferior frontolateral cortex: Figure 6.1d shows single-subject dipole solutions (indicated by striped disks), and the grand average of these source reconstructions (white dipoles); the grand average data show that sources of the ERAN are located bilaterally in inferior Brodmann area (BA) 44 (see also Maess et al. 2001); the dipole strength was nominally stronger in the right hemisphere, but this hemispheric difference was not statistically significant. This region is in the left hemisphere and is usually referred to as part of "Broca's area," although it is presumed that music-syntactic processing also receives additional contributions from the ventrolateral premotor cortex and the anterior superior temporal gyrus (i.e., the planum polare) (discussed below; see also Koelsch 2006).

Results of the MEG study (Koelsch 2000; Maess et al. 2001) were supported either by functional neuroimaging studies using chord sequence paradigms reminiscent of that shown in Figure 6.1b (Koelsch et al. 2002a, 2005; Tillmann et al. 2006) or studies that used "real," multipart music (Janata et al. 2002b) and melodies. These studies showed activations of inferior frontolateral cortex at coordinates highly similar to those reported in the MEG study (Figure 6.1e). Particularly the fMRI study by Koelsch et al. (2005; Janata et al. 2002a) supported the assumption of neural generators of the ERAN in inferior BA 44: In addition, ERAN has been shown to be larger in musicians than in nonmusicians (Koelsch et al. 2002b), and, in the fMRI study by Koelsch et al. (2005), effects of musical training were correlated with activations of inferior BA 44, both in adults as well as children.

Moreover, data recorded from intracranial grid electrodes from patients with epilepsy identified two ERAN sources: one in the inferior frontolateral cortex and one in the superior temporal gyrus (Sammler 2008). The latter was inconsistently located in anterior, middle, and posterior-superior temporal gyrus.

Finally, it is important to note that inferior BA 44 (part of Broca's area) is involved in the processing of syntactic information during language perception (e.g., Friederici 2002), in the hierarchical processing of action sequences (e.g., Koechlin and Jubault 2006), and in the processing of hierarchically organized mathematical formulas (Friedrich and Friederici 2009). Thus, Broca's area appears to play a role in the hierarchical processing of sequences that are arranged according to complex regularities. On a more abstract level, it is highly likely that Broca's area is involved in the processing of hierarchically organized sequences in general, be they musical, linguistic, action-related, or mathematical.

In contrast, the processing of musical structure with finite-state complexity does not appear to require BA 44. Instead, it appears to receive main contributions from the ventral premotor cortex (PMCv). Activations of PMCv have been reported in a variety of functional imaging studies on auditory processing—using musical stimuli, linguistic stimuli, auditory oddball paradigms, pitch discrimination tasks, and serial prediction tasks—which underlines the importance of these structures for the sequencing of structural information, the recognition of structure, and the prediction of sequential information (Janata and Grafton 2003; Schubotz 2007) (Figure 6.1d). With regard to language, Friederici (2004) reports that activation foci of functional neuroimaging studies on the processing of long-distance hierarchies and transformations are located in the posterior inferior frontal gyrus (with the mean of the coordinates reported in that article being located in the inferior pars opercularis), whereas activation foci of functional neuroimaging studies on the processing of local structural violations are located in the PMCv (see also Friederici et al. 2006; Makuuchi et al. 2009; Opitz and Kotz 2011). Moreover, patients with a lesion in the PMCv show disruption of the processing of finite-state, but not phrase-structure, grammar (Opitz and Kotz 2011).

In terms of the cognitive processes involved in music-syntactic processing (see above), the ERAN elicited in the studies mentioned was probably due to a disruption of musical structure building as well as the violation of a local prediction based on the formation of musical expectancy. That is, it seems likely that, in the studies reported, processing of local and (hierarchically organized) long-distance dependencies elicited early negative potentials, and that the observed ERAN effect was a conglomerate of these potentials. The electrophysiological correlates of the formation of musical expectancy, on one hand, and the building of hierarchical structures, on the other, have not yet been separated. It seems likely, however, that the former may primarily involve the PMCv, whereas Broca's area may be involved in the latter.

Interactions between the Syntactic Processing of Language and Music

The strongest evidence for shared neural resources in the syntactic processing of music and language stems from experiments that show interactions between both (Koelsch et al. 2005; Steinbeis and Koelsch 2008b; for behavioral studies see Slevc et al. 2009; Fedorenko et al. 2009; see also Patel, this volume).[1] In these studies, chord sequences were presented simultaneously with visually presented sentences while participants were asked to focus on the language-syntactic information, and to ignore the music-syntactic information (Figure 6.2).

Using EEG and chord sequence paradigms reminiscent of those described in Figure 6.1b, two studies showed that the ERAN elicited by irregular chords interacts with the left anterior negativity (LAN), a component of an event-related

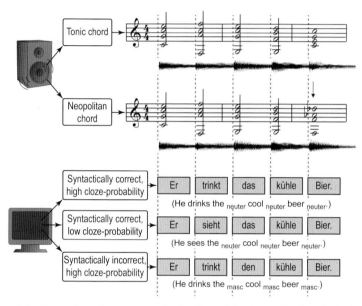

Figure 6.2 Examples of experimental stimuli used in the studies by Koelsch et al. (2005) and Steinbeis and Koelsch (2008b). (a) Examples of two chord sequences in C major, ending on a regular (upper row: the tonic) and an irregular chord (lower row: the irregular chord, a Neapolitan, is indicated by the arrow). (b) Examples of the three different sentence types (English translations of the German sentences used in the experiment). Onsets of chords (presented auditorily) and words (presented visually) were synchronous. Reprinted with permission from Steinbeis and Koelsch (2008b).

---

[1] It is a logical fallacy to assume that one could provide empirical evidence for resources that are "distinctively musical" vs. "distinctively linguistic," because it is not possible to know with certainty what the musical analog for a linguistic phenomenon is (and vice versa). For example, if I do not know the musical analog of a verb inflection, then I can only arbitrarily pick, out of an almost infinite number of musical phenomena, one by which to compare the processing of such musical information with the processing of verb inflections. If the data point to different processes, then there is always the possibility that I just picked the wrong musical analog (and numerous studies, published over the last two decades, have simply made the wrong comparisons between musical and linguistic processes, ostensibly showing distinct resources for music and language). Only if an *interaction* between processes is observed, in the absence of an interaction with a control stimulus, can one reasonably assume that both music and language share a processing resource.

potential (ERP) elicited by morphosyntactic violations during language perception. Using German sentences, Koelsch et al. (2005) and Steinbeis and Koelsch (2008b) showed that morphosyntactically irregular (gender disagreement) words elicited an LAN (compared to syntactically regular words; Figure 6.3a). In addition, the LAN was reduced when the irregular word was presented simultaneously with a music-syntactically irregular chord (compared to when the irregular word was presented with a regular chord, Figure 6.3b).

No such effects of music-syntactically irregular chords were observed for the N400 ERP, when words that were syntactically correct, but which had a low semantic cloze probability (e.g., "He sees the cold beer"), were elicited compared to words with a high semantic cloze probability (e.g., "He drinks the cold beer")[2] (Figure 6.3c). The final words of sentences with low semantic cloze probability elicited a larger N400 than words with high semantic cloze probability. This N400 effect was not influenced, however, by the syntactic regularity of chords; that is, the music-syntactic regularity of chords specifically affected the syntactic (not the semantic) processing of words (as indicated by the interaction with the LAN).

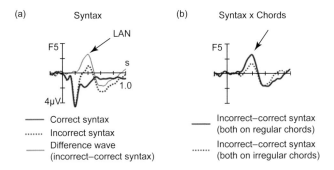

Figure 6.3 Total average of ERPs elicited by the stimuli shown in Figure 6.2. Participants ignored the musical stimulus, concentrated on the words, and, in 10% of the trials, answered whether the last sentence was (syntactically or semantically) correct or incorrect. (a) Compared to regular words, morphosyntactically irregular words elicit a LAN, best seen in the difference wave (thin line, indicated by the arrow). The LAN had a left anterior scalp distribution and was maximal at the electrode F5. All words were elicited on regular chords. (b) LAN effects (difference waves) are shown for words presented on regular chords (thick line is identical to the thin difference wave in (a) and irregular chords (dotted line). The data show that the morphosyntactic processing (as reflected in the LAN) is reduced when words have to be processed simultaneously with a syntactically irregular chord.

---

[2] The semantic cloze probability of the sentence "He drinks to cold beer" is higher than the cloze probability of the sentence "He sees the cold beer": after the words "He sees the cold..." anything that is cold and can be seen is able to close the sentence, whereas after the words "He drinks the cold..." only things that are cold and that one can drink are able to close the sentence.

In the study by Koelsch et al. (2005), a control experiment was conducted in which the same sentences were presented simultaneously with sequences of single tones. The tone sequences ended either on a standard tone or on a frequency deviant. The physical MMN elicited by the frequency deviants did not interact with the LAN (in contrast to the ERAN), indicating that the processing of auditory oddballs (as reflected in the physical MMN) does not consume resources related to syntactic processing (Figure 6.3d). These ERP studies indicate that the ERAN reflects syntactic processing, rather than detection and integration of intersound relationships inherent in the sequential presentation of discrete events into a model of the acoustic environment. The finding that language-syntactic deviances—but not language-semantic deviances or acoustic deviances—interacted with music-syntactic information suggests shared resources for the processing of music- and language-syntactic information.

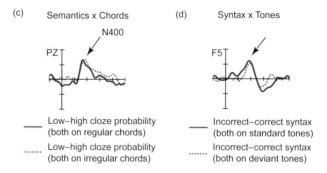

Figure 6.3 (cont'd) (c) shows the analogous difference waves for the conditions in which all words were syntactically correct, but in which ERPs elicited by words with high semantic cloze probability (e.g., "He drinks the cold beer") were subtracted from ERPs elicited by words with low semantic cloze probability (e.g., "He sees the cold beer"). The solid line represents the condition in which words were presented on regular chords, the dotted line represents the condition in which words were presented on irregular chords. In both conditions, semantically irregular (low-cloze probability) words elicited an N400 effect. The N400 had a bilateral centroparietal scalp distribution and was maximal at the electrode PZ. Importantly, the N400 was not influenced by the syntactic irregularity of chords (both difference waves elicit the same N400 response). (d) ERP waves analogous to those shown in (b) are shown. Here, however, tones were presented (instead of chords) in an auditory oddball paradigm (tones presented at positions 1–4 were standard tones, and the tone at the fifth position was either a standard or a deviant tone, analogous to the chord sequences). As in the chord condition, morphosyntactically irregular words elicit a clear LAN effect (thick difference wave). In contrast to the chord condition, virtually the same LAN effect was elicited when words were presented on deviant tones. Morphosyntactic processing (as reflected in the LAN) is not influenced when words have to be processed simultaneously with an acoustically deviant tone. Thus, the interaction between language- and music-syntactic processing shown in (b) is not due to any acoustic irregularity, but rather to specific syntactic irregularities. The scale in (b) to (d) is identical to the scale in (a). Data are presented in Koelsch et al. (2005). Reprinted with permission from Koelsch et al. (2000).

These ERP findings have been corroborated by behavioral studies: In a study by Slevc et al. (2009), participants performed a self-paced reading of "garden-path" sentences. Words (presented visually) occurred simultaneously with chords (presented auditorily). When a syntactically unexpected word occurred together with a music-syntactically irregular (out-of-key) chord, participants needed more time to read the word (i.e., participants showed stronger garden-path effect). No such interaction between language- and music-syntactic processing was observed when words were semantically unexpected, or when the chord presented with the unexpected word had an unexpected timbre (but was harmonically correct). Similar results were reported in a study in which sentences were sung (Fedorenko et al. 2009). Sentences were either subject-extracted or object-extracted relative clauses, and the note sung on the critical word of a sentence was either in-key or out-of-key. Participants were less accurate in their understanding of object-related extractions compared to subject-extracted extractions (as expected), because the object-extracted sentence constructions required more syntactic integration compared to subject-extracted constructions. Importantly, the difference between the comprehension accuracies of these two sentence types was larger when the critical word (the last word of a relative clause) was sung on an out-of-key note. No such interaction was observed when the critical word was sung with greater loudness. Thus, both of these studies (Fedorenko et al. 2009; Slevc et al. 2009) show that music- and language-syntactic processing specifically interact with each other, presumably because they both rely on common processing resources.

The findings of these EEG and behavioral studies, showing interactions between language- and music-syntactic processing, have been corroborated by a recent patient study (Patel 2008). This study showed that individuals with Broca's aphasia also show impaired music-syntactic processing in response to out-of-key chords occurring in harmonic progressions. (Note that all patients had Broca's aphasia, but only some of them had a lesion that included Broca's area.)

In conclusion, neurophysiological studies show that music- and language-syntactic processes engage overlapping resources (in the frontolateral cortex). The strongest evidence that show these resources underlie music- and language-syntactic processing stems from experiments that demonstrate interactions between ERP components reflecting music- and language-syntactic processing (LAN and ERAN). Importantly, such interactions are observed (a) in the absence of interactions between LAN and MMN (i.e., in the absence of interactions between language-syntactic and acoustic deviance processing, reflected in the MMN) and (b) in the absence of interactions between the ERAN and the N400 (i.e., in the absence of interactions between music-syntactic and language-semantic processing). Therefore, the reported interactions between LAN and ERAN are syntax specific and cannot be observed in response to any kind of irregularity. However, whether the interaction between ERAN and

LAN is due to the processing of local or long-distance dependencies (or both) remains to be determined.

Musical Meaning

To communicate, an individual must utter information that can be interpreted and understood by another individual. This section discusses neural correlates of the processing of meaning that emerge from the interpretation of musical information by an individual. Seven dimensions of musical meaning are described, divided into the following three classes of musical meaning:

1. Extramusical meaning can emerge from the act of referencing a musical sign to a (extramusical) referent by virtue of three different types of sign quality: iconic, indexical, and symbolic.
2. Intramusical meaning emerges from the act of referencing a structural musical element to another structural musical element.
3. Musicogenic meaning emerges from the physical processes (such as actions), emotions, and personality-related responses (including preferences) evoked by music.

Thus, in contrast to how the term *meaning* is used in linguistics, *musical meaning* as considered here is not confined to conceptual meaning; it can also refer to nonconceptual meaning. In language, such nonconceptual meaning may arise, for example, from the perception of affective prosody. Moreover, I use the term *musical semantics* in this chapter (instead of simply using the terms "musical meaning" or "musical semiotics") to emphasize that musical meaning extends beyond musical sign qualities: For example, with regard to intramusical meaning, musical meaning can emerge from the structural relations between successive elements. Another example, in terms of extramusical meaning, is that during the listening of program music, the processing of extramusical meaning usually involves integration of meaningful information into a semantic context. Note, however, that the term *musical semantics* does not refer to binary (true-false) truth conditions. I agree with Reich (2011): no musical tradition makes use of quantifiers (e.g., "all," "some," "none," "always"), modals (e.g., "must," "may," "necessary"), or connectives (e.g., "and," "if...then," "if and only if," "neither...nor") unless music imitates language (such as drum and whistle languages; Stern 1957). Hence, the term "musical semantics" should not be equated with the term "propositional semantics" as it is used in linguistics. Also note that during music listening or music performance, meaning can emerge from several sources simultaneously. For example, while listening to a symphonic poem, meaning may emerge from the interpretation of extramusical sign qualities, from the processing of the intramusical structure, as well as from music-evoked (musicogenic) emotions.

Extramusical Meaning

Extramusical meaning emerges from a reference to the extramusical world. It comprises three categories:

1. Iconic musical meaning: Meaning that emerges from common patterns or forms, such as musical sound patterns, that resemble sounds or qualities of objects. This sign quality is reminiscent of Peirce's "iconic" sign quality (Peirce 1931/1958); in language, sign quality is also referred to as onomatopoeic.
2. Indexical musical meaning: Meaning that arises from the suggestion of a particular psychological state due to its resemblance to action-related patterns (such as movements and prosody) that are typical for an emotion or intention (e.g., happiness). This sign quality is reminiscent of Peirce's "indexical" sign quality (for a meta-analysis comparing the acoustical signs of emotional expression in music and speech, see Juslin and Laukka 2003). Cross and Morley (2008) refer to this dimension of musical meaning as "motivational-structural" due to the relationship between affective-motivational states of individuals and the structural-acoustical characteristics of (species-specific) vocalizations. With regard to intentions, an fMRI study (Steinbeis and Koelsch 2008b), which will be discussed later in detail, showed that listeners automatically engage social cognition as they listen to music, in an attempt to decode the intentions of the composer or performer (as indicated by activation of the cortical theory of mind network). That study also reported activations of posterior-temporal regions implicated in semantic processing, presumably because the decoding of intentions has meaning quality.
3. Symbolic musical meaning: Meaning due to explicit (or conventional) extramusical associations (e.g., any national anthem). Peirce (1931/1958) denoted this sign quality as symbolic (note that the meaning of the majority of words is due to symbolic meaning). Symbolic musical meaning also includes social associations such as between music and social or ethnic groups (for the influence of such associations on behavior, see Patel 2008). Wagner's leitmotifs are another example of symbolic extramusical sign quality.

Extramusical Meaning and the N400

The processing of extramusical meaning is reflected in the N400. As described earlier in this chapter, the N400 component is an electrophysiological index of the processing of meaning information, particularly conceptual/semantic processing or lexical access, and/or post-lexical semantic integration. Koelsch et al. (2004) showed that the N400 elicited by a word can be modulated by the meaning of musical information preceding that word (see Figure 6.4). Further studies have revealed that short musical excerpts (duration ~ 1 s) can also elicit N400 responses when presented as a target stimulus following meaningfully unrelated words (Daltrozzo and Schön 2009). Even single chords can elicit N400 responses, as shown in affective priming paradigms using chords as targets (and words as prime stimuli) or chords as primes (and words as targets; Steinbeis and Koelsch 2008a). Finally, even single musical sounds can elicit N400 responses due to meaningful timbre associations (e.g., "colorful," "sharp"; Grieser Painter and Koelsch 2011).

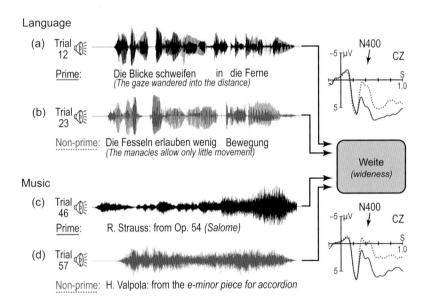

Figure 6.4 Examples of the four experimental conditions preceding a visually presented target word. Top left: sentence priming (a) and not priming (b) the target word *Weite* (wideness). Top right: Total-averaged brain electric responses elicited by target words after the presentation of semantically related (solid line) and unrelated prime sentences (dotted line), recorded from a central electrode. Compared to the primed target words, unprimed target words elicited a clear N400 component in the ERP. Bottom left: musical excerpt priming (c) and not priming (d) the same target word (excerpts had similar durations as sentences). Bottom right: Total-averaged ERPs elicited by primed (solid line) and non-primed (dotted line) target words after the presentation of musical excerpts. After the presentation of sentences, target words presented after unrelated musical excerpts elicited a clear N400 component compared to target words presented after related excerpts. Each trial was presented one time; conditions were distributed randomly, but in counterbalanced order across the experiment. Each prime was used in another trial as a non-prime for a different target word (and vice versa); thus, each sentence or musical excerpt was presented twice (half were first presented as primes, the other half as non-primes). Audio examples are available on www.stefan-koelsch.de. Reprinted with permission from Koelsch et al. (2004).

Intramusical Meaning

Musical meaning can also emerge from intramusical references; that is, from the reference of one musical element to at least one other musical element (e.g., a G major chord is usually perceived as the tonic in G major, as the dominant in C major, and, in its first inversion, possibly as a Neapolitan sixth chord in F# minor). The following will illustrate that the so-called N5 appears to be an electrophysiological correlate of such processing of intramusical meaning.

N5 was described first in reports of experiments using chord sequence paradigms with music-syntactically regular and irregular chord functions (see also Figure 6.1b). As described above, such irregular chord functions typically elicit two brain-electric responses: an early right anterior negativity (the ERAN, which is taken to reflect neural mechanisms related to syntactic processing) and a late negativity, the N5. Initially, N5 was proposed to reflect processes of harmonic integration, reminiscent of the N400 reflecting semantic integration of words. N5 was therefore proposed to be related to the processing of musical meaning, or semantics (Koelsch et al. 2000). N5 also shows some remarkable similarities with the N400.

Harmonic Context Buildup

N5 was first observed in experiments using paradigms in which chord sequences (each consisting of five chords) ended either on a music-syntactically regular or irregular chord function. Figure 6.5 shows ERPs elicited by regular chords at positions 1 to 5; each chord elicits an N5 (see arrow in Figure 6.5), with the amplitude of the N5 declining toward the end of the chord sequence. Amplitude decline is taken to reflect the decreasing amount of harmonic integration required with progressing chord functions during the course of the cadence. The small N5 elicited by the (expected) final tonic chord presumably reflects the small amount of harmonic integration required at this position of a chord sequence. This phenomenology of the N5 is similar to that of the N400 elicited by open-class words (e.g., nouns, verbs): as the position of words in a sentence progresses, N400 amplitude declines toward the end of a sentence (Van Petten and Kutas 1990). In other words, during sentence processing, a semantically correct final open-class word usually elicits a rather small N400, whereas the open-class words preceding this word elicit larger N400 potentials. This is due to the semantic expectedness of words, which is rather unspecific at the beginning of a sentence, and which becomes more and more specific toward the end of the sentence (when readers can already guess what the last word will be). Thus, a smaller amount of semantic integration is required at the end of a sentence, reflected in a smaller N400. If the last word is semantically unexpected, then a large amount of semantic processing is required—reflected in a larger amplitude of N400.

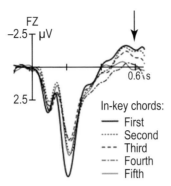

Figure 6.5 Total-averaged ERPs elicited by regular chords, separately for each of the regular chords (first to fifth position) of a five-chord cadence. Amplitude of the N5 (indicated by the arrow) is dependent on the position of the chords in the cadence: amplitude of the N5 decreases with increasing harmonic context buildup. Reprinted with permission from Koelsch (2012).

Harmonic Incongruity

Compared to regular chord functions, irregular chord functions typically elicit an ERAN, which is taken to reflect neural mechanisms related to syntactic processing. In addition, irregular chords elicit an N5 with a larger amplitude than the N5 elicited by regular chord functions (see Figure 6.2c) (for studies on N5 effects for melodies, see Miranda and Ullman 2007; Koelsch and Jentschke 2010). This increase of the N5 amplitude is taken to reflect the increased amount of harmonic integration, reminiscent of the N400 reflecting semantic integration of words (Figure 6.4). That is, at the same position within a chord sequence, N5 is modulated by the degree of fit with regard to the previous harmonic context, analogous to the N400 (elicited at the same position within a sentence), which is modulated by the degree of fit with regard to the previous semantic context (see also Figure 6.4). Therefore, N5 is proposed to be related to the processing of musical meaning, or semantics, although the type of musical meaning is unclear (Koelsch et al. 2000). Further evidence for the notion that the ERAN reflects music-syntactic and N5 music-semantic processes is reported next.

N5 and N400

We have seen (Figure 6.2) that LAN elicited by morphosyntactic violations in language is influenced by the music-syntactically irregular chord functions, whereas the N400 is not affected by such irregularities. In addition, Steinbeis and Koelsch (2008a) found that ERAN was smaller when elicited on syntactically wrong words compared to the ERAN elicited on syntactically correct words (cf. Figure 6.3a). This lends strong support to the notion that ERAN reflects syntactic processing.[3]

Moreover, and most importantly with regard to the processing of musical meaning, results of Steinbeis and Koelsch (2008a) study also show an interaction between the N5 and the semantic cloze probability of words (in the absence of an interaction between the N5 and the syntactic regularity of words; Figure 6.3b, c): N5 was smaller when elicited on words with a semantic low cloze probability (e.g., "He sees the cold beer") than on words with a semantic high cloze probability (e.g., "He drinks the cold beer"). Importantly, N5 did not interact with the syntactic processing of words, indicating the N5 potential can be modulated specifically by semantic processes; namely by the activation of lexical representations of words with different semantic fit to a previous context. This modulation indicates that N5 is related to the processing of meaning information. Note that the harmonic relation between the chord functions of a harmonic sequence is an intramusical reference (i.e., a reference of one musical element to another musical element, but not a reference to anything belonging to the extramusical world). Therefore, there is reason to believe that N5 reflects the processing of intramusical meaning. The fact that irregular chord functions usually elicit both ERAN and N5 potentials suggests that irregular chord functions evoke syntactic processes (as reflected in the ERAN) as well as semantic processes (as reflected in the N5). In addition, as will be discussed later, irregular chord functions may also evoke emotional effects.[4]

Musicogenic Meaning

In musicogenic meaning (Koelsch 2011b), listeners not only perceive the meaning expressed by the music, they process the musical information individually in terms of (a) physical activity, (b) emotional effects, and (c) personality-related effects. This, in turn, adds quality to the meaning for the perceiver.

Physical

Individuals tend to move to music. For example, they may sing, play an instrument, dance, clap, conduct, nod their heads, tap, or sway to music. In short, individuals tend to display some sort of physical activity in response to, and

---

[3] The fact that an interaction between language-syntactic irregularity and the ERAN was observed in the study by Steinbeis and Koelsch (2008a) but not in the study by Koelsch et al. (2005) is probably due to the task: participants in the study by Steinbeis and Koelsch (2008a), but not in the study by Koelsch et al. (2005), were required to attend to the chords. For details see Koelsch (2012).

[4] See Koelsch (2012) for further intramusical phenomena that give rise to musical meaning, including meaning that emerges from the buildup of structure, the stability and extent of a structure, the structure following a structural breach, resolution of a structural breach, and meaning emerging from large-scale musical structures.

in synchrony with, music. The mere fact that an individual shows such activity carries meaning for the individual. In addition, the way in which an individual moves is in itself an expression of meaning information. Movements are "composed" by the individual. Thus they need to be differentiated from motor effects that may result as an emotional effect of music, such as smiling during emotional contagion when listening to joyful music.

In a social situation, that is, when more than one individual moves to or plays music, meaning also emerges from joint, coordinated activity. For example, an action-related effect that becomes apparent in music based on an isochronous pulse—a pulse to which we can easily clap, sing, and dance—is that individuals synchronize their movements to the external musical pulse. In effect, in a group of individuals, this leads to coordinated physical activity. Notably, humans are one of the few species that are capable of synchronizing their movements to an external beat (nonhuman primates apparently do not have this capability, although some other species do; for a detailed account, see Fitch and Jarvis, this volume). In addition, humans are unique in that they can understand other individuals as intentional agents, share their intentionality, and act jointly to achieve a shared goal. In this regard, communicating and understanding intentions as well as interindividual coordination of movements is a prerequisite for cooperation. Cross stated that, in a social context, musical meaning can emerge from such joint performative actions and referred to this dimension as "socio-intentional" (Cross 2008:6).

Emotional

Musicogenic meaning can also emerge from emotions evoked by music. This view considers that feeling one's own emotions is different from the recognition of emotion expressed by the music (Gabrielson and Juslin 2003), the latter usually being due to indexical sign quality of music. The different principles by which music may evoke emotions are discussed elsewhere (e.g., Gabrielson and Juslin 2003; Koelsch 2012); here, the meaning that emerges from (music-evoked) emotions is discussed.

The evocation of emotions with music has important implications for the specificity of meaning conveyed by music as opposed to language. In communicating emotions, language faces several problems: In his discussion about rule following and in his argument against the idea of a "private language," Wittgenstein (1984) demonstrates that "inner" states (like feelings) cannot be directly observed and verbally denoted by the subject who has these states. His argument shows that the language about feelings functions in a different mode to the grammar of words and things. Wittgenstein argues that it is not possible (a) to identify correctly an inner state and (b) to guarantee the correct language use that is not controlled by other speakers. This means (c) that it is impossible for the speaker to know whether his or her use corresponds to the rules of the linguistic community and (d) whether his or her use is the same in different situations. According to Wittgenstein, correct use of the feeling vocabulary is only possible in specific language games. Instead of assuming a direct interaction of subjective feelings and language, Gebauer (2012) proposed that feeling sensations (Wittgenstein's *Empfindungen*) are reconfigured by linguistic expressions (although reconfiguration is not obligatory for subjective feeling). This means that there is no (direct) link or translation between feelings and words, thus posing fundamental problems for any assumption of a specificity of verbal communication about emotions. However, affective prosody, and perhaps even more so music, can evoke feeling sensations (*Empfindungen*) which, before they are reconfigured into words, bear greater interindividual correspondence than the words that individuals use to describe these sensations. In other words, although music seems semantically less specific than language (e.g., Slevc and Patel 2011; Fitch and Gringas 2011), music can be more specific when it conveys information about feeling sensations that are difficult to express in words, because music can operate prior to the reconfiguration of feeling sensations into words. Note that in spoken language, affective prosody also operates in part on this level, because it elicits sensational processes in a perceiver that bear resemblance to those that occur in the producer. I refer to this meaning quality as a priori musical meaning. The reconfiguration of a feeling sensation into language involves the activation of representations of a meaningful concept, such as "joy," "fear," etc. Zentner et al. (2008) report a list of 40 emotion words typically used by Western listeners to describe their music-evoked feelings. Such activation presumably happens without conscious deliberation, and even without conscious (overt or covert) verbalization, similar to the activations of concepts by extramusical sign qualities, of which individuals are often not consciously aware.

Personal

Feeling sensations evoked by a particular piece of music, or music of a particular composer, can have a personal relevance, and thus meaning, for an individual in that they touch or move the individual more than feeling sensations evoked by other pieces of music, or music of another composer. This is in part due to interindividual differences in personality (both on the side of the recipient and on the side of the producer). Because an individual has a personality (be it a receiver or producer of music), and personalities differ between individuals, there are also interindividual differences among receivers in the preference for, or connection with, a particular producer of music. For example, one individual may be moved more by Beethoven than by Mozart, while the opposite may be true for another. Music-evoked emotions may also be related to one's inner self, sometimes leading to the experience that one recognizes oneself in the music in a particular, personal way.

Music-Evoked Emotions

Unexpected Harmonies and Emotional Responses

A study by Steinbeis et al. (2006) tested the hypothesis that emotional responses can be evoked by music-syntactically unexpected chords. In this study, physiological measures including EEG, electrodermal activity (EDA, also referred to as galvanic skin response), and heart rate were recorded while subjects listened to three versions of Bach chorales. One version was the original version composed by Bach with a harmonic sequence that ended on an irregular chord function (e.g., a submediant). The same chord was also rendered expected (using a tonic chord) and very unexpected (a Neapolitan sixth chord). The EDA to these three different chord types showed clear differences between the expected and the unexpected (as well as between expected and very unexpected) chords. Because the EDA reflects activity of the sympathetic nervous system, and because this system is intimately linked to emotional experiences, these data corroborate the assumption that unexpected harmonies elicit emotional responses. The findings from this study were later replicated in another study (Koelsch et al. 2008), which also obtained behavioral data showing that irregular chords were perceived by listeners as surprising, and less pleasant, than regular chords.

Corroborating these findings, functional neuroimaging experiments using chord sequences with unexpected harmonies (originally designed to investigate music-syntactic processing; Koelsch et al. 2005; Tillmann et al. 2006) showed activations of the amygdala (Koelsch et al. 2008), as well as of orbital frontal (Tillmann et al. 2006), and orbital frontolateral cortex (Koelsch et al. 2005) in response to the unexpected chords. The orbital frontolateral cortex (OFLC), comprising the lateral part of the orbital gyrus (BA 11) as well as the medial part of the inferior frontal gyrus (BA 47 and 10), is a paralimbic structure that plays an important role in emotional processing: OFLC has been implicated in the evaluation of the emotional significance of sensory stimuli and is considered a gateway for preprocessed sensory information into the medial orbitofrontal paralimbic division, which is also involved in emotional processing (see Mega et al. 1997; Rolls and Grabenhorst 2008). As mentioned above, the violation of musical expectancies has been regarded as an important aspect of generating emotions when listening to music (Meyer 1956), and "breaches of expectations" have been shown to activate the lateral OFC (Nobre et al. 1999). Moreover, the perception of irregular chord functions has been shown to lead to an increase of perceived tension (Bigand et al. 1996), and perception of tension has been linked to emotional experience during music listening (Krumhansl 1997).

These findings show that unexpected musical events do not only elicit responses related to the processing of the structure of the music, but also emotional responses. (This presumably also holds for unexpected words in sentences and any other stimulus which is perceived as more or less expected.) Thus, research using stimuli that are systematically more or less expected should ideally assess the valence and arousal experience of the listener (even if an experiment is not originally designed to investigate emotion), so that these variables can potentially be taken into account when discussing neurophysiological effects.

Limbic and Paralimbic Correlates

Functional neuroimaging and lesion studies have shown that music-evoked emotions can modulate activity in virtually all limbic/paralimbic brain structures—the core structures of the generation of emotions (see Figure 6.6 for an illustration). Because emotions include changes in endocrine and autonomic system activity, and because such changes interact with immune system function (Dantzer et al. 2008), music-evoked emotions form an important basis for beneficial biological effects of music as well as for possible interventions using music in the treatment of disorders related to autonomic, endocrine, and immune system dysfunction (Koelsch 2010; Koelsch and Siebel 2005; Quiroga Murcia et al. 2011).

Using PET, Blood et al. (1999) investigated brain responses related to the valence of musical stimuli. The stimuli varied in their degree of (continuous) dissonance and were perceived as less or more unpleasant (stimuli with the highest degree of continuous dissonance were rated as the most unpleasant). Increasing unpleasantness correlated with regional cerebral blood flow (rCBF) in the (right) parahippocampal gyrus, while decreasing unpleasantness correlated with rCBF in the frontopolar and orbitofrontal cortex, as well as in the (posterior) subcallosal cingulate cortex. No rCBF changes were observed in central limbic structures such as the amygdala, perhaps because the stimuli were presented under computerized control without musical expression (which somewhat limits the power of music to evoke emotions).

However, in another PET experiment, Blood and Zatorre (2001) used naturalistic music to evoke strongly pleasurable experiences involving "chills" or "shivers down the spine." Participants were presented with a piece of their own favorite music using normal CD recordings; as a control condition, participants listened to the favorite piece of another subject. Increasing chills intensity correlated with rCBF in brain regions thought to be involved in reward and emotion, including the ventral striatum (presumably the nucleus accumbens, NAc; see also next section), the insula, anterior cingulate cortex (ACC), orbitofrontal cortex, and ventral medial prefrontal cortex. Blood and Zatorre also found decreases of rCBF in the amygdala as well as in the anterior hippocampal formation with increasing chills intensity. Thus, activity changes were observed in central structures of the limbic/paralimbic system (e.g., amygdala, NAc, ACC, and hippocampal formation). This was the first study to show modulation of amygdalar activity with music, and is important for two reasons: First,

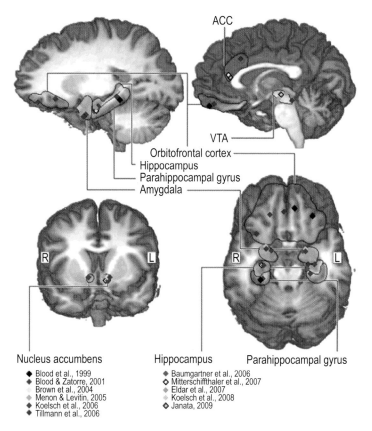

Figure 6.6 Illustration of limbic and paralimbic structures. The diamonds represent music-evoked activity changes in these structures (see legend for references, and main text for details). Note the repeatedly reported activations of amygdala, nucleus accumbens, and hippocampus, which reflect that music is capable of modulating activity in core structures of emotion (see text for details). Top left: view of the right hemisphere; top right: medial view; bottom left: anterior view; bottom right: bottom view. Reprinted with permission from Koelsch (2010).

the activity of core structures of emotion processing was modulated by music, which supports the assumption that music can induce "real" emotions, and not merely illusions of emotions (for details, see Koelsch 2010). Second, it strengthened the empirical basis for music-therapeutic approaches for the treatment of affective disorders, such as depression and pathologic anxiety, because these disorders are partly related to dysfunction of the amygdala. In addition, depression has been related to dysfunction of the hippocampus and the NAc (Drevets et al. 2002; Stein et al. 2007).

An fMRI study conducted by Koelsch et al. (2006) showed that activity changes in the amygdala, ventral striatum, and hippocampal formation can be evoked by music even when an individual does not have an intense "chill" experiences. This study compared brain responses to joyful instrumental tunes (played by professional musicians) with responses to electronically manipulated, continuously dissonant counterparts of these tunes. Unpleasant music elicited increases in blood oxygenation level-dependent (BOLD) signals in the amygdala, hippocampus, parahippocampal gyrus, and temporal poles; decreases of BOLD signals were observed in these structures in response to the pleasant music. During the presentation of the pleasant music, increases of BOLD signals were observed in the ventral striatum (presumably the NAc) and insula (in addition to some cortical structures not belonging to limbic or paralimbic circuits, which will not be further reported here). In addition to the studies from Blood and Zatorre (2001) and Koelsch et al. (2006), several other functional neuroimaging studies (for reviews, see Koelsch 2010; Koelsch et al. 2010) and lesion studies (Gosselin et al. 2007) have showed involvement of the amygdala in emotional responses to music. Most of these studies reported activity changes in the amygdala in response to fearful musical stimuli, but it is important to note that the amygdala is not only a "fear center" in the brain. The amygdala also plays a role for emotions that we perceive as pleasant (for further details, see Koelsch et al. 2010).

Compared to studies investigating neural correlates of emotion with stimuli other than music (e.g., photographs with emotional valence, or stimuli that reward or punish the subject), the picture provided by functional neuroimaging studies on music and emotion is particularly striking: The number of studies reporting activity changes within the (anterior) hippocampal formation is remarkably high (for reviews, see Koelsch 2010; Koelsch et al. 2010). Previously it was argued (Koelsch et al. 2010) that the hippocampus plays an important role in the generation of tender positive emotions (e.g., joy and happiness), and that one of the great powers of music is to evoke hippocampal activity related to such emotions. The activity changes in the (anterior) hippocampal formation evoked by listening to music are relevant for music therapy because patients with depression or posttraumatic stress disorder show a volume reduction of the hippocampal formation, associated with a loss of hippocampal neurons and blockage of neurogenesis in the hippocampus (Warner-Schmidt and Duman 2006), and individuals with flattened affectivity (i.e., a reduced capability of producing tender positive emotions) show reduced activity changes in the anterior hippocampal formation in response to music (Koelsch et al. 2007). Therefore, it is reasonable to assume that music can be used therapeutically to: (a) reestablish neural activity (related to positive emotion) in the hippocampus, (b) prevent death of hippocampal neurons, and (c) stimulate hippocampal neurogenesis.

Similarly, because the amygdala and the NAc function abnormally in patients with depression, studies showing modulation of activity within these structures motivate the hypothesis that music can be used to modulate activity of these structures (either by listening to or by making music), and thus ameliorate symptoms of depression. However, the scientific evidence for the effectiveness of music therapy on depression is surprisingly weak, perhaps due to the lack of high-quality studies and the small number of studies with randomized, controlled trials (e.g., Maratos et al. 2008).

An Evolutionary Perspective: From Social Contact to Spirituality—The Seven Cs

As discussed, music can evoke activity in brain structures involved in reward and pleasure (e.g., the NAc) as well as changes in the hippocampus, possibly related to experiences of joy and happiness. Here I will summarize why music is so powerful in evoking such emotions and relate these explanations to the adaptive value of music.

Music making is an activity involving several social functions. These functions can be divided into seven different areas (see also Koelsch 2010). The ability and the need to practice these social functions is part of what makes us human, and the emotional effects of engaging in these functions include experiences of reward, joy, and happiness; such effects have important implications for music therapy. Disengagement from these functions represents an emotional stressor and has deleterious effects on health (e.g., Cacioppo and Hawkley 2003). Therefore, engaging in these social functions is important for the survival of the individual, and thus for the human species. Below, I provide an outline of the seven different dimensions of social functions:

1. When we make music, we make *contact* with other individuals. Being in contact with others is a basic need of humans, as well as of numerous other species (Harlow 1958). Social isolation is a major risk factor for morbidity as well as mortality (e.g., Cacioppo and Hawkley 2003). Although no empirical evidence is yet available, I hypothesize that social isolation will result in hippocampal damage and that contact with other individuals promotes hippocampal integrity.
2. Music automatically engages social *cognition* (Steinbeis and Koelsch 2008a). As individuals listen to music, they automatically engage processes of mental state attribution ("mentalizing" or "adopting an intentional stance"), in an attempt to understand the intentions, desires, and beliefs of those who actually created the music. This is often referred to as establishing a "theory of mind" (TOM). A recent fMRI study (Steinbeis and Koelsch 2008a) investigated whether listening to music would automatically engage a TOM network (typically comprising anterior frontomedian cortex, temporal poles, and the superior temporal sulcus). In this study, we presented nontonal music from Arnold Schönberg and Anton Webern to nonmusicians, either with the cue that they were written by a composer or with the cue that they were generated by a computer. Participants were not informed about the

experimental manipulation, and the task was to rate after each excerpt how pleasant or unpleasant they found each piece to be. A post-imaging questionnaire revealed that during the composer condition, participants felt more strongly that intentions were expressed by the music (compared to the computer condition). Correspondingly, fMRI data showed during the composer condition (in contrast to the computer condition) a strong increase of BOLD signals in precisely the neuroanatomical network dedicated to mental state attribution; namely, the anterior medial frontal cortex (aMFC), the left and right superior temporal sulcus, as well as left and right temporal poles. Notably, the brain activity in the aMFC correlated with the degree to which participants thought that an intention was expressed in the composed pieces of music. This study thus showed that listening to music automatically engages areas dedicated to social cognition (i.e., a network dedicated to mental state attribution in the attempt to understand the composer's intentions).

3. Music making can engage "*co-pathy*" in the sense that interindividual emotional states become more homogenous (e.g., reducing anger in one individual, and depression or anxiety in another), thus decreasing conflicts and promoting cohesion of a group (e.g., Huron 2001). With regard to positive emotions, for example, co-pathy can increase the well-being of individuals during music making or during listening to music. I use the term co-pathy (instead of empathy) because empathy has many different connotations and definitions. By using the term co-pathy we not only refer to the phenomenon of thinking what one would feel if one were in someone else's position, we also refer to the phenomenon that one's own emotional state is actually affected in the sense that co-pathy occurs when one perceives (e.g., observes or hears), or imagines, someone else's affect, and this evokes a feeling in the perceiver which strongly reflects what the other individual is feeling (see also Singer and Lamm 2009). Co-pathy should be differentiated from:

- Mimicry, which is a low-level perception action mechanism that may contribute to empathy.
- Emotional contagion, which is a precursor of empathy (e.g., children laughing because other children laugh); both mimicry and emotional contagion may occur outside of awareness and do not require a self/other concept.
- Sympathy, empathic concern, and compassion, which do not necessarily involve shared feelings (e.g., feeling pitiful for a jealous person, without feeling jealous oneself) (see Singer and Lamm 2009).

Thus, co-pathy requires self-awareness and self/other distinction (i.e., the capability to make oneself aware that the affect may have been evoked by music made by others, although the actual source of one's emotion lies within oneself).

4. Music always involves *communication* (notably, for infants and young children, musical communication during parent–child singing of lullabies and play songs is important for social and emotional regulation, as well as for social, emotional, and cognitive development; Fitch 2006a; Trehub 2003; see also Trehub, this volume). Neuroscience and behavioral studies have revealed considerable overlap between the neural substrates and cognitive mechanisms that underlie the processing of musical syntax and language syntax (Koelsch 2005; Steinbeis and Koelsch 2008b). Moreover, musical information can systematically influence semantic processing of language (Koelsch et al. 2004; Steinbeis and Koelsch 2008b). It is also worth noting that the neural substrates engaged in speech and song strongly overlap (Callan et al. 2006). Because music is a means of communication, active music therapy (in which patients make music) can be used to train skills of (nonverbal) communication (Hillecke et al. 2005).

5. Music making also involves *coordination* of actions. This requires synchronizing to a beat and keeping a beat, a human capability that is unique among primates (Patel et al. 2009). The coordination of movements in a group of individuals appears to be associated with pleasure (e.g., when dancing together), even in the absence of a shared goal (apart from deriving pleasure from concerted movements; see also Huron 2001). Interestingly, a recent study from Kirschner and Tomasello (2009) reported that children as young as two and a half years of age synchronized more accurately to an external drum beat in a social situation (i.e., when the drum beat was presented by a human play partner) compared to nonsocial situations (i.e., when the drum beat was presented by a drumming machine, or when the drum sounds were presented via a loudspeaker). This effect might originate from the pleasure that emerges when humans coordinate movements between individuals (Overy and Molnar-Szakacs 2009; Wiltermuth and Heath 2009). The capacity to synchronize movements to an external beat appears to be uniquely human among primates, although other mammals and birds might also possess this capacity. A current hypothesis (e.g., Patel 2006) is that this capacity is related to the capacity of vocal learning (e.g., as present in humans, seals, some song birds, but not in nonhuman primates), which depends (in mammals) on a direct neural connection between the motor cortex and the nucleus ambiguus, which is located in the brainstem and which contains motor neurons that innervate the larynx; the motor cortex also projects directly to brainstem nuclei innervating the tongue, jaw, palate, and lips (e.g., Jürgens 2002).

6. A sound musical performance by multiple players is only possible if it also involves *cooperation* between players. Cooperation implies a shared goal, and engaging in cooperative behavior is an important potential source of pleasure. For example, Rilling et al. (2002) reported an association between cooperative behavior and activation of a reward network including the NAc. Cooperation between individuals increases interindividual trust as well as the likelihood of further cooperation between these individuals. It is worth noting that only humans have the capability to communicate about coordinated activities during cooperation to achieve a joint goal (Tomasello et al. 2005).
7. Music leads to increased social *cohesion* of a group (Cross and Morley 2008; Lewis, this volume). Many studies have shown that humans have a "need to belong" and a strong motivation to form and maintain enduring interpersonal attachments (Baumeister and Leary 1995). Meeting this need increases health and life expectancy (Cacioppo and Hawkley 2003). Social cohesion also strengthens the confidence in reciprocal care (see also the caregiver hypothesis; Fitch 2005) and the confidence that opportunities to engage with others in the mentioned social functions will also emerge in the future.

Although it should clearly be noted that music can be used to manipulate other individuals as well as to support nonsocial behavior (e.g., Brown and Volgsten 2006), music is still special—although not unique—in that it can engage all of these social functions at the same time. This is presumably one explanation for the emotional power of music (for a discussion on the role of other factors, such as sexual selection, in the evolution of music, see Huron 2001; Fitch 2005). Music, therefore, serves the goal of fulfilling social needs (e.g., our need to be in contact with others, to belong, to communicate). In addition, music-evoked emotions are related to survival functions and other functions that are of vital importance for the individual.

It is worth mentioning that the experience of engaging in these social functions, along with the experience of the emotions evoked by such engagements, can be a spiritual experience (e.g., the experience of communion; the use of "spiritual" and "communion" is not intended to infer religious context). This may explain why many religious practices usually involve music.

Engaging in social functions during music making evokes activity of the "reward circuit" (from the lateral hypothalamus via the medial forebrain bundle to the mesolimbic dopamine pathway, and involving the ventral tegmental area with projection to the NAc) and is immediately perceived as fun. Interestingly, in addition to experiences of mere fun, music making can also evoke attachment-related emotions (due to the engagement in the mentioned social functions), such as joy and happiness. This capacity of music is an important basis for beneficial biological effects of music, and thus for the use of music in therapy.

A Multilevel View of Language and Music: The Music–Language Continuum

The above discussion has illustrated some overlaps of the cognitive operations (and neural mechanisms) that underlie music- and language-syntactic processing, as well as the processing of meaning in music and language. These overlaps indicate that "music" and "language" are different aspects, or two poles, of a single continuous domain. I refer to this domain as the music–language continuum. Several design features (Fitch 2006a; Hockett 1960a) of "music" and "language" are identical within this continuum: complexity, generativity, cultural transmission, and transposability. Complexity means that "musical signals (like linguistic signals) are more complex than the various innate vocalizations available in our species (groans, sobs, laughter and shouts)" (Fitch 2006a:178). Generativity means that both "music" and "language" are structured according to a syntactic system (usually involving long-distance dependencies). Cultural transmission means that music, like language, is learned by experience and culturally transmitted. Transposability means that both "music" and "speech" can be produced in different keys, or with different "starting tones," without their recognition being distorted.

Two additional design features that could be added to this list are universality (all human cultures that we know of have music as well as language) and the human innate learning capabilities for the acquisition of music and language. Even individuals without formal musical training show sophisticated abilities with regard to the decoding of musical information, the acquisition of knowledge about musical syntax, the processing of musical information according to that knowledge, and the understanding of music. The innate learning capability for the acquisition of music indicates that musicality is a natural ability of the human brain; it parallels the natural human ability to acquire language. Perhaps this natural human ability is a prerequisite for language acquisition, because it appears that an infant's first step into language is based on prosodic features: Infants acquire considerable information about word and phrase boundaries (possibly even about word meaning) through different types of prosodic cues (i.e., of the musical cues of language such as speech melody, meter, rhythm, and timbre). With regard to production, many cultures do not have concepts such as "musician" and "nonmusician" let alone "musical" and "unmusical" (Cross 2008). This indicates that, at least in some cultures, it is natural for everyone to participate actively in cultural practices involving music making. Even in Western cultures, which strongly distinguishes between "musicians" and "nonmusicians," it is natural for everyone to participate in singing (e.g., during religious practices or while at rock concerts).

Beyond these design features, there are also design features that are typical for either "music," at one end of the continuum, or "language," at the other, but which overlap between language and music in transitional zones, rather than being clear-cut distinctive features for "music" or "language" in general. These features are scale-organized discrete pitch, isochrony, and propositional semantics.

Pitch information is essential for both music and speech. With regard to language, tone languages rely on a meticulous decoding of pitch information (due to tones coding lexical or grammatical meaning), and both tonal and nontonal languages use suprasegmental variations in F_0 contour (intonation) to code structure and meaning conveyed by speech (e.g., phrase boundaries, questions, imperatives, moods, and emotions). Music often uses sets of discrete pitches, whereas discrete pitches are not used for speech. However, pitches in music are often less discrete than one might think (e.g., glissandos, pitch bending), and many kinds of drum music do not use any scale-organized discrete pitches. On the other hand, the pitch height of different pitches produced during speaking appears not to be arbitrary, but rather to follow principles of the overtone series, which is also the basis of the pitches of many musical scales (e.g., Ross et al. 2007). In particular, emphatic speech (which borders on being song) often uses discrete scalelike pitches (in addition to more isochronous timing of syllables). This illustrates that discrete pitches (such as piano tones) are at the musical end of the music–language continuum and that a transitional zone of discrete pitch usage exists in both "music" and "language" (for the use of pitch in music and language, see Ladd, this volume).

The isochronous tactus (or "beat"), on which the musical signals are built in time, is at the "musical" end of the continuum. Though such an isochronous pulse does not appear to be a characteristic feature of spoken language, it can be found in poetry (Lerdahl 2001a), ritualistic speech, and emphatic speech. Still, not all kinds of music are based on a tactus (in particular, pieces of contemporary music). Thus, like discrete pitches, isochronous signals are more characteristic of the musical end of the music–language continuum, and the transitional zone from isochronous to non-isochronous signals is found in both music and speech. As mentioned above, emphatic speech borders on song and often uses discrete scale-like pitches as well as more isochronous timing of syllables. Art forms such as poetry, rap music, or recitatives represent transitional zones from speech to song.

With regard to the language end of the continuum, I have already mentioned in the discussion on musical meaning that no musical tradition makes use of propositional semantics unless music imitates language, as in drum and whistle languages (Stern 1957). Nevertheless, music can prime representations of quantifiers (such as "some" and "all") and possibly also evoke at least vague associations of some modals (such as "must" in passages conveying strong intentionality) or connectives (by establishing dependency relations between musical elements, such as the pivot chord in a tonal modulation that either belongs to one or to another key). In Western music, such capabilities can be used to convey narrative content, but clearly there is no existence of (or necessity for) a full-blown vocabulary of propositional semantics of language. On the other hand, quantifiers, modals, or connectives are often used imprecisely in everyday language (think of the "logical and," or the "logical or"). The mere existence of the two words "propositional" and "nonpropositional" leads easily to the illusion that there is a clear border between these two concepts or that one is the opposite of the other. However, although propositional semantics is characteristic of the language pole of the music–language continuum, a transitional zone of propositionality overlaps both language and music. Fitch noted that "lyrical music, which because it incorporates language, thus automatically inherits any linguistic design features (Fitch 2006a:176). Therefore, anyone interested in what listening to music with propositional semantics feels like, just has to listen to a song with lyrics containing propositional semantics (for the hypothesis that true-false conditions are not required in music to be made mutually explicit, see Cross 2011; Koelsch 2011a).

Meaning specificity is another design feature that is often taken as characteristic for language. Language appears to be more suitable to refer to objects of the extra-individual world; that is, objects which can be perceived by different individuals, and whose existence and qualities can thus be verified, or falsified, by others. However, although they possess a limited vocabulary, musical cultures have extramusical sign qualities that can also convey such meaning; the symbolic sign quality of music, for example is, by definition, just as specific as the symbolic sign quality of words (although such a lexicon is not comparable quantitatively to the lexicon of language). Similarly to the terms "propositional" and "nonpropositional," the terms "communication" (in the sense of conveying specific, unambiguous information with language) and "expression" (in the sense of conveying rather unspecific, ambiguous information with music) are normally used as if there were two separate realms of conveying meaningful information (communication and expression) with a clear border between them. However, this notion is not accurate, because there is a continuous degree of specificity of meaning information, with "expression" being located toward one end and "communication" toward the other.

More importantly, music can communicate states of the intra-individual world; that is, states which cannot be perceived by different individuals, and whose existence and qualities can thus not be falsified by others; they can, however, be expressed in part by facial expression, voice quality, and prosody (Scherer 1995; Ekman 1999): Music can evoke feeling sensations which bear greater interindividual correspondence with the feeling sensations of the producer than the words that an individual can use to describe these sensations (see discussion of a priori musical meaning in the section on musical meaning). In this sense, music has the advantage of defining a sensation without this definition being biased by the use of words. Although music might seem to be "far less specific" (Slevc and Patel 2011) than language (in terms of semantics), music can be more specific when it conveys sensations that are problematic to express in words. Importantly, in spoken language, affective prosody operates in part on this level because it elicits sensational phenomena in a perceiver that resemble those that occur in the producer. This notion is supported by the observation that affective information is coded with virtually identical acoustical features in speech and music (Scherer 1995; Juslin and Laukka 2003). For the design feature of translatability, see Patel (2008); for the design features of performative contexts and repertoire, see Fitch (2006a).

The description of the design features illustrates that the notion of clear-cut dichotomies of these features, and thus of clear-cut boundaries between music and language, is too simplistic. Therefore, any clear-cut distinction between music and language (and thus also any pair of separate definitions for language and music) is likely to be inadequate, or incomplete, and a rather artificial construct. Due to our "language games" (Wittgenstein), the meaning of "music" and "language" is sufficiently precise for an adequate use in everyday language. For scientific language, however, it is more accurate to consider the transitional nature of the design features and to distinguish a scientific use of the words "music" and "language" from the use of these words in everyday language. Using the term music–language continuum acknowledges both the commonalities between music and language and the transitional nature of the design features of music and language.

7

Film Music and the Unfolding Narrative

Annabel J. Cohen

Abstract

This chapter focuses on the role of music in narrative film. Unlike most other sensory information in a film (i.e., the visual scenes, sound effects, dialog, and text), music is typically directed to the audience and not to the characters in the film. Several examples will familiarize the reader with some of the subtleties of film music phenomena. Two aspects of film music are introduced: *congruence*, which focuses on purely structural aspects, and *association*, which focuses on the associative meaning of the music. The nature of and interplay between the emotional experience of the audience (referred to as internal semantics) and the external "reality" of the film (referred to as external semantics) are discussed, and an assessment is made as to where music (in particular, film music) resides with respect to these two domains. Because the two dimensions of structure and association are orthogonal to the internal–external semantic dimensions, they define four quadrants for describing the relation between music (structure and associations) and film narrative's internal and external semantics. Finally, the concept of a *working narrative* (WN) is introduced as the audience's solution to the task of integrating and making sense out of the two sources of information provided in the film situation: sensory information (including the acoustic information of music) as well as information based on experience including a story grammar. The author's *congruence-association model with the working narrative construct* (CAM-WN) accommodates the multimodal context of film, while giving music its place.

The Curious Phenomenon of Film Music: What Is It Doing There?

Music serves many roles in film (Copland 1939; Cohen 1999; Cross 2008; Levinson 1996), from masking extraneous noise in the theater (in the days of silent film, the noise of the film projector), to establishing mood, interpreting ambiguity, providing continuity across discontinuous clips within a montage, furnishing music that would naturally appear in the film's diegesis or story, directing attention, and engaging the viewer. Some of these roles have been submitted to empirical investigation, but most have not.

During the silent film era, directors hired composers to create original music (e.g., Eisenstein worked with Prokofiev), although more rarely, some directors, like Charles Chaplin, composed their own film music. Creating original music is standard practice in modern film-making (e.g., John Williams composed for *Star Wars*); however, directors often choose to use precomposed music as well. For example, in the film *The Social Network* (Fincher and Spacey 2010), composers Trent Reznor and Atticus Ross adapted Edvard Grieg's *Hall of the Mountain King* for the Henley Race rowing competition scene. Many films, including *The Social Network*, employ a mixture of appropriated and original music.

Sometimes a film score has large segments that are coherent in the absence of visuals and can well stand alone on a soundtrack album. Many film scores employ a musical phrase or repeating motif.[1] Others may sometimes simply employ a long tone, which by itself could have little appeal. Using *The Social Network* as an example (Sony Pictures Digital 2010),[2] the opening titles entail the drone of a low note slowly and unobtrusively embellished by its higher or lower octave over a four-minute period; a simple, otherwise unaccompanied melody unfolds intermittently. In contrast to these uncomplicated configurations and instrumentation, complex orchestral works (e.g., those of John Williams or Michael Kamen) are also used in film. Simple or complex, sound patterns all fall into the category of music—music that serves the film audience, not the characters in the film drama. Imagine a (nonhuman) computer trying to determine what acoustical information is meant for the audience and what is intended as part of the on-screen action. Such a thought experiment reflects the remarkable human ability to integrate mentally the musical information in a film with the other information of the film so as to arrive at an engaging film narrative.

The successful use of original music composed or adapted for a particular film relies, at least in part, on the composer's understanding of the director's goals in conveying a message and engaging the audience. Directors communicate their guidelines to a composer in various ways: they may provide a *temp track* (examples that the director has chosen from songs or compositions familiar to him or her), talk with the composer, or request examples from the composer for consideration. Temp tracks sometimes end up in the film. For example, composer Alex North's complete score for *2001: A Space Odyssey* (1968) was rejected in favor of the classical selections that director Stanley Kubrick had originally chosen as temp tracks (Karlin and Wright 2004:30). A similar fate befell composer Wendy Carlos' work for Kubrick's (1989) *The Shining*. Sometimes however, composers, such as Aaron Copland for the film *Of Mice and Men* (Milestone 1939), are given freer rein. The significance of a

---

[1] A motif is analogous to a meaningful phrase in language—a musical concept distinct from other musical concepts arising from a small number of sounded elements, distinctively different from other sequences of notes.

[2] The original score is available online: http://www.thesocialnetwork-movie.com/awards/#/music.

meeting of minds between the director and composer is underlined by a number of director–composer partnerships: Sergei Eisenstein and Sergei Prokofiev, Alfred Hitchcock and Bernard Hermann, James Cameron and James Horner, Steven Spielberg and John Williams, David Cronenberg and Howard Shore, David Fincher and Trent Reznor. Seasoned composers would perhaps argue that what is essential about film scoring is an innate sense of what is right for the film.

In terms of "what is right for the film," a first thought might be that the music sets the mood of the film. However, music serves many purposes in a film which collectively bridge the gap between the screen and the audience. Decades ago, Aaron Copland (1940:158) quoted composer Virgil Thompson as saying it best:

> The score of a motion picture supplies a bit of human warmth to the black-and-white, two-dimensional figures on the screen, giving them a communicable sympathy that they otherwise would not have, bridging the gap between the screen and the audience. The quickest way to a person's brain is through his eye but even in the movies the quickest way to his heart and feelings is still through the ear.

This implies that music quickly and effectively adds an emotional dimension to film, allowing an audience to relate more fully to it and engage in it, linking the internal world of the audience to the external world represented by two-dimensional moving images and a soundtrack. Emotional information can, of course, be conveyed without music via other visual and auditory media channels; consider, for example, a *visual image* of children playing in a garden on a sunny day, *words* in a subtitle description "happy children at play," the *sound effects* of laughter of children playing, or their fanciful spoken *dialog*. Film music, unlike these other four media channels—scenographic, textual, sound effects, or speech—has a direct route to the audience's "heart and feelings," or so it was thought by Copland and Thompson (see above quotation) and no doubt by many directors and audiences alike. Such a description is, of course, hardly scientific (for a model of various component processes that interact in "direct" feelings of emotion, see Scherer, this volume). Nonetheless, the notion of supplying *warmth* to a two-dimensional figure implies that music makes the film more compelling and engages empathy (though perhaps by overlapping with mirror mechanisms; see Fogassi, this volume). This further implies that when music accompanies film, it prompts emotional associations and responses of listeners and adds emotional dimensions. Some examples are provided by horror films like *Psycho* (Hitchcock 1960) and *Jaws* (Spielberg et al. 1988), where a short musical motif played at particular scary times in the film becomes associated with a particular kind of fear (Cross and Woodruff 2009). Excerpts that take on meaning through their association with events or characters in a drama are referred to as *leitmotifs*, a term applied for similar use in the operas of Wagner (Cooke 2008:80–83). The explanation of the leitmotif rests on developing a conditioned reflex, although understanding what exactly is conditioned, and the time course involved, is not a simple matter (Cohen 1993). Focusing on these leitmotifs alone belies the complexity of music perception and cognition as well as the fact that music engages so much of the brain (Overy and Molnar-Szakacs 2009; Koelsch, this volume; Janata and Parsons, this volume). The perceived unity of a piece of music suggests that music is a singular thing. However, music is comprised of many components: from a single pitch, to relations between pairs of notes and among groups of notes, timbre, loudness and timing patterns, harmony, and patterning on various hierarchical levels, etc.

When discussing the role music in film, I find it useful to distinguish two aspects of music: congruence and association. *Congruence* focuses on the structure of music, which can overlap with structures in other sensory domains. It is exemplified by cartoon music, which matches music to an action. Known somewhat pejoratively as "Mickey Mousing," congruence is readily employed in narrative film. Kim and Iwamiya (2008) studied perception of auditory patterns and moving letter patterns (telops) as might be seen in television commercials. Their studies, which involved rating scales, revealed sensitivity to similar patterns of motion across visual and audio modalities (see also Lipscomb 2005; Kendall 2008). Distinct from congruence (i.e., the structural properties of film music), *association* focuses on the meanings that music can bring to mind. The concept of a leitmotif belongs here as one example.

To date, research on film music has focused primarily on association; that is, how music contributes to the meaning of a film (Cohen 2010). Studies have shown that music influences meaning regardless of whether it precedes or foreshadows (Boltz et al. 1991), accompanies (Boltz 2004; Shevy 2007), or follows a film event (Tan et al. 2008). Several experiments by Thompson, Russo, and Sinclair (1994) indicate that music can evoke a sense of closure. This interpretative or modifying role arises primarily when a scene is ambiguous (Cohen 1993). The direction of the modifier seems to be typically audio to visual, and is consistent with the view that vision is the dominant sense. An unambiguous visual image can, however, enforce interpretation of music that has neutral meaning (Boltz et al. 2009; Spreckelmeyer et al. 2006). Film music has also been shown to affect memory of a film (Boltz 2004; Boltz et al. 1991). However, the complex issue of explaining the role of music considered part of the scene versus the same music intended solely for the audience has only been the focus of two investigations (Fujiyama et al. 2012; Tan et al. 2008).

The Diegetic World of the Film

Whether for a feature film of several hours or for a commercial as short as 15 seconds, twenty-first century audiences typically take the presence of film music for granted (i.e., they find it acceptable and ordinary). Yet, even if all other aspects of the film scene evince utmost realism, most music that accompanies a depicted scene would not belong in the film world's acoustic reality. The life and world of film characters is referred to, in film theory, as the *diegesis*. Information outside the world of the film is referred to as *nondiegetic*. Typically, film music plays no part in the diegesis. Consider a scene from the film *Road House* (Herrington and Silver 1989), where the villain (played by Ben Gazzara) displays his "king of the road" effrontery in a comic light, swerving down the road in his convertible to the accompaniment of the 1950s hit rock and roll song *Sh-Boom*. The music, if noticed at all, seems nondiegetic: its regular rhythm and undulating contour of the melody mirrors the almost sine-wave contour of the car's back-and-forth trajectory across the highway's midline. However, as the scene continues, we discover that Gazzara is singing along to the music (Figure 7.1), thus conferring a diegetic role for the music.

Most film music is, however, nondiegetic; that is, it is not meant to be heard by the film characters, unless for comic effect. In *Blazing Saddles* (Brooks and Hertzberg 1974), for example, the background big band music *April in Paris* is heard nondiegetically as the jazzy sheriff begins his gallant ride on horseback across the desert (Figure 7.2). However, when the Sheriff surprisingly comes across Count Basie and his band performing visibly in the desert, the music immediately becomes diegetic. The jazzy associations of the music, consistent with the super-cool personality of the sheriff, fool the audience into ignoring the music as an audible phenomenon. The appearance of Count Basie and his band in the film summons attention to the music as a performance (e.g., that the sounds are played by musicians, and that one cannot hear this music played any better). It is generally agreed by film theorists that much film music is largely unheard, as captured by the title of Gorbman's (1987) book *Unheard Melodies: Narrative Film Music*; that is, film music is not heard as music. Instead, the audience attends to second-order information provided by the structure of the music or the ideas and emotions associated with the music. This second-order information, in turn, directs attention either to other visual or auditory patterns in the film or to the meanings conveyed by the music which help interpret the meaning of the

Figure 7.1 Scene from the *Road House* (Herrington and Silver 1989): the antagonist (played by Ben Gazzara) careens down the road singing to *Sh-Boom*, demonstrating the diegetic nature of the music. (http://www.youtube.com/watch?v=xLdyuwqik4Q)

Figure 7.2 In this scene from *Blazing Saddles* (Brooks and Hertzberg 1974), the nondiegetic role of the film music (*April in Paris*) changes abruptly when the main character, Sheriff Bart, encounters Count Basie and his band in the desert playing the very tune being heard by the audience. The joke is on the audience, who most likely assumed (if they assumed anything at all) that what they were hearing was "just film music." (http://www.youtube.com/watch?v=7cLDmgU2Alw&feature=player_detailpage)

film. Characters in the film are unaware of the music, unless, for example, the narrative includes such events as a live concert, or someone practicing saxophone, or a band visibly playing in the desert, as in *Blazing Saddles*. The audience is often unaware of the music as well but, unlike the film characters, it is affected by the music. In a pilot study in which film excerpts were shown without music, Tan et al. (2007:171) report that approximately two-thirds of the participants appeared to believe "that the images were accompanied by a film score even when there was no music at all." Just as people can speak without explicitly knowing grammar, so can they enjoy a film without appreciating all the components of its narrative structure.

In the silent film era, music was primarily nondiegetic, serving to establish moods or interpretive contexts. After 1915, film distributors often included guidelines for music selections. Collections of music suited for various film contexts were specifically developed and published for use in silent film. There were exceptions: the Eisenstein films in Russia had scores composed by Prokofiev (*Alexander Nefsky*, 1938; *Ivan the Terrible*, 1944/1958) and by Edmund Meisel (*Battleship Potemkim*, 1925); shorter avant works, such *Entra'acte* (1924), directed by René Clair in France commissioned music by Camille Saint-Saens. Some of these full scores have survived or been restored. Occasionally live performances of the film and score are mounted in concert halls around the world.

When the talkies arrived in 1926, film directors temporarily retired the nondiegetic use of music, reclaiming it gradually by the mid 1940s. From the late 1920s to the early 1940s, music typically was only included in films when the plot called for it (Prendergast 1992:23). Directors, however, liked to have music in the film, and thus plots were contrived to include it. Eventually, such forced plots were found to be less than effective. Similarly, the capacity for sound recording led to the birth of the movie musical, which became very popular during this time. That the majority of directors, even Hitchcock, were long in rediscovering the nondiegetic function of film music underlines that the presence of music in a realistic film is counterintuitive. Why would film audiences need music to make the film more compelling, when the music is not part of the depicted action and all the actual sound of the scene can be represented through sound recordings?

Based on her comparisons of deaf and hearing audiences as they viewed films from the silent era, Raynauld (2001) emphasizes that sound has always played a role in films, even before the advent of talkies. It was just that silent film audiences had to imagine the sounds. Thus, although films produced between 1895 and 1929 were silent from a technical standpoint, they were not silent "from the vantage point of narrative" (Raynauld 2001:69). Actors and actresses spoke the words in the script even though they could not be heard, and title slides bridged some of the gaps. Consistent with the audience's "hearing" of the speech track of silent films, the first talking pictures often led to audience disappointment when the speech of the characters did not match how the audience had imagined them. This disappointment provided a key plot point in the twenty-first century "silent film" *The Artist* (Hazanavicius and Langmann 2011). Here, the protagonist cannot make the transition to the talkies due to his accent, a point revealed only in the last few minutes of the film. In retrospect, as nondiegetic music had a place in these "silent" films, it might follow that its role would be retained in talking films. However, the 1920s invention of the talking film heralded the demise of the robust silent film music industry, an industry that had supplied work to the vast majority of all performing musicians. In the United Kingdom, for example, 80% of all professional musicians lost their livelihoods at that time (Cooke 2008:46). As mentioned, directors now assumed that music for the talkies must be part of the diegesis; however, their contrived plots or films without music soundtracks were found lacking. Eventually the industry of film music returned under the new guise of the full prerecorded film score to replace the live music performance of the nondiegetic music of the silent film era.

The acceptance of and need for film music in talkies (i.e., the sound film) can be explained by considering the audience member as an active rather than passive receiver of the sensory input of the film (Bordwell 1996:3; see also the approach to narrative by Abbott 2008). While narrative approaches have been applied to the analysis of music on its own (Levinson 1997:167–169), the film as story obviously demands such consideration (Bordwell 2006). Levinson (1996:252) argues for three levels of narrative within a film: (a) the fictional story, (b) the perceptual enabler, who presents "the story's sights and sounds," and (c) the filmmaker, who is not on the same plane as the diegesis. He suggests that film music can be revealing on each of these levels. In his view, the contribution of music to the fiction is fairly obvious, for example, in the case of disambiguating context. The setting of mood in the composed film score may be linked more with the enabler, whereas the establishment of specific association to borrowed music (e.g., known popular music) reveals much about the filmmaker. Levinson challenges the accepted view that film music is unheard (Gorbman 1987) and argues instead that it is by necessity heard so as to help the audience understand the position of the narrator and the filmmaker, although one might still argue that music may be heard more or less unconsciously. He does not believe that the audience is responsible for creating the film, but rather that it is the job of the audience to understand how the narrative, as created by the narrator and filmmaker, is to be understood. It is important to keep Levinson's view in mind while considering the simpler view of the audience as creator of the narrative (and of the higher-order levels of narrator, and filmmaker, should information in the film be sufficient to allow this aspect).

The very presence of nondiegetic music in a film challenges cognitive science, film theory, and musicology for an explanation. I have addressed this issue in previous articles (e.g., Cohen 1990, 2013). The phenomena, however, demand continuing attention and exploration. Robin Stilwell (2007) has referred to the "fantastical gap" between the world of the film and the world of the audience. When we read a book, we construct the story; we do not construct music imagery to accompany the narrative. No music score attaches itself to our day-to-day existence, although some of us may imagine music throughout the day, and such music may be appropriate to our moods. Sometimes such music is merely an earworm that we cannot get out of our heads. When we dream, a music soundtrack is not usually part of our dreams. In contrast to these examples of narrative in our lives, in which music soundtracks play no role, in film, a director uses music to make a scene more compelling. Realistic details down to a car's license plate command a film director's attention, and music is used when it would not normally occur as part of such details. Why, then, is there a place for music? The situation is puzzling. The brain needs music in order for film to work, in order to engage fully in a scene. What gap, then, does music fill? What part of music fills the gap? Is all of the music necessary? Can the brain choose what part it needs and ignore the rest?

Film Score Examples: Opening Runs in *Chariots of Fire*, *The Social Network*, and *Jaws*

Film music does more than create an auditory shortcut to a mental emotion center. The power of music in film can be attributed to its unique *musical* features; that is, the capacity of patterns of notes in melodies and harmonies to connect to emotions via musically induced tension and relief (Meyer 1956), musical implications and realizations (Narmour 1991), and musical motifs and meanings. In addition, music contributes to film through *structural* features shared with or in counterpoint to other aspects of film: visual scenes and objects, sound effects, speech, or even visual text (Cohen 2009; Fahlenbrach 2008). Whereas the contribution of music to film has been appreciated since the time of Copland and even earlier by Hugo Münsterberg (1916/1970)—the first psychologist to direct attention to film—the scientific knowledge acquired in recent years about brain function, music perception, and cognition may help reveal music's contribution in not one but in many different ways to an audio-visual presentation.

Let us consider examples of music from the openings of three well-known films. Each scene involves one or more characters in relentless motion toward a goal that is significant to the film. *Chariots of Fire* (Hudson and Putnam 1981) opens with a group of runners training on a seacoast in England (Figure 7.3). The music begins with quickly repeating (sixteenth) notes in the bass, which establish a reference note (keynote, main note, or tonic) for the entire piece. The persistent repeating note might be regarded as reflecting the determined action of the runners. The higher-pitched soprano line introduces a two-note motif reminiscent of a military bugle call: the second note is higher than the first (the distance from *do* to *sol*, also known as a rising *perfect fifth*). Soon a simple melody that entails triplets (three notes to a beat, as in the word *won-der-ful*) contrasts with the low repeating sixteenth notes (four notes to a beat, as in *mo-tion-pic-ture*) and imparts an additional layer of meaning to the characters and situation. As with the different temporal patterns in the music, there are different temporal patterns in the visual scene: the rolling waves and the runners. For some audience members, the triplet patterns may relate to the rolling waves while the sixteenth notes connect with the runners. The music seems to add to the film but, of course, the music is absent in the diegetic world of the actors: the runners do not hear the music.

Let us compare this to the opening of *The Social Network* (Fincher and Spacey 2010). After being rejected by his girlfriend at the Thirsty Scholar pub, the main character, Mark Zuckerberg, emerges from a subway stop and pensively navigates across the Harvard campus to his dorm (Figure 7.4). The accompanying music signifies that this is no ordinary "walk in the park": an unrelenting low bass note (actually just a semitone away from the keynote of *Chariots of Fire*) is introduced, which segues into its octave (its nearest harmonic neighbor, from the point of view of physical acoustics) and then back

Figure 7.3 Opening scene from the *Chariots of Fire* (Hudson and Putnam 1981).

Figure 7.4 Opening scene from *The Social Network* (Fincher and Spacey 2010).

again to the bass note. The change might metaphorically signify that something is unstable; something is brewing. This unrelenting bass note line almost becomes attached to the central moving Zuckerberg character: sometimes he is a speck far less visible than other passersby on his route, but we do not lose sight of him. As in the opening of *Chariots of Fire*, a simple higher melody is presented. The path of Zuckerberg can be likened to the audiences' melody line, the figure against the background harmony of the visual scene. His footsteps contribute to the sound effects. Sometimes there are other sound effects: rustling belonging to a passerby, or music played by a street-musician violinist. The violin music, however, sounds like the violinist is tuning up, though visually the violinist appears to be engaged in a demanding, presumably classical, work. The audible violin music and the violinist that is visualized do not fit together in any realistic way, but the audience likely ignores this due to the focus placed on Mark, the main character—a character who will design the social network through his superior quantitative and leadership skills, and who will suffer through a lack of social intelligence. These contrasting aspects of his character are made clear in the first scene at the Thirsty Scholar Pub, and the music, with its contrasting soprano and bass lines, help reinforce this: the simple childlike melody and the brooding baseline of cold pragmatic reality (as described by the director, David Fincher, on the DVD additional tracks). Again, the music in no way distracts from the storyline and is not part of Mark Zuckerberg's world—other than the violinist, whose sound is not faithfully represented.

Now consider the opening scene from *Jaws*: a hippy beach party. Faint realistic music comes from a harmonica here, a guitar player there. The camera focuses on a male eyeing a girl at a distance. The music and sounds of the party are almost inaudible. He approaches the girl; she is delighted and wordlessly beckons him to follow her. She, like the protagonists of the two previous films, begins a determined run along the beach, with the boy following, somewhat in a drunken stupor (Figure 7.5). She gallops ahead, and her silhouette depicts her disrobing piece by piece, until she arrives presumably naked at the water's edge and plunges in. What is different in this example is that there is no music

Figure 7.5 Opening scene from *Jaws* (Spielberg et al. 1988).

during the determined run. In fact, there is no music at all until she enters the water. At that point, a repeating bass rhythm of the *Jaws* theme begins and develops in conjunction with the screams and thrashing motions of the woman fighting against the shark, depicted by the music.

Each of these films received awards for their film scores. In general, they all open with a repeating low pitch, signifying a protagonist with a driven nature, but there are differences. *Chariots of Fire* and *The Social Network* use music to underscore the act of running; in *The Social Network*, music defines the protagonist whereas in *Chariots of Fire* it depicts a group of runners. In *Jaws*, however, the onset of music foreshadows the entry of the shark, the primary protagonist of the film; the preceding scene of running lacks music altogether. In none of these examples is the music complex, although many films scores do use more complex music from the start, and these three films all use more complex music elsewhere. The simplicity of these backgrounds helps explain the influence of the congruence and associationist aspects of the music on the audience's interpretation of and engagement in the film narrative. Before discussing this further, let us first consider the internal and external worlds of film.

Internal and External Semantics and Unfolding the Narrative of the Film

Film music provides a unique perspective on the semantics of internal and external worlds (see Seifert et al., this volume). By *internal semantics*, I am referring to the emotional perspective of the perceiver: how the audience member feels while watching a film, and how the audience member feels about the characters and events depicted in the film. By *external semantics*, I mean the regularities of nature and cultural conventions of human interaction and human–environment interaction: the rules of language grammar and the pragmatics of discourse, the natural phenomena of climate, geography, and the rise and fall of the sun, the customs of dress and behavior, as well as the systems of social value of various historical times. All of these events can be learned from experience of the physical and social worlds. As the quote earlier from

Copland and Virgil Thompson implies, film music bridges the internal world of the audience member and the information provided by the screen. Film music thus provides an opportunity to study the relation between internal and external semantics, potentially contributing to the cognitive scientific understanding of how individuals make sense and personal meaning out of the cues that nature and social reality provide. Film music helps with this complex problem because of some essential differences from, as well as some less obvious similarities with, other media channels of the multimodal display. As pointed out earlier, film music, in contrast to the speech and sound effects in a film, is typically to be heard only by the audience, not by the film characters; yet most of the other sounds and sights in the film are audible or visible to both audience and actor.

Consider the audience member in a public or home theater presented for about two hours with a large rectangle of moving images, and sometimes printed text, as well as sounds from audio speakers, typically while sitting in a darkened room, often in the midst of other viewer-listeners. Each viewer-listener faces a two-hour task of organizing the presented information in an engaging and enjoyable manner, justifying the time spent and the financial costs of the movie ticket or, if in the case of home theater, the investment in audiovisual equipment. How does the music facilitate the task? A clue lies in the distinction between internal and external worlds in the film music context because, unlike the other light and acoustic media components of film which depict the external world (i.e., the scenography, the printed text, the sound effects and the spoken dialog), music typically exists in both inner and outer worlds, although, as will be suggested, only a part of the music functions within the internal semantics. As will be shown, music can play on emotions (the internal world) while directing attention to the information in the external world. Music may add unity to the experience of film, because music, even in its simplest form, operates in both amodal structural (i.e., not specific to sound or vision) and associative (meaning) realms, both of which connect to external and internal semantics (see Table 7.1).

Table 7.1 aims to represent how the structural and associationist aspects of various musical elements affect the internal and external meaning of a film. Elements of the film score (left-hand column) are deconstructed into structural congruence and association. The top panel focuses on the structural aspects of these musical elements: how they change over time (e.g., increasing or decreasing intensity or loudness), or how two aspects relate to each other (e.g., the consonant or dissonant relation of tones). The bottom left panel focuses on associations of various musical elements (e.g., what is associated with a low tone as opposed to a high tone, or consonance as opposed to dissonance). The musical elements in the upper and lower panels overlap but their structural and association aspects differ. Thus we find in the list the analysis of the pitch, tempo, and metric relations (temporal patterning), direction of the pitch (whether it ascends or descends), absolute intensity (soft/loud), pattern of intensity change

Table 7.1 Representation of how music structure and music associations impact the interpretation of the unfolding film narrative with respect to internal and external semantics.

	Internal Semantics of Film	External Semantics of Film
Music Structure (Congruence)	• Cross-domain structural congruencies produce an aesthetic response (e.g., musical triplets to the ocean waves in *Chariots of Fire*) • A low repeating pulsating pitch (e.g., the drone in *The Social Network* or *Chariots of Fire*) engages subcortical processing and higher brain levels • Parallel music/scenographic figure/ground direction of attention (e.g., Zuckerberg figure as a musical stream across the Harvard campus) • Parallel temporal structures may compete for and share syntactic resources (Broca's area) • Bypass acoustics (sounds)	• Congruencies between real (diegetic) music heard by a film character and activities of the character provide information about the character (musicality, human-ness of the character; e.g., Gazzara singing and driving in *Road House*) • Degree of synchrony between music and some other film modality conveys information about the real world of the film (e.g., oscillating path of the car in *Road House*) • Acoustics included in diegetic music • Acoustics audible
Music Meaning (Associations)	• Metaphoric, embodied parallels (e.g., rising pitch) give meaning to events eliciting a mirror neuron system (e.g., motor system to raise pitch of the voice, or raise a limb) • Sounds pulsed at the heart rate engage deep emotions (e.g., *Jaws* semitone motif) • Tension release and increase support personal emotional interpretation and involvement • Bypass acoustics	• Learned associations and conventions (national anthems, funerals, weddings) convey information about era and other context and establish mood • Acoustics may be bypassed if only associations contribute (specific music is not recognized, as it plays no role) • Repeated music-film contingencies create leitmotifs and further plot development • Diegetic music that fits the story is heard as music (e.g., dance music; a performance; Count Basie's band in *Blazing Saddles*); acoustics enters consciousness

(increasing/decreasing), triad relations (major, minor, diminished, augmented triads), chord analysis, chord progression, and phrase repetition. The list is not meant to be exhaustive; the various dimensions and elements identified are amenable to variation on hierarchical levels not represented here and would connect with various psychomusicological theoretical stances (the generative theory of tonal music by Lerdahl and Jackendoff 1983; see also Lerdahl, this volume; Huron 2006; Narmour 1991). Lipscomb (2005) has identified timing pattern (rhythm) as a structural feature whereas Cohen (2005) focuses on the highness and lowness of pitch and slow or fast tempo for its impact on happy–sad meaning. To the best of my knowledge, no empirical study has varied both structural and association aspects together; such a study begs to be conducted.

In Table 7.1, the two right-hand columns refer to the internal and external semantics of the film. The former focuses on the audience's feelings about the characters or situations and engagement in the plot, whereas the latter refers to acquisition of knowledge portrayed by the film: information about the characters, the time period, and setting.

Below I discuss each quadrant of Table 7.1 in turn so as to explain how musical structure and musically inspired multimodal structural congruence can affect the internal and external semantics in the listening/viewing experience of a film and, likewise, how the associations of, or perhaps to, music can affect the internal and external semantics of a film.

Music Structure (Congruence) and Internal Semantics of Film

Music can create attentional focus to aspects of the scenography with which it shares structure (i.e., temporal patterning). To provide a mechanism for this, I propose that aspects of the structure of the music which match the structure of other nonmusical elements of the film produce similar neural excitation patterns of system-specific (i.e., auditory or visual system) cell networks in the brain. The resonance of these similar excitation patterns brings to attention (i.e., to consciousness) those components in the film that embody these shared characteristics. This view is consistent with a recent review by Kubovy and Yu (2012:254), who argue that "the strongest relation between perceptual organization in vision and perceptual organization in audition is likely to be by way of analogous Gestalt laws." As an example, consider the motion pattern of the runners in *Chariots of Fire*, which matches the low-note rhythm of the music, or the melodic motif of *The Social Network* theme that depicts a moving speck (to which the audience's attention is drawn), which we know to be Mark Zuckerberg wending his way back to his dorm. It can be argued (though yet to be empirically tested) that similar music–visual structure draws attention to the visual information which gives rise to that shared structure. Although Gestalt psychology does not refer to a principle of multimodal similarity, the notion is a simple extension of the rather amodal principle of grouping by similarity within a sensory modality. Also consistent with Gestalt theoretic notions is the aesthetic satisfaction that arises from the registration of good form—an aspect of the internal semantics of the film. Such a response, which engages an audience member from an aesthetic standpoint, would also provide a focus of attention.

Even pitch analysis admits to consideration from the perspectives of structure, though in a completely auditory realm. All melody and harmony require pitch. The study of psychoacoustics tells us that the representation of pitch itself entails complex pattern recognition. The physical basis of pitch is a repeating pattern of air pressure. At a rate of between 30–20,000 times per second, any repeating air pressure pattern will produce the perception of pitch in the normal human auditory system. The sensation of pitch, as distinct from noise, requires the mental registration of approximately six cycles (Boomsliter et al. 1970). Recent evidence provided by Bidelman and Krishnan (2011) shows that all levels of the auditory system follow individual cycles of the waveform beginning with lower brain centers. When two tones are presented simultaneously, simpler ratio relations (like the octave 2:1) are represented most clearly, and the rank ordering of the faithfulness of the representation of these relations matches the ordering of the perceived consonance. While Western tonal music scales generally exploit small ratios or their approximation, certain music exploits the simpler relations more than others. Many so-called good soundtracks (e.g., *Chariots of Fire*, *The Social Network*, and *Jaws*) exploit these simple relations, thus giving processing precedence in both lower and higher brain centers, as compared to music that exploits more complex (more dissonant) musical relations. The engagement of the sensation of pitch (as opposed to noise) may serve to engage the audience member due to the establishment of (possibly large-scale) coherent brain activity entailed by pitch representation.

Low pitches characterize the music of the three award-winning scores highlighted above. My speculation is that low sounds may have special relevance for their ability to command engagement without calling attention to themselves as music per se. Interestingly, low pitch (40–130 Hz) in the audible frequency range is high frequency in brain wave EEG patterns, and it has been suggested that this gamma range of EEG may be significant in the establishment of consciousness (John 2003; Edelman and Tononi 2000). John (2003) has reviewed research that indicates how endogenously (self-) generated rhythms can suppress exogenous (automatically evoked) information. Such a principle may account for the many anecdotes and reports that the acoustic information of film music does not reach consciousness as it normally would outside the film context. Boomsliter et al. (1970) likened John's basic notion of endogenous processing of flicker to the processing of pitch versus noise. It is not suggested that electrical gamma waves would simply arise from physical acoustic energy waves of the same frequency; however, a possible relation between the two and the effect on consciousness necessitates further study (cf. John 2002:5).

A good example of structural congruence can be found in Kubrick's *The Shining*. Toward the climax of the film, the child, Danny, retrieves a butcher knife from the bedside table near his sleeping mother (Figure 7.6). With the knife held up in his left hand, the child's right-hand fingers move up toward the tip of the blade just as an ascending dissonant sequence of three tones sound from Penderecki's *De Natura Sonoris No. 2*; the audience's attention is drawn to the blade, its sharpness and danger. When Danny's fingers descend, the movement is accompanied by two more notes which add further emphasis; as Danny places his fingers lower on the blade, a lower note sounds

Figure 7.6 Scene from *The Shining* (Kubrick 1989), where Danny senses the sharpness of the knife. His finger motion up and down the blade parallels the direction and timing of the accompanying music—a dissonant melody from Penderecki's *de Natura Sonoris No. 2*. (http://www.youtube.com/watch?v=4iwK2jwqFfk)

and becomes a drone. In addition, immediately prior to and after this short episode, Danny repeats, in a low croaky voice, the word "redrum"—murder spelled backward—which functions as a type of musical ostinato (a repeating melodic fragment), in counterpoint with the discordant instrumental music of Penderecki. This functions much like the background motif of the *Jaws* semitone theme, the repeating rhythm of the *Chariots of Fire* theme, and the repeating bass note of *The Social Network* introduction.

Assuming that watching an action elicits the same neural activity as enacting the action oneself, the engagement of the mirror neuron system is likely to be a major contribution to the internal semantics of the film experience. Watching Danny move his fingers up and down the knife blade may engage an audience member in representing this same activity personally. Because the direction of the melody can also engage the same mirror neuron system (Overy and Molnar-Szakacs 2009), film music can reinforce the empathetic effect. Any ascending pattern of notes will parallel Danny's fingers moving up the knife blade, allowing the audience member more readily to identify directly with the action and empathize with the character. Once attention is directed by visual information, which shares structure with the music, other associative and metaphorical components of the music can connect to the internal semantics, as will next be described. It is important to note, however, that none of the music phenomenon above requires that sounds be heard. The impact of the sound arises from its structure not the sound per se.

Music Meaning (Associations) and Internal Semantics of Film

In addition to the motoric, gestural information described in the example of Danny from *The Shining*, music brings with it emotions that can be ascribed to the experience. Blood et al. (1999) have shown that different regions of the brain are affected by consonant and dissonant music passages. Thus, we would expect the discordant notes of Penderecki in *The Shining* example to increase a personal sense of unpleasantness and danger associated with the sharp object, as compared to the associations that an ascending consonant passage from a Mozart concerto might bring. In the example of Ben Gazzara driving his convertible to the music of *Sh-Boom*, the jaunty melody brings with it carefree associations and old-fashioned coolness. The music helps the audience to become Ben Gazzara, the King of the Road, or to perceive the sharpness of the knife, as Danny's fingers run up and down the knife's blade, like a discordant musical arpeggio. These are no longer cardboard characters; the amodal musical properties of ascending and temporal pitch patterns coupled with embodied motion plus the associations that come with the music enable an audience member to identify more easily with or, in a sense, become the characters on the screen.

Sounds pulsing at the heart beat or breath rate (as in *Jaws*) may bring associations of life and death. The random modulations of the repeating tone in *The Social Network* mirror anxiety and uncertainty—something is about to happen. These associations support personal interpretation and involvement in the drama. As in the structural analysis described earlier, these association effects are independent of actually registering the music as sound. By analogy, readers are typically oblivious to the letters on the page as they extract the meaning of the words.

Music Structure (Congruence): External Semantics of Film

Mental analysis of film music, carried out on various hierarchical levels, may lead to the identification of the music, if it is a known piece, or to the recognition of its style. As discussed, film music may be, or can become, part of the story, the diegesis. In the *Road House* example discussed earlier, the music begins in the seemingly nondiegesis, but as the camera pans in for a close-up of Ben Gazzara, we realize that he is singing along with the music—the music has thus crossed the "fantastical gap" (Stilwell 2007). In this example, the source of music can be attributed to the convertible's radio. Of particular interest to the consideration of music-visual congruence is the synchrony of the music structure and actions of Ben Gazzara, the driver. He sings along with the music. The congruence indicates his musicality. His bass voice is distinct from that of the tenor range of the recording. His voice and mouth are synchronized with the music, as are the movements of his head. He embodies the music, and he imparts that embodiment to his steering wheel and the undulating motion of his vehicle. This music-visual structural congruence contributes information about the external reality of the film narrative, in particular about the Ben Gazzara character. The audience does not give this a second thought, but a very different personality would be portrayed by a driver who sings off pitch, out of synchrony, without embodying the music. Synchrony adds a dimension to this character. It is also a source of humor and drama in the film as the world of abstract structure—that of a dance with a car—clashes with an object that also inhabits the real world; specifically, a driver coming in the opposite direction, quite out of phase with the ongoing congruent audiovisual pattern. This addition to the communication of the external reality of the film world arises by virtue of the structural congruence of the music with other nonmusical structure in the film. The phenomenon requires that music be part of the diegesis and that the music is heard.

Music Meaning (Associations): External Semantics of Film

The style of the music provides associations that contribute to the external semantics. Diegetic music provides information about the world of the film in a very obvious way. It is the way in which directors, in the early days of talkies, thought film music should work. The old rock and roll tune *Sh-Boom* helps represent the character of Ben Gazzara, telling the viewer his tastes in music, characterizing his age, providing a sense of the complexities of his personality, and appearing in a fairly human light in some respects.

Some films incorporate music performances within the drama to showcase a potential hit song for the film. Some background music may be used to set the scene and provide information about the context of the film through association. If new associations are formed between the music and events or characters in the film (i.e., if leitmotifs are created by the pairing of music and film) and these serve to further the plot and provide information about the reality of the film (e.g., a shark is nearby), this can also be considered as a musical contribution to the external semantics of the film. At the same time, the emotional impact of this cue could be regarded as an associative effect with implications on the internal semantics of the film. In some cases, the music contributing to the external semantics of film may be the focus of consciousness, but this is not necessary, as in the case when music associations set the historical period or general setting.

Summary

The relationships between the structural and associationist dimensions and semantic (internal and external) dimensions of a film, outlined in Table 7.1, show how the music from a film score can contribute to the impact of the film, from both personal emotional and external meaning perspectives. This structural–associationist analysis can also be applied to any physical dimension of the film: scenography, the text, the sound effects, the spoken dialog, or music. For example, Danny's repetition of the pseudo-word "redrum" (murder backwards) has a structural component quite apart from any other meaning that it might have: this repeating pattern, spoken in a croaky low pitch, draws attention to Danny and his eerie extrasensory perception.

Film Music and the Unfolding Narrative 191

The Structure of Film

A few general observations about narrative are in order. Let us first note its roots in our embodied interactions with the physical and social world. It "is found in all activities that involve the representation of events in time" (Abbott 2008:xii). Narrative or story appears in many guises. The most obvious is through written text: fiction books and stories that are a natural part of lives from birth through to old age. Developmental psychologist Katherine Nelson (2005) identifies the development of narrative understanding and generation as the fifth stage in the development of consciousness, and states that it occurs between the ages of three and six years.

Historically, stories have a much earlier origin than texts; stories have been told since the earliest civilizations and are exemplified by myths. Storytelling remains an everyday skill—one that can be developed to an extraordinary level of expertise. In film, stories are told visually through scenography and typically, due to technical advances in cinema technology, with added text and talking voices. Sound effects help to tell stories and so, curiously, does music. Each of these domains makes specific demands on the brain, and it is a puzzle to determine whether additional domains (language, sound effects, and music) make it easier or harder for the brain to process the narrative.

Regardless of the medium of presentation, as in everyday perception and cognition, audiences of a film have two sources of information (Figure 7.7): (a) the lower-order multimodal physical/sensory input (i.e., the patterns of light that hit the retina as well as the patterns of sound waves received at the ear) and (b) their higher-order knowledge, based on their own remembered or stored experience. The latter includes the rules and conventions about social norms, the grammar of story construction, and memory of one's own reactions to such information. Long-term memory includes all of the information that an individual can draw on to make sense of the world based on experience. From this arises the capacity for narrative, developed early in life. Both of these sources might be regarded as coming from the external world, but in different ways. Multimodal physical information (Figure 7.7a), which impacts the sensory receptors of the eye and ear, leads to automatic (exogenous) sensory processing, whereas memory (Figure 7.7b), prompted by fast preprocessing of the sensory information, leads to elicited (endogenous) generation of higher-order hypotheses and expectations about the reality represented at the sensory receptors. The output of the top-down endogenous processes (Figure 7.7b) may or may not find a match in the slower bottom-up exogenous processes (Figure 7.7a). Following the concepts of many theorists (Edelman and Tononi 2000; Grossberg 1995, 2007; John 2003; Kintsch 1998a), the best match leads to consciousness and, in my model for the role of music in film (Cohen 2005, 2009, 2010, 2013), to the *working narrative* (Figure 7.7c).

The effect of music on the interpretation of film has often been shown to be additive; for example, happy music plus neutral scene leads to happier scene

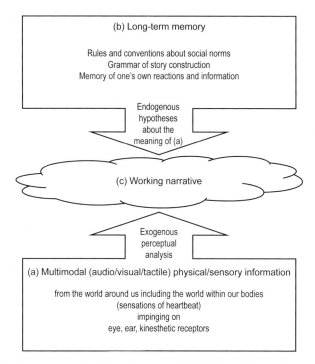

Figure 7.7 Two sources of information are available to the film audience member and lead to the working narrative.

(Cohen 2005). In addition, if camera technique directs attention to a particular visual focus (e.g., to highlight the main character in a film), the associations from the music would also be so focused. However, music itself may offer its own means of directing attention (Table 7.1). The role of congruent music-visual structure on visual attention may modulate an otherwise general associative or semantic effect of film music such that the meaning of the music is ascribed to a *particular* visual focus of attention (Cook 1998; Marshall and Cohen 1988; Cohen 2010). Thus the emotional associations of the music plus the other narrative elements internal to the scene would link together for a cumulative effect. Recall, for example, the scene from *The Shining*, where Danny retrieves the butcher knife that his mother has placed on her bedside stand as she naps. Emphasizing how sharp it is, he glides his fingers up its edge. As his fingers ascend toward the tip, the background music runs through a discordant high pitch arpeggio. The association of discordance and the ascension of the arpeggio add to the ominous nature and danger of the scene. The shared amodal property of ascension, common to the visual action and the music pattern, form a hinge that binds together an audiovisual Gestalt to which now additional associations of the music and the scene can be integrated. The audience member can empathize with this up and down motion along the blade, and the discordance can be ascribed to this focus of attention. This particular motif, or simply the last sustaining low note of the passage, may be linked to the scene or to the knife.

The process described above may be further considered as two stages: (a) focus attention through temporal congruence and (b) ascribe meaning of music to the focus. While there is no evidence that these processes occur in this order, there is evidence for both independently, but not yet in the same study. The anecdotal impression of music and film "going well together" is compelling, as in the example of the knife clip in *The Shining*. The scene from *Road House* previously described provides another. Here the evil city tycoon (Ben Gazzara) drives from his mansion in his convertible singing *Sh-Boom*. As far as he is concerned, he owns not only the city but the highway too, and as the contour of the melody undulates up and down, his vehicle outlines an undulating contour above and below the highway midline. The audio and visual are bound by their amodal similarity, and the jauntiness of the music funnels through to add to the personality of the tycoon. However, the visual and musical trajectories (i.e., their up and down patterns) are, in fact, not exactly in phase, and thus the phenomenon under discussion requires a variety of conditions for testing if we are to understand the relative contributions of the coordinated melodic and visual motion and the ascription of the meaning of the music to the visual focus of attention.

The narrative and multimedia context of film emphasize the ability of the brain to integrate multiple sources of information. This ability, however, is simply what the brain does in most waking minutes. Hasson et al. (2008b) have been exploring the intersubject correlation in brain activity as measured with fMRI across groups of viewers of the same film for more than ten minutes. Their study shows a systematic increase in correlated brain activity as a function of the degree of structure of the film: from a (low-structure) segment of reality video of New York City, to a sophisticated TV comedy *Curb Your Enthusiasm* by Larry David, to *The Good, the Bad and the Ugly* directed by Serge Leone, and finally to a TV thriller directed by Hitchcock, representing increasing levels of aesthetic control. The correlations or cortical activity across audiences (different for the four presentations) range from less than .05 to over .65, respectively. The technique looks promising for application to music variables, provided that the noise of the magnet is not so loud as to disrupt the impact of the music. Hasson et al. (2008b) also review related work, comparing responses to video (silent film) and audio story, which show strong differences in response. In addition, repeated viewings of intact versus scrambled versions of films led to different degrees of correlation.

Film places a boundary on audiovisual stimulus conditions, and this control is helpful in understanding how music and language work within the larger sensory context. With the additional context from other sources of information of visual scenes, text, and sound effects, individual and overlapping contributions of music and speech become apparent. The concept of language and grammar may apply similarly to all five of the domains of visual text, visual scenes, music, sound effects, and speech. By viewing the task of the brain exposed to film as one of integrating information from many sources, it may be suggested that the brain (at least sometimes) does not care where it gets information. It will take it from anywhere as long as it helps to get the story right. Getting the story right is the topic of a larger discussion, but for the present, the criteria would assume Grice's (1975) cooperative principle that information is narrated with the aim of creating meaning and entertainment. However, Grice's principles apply to discourse not to art, and thus additional principles are at stake (e.g., the audience is owed not simply a message but a nice experience, one which might take the scenic rather than the most direct route). There is a host of features on screen at any given time which may reward the viewer's attention, but which may or may not be essential to the story.

It would be parsimonious to apply the same code to information from different sensory sources; information from all source modalities would share the same code. Moreover, there is no reason to tag each piece of information as to its origin; that is, it does not matter whether an association of sadness came from music, or a sound effect, or from text. Instead, it is important to code for external and internal semantics. Structural and associative information apply to both external and personal realms.

Because music adds so much to the film narrative experience (or conversely, because much is lost to film narrative experience in the absence of music), and because how music functions in film is by no means unidimensional, understanding how music operates in film may hold a clue to how all film information is bound together in consciousness, and to how narrative works. Once we figure out how music works in film, it may be easier to understand how multimodal scenes are interpreted in the real world when a music soundtrack is typically absent as well as when, instead, physical and the social information are more strongly integrated (e.g., sensory information is more prominent).

The Congruence-Association Model and Its Extension to Working Narrative

Expanding further on Figure 7.7, my view is that the brain encodes associations and structure from all five channels of sensory information; that is, two visual (scenes and text) and three auditory (music, speech, and sound effects) channels. This five-channel classification was introduced by the film theorist Metz (1974) and promulgated by film scholar Stam (2000; see also Green 2010). In the model presented here, information in these five channels leads first to a passive, automatic, exogenous activity arising from the impact of energy from the real world undergoing electrochemical transduction by sensory neurophysiological processes. Faster and slower bottom-up processes co-occur; the faster processes (or preprocesses) signal to the higher structures in long-term

Film Music and the Unfolding Narrative

memory, consequently initiating top-down endogenous processes. For a recent discussion of the similar significance of top-down processes, see Gazzaley and Nobre's (2012) article on bridging selective attention and working memory.

Preprocessing of some information primes long-term memory and narrative proclivities for establishing the bottom-meets-top matching process, whereby the best match of top-down inferences to bottom-up sensory input wins out as the *working narrative* in consciousness. In general, diverse processes informed by working memory of recent episodes, as well as by long-term knowledge of the social and physical world, forge a coherent interpretation of (a portion of) the original sensory data. This approach resembles a schema theory (John 2002; Kintsch 1998a; Arbib and Hesse 1986), but some unique features, found only in the *congruence association model with working narrative* (CAM-WN), accommodate film music conundrums (see Figure 7.8). In addition to the five channels previously mentioned, an additional kinesthetic channel is included to accommodate the mirror system activity and extend to the nonfilm (real-world) situation. The nonfilm situation would have a reduced role for the music channel.

The CAM-WN model presents two sources of information available to an audience: *surface information*, which is sensory information in six channels

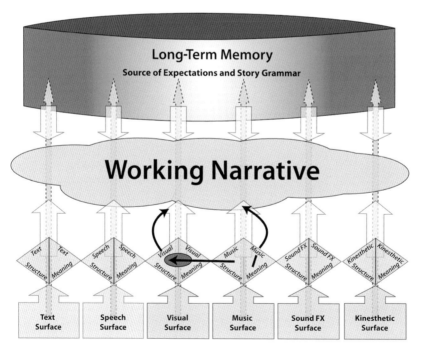

Figure 7.8 Congruence-association model with working narrative (CAM-WN). See text for explanation.

(text, speech, visual, music, sound effects, and kinesthetic), and *long-term memory*, which includes expectations and story grammar. Surface information entails bottom-up processing—both fast preprocessing (as represented by the dashed lines) and slow, more accurate processing. Long-term memory processes are top down and are elicited by the arrival of activation from the fast preprocessing. The surface information from all six sources is coded similarly into structural features and meaning (associations). The shared code (of structure and meaning) allows for structural congruencies and cumulative meanings, and assists in targeting elements that are relevant to the story. In Figure 7.8, a horizontal arrow from music structure to an oval in the visual channel represents shared music-visual structure in this particular instance. The structural congruence directs attention to the congruent visual pattern which becomes the attended target. These attended elements are the target in a contest to find the best match with top-down processes. The best match enters into consciousness in the working narrative, which is the audience's conscious unfolding narrative. Only matched items achieve consciousness. Thus meanings (associations) of the music and music structures may contribute to the narrative without the actual sounds of the music entering consciousness, because those sounds are not matched by endogenous generation.

The music channel, in part, distinguishes this model from "just a theory of perception" or "just a theory of information processing." The translation of these elements into neural equivalents may well be possible with increasing numbers of studies of brain imaging; however, very few brain imaging studies have focused on music and moving images to date, and more are needed. The CAM-WN framework is helpful because of the focus it gives to music; it also draws attention to the fact that music is not one thing but is comprised of many separate components, each of which can be analyzed in more than one way. Like each of the other channels of film information, music is analyzed for associations and for structure. This approach can inspire studies that isolate particular conditions within and between different modalities (e.g., the effects of different tempos, rhythms, or intensity changes, etc. on meaning within two modalities and the effect of the combination of tempos on meaning when the modalities are combined).

The CAM-WN model is general enough to incorporate other proposals which are relevant to the question at hand but are descriptive and not predictive. Examples at two ends of the spectrum are provided by E. Roy John's neural theory of consciousness (John 2003), as discussed above, and Kathrin Fahlenbrach's insightful approach to explaining the contribution to film meaning through the components of the soundtrack. Fahlenbrach argues that audiovisual scenes refer to presymbolic metaphorical structures in our cognitive system that offer rich material for visual and acoustic concretization of both very complex meanings and the representation of bodily and emotional experience (Fahlenbrach 2008). For Fahlenbrach, audiovisual metaphors function, with reference to emotional metaphors, as emotion markers in empathetic scenes, which represent culminating points of a film. She stresses, however, the need to test her model empirically.

John's focus on the match between exogenous and endogenous processes and Fahlenbrach's focus on the amodal component of media are encompassed by the CAM-WN model. Thus the CAM-WN model allows application of concepts from various levels of discourse within the same framework and specifically incorporates music. Fahlenbrach (2008) does not make reference to specific music features, and John (2003) does not even mention music, although Boomsliter et al. (1970) related his empirical research on exogenously and endogenously produced rhythms to pitch processing. The incorporation of the schema theory proposed by Herrmann et al. (2004) accommodates the matching component necessary for recognition and consciousness and adds the mechanism of the gamma band, consistent with the earlier proposal of John (2003), the function proposed by Fries (2009), and the evidence for frequency following at lower and higher brain levels for pitch analysis by Bidelman and Krishnan (2011).

CAM-WN accounts for the tolerance and prevalence of nondiegetic music that is artificial with respect to the ongoing world of the film (Boltz 2004; Cohen 2010). In CAM-WN, the top-down inference process (from long-term memory) can match the *meanings* of the music without matching the acoustical information (sounds) of the music. CAM-WN proposes something like an illusory conjunction (Treisman and Schmidt 1982) that forms between music meanings and visual object information, leaving behind the acoustic shell or vehicle that is never to appear in the working narrative (the story), though potentially preserved in an independent music memory simultaneously being updated (and without which there would be fewer CD album downloads and sales). John (2003) refers to data that shows how endogenous generation can suppress exogenous information. Endogenous information leads to perception although the exogenous information is in fact registered (without reaching awareness). CAM is descriptive and to some extent predictive, but twenty-first century cognitive science is, or will soon be, able to do better by explaining more specifically how brain mechanisms might carry out and elucidate CAM's functions. Within the last decade there has been progress to this end; however, only one brain imaging study has focused on both the moving image and the music track (Eldar et al. 2007).

Reference was made earlier to the notion of film music contradicting the diegesis. By this I mean that the music would not fit the fictional world depicted: Count Basie and his big band have no place on the desert (*Blazing Saddles*); a violinist happens to be along Mark Zuckerberg's path (*The Social Network*), not three cellos, an electric and a nonelectric piano, and the violinist is playing a piece not tuning up. The acoustics contradict the scene depicted. The emotion conveyed by the music does not contradict the scene, supporting both its internal and external semantics. In CAM-WN the lack of a match between exogenous and endogenous activity associated with the actual sound of the music means that the sound is outside of consciousness. No other current model or framework for narrative deals explicitly with this particular and anecdotally prevalent case.

Using fMRI, Eldar et al. (2007) showed that combining music which evoked emotional meaning with neutral films increased activity in the amygdala, hippocampus, and lateral prefrontal regions, as compared to a condition in which the film was absent. A similar finding was earlier reported by Baumgartner et al. (2006) who, using affective still pictures rather than motion pictures, showed activation in the amygdala and hippocampus only in the presence of music; this led the authors to propose that music engages the emotional system whereas so-called emotional pictures engage a cognitive system (for a discussion of the neural basis of music-evoked emotions, see Koelsch 2010). The work by Eldar et al. and Baumgartner et al. suggest that music, rather than vision alone, provides the link to an emotional feeling state in film. Their data were also supported by subjective responses which showed additivity of emotional meaning from auditory and visual sources. The results concur with musicologist Nicholas Cook's (1998:23) description of the role of music in the multimedia context: "a bundle of generic attributes in search of an object....a structured semantic space, a privileged site for the negotiation of meaning." The level of rigor required or that can be expected is represented by the approach taken by Arbib (2010) and Arbib and Lee (2008) in their attempt to explain how a verbal description arises when information is visually presented. Details remain, however, to be resolved. Arbib and Lee (2008:151) show, for example, how the same visual image will lead to different grammatical interpretations depending on context. Moving this example to the present discussion, such context can be provided by musical structure arising from the first level of CAM analysis, or it can be provided by endogenous generation arising from the flow from long-term memory that has been primed by preliminary processes. The information is available to construct the theory within the CAM-WN framework.

Audience engagement or absorption in a film, particularly when facilitated by music, begs for consideration of the topic of consciousness. Revonsuo (1998) has focused on neural synchrony that leads to binding, which brings about consciousness (see also Crick and Koch 1990). Here I suggest that music input facilitates creation of such synchrony. Luo et al. (2010) obtained magnetoencephalography (MEG) recordings from participants who watched film clips with matched or mismatched audio tracks. Their analysis revealed that a "cortical mechanism, delta-theta modulation, across early sensory areas, plays an important 'active' role in continuously tracking naturalistic audio-video streams, carrying dynamic multi-sensory...information....Continuous cross-modal phase modulation may permit the internal construction of behaviorally relevant stimuli" (Luo et al. 2010:1–2). In particular, the auditory cortex was found to track not only auditory stimulus dynamics but also the visual dynamics as well. The converse was true of the visual cortex. The system favors congruent as opposed to incongruent audiovisual stimuli. "The phase patterns of the ongoing rhythmic activity in early sensory areas help construct a temporal framework that reflects both unimodal information and multimodal context from which the unified multisensory perception is actively constructed" (Luo et al. 2010:10). It might follow that film music produces a source for electrical brain wave patterns that can bind with accent patterns from visual sources (e.g., runners in *Chariots of Fire*, the motion of Mark Zuckerberg's image in *The Social Network* as he traverses the Harvard campus, or Danny's finger motion along the knife blade in *The Shining*) but not necessarily only these visual patterns, perhaps patterns from sound effects (e.g., the sound of motion on the pavement, the sounds of speech patterns) or the visual patterns of articulation.

Hermann et al. (2004:347) describe gamma-band oscillations (30 Hz) that enable "the comparison of memory contents with stimulus-related information and the utilization of signals derived from this comparison." Their model attempts to explain early gamma-band responses in terms of the match between bottom-up and top-down information. Furthermore, it assumes that late gamma-band activity reflects "the readout and utilization of the information resulting from this match." This mechanism, which they describe as the *match and utilization model*, seems like that which is required by the CAM-WN model in preprocessing information from the initial sensory channels, activating relevant parts of long-term memory so that appropriate inferences can be generated to make the best match with the slower and more detailed processing of the initial encoding. Commonalities with the system proposed by John (2003) deserve mention, although John's reference to gamma band begins at 40 Hz, not 30 Hz.

A remaining issue is that of the mirror neuron system. Film is an ideal context for such study because characters in the film, with whom the audience identifies, provide images to be mirrored. In addition, music also leads to mirror activities in terms of dance and other movement (Overy and Molnar-Szakacs 2009; see also Calvo-Merino et al. 2006, 2008). Speech may engage the mirror neuron system (Arbib 2005a; Rizzolatti and Arbib 1998). The prefrontal and parietal location of the mirror neuron system may overlap with a common area of linguistic and music semantic processing. The CAM-WN model accommodates motoric and kinesthetic information by means of the kinesthetic channel. Information in the working narrative normally provided by the kinesthetic surface could be provided by musical information, following Overy and Molnar-Szakacs (2009).

CAM-WN makes sense of Hasson et al.'s (2008b) fMRI correlations arising from viewing film. All audience members received the same four conditions of stimulus input that differed in aesthetic control. Hitchcock's cinematic techniques were found to control processing to a greater extent than any of the techniques used by the other directors in the study. Presumably, Hitchcock is a master of directing audience attention to particular information in the visual scene. He provides sufficient information that passes quickly to long-term memory, engaging narrative processes and hypothesis generation to match or not match information arising from the sensory bottom-up processing. Only some of the parsed and processed information is best matched and integrated into the ongoing working narrative. In contrast to the Hitchcock film excerpt, a video of New York produced few common hypotheses arising from the story grammar and long-term memory. Every viewer's experience or imagined experience of New York differs. Thus, each viewer would create a different working narrative. Adding music to this videography, however, might introduce constraints at several levels and lead to higher correlations among viewers. Future research is needed to test this.

The CAM-WN approach also accommodates Levinson's (1996) challenge of the much accepted view that film music is unheard (Gorbman 1987). He has argued that film music is by necessity heard so as to help the audience understand the position of the narrator and the filmmaker. He believes that the job of the audience is to understand how the narrative, as created by the narrator and filmmaker, is to be understood. CAM-WN certainly accommodates the simpler view of the audience as the creator of the narrative, but it also accommodates the higher-order levels of narrator and filmmaker, should information in the film be sufficient to allow this aspect. This allowance, in part, depends on the audience member's prior knowledge. Clearly an audience member steeped in philosophy, aesthetics, film, and music will have more knowledge to draw on than a member who lacks this experience. To test predictions from CAM-WN, group differences in consciousness of film music, narrator, director, and so on clearly require further psychological research.

With its three auditory channels, the CAM-WN model helps show the subtle balance between speech, sound effects, and music in film. Loud music will mask speech and sound effects, the two primary auditory contributors of diegetic information. The film will then become increasingly ambiguous and more easily influenced by the music because loud music draws attention (still potentially unconscious) to itself and diminishes diegetic information due to masking. Cognitive science and psychoacoustics can provide insight regarding the relative contributions of the sound effects, speech, and music, each of which can convey structural information and semantic meaning. Each of these acoustical modalities is a different kind of vehicle designed to transport some kinds of information better than others. To date, however, no psychological study has investigated the signal-to-noise ratio relations in the soundtrack mix of sound effects, speech, and music. Future studies are needed to address questions about how film music works in film, how the brain handles these multiple sources of acoustic information, and how it integrates them with the visual information.

The inclusion of music in film might be regarded as one of nature's gifts to cognitive science, because, as has been emphasized, the presence of music in film is illogical and paradoxical on several grounds. Other illusions that have dropped in our path, such as Treisman and Schmidt's (1982) illusory conjunctions in vision or Roger Shepard's ascending staircase in the auditory realm, have provided insight into how the mind works. Likewise, the study of film music—from its paradoxes to its most obvious features—may help us unpack the mysterious relations of music, language, and the brain.

Acknowledgments

I wish to thank the Ernst Strüngmann Forum for the opportunity to participate. This chapter benefitted from the insight, inspiration, and encouragement of Michael Arbib, and the conscientious editorial work of Julia Lupp and her team. The Social Sciences and Humanities Research Council of Canada is acknowledged for the generous support of my research in film music for over two decades, and the Canada Foundation for Innovation is acknowledged for providing infrastructure in the CMTC Facility at the University of Prince Edward Island.

First column (top to bottom):   Uwe Seifert, Jônatas Manzolli, Tom Fritz, Paul Verschure, Uwe Seifert, Michael Arbib, Leo Fogassi
Second column: Paul Verschure, Nikki Rickard, Leo Fogassi, Nikki Rickard, Jônatas Manzolli, Annabel Cohen
Third column: Michael Arbib, Annabel Cohen, Gina Kuperberg, Tom Fritz, Klaus Scherer, Gina Kuperberg, Klaus Scherer

8

Semantics of Internal and External Worlds

Uwe Seifert, Paul F. M. J. Verschure, Michael A. Arbib,
Annabel J. Cohen, Leonardo Fogassi, Thomas Fritz,
Gina Kuperberg, Jônatas Manzolli, and Nikki Rickard

Abstract

This chapter analyzes the similarities and differences of meaning in language and music, with a special focus on the neural underpinning of meaning. In particular, factors (e.g., emotion) that are internal to an agent are differentiated from factors that arise from the interaction with the external environment and other agents (e.g., sociality and discourse). This "world axis" (from internal to external worlds) is complemented by three other axes: the "affective–propositional axis," the "sensorimotor–symbolic axis," and the "structure axis" (from small- to large-scale structure). Common structure–function relationships in music and language and their neuronal substrate are addressed, with emphasis on expectation and prediction. A special focus has been put on how the factors discourse or narrative relate to emotion and appraisal. Neurocinematics is studied for its focus on large-scale structure where music and language strongly interact.

Four Axes for the Study of Meaning

Our discussion can be organized along four general axes (Figure 8.1). The first (world axis) describes a continuum between involvement with the "internal world" concerned with the grounding of language and music in the self's emotions and drives (and, perhaps, its embodiment, perception, and cognition) and the "external world" of the physical and social environment. The second (affective–propositional axis) describes a continuum from affective to propositional meaning, whereas the third (sensorimotor–symbolic axis) is based on levels ranging from sensorimotor to conceptual or symbolic processing. The fourth (structure) axis considers structure ranging from small scale (such as a sentence in language) to large scale (such as a narrative or discourse). These

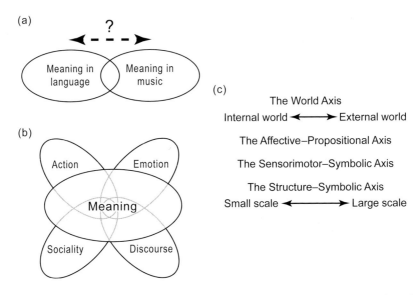

Figure 8.1 (a) We seek to understand to what extent meanings in language and music engage different or shared brain mechanisms. (b) In seeking to understand meaning, we take action (and perception), emotion, discourse, and sociality into account. (c) We structure our discussion in terms of four axes which, with some overlap, help us explore similarities and differences between music and language. Left versus right on the four axes does not, in general, correlate with the differences between music and language.

axes are neither all-inclusive nor mutually exclusive of one another, but provide a fruitful framework for our discussion.

Although this will be pursued only lightly in this chapter, we note another important question: To what extent are the perception and production of language and music autonomous activities of the individual, and to what extent are they emergent properties of group interactions? Music is used in the external world to promote social bonds, group synchronization, and coordination of emotional states/consciousness as in hunting, work, ceremony, ritual, and dance (see Lewis, this volume, as well as Levinson, this volume, who contrasts turn-taking in conversation with social coordination in group performance of music or dance). Musical discourse driven by collective performance can be found in Balinese Gamelan music, the percussive music of Ghana West Africa, or Brazilian drum music as well as in some contemporary interactive music systems (see Verschure and Manzolli, this volume). In each of these cases, music emerges from the interaction between individuals; that is, without a score and usually without a conductor. These examples challenge that part of the Western musical tradition which relies on a composer and a conductor to define and guide, respectively, the production of a musical piece and its meaning.

The World Axis

What constitute the "internal" and "external worlds" that define our first axis? At the start of our discussion, the "external world" was seen primarily as the physical and social world that surrounds the individual, whereas the "internal world" was the realm of feelings and emotions. As the discussion continued, some sought to extend the notion of "internal world" to include any thoughts one might entertain: one's inner feelings, factual and counterfactual states of the world, or wishes, dreams, and desires. In either case, it is clear that language allows us to communicate about states of our internal world as well as states of the external world. It is not clear, however, to what extent music fulfills a similar role. Traditionally, music has been more strongly related to the internal world of emotions or affective response, also referred to as extramusical semantics or musicogenic meaning (Koelsch 2011c, this volume). Thus, understanding the relationship of music and language to the world axis provided a central challenge for our discussions.

The Affective–Propositional Axis

We may use language to say "John is sad" or put together several sentences that let one infer that what John experienced made John sad; a story may be so evocative that, in hearing it, we not only come to know of John's sadness but come to share it, as the story makes us sad. Neither an explicit proposition nor the sort of inferences that language supports seems apropos to music. Perhaps, however, the mechanisms whereby a story may move us to share an emotion may be operative as we experience music in certain ways. Aesthetically, a satisfying piece of writing and a satisfying piece of music might engage our emotions via shared resources in the brain, even though accessed differently.

The Sensorimotor–Symbolic Axis

We share many basic sensorimotor mechanisms with nonhuman animals, but humans are (almost) unique in their ability to use symbols as distinct from limited call systems or "nonsemantic" song for communication. We seek to understand similarities and differences between "basic" processes of the action–perception cycle, and the unique forms of human social interaction rooted in music and language.

The Structure Axis

Much of linguistics has focused on the word and the single sentence as objects of study; these are elements of small-scale structure. We may also consider large-scale structure, as in an extended narrative or conversation, or in a complete musical performance, whether emergent or based on a prior composition

206 U. Seifert et al.

as for a symphony. Clearly, large-scale structure places different demands on memory from that which is required for production or perception of, say, a single sentence in isolation.

The Way Forward

In what follows, we discuss the meanings of language and music in relation to their differential engagement along the four axes, as well as how these processes involve regions of the human brain in different patterns of cooperative computation. Specific studies are included that seek neural correlates for processes related to meaning. Figure 8.2 is designed as a reference point for a number of the key regions of the human brain implicated in the studies. Figure 8.2a shows a lateral view of the cerebral cortex. Figure 8.2b singles out four cortical regions located medially (where the left and right hemispheres face each other), as well as four important subcortical systems implicated in reward, emotion and memory. Another relevant subcortical system, the basal ganglia, is not shown. As is well known, both the size and functionality of cerebral regions may vary between the left and right cortical hemispheres.

Is the semantic system, however, connected to specialized meaning systems, or is there a single semantic system underlying the four basic axes in music,

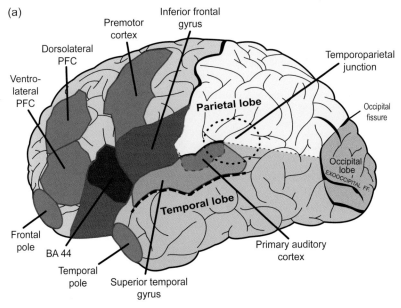

Figure 8.2 Lateral (a) and medial (b) view of key brain regions identified in this chapter as being critical to music or language. Cortical structures include prefrontal cortex (PFC), temporal pole, primary auditory cortex, Brodmann area 44 (BA 44), and frontal pole. Subcortical structures include amygdala, hippocampus, nucleus accumbens, and parahippocampal gyrus.

language, and action? Many of the neural data seem to support the former view (for a discussion of shared resources, see Patel, this volume). On the other hand, those arguing in favor of a unified semantic system hold that meaning is the result of evolutionary processes which capitalize on unifying principles (see Fitch and Jarvis, as well as Cross et al., this volume). Certainly, animals respond to some of the basic features of sound patterns that humans associate with music (Rickard et al. 2005). For instance, rhesus monkeys perceive octave equivalence (Wright et al. 2000), cockatoos synchronize relatively well to a beat (Patel et al. 2009), and chicks are aroused by rhythmic auditory patterns (Toukhsati et al. 2004). However, none of this means that these species "have" music, let alone language. In this chapter, we will advance this debate but not resolve it.

Meaning before Language and Music: The World Axis

Throughout this report, the world axis will be a continuing, often implicit, theme. Here we set a baseline by assessing the world axis as a property of organisms, generally, before turning to the uniquely human attributes of language and music. We characterize the organism's interaction with the external world in terms of the action–perception cycle and the way this is influenced by the internal world through the role of emotion, and briefly discuss levels of processing. Thereafter we address the other axes: the affective–propositional axis, the sensorimotor–symbolic axis, and the structure axis.

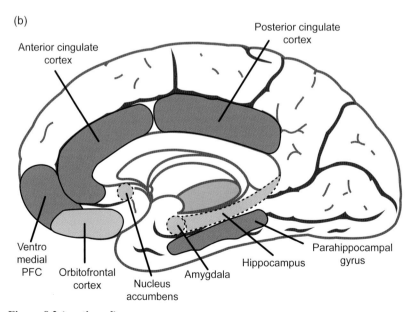

Figure 8.2 (continued)

The Action–Perception Cycle

Action and perception are integrated, whether in intelligent human behavior, the activity of an animal, or the smooth functioning of an adaptable robot. An organism is not passive, receiving a series of "sensory samples" to be classified into one of some small number of categories. Rather, it is constantly and actively seeking information relevant to its actions. If it has miscomputed what is relevant to it, it will ignore much relevant information. Here is an example of the dependence of perception on one's current goals: If you are walking down the street to go to the store, you may avoid colliding with people without seeing who they are, much to your embarrassment if you pass a friend who may feel snubbed. However, if you are walking down the street looking for someone, you will most likely recognize any friend you encounter, even those for whom you were not looking. Consider another example: the way we perceive a chair will be quite different depending on whether we intend to sit on it or to paint it.

Both examples help establish the basic notion of the action–perception cycle (Neisser 1976; Arbib 1989) which, being a cycle, is also known as the perception–action cycle (Fuster 2004; Janata and Parsons, this volume). The subject's exploration of the world is directed by *anticipatory schemas*, plans for perceptual action as well as readiness for particular kinds of sensory structure. The information that is picked up modifies the perceiver's anticipations of certain kinds of information which, thus modified, direct further exploration. For example, to tell whether any coffee is left in a cup, we may take the cup in hand and tilt it to make the interior visible, tilting the cup further, when we fail to see any coffee, until we see some or discover that the cup is empty. Thus, the organism is perceiving as it is acting; that is, as it is making, executing, and updating plans. It must stay tuned to its spatial relationship with its immediate environment, anticipating facets of the environment before they come into view, so that expectation and prediction are essential aspects of behavior.

We perceive the environment to the extent that we are *prepared to interact* with it in some reasonably structured fashion. We can show that we have perceived a cat by naming it; however, our perception of it often involves no conscious awareness of its being a cat per se; for example, when it jumps on our lap while we are reading and we simply classify it by the action we take as "something to be stroked" or "something to be pushed off." Arbib (1981) posited that the world is perceived in terms of assemblages of dynamic, adaptive units, *perceptual schemas*, each of which roughly corresponds to a *domain of interaction*; this may be an object in the usual sense, an attention-riveting detail of an object, or some domain of social interaction. Of course, the available schemas, and thus perception itself, will vary greatly from species to species— both people and dogs perceive telegraph poles in terms of communication systems, yet the perceived means of communication are very different—as well as from individual to individual. A schema can be "tuned" to the current situation, and in many cases, this will include updating parameters for size, orientation, and location. In general, the activation of a perceptual schema gives access to various *motor schemas* (the same object may be acted upon in diverse ways), which encode knowledge of how to interact with that which the perceptual schema represents.

The reader may object that much of human behavior is verbal rather than involving activity of the body and limbs, and regret that our emphasis on "perception as preparation to interact" would seem to exclude most of people's more intelligent behavior from study. But human brains have evolved from the brains of animals which interacted in a complex fashion with their environments without the aid of language. Indeed, it was the consensus of the Forum that language (and music) can best be understood in relation to already complex systems that we share with other species (Cross et al., this volume). Although conscious planning may certainly play a role in directing our behavior, sophisticated sensorimotor strategies exist that do not require an arch controller. Verbal mediation may sometimes be paramount, but in many cases it will simply bias or only monitor the underlying sensorimotor dynamics.

Piaget was perhaps the most influential user of the word *schema*, who defined the "schema of an action" as the structure of "the generalizable characteristics of this action, that is, those which allow the repetition of the same action or its application to a new content" (Beth and Piaget 1966:235). He offers a constructivist theory of knowledge (Piaget 1954, 1971) in which the child builds up a basic repertoire of schemas (e.g., for grasping) through sensorimotor interactions with the world. These develop during various stages until the child has schemas for abstract thought that are no longer rooted in sensorimotor particularities. This notion of schema is also related to Peirce's notion of a "habit," with its connections between the individual and the social. Peirce (1931/1958) held that there is a strong analogy between the evolution of species and the evolution of science. He used the term "habit" in a sense sufficiently general to cover the hereditary action patterns of the individuals of a species, learned habits of individuals, as well as the concepts, rules, methods, and procedures of social institutions.

For Piaget, the process of adaptation in cognitive development is composed of the inextricably intertwined processes of assimilation and accommodation. Assimilation is the process whereby the data of the world are assimilated to currently available schemas, forming a schema assemblage which is not merely a passive representation but an active process which leads to activity and exploration within a structure of anticipations. To the extent that the structure of anticipations proves dissonant with consequent experience, there is grist for accommodation—the process whereby the individual's repertoire of schemas changes over time to better reflect other aspects of the world beyond those assimilable to current schemas.

All this frames the language- and music-free notion of "behavioral meaning" of an event or object; namely its implications for further action and perception (as in the case of "a thunderstorm means rain" and "the expectation of rain means you should take an umbrella") and the possible updating of memory structures. As in the case of a chair, the "meaning" may be highly context dependent (a chair is to sit; a chair is to paint). For some "meanings," many people will employ similar perceptual or motor schemas to engage with an object or respond to an event, thus establishing a set of core meanings which, in turn, can anchor the meanings of words in a language. To proceed further we need to assess more explicitly not only our interactions with the external world but also those internal changes known as feelings or emotions.

Emotion

What is the difference between the evaluations of a stimulus event that give rise to an emotion process and those that do not? This question is difficult to answer because the transition is continuous, having to do primarily with the relevance or pertinence of the event to that about which an individual cares most: personal well-being, important aims, intimate relationships, cultural values. This type of emotion–antecedent evaluation has come to be called *appraisal*, after the pioneering work of Arnold and Lazarus in the 1960s, followed by a relatively large group of appraisal theorists (see Ellsworth and Scherer 2003). The emotion–antecedent appraisal consists of a complex recursive process of sequential cumulative checking of factors such as novelty, intrinsic pleasantness, goal conduciveness, coping potential, and self/norm compatibility.

This description of appraisal emphasizes aspects that can (but need not) be brought to conscious awareness. They are complemented by a host of processes that link from basic drives to cortical systems via subcortical reward and conditioning processes which can be studied in nonhumans. Writing as a neurophysiologist, LeDoux (2012) downplays questions about whether emotions that humans consciously feel are also present in other animals, focusing instead on the extent to which so-called survival circuits and corresponding functions that are present in other animals are also present in humans. Though such functions are not causally related to emotional feelings, they contribute to these, at least indirectly. The interaction of the appraisal process with components of the emotion process (e.g., action tendencies, physiological changes, motor expressions, and feelings that integrate all of these components) does not need to reach a static endpoint. We may talk of an *action–emotion–perception* cycle, where the efferent effects occur continuously in the appraisal process as soon as there is sufficient stability of an appraisal on a specific criterion to warrant the energy invested in an action (see Scherer, this volume). Further, emotions play a role in social interactions. Arbib and Fellous (2004) distinguish the role of emotions in modulating the prioritizing or planning (possibly unconscious) of one's actions from the facial and postural expressions that convey something of one's emotional state to others (Darwin 1872/1998). The latter plays a crucial role in coordinating behavior (e.g., when you observe that someone is angry, you will interact very differently with them than if you perceive them to be sad).

One can distinguish between *utilitarian*, *aesthetic*, and *epistemic* emotions. Fear and disgust are examples of utilitarian emotions, which are also called basic, modal, or primordial emotions. Utilitarian emotions, like drives, serve the most elementary basic needs of an animal. Sublimity, awe, and beauty might serve as examples of aesthetic emotions. Aesthetic emotions are related to enjoyment (Scherer, this volume). Interest and curiosity exemplify epistemic emotions, which are related to the pleasure elicited through intellectual play and understanding. Epistemic emotions are related to conceptual thinking. States of our affective internal world range from drives to emotions and shade our personal and interpersonal utilitarian emotions to epistemic and aesthetic emotions. Humans evaluate world phenomena according to epistemic and aesthetic emotions. These evaluation types are not exclusive and may overlap in terms of the underlying mechanisms.

Levels of Processing

We discussed the possibility that processes which underlie the action–perception cycle and emotion–antecedent appraisal most likely occur on different levels of information processing (e.g., sensorimotor, schematic, associative, conceptual) and that some appraisal criteria, mostly driven by intrinsic stimulus characteristics, might be processed more easily and more rapidly at lower levels, whereas checks involving external inference require processing at higher levels. Specifically, Scherer (2009) postulates four such levels (entailing different neural structures and circuits):

1. A low sensorimotor level with a pattern-matching mechanism that is largely genetically determined, using criteria consisting of appropriate templates.
2. A schematic level, based on memory traces from individual learning processes, in the form of uniform, holistic templates and occurring in a fairly automatic, unconscious fashion.
3. An association level, based on multiform associations spreading and involving various cortical association areas. This may either occur automatically and unconsciously or in a deliberate, conscious fashion, depending on attentional investment.
4. A conceptual level, involving propositional knowledge and underlying culturally shaped representational meaning systems, probably requiring consciousness and effortful calculations in prefrontal cortical areas.

Similar hierarchies have been offered by Ortony et al. (2005) and Sloman et al. (2005). Verschure (2012) has advanced a related multilevel perspective in a recent theory, the *distributed adaptive control* architecture (for a description, see Verschure and Manzolli, this volume), which proposes that the brain can be seen as a layered architecture comprising reactive, adaptive and contextual layers. For example, the above four levels can be mapped to the role of different brain structures in an integrated model of the sensory and motor learning components of classical conditioning (Hofstötter et al. 2002; Inderbitzin et al. 2010; see Figure 15.4 in Arbib et al., this volume) as follows:

1. The low sensorimotor level would match the unconditioned stimulus to an unconditioned response, coupling actions such as an air puff to the cornea and the closure of the eyelid. This is essentially a brainstem process.
2. The schematic level would be constructed from the association of the conditioned stimulus (e.g., a tone) to the unconditioned stimulus, and involves the amygdala and cerebellum.
3. The association level can be the integration of the now salient conditioned stimulus into a more elaborate behavioral plan that would rely on cortical mechanisms.
4. At the conceptual level, advanced planning and meta-representational systems of the prefrontal cortex (PFC) would come into play.

The cited model offers a detailed analysis of levels 1 and 2. Further modeling to link action and emotion requires that reinforcement learning and other mechanisms be brought into play (Arbib et al., this volume). Fuster (2004) explicitly discusses upper processing stages of what he calls the "perception–action cycle."

Meaning in Language and Music: The Affective–Propositional Axis

Semiotics

In his semiotics (i.e., the study of signs and sign processes, designation, likeness, analogy, metaphor, symbolism, signification, and communication), Peirce (1931/1958) distinguished at least 64 different classes of signs, including his well-known fundamental distinction between index, icon, and symbol. An icon or iconic sign is based on a similarity relation; an indexical sign or index is based on a causal relation (e.g., "smoke as a sign for fire"), or at least a spatiotemporal correlation (Deacon 1997); the relation between a symbol or symbolic sign and its referent is established by convention. Peirce's definition of a sign refers to a ternary relation as a sign process called "semiosis." However, it must be admitted that in most discussion on language (and in this chapter), the main function of language is considered to be *signification*, conveying the meaning that a term, symbol, or character is intended to convey (cf. Morris 1964). Its signs are symbolic, conventional, and arbitrary, even if some spoken words (as distinct from written words) resemble iconic (as in the case of onomatopoeia) or indexical forms. Many signs in sign language appear iconic, though a conventionalized form is neurally distinguishable from the pantomime it may resemble (Corina et al. 1992; Marshall et al. 2004).

At the core of Peirce's semiotics is the notion that meaning generation is based on induction, deduction, and abduction. Where the first two processes bring us from particulars to general observations and vice versa, the latter notion involves the insight that the premises on which deduction is based can, in turn, be subject to change. Following Peirce's semiotics and combining logical positivism and pragmatism, Morris (1938) introduced the distinction syntactics, semantics, and pragmatics:

- Syntactics: the relations among signs in formal structures.
- Semantics: the relation between signs and the things to which they refer (i.e., their meaning).
- Pragmatics: the relation between signs and the effects they have on the people who use them.

This terminology—with a slight change from "syntactic" to "syntax"—is still used today and is widely accepted in linguistics, logic, and philosophy. Can it be applied to the meaning of action and music as well as language?

Frey et al. (2009) carried out an event-related potential (ERP) study that investigated music processing as sign processing and discussed temporal semiotic units as a basis for musical meaning. In discussing research on musical semantics in cognitive neuroscience of music and neuromusicology in which semiotic sign relations also come into play, Reich (2011) argues that Koelsch's (2011a) results indicate that the brain processes semiotic relations (for commentaries, see Slevc and Patel 2011; Seifert 2011; for a reply, Koelsch 2011c). An issue in this discussion is the interpretation of the N400-evoked brain potential as an indicator of a restricted semantics with a symbolic signification relation (Besson et al. 2011) as opposed, for example, to the P600. However, the relation of combinatoriality (i.e., pure syntax) and semantics is tricky, as a recent result from ERP studies on *language* processing indicates. Kuperberg (2007) reports that the P600 (usually associated with syntactic processing) appeared in a semantic task for which there was no change in N400 (usually associated with semantic complexity) (see Hagoort and Poeppel, this volume; Caplan 2009; Patel et al. 1998). Note, however, that the 400 and 600 ms time stamps do not guarantee that the same processes are being manifest in different experiments. Teasing apart which processes are shared and which are different remains a major challenge.

Propositional Meaning of Language

We are able to describe states of the world to others using language: we ask people to inform us about the state of the world, their intentions, and/or the intentions of others as well as ask them to achieve or help us achieve new goals. The "core semantics" of language is the matching of spoken or signed speech to actual (i.e., remembered or present), desired, and hypothetical states of the world; this is *propositional meaning*. As such, it supports not only statements and questions but also *inference*. When spoken, a sentence can also convey other types of meaning; for example, affective meaning through its prosody means that there is more than just propositional content to an utterance. We will have relatively little to say about propositional meaning in what follows, since we regard the notion of "meaning" for music as being, for now, far more mysterious and thus in need of greater discussion.

Dimensions of Musical Meaning

Whereas propositional meaning might be the dominating symbolic sign relation of language, musical meaning uses iconic, indexical, and symbolic sign relations in a less conventional way. For example, the call of a cuckoo (in music reproduced by a descending third, a so-called "cuckoo third") might signify, if interpreted as an icon of the cuckoo's call, an index of spring or a symbol of nature. The interplay of the different sign relations may be experienced while listening to the second movement of Ludwig van Beethoven's *Pastorale* Symphony (Karbusicky 1986:275).

Meaning is usually tied to the notion of the subject for whom the sign has *significance*. It may also be considered a systemic property of agent–environment interaction. In music, we have already noted examples of musical discourse in various cultures that are driven by collective performance. These examples raise the question of whether language, music, and action can be considered emergent phenomena with no prime movers. Are music, language, and action emergent and situated, or are they predefined and particular? It would seem that either aspect may hold true on certain occasions, which leads us to question how they might interact, and how autonomous meaning construction translates into future automatic and predefined processing.

The Issue of Shared Semantic Resources

A general theme of the Forum was the extent to which language and music should be considered simply as terms for different regions of a language–music continuum. Most would accept that, whatever the existence of intermediate forms, one can distinguish, for example, language defined for its propositional content from the music as exemplified by Mozart (and many other genres from different cultures) as lying outside the overlap depicted in Figure 8.1a. Given this, many discussants held that distinct neural systems are engaged in different forms of meaning (e.g., propositional vs. affective). Others supported the assumption of a single meaning system mapped to language, music, and action, and they offered semiotics as a framework to define such a common system, noting Deacon's work on the interpretative capabilities of the human brain as a starting point for linking semiotic reasoning to brain research (Deacon 1997, 1998, 2006). Deacon draws attention to changes in the structure of the brain's motivational systems as an essential evolutionary change which made the interpretative function of the human brain possible (see Hagoort and Poeppel, this volume), an important dimension underplayed in the mirror-system based approach to language evolution (see Arbib and Iriki, this volume).

Meaning, Emotion, and Appraisal in Language and Music

Bühler (1934/1988) made an important contribution to the study of meaning by pointing out that language fulfills three functions: (a) as a symptom of the speaker's state, (b) as an appeal to produce a reaction by the listener, and (c) as a symbol in the culturally shared communication code. This approach to meaning can easily be extended to other communication systems, such as nonverbal expressions and music (see Scherer, this volume). The notion of an appeal function is extremely useful when considering the role of emotion in constituting meaning for the individual.

Meaning in music has traditionally been tied to emotions (Scherer, this volume), and several basic emotional expressions in Western music have been identified in an African culture insulated from Western musical influences, suggesting that the linkage of these musical forms to emotions may be universal (Fritz et al. 2009). Music, however, does not need to be seen in these terms alone; emotions may figure as only one dimension in which musical experience and expression is organized. Others could be cognition, culture, and action (Morris 1938). In our discussion of the processing of emotions, we referred primarily to the *multicomponent process model* as a starting point (for details, see Scherer, this volume). Our earlier discussion (see section on "Levels of Processing") suggests how this process-level model might be mapped to the neuronal substrate.

Emotional effects of language have mainly been studied in connection with prosody (see Thompson-Schill et al., this volume). In our discussions, we focused on the elicitation of emotion through music, often with the aforementioned levels continuously interacting. While music has often been called the "language of emotion," researchers and laymen usually find it difficult to pinpoint which emotions are thereby produced, as the responses are often quite subtle and may involve blends of different affects. Scherer (this volume) argues that while music may produce basic or utilitarian emotions, aesthetic or epistemic emotions are more often produced. Scherer and Zentner (2001) identified a number of "routes" that music can induce emotions: appraisal, memory, entrainment, emotional contagion, and empathy (for a more detailed account, see Scherer and Coutinho 2013).

Juslin and colleagues offer another view on processes or mechanisms underlying music-induced emotions which connects to the processing of "normal" emotions (Juslin 2009; Juslin and Västfjäll 2008b; Juslin et al. 2010), where several utilitarian emotions (e.g., empathy, emotional contagion, synchronization, and entrainment) can be evoked. These processes might be grounded in neural systems that imply mirror mechanisms (cf. Fogassi, this volume), but just how they are involved in musical emotion processing remains an open question. Chapin et al. (2010) found that subjects listening to expressive music of pieces they had already learned to play, in contrast to listening to pieces which they had not yet played, activated a human mirror system, including bilateral BA 44/45, superior temporal sulcus, ventral PMC, and inferior parietal cortex, along with other motor-related areas and the insula (Figure 8.2). A mirror mechanism may be involved in rhythmic coordination (i.e., synchronization and entrainment) and form the basis for shared understanding of people at a level of pre-reflective consciousness (McGuiness and Overy 2011; Molnar-Szakacs and Overy 2006; Overy and Molnar-Szakacs 2009). Zatorre et al. (2007) report on interactions of the auditory and motor system during music making and cognition which involve the posterior superior temporal gyrus, BA 9/4, BA 8, BA 44, BA 45, pre-supplementary motor area and supplementary motor area, rostral and caudal dorsal premotor cortex, ventral premotor cortex and primary motor cortex. Other studies indicate the involvement of the right homotope of BA 44 in connection with studies indicating the involvement of Broca's area and the premotor cortex in music processing (e.g., Fadiga et al. 2009; Koelsch 2006, 2011c; Maess et al. 2001). In connection with the role of the motor system for prediction, Bubic et al. (2010) point to an important role of mirror mechanisms in meaning formation in music. (For detailed overviews of recent findings on brain structures involved in processing of musical emotions, see Koelsch 2011a and this volume; Panksepp and Trevarthen 2009; Peretz 2010).

An old dichotomy between language and music is the notion that language was perceived to be more related to logic and the human mind whereas music was grounded in emotion and the human body. This view, however, is controversial. At the same time, we have the impression that little research has been done to clarify the matter. Especially in language research, the relation to emotional processing—to ground meaning at the word, sentence, and discourse/text levels—has been little studied.

Duration and time course enable *affect*, *emotion*, and *mood* to be distinguished. Affects might be conceived of as short emotional events. Distinguishing mood and emotions, one can say that emotions are object specific and shorter in duration than moods. Despite the extended duration of much music, we can still argue that emotions are more relevant here. If continuous measures are used (Schubert 2001, 2010), we can view the dynamic nature of music as a set of discrete emotional events of varying duration. The entire form of a piece can induce an aesthetic appreciation and a cognitive recognition of a theme that is accompanied by an aesthetic emotion.

Expectation in Music

Meyer's model of musical meaning (Meyer 1956) is composed of three distinct instances (or processes): *hypothetical meaning*, *evident meaning*, and *determinate meaning*. Hypothetical meaning is the unconscious generation of expectations related to and specific to a stimulus that could be described by probabilistic relationships between antecedents and consequents. Evident meaning occurs when the consequent becomes actualized as a concrete musical event, reaching a new stage of meaning. It appears when that relation between antecedent and consequent is actually perceived. Narmour (1990) developed the related implication–realization theory in which an unrealized implication results in a surprise. Determinate meaning refers to meaning that arises "only after the experience of the work is timeless in memory, only when all the meanings which the stimulus has had in the particular experience are realized and their relationships to one another comprehended as fully as possible" (Meyer 1956:58).

Biologically speaking, "accurate expectations are adaptive mental functions that allow organisms to prepare for appropriate action and perception" (Huron 2006:3). Huron offers an account of musical expectation in which two different neural pathways operate concomitantly and are correlated with a feeling of surprise. According to the neurophysiological (nonmusical) studies (LeDoux 1996; Rolls 2005a, 2011) that ground the dual path model: (a) In the case of dangerous stimuli, the fast limbic track results in negative emotional states and prepares the organism for quick action while it is in this unpredictable situation: being surprised means previously having predicted wrongly. (b) The slow track involves cortical areas that are responsible for providing a contextualized but time-consuming appraisal. Slow appraisal can result in a valence in contrast to the outcome of the fast track (e.g., when the situation analysis reveals that the event, besides surprising, was not dangerous to the organism). The contrast between the negative and the potentially positive appraisal reinforces the final positive state. For Huron (2006), positive emotions can come from two possibilities: when the anticipation is correct (limbic reward) and when the anticipation is not correct but is also not dangerous (contrastive valence).

Meaning in Language and Music: The Sensorimotor–Symbolic Axis

The challenge in relating music, language, and action is to discover possible mappings from features that are apparently unique to one or another domain. In language, for example, compositionality and recursion are key features that permit a speaker to produce a practically infinite set of syntactically correct sentences while also exploiting this structure to produce and understand sentences which convey an open-ended repertoire of novel meanings (compositional semantics). It can be argued that the ability for tool making exhibits more basic structures for action that can also be manifested in language (Stout 2011; Stout and Chaminade 2009), but this does not imply that the circuits for praxis are the same as those for language; there may simply be an evolutionary relation between them (Arbib 2006a). When comparing language to music and action, the question then arises: To what extent do music and action exploit such computational features? Here, we may ask whether similarities in structure extend to mechanisms that support compositional semantics, since when Fitch (2006a) and Mithen (2005) chart the possible evolution of language from a musical protolanguage, they have to posit evolutionary changes in the brain to support the emergence of propositional semantics.

The discovery of mirror mechanisms for action understanding in the 1990s (see Fogassi, this volume) opened up new ways to think about social neuroscience, focusing on shared brain mechanisms for generating the actions of self and observing the actions of others. The notion of a "mirror neuron" is mainly used in connection with (single) cell recording studies (especially in macaques), whereas "mirror system" refers to the activation of networks in human brains as evidenced by imaging studies. More generally, "mirror mechanism" is used and defined by Rizzolatti and Sinigaglia (2010:264) as "the mechanism that unifies perception and action, transforming sensory representations of the behavior of others into motor representations of the same behavior in the observer's brain."

Thus, certain motor systems in the brain that are involved in planning and execution of actions (e.g., areas of parietal cortex and premotor cortex) are also involved in the perception of action. Assessment of evolution (Arbib and Iriki, Fitch and Jarvis, Cross et al., all this volume) seeks, in part, to understand how general sensorimotor processes ground the symbolic functions of the human brain, and to what extent brain mechanisms that support language and music evolved separately, or whether the evolution of one provided the scaffolding for the other (Patel 2008). This suggestion of the evolutionary grounding of music and language in the action–perception cycle raises the possibility that the action of the self, as experienced in the internal world, provides a foundation for meaning in language and music. Learning how understanding the actions of others may be mediated (in part, at least) through mirror mechanisms was a first step in extending research on how the motor system is engaged in understanding the external world to develop a neuroscience of social interaction. As a result, one must consider that the generation and assignment of meaning by the brain may involve a wide range of systems. In any case, the generation of meaning is closely tied to the sociality of the agent, thus grounding meaning in social interaction.

A word of caution is in order: some authors talk as if mirror neurons or mirror systems alone bear the whole weight of recognizing the actions of others as a basis for social interaction. Their contribution can, however, be better understood within the context of a much larger brain system. Buccino et al.

(2004a) used fMRI to study subjects viewing a video, without sound, in which individuals of different species (man, monkey, and dog) performed ingestive (biting) or communicative (talking, lip smacking, barking) acts. In the case of biting there was a clear overlap of the cortical areas that became active in watching man, monkey, and dog, including activation in areas considered to be mirror systems. However, although the sight of a man moving his lips as if he were talking induced strong "mirror system" activation in a region that corresponds to Broca's area, the activation was weak when the subjects watched the monkey lip smacking, and it disappeared completely when they watched the dog barking. Buccino et al. (2004a) concluded that actions belonging to the motor repertoire of the observer (e.g., biting and speech reading) are mapped on the observer's motor system via mirror neurons, whereas actions that do not belong to this repertoire (e.g., barking) are recognized without such mapping. In view of the distributed nature of brain function, it would seem that the understanding of *all* actions involves general mechanisms that need not involve the mirror system strongly, but that for actions that are in the observer's repertoire, these general mechanisms may be complemented by activity in the mirror system, which enriches that understanding by access to a network of associations linked to the observer's own performance of such actions.

In social cognition, the mirror systems of the brain assist it in inferring the intentions, feelings, and goals of conspecifics by interpreting their overt behavior in terms of the internal models of one's own motor system. If we would like to understand meaning in music or otherwise (action, language), there is no way around conceptualizing a transfer of meaning from one individual to another. This must rest on the ability of brain systems in the "social brain" (Frith 2007, 2008) to learn patterns of linkage between the goals of others and their overt behavior (Arbib 2012, chapter 12). Both music and language constitute an interactive social behavior (Clayton 2009; Cross 2010, 2012; Cross and Woodruff 2009; Lewis, this volume; Levinson, this volume). Both play a role, for example, in entrainment, bonding, lullabies, and ritual (Cross et al., this volume). This poses a methodological constraint on the investigation of how musical meaning is perceived in an ecologically valid situation. This is, of course, the case for many mechanisms in psychology. However, because sociality is a crucial component in music, this constraint is not to be underestimated.

How, then, can collective, external representations of interacting brains be investigated? Little is known today to answer this question but work has begun. For example, Sänger et al. (2012) simultaneously recorded the electroencephalogram (EEG) from the brains of each of 12 guitar duos repeatedly playing a modified Rondo in two voices. They found that phase locking as well as within-brain and between-brain phase coherence connection strengths were enhanced at frontal and central electrodes during periods that put particularly high demands on musical coordination. Phase locking was modulated in relation to the experimentally assigned musical roles of leader and follower, which they saw as supporting the notion that synchronous oscillations have functional significance in dyadic music performance. They state that "we expected frontal and central electrodes to be predominantly involved, as they cover the PFC, which has been associated with theory of mind activity, the premotor cortex, where the human mirror neuron system is suspected, and the motor and somatosensory cortices" (Sänger et al. 2012:312). Thus, while the results offer an interesting step in assessing the neural basis of musical coordination, they offer no insight into whether mirror systems play a distinctive role.

Situated interaction in groups forms "group minds" for a short period of time. These must be distinguished from longer-lasting consolidation of these in institutions (i.e., habits governed by social rules). Specific languages and musical genres may be conceived of as such institutions (cf. the "social schemas" of Arbib and Hesse 1986; the "memes" of Dawkins 1976; and the "collective representations" of Durkheim 1938; here the emphasis is on how social interactions build neural representations that underlie related patterns of behavior). Mirror mechanisms are only one example of the interpretative and socially coordinative capabilities of the brain. One could argue that they all could be deciphered by virtue of their link to action, as praxis is generalized to include the special kinds of social interactions and embodied manifestations that language and music make possible, singly and in combination.

Meaning in Language and Music: The Structure Axis

Much of the research on action, music, and language looks at structure on a small scale: How do a few actions combine to reach a well-defined goal? How are words combined to endow an utterance with a desired meaning? How do notes in a short passage combine to build and release tension? A major goal of our discussions was to assess how the neuroscience of language and music might be extended to cover large-scale structure, as exemplified in the cumulative impact of a narrative when expectations are created and then met or deflated. For example, some research in music cognition has distinguished between the local phrase level and a listener's global sense of an entire piece of music (Deliège et al. 1996; Tan and Spackman 2005). One anchor for this discussion was Cohen's analysis (this volume) of how language, music, and action come together in film. Of course, a film is a composed work whose intended meanings may or may not be realized in the brains of viewers (for movies whose unfolding is controlled by the viewer's emotional response, see Tikka 2008). By contrast, whether a conversation or a group performance that responds to spontaneous interpolations of performers, though usually within an assigned framework, we see the challenge of understanding brains in social interaction, where the actions of each participant change the dynamics of the action–perception cycles of the others in a dynamic shaping and reshaping.

Meaning on a Larger Scale: Film and Narrative

Most psycholinguistic models of discourse/text comprehension make reference to a so-called situation or mental model: the evolving representation of meaning. Unlike the surface form, or individual propositions, the situation model is retained in memory, but the nature and stability of that memory, and its relation to brain processes, remains very much an open question. Two general approaches have been taken to describe the structure of the situation model within language processing. The first seeks to describe the individual events and relationships between them along causal, spatial, temporal, and referential dimensions. Most of this work has used Zwaan's event-indexing model as a general framework (Zwaan and Radvansky 1998). More recently, a similar approach was used to describe the structure of film (Zacks and Magliano 2011; Zacks et al. 2010). The second approach appeals to the idea of a narrative structure where, for example, readers identify narrative episodes and subepisodes of a text (Gernsbacher 1990). Early "story grammars" attempted to describe narrative structure at a theoretical level. More recently this has been generalized to visual images based on an analysis of comic strip comprehension (Cohn et al. 2012).

Stories can be transmitted by a variety of media: through face-to-face conversation, over the telephone, radio programs, printed media, comic books, theater, and moving images (film, cinema). Initially, the function of the moving image was thought to be to present reality (e.g., to understand how a horse gallops). Gradually, directors came to realize that a moving image could be used to tell stories. The successful narrative has sentence-by-sentence (scene-by-scene) meaning, cumulative meaning (a trajectory), and structural meaning on both small and large scales. A narrative's meaning is not solely constructed from "propositional" knowledge; emotional processing is also involved at different levels. Generally, people become emotionally involved in a film. Tan et al. (2007) argue that film creates real emotion in the sense of Frijda (1986; cf. Cohen 2010, this volume), connecting with the semantics of the internal (personal) world. There is also a buildup of like/dislike, trust/distrust, love/revulsion concerning different characters; this strongly affects what we grow to expect and how we will react to their behavior as the story unfolds. We both observe the emotions of others and have emotions about others.

The attribution of meaning to a narrative raises the question of how we, as observers, link to its components. Here one would expect that mirror mechanisms play a key role in relation to empathy and emotion as well as its role in relation to actions; both feed into judging the intentions of others and require a system of which mirror mechanisms form a part (see below). In social cognition, "empathy" explains such an unmediated understanding of others. Jeannerod (2005) tied the simulation theory of mind to empathy and its origins in Theodor Lipps' concept of *Einfühlung* (for a proposal relating musical meaning, empathy, imitation, and evolution from a mirror mechanism perspective, see Seifert and Kim 2006). Briefly, empathy is conceived of as a projection mechanism of one's own internal states onto objects or others. Therefore, empathy in Lipps' psychological aesthetic theory is an operation, which serves as a basis (Gregory 2004) for "aesthetic enjoyment of our own activity in an object." Such a process referring to objects, rather than conspecifics, might be termed "aesthetic empathy." Irrespective of these roots, a fundamental question involves how we extract meaning about the "internal world" of other agents and aesthetic meaning through the observation of their surface properties.

Gallese (2011) refers to Lipps' ideas on empathy in developing his view of a neuroaesthetics based on mirror mechanisms and simulation theory. Calvo-Merino and colleagues are concerned with neuroaesthetics of performing art and mainly investigate observation of dance movements (Calvo-Merino et al. 2005; Calvo-Merino and Haggard 2011). They start their considerations concerning neuroaesthetics with mirror mechanisms for action understanding and report that the intraparietal sulcus, the dorsal premotor cortex, the superior parietal lobe, the superior temporal sulcus, and the ventral premotor cortex are involved in an action simulation network for perceiving whole body dance movements. One might speculate that dance observation and aesthetic experience might be based on some special kind of empathy, which some call "kinesthetic empathy" (Foster 2010; Reason and Reynolds 2010; Reynolds and Reason 2012) as the basis for aesthetic empathy. Their work, however, is related to the differential activation of a mirror system depending on whether or not the observer is expert in the observed dance genre, reminding us that empathy can involve understanding the intentions underlying another's actions, not just recognition of emotional state (Gazzola et al. 2006). Of course, one can be a connoisseur of a dance form without being a performer. The same holds for sports. Aglioti et al. (2008) investigate the dynamics of action anticipation and its underlying neural correlates for subjects observing free shots at a basket in basketball. Professional basketball players predicted the success of free shots at a basket earlier and more accurately than did individuals with comparable visual experience (coaches or sports journalists) and novices. Moreover, performance between athletes and the other groups differed before the ball was seen to leave the model's hands, suggesting that athletes predicted the basket shot's fate by reading body kinematics. Both visuomotor and visual experts showed a selective increase of motor-evoked potentials during observation of basket shots. However, only athletes showed a time-specific motor activation during observation of erroneous basket throws. Aglioti et al. (2008) suggest that achieving excellence in sports may be related to the fine-tuning of specific anticipatory mirror mechanisms which support the ability to predict the actions of others ahead of their realization.

The role of music in contributing to the interpretation of film has been relatively well studied (Boltz 2001, 2004; Boltz et al. 1991, 2009). Perhaps the most significant role of music, however, is one associated with the semantics of the internal world: it binds the audience to the screen through the provision of felt emotion, thus connecting the internal and external worlds. For some reason, music seems to provide emotion better than the other domains of information (e.g., speech, text, sound effects, and visual images).

Little is known about markers for neural processes involved in story comprehension as well as the neural structures involved and levels of processing. The current state of the neural basis of discourse and text comprehension has been summarized by Perfetti and Frishkoff (2008), who also stress that it is important to distinguish between the processing of surface form and the processing of more abstract situation models. The temporal lobe and PFC (inferior frontal gyrus) appear to be involved in sentence and text processing, the ventral and dorsomedial PFC in certain inferencing processes, and the dorsomedial PFC in inferring coherence across sentences. Perfetti and Frishkoff cite evidence that the right hemisphere contributes to discourse coherence through inference making. Further, the right hemisphere seems to be involved in non-literal (e.g., metaphorical) and emotive meaning processing and, in general, it is thought that a left-lateralized language network (which includes frontal regions) is engaged in such evaluative and inferencing processes. The left hemisphere is involved in integration meaning across sentences at two levels: (a) coherent semantic representations at successive clauses and sentences and (b) a situation model that is continuously updated. Text comprehension seems to be supported by a large network consisting of the lateral PFC (including the inferior frontal gyrus and the dorsolateral PFC), the anterior temporal lobes (including the temporal pole), and the dorsomedial PFC (including the anterior and posterior cingulate cortex).

There is no simple mapping of either level to neuronal structure (see, however, our earlier discussion in the section on "Levels of Processing"). The text-base (i.e., understanding of words and sentences) maps onto regions in the ventral PFC as well as dorsolateral PFC and anterior temporal cortex. The general conclusion drawn by Perfetti and Frishkoff (2008) is that the only special structures involved in text processing are those specialized for language processing, in particular for syntax processing. However, text comprehension, in general, relies on brain regions for cognitive, affective, and social processing and memory updating. The anterior and posterior cingulate cortex are, for example, involved in such processing. We speculate that the distinction between surface form and situation model might apply to film and music as well, and that it is important to investigate how such situation models are built up from the surface form in processing large-scale structures.

The cognitive neuroscience of music has just started to investigate large-scale music processing. In one study, Sridharan et al. (2007) investigated the neural dynamics of event segmentation using entire symphonies under real listening conditions. Transition between movements was especially scrutinized. They found that a ventral frontotemporal network was involved in detecting the early part of a salient event whereas a dorsal frontoparietal network associated with attention and working memory was activated in the latter part of the salient event. Since film entails not only moving images but also speech discourse, sound effects, music, and text, research in these areas can be expected to apply to film. Studies of the event segmentation of film (Zacks et al. 2007; Zacks and Swallow 2007; Zacks and Magliano 2011) are finding systematic brain imaging responses involving the PFC. This is consistent with the neuronal substrate involved in the segmentation of large-scale musical works (Sridharan et al. 2007).

Using movies as stimuli can aid cognitive neuroscience in investigating the brain's mechanisms (e.g., memory, attention, prediction, and emotion) involved in the processing of large-scale structures. Movies provide ecologically valid stimuli that can be close to "real-world" conditions (Spiers and Maguire 2006, 2007). Indeed, phenomena such as change blindness (i.e., the state of being oblivious to a visual anomaly; Simons and Levin 1997; Simons and Rensink 2005) are as present in real-world perception as well as in film change blindness. Spiers and Maguire (2006) focused on "mentalizing," in the sense of considering the thoughts and intentions of other individuals, in order to act in a socially appropriate manner. This has been linked with the activation of a cortical network, including the posterior superior temporal sulcus, the temporoparietal junction, the medial PFC, and the temporal poles. Spiers and Maguire (2006) found the right posterior superior temporal sulcus, the medial PFC, and the right temporal pole active with spontaneous mentalizing. The right posterior superior temporal sulcus was active during several different subtypes of mentalizing events. The medial PFC seemed to be particularly involved in thinking about agents that were visible in the environment.

Inferring the intention of agents has also been studied using narratives. This spontaneous mentalizing showed to some extent activations related to agent understanding in story, cartoon, and film comprehension tasks. Although the temporoparietal junction has been reported to be active during the understanding of an agent's intentions during story comprehension, a relevant activation could not be observed in the spontaneous mentalizing task. Spiers and Maguire explain this low activation by suggesting that story comprehension might involve more semantic processing than spontaneous mind reading. Nevertheless, it may be that spontaneous mentalizing is some kind of story processing mode that brings together themes, episodes, events, and subevents.

But how can one study the neural basis of something that occurs spontaneously as we go about our daily lives? Combining fMRI, a detailed interactive virtual reality simulation of a bustling familiar city, and a retrospective verbal report protocol, Spiers and Maguire found increased activity in the right posterior superior temporal sulcus, the medial PFC, and the right temporal pole associated with spontaneous mentalizing. In addition, the right posterior superior temporal sulcus was consistently active during several different subtypes of mentalizing events. By contrast, medial PFC seemed to be particularly involved in thinking about agents that were visible in the environment.

Some teams are beginning to apply intersubjective correlation and fMRI studies in the field of cinema (also referred to as "neurocinematics") as a tool to assess the extent of similarities elicited by related narratives and soundscapes in the human brain. Hasson et al. (2004), for example, had subjects freely view a popular movie for half an hour of while undergoing functional brain imaging. By letting the brain signals "pick up" the optimal stimuli for each specialized cortical area, they found a voxel-by-voxel response to the moving image that was strikingly correlated between individuals, not only in primary and secondary visual and auditory areas but also in higher-order association cortical areas. These took the form of a widespread cortical activation pattern correlated with emotionally arousing scenes and regionally selective components. In a later study, Hasson et al. (2008a) assessed the neural bases of episodic encoding of real-world events by measuring fMRI activity while observers viewed a novel TV sitcom. Three weeks later, they tested subjects' subsequent memory for the narrative content of movie events and found a set of brain regions, whose activation was significantly more correlated across subjects during portions of the movie, that were successfully recalled as compared to periods whose episodes were unsuccessfully encoded. These regions include the parahippocampal gyrus, superior temporal gyrus, anterior temporal poles, and the temporoparietal junction.

In one fMRI study with music and film, Eldar et al. presented emotional music with neutral film clips and found activation of key limbic system structures (amygdala and hippocampus) and lateral PFC regions only when the music and film were combined, leading them to say: "Thus, it is possible that lateral temporal regions, such as the superior temporal sulcus and the temporal pole, mediate the tying of emotion to representations of particular objects, people, or situations" (Eldar et al. 2007:2838). It seems that some kind of binding between cognitive domains took place, in which shared information structures were involved. A similar mechanism may recently have been shown by Willems et al. (2011)—not with music, but instead with sentences which expressed or did not express fear presented with or without a neutral visual image. In the presence of the neutral visual image, the sentence expressing fear activated the right temporal pole in comparison to conditions in which the sentence did not express the fear emotion or the sentence was absent. They conclude that "the right anterior temporal pole serves a binding function of emotional information across domains such as visual and linguistic information" (Willems et al. 2011:404). By "information structure in discourse processing," linguists address how prosodic stress can indicate the topic of a sentence; this helps to clarify what new information it adds to what has gone before (cf. Poeppel and Hagoort, this volume). Research on information structure seems to be important to explore bindings in cognitive domains such as language or music. There is a clear hierarchy in language: one begins with phonemes or syllables, then words; thereafter, phrases and sentences are constructed from which discourse emerges. Such a well-defined a hierarchy for structure building in music is difficult to find. A piece of music, such as a symphony, seems intuitively to be better conceived of as more akin to discourse than to a sentence, but we currently lack a firm theoretical framework to assess this claim.

Unfortunately, to date, almost all research on musical form has been taxonomic (Randel 2003). For example, single and compound forms are distinguished. (A compound form consists of more than one single form, as in, e.g., a suite, sonata, symphony, and string quartet.) A starting point at a conceptual mid-level for further research on large-scale musical structures might be to investigate spinning out (*Fortspinnung*) of short melodic figures by means of sequences. Richard Wagner used sequential techniques for addressing the leitmotifs in his musical dramas. The interesting point is that leitmotifs also refer to extramusical entities. Thus, spinning out and leitmotifs might serve as a conceptual bridge to investigate principles and mechanisms for coherence of "musical narratives" in connection with neurocinematics and to study more general processes involved in multimodal meaning formation, including emotion. Since these concepts are derived from a specific Western musical tradition, they will also need to be tested using non-Western musics.

The *generative theory of tonal music* (GTTM; Lerdahl, this volume) seems to conceive of a musical piece partially in terms that are analogous to the description of linguistic sentences structured at multiple levels, although important differences are noted. GTTM does not, however, address the sort of "semantics" exemplified by music's role in film but rather focuses more on sound pattern and the building and release of "tension" in a musicological sense. This approach is rooted in the Western musical tradition (Verschure and Manzolli, this volume), though other studies have begun to assess the relevance of syntactic methods to the study of non-Western musical genres. If musical pieces are, at least partially, conceived of as texts, then the question becomes: Which semantic or syntactic operations and mechanisms establish the coherence of such large-scale musical structures, and to what extent do they implicate sharing of brain resources with the processing of the (apparently somewhat different) semantic or syntactic operations of language? In neurocognitive research on discourse comprehension, linguistic cues (e.g., repetition and synonyms) are distinguished from nonlinguistic cues (e.g., pragmatic or world knowledge).

Structure Building: The Hypothesis of Intermediate-Term Memory

The large-scale structure processing involved in dialog, text, film, music, and situation comprehension can be viewed as involving different memory processes: predictions and expectations about future events and episodes, based on actual sensory input and past experiences of events and episodes. All are used to construct a situation model. These processes seem to operate on different levels and timescales and involve conscious as well as unconscious processing. The "stream of consciousness" created by the brain's processing is constituted in a "specious present," which acts as an interface between the near future and the near past. Edmund Husserl provided an introspective phenomenological analysis of time consciousness, using melody perception as one example (Clarke 2011; Montague 2011).

Recent research indicates that it might be necessary to hypothesize an *intermediate-term memory* between working memory and long-term memory (Baddeley 2000, 2007; Donald 2001; Ericsson and Kintsch 1995). For example, as we read a chapter, the exact words of a sentence might be quickly lost from working memory, while the contribution of a sentence to the plot might be retained in intermediate working memory. Nonetheless, only those retained in a longer-term "narrative memory" will guide our reading of the book a day or two later. An intermediate-term memory, existing between short- and longer-term memory stages, has long been recognized by animal researchers (e.g., McGaugh 1966; Gibbs and Ng 1979); it is during this stage that memory is labile and sensitive to neuromodulation from arousal sources, including those associated with emotional significance (Crowe et al. 1990; McGaugh 2000). Recent studies have demonstrated that music is capable of modifying the strength of memory (e.g., of a slideshow with narrative or a word list) for a limited period after learning, consistent with emotional neuromodulation of an intermediate stage of memory consolidation (Judde and Rickard 2010; Rickard et al. 2012). However, such studies do not address the way in which the details of changing working memories over the course of hours may become compacted into a structure of key episodes of, perhaps, exceptional epistemic or emotional significance. Indeed, the relevance of the hypothetical construct of intermediate-term memory seems to be overlooked in current research on musical memory, as an inspection of recent work on musical memory indicates (Snyder 2000, 2009; Tillmann et al. 2011).

The *congruence association model with working narrative* (CAM-WN; Cohen, this volume) offers a film-centered starting point for investigating the structuring of events in time as well as the formation of multimodal meaning that depends on music, language, vision, and emotion. CAM-WN will need to be linked, however, with current cognitive neuroscience results of large-scale structures. In more general terms, one could interpret the narrative mode of a brain as exploiting mechanisms of prediction (Bar 2009; Friston 2010; Schacter et al. 2007) while noting the importance of emotional memory and the interaction of cognition and emotion (LeDoux 1996, 2000, 2004; Rolls 2011) at the interface of conscious and unconscious processing (Baddeley 2000, 2007). As yet, however, we do not have data that can tease apart the contributions of music and language to a complete experience and relate these contributions to brain processes.

Cognitive neuroscience of music has just begun to investigate rigorously the brain's processing of surprise and prediction on large-scale structures as well as the relation between music and reward (Salimpoor et al. 2011; Zald and Zatorre 2011; Arbib et al., this volume). Tone semantics might serve as a starting point for exploring whether it is possible to relate concepts of the *component process model* of emotion processing (Scherer, this volume), the CAM-WN framework, and the distributed adaptive control architecture (Verschure 2012; Verschure and Manzolli, this volume). It is possible to relate processes and levels involved in tone semantics to processing large-scale structures (e.g., film scenes and their music). Low-level processes involved in tone semantics might be involved both in establishing a surface form and in building up a situation model. Traditional concepts from psychoacoustics, such as frequency (pitch), intensity (loudness), and timbre (frequencies, their intensities, and phase), can be linked to large-scale processing (for empirical evidence concerning timbre, see Alluri et al. 2012). As evidenced by the work of Koelsch (2011c), it is possible to relate specific processes to specific brain regions. The question is: Which processes and levels are involved in establishing the meaning of tones, their significance, and signification?

Conclusion: A Comparative Approach for Future Research

In our discussions, we compared language and music with respect to semantics. We considered how an action-oriented embodied approach to a comparative study of semantics in language and music might take emotion and mirror mechanisms into account. In action, language, and music processing, syntactic or structural aspects of processing might be carried out by specific shared neural networks, though limited sharing seemed more likely to many of the discussants. We discussed additional changes in assessing comprehension of large-scale structures (e.g., situations, narratives, or musical form), but said too little about production, and thus the challenges of understanding the interaction between multiple brains when people talk or make music together. We addressed the notion of intermediate-term memory, integrating episodic, emotional, and semantic information from successive working memories as a basis for anticipating and making sense of new events within a narrative or, indeed, the rigors of daily life. This form of memory continuously integrates information from the past and present to enable the construction of a situation model, which will be used to predict upcoming events. This kind of "narrative" memory is related to both conscious and unconscious processing. The investigation of these processes as well as the relationship between unconscious and conscious processing of signification and significance constitute important future avenues for research.

We have stressed, though too briefly, the social types of communication formed by music and language and the mirror mechanisms that may be involved. In music, mirror mechanisms appear to be part of the neural systems that establish a feeling of communion and prereflective shared meaning. In social interaction, mirror mechanisms (in concert with other brain regions) might be involved in emotional contagion, aesthetic empathy, synchronization, and entrainment; they may also serve to regulate mood and establish group identity. Social interactions can (for good or ill) temporarily establish a "group mind" during these interactions and, in due course, can yield longer-lasting shared habits, which may contribute to the formation of institutions. Brain research and cognitive science must take these phenomena into account if a greater understanding of the (human) mind, consciousness, and the brain's interpretational capability is to be achieved and a fuller account of the brain's most important neurocognitive higher-level functions (such as language and music) is to be given.

For progress to be made in comparing action, language, and music in cognitive neuroscience, mechanisms for data sharing must be established (Eckersley et al. 2003). Moreover, for a top-down approach, which links psychological concepts onto brain structures and processes, a cognitive ontology (including, at a minimum, concepts from action, music, language, memory, and emotion/motivation) is needed to address conceptual problems and to develop new experimental designs for a comparative approach (Arbib et al. 2013). Promising topics for future research include:

- How the prospective brain constructs narratives.
- The role of cognitive–emotional interactions.
- The interface of conscious and preconscious processing.
- The neural dynamics of shared experience.
- The neural correlates of culture-building and institutions.
- The role of emotional memory in constituting conscious experiences.

A key challenge in our analysis of music, language, and the brain is to elevate this analysis to a system-level description, where basic elements are clarified within the overall structures of action, perception, and memory. In this analysis, an important role can and should be played by computational approaches that are attuned to defining and manipulating system-level features, moving from syllables and pitches to musical form and narrative. This could open up new ways to understand the brain's interaction with the world through action and how meaning and art emerge as a result.

Acknowledgment

We gratefully acknowledge the contributions of Klaus Scherer to our discussions.

Part 3: Structure

9

The Infrastructure of the Language-Ready Brain

Peter Hagoort and David Poeppel

Abstract

This chapter sketches in very general terms the cognitive architecture of both language comprehension and production, as well as the neurobiological infrastructure that makes the human brain ready for language. Focus is on spoken language, since that compares most directly to processing music. It is worth bearing in mind that humans can also interface with language as a cognitive system using sign and text (visual) as well as Braille (tactile); that is to say, the system can connect with input/output processes in any sensory modality. Language processing consists of a complex and nested set of subroutines to get from sound to meaning (in comprehension) or meaning to sound (in production), with remarkable speed and accuracy. The first section outlines a selection of the major constituent operations, from fractionating the input into manageable units to combining and unifying information in the construction of meaning. The next section addresses the neurobiological infrastructure hypothesized to form the basis for language processing. Principal insights are summarized by building on the notion of "brain networks" for speech–sound processing, syntactic processing, and the construction of meaning, bearing in mind that such a neat three-way subdivision overlooks important overlap and shared mechanisms in the neural architecture subserving language processing. Finally, in keeping with the spirit of the volume, some possible relations are highlighted between language and music that arise from the infrastructure developed here. Our characterization of language and its neurobiological foundations is necessarily selective and brief. Our aim is to identify for the reader critical questions that require an answer to have a plausible cognitive neuroscience of language processing.

The Cognitive Architecture of Language Comprehension and Production

The Comprehension of Spoken Language

When listening to speech, the first requirement is that the continuous speech input is perceptually segmented into discrete entities (features, segments, syllables) that can be mapped onto, and will activate, abstract phonological representations that are stored in long-term memory. It is a common claim in state-of-the-art models of word recognition (the cohort model: Marslen-Wilson 1984; TRACE: McClelland and Elman 1986; the shortlist model: Norris 1994) that the incoming and unfolding acoustic input (e.g., the word-initial segment *ca...*) activates, in parallel, not only one but a whole set of lexical candidates (e.g., *captain, capture, captivate, capricious...*). This set of candidates is reduced, based on further incoming acoustic input and contextually based predictions, to the one that fits best (for a review, see Poeppel et al. 2008). This word recognition process happens extremely fast, and is completed within a few hundred milliseconds, whereby the exact duration is co-determined by the moment at which a particular word form deviates from all others in the mental lexicon of the listener (the so-called recognition point). Given the rate of typical speech (~4–6 syllables per second), we can deduce that word recognition is extremely fast and efficient, taking no more than 200–300 ms.

Importantly, achieving the mapping from acoustics to stored abstract representation is not the only subroutine in lexical processing. For example, words are not processed as unstructured, monolithic entities. Based on the morphophonological characteristics of a given word, a process of lexical decomposition takes place in which stems and affixes are separated. For spoken words, the trigger for decomposition can be something as simple as the inflectional rhyme pattern, which is a phonological pattern signaling the potential presence of an affix (Bozic et al. 2010). Interestingly, words seem to be decomposed by rule; that is, the decompositional, analytic processes are triggered for words with obvious parts (e.g., teacup = tea-cup; uninteresting = un-inter-est-ing) but also for semantically opaque words (e.g., bell-hop), and even nonwords with putative parts (e.g., blicket-s, blicket-ed). Decomposing lexical input appears to be a ubiquitous and mandatory perceptual strategy (e.g., Fiorentino and Poeppel 2007; Solomyak and Marantz 2010; and classic behavioral studies by Forster, Zwitserlood, Semenza, and others). Many relevant studies, especially with a view toward neurocognitive models, are reviewed by Marslen-Wilson (2007).

Recognizing word forms is an entrance point for the retrieval of syntactic (lemma) and semantic (conceptual) information. Here, too, the process is cascaded in nature. That is, based on partial phonological input, meanings of multiple lexical candidates are co-activated (Zwitserlood 1989). Multiple activation is less clear for lemma information that specifies the syntactic features (e.g., word class, grammatical gender) of a lexical entry. In most cases, the phrase structure context generates strong predictions about the syntactic slot (say, a noun or a verb) that will be filled by the current lexical item (Lau et al. 2006). To what degree lemma and concept retrieval are sequential or parallel in nature during online comprehension, is not clear. Results from electrophysiological recordings (event-related brain potential, ERP), however, indicate that most of the retrieval and integration processes are completed within 500 ms (Kutas and Federmeier 2011; see also below).

Thus far, the processes discussed all relate to the retrieval of information from what is referred to in psycholinguistics as the mental lexicon. This is the information that in the course of language acquisition gets encoded and consolidated in neocortical memory structures, mainly located in the temporal lobes. However, language processing is (a) more than memory retrieval and (b) more than the simple concatenation of retrieved lexical items. The expressive power of human language (its generative capacity) derives from being able to combine elements from memory in endless, often novel ways. This process of deriving complex meaning from lexical building blocks (often called composition) will be referred to as *unification* (Hagoort 2005). As we will see later, (left) frontal cortex structures are implicated in unification.

In short, the cognitive architecture necessary to realize the expressive power of language is tripartite in nature, with levels of form (speech sounds, graphemes in text, or manual gestures in sign language), syntactic structure, and meaning as the core components of our language faculty (Chomsky 1965; Jackendoff 1999; Levelt 1999). These three levels are domain specific but, at the same time, they interact during incremental language processing. The principle of compositionality is often invoked to characterize the expressive power of language at the level of meaning. A strict account of compositionality states that the meaning of an expression is a function of the meanings of its parts and the way they are syntactically combined (Fodor and Lepore 2002; Heim and Kratzer 1998; Partee 1984). In this account, complex meanings are assembled bottom-up from the meanings of the lexical building blocks via the combinatorial machinery of syntax. This is sometimes referred to as simple composition (Jackendoff 1997). That some operations of this type are required is illustrated by the obvious fact that the same lexical items can be combined to yield different meanings: *dog bites man* is not the same as *man bites dog*. Syntax matters. It matters, however, not for its own sake but in the interest of mapping grammatical roles (subject, object) onto thematic roles (agent, patient) in comprehension, and in the reverse order in production. The thematic roles will fill the slots in the situation model (specifying states and events) representing the intended message.

That this account is not sufficient can be seen in adjective–noun expressions such as *flat tire, flat beer, flat note*, etc. (Keenan 1979). In all these cases, the meaning of "flat" is quite different and strongly context dependent. Thus, structural information alone will need to be supplemented. On its own, it does not suffice for constructing complex meaning on the basis of lexical-semantic building blocks. Moreover, ERP (and behavioral) studies have found that nonlinguistic information which accompanies the speech signal (such as information about the visual environment, about the speaker, or about co-speech gestures; Van Berkum et al. 2008; Willems et al. 2007; Willems et al. 2008) are unified in parallel with linguistic sources of information. Linguistic and nonlinguistic information conspire to determine the interpretation of an utterance on the fly. This all happens extremely fast, usually in less than half a second. For this and other reasons, simple (or strict) composition seems not to hold across all possible expressions in the language (see Baggio and Hagoort 2011).

We have made a distinction between memory retrieval and unification operations. Here we sketch in more detail the nature of unification in interaction with memory retrieval. Classically, psycholinguistic studies of unification have focused on syntactic analysis. However, as we saw above, unification operations take place not only at the syntactic processing level. Combinatoriality is a hallmark of language across representational domains (cf. Jackendoff 2002). Thus, at the semantic and phonological levels, too, lexical elements are argued to be combined and integrated into larger structures (cf. Hagoort 2005). Nevertheless, models of unification are most explicit for syntactic processing. For this level of analysis, we can illustrate the distinction between memory retrieval and unification most clearly. According to the *memory, unification, and control* (MUC) model (Hagoort 2005), each word form in the mental lexicon is associated with a structural frame (Vosse and Kempen 2000). This structural frame consists of a three-tiered unordered tree, specifying the possible structural environment of the particular lexical item (see Figure 9.1).

The top layer of the frame consists of a single phrasal node (e.g., noun phrase, NP). This so-called root node is connected to one or more functional nodes (e.g., subject, S; head, hd; direct object, dobj) in the second layer of the frame. The third layer again contains phrasal nodes to which lexical items or other frames can be attached.

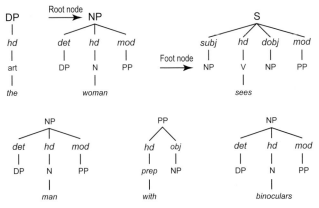

Figure 9.1 Syntactic frames in memory. Frames such as these are retrieved on the basis of incoming word form information (*the, woman,* etc). DP: determiner phrase; NP: noun phrase; S: sentence; PP: prepositional phrase; art: article; hd: head; det: determiner; mod: modifier; subj: subject; dobj: direct object. The head of a phrase determines the syntactic type of the frame (e.g., noun for a noun phrase, preposition for a prepositional phrase)

This parsing account is "lexicalist" in the sense that all syntactic nodes—S, NP, VP (verb phrase), N, V—are retrieved from the mental lexicon. In other words, chunks of syntactic structure are stored in memory. There are no syntactic rules that introduce additional nodes, such as in classical rewrite rules in linguistics (S → NP VP). In the online comprehension process, structural frames associated with the individual word forms incrementally enter the unification workspace. In this workspace, constituent structures spanning the whole utterance are formed by a unification operation (see Figure 9.2). This operation consists of linking up lexical frames with identical root and foot nodes, and checking agreement features (number, gender, person, etc.). Although the lexical-syntactic frames might differ between languages, as well as the ordering of the trees, what is claimed to be universal is the combination of lexically specified syntactic templates and unification procedures. Moreover, across language the same distribution of labor is predicted between brain areas involved in memory and brain areas that are crucial for unification.

The resulting unification links between lexical frames are formed dynamically, which implies that the strength of the unification links varies over time until a state of equilibrium is reached. Due to the inherent ambiguity in natural language, alternative unification candidates will usually be available at any point in the parsing process. That is, a particular root node (e.g., prepositional phrase, PP) often finds more than one matching foot node (i.e., PP) (see Figure 9.2) with which it can form a unification link (for examples, see Hagoort 2003).

Ultimately, at least for sentences which do not tax the processing resources very strongly, one phrasal configuration results. This requires that among the alternative binding candidates, only one remains active. The required state of equilibrium is reached through a process of lateral inhibition between two or

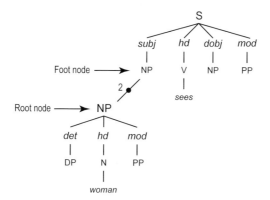

Figure 9.2 The unification operation of two lexically specified syntactic frames. Unification takes place by linking the root node NP to an available foot node of the same category. The number 2 indicates that this is the second link that is formed during online processing of the sentence, *The woman sees the man with the binoculars.*

more alternative unification links. In general, due to gradual decay of activation, more recent foot nodes will have a higher level of activation than the ones that entered the unification space earlier. In addition, strength levels of the unification links can vary as a function of plausibility (semantic) effects. For instance, if instrumental modifiers under S-nodes have a slightly higher default activation than instrumental modifiers under an NP-node, lateral inhibition can result in overriding a recency effect.

The picture that we sketched above is based on the assumption that we always create a fully unified structure. This is, however, unlikely. In our actual online processing of life in a noisy world, the comprehension system will often work with just bits and pieces (e.g., syntactic frames) that are not all unified into one fully unified phrasal configuration. Given both extralinguistic and language-internal contextual prediction and redundancy, in the majority of cases this is still good enough to derive the intended message (see below).

The *unification model*, as formalized in Vosse and Kempen (2000), has nevertheless a certain psychological plausibility. It accounts for sentence complexity effects known from behavioral measures, such as reading times. In general, sentences are harder to analyze syntactically when more potential unification links of similar strength enter into competition with each other. Sentences are easy when the number of U-links is small and of unequal strength. In addition, the model accounts for a number of other experimental findings in psycholinguistic research on sentence processing, including syntactic ambiguity (attachment preferences; frequency differences between attachment alternatives), and lexical ambiguity effects. Moreover, it accounts for breakdown patterns in agrammatic sentence analysis (for details, see Vosse and Kempen 2000).

So far we have specified the memory and retrieval operations that are triggered by the orthographic or acoustic input. Similar considerations apply to sign language. In our specification of the processing steps involved, we have implicitly assumed that ultimately decoding the meaning is what language comprehension is about. However, while this might be a necessary aspect, it cannot be the whole story. Communication goes further than the exchange of explicit propositions. In essence, it is a way to either change the mind of the listener, or to commit the addressee to the execution of certain actions, such as closing the window in reply to the statement *It is cold in here*. In other words, a theory of speech acts is required to understand how we get from coded meaning to inferred speaker meaning (cf. Levinson, this volume; Grice 1989).

Another assumption that we made, but which might be incorrect, relates to how much of the input the listener/reader analyzes. This is what we alluded to briefly in the context of unification. In classical models of sentence comprehension—of either the syntactic structure-driven variety (Frazier 1987) or in a constraint-based framework (Tanenhaus et al. 1995)—the implicit assumption is usually that a full phrasal configuration results and a complete interpretation of the input string is achieved. However, oftentimes the listener interprets the input on the basis of bits and pieces that are only partially analyzed. As a consequence, the listener might overhear semantic information (cf. the Moses illusion; Erickson and Mattson 1981) or syntactic information (cf. the Chomsky illusion; Wang et al. 2012). In the question *How many animals of each kind did Moses take on the ark?*, people often answer "two," without noticing that it was Noah who was the guy with an ark, and not Moses. Likewise, we found that syntactic violations might go unnoticed if they are in a sentence constituent that provides no new information (Wang et al. 2012). Ferreira et al. (2002) introduced the phrase *good-enough processing* to refer to the listeners' and readers' interpretation strategies. In a good-enough processing context, linguistic devices that highlight the most relevant parts of the input might help the listener/reader in allocating processing resources optimally. This aspect of linguistic meaning is known as *information structure* (Büring 2007; Halliday 1967). The information structure of an utterance essentially focuses the listener's attention on the crucial (new) information in it. In languages such as English and Dutch, prosody plays a crucial role in marking information structure. For instance, in question–answer pairs, the new or relevant information in the answer will typically be pitch accented. After a question like *What did Mary buy at the market?* the answer might be *Mary bought VEGETABLES* (accented word in capitals). In this case, the word "vegetables" is the focus constituent, which corresponds to the information provided for the Wh-element in the question. There is no linguistic universal for signaling information structure. The way information structure is expressed varies within and across languages. In some languages it may impose syntactic locations for the focus constituent; in others focus-marking particles are used, or prosodic features like phrasing and accentuation (Kotschi 2006; Miller 2006).

In summary, language comprehension requires an analysis of the input that allows the retrieval of relevant information from memory (the mental lexicon). The lexical building blocks are unified into larger structures decoding the propositional content. Further inferential steps are required to derive the intended message of the speaker from the coded meaning. Based on the listener's comprehension goals, the input is analyzed to a lesser or greater degree. Linguistic marking of information structure co-determines the depth of processing of the linguistic input. In addition, nonlinguistic input (e.g., co-speech gestures, visual context) is immediately integrated into the situation model that results from processing language in context.

Producing Language

While speech comprehension can be described as the mapping from sound (or sign) to meaning, in speaking we travel the processing space in the reverse order. In speaking, a preverbal message is transformed in a series of steps into a linearized sequence of speech sounds (for details, see Levelt 1989, 1999). This again requires the retrieval of building blocks from memory and their unification at multiple levels. Most research on speaking has focused on single word production, as in picture naming. The whole cascade of processes, from the stage of conceptual preparation to the final articulation, happens in about 600 ms (Indefrey and Levelt 2004). Since we perform this process in an incremental fashion, we can easily utter 2–4 words per second. Moreover, this is done with amazing efficiency; on average, a speech error occurs only once in a thousand words (Bock 2011; Deese 1984). The whole cascade of processes starts with the preverbal message, which triggers the selection of the required lexical concepts (i.e., the concepts for which a word form is available in the mental lexicon). The activation of a lexical concept leads to the retrieval of multiple lemmas and a selection of the target lemma, which gets phonologically encoded. At the stage of lemma selection, morphological unification of, for instance, stem and affix takes place. Recent intracranial recordings in humans indicate that certain parts of Broca's region are involved in this unification process (Sahin et al. 2009). Once the phonological word forms are retrieved, they will result in the retrieval and unification of the syllables that compose a phonological word in its current speech context.

Although speech comprehension and speaking recruit many of the same brain areas during sentence-level semantic processes, syntactic operations, and lexical retrieval (Menenti et al. 2011), there are still important differences. The most important difference is that although speakers pause, repair, etc., they nevertheless cannot bypass syntactic and phonological encoding of the utterance that they intend to produce. What is good enough for the listener is often not good enough for the speaker. Here, the analogy between perceiving and producing music seems obvious. It may well be that the interconnectedness of the cognitive and neural architectures for language comprehension and production enables the production system to participate in generating internal predictions while in the business of comprehending linguistic input. This prediction-is-production account, however, may not be as easy in relation to the perception of music, at least for instrumental music. With few exceptions, all of humankind are expert speakers. However, for music, there seems to be a stronger asymmetry between perception and production. This, then, results in two questions: Does prediction play an equally strong role in language comprehension and the perception of music? If so, what might generate the predictions in music perception?

The Neurobiological Infrastructure

Classically, and based primarily on evidence from deficits in aphasic patients, the perisylvian cortex in the left hemisphere has been seen as the crucial network for supporting the processing of language. The critical components were assumed to be Broca's area in the left inferior frontal cortex (LIFC) and Wernicke's area in the left superior temporal cortex, with these areas mutually connected by the arcuate fasciculus. These areas, and their roles in language comprehension and production, are often still described as the core language nodes in handbooks on brain function (see Figure 9.3).

However, later patient studies, and especially recent neuroimaging studies in healthy subjects, have revealed that (a) the distribution of labor between Broca's and Wernicke's areas is different than proposed in the classical model, and (b) a much more extended network of areas is involved, not only in the left hemisphere, but also involving homologous areas in the right hemisphere. One alternative proposal is the MUC model proposed by Hagoort (2005). In this model, the distribution of labor is as follows (see Figure 9.4): Areas in the temporal cortex (in yellow) subserve the knowledge representations that have been laid down in memory during acquisition. These areas store information about word form, word meanings, and the syntactic templates that we discussed above. Dependent on information type, different parts of temporal cortex are involved. Frontal cortex areas (Broca's area and adjacent cortex, in blue) are crucial for the unification operations. These operations generate larger structures from the building blocks that are retrieved from memory. In addition, executive control needs to be exerted, such that the correct target

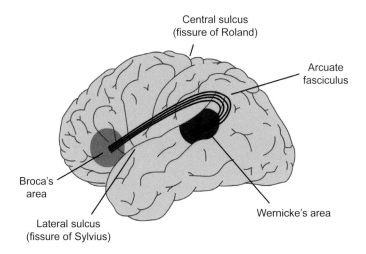

Figure 9.3 The classical Wernicke–Lichtheim–Geschwind model of the neurobiology of language. In this model Broca's area is crucial for language production, Wernicke's area subserves language comprehension, and the necessary information exchange between these areas (such as in reading aloud) is done via the arcuate fasciculus, a major fiber bundle connecting the language areas in temporal cortex (Wernicke's area) and frontal cortex (Broca's area). The language areas border one of the major fissures in the brain, the so-called Sylvian fissure. Collectively, this part of the brain is often referred to as perisylvian cortex.

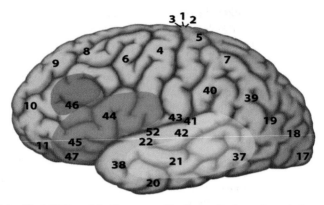

Figure 9.4 The MUC model of language. The figure displays a lateral view of the left hemisphere. The numbers indicate Brodmann areas. These are areas with differences in the cytoarchitectonics (i.e., composition of cell types). The memory areas are in the temporal cortex (in yellow). Unification requires the contribution of Broca's area (Brodmann areas 44 and 45) and adjacent cortex (Brodmann areas 47 and 6) in the frontal lobe. Control operations recruit another part of the frontal lobe (in pink) and the anterior cingulate cortex (not shown in the figure).

language is selected, turn-taking in conversation is orchestrated, etc. Control areas involve dorsolateral prefrontal cortex (in pink) and a midline structure known as the anterior cingulate cortex (not shown in Figure 9.4).

In the following sections we discuss in more detail the brain networks which support the different types of information that are crucial for language. We briefly describe the neurobiological infrastructure underlying the tripartite architecture of the human language system. For the three core types of information (phonological, syntactic, and semantic), we make the same general distinction between retrieval operations and unification: Retrieval refers to accessing language-specific information in memory. Unification is the (de)composition of larger structures from the building blocks that are retrieved from memory. As we will see below, a similar distinction has been proposed for music, with a striking overlap in the recruitment of the neural unification network for language and music (Patel 2003 and this volume).

The Speech and Phonological Processing Network

As we noted at the outset, speech perception is not an unstructured, monolithic cognitive function. Mapping from sounds to words involves multiple steps, including operations that depend on what one is expected to do as a listener: remain silent (passive listening), repeat the input, write it down, etc. The different tasks will play a critical role in the perception process. Accordingly, it is now well established that there is no single brain area that is responsible for speech perception and the activation/recruitment of phonological knowledge.

Rather, several brain regions in different parts of the cerebral cortex interact in systematic ways in speech perception. The overall network, which also includes subcortical contributions (see recent work by Kotz and Schwartze 2010), has been established by detailed consideration of brain injury and functional imaging data (for reviews and perspectives on this, see Binder 2000; Hickok and Poeppel 2000, 2004, 2007; Poeppel et al. 2008; Scott and Johnsrude 2003). Figure 9.5, from Hickok and Poeppel (2007), summarizes one such perspective, emphasizing concurrent processing pathways.

Areas in the temporal lobe, parietal areas, and several frontal regions conspire to form the network for speech recognition. The functional anatomy underlying speech–sound processing is comprised of a distributed cortical system that encompasses regions along at least two processing streams. A ventral, temporal lobe pathway (see Figure 9.5b) primarily mediates the mapping from sound input to meaning/words (lower pathway in Figure 9.5a). A dorsal path incorporating parietal and frontal lobes enables the sensorimotor transformations that underlie

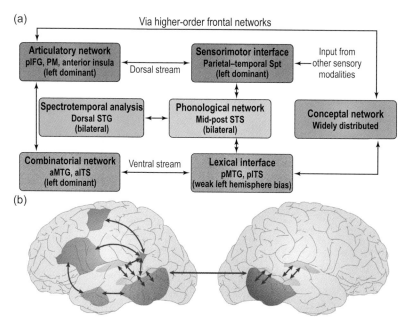

Figure 9.5 A model of the speech and phonological processing network. The earliest stages of cortical speech processing involve some form of spectrotemporal analysis, which is carried out in auditory cortices bilaterally in the supratemporal plane. Phonological-level processing and representation involves the middle to posterior portions of the superior temporal sulcus (STS) bilaterally, although there might be a left-hemisphere bias at this level of processing. A dorsal pathway (blue) maps sensory or phonological representations onto articulatory motor representations. A ventral pathway (pink) provides the interface with memory representations of lexical syntax and lexical concepts (reprinted with permission from Hickok and Poeppel 2007).

mapping to output representations (upper pathway in Figure 9.5a). This anatomic fractionation suggests that hypothesized subroutines and representations of speech processing have their own neural realization, as indicated in the boxes, and supports models which posit a componential architecture (e.g., this dual pathway model). This distributed functional anatomy for speech recognition contrasts with other systems. For example, in the study of face recognition, one brain region plays a disproportionately large role (the fusiform face area). However, the functional anatomic models that have been developed for speech recognition and phonological processing are much more extended and bear a resemblance to the organization of the visual system. In the parallel pathways in the visual system, we contrast a where/how (dorsal) and a what (ventral) system (Kravitz et al. 2011).

One way to carve up the issue—admittedly superficial, but mnemonically useful—is purely by anatomy: temporal lobe–memory; parietal lobe–analysis/coordinate transformation; frontal lobe–synthesis/unification (Ben Shalom and Poeppel 2008). The areas in the temporal lobe (in addition to sensory/perceptual analysis in the superior temporal lobe) have a principal role in storage and retrieval of speech sounds and words. These areas underlie the required memory functions. One region in the temporal lobe of special interest in the mapping from sound form to lexical representation is the superior temporal sulcus (STS); it receives inputs from many areas, including core auditory fields, visual areas, etc., and it sits adjacent to middle temporal gyrus (MTG), the putative site of lexical representations proper (Hickok and Poeppel 2004; Indefrey and Levelt 2004; Lau et al. 2006; Snijders et al. 2009). The areas in the parietal cortex (SPT, SMG, angular gyrus, intraparietal sulcus) are implicated in analytic functions (e.g., sublexical phonological decisions; sensorimotor transformations). The areas in frontal cortex (various areas in the inferior frontal cortex and dorsomedial frontal cortex) play an obvious role in setting up motor output programming, but, more critically, underlie unification operations.

The Syntactic Network

In comparison with phonological and semantic processing, which have compelling bilateral contributions (in contrast to the classical left-hemisphere-only model), syntactic processing seems strongly lateralized to the left hemisphere perisylvian regions. Indirect support for a distinction between a memory component (i.e., the mental lexicon) and a unification component in syntactic processing comes from neuroimaging studies on syntactic processing. In a meta-analysis of 28 neuroimaging studies, Indefrey (2004) found two areas that were critical for syntactic processing, independent of the input modality (visual in reading, auditory in speech). These two supramodal areas for syntactic processing were the left posterior STG/MTG and the LIFC. The left posterior temporal cortex is known to be involved in lexical processing (Hickok and Poeppel 2004, 2007; Indefrey and Cutler 2004; Lau et al. 2006). In connection to the *unification model*, this part of the brain might be important for the retrieval of the syntactic frames that are stored in the lexicon. The *unification space*, where individual frames are connected into a phrasal configuration for the whole utterance, might recruit the contribution of Broca's area (LIFC).

Direct empirical support for this distribution of labor between LIFC (Broca's area) and temporal cortex was recently found in a study of Snijders et al. (2009). These authors performed an fMRI study in which participants read sentences and word sequences containing word–category (noun–verb) ambiguous words at critical position (e.g., "watch"). Regions contributing to the syntactic unification process should show enhanced activation for sentences compared with words, and only within sentences display a larger signal for ambiguous than unambiguous conditions. The posterior LIFC showed exactly this predicted pattern, confirming the hypothesis that LIFC contributes to syntactic unification. The left posterior MTG was activated more for ambiguous than unambiguous conditions, as predicted for regions subserving the retrieval of lexical-syntactic information from memory. It thus seems that the LIFC is crucial for syntactic processing in conjunction with the left posterior MTG, a finding supported by patient studies with lesions in these very same areas (Caplan and Waters 1996; Rodd et al. 2010; Tyler et al. 2011).

In the domain of music perception, a similar model has been proposed by Patel (2003). Although in the past, perspectives on language and music often stressed the differences, Patel has introduced and strongly promotes an alternative view: that at many levels, the similarities between music and language are more striking than the differences. Clearly, the differences are undeniable. For instance, there are pitch intervals in music that we do not have in language; on the other hand, nouns and verbs are part of the linguistic system without a concomitant in music. These examples point to differences in the representational structures that are domain specific and laid down in memory during acquisition. However, the processing mechanisms (algorithms) and the neurobiological infrastructure to retrieve and combine domain-specific representations might be shared to a large extent. This idea has been made explicit in Patel's *shared syntactic integration resource hypothesis* (SSIRH in short; see Patel, this volume). According to this hypothesis, linguistic and musical syntax have mechanisms of sequencing in common, which are instantiated in overlapping frontal brain areas that operate on different domain-specific syntactic representations in posterior brain regions. Patel's account predicts that lesions affecting the unification network in patients with Broca's aphasia should also impair their unification capacity for music. In fact, this is exactly what a collaborative research project between Patel's and Hagoort's research groups has found (Patel et al. 2008a).

The Semantic Network

In recent years, there has been growing interest in investigating the cognitive neuroscience of semantic processing (for a review of a number of different approaches, see Hinzen and Poeppel 2011). A series of fMRI studies has aimed at identifying the semantic processing network. These studies either compared sentences containing semantic/pragmatic anomalies with their correct counterparts (e.g., Friederici et al. 2003; Hagoort et al. 2004; Kiehl et al. 2002; Ruschemeyer et al. 2006) or sentences with and without semantic ambiguities (Davis et al. 2007; Hoenig and Scheef 2005; Rodd et al. 2005). The most consistent finding across all of these studies is the activation of the LIFC, in particular BA 47 and BA 45. In addition, the left superior and middle temporal cortices are often found to be activated, as well as left inferior parietal cortex. For instance, Rodd and colleagues had subjects listen to English sentences such as *There were dates and pears in the fruit bowl* and compared the fMRI response of these sentences to the fMRI response of sentences such as *There was beer and cider on the kitchen shelf*. The crucial difference between these sentences is that the former contains two homophones, i.e., "dates" and "pears," which, when presented auditorily, have more than one meaning. This is not the case for the words in the second sentence. The sentences with the lexical ambiguities led to increased activations in LIFC and in the left posterior middle/inferior temporal gyrus. In this experiment all materials were well-formed English sentences in which the ambiguity usually goes unnoticed. Nevertheless, very similar results were obtained in experiments that used semantic anomalies. Areas involved in semantic unification were found to be sensitive to the increase in semantic unification load due to the ambiguous words.

Semantic unification could be seen as filling the slots in an abstract event schema, where in the case of multiple word meanings for a given lexical item competition and selection increase in relation to filling a particular slot in the event scheme. As with syntactic unification, the availability of multiple candidates for a slot will increase the unification load. In the case of the lexical ambiguities there is no syntactic competition, since both readings activate the same syntactic template (in this case the NP-template). Increased processing is hence due to integration of meaning instead of syntax.

In short, the semantic processing network seems to include at least LIFC, left superior/middle temporal cortex, and the (left) inferior parietal cortex. To some degree, the right hemisphere homologs of these areas are also found to be activated. Below we will discuss the possible contributions of these regions to semantic processing.

An indication for the respective functional roles of the left frontal and temporal cortices in semantic unification comes from a few studies investigating semantic unification of multimodal information with language. Using fMRI, Willems and colleagues assessed the neural integration of semantic information from spoken words and from co-speech gestures into a preceding sentence context (Willems et al. 2007). Spoken sentences were presented in which a critical word was accompanied by a co-speech gesture. Either the word or the gesture could be semantically incongruous with respect to the previous sentence context. Both an incongruous word as well as an incongruous gesture led to increased activation in LIFC as compared to congruous words and gestures (for a similar finding with pictures of objects, see Willems et al. 2008). Interestingly, the activation of the left posterior STS was increased by an incongruous spoken word, but not by an incongruous hand gesture. The latter resulted in a specific increase in dorsal premotor cortex (Willems et al. 2007). This suggests that activation increases in left posterior temporal cortex are triggered most strongly by processes involving the retrieval of lexical-semantic information. LIFC, on the other hand, is a key node in the semantic unification network, unifying semantic information from different modalities. From these findings it seems that semantic unification is realized in a dynamic interplay between LIFC as a multimodal unification site, on the one hand, and modality-specific areas on the other.

Although LIFC (including Broca's area) has traditionally been construed as a language area, a wealth of recent neuroimaging data suggests that its role extends beyond the language domain. Several authors have thus argued that LIFC function is best characterized as "controlled retrieval" or "(semantic) selection" (Badre et al. 2005; Moss et al. 2005; Thompson-Schill et al. 2005; Thompson-Schill et al. 1997; Wagner et al. 2001). How does the selection account of LIFC function relate to the unification account? As discussed elsewhere, unification often implies selection (Hagoort 2005). For instance, in the study by Rodd and colleagues described above, increased activation in LIFC is most likely due to increased selection demands in reaction to sentences with ambiguous words. Selection is often, but not always, a prerequisite for unification. Unification with or without selection is a core feature of language processing. During natural language comprehension, information has to be kept in working memory for a certain period of time, and incoming information is integrated and combined with previous information. The combinatorial nature of language necessitates that a representation is constructed online, without the availability of an existing representation of the utterance in long-term memory. In addition, some information sources that are integrated with language do not have a stable representation in long-term memory such that they can be selected. For instance, there is no stable representation of the meaning of co-speech gestures, which are highly ambiguous outside of a language context. Still, in all these cases increased activation is observed in LIFC, such as when the integration load of information from co-speech gestures is high (Willems et al. 2007). Therefore, unification is a more general account of LIFC function. It implies selection, but covers additional integration processes as well.

Importantly, semantic processing is more than the concatenation of lexical meanings. Over and above the retrieval of individual word meanings, sentence and discourse processing requires combinatorial operations that result in a coherent interpretation of multi-word utterances. These operations do not adhere to a simple principle of compositionality alone. World knowledge, information about the speaker, co-occurring visual input, and discourse information all trigger similar electrophysiological responses as sentence-internal semantic information. A network of brain areas, including the LIFC, the left superior/middle/inferior temporal cortex, the left inferior parietal cortex and, to a lesser extent, their right hemisphere homologs, are recruited to perform semantic unification. The general finding is that semantic unification operations are under top-down control of the left and, in the case of discourse, also the right inferior frontal cortex. This contribution modulates activations of lexical information in memory as represented by the left superior and middle temporal cortex, with presumably additional support for unification operations in left inferior parietal areas (e.g., angular gyrus).

The Network Topology of the Language-Ready Brain

We have seen that the language network in the brain is much more extended than was thought for a long time and includes areas in the left hemisphere as well as right hemisphere. However, the evidence of additional activations in the right hemisphere and areas other than Broca and Wernicke does not take away the strong bias in favor of left perisylvian cortex. In a recent meta-analysis based on 128 neuroimaging papers, Vigneau et al. (2010) compared left and right hemisphere activations that were observed in relation to language processing. On the whole, for phonological processing, lexical-semantic processing, and sentence or text processing, the activation peaks in the right hemisphere comprised less than one-third of the activation peaks in the left hemisphere. Moreover, in the large majority of cases, right hemisphere activations were in homotopic areas, suggesting strong interhemispheric influence. It is therefore justified to think that for the large majority of the population (with the exception of some portion of left-handers, cases of left hemispherectomy, etc.), the language readiness of the human brain resides to a large extent in the organization of the left perisylvian cortex. One emerging generalization is that the network of cortical regions subserving output processing (production) is very strongly (left) lateralized; in contrast, the computational subroutines underlying comprehension appear to recruit both hemispheres rather extensively, even though there also exists compelling lateralization, especially for syntax (Menenti et al. 2011).

Moreover, the network organization of the left perisylvian cortex has been found to show characteristics that distinguishes it from the right perisylvian cortex and from homolog areas in other primates. A recent technique for tracing fiber bundles in the living brain is diffusion tensor imaging (DTI). Using DTI, Rilling et al. (2008) tracked the arcuate fasciculus in humans, chimpanzees, and macaques and found a prominent temporal lobe projection of the arcuate fasciculus in humans that is much smaller or absent in nonhuman primates (see Figure 9.6). Moreover, connectivity with the MTG was more widespread and of higher probability in the left than in the right hemisphere. This human specialization may be relevant for the evolution of language. Catani et al. (2007) found that the human arcuate fasciculus is strongly lateralized to the left, with quite some variation on the right. On the right, some people lack an arcuate fasciculus, in others it is smaller in size, and in a minority of the population this fiber bundle is of equal size in both hemispheres. The presence of the arcuate fasciculus in the right hemisphere correlated with a better verbal memory. This pattern of lateralization was confirmed in a study on 183 healthy right-handed volunteers aged 5–30 years (Lebel and Beaulieu 2009). In this study the lateralization pattern did not differ with age or gender. The arcuate fasciculus lateralization is present at five years of age and remains constant

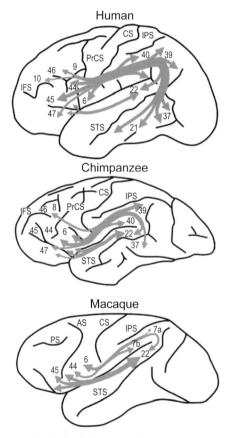

Figure 9.6 The arcuate fasciculus in human, chimpanzee, and macaque in a schematic lateral view of the left hemisphere. Reprinted from Rilling et al. (2008) with permission from Macmillan Publishers Ltd.

throughout adolescence into adulthood. However, another recent study comparing seven-year-olds with adults (Brauer et al. 2011b) shows that the arcuate fasciculus is still relatively immature in the children.

In addition to the arcuate fasciculus, which can be viewed as part of a dorsal processing stream, other fiber bundles are also important in connecting frontal with temporoparietal language areas (see Figure 9.7). These include the superior longitudinal fasciculus (adjacent to the arcuate fasciculus) and the extreme capsule fasciculus as well as the uncinate fasciculus, connecting Broca's area with superior and middle temporal cortex along a ventral path (Anwander et al. 2007; Friederici 2009a; Kelly et al. 2010).

DTI is not the only way to trace brain connectivity. It has been found that imaging the brain during rest reveals low-frequency (<0.1 Hz) fluctuations in the fMRI signal. It turns out that these fluctuations are correlated across areas that are functionally related (Biswal et al. 1995; Biswal and Kannurpatti 2009). This so-called resting state fMRI can thus be used as an index of functional connectivity. Although both DTI and resting state fMRI measure connectivity, in the case of DTI the connectivity can often be related to anatomically identifiable fiber bundles. Resting state connectivity measures the functional correlations between areas without providing a correlate in terms of an anatomical tract. Using the resting state method, Xiang et al. (2010) found a clear topographical functional connectivity pattern in the left inferior frontal, parietal, and temporal areas. In the left but not the right perisylvian cortex, functional connectivity patterns obeyed the tripartite nature

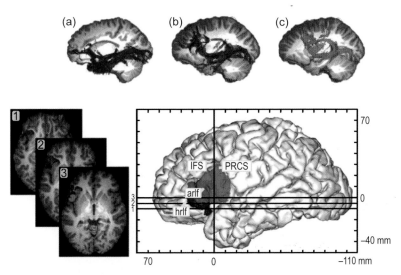

Figure 9.7 Connectivity patterns between parts of frontal cortex (in red, blue, and green) with parietal and temporal areas. Colored areas in left frontal cortex are connected via fiber bundles with the same color (a, b, c). Reprinted from Friederici (2009a) with permission from Elsevier.

of language processing (phonology, syntax, and semantics). These results support the assumption of the functional division for phonology, syntax, and semantics of the LIFC, including Broca's area, as proposed by the MUC model (Hagoort 2005), and revealed a topographical functional organization in the left perisylvian language network, in which areas are most strongly connected according to information type (i.e., phonological, syntactic, and semantic).

In summary, despite increasing evidence of right hemisphere involvement in language processing, it still seems clear that the left perisylvian cortex has certain network features that stand out in comparison to other species, making it especially suited for supporting the tripartite architecture of human language.

Neurophysiology and Timing

Although we have thus far emphasized functional neuroanatomy and the insights from imaging, it is worth bearing in mind what electrophysiological data add to the functional interpretations we must entertain. As discussed at the outset, one of the most remarkable characteristics of speaking and listening is the speed at which it occurs. Speakers produce easily between two and five words per second; information that has to be decoded by the listener within roughly the same time frame. Considering that the acoustic duration of many words is in the order of a few hundred milliseconds, the immediacy of the electrophysiological language-related effects is remarkable. For instance, the early left anterior negativity (ELAN), a syntax-related effect (Friederici et al. 2003), has an onset on the order 100–150 ms after the acoustic word onset. The onset of the N400 is approximately at 250 ms, and another language relevant ERP, the so-called P600, usually starts at about 500 ms. Thus the majority of these effects happen well before the end of a spoken word. Classifying visual input (e.g., a picture) as depicting an animate or inanimate entity takes the brain approximately 150 ms (Thorpe et al. 1996). Roughly the same amount of time is needed to classify orthographic input as a letter (Grainger et al. 2008). If we take this as our reference time, the early appearance of an ELAN response to a spoken word is remarkable, to say the least. In physiological terms, it might be just too fast for long-range recurrent feedback to have its effect on parts of primary and secondary auditory cortex involved in first-pass acoustic and phonological analysis. Recent modeling work suggests that early ERP effects are best explained by a model with feed-forward connections only. Backward connections become essential only after 220 ms (Garrido et al. 2007). The effects of backward connections are, therefore, not manifest in the latency range of at least the ELAN, since not enough time has passed for return activity from higher levels. However, in the case of speech, the N400 follows the word recognition points closely in time. This suggests that what is happening in online language comprehension is presumably, for a substantial part, based on predictive processing. Under most circumstances, there is simply not enough time for top-down feedback to exert control over a preceding bottom-up analysis. Very likely, lexical, semantic, and syntactic cues conspire to predict very detailed characteristics of the next anticipated word, including its syntactic and semantic makeup. A mismatch between contextual prediction and the output of bottom-up analysis results in an immediate brain response recruiting additional processing resources for the sake of salvaging the online interpretation process. Recent ERP studies have provided evidence that context can indeed result in predictions about a next word's syntactic features (i.e., gender; Van Berkum et al. 2005) and word form (DeLong et al. 2005). Lau et al. (2006) provided evidence that the ELAN elicited by a word category violation was modulated by the strength of the expectation for a particular word category in the relevant syntactic slot. In summary, we conclude that predictive coding is likely a central feature of the neurocognitive infrastructure.

Neural Rhythms and the Structure of Speech

A final issue relates to the convergence of intrinsic aspects of brain function and temporal characteristics of the speech signal. It is known that the brain generates intrinsic oscillatory rhythms which can be characterized by their frequency bands (for an extended discussion of the neural underpinnings, see Buzsáki 2006). For instance, theta oscillations are defined as activity between ~4 and 8 Hz, the alpha rhythm has its center peak at about 10 Hz (~9–12 Hz), and beta oscillations are found at around 20 Hz. Finally, gamma oscillations are characterized by frequencies above 40 Hz (see also Arbib, Verschure, and Seifert, this volume.) A recent, and admittedly still speculative, hypothesis suggests the intriguing possibility that some of these neuronal oscillations have temporal properties that make them ideally suited to be the carrier waves for processing aspects of language that are characterized by the different timescales at which they occur (e.g., Giraud et al. 2007; Luo and Poeppel 2007; Schroeder et al. 2008; Giraud and Poeppel 2012).

Naturalistic, connected speech is aperiodic, but nevertheless quasi-rhythmic as an acoustic signal. This temporal regularity in speech occurs at multiple timescales; each of these scales is associated with different types of perceptual information in the signal. Very rapidly modulated information, say 30–40 Hz or above (low gamma band), is associated with the spectrotemporal fine structure of a signal and is critical for establishing the order of rapid events. Modulation at the rate of 4–8 Hz (the so-called theta band) is associated with envelope fluctuations, discussed below. Modulations at slow rates, say 1–3 Hz, typically signal prosodic aspects of utterances, including intonation contour and phrasal attributes. We briefly elaborate on one of these scales: the intermediate scale.

There exists one pronounced temporal regularity in the speech signal at relatively low modulation frequencies. These modulations of signal energy (in reality, spread out across a filter bank) are well below 20 Hz, typically peaking roughly at a rate of 4–6 Hz. From the perspective of what the auditory cortex receives as input, namely the modulations at the output of each filter/channel of the filter bank that constitutes the auditory periphery, these energy fluctuations can be characterized by the "modulation spectrum" (Greenberg 2005; Kanedera et al. 1999). For speech produced at a natural rate, the modulation spectrum across languages peaks between 4–6 Hz (e.g., Elliott and Theunissen 2009). Critically, these energy modulations correspond in time roughly to the syllabic structure (or syllabic "chunking") of speech. The syllabic structure, as reflected by the energy envelope over time, in turn, is perceptually critical because (a) it signals the speaking rate, (b) it carries stress and tonal contrasts, and (c) cross-linguistically the syllable can be viewed as the carrier of the linguistic (question, statement, etc.) or affective (happy, sad, etc.) prosody of an utterance. As a consequence, a special sensitivity to envelope structure and envelope dynamics is critical for successful auditory speech perception.

One hypothesis about a potential mechanism for chunking speech (and other sounds) is based on the existent neuronal infrastructure for dealing with temporal processing in general. In particular, *cortical oscillations* could be efficient instruments of auditory cortex output discretization/chunking/ sampling. Neuronal oscillations reflect synchronous activity of neuronal assemblies (either intrinsically coupled or coupled by a common input). Importantly, cortical oscillations are argued to shape and modulate neuronal spiking by imposing phases of high and low neuronal excitability (e.g., Fries 2005; Schroeder et al. 2008). The assumption that oscillations cause spiking to be temporally clustered derives from the observation that spiking tends to occur in the troughs of oscillatory activity (Womelsdorf et al. 2007). It is also assumed that spiking and oscillations do not reflect the same aspect of information processing. While spiking reflects axonal activity, oscillations are said to reflect mostly dendritic postsynaptic activity (Wang et al. 2012).

Neuronal oscillations are ubiquitous in cerebral cortex and other brain regions (e.g., hippocampus), but they vary in strength and frequency depending on their location as well as the exact nature of their generators. In human auditory cortex, at rest (i.e., no input), ~40 Hz activity (low gamma band activity) can be detected (using concurrent EEG and fMRI) in the medial part of Heschl's gyrus, a region that is situated just next to core primary auditory cortex. In response to linguistic input, gamma oscillations spread to the whole auditory cortex as well as to classical language regions, where they cannot be detected at rest (Morillon et al. 2010).

If there exists a principled relation between the temporal properties of neuronal oscillations and the temporal properties of speech (i.e., delta band/intonation contour, theta band/syllabic rate, gamma band/segmental modulation), it stands to reason that these correspondences are not accidental. The speech processing system is exploiting the neuronal, biophysical infrastructure and yielding speech phenomena at timescales provided. In this context, it is worth remembering that the observed neuronal oscillations are not merely "driven in" to the system by external signal properties but are rather endogenous aspects of brain activity. Indeed, experimental data from many animal studies as well as some recent human data show that neuronal oscillations in these ranges are endogenous and evident in auditory and motor areas (Giraud et al. 2007; Morillon et al. 2010).

Such data suggest an intriguing evolutionary scenario in which neuronal processing timescales follow from purely biophysical constraints (and therefore will also be visible in other primates) for the basis for timing phenomena in speech processing. The cognitive system is grafted on top of structures that provide hardware constraints, setting the stage for potential coevolutionary scenarios of brain and speech. (Fogassi, this volume, offers a complementary perspective on the evolution of speech; Arbib and Iriki, this volume, place more emphasis on the role of gesture in the evolution of the language-ready brain.)

Final Remarks

The data from neurobiology, cognitive neuroscience, psycholinguistics, and linguistics lead to a similar conclusion across domains: there is no single computational entity called "syntax" and no unstructured operation called "semantics," just as there is no single brain area for words or sounds. Because these are structured domains with considerable internal complexity, unification, or linking operations as outlined in the MUC perspective above, is necessary. Cognitive science research, in particular linguistic and psycholinguistic research, shows convincingly that these domains of processing are collections of computational subroutines. Therefore it is not surprising that the functional anatomy is not a one-to-one mapping from putative language operation to parts of brain. In short, there is no straightforward mapping from syntax to brain area X, semantics to brain area Y, phonology to brain areas Z, etc. Just as cognitive science research reveals complexity and structure, so the neurobiological research reveals fractionated, complex, and distributed anatomical organization. Moreover, this fractionation is not just in space (anatomy) but also in time: different computational subroutines act at different points in the time course of language processing. When processing a spoken sentence, multiple operations occur simultaneously at multiple timescales and, unsurprisingly, many brain areas are implicated in supporting these concurrent operations. The brain mechanisms that form the basis for the representation and processing of language are fractionated both in space and in time, necessitating theories of unification that underpin how we use language to arrive at putatively unified interpretations.

Music is in many ways like language. Although it is not very helpful to try to make direct comparisons between building blocks of music and language (e.g., to claim that words correspond to notes), music is almost certainly another complex faculty that has to be decomposed in multiple subroutines, each recruiting different nodes in a complex neuronal network. It is likely that some of the nodes in the neuronal networks that support the perception and production of music are shared with language. In both cases, meticulous analyses is required to determine what the primitives are (for a discussion of this approach and an attempt to make explicit what is shared and what is different, see Fritz et al., this volume); that is, what the "parts list" is (e.g., features, segments, phonemes, syllables, notes, motifs, intervals). This will enable us to meet the challenge of mapping the list of primitives for language and music to the computations executed in the appropriate brain areas.

10

Musical Syntax and Its Relation to Linguistic Syntax

Fred Lerdahl

Abstract

Music is meaningful, but there is no musical counterpart to the lexicon or semantics of language, nor are there analogs of parts of speech or syntactic phrases. This chapter seeks to establish a notion of musical syntax at a more fundamental level, starting from the view that syntax can be broadly defined as the hierarchical organization of discrete sequential objects which generate a potentially infinite set of combinations from a relatively small number of elements and principles (thereby extending not only to linguistic syntax in the usual sense but also to a syntax of phonology). The elementary musical objects in this approach are perceived pitches, chords, and rhythms. Sequences of musical events receive three types of structure: groupings, grids, and trees. Using a Beatles song as illustration, the formation of successive structural levels is described and issues of sequential ordering, the status of global structural levels, contour, and the question of psychological musical universals are discussed. The strongest correspondences between music and language appear to be between musical syntax and linguistic phonology, not musical syntax and linguistic syntax.

Background on Musical Syntax

Music has always been the most theoretically laden of the arts. In the Western tradition, early theorizing largely focused on details of tuning, scales (modes), and rhythmic proportion. In the Renaissance, theorists developed principles to control horizontal and vertical intervallic relations (Zarlino 1558), which gradually coalesced into the pedagogies of counterpoint and harmony still taught to undergraduate music majors today. The specification of scale and chord type and the treatment of dissonance constitute a sort of morphology of a musical style.

In the eighteenth century, Jean-Philippe Rameau (1726) proposed a syntactic treatment of harmonic progression linked to the newly discovered overtone series. Other theorists addressed larger levels of musical form, pursuing an analogy to Classical rhetoric (Mattheson 1739) and articulating phrasal forms (Koch 1793). The nineteenth century focused increasingly on chromatic harmony, culminating in Hugo Riemann's (1893) theory of harmonic function, which bears the seeds of a syntactic constraint system. In the early twentieth century, Heinrich Schenker (1935) developed a comprehensive analytic method that generates structure in a series of self-similar hierarchical levels, starting from a simple underlying form and yielding, through elaborative and transformational operations, the surface variety of a given piece. As such, he anticipated generative linguistics.

Recent music theory has often taken a psychological turn, beginning with Leonard Meyer's (1956) reliance on Gestalt principles and probabilistic methods to account for melodic expectation. His argument that emotion in music arises from denied expectation remains a touchstone for research on musical emotion. Jackendoff and I adopted the methodological framework of generative linguistics to develop a theory of musical cognition in entirely musical terms (generative theory of tonal music, GTTM; Lerdahl and Jackendoff 1983). We proposed four interacting hierarchical components: the rhythmic components of (a) grouping and (b) meter and two kinds of pitch hierarchy, (c) time-span reduction, and (d) prolongational reduction. Time-span reduction provides an interface between rhythm and pitch, whereas prolongational reduction describes nested patterns of departure and return that are experienced as waves of tension and relaxation. Each component is regulated by well-formedness rules, which stipulate possible structures, and preference rules, which assign to given musical passages, in gradient fashion, specific structures that the theory predicts are cognized (see also Jackendoff and Lerdahl 2006).

The growing cognitive science of music has bred fruitful interdisciplinary work, notably Krumhansl's (1990) theoretically informed experiments on the cognitive schematic organization of pitches, chords, and keys. These results underlie the post-GTTM construction of a unified and quantitative hierarchy of pitch relations (tonal pitch space, TPS; Lerdahl 2001b; see also Lerdahl and Krumhansl 2007). In other work, Huron (2006) explored expectation from the perspectives of statistical learning and evolutionary psychology. Tymoczko's (2006) geometrical model of all possible chords, while mathematically sophisticated, may be of limited relevance to music cognition since it sets scales aside and assumes perceptual equivalence of all members of a chord. Most musical idioms, if they have chords at all, build them out of pitches of a musical scale, and chords in most idioms have perceptually salient roots which must be factored into measurements of distance (where spatial distance correlates with cognitive distance).

Current interest in the relationship between music and language has been fueled by the idea that these two uniquely human capacities evolved from protomusical, protolinguistic expressive utterances (Brown 2000; Darwin 1876; Fitch 2010; Rousseau 1760/1852; see also Arbib and Iriki, this volume). Patel

(2008 and this volume) provides a thorough review from a neuroscientific perspective of what music and language do and do not share. He advances the hypothesis that while linguistic and musical structures may have different storage areas in the brain, their processing shares brain resources. He approaches the two capacities in areas where they do not mix, instrumental music and nonpoetic speech, so as to maintain a clear comparison. I take the opposite approach by analyzing the sounds of poetry using GTTM's components (Lerdahl 2001a). This analysis presents evidence for where musical and linguistic structures do and do not overlap, and it provides an account of the variables in setting text to music.

Rohrmeier (2011; see also Koelsch, this volume) has implemented a tree-structure version of Riemann's functional theory of harmony in ways that resemble early generative grammar (Chomsky 1965). His trees decompose into the grammatical categories of tonic region, dominant region, and subdominant region, which in turn are spelled out in terms of tonic-functioning, dominant-functioning, and subdominant-functioning chords. He does not, however, address rhythm or melody. GTTM rejected an early and less-developed version of this approach (Keiler 1977), if only because it does generalize to other musical styles. Indeed, until recently, most of the world's music did not have harmonic progressions. A second issue concerns the status of Riemannian harmonic functions, which derive from Hegelian philosophy via Hauptmann (1853). It is unclear what cognitive claim could lie behind the tripartite functional classification. In contrast to Rohrmeier's approach, GTTM and TPS seek a theoretical framework that can be applied, with suitable modifications, to any type of music; hence their tree structures and treatment of functionality are unencumbered by stylistic restrictions or a priori categories. Rohrmeier's work, however, succeeds well within its self-imposed limits.

Katz and Pesetsky (2011) seek unity between music theory and the minimalist program in current generative linguistics. They reinterpret GTTM's components in pursuit of the claim that the two cognitive systems are identical except for their inputs: phonemes and words in one, pitches and rhythms in the other. It is a highly suggestive approach, although they are forced into the uncomfortable position of positing a single entity, the cadence, as the underlying generative source of all musical structures, including rhythm. Here they follow the Chomsky paradigm in which syntax is the centerpiece from which phonological and semantic structures are derived. Jackendoff's (2002) parallel-architecture linguistic theory is more like GTTM's organization, with its equal and interactive components.

Syntax Abstractly Considered

A recurring pitfall in discussions of musical syntax is the search for musical counterparts of the tripartite division of linguistic theory into syntax, semantics, and phonology (Bernstein 1976). There is no reason to suppose that this division transfers in any straightforward way to music. Music is meaningful, but there is no musical counterpart to the lexicon or semantics, just as there is no analogy to parts of speech and syntactic phrases, or binary phonemic oppositions and moras. Comparisons of musical and linguistic organization must thus begin at a more fundamental level.

If we broadly define syntax here as the hierarchical organization of discrete sequential objects, we can speak not only of linguistic syntax in the usual sense but also of the syntax of phonology. Any syntax constitutes a "Humboldt system" capable of generating a potentially infinite set of outputs from a relatively small number of elements and principles (Merker 2002). In both language and music, there are four abstract aspects in a syntactic hierarchy: a string of objects, nested groupings of these objects, a prominence grid assigned to the objects, and a headed hierarchy of the grouped objects.

Strings and Groupings

The constitution of a string of objects varies according to component and level within a component. In linguistic phonology, the object at the segmental level is a phoneme; at suprasegmental levels, it is the syllable, then the word, then a prosodic unit. In linguistic syntax, the low-level object is a lexical item with discrete features; at higher levels it is an X-bar phrase (i.e., a syntactic phrase whose principal constituent, or head, is an X; e.g., a verb phrase is a constituent headed by a verb), then a sentence. Music has different objects. Music theory tends to ignore the psychoacoustic level, which corresponds more or less to that of phonetics in linguistic studies, and treats perceived pitches, chords, and rhythms as its elementary objects. These can be referred to as "(pitch) events." At larger levels, units consist of groupings of events.

How do linguistic and musical objects relate? Perhaps the most basic correspondence is between syllable and note. In a text setting, a single syllable is usually set to a single note (the less frequent case is melisma, in which a syllable continues over several notes). At a sub-object level, syllables typically break down into a consonant plus a vowel, corresponding roughly to the attack and sustained pitch of a note. Syllables group into polysyllabic words and clitic phrases, which group into phonological and intonational phrases. These levels correspond more or less to the musical levels of motive, subphrase, and phrase, respectively. Possibly of deeper import than these broad correspondences is that they are made between music and phonology, not music and linguistic syntax.

In both domains, groupings apply to contiguous objects. The general form of grouping structure is illustrated in Figure 10.1. Higher-level groups are made up of contiguous groups at the next lower level. These strictures apply only within a component. Grouping boundaries in one component of language or music often do not coincide with those in another. For example, prosodic

Musical Syntax and Its Relation to Linguistic Syntax 261

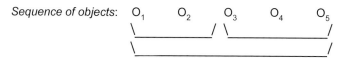

Figure 10.1 Abstract form of grouping structure.

and linguistic-syntactic boundaries are often different, as are phrasal and prolongational boundaries in music.

Generative linguistics has long posited movement transformations that reorder syntactic strings in specified ways to account for certain syntactic phenomena. Lately, movement has become "internal merge" (Chomsky 1995). Such phenomena appear not to exist in music and, except for Katz and Pesetsky (2011), music theory has never considered them.

Grids

Here we consider "grids" as they arise in phonology and in the metrical components of poetry and music. An instance of the general form of a grid is illustrated in Figure 10.2, where an X represents stresses or beats. If an X is prominent or strong at a given level, it is also an X at the next larger level. The X's do not group in the grid per se; that is, a grid is not a tree structure.

There are two kinds of grids in language and music: a stress grid and a metrical grid. A stress grid in phonology represents relative syllabic stress, confusingly called a "metrical grid" in the literature (e.g., Liberman and Prince 1977). A linguistic metrical grid, in contrast, represents strong and weak periodicities in a poetic line against which stresses do or do not align. Stresses in ordinary speech are usually too irregular to project meter (Patel 2008). The cues for beats in musical meter are more periodic and mutually reinforcing than those of most spoken poetry. Consequently a musical metrical grid often has many levels.

Stress (or psychoacoustic prominence) in music is less rule-governed than in phonology and plays little role in music theory. Much more important in music is another kind of prominence—pitch-space stability—for which there is no linguistic equivalent. The most stable event in music is the tonic: a pitch or chord that is the point of orientation in a piece or section of a piece. Other pitches and chords are relatively unstable in relation to the tonic. The degree of instability of nontonic pitches and chords has been well established empirically and theoretically (Krumhansl 1990; Lerdahl 2001b).

```
        X           X           X
    X       X   X       X       X
    X   X   X   X   X   X   X   X
```

Figure 10.2 Abstract form of a grid.

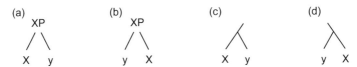

Figure 10.3 Abstract form of an X-bar syntactic tree and an equivalent musical tree. X's dominance is represented symbolically in (a) and (b) by "XP"; in (c) and (d), it is depicted by the longer branch stemmed to X.

Headed Hierarchies

Groupings and grids are nonheaded hierarchies; that is, adjacent groups or X's do not form dominating-subordinating constituencies. Strings of words and musical events, however, form headed hierarchies. The objects of headed hierarchies in linguistic syntax are grammatical categories, represented in simplified form in Figure 10.3a, b, where XP stands for X-bar phrase, X for the head constituent (noun in a noun phrase, verb in a verb phrase, etc.), and Y for any nonhead grammatical constituent within the phrase, either before or after X. The same hierarchy is conveyed in Figure 10.3c, d using GTTM's musical tree notation. Musical trees, however, do not represent X-bar categories but rather noncategorical elaborative relations. In time-span reduction, branching occurs between events in the nested rhythmic structure, with the more stable event dominating at each level. In prolongational reduction, the tree shows tensing-relaxing relations between stable and unstable events, with right branching (Figure 10.3c) for a tensing motion and left branching (Figure 10.3d) for a relaxing motion.

Syntax Illustrated

Some short examples from the Beatles song *Yesterday* will illustrate these abstract structures, moving from lyrics to music and beginning with phonology instead of linguistic syntax. The focus will be on representations rather than derivations. Figure 10.4a illustrates a prosodic analysis of the word "yesterday," employing a tree notation with strong (S) and weak (W) nodes (Liberman and Prince 1977). Figure 10.4b conveys the same information using a combination of prosodic grouping and stress grid (Selkirk 1984; Hayes 1989). Figure 10.4c translates Figure 10.4a, b into musical tree notation, in which domination is represented by branching length.

  A prosodic analysis of the entire first line of the lyric appears in Figure 10.5a, showing a stress grid, an inverted metrical grid, and prosodic grouping. There is a silent metrical beat between "-day" and "all," as suggested by the comma. The result, based on prosodic criteria independent of the Beatles' setting, is triple meter at the largest level, with strong beats on "Yes-," "troub-," and "-way." Somewhat unusually, the stress and metrical grids fully

Musical Syntax and Its Relation to Linguistic Syntax   263

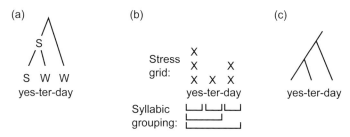

Figure 10.4 Prosodic analysis of the word "yesterday" with mostly equivalent notations: (a) strong–weak (S–W) tree; (b) prosodic grouping and stress grid; (c) timespan tree.

align. Figure 10.5b converts the metrical pattern in Figure 10.5a into standard musical notation. The notes on "Yes-," "troub-," and "-way" are raised slightly to convey the heavy stresses on these syllables (pitch height is a contributor to the perception of stress).

Figure 10.6 shows the Beatles' musical setting, accompanied by a musical grouping analysis and metrical grid. To the left of the grid are the note durations of the metrical levels. The setting departs in a few respects from the metrical analysis in Figure 10.5. The first two syllables of "yesterday" are shortened, causing syncopation (relative stress on a weak beat) and lengthening on "-day." The stress-metrical pattern of "yesterday" essentially repeats in the rhymed "far away," whereby "far" receives the major metrical accent

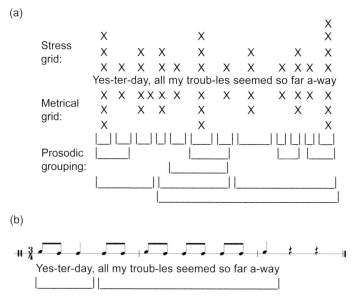

Figure 10.5 (a) Prosodic analysis of the first poetic line of *Yesterday*; (b) conversion of the metrical pattern in (a) into musical notation.

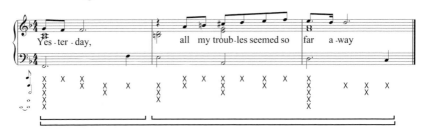

Figure 10.6 The first phrase from the song *Yesterday*, with the metrical grid and global grouping added below.

instead of "-way." This emphasis shifts the semantic focus of the second part of the phrase, conveying distance or "farness" between a "happy yesterday" and "troubled today."

Figure 10.7 displays one interpretation of the syntactic tree of the sentence (Jackendoff, pers. comm.). Figure 10.8 shows the prolongational tree of the corresponding music. Derivational levels are labeled in the tree, and a conventional Roman numeral harmonic analysis appears beneath the music. The linguistically dominating words, "troubles" and "seemed," which are the main noun and verb of the sentence in Figure 10.7, are rather embedded in the musical tree in Figure 10.8. The musically dominating words are "yesterday" and "far away," specifically the rhyming syllables "-day" and "-way."

More basic than these particular divergences, however, is the dissimilarity of the trees themselves: a linguistic-syntactic tree consists of parts of speech and syntactic phrases, but a musical tree assigns a hierarchy to events without grammatical categories. In this respect, musical trees are like phonological

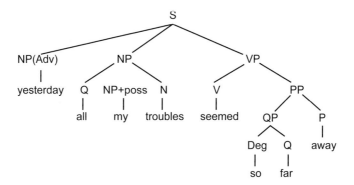

Figure 10.7 Syntactic tree for the first poetic line of *Yesterday*. S = sentence, NP(Adv) = noun phrase with adverbial function, NP = noun phrase, VP = verb phrase, PP = prepositional phrase, Q = quantifier, NP+ poss = possessive pronoun, N = noun, V = verb, QP = quantifier phrase, P = preposition, and Deg = degree.

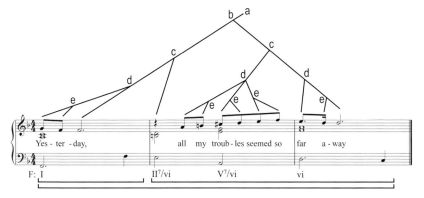

Figure 10.8 Prolongational tree for the first musical phrase of *Yesterday* (bars 1–3).

trees, as in Figure 10.4c, instead of linguistic-syntactic trees. Furthermore, the phrase categories in Figure 10.7 include word groupings (e.g., the noun phrase "all my troubles") whereas all the leaves of a prolongational tree are single events. (For a contrasting approach, see Rohrmeier 2011).

In bars 2–3 (Figure 10.8), the music that portrays "far away" progresses melodically to a higher register and harmonically from the tonic (I) F major to D minor. In the next phrase (Figure 10.9), melody and harmony return in bars 4–5, reflecting the sense of the words "Now it looks as though they're here to stay." The tree represents the return by its highest branch at level *a*, which attaches to the highest branch at level *a* in Figure 10.8. Thus the tonic F major in bar 5 "prolongs" the tonic in bar 1. The continuation in bars 5–6 in turn prolongs the tonic of bar 5.

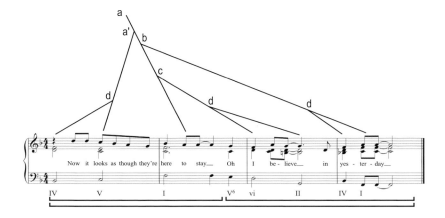

Figure 10.9 Prolongational tree for the second musical phrase of *Yesterday* (bars 4–7).

Sequential Ordering in Music

Any consideration of syntax involves not only a hierarchy of elements but also their sequential ordering. Word order is crucial in English, a comparatively uninflected language. In the syntax of various other languages, case markers often take the place of word order. There appears to be no musical equivalent to the syntactic role of case markers.

Are there constraints on the order of musical events? For Western tonal music, the answer is yes, although there is more freedom than in English syntax. At a very local level, dissonance treatment requires specific continuations. For example, the dissonances on "Yes-" and "far" in Figure 10.8 demand and receive resolution on adjacent and consonant scale degrees. At the phrase or double-phrase level, the most restrictive constraint is the cadence, a formulaic way of establishing closure at a phrase ending. In Western tonality, the standard cadence is a two-event process: a melodic step to the tonic pitch over an accompanying dominant-to-tonic harmonic progression. Cadential preparation, for which there are only a few options, usually precedes a cadence. At the onset of a phrase, the norm is to start on the tonic or member of the tonic chord. Hence a phrase ordinarily begins and ends on a tonic, the point of stability or relaxation. After the beginning, ordering is relatively unrestricted, with the proviso that the continuation induces tension as the music departs from the tonic. The high point of tension occurs somewhere in the middle of the phrase, followed by relaxation into the cadence. In sum, ordering constraints are generally strongest at the phrasal boundaries and weakest in the middle. Figure 10.10 sketches this pattern of stability from tension to closure.

Yesterday manifests this pattern at the double phrase more clearly than at the single phrase level. In the opening phrase (Figure 10.8), the tonic (I) in bar 1 departs and tenses to D minor (vi) in bar 3. In the relaxing answering phrase (Figure 10.9), the subdominant (IV) on the downbeat of bar 4 prepares the dominant-tonic (V–I) cadence in bars 4–5. Bars 6–7 function as a tag, a confirmation of this main action.

The trees in Figures 10.8–10.10 refer to musical features—tonic, key, cadence, dissonance, tension, stability—which do not translate into linguistic-syntactic terms. The only area where pitch is shared between the two media

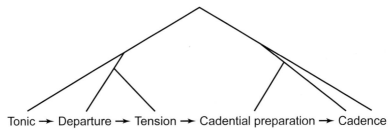

Figure 10.10 Normative prolongational pattern of tension and relaxation in phrases.

is the rise and fall of melodic and speech contour. In other respects, pitch relations are unique to music. A robust tradition represents pitch relations in multidimensional spaces that have no linguistic counterpart (Krumhansl 1990; Lerdahl 2001b; Tymoczko 2011). There is increasing evidence that spatial pitch representations have neural correlates, not only logarithmic pitch height on the basilar membrane and in the auditory cortex (Weinberger 1999) but also cycle-of-fifths key relations among major keys (Janata et al. 2002a).

Figure 10.10 represents a schema applicable to many musical styles. Ordering constraints can also arise from schemas specific to a single style. In the Classical period, there are a number of stock subphrase melodic and harmonic patterns (Gjerdingen 1996). These patterns slot into beginning, middle, or end positions to form functionally ordered units (TPS). Similarly, classical schemas within and above the phrase level coalesce into intermediate levels of form (Caplin 1998). These larger units, comprised of phrase groups with characteristic tree patterns, are also susceptible to ordering constraints.

The tightness or laxness of ordering at various musical levels corresponds to the expectation, or predictability, of future events: the tighter the constraint at a given point, the higher the probability of a particular outcome. Huron (2006) argues that predictability is largely a result of exposure to statistical distributions. Yet while distributions are undoubtedly important in learning the syntax of a musical idiom, a statistical account alone is insufficiently explanatory. Why do dissonances resolve to adjacent pitches? Why is dominant-to-tonic the standard cadence in the classical style? Why, and how, are some chords and keys closely related while others are only distantly related? Why does the normative phrase go from relaxation to tension to relaxation? Why do certain style-specific schemas occur at certain positions in a phrase or section? There is no space here to resolve such questions except to say that the answers go beyond statistics to issues involving mental representation, psychoacoustics, and biology.

Structure at Global Musical Levels

Linguistic-syntactic trees apply only up to the level of the sentence. Larger linguistic levels express discourse or information structure. A tradition in music theory, however, carries the logic of prolongational syntax from the smallest detail up to a large movement such as sonata form (Schenker 1935; GTTM). This view is too uniform and needs to be supplemented by methods that incorporate aspects of discourse structure. Music theory has yet to develop such an approach in any detail.

Three related questions arise in this connection. First, to what extent do ordinary listeners hear prolonged pitches or chords, especially the tonic, over long time spans? The empirical literature on this matter, technical flaws aside, is not very encouraging (e.g., Cook 1987b). TPS offers a way to resolve the issue by positing prolongational functions instead of key identity as the operative factor in the perception of long-range connections. A piece may begin in one key and end in another, but if the ending key is well established it can function as a global tonic and thereby provide closure.

Second, what are the units of analysis at global levels? TPS proposes feature reduction of events at underlying levels and a syntactic ordering of small schematic groups, of the kind discussed in Gjerdingen (1996) and Caplin (1998), as the units of analysis at larger levels. Metaphorically, if pitch events are atoms of analysis, in a larger view small schematic groups act as molecules of analysis.

Third, how deep does hierarchical embedding in prolongational structure extend? In principle, embedding could be of indefinite depth, but in practice there are limits to its cognition. (The same holds true in language.) Larger units of analysis will alleviate this problem, for at global levels there are fewer objects to embed.

It is perhaps unnecessary to add that the hypothesis that only the narrow language faculty possesses recursion is not tenable (Hauser et al. 2002). (Recursion is meant as hierarchical self-embedding, as when a sentence includes one or more sentential subordinate clauses.) There is empirical evidence that music is also cognized hierarchically and recursively (Dibben 1994; Lerdahl and Krumhansl 2007). For example, the Bach chorale tested in Lerdahl and Krumhansl (2007) shows recursion of the harmonic progression I → V–I at multiple levels.

Contour

Intonation in speech and melody in music share the feature of pitch contour. In speech, pitch rises and falls in continuous glides; in music, pitch height is ordinarily steady from one pitch to the next. In most musical systems, intervals between pitches are constant, whereas in speech intervals they are in constant flux. Slight variations in musical tuning do not undermine interval constancy as long as the intervals are perceived categorically. Fixed intervals and their relative psychoacoustic consonance permit a melody to project degrees of stability and instability for which there is no linguistic analog.

This point holds for tone languages as well. Speech tones are not fixed but relative to each other, the voice of speaker, and the natural declination of speech utterances (Ladd 2008). Tone languages as well as nontone languages lack fixed pitch intervals, scales, tonics, and other fundamental features of music.

The question to pose in this context is whether speech contour is syntactic. If it is, it is not merely continuous in rise and fall but must consist of semistable objects connected hierarchically. The topic of intonational units has been much debated (Ladd 2008). On one side is the view that intonation is fully continuous and definable only by its shapes (Bolinger 1986). Autosegmental-metrical theory, in contrast, posits intonational objects at the highest peak and phrase boundaries of an utterance. Movement between these points is unconstrained, and the objects are related sequentially but not hierarchically (Pierrehumbert 1980).

In adapting GTTM to the analysis of the sounds of poetry, I have developed a derivational model that assigns each syllable of an utterance to one of four tiered but relative (not absolute) levels of pitch height (Lerdahl 2001a). The syllables are organized hierarchically, using the factors of prosodic grouping, stress, and a few canonical contour shapes. In this view, speech intonation is syntactic. The approach finds provisional backing in the recent technique of prosogram analysis of speech contour, which assigns pitch and timing to each syllable in a sequence (Mertens 2004; Patel 2008). Tierney and Patel (pers. comm.) have applied prosogram analysis to Robert Frost's reading of one of his poems analyzed in Lerdahl (2001a), lending indirect support for the theoretical analysis.

The complementary question, whether contour in a musical line is syntactic, has hardly been raised, no doubt because for tonal music, the syntax of pitch relations—scales, counterpoint, harmony, tonality, relative stability, and event hierarchies—has always been more central. Instead, musical contour theories have developed in the context of nontonal contemporary Western music. With the exception of Morris (1993), these theories are not hierarchical in orientation.

Psychological Universals in Music

Generative linguistics seeks to define universal grammar beyond the particularities of this or that language (Chomsky 1965). The term universal is intended not in a cultural but in a psychological sense. In this view, a feature of universal grammar need not appear in every language; rather, it describes a part of the organization of the language capacity itself. Although far less rigorous comparative work has been done in music than in language, the musical situation is comparable. The structures of grouping, grid, and tree apply to all of music cognition, and the particulars of given musical idioms vary in systematic ways within that framework.

To take a relatively simple case, the stresses in some musical styles are irregular enough that no meter is inferred (e.g., the beginning of a North Indian raga, some Japanese gagaku music, some contemporary Western art music). However, if stresses are sufficiently regular, a mentally inferred metrical grid of beats comes into play against which events are heard and measured. There are only a few types of metrical grid:

1. Multiple levels of beats, with equidistant beats at each level and with beats two or three beats apart at the next larger level (as in Western tonal music; see Figure 10.11a).

2. Multiple levels of beats, with equidistant beats at the smallest level, beats two or three beats apart within the next level, and often equidistant beats at a still larger level (as in much Balkan music; see Figure 10.11b).
3. Multiple beat streams, each with equidistant but often noncoincident beats at the lowest metrical level and equidistant beats at a larger level (as in some sub-Saharan African music, some Indian music, and some Western art music; see Figure 10.11c).

Not all kinds of music invoke metrical grids, so grids are not culturally universal; however, if a grid is inferred, its form is formally constrained as described. Each of the three grid types can combine in different ways within its own type (e.g., different combinations of two and three in type Figure 10.11a), and each can combine with the other types. Depending on the regularity of stresses, there can be as few as two and as many as eight metrical levels. The result is a small combinatorial explosion of possible (or well-formed) grids. A given musical style typically utilizes a small subset of possible grids.

Beyond well-formedness, metrical grids are constrained by perceptual limits of tempo. Six hundred beats per minute (100 ms) are too fast to distinguish clearly, whereas 10 beats per minute (6 s) are too slow enough to gauge accurately (London 2004). A middle tempo, from about 70–100 beats per minute (857–600 ms), is perceptually the most salient, and integer multiples or divisions of the tactus tend to be heard in relation to it (GTTM). It has long been hypothesized that the tactus has a biological basis in the human heart rate.

When listening to music, listeners infer particular metrical grids by finding the best match, or fewest violations, between stress patterns in the musical signal and the repertory of possible grids in the style in question. This process happens automatically and quickly. (Next time you turn on the radio and hear music, observe that it takes a moment to find the beat.) GTTM lays out, through its interacting metrical preference rules, how this process happens. Some of the rules appear to be psychologically universal, especially those that incorporate Gestalt principles, whereas others are style specific. No doubt the list of factors is incomplete, if only because of GTTM's orientation to Western tonal music. Comparative study of music from around the globe promises to enrich and correct claims of psychological musical universality with respect not only to meter but also to other musical components.

(a)
Level n+2: x   x   x x
Level n+1: x x x x x x x x
Level n: xxxxxxxxxxxxxx

(b)
x   x
x x x   x x x   x
xxxxxxxxxxxxxxx

(c)
n+1: x   x   x
n: [x x x x x x x x
    xxxxxxxxxxx

Figure 10.11 Types of metrical grids: (a) beats at level n are two beats apart at level n + 1 and three beats apart at level n + 2; (b) beats at level n that are 2 + 2 + 3 beats apart at level n + 1 and three beats apart at level n + 2; (c) two streams of beats at level n, dividing level n + 1 into five and three parts.

Shared Structures and Evolution

This review of musical syntax and its relation to linguistic syntax has taken a broad view of syntax to include not just linguistic syntax but any hierarchical ordering of sequential objects. From an abstract perspective, music and language share three kinds of syntactic structures: groupings, grids, and trees. Groupings and grids are nonheaded hierarchies whereas trees are headed hierarchies.

The broad definition of syntax permits parts of phonology as well as music to be viewed in syntactic terms. Indeed, phonological trees representing syllabic prominence are formally equivalent to musical trees (Figure 10.4c, 10.8, and 10.9), even though the leaves of the trees are different: syllables on one hand, pitch events on the other. This similarity stands in contrast to linguistic-syntactic trees, which are built of syntactic categories.

Further connections between phonology and music emerge when poetry is considered instead of ordinary speech (Lerdahl 2001a). The inference of a poetic meter from patterns of syllabic stress operates in the same way as the best-match process between musical stress and grid. The departure and return of events, especially of the tonic, which is so important to the sense of tension and relaxation in music, is comparable at local levels of poetry in recurrent sound patterns in alliteration, assonance, and especially rhyme. The intuition in both cases is of a return and connection to the same or similar object: a pitch event in music and a syllable in language.

Table 10.1 suggests a taxonomy of shared and unshared musical and linguistic structures. Fixed pitches and the pitch structures that arise from them belong exclusively to music. The lexicon, which specifies word meanings and parts of speech, belongs only to language. From combinations of lexical items come semantic structures such as truth conditions and reference, for which there is no musical counterpart. Also in the linguistic category are linguistic-syntactic relations and various phonological structures such as distinctive features (Jakobson et al. 1952). Shared structures are mostly in the domain of rhythm, broadly conceived: the relative duration of pitches and syllables; grouping into phrases and sections in music and prosodic phrases in language; patterns of stress (or contextual psychoacoustic salience); and metrical grids. Nonrhythmic features shared by both domains are pitch contour and recurrent patterns of timbre (sound quality). "Metrical grids" and "recurrent sound patterns" are given asterisks because these features, while common in music, appear in language mainly in poetry but not in normal speech.

This classification broadly fits with Peretz and Coltheart's (2003) modular model of music processing based on neuropsychological evidence. They place pitch and rhythm in different brain modules and view contour as being processed prior to fixed-pitch relations.

Table 10.1 can also be considered in an evolutionary light. As mentioned, there is a long-standing view that expressive animal utterances preceded the

Table 10.1 Hypothesized organization of musical and linguistic structures (adapted from Lerdahl 2001a).

Exclusively musical structures	Shared structures	Exclusively linguistic structures
Fixed pitches, intervals, and scales	Durational patterns	Lexicon (word meaning and parts of speech)
Harmony	Grouping (prosodic hierarchy)	Semantic structures (truth conditions, reference, entailment)
Counterpoint	Stress (psychoacoustic salience)	Syntactic units
Tonality	*Metrical grids	Phonological distinctive features (and other phonological structures)
Pitch prolongations	Contour	
Tonal tension and attraction	*Recurrent sound patterns	

* Common features in music, but which in language appear primarily in poetry, not in normal speech.

emergence of the separate faculties of music and language. The structural features that cries and calls display are the shared structures in Table 10.1. Animal sounds consist of long and short elements that group into larger units, albeit with comparatively shallow nesting. Relative stress (contextual salience) is a psychoacoustic feature of animal calls, as is pitch contour. Animal utterances show cyclic patterns of recurrent features, especially birdcalls—a case of convergent evolution rather than direct descent. The exception under "shared structures" is metrical grids; only humans and some songbirds engage in behavior with multiple periodicities (Fitch 2006a). These shared structures give rise to the possibility of a pared-down syntactic analysis of animal utterances using GTTM's components, much as in the analysis of spoken poetry.

The shared structures formed a syntactic foundation for the subsequent specializations of music and language. Fitch (2006a) makes a useful distinction in this regard between unlearned animal signals (e.g., ape or birdcalls) and learned complex signals (e.g., bird or whale song). In this scenario, learned complex signals acted as a stage on the way to the full development of the separate faculties of music and language. Music grew in the direction of fixed pitches and intervals and consequent complex pitch structures, and language grew in the direction of word meanings and their combinations.

11

An Integrated View of Phonetics, Phonology, and Prosody

D. Robert Ladd

Abstract

"Phonetics, phonology, and prosody" do not, as might be thought, constitute a set of three separate subsystems of language: linguistic sound systems have both phonetic and phonological aspects, and this applies as much to "prosody" as to other areas of language. The distinction between phonetics and phonology is most often applied to segmental sounds (i.e., sounds that are typically represented by individual letters in alphabetic writing systems), and "prosody" is often used to refer to any nonsegmental phenomena. However, there is little justification for defining prosody as some kind of separate channel that accompanies segmental sounds; prosody so conceived is no more than a loose collection of theoretical leftovers. Moreover, it is easy to identify phonetic and phonological aspects of at least some phenomena that are often thought of as prosodic (e.g., lexical tone).

Nevertheless, various properties might motivate talking about a separate subsystem "prosody." In particular, there are good reasons to think that the essence of prosody is the structuring of the stream of speech into syllables, phrases, and other constituents of various sizes, which may have internal structure, e.g., a head or nucleus of some sort. Rather than looking for specific local acoustic cues that mark a boundary or a stressed syllable—a quest motivated by a linear view of sound structure—we should be looking for cues that lead the perceiver to infer structures in which a boundary or a given stressed syllable are present. Clear analogs in music abound (e.g., harmonic cues to meter mean that a note can be structurally prominent without being louder or longer or otherwise acoustically salient). It is thus likely that our understanding of music should inform research on linguistic prosody rather than the reverse. The existence of abstract hierarchical structure in what is superficially a linear acoustic signal unfolding in time is a key aspect of what music and language share. Much the same is true of signed languages, though of course they are based on a stream of visible movements rather than an acoustic signal.

These issues are relevant to the notion of duality of patterning: the building up of meaningful units (e.g., words) out of meaningless ones (e.g., phonemes). This is said to be a central design feature of language and may be absent from music and, for example, birdsong. However, the division between meaningful and meaningless elements is less sharp than it appears, and the fact that words are composed of phonemes is arguably just a special case of the pervasive abstract hierarchical structure of language. If this view can be upheld, the issue of whether music exhibits duality of patterning can be seen as the wrong question, along with the question of whether birdsong is more like phonology or more like syntax. Instead, it suggests that, evolutionarily, music and language are both built on the ability to assemble elements of sound into complex patterns, and that what is unique about human language is that this elaborate combinatoric system incorporates compositional referential semantics.

Introduction

The title of this chapter was assigned to me in the organizers' brief in advance of the Forum. One possible interpretation of such a title would be that it takes phonetics, phonology, and prosody to be three separate and parallel subsystems of language, which need to be considered in a broader, unified perspective. I do not believe this is a productive starting point. Unquestionably, there are people who treat the three terms as referring to three distinct *areas of study*, but the only one that is sometimes thought of as an actual subsystem is prosody; phonology and phonetics are *aspects* of the sound system of language, and if prosody is a subsystem we can reasonably speak of phonetic and phonological aspects of prosody as well. In this overview I first make clear what it means to distinguish between the phonetic and phonological aspects of sound structure; thereafter I tackle the question of whether it makes sense to regard prosody as a well-defined subsystem.

Wherever possible I have drawn analogies to music and discuss at some length the ways in which hierarchical structure plays a role in both music and language. However, the main goal of this chapter is to lay out issues and findings about the function and structure of sound in language.

Phonetics and Phonology

Phonetics and phonology both conventionally refer to different aspects of the sounds of language, and by extension to analogous properties of signed (manual/visual) languages.

Phonetics

Phonetics deals with the production (*articulatory phonetics*), transmission (*acoustic phonetics*), and perception (sometimes *auditory phonetics*) of speech sounds. Somewhat more narrowly, it is often seen as being centrally concerned with delimiting the features that can serve as the basis of linguistic distinctions.

An Integrated View of Phonetics, Phonology, and Prosody   275

(For example, no known language uses a distinction between the normal pulmonic airstream and the esophageal airstream—belching—to distinguish different meanings.) This more specific concern means that, as a field, phonetics may not be very coherent theoretically. Since by definition it concerns the physical and perceptual nature of speech sounds, functional considerations are never far from the surface; function is the only reason whistles and belches lie outside phonetics but clicks and trills do not (for further discussion, see Pike 1943). Furthermore, our understanding of phonetics is probably distorted by long-standing scientific divisions of labor. Auditory phonetics tends to be neglected by linguistics and typically falls into the realm of clinical audiology and perceptual psychology. Acoustic phonetics, though increasingly relevant to linguistics, is also of concern to electrical engineers and speech technologists. Only articulatory phonetics is solidly a part of linguistics, though it is also important to clinicians.

To some extent, this division of labor is probably due to the anchoring of articulatory phonetics in language teaching and traditional philology: since the nineteenth century, an important focus of phonetic research has been on devising a transcription system applicable to all languages. This emphasis has entrenched an alphabetic bias and given rise to the notion of "phone"; that is, the idea that we can identify sound segments like [d] or [p] in speech independently of the language being spoken (see Pike 1943). This notion is probably spurious (discussed at length by Ladd 2011): modern acoustic investigation reveals the full extent of subtle language-specific differences even in occurrences of "the same" sound.

As for what we might call the "phonetics" of music (e.g., the spectra of different musical instruments, the physical nature of *bel canto* singing, the perception of pitch and melodic structure), this has been investigated by physicists, phoneticians, and psychologists, with varying interests, backgrounds, and theoretical concerns. A unified understanding of the physical side of language and music is still a long way off, though many of the elements are in place.

Phonology

Phonology deals with the way sound is structured in language. The most basic aspect of this structure is the *phonemic principle*. Since the late nineteenth century it has been clearly understood that one language may signal semantic distinctions by means of phonetic differences that are irrelevant in another language (e.g., the distinction between back- and front-rounded vowels in French *cou* /ku/ "neck" and *cul* /ky/ "arse" poses difficulties for speakers of, say, English or Spanish). This means that knowing a language involves knowing both a system of phonetic *categories* and the abstract structural relations between the categories. In turn, this means that we must define "the vowel sound of *cou*" not just by its physical (phonetic) properties, but also (perhaps primarily?) in abstract phonological terms based on the set of contrasts in French vowels: part of the meaning of *cou* is that it is not *cul*. No occurrence of a given vowel is exactly like any other in every physical detail, so the occurrence of one phoneme or the other—and hence one word or the other—is partly a cognitive fact, not a physical one.

Something like the phonemic principle applies in signed languages as well: no two occurrences of a given sign are physically identical, but the physical variability is organized into categories of handshape, location, and movement specific to each signed language. A plausible musical analog involves scale notes: different musical traditions may have different note categories, but within a musical culture, though various productions of a given note may differ on many acoustic dimensions, listeners will accept a wide variety of signals as instances of the same note. This is, for example, what made it possible to explore different tuning systems during the development of Western art music (e.g., Rasch 2002): the underlying "phonological" system of notes remained the same, and the adjustments to interval sizes were strictly on the level of "phonetic" detail.

Phonology also covers a number of language-internal structural regularities. The most important of these are *morphophonemics* (systematic alternations between otherwise distinct phonemes, e.g., the relation between /k/ and /s/ in *cynic*/*cynicism* or *opaque*/*opacity*) and *phonotactics* (constraints on arrangements of phonemes, whereby, e.g., *bling* is a well-formed English monosyllabic word but *dling* or *lbing* are not). Both morphophonomics and phonotactics require reference to other structures. For example, any speaker of English adding the suffix *–ish* to opaque would certainly produce *opaque-ish*, not *opacish*, even though the *–ish* suffix begins with the same vowel as *–ity*. Similarly, even though the sequence *dling* is ill-formed as an English monosyllable, it is perfectly acceptable in a word like *foundling*: only the sequence *ling* constitutes a syllable; the *d* belongs to the end of the preceding syllable. Structural regularities are not purely a matter of sound sequences; they depend on the organization of sounds into larger units like syllables and suffixes. Here again there are obvious musical analogs: a chord progression that makes musical sense in one key may be quite jarring in another.

Linking Phonetics and Phonology

Any complete description of the sounds of a language involves both phonetic and phonological facts. It is a phonetic fact that English /t/ and /d/ are typically produced with the tip of the tongue against the alveolar ridge, unlike the roughly corresponding sounds of French, which are produced with the tip of the tongue against the back of the upper teeth. It is a phonetic fact that English /t/ at the beginning of a syllable is produced with a voice onset time (VOT)—the gap between the release of the tongue-tip closure and the beginning of vocal fold vibration—of approximately 50 milliseconds, whereas the corresponding sound in Chinese is produced with a somewhat longer VOT and in French with a considerably shorter VOT.

It is a phonological fact about English that /t/ and /d/ are categorically distinct phonemes (as shown by *minimal pairs* like *town/down* or *bet/bed*); the phonetic basis of the distinction lies primarily in the VOT, with /t/, as just noted, having a VOT of some 50 ms in syllable initial position and /d/ being either "prevoiced" (i.e., with voicing onset before the release of the closure) or having roughly simultaneous closure release and voice onset (VOT of 0). There is nothing natural or inevitable about this: some languages (e.g., many of the Australian languages) have only a single phoneme in the general area of /t/ and /d/, with widely variable VOT, whereas others (e.g., Thai) have three (phonetically prevoiced, voiceless, and "voiceless aspirated," i.e., with long VOT). It is also a phonological fact about English that this distinction is not found after a syllable-initial /s/ (there are words like *stuff* and *stare*, but nothing like **sduff* or **sdare* corresponding to distinctions like *tough/duff* and *tear/dare*).

At the boundary between phonetics and phonology are at least two kinds of facts. One is allophony: systematic variation in the phonetic realization of phonemes in different phonological contexts. For example, the sound spelled *t* in *stuff* and *stare* has a much shorter VOT than the sound spelled *t* in *tough* and *tear* and is phonetically much closer to syllable initial /d/ than to syllable-initial /t/, yet it counts phonologically as /t/. The other is neutralization: the suppression of a distinction in specific phonological contexts by virtue of the overlapping phonetic realization of otherwise distinct phonemes. For example, in many varieties of English, /t/ and /d/ occurring between vowels (as in *betting* and *bedding*) may be phonetically indistinguishable, even though the distinction is maintained in other phonological contexts (as in *bet* and *bed*).

Prosody

Beyond the Segment

The preceding section gives an idea of the kinds of phenomena that have traditionally occupied phoneticians and phonologists. All of them relate primarily to segmental sounds: sounds indicated by separate symbols in alphabetic writing, which in some sense succeed one another in time. Yet two things have long been clear. First, speech involves continuous movement of the articulators, with few objective boundaries between segments. Second, there are many linguistically meaningful distinctions that are difficult or impossible to represent segmentally, either because they involve phonetic events that are simultaneously rather than sequentially ordered (e.g., emphasis on one word rather than another), or because they seem to involve subtle modifications of the segmental sounds over longer stretches of speech (e.g., slowing of speech rate in the final one or two seconds of an utterance).

We can probably ignore the first issue—the fact that speech is physically continuous—because segments might still be part of an appropriate idealization at some level of description. Evidence in favor of including segment-sized units in our theories of speech includes transposition errors such as spoonerisms, the widespread use of rhyme and alliteration in poetry, and the invention of alphabetic writing. It is probably true that syllables are a psychologically more salient unit than segments, and that learning to read an alphabetic writing system enhances phonological awareness of segments, but there is no reason to doubt that both types of units exist, in some sort of hierarchical relationship. Note that even Japanese *kana* syllabaries, which provide unanalyzable symbols for syllables like /ka/ and /mi/, are nevertheless traditionally arranged in a way that betrays an awareness of segments: /a/ /i/ /u/ /e/ /o/, then /ka/ /ki/ /ku/ /ke/ /ko/, and so on through sets beginning /sa/ /ta/ /na/ /ha/ /ma/ /ya/ /ra/, and /wa/.

The second issue, however—the existence of distinctions that do not seem to be based on segmental sounds—is fundamental to our concerns here, because the list of nonsegmental features that are linguistically meaningful is extremely diverse. At the word level this includes stress and accent (e.g., the difference between the noun *object* and the verb *object*) and tone (e.g., the difference in Mandarin Chinese between *tāng* [with high-level pitch] "soup" and *táng* [with rising pitch] "sugar"). At the sentence or utterance level, there are meaningful phonetic features that do not involve distinctions between one word and another, and which do not lend themselves to being represented in terms of different segments. Some of these involve:

- Different phrasing or ways of grouping the same words: *This is my sister, Helen* can either be addressed to Helen and not mention the sister's name or to some unnamed person specifying that the sister is called Helen.
- Different ways of pronouncing the same words so as to signal different interpersonal or sentence-level meanings: *Okay.* vs. *Okay?* This is the core of what is often called *intonation*.
- Different ways of pronouncing the same words so as to signal differences of what has often been called *information structure*: differences of focus and emphasis like *They saw me* vs. *They saw me*. This issue figures prominently in Hagoort and Poeppel (this volume) and has been addressed in a great deal of literature going back many decades.

The Temptations of "Prosody" As a Unity

Because of the power of the segmental/alphabetic model for phonetics and phonology, all of the functionally disparate phenomena just listed are frequently lumped together under the rubric "prosody." This loose use of the term is perhaps especially tempting in the context of discussing the relation between language and music, because it is easy to think of prosody as "the music of language." However, I do not believe it will be helpful in the long run to treat prosody as a catchall category for everything in the sound structure of language that cannot easily be idealized as a string of consonants and vowels. We need to recognize at least three things about "prosody" if we are to make any progress at all.

Eroding the Segmental/Suprasegmental Distinction

First, it seems clear that at least some "suprasegmental" phonemes are functionally identical to segmental ones. The most important case is that of tone: the difference between *tāng* and *táng* in Chinese is entirely comparable to the difference between *town* and *down* in English, hard though this may be for native speakers of English to believe. Indeed, there is plenty of evidence from various languages that phonological distinctions based on VOT (as in *town/down*) can sometimes be reinterpreted, in historical language change, as phonological distinctions based on pitch (as in *tāng/táng*), and vice versa; such a reinterpretation appears to be going on right now in Standard Korean (Silva 2006). Moreover, the kinds of phonological and phonetic phenomena familiar from the study of segmental phonemes—things like neutralization, allophony, morphophonological alternation, and phonotactic regularities—are routinely encountered in tone systems.

Conversely, any useful theory of phonological structure must allow for the fact that phonemic distinctions can occur simultaneously, not just in ordered strings. This also becomes clear in the study of signed languages, where the internal structure of signs often involves the simultaneous occurrence of meaningful movements in different dimensions. Readers with a background in phonology will recognize that this raises the issue of distinctive features, which are often conceived of as the simultaneous components of a phoneme (e.g., voicelessness, labiality). Unfortunately any discussion of feature theory would require another article at least as long as this one. Especially since the relevance of feature theory to music is doubtful, I cannot pursue these issues here.

Beyond the Ordinary Phonemic Distinctions

Second, certain properties of sound structure, including some not traditionally classed as suprasegmental or prosodic, really do need to be regarded as different from ordinary phonemic distinctions. I believe that the key to understanding these features is to focus on the fact that they involve syntagmatic relations between different constituents of a phonological string rather than the paradigmatic oppositions between phonemes. The distinction between the paradigmatic and syntagmatic "axes" of structure was first discussed explicitly by Saussure: a paradigmatic opposition is one between an element that is present in the signal and others that could be present but are not, whereas syntagmatic relations are those between elements standing in construction with one another. For example, in the sentence *I met the man called "the Enforcer"* there is a paradigmatic relation between *"the Enforcer"* and a huge range of other possible names and nicknames the man might have, such as *"Scarface,"* *"the Boss," "Harry,"* and so on. At the same time, there is a syntagmatic relation between *the man* and *"the Enforcer,"* which changes abruptly if we place the same sequence of words in the sentence *The man called "the Enforcer," but received no reply.* Similarly, when we hear *that's tough*, we are in some sense hearing the same string as in *that stuff* (i.e., the same paradigmatic selection of phonemes is involved in both cases), but there are subtle phonetic cues to the fact that the internal organization—the pattern of syntagmatic relations—is different. The classical phonemic approach to such cases was to posit a "juncture" phoneme that took its place in the string at the boundary between the larger units; modern "metrical" and "prosodic" phonology assume rather that the string has a hierarchical structure which affects its realization, though there is little agreement on the details.

In the same way, it seems likely that stress and accent should be regarded as involving syntagmatic relations of strength or prominence between syllables in a structure (Liberman and Prince 1977), rather than absolute (paradigmatic) features such as "secondary stress" that are properties of a given syllable without regard to the structure. Lumping stress together with, for example, tone as "suprasegmental" or "prosodic" obscures this difference (Himmelmann and Ladd 2008; Hyman 2006). However, there is little agreement on how to think about these linguistic phenomena. In music, the distinction can be illustrated by the ways in which the names *do-re-mi* are used to label notes in the Western musical scale: in French or Italian, *do* (or *ut*) refers to the note called C in English or German (approximately 261 Hz for "middle C," and corresponding notes in other octaves). In English or German, on the other hand, *do* refers to the tonic note in a given key (i.e., C or any other note when it serves as tonic). The French/Italian use refers to the paradigmatic value of a note (C is C regardless of what structural role it plays in a given context); the English/German use refers to its syntagmatic value.

I return to the whole issue of syntagmatic structure below.

Around the Edge of Language

Third, and finally, it is important *not* to assign the label "prosody" to *all* peripheral or "paralinguistic" aspects of communication. There is unquestionably good reason to think of certain features of speech as being "around the edge of language" (Bolinger 1964). For example, it is a commonplace observation that we can say "the same" sentence in different ways, and that it is often possible to infer whether a speaker is angry or sad even if they are speaking a language we do not understand. Cultural differences and "display rules" (Ekman and Friesen 1969) notwithstanding, there do appear to be communicative behaviors and displays, such as "raising the voice," that can be regarded as characteristic of the human species and thus treated as accompaniments of language in some narrow sense of that term (e.g., Ohala 1984; Scherer 1995). Some of these (e.g., facial expression) are physically distinct from the speech signal, and we can reasonably idealize away from some (e.g., overall volume or pitch range) in characterizing what is essential about language. Moreover, brain imaging studies (e.g., Belin et al. 2000) have established some of the neurocognitive bases of a distinction between what Belin and his colleagues call "speech" and "voice"; a bilateral "temporal voice area," close to but distinct from the language-related centers in the left hemisphere, is apparently involved in recognizing individual voices and processing nonlinguistic vocalizations like screams, coughs, and laughter independently of any linguistic message that may be involved. There are, thus, sound empirical and theoretical reasons for distinguishing a core phenomenon "language" from associated communicative behavior (see also Scherer, this volume). Nevertheless, many important definitional and theoretical issues remain, in part because it is extremely difficult to draw a clear boundary between the "associated communicative behavior" and features with clear linguistic functions that we might nevertheless want to call "prosodic." Voice pitch is the most obvious problematic feature here, subserving as it does functions ranging from phonemic tone to overall signals of, for example, dominance and emotional arousal.

The point that I would emphasize most strongly is that we should not equate specific linguistic or paralinguistic functions with specific *physically distinct* aspects of the signal. This assumption underlies much past and current research, such as studies that attempt to get at the emotional content of speech by low-pass filtering or other means that render the words unintelligible. Procedures like these presuppose that speech and voice are functionally *and physically* separable, even though appropriately designed experiments make it clear that the affective (emotional, interpersonal, etc.) message of intonation patterns is not independent of linguistic structure. For example, the speaker attitude signaled by rising or falling final pitch in questions depends on whether the question is a yes/no question or a "WH-question" (i.e., a question with a word like *who* or *how*); with yes/no questions in German, final rises are heard as more pleasant and polite than final falls, whereas exactly the opposite is true for WH-questions (Scherer et al. 1984). Even aspects of the signal that are physically distinct (e.g., vocal and facial aspects of sign language) may still be interpreted as part of an integrated whole.

The idea that multiple types of meaning can be carried by a single signal recalls the insight from modern sociolinguistics that "indexical" meanings of astonishing precision can be conveyed by phonetic variation in the realization of phonological categories. The classic case is that attitudes to "off-island" people are reflected in the phonetic details of how natives of Martha's Vineyard pronounce the vowel of *right* and *wide* (Labov 1963). Here there can be no question of parallel channels in the signal; rather, the signal—specifically the phonetics of the vowel—is multifunctional, conveying both lexical choices and speaker attitudes. In some way, the whole signal must therefore be processed via multiple interacting pathways to yield the multifaceted percept; it is not a matter of decomposing the signal into physically separate streams and evaluating the separate streams independently. Recent work by Bestelmeyer et al. (2010) on the neural processing of sociolinguistically meaningful features of speech provides evidence for the idea that multiple brain pathways may be involved: Bestelmeyer et al. showed that various nonlinguistic areas of the brain, including the temporal voice area, were active when British listeners detected the differences between Scottish and Southern English accents in semantically neutral utterances. Yet some sign language research still attempts to analyze the semantic contributions of manual, facial, and whole-body movements separately (e.g., Napoli and Sutton-Spence 2010).

Further evidence against treating prosody or voice as a phonetically separable parallel channel is based on Bregman's findings on "auditory scene analysis" (Bregman 1990). Bregman showed that certain sequences of tones may be partitioned perceptually, so that they are not heard as a single sequence, but as two separate sequences or "streams" occurring in parallel; the "parallel channel" conception of prosody lends itself readily to being thought of in terms of Bregman-style streams. However, Bregman's work consistently shows that when a listener perceives an incoming signal as consisting of separate auditory "streams," it becomes difficult to detect temporal coordination between streams. If prosodic features (or facial features in signed languages) do form a parallel channel, then Bregman's findings are hard to reconcile with a consistent finding of research on both signed and spoken languages; namely that features belonging to what might be thought of as separate streams—such as pitch and segments—are extremely tightly coordinated in time in speech production, and that listeners are extremely sensitive to small differences in temporal coordination (e.g., Arvaniti et al. 1998).

Theoretical Attempts to Address "Prosody" in Linguistics, and Links to Music

In the 1970s and 1980s there was some attempt to incorporate some of the phenomena loosely labeled "prosodic" into phonological theory. "Autosegmental" phonology (Goldsmith 1990) sought to deal with the theoretical problems posed by tone (i.e., by phoneme-like elements that are not strictly linearly ordered but occur simultaneously with others), whereas "metrical" and "prosodic" phonology (Selkirk 1984; Nespor and Vogel 1986) tackled some of the problems of syntagmatic structure and the internal organization of phonological strings. (The fact that autosegmental and metrical/prosodic phonology underwent largely separate developments almost certainly reflects the fundamental difference between syntagmatic and paradigmatic aspects of language structure [cf. the discussion above, "Beyond the Ordinary Phonemic Distinctions"]; in this connection it is relevant to note that rhythm and melody are treated quite separately by Patel [this volume], for what I believe are analogous reasons. For more on the idea of metrical phonology as a theory of syntagmatic structure in phonology, see Beckman [1986, especially chapter 3]). The outcome of these theoretical efforts was rather inconclusive, for reasons I explore in more detail in Ladd (2011). Essentially, both remained wedded to an excessively categorical, segment-based conception of phonetics inherited from the early days of the International Phonetic Association in the late nineteenth century.

Tonal Distinctions and Singing

Much of what has been addressed under the heading of autosegmental phonology concerns the phonology of segments and as such is probably of little relevance to the relation between language and music. This is so despite the fact that this general line of research began with the study of phonological tone. I have already made clear above that tone phonemes are functionally entirely equivalent to segmental phonemes and that it is inappropriate to include tone as part of some peripheral subsystem called prosody—the "music of language," as it were. However, one point is worth comment in this connection: the fact that both tonal distinctions and singing depend on pitch frequently leads to puzzlement over how it is possible to sing in a tone language. For the most part, the solution to this puzzle is that it is no more mysterious than the fact that it is possible to whisper in a language (like English) with voicing contrasts: language is highly redundant, and most phonemic distinctions—including tone and voicing distinctions—have multiple phonetic cues. I am investigating this question together with colleagues as part of an ongoing research project on song in Dinka, a major language of South Sudan (http://www.lel.ed.ac.uk/nilotic/). Dinka has phonemic distinctions of tone, voice quality, and quantity (vowel duration), all of which must interact in singing with the phonetic demands of the music, without seriously endangering the intelligibility of the song texts. It appears that this accommodation involves both the preservation of redundant phonetic cues (e.g., breathy vowels are slightly shorter than nonbreathy vowels, exactly as in speech) and constraints on matches between words and music (e.g., musically short notes are more likely to occur with phonologically short vowels, and melodic rises and falls are more likely to match the linguistically specified pitch movement from one syllable to the next than to contradict it). This last constraint has been documented for Cantonese (Wong and Diehl 2002) and for Shona (Schellenberg 2009).

Hierarchical Structures in Phonology

The original work on metrical phonology (Liberman 1979; Liberman and Prince 1977) drew heavily on ideas from music. Liberman propounded a view in which the essence of prosody is the structuring of the stream of speech into syllables, phrases, and other constituents of various sizes which may have internal structure (e.g., a head or nucleus of some sort). Rather than looking for specific local acoustic cues that mark a boundary or a stressed syllable—a quest motivated by a linear paradigmatic view of sound structure—we should be looking for cues that lead the perceiver to infer structures in which a boundary or a given stressed syllable are present. Clear analogs in music abound (e.g., harmonic cues to meter mean that a note can be structurally prominent without being louder or longer or otherwise acoustically salient), and Liberman explicitly related his conception that stress is a relation between two elements in a structure to traditional notions from the study of music and poetry, such as tension and relaxation (see Lerdahl, this volume).

The idea of metrical phonology gave rise to a flurry of theoretical discussion in the 1980s and 1990s about hierarchical structure in phonology, all of which in one way or another assumed that utterances can be subdivided into phrases which, in turn, can be divided into smaller units (e.g., "clitic groups," "prosodic words," "feet") which ultimately consist of syllables that themselves consist of segments. This discussion went under the rubric "prosodic phonology." The inclusion of "prosodic words" in this list of proposed elements of phonological constituent structure draws attention to a fact that may be unfamiliar to many readers: an important line of evidence for the independence of phonological and syntactic structure has nothing to do with what is conventionally thought of as "prosody" but comes from the word-like phonological status of elements that straddle syntactic boundaries. These include combinations of preposition and article like French *au* /o/ "to the" or combinations of subject pronoun and auxiliary like English *he's* and *it's*. In the latter case, the voicing of the *s* is determined by the voicing of the preceding segment in accordance with ordinary word-internal phonological principles.

The details of the prosodic hierarchy have never been settled and, in my view, the whole line of work remains somewhat inconclusive. There is now general agreement, however, that phonology involves some sort of hierarchical structure *that is distinct from the hierarchical structures of morphology and syntax*. This idea is, of course, a long-standing assumption in traditional metrics—"prosody" in the original sense—whereby poetic lines consist of feet that consist of syllables and can be grouped into couplets, verses, etc.:

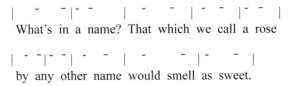

Units like feet and lines are clearly units of sound structure, not grammatical structure. Some feet are grammatical constituents (*a name, a rose*), but others are not (*What's in, -ny o-*). Some line boundaries (e.g., the end of the second line) are sentence boundaries, whereas others (e.g., the end of the first line) are not.

In the 1980s much attention was devoted to matches and mismatches between hierarchical phonological structure and hierarchical syntactic structure, again without any very clear conclusions. One line of research (Selkirk 1984; Ladd 2008) suggests that syntactic structure is recursive while prosodic structure is not and that mismatches may be explained partly by the need to reconcile structures with different formal properties. Another view is Steedman's idea that, in effect, syntactic structure *is* prosodic structure and that much of what is often considered syntactic involves semantics (Steedman 2000). Whatever turns out to be the most useful way of discussing this issue, it is worth noting that many poetic traditions place constraints on the relation between grammatical and prosodic structure. For example, in Latin hexameter a caesura—a grammatically determined pause—was supposed to occur in the middle of the third foot of the line, as in the first line of Virgil's *Aeneid*:

$$- \smile \smile \;|\; - \quad \smile \smile \;|\; - \quad -\;|\; - \quad -\;|\; - \quad \smile \smile \;|\; - \; -$$
Arma virumque cano / Troiae qui primus ab oris

Moreover, traditional metrics has terminology (e.g., *enjambment*) for specific types of mismatches between the two structures. Many song traditions have conventions about how texts are to be set to a melody, which involve similar considerations. I do not believe that the details of the mappings between phonological and syntactic hierarchies are especially relevant to understanding the relation between language and music, but it seems clear that the *existence* of abstract hierarchical structure, in what is superficially a linear acoustic signal unfolding in time, is a key aspect of what music and language share (see Lerdahl, this volume).

I have said nothing about how this structure is processed by listeners, in part because this topic lies well outside my expertise. Mainstream psycholinguistics provides an extensive literature on the processing of *syntactic* structure, including some discussions of how prosodic features affect processing (for a review, see Cutler et al. 1997); there is also literature from within music psychology on how musical structure is apprehended. To my knowledge there is little analogous research on the processing of phonological structure (e.g., poetic meter) in spoken language, in part because it would be difficult to distinguish operationally between phonological and syntactic processing. In any case, I emphasize that the key point of interest here is the fact that both musical and linguistic signals have hierarchical structure, not that there are detailed similarities and differences between them.

Coda: On Duality of Patterning

The question of whether the structures of phonology and syntax are separate and distinct is relevant to the notion of duality of patterning. Duality of patterning is said to be a central design feature of language (Hockett and Ascher

1964) that may be absent from music and, for example, birdsong (see Fitch and Jarvis, this volume), and is thus potentially central to the larger question under consideration in this book. In Ladd (2012), I explore the notion that duality of patterning is not in itself fundamental, but is rather epiphenomenal on other more basic properties of language. The division between meaningful and meaningless elements is less sharp than it appears, and the fact that words are composed of phonemes is arguably just a special case of the pervasive abstract hierarchical structure that is present in language. The argument can be summarized as follows.

Of all Hockett's design features, duality of patterning is the most misrepresented and misunderstood; in particular, it is frequently conflated with or linked to productivity (Fitch 2010). Hockett seems to have regarded duality of patterning as the single most important breakthrough in the evolution of language (Hockett 1973:414), yet he himself was unsure whether to ascribe duality of patterning to the dance of the honeybee (Hockett 1958:574). The basis of the duality idea is Hjelmslev's work on language structure (e.g., Hjemlslev 1953), which assumed the utter separateness of the patterns of organization of the Saussurean signifier (Hjelmslev's "expression plane") and the signified (his "content plane"). Hockett, for his part, emphasized the fact that the elements of phonology are meaningless but are built up into the meaningful elements of morphosyntax. Either way, the distinction appears less sharp when carefully scrutinized. As we just saw, there are many different views of the "syntax-prosody interface," including Steedman's view that prosodic structure *is* syntactic structure; as we also just saw, traditional poetic metrics draw no clear line between prosodic structure (e.g., feet, lines) and syntactic structure (e.g., caesura, enjambment). As for the boundary between the meaningless and the meaningful, it too is blurred in several important ways, including (a) systems of "ideophones" or "mimetics" in many languages (Voeltz and Kilian-Hatz 2001); (b) non-morpheme-based sound-symbolic effects of the sort found in English *glow*, *gleam*, *glisten*, etc.; and (c) morpheme-like formatives with no obvious meaning, such as the Latinate English prefixes *con-* and *pro-*. It is extremely relevant in this connection that recent research on Bedouin sign language, a new language now in its fifth generation, by Aronoff and colleagues has concluded that phonological structure—and hence duality of patterning—has emerged as the language's overall structure has become more complex (Aronoff 2007). This is consistent with the fact that productivity and duality of patterning have regularly been conflated or even confused in the literature.

If this view can be upheld, it begins to make the issue of whether music exhibits duality of patterning look like the wrong question, along with the issue of whether music and birdsong are more like phonology or more like syntax (Yip 2006, 2010; see also chapters by Fitch and Jarvis, Patel, and Lerdahl, this volume). Instead, it suggests that both music and language are evolutionarily built on the ability to assemble elements of sound into complex patterns, and that what is unique about human language is that this elaborate combinatoric system incorporates compositional referential semantics (see Trehub, this volume).

First column (top to bottom):   Sharon Thompson-Schill, Bob Ladd, Peter Hagoort, Peter Dominey, Henkjan Honing, Mark Steedman, Stephen Levinson
Second column: Stephen Levinson, Stefan Koelsch, Peter Dominey, group discussion, Henkjan Honing, Bob Ladd, Sharon Thompson-Schill
Third column: Peter Hagoort, Fred Lerdahl, Mark Steedman, Henkjan Honing, Fred Lerdahl, Stephen Levinson

12

Multiple Levels of Structure in Language and Music

Sharon Thompson-Schill, Peter Hagoort, Peter Ford Dominey,
Henkjan Honing, Stefan Koelsch, D. Robert Ladd,
Fred Lerdahl, Stephen C. Levinson, and Mark Steedman

Abstract

A forum devoted to the relationship between music and language begins with an implicit assumption: There is at least one common principle that is central to all human musical systems and all languages, but that is not characteristic of (most) other domains. Why else should these two categories be paired together for analysis? We propose that one candidate for a common principle is their structure. In this chapter, we explore the nature of that structure—and its consequences for psychological and neurological processing mechanisms—within and across these two domains.

The Syntax of Music and Language

A Cautionary Prelude

A theme which runs throughout this book is the importance of recognizing the diversity of forms that are called "music" or "language," and the dangers of an overly narrow focus. Unfortunately, at this stage in the development of these fields, there are limitations in the data that are available for analysis. Therefore, although we have tried to focus on general principles, much of what we have to say about the structure of language is based on written English language and much of what we have to say about the structure of music is based on scores of Western tonal music. The consequences of this limitation for our understanding of structure can be illustrated with one example from linguistics: In contrast to English, Czech is usually described as a language with "free word order." However, word order in Czech acts as a marker for discourse "information structure," marking topic and comment, given and new—a function which intonation performs in English. English employs free order with respect to information structure and is rigid with respect to argument structure, whereas Czech is the reverse. If modern syntactic theory had started with Czech then we might have called English free word order and Czech rigid. In other words, the mere inclusion of cross-cultural comparisons does not ensure a non-ethnocentric approach to the study of language and music structure, in the same way that the study of spoken language comprehension has been completely shaped by the influence of our alphabet-focused written language system.

Hierarchical Structure

Language and music arrive through the ear and exit through the mouth (or fingers or feet) as actions in time; that is, as a continuous stream of (primarily) acoustic or motor information. But that is *not* the end of the story: we can process an acoustic input by grouping one sound with the next, in the same linear order in which they arrive, with no need to restructure the input. But the situation is quite different. A given linguistic or musical string is best described by a hierarchy, assembled out of elements in the string, in a way that captures meaning relations among the elements beyond their temporal order. Many of the details about the hierarchical organization of elements in music and language (i.e., of syntax) are reviewed by Lerdahl (this volume).

It is important to understand that when linguists talk about hierarchical structure, they distinguish two levels of structure. The most important level of hierarchical structure is the level of *meaning representation*. Such representations are sometimes called "logical forms," because the way linguists write them down often looks like some version of first-order logic, with which it shares such properties as recursivity and compositionality. (This is not to claim that the psychologically real meaning representations look anything like a standard logic.) Such representations are closely related to underlying conceptual relations, standing in a subsumption relation to them, according to which, and at some level, they must be the same for all languages. (The reason for believing this is that it is hard to see how children could learn the language of their culture without access to meaning representations. Since languages differ in their surface forms, and children can learn any of them, this meaning representation must be the same for all.) Since languages differ, in particular, in the order of elements like verbs and noun phrases, we should probably think of logical forms as unordered structures, although of course when we write them down, we will have to choose an ordering on the page. (The fact that so many distinct notations are on offer for essentially the same purpose strongly suggests that the linguists do not have a very clear idea of what the universal logical language really looks like.)

The second kind of structure which linguists talk about is sometimes referred to as *surface structure*. Such structure is a grouping of the elements of the sentence into structural constituents like the English noun phrase and verb phrase. In English and some other relatively rigid word-order languages, such constituents are closely related to such elements of the logical form as predicates and arguments, and can even reasonably be claimed to exhibit a similarly recursive structure. However, other free word-order languages, like Turkish or Latin, do not exhibit any obvious surface constituency and allow considerable freedom for elements that are related at the level of logical form to be nonadjacent in the string. While there is some evidence from phenomena like coordination of some kind of structure, such structure seems to be related to the process of derivation of logical form, rather than to interpretable syntactic structure. In this connection, it is interesting to note that it is commonplace in computational linguistics to regard all surface structure, including that attributed to English, as being epiphenomenal on the process of deriving logical form.

Both in language and in music, elements that are nonadjacent in the sentence may be grouped in the hierarchical meaning representation, as in the case of the "right node raising" construction in (a) language and (b) "interrupted cadences" in music:

(a) I grow, and you sell, beans. = I grow beans, and you sell beans.
(b) II^7 V^7, II^7 V^7, I = II^7 V^7 I, II^7 V^7 I.

One point about these groupings is worth making explicit here: The fact that *grow* belongs with *beans* (and V^7 with I) derives from their interpretation. The interest of long-range dependencies is that they show a similarity in the way in which sentences and music are mapped onto meaning; more specifically, sentence-internal semantics and intramusical meaning (see Koelsch, this volume). Whether or not there are other types of meaning is reviewed by Seifert et al. (this volume). Here, we simply make the point that there is a broad similarity between language and music in the way syntax maps strings onto hierarchical meaning representations.

The fact that language and music have structures with nonadjacent, long-distance dependencies reflects a fundamental property of the domain, not some peculiar quirk of each system: We can consider that the semantic content of language is a form of high-dimensional representation. Language production must find a way to transform this high-dimension representation into a linear or near-linear sequence. In making this transformation, items that were "adjacent" (i.e., related) in the high-dimensional space will become separated, thus creating long-distance dependencies. In other words, dimensional reduction (in this case from many to just one) requires a distortion in some of the relations (e.g., compare a map to a globe). Establishing dependencies, including the long-range variety, is the job of syntax.

There may be differences in the types of hierarchical structures (e.g., how deep vs. how flat they are) between music and language, as well as within music and within language. The organization of the levels in these hierarchies is highly culture-dependent: meaning might be expressed in tree structures in a rigid word-order language, like English, and in non-ordered dependency relations in another. Music always seems to use linear order as the fundamental organizing principle, but there are nonetheless examples of nonadjacent dependencies in music.

A related problem that immediately arises in any structural analysis is how to define a constituent; that is, how to extract a discrete object from a continuous signal (i.e., the leaves of the tree). In language, the smallest object (or event) is usually taken to be the phoneme. In music, one can define an event as a separately sounding drum beat, pitch, or chord. It makes sense to simplify the surface of a musical texture to arrive at syntactically useful events without unnecessary clutter, but defining "syntactically useful" is not without its challenges. Consider the "Ooh-ooh-ooh" in the 1935 recording by Billie Holiday that is followed by the beguilingly rhyming "what a little moonlight can do-oo-oo." If you look at the transcription of this opening text, you will see little more than three "ooh's." However, it seems virtually impossible to reduce them meaningfully to individual sounding notes. Where does one note begin and the other end? By contrast, the meaning of "ooh-ooh-ooh," as sung by Billie Holiday, seems to be carried by the (indivisible) whole rather than in the individual sounds and notes. This example demonstrates how letters, and language in general, fall short. Jackendoff and a number of others dismiss melodic utterances like "oh," "wow," "hmm," and "hey" as relics from the "one-word stage" of language. Others emphasize, more plausibly, that such expressions represent a fundamental and very old aspect of language and music.

A second problem that arises concerns the determination of the maximal domain of the hierarchy: For example, in syntax, it is usually presumed to be the clause, but in Western orchestral music it may extend (with perhaps loss of perceptual relevance) to the movement. At the level of maximal domains, one may have a shift of "currency" as it were: a sentence in conversation may constitute part or the whole of a turn; a turn delivers a speech act, which is something more than a chunk of propositional meaning; it counts as an action, so that it can be responded to by an action, verbal or otherwise (cf. requesting the wine, responded to by providing it).

A further set of problems that one might address, but which in our opinion is fruitless, is the effort to relate specific structural elements of music to those of language. There is little to be gained by endless discussion of whether words correspond to notes or something else. Instead efforts should be directed toward clearly formulating a testable hypotheses (e.g., about brain activity; see Koelsch and Patel, both this volume) based on the assumption that there is hierarchical grouping structure in both music and language.

The Ambiguity Problem

The average length of a sentence in the *Wall Street Journal* is 25 words. If one attempts to compute the meaning of such a sentence from a parser that is drawing from an annotated database of over a million words, one discovers that there are hundreds (if not thousands) of syntactically valid analyses from which to choose. Yet, we can read the *Wall Street Journal* without any difficulty (at least with regard to the parsing problem). The question, therefore, is how humans do this, not only with language but also with music. Here we should distinguish between global ambiguity, as in the sentence *Visiting relatives can be a nuisance*, and local ambiguity, as in a sentence which begins *Have the officers...* (which is locally ambiguous at the beginning of the question or a command) but resolves it at the end, e.g., with *arrived?* or *dismissed!*.

Ambiguity in Music

The mapping between the linearly ordered event sequence and the hierarchically organized structure of music is not one-to-one. There is both local and global ambiguity. A diminished chord that contains the note C, as played on the piano, is locally ambiguous in terms of notation: you can write the minor third above the C as an E flat or as a D sharp. If the next chord is G, then it is a "C diminished" chord and you express the tone as an E Flat. If, however, the piece is in E minor, then it is an "A diminished" chord and you write the tone as a D sharp. In the "whole-tone" scale used by Debussy, music is based on the augmented chord, which is similarly ambiguous to the diminished, but in whole-tone music, it never gets disambiguated. As a result, many of Debussy's pieces are in a sense globally ambiguous as to any tonal center or key. Thus, just as the reader of the *Wall Street Journal* needs to extract one interpretation from many possibilities, so too does the listener to music.

In light of the pervasive ambiguity in music, several principles describe how one interpretation comes to be favored over another. These principles describe transitions in a multidimensional tonal space that is crystalline in its multidimensional regularity and beauty. We briefly digress from the topic of ambiguity to describe this space (another "structure" of music).

The development of pitch space models began in pedagogical eighteenth-century music theory treatises. Part of the cognitively valid solution was intuitively achieved already in the early eighteenth century (Euler 1739), and was developed in computational terms by Longuet-Higgins and Steedman (1971), using a Manhattan city-block distance metric for harmonic distance. In the early 1980s, the experimental psychologist Krumhansl and collaborators established empirically the shape of tonal spaces at three levels of organization: pitch, chord, and key (Krumhansl 1990). In Lerdahl's *tonal pitch space*, he develops a quantitative music theoretic model that correlates with the data and unifies the three levels formally (Lerdahl 2001b). Lerdahl and Krumhansl (2007) successfully tested the tension model that relies in part on the pitch space theory.

When selecting an interpretation of an ambiguous musical event, the principle of the shortest path ranks and selects the most efficient, most probable solution to both tonic-finding (finding the tonal point of reference) and tree-building (forming a hierarchical representation of events). It does this by measuring distances in the tonal space. The attraction component treats the variable tendencies of pitches to move to other pitches. For example, the leading tone is strongly attracted to the tonic, which is more stable and very proximate. The relation is asymmetric: the tonic is only weakly attracted to the leading tone. Attractions are calculated for each voice in a harmonic progression, and the shortest path also enters into the equation. The overall picture is a kind of force field within which pitches and chords behave: the stronger the attraction, the stronger the expectation. As with the principle of shortest path, this procedure can be cast in terms of probabilities. All but the strongest probabilities are pruned away quickly. If an improbable event follows, the experience is a surprise or jolt. The principle of prolongational good form supplements the principle of the shortest path in building a tree structure for event sequences. Prolongational good form encourages, among other things, the characteristic tonic–dominant–tonic (I–V–I) relationship that is at the heart of classical tonal music. Typically, this relationship occurs recursively in the course of a piece.

Thus far we have talked only about pitches and chords, but as Lerdahl reviews (this volume), there are also hierarchical rhythmic structures (although in the case of rhythm, groups do not form dominating-subordinating constituencies). The mapping of rhythm events onto a hierarchy is also ambiguous. As an example, consider a rhythm of three time intervals: 0.26 s to 0.42 s to 0.32 s (i.e., which occur on a continuous timescale). If that rhythm is primed (preceded) by a rhythm in duple meter, it will be perceived by the majority of (Western) listeners as 1:2:1. However, when the same rhythm is preceded by a fragment of music in triple meter, then the majority of participants will perceive it as 1:3:2. Physically, the musical events are identical. However, perceptually and cognitively they are distinct; this turns out to be common in the space of all possible rhythms of a certain duration (Desain and Honing 2003). Research in rhythmic categorization has shown that this process remains open to top-down cognitive influences, either influenced by the preceding musical context (veridical expectation) or by expectations constructed from earlier exposure to music (schematic expectation) (Bharucha 1994; Huron 2006). A consequence of this is that hierarchical analysis based on categorized rhythm (e.g., 16th–8th–16th notes or 1:2:1; cf. Lerdahl, this volume) is dependent on the outcome of the analysis of which it is actually the input.

Ambiguity in Language

As mentioned above, a computational model of parsing based on corpus data shows a remarkable degree of syntactic ambiguity. As Steedman has noted, "the reason human language processing can tolerate this astonishing degree of ambiguity is that almost all of those syntactic analyses are semantically completely anomalous." Thus, the resolution has to come from the interfaces with discourse semantics and world knowledge, but how these interface operations are computationally handled in an incremental processing system is unsolved. (We return to a longer discussion of this unsolved problem at the end of this chapter.)

It is interesting to consider the use in language of "least-effort" heuristics, of the kind that were applied to musical disambiguation in the last section. Similar "shortest move" principles have frequently been proposed in linguistics since Rosenbaum's Minimal Link Condition on control (Rosenbaum 1967), most recently in the form of the economy principles of Chomsky (1995). Such principles were proposed in answer to the question, "Why does the long-distance dependency upon the subject of the infinitival *to go* in the sentence *Mary wants John to go* refer to John's departure, not Mary's, as it would be in the sentence *Mary wants to go*? They claim that it is because, in both cases, the infinitival must choose the closest antecedent noun phrase. The trouble is that there is a small class of verbs, like promise, whose infinitivals target the nonproximal noun phrase, as in *Mary promised John to go*, in which it is Mary's departure that is at stake.

Earlier we said the sole *raison d'etre* of syntax is to build structural meaning representations, and that surface structure should be viewed as a record of the process by which the meanings get built. It follows that the operations of surface syntax give us something on which to hang the Bayesian priors of a parsing model. Such parsing models have to disambiguate quickly, as we do not have the luxury of contemplating thousands of possible structures before we select the most likely one. As Levinson's turn-taking work illustrates (this volume), we have to have disambiguated an utterance before it is finished (in order to plan our own).

However, if the point of a parsing model is to disambiguate, why are both music and language not merely ambiguous systems, but are designed to yield massive numbers of irrelevant parses? Some have argued that grammar-based parsing models really play a very limited role in comprehension and exist primarily to regularize production (Ferreira 2007). That is, syntax might exist for production but be relatively useless for comprehension. Of course, this creates a new puzzle; namely, how do we get to semantics without syntax and what analysis is "good enough" for comprehension, without requiring a full or correct parse?

Instead of questioning whether or not comprehension requires parsing, another approach to this puzzle is to question some of the assumptions that go into the models. The syntax–semantics interface seems to work quite well in human language comprehension, but appears to raise severe problems for machine processing of language. This is partly because extrasyntactic sources of information (e.g., context, world knowledge) are known to play an important part in disambiguation. Still, it is currently difficult to exploit such knowledge in computational systems, due to the lack of adequate semantic representations.

This suggests that the current separation and representation of syntax and semantics may have some fundamental problems.

Ambiguity between Music and Language

An initial question was posed at this Forum: If an archeologist from another era were to come across several music scores, how could this be determined to be musical notation and not fragments of a written language? Alternatively, how could samples of written language be classified as text and not musical notation?

We asked a slightly different question: When a listener hears an acoustic signal (with a certain set of spectral properties), how does the listener decide whether it is language or music? This represents an additional type of ambiguity, which occurs not at the level of mapping events onto objects in a hierarchical structure, but at the level of constructing the events in the first place.

One might object, at this point, and argue that this is not a type of ambiguity that occurs unless one is trying to discriminate between forms of language and music that are closer to the center of the music–language continuum. However, to illustrate that even an English speaker familiar with Western tonal music can have two incompatible interpretations of a single acoustic input, we describe an illusion first observed by Diana Deutsch (1995): A sentence containing the phrase "sometimes behave so strangely" is perceived by a listener as intended (i.e., as a sentence). If this snippet of the phrase is looped repeatedly, perception changes. As semantic satiation takes hold, the phrase begins to sound like music. Indeed, when the phrase is heard again, replaced in the middle of the sentence from which it was removed, it continues to be perceived as song in the middle of an otherwise normal sentence.

This illusion illustrates a point that is both obvious and profound. The obvious part is that both language and music share a sensory channel, which begins at the cochlea. As such, this allows for the possibility of competition between interpreting the signal as speech versus music. Just as a Necker cube cannot be interpreted as being in two orientations at once, so too must the listener select a single interpretation of "sometimes behave so strangely" (and other such phrases that produce this illusion). Moreover, if the interpretation is as language, high pitch is heard as stress or accent, but if the interpretation is as music, high pitch is heard as unaccented pitch (Ladd, pers. comm.). The profound part is that the illusion reveals that there is not sufficient information in the signal itself to discriminate unambiguously between these two interpretations. The "decision" appears to be made on the basis of something that is not acoustic at all; namely, semantics. This is not to say that, on average, the acoustic signal between music and language does not differ (which of course it does) but rather that there is overlap between their distributions. Information is extracted from the acoustic signal that allows one to select the most likely interpretation. Next we consider the streams of information, extracted from an acoustic input, that are used to construct the objects of analysis in linguistic and musical structures.

Streams of Information

Our discussion of structure has thus focused on one particular structural description: the hierarchies that are constructed out of a linearly ordered sequence of events, such as phonemes/words or notes/chords. Here we turn to a different type of structural analysis: a *description of the system* (i.e., the types of information in the signal that are represented) as opposed to a description of the content itself (i.e., the structure of a phrase). This is a nonhierarchical structure that we will henceforth refer to as a *set of streams*, rather than by hierarchically ordered levels (as with syntax). These streams can be partially independent (unlike syntactic trees) and can be used for different functions. In the simple case, a stream of information can be thought of as a distinctive modality or medium of transmission, like manual gesture which accompanies speech. However, even within a modality, there are different types of information that must be extracted from a single acoustic signal.

The Big Three

When linguists refer to types of information, or representations, in language, they are often referring to *semantics*, *syntax*, and *phonology*. We, too, could have approached the question of the structure of the language system with these terms but chose not to for several reasons. First, the distinctions between these three domains in language are not entirely clear. Although they are necessary constructs for linguistic theory, it is not clear that they are distinct kinds of representations or processes; this may explain why efforts to localize "syntax" or "semantics" to a discrete cortical module have by and large been unsuccessful. Second, in the context of this Forum (the relation between language and music), analyzing the semantics, syntax, and phonology of language immediately invites comparisons of each of these subsystems to some counterpart subsystem in music. Just as trying to relate notes to words is fruitless, so too is the attempt to match parts of the language system to parts of the music system. Semantics, syntax, and phonology are functional descriptions, and because the functions of language and music are different, it is hopeless to impose one system's labels on the other system. As we reviewed above, there is, however, overlap in the processing of language and music at levels of analysis where we can make comparisons; namely, the content of the signal. One can usefully ask what kind of information can be extracted from the acoustic signal, what functions each type of information supports (within language and music), and what similarities and differences exist between the types of information used—and the means by which they are integrated—in these two domains.

The Problem of Prosody

One example that illustrates the difference between using a functional label and a description of the information that a function requires is in the domain of prosody. Ladd (this volume) discusses the confusion that has been created by the catchall use of the word prosody to mean "everything left in language when you remove the words." This conventional (albeit recent) use of the word prosody as a functional label has the effect of implying the same function for a whole host of different "suprasegmental" signals. As reviewed in his chapter, the clearest case where this coarse functional grouping is inappropriate is that of lexical tone, which plays a purely phonological role; that is, tone variation in Mandarin has the same function as voice onset time variation in English. (Rather than attributing the error of calling lexical tone prosody to sloppiness, we suspect this is another example of the English-centric influence on linguistics.)

We believe that we can clear up these muddy waters by shifting the emphasis from theoretically laden functional labels to more neutral informational descriptors. There is something in common to "everything but the words"; namely, that unlike the words (i.e., unlike consonants and vowels which require discrimination of very rapid transitions in the acoustic signal), lower-frequency information is used to discriminate lexical tone (in a tonal language), accent (in a nontonal language), intonation and phrasing (at the sentence level), and emotional content. In turn, these discriminations can affect phonology, syntax, and semantics.

We also believe that descriptions of the content of the acoustic signal aids in the interpretation of observed similarities and differences between language and speech. Below, we discuss some of these comparisons (e.g., lateralization differences).

Decomposing the Signal

If we are to make any progress in understanding the different kinds of information which are present in an acoustic signal that support the functions necessary for language and music processing, we must solve the problem of how a single signal is decomposed into its constituent parts. If the parts are, as suggested in the discussion of prosody, distinguished from each other based on their temporal frequency in the signal, all that is needed is a filter with a specific bandwidth. Hagoort and Poeppel (this volume) describe a candidate mechanism for implementing such a system. This mechanism for chunking speech (and other sounds) is based on a neuronal infrastructure that permits temporal processing in general. In particular, intrinsic cortical oscillations at different frequency bands (e.g., theta between 4–7 Hz, gamma > 40 Hz) could be efficient instruments for decomposing the input signal at multiple timescales. Neuronal oscillations reflect synchronous activity of neuronal assemblies. Importantly, cortical oscillations can shape and modulate neuronal spiking by imposing phases of high and low neuronal excitability (cf. Schroeder et al. 2008). The assumption is that oscillations cause spiking to be temporally clustered. Oscillations at different frequency bands are then suggested to sample the speech signal at different temporal intervals. There is evidence that left temporal cortex auditory areas contain more high-frequency oscillations (closely corresponding to the length of the rapid transitions of consonants and vowels) and that right temporal cortex auditory areas are more strongly dominated by more low-frequency oscillations (closely corresponding to the length of syllables). In addition, we know that auditory signals with high-frequency patterns produce more activation in the left temporal cortex, whereas low-frequency patterns produce more activation in the right temporal cortex.

According to this account, information from the acoustic signal (for speech or music) is decomposed into (simplifying here, to only two streams) high- and low-frequency information. If the acoustic stream is speech, high-frequency information can be used to discriminate phonemes, whereas low-frequency information can be used to calculate stress, accent, or (in a tonal language) tone. If the acoustic stream is song or music, there is less information present at high frequencies (music, including song, is slower than speech on average), which might explain the relative prominence of right over left temporal activation during music compared to speech perception. In effect, Poeppel's suggestion is that that the biological constraints (i.e., the oscillations that are part of the "hardware") on speech comprehension may have shaped the properties of our language, capitalizing on these naturally occurring oscillation frequencies to split the signal into what we now call phonemes and syllables.

Music, of course, would be analyzed with the same streams, segregated by the same oscillatory mechanisms. However, in music, quite different timescales operate: low-frequency scales can be associated with the pulse (or tactus) of the music (in the order of 400–600 ms). A cognitive phenomenon named beat induction is commonly associated with brain regions such as the basal ganglia (Grahn and Brett 2007). There are also faster timescales associated with variable durations (i.e., rhythm, associated with activity in the cerebellum; Grube et al. 2010) and expressive timing (minute intentional variations in the order of 50–100 ms). Finally, on a comparable timescale, there are timbral aspects of music, such as the information that human use to distinguish between instruments from the attack of the acoustic signal.

If spreading information across the channels is useful for language comprehension, why then does music not capitalize on this split? One provocative idea is that it used to, under theories that music originated from song (so the high-frequency channel would have used for phoneme discrimination too). Physical constraints may also factor into how long it takes to produce a specific pitch (when, as in music, a rough approximation is not acceptable).

Although this whole argument sounds very tidy, if not simplistic, it has problems. First, the role of intonational marking of phrase boundaries is not immediately clear. A lot of the evidence from recent work on intonation indicates that, to a very considerable extent, intonation works in terms of local pitch events, not holistic contours. How those get produced and interpreted does not seem to fall out from thinking in terms of smaller and larger domains. (Perhaps an intonational boundary marker is similar to a chord change: the new chord itself is a local event, but it defines a new larger stretch of the harmonic structure.)

One potential consequence of decomposing a signal into separate streams so early in processing is that there has to be some mechanism for maintaining coordination between the streams in tight temporal alignment. It is not clear how best to think about this. It seems inappropriate to treat the separate streams as separate in the sense of Bregman (1990): it is well established that syllable-level pitch movements are very precisely aligned relative to the articulatory gestures for consonants and vowels (e.g., Arvaniti et al. 1998), and that (unlike in Bregman's research) listeners are very sensitive to differences in alignment. Further research is required to reconcile the apparent separateness of the streams from the equally apparent unity of the whole signal.

Cross-Modal Streams

Our discussion of streams of information has thus far focused on decomposing the acoustic signal. However, the coordination problem just raised extends to other modalities as well. We will use as our case study of cross-modal integration the case of gesture. Of course, there is one population in which gesture and language occupy a single modality; namely, users of sign language. Interestingly, some of the issues raised above are pertinent to studies of sign language as well, such as the location of information differing in functional relevance on the face.[1]

There are a number of different categories of gestures, including some that are tightly locked to the onset of a word (e.g., indexical gestures, such as pointing to accompany "this one"). Others precede the onset of a word by only a fractional period of time (e.g., an iconic gesture, such as a hammering

---

[1] When thinking about the relation between emotion and language and speech, it may be worth considering whether emotion is more like a modality (e.g., hand movements) or a functional system (e.g., like prosody).

motion when saying "hammer"). Despite these slight timing differences, both the speech signal and the co-speech gestures result in a common representation, and hence have to be integrated in comprehension and jointly planned in production. Frontal cortex seems to play a role in this integration process (Willems et al. 2007). In ordinary conversational settings, even when speakers talk on the phone, speech does not occur in isolation but is embedded in a multimodal communicative arrangement, including the use of hand gestures.

Where Next?

Linking Psycho- and Neurolinguistics with Computational Linguistics

Currently, syntactic processing in computational linguistics and psycholinguistics or neurocomputational models of human sentence processing have almost entirely diverged, and pay almost no attention to each other in the literature. The reason is that parsing on the scale that is required to read the newspaper requires very large grammars, with thousands of rules, and that very large grammars engender huge ambiguity, with hundreds and even thousands of valid derivations for each sentence. Accordingly, state-of-the-art parsers use statistical models of derivations, which allow a probability to be assigned to any analysis of a given sentence. Such statistical models are derived from human-labeled sentences in a "treebank."

These models are rightly despised by psychologists and, in fact, are quite weak, working at about 90% accuracy on a number of measures. They are weak because we do not have enough labeled data on which to train them (and we never will). Psychologists know that the parser draws on all levels of linguistic representation for disambiguation, incrementally and at high speed, including semantics, and even referential context and logical inference. One might expect that they would be able to offer an alternative to the computationalists.

Unfortunately, the models that psycholinguists currently embrace seem to be predicated on the assumption that you sometimes have at most two alternatives, and propose strategies such as "best-first" which have no chance at all of coping with realistically large levels of ambiguity. Moreover, all semantic theories on offer from linguists exhibit highly complex mappings to syntactic structure, involving processes like "covert movement," with which it is very hard to do effective inference. Part of the problem, as Levinson's work shows, is that whatever the real semantics is, the markers found in real languages do not seem to be transparent to the primitive concepts of the presumed universal semantics. Indeed, it seems possible that there is no such primitive that is transparently marked in any attested language.

The open problem that our discussions raised is this: Can we provide psychologically plausible parsing mechanisms that will work at the scale of real human language processing, and can we identify a "natural" semantics and conceptual system which supports inference that can be smoothly integrated with them?

Musical Dialog

Levinson (this volume) stresses that a central aspect of language is the online dialogic interaction: while one is listening to a partner, one is simultaneously predicting the partner's upcoming words and preparing one's own subsequent utterance. This requires three concurrent but distinct linguistic representations to be managed at one time: (a) the representation of what the speaker is saying, (b) the representation of what the listener believes this speaker will say (which we know from Levinson is often the same as the former but which still must be tracked), and (c) the representation of the listener's planned utterance which is being prepared. These representations, of course, are being crafted in the face of all of the ambiguity just discussed. From a processing perspective, this parallel comprehension and production is very distinct from "passive" listening to a narrative. There are conversations in some musical forms, although there may be some differences between the demands they create in music and in language, depending on whether one musician is creating a plan in response to the music of the other musician or not. Just as conversational turn-taking has processing implications in language, the study of musical dialog may constrain hypotheses about musical production and comprehension. It may also be fruitful to examine the extent to which there are common mechanisms in language and music for managing these interactions.

Physical Constraints

Above we suggested that properties of the events that compose language and music reflect biological constraints, such as the proposed correspondence between the length of a syllable and the theta oscillation. There are other kinds of biological constraints that one might also usefully consider. For example, in language, the need to breathe constrains the length of a prosodic utterance. The musical analog is the "phrase," which has about the same length as a prosodic utterance. A typical phrase ends in a cadence (formulaic way of achieving closure within a given musical style). The analog is close because much, if not most, music is sung. The breathing constraint applies not only to the voice but also to wind instruments (woodwinds or brass), though not to string or percussion instruments. Nevertheless, music played by winds usually follows the same phrase lengths as vocal music.

Of course, a tempting correspondence such as this may prove to be misleading. Although there may be some evolutionary link between breath groups and linguistic phrases, as between opening and closing jaw gestures and syllables, there is a lot you can say about syllable structure for which basic oscillatory jaw movement is essentially irrelevant. The same may be true of phrases and breath groups.

Action

Cross et al. (this volume) discuss the problem of the evolution of language and music. To their thoughts, we add an insight that comes from our consideration of the structure of language and music. As described above, both language and music require mapping between a linear sequence and a hierarchical structure, which may involve grouping events that are nonadjacent but which are, instead, connected by their meaning. So, too, elementary actions can be sequenced to form compound actions or plans that have a hierarchical structure. Sensorimotor planning of this kind, including planning that involves tools, is not—unlike language—confined to humans. The mastery of the relevant action representation, including tool use and effects on other minds, also immediately precedes the onset of language in children.

Planning in nonlinguistic and prelinguistic animals is striking for two reasons: (a) the ability to sequence actions toward a goal in abstraction from their actual performance and (b) the fact that this ability is strongly dependent on an affordance-like association between the immediate situation and the objects that it includes, and the actions they make possible. The close relation between planning with tools and other minds and language suggests that this kind of planning provides the substrate onto which language can be rather directly attached, both in evolutionary and developmental terms.

The field of artificial intelligence has created computationally practical representations of actions and planning. It might be interesting to consider how linguistic syntax and semantics (as well as aspects of nonsyntactic speech acts related to discourse, discussed by Levinson, this volume) could be derived from such representations.

Part 4: Integration

13

Neural Mechanisms of Music, Singing, and Dancing

Petr Janata and Lawrence M. Parsons

Abstract

Song and dance have been important components of human culture for millennia. Although song and dance, and closely related music psychological processes, rely on and support a broad range of behavioral and neural functions, recent work by cognitive neuroscientists has started to shed light on the brain mechanisms that underlie the perception and production of song and dance. This chapter enumerates processes and functions of song that span multiple timescales, ranging from articulatory processes at the scale of tens and hundreds of milliseconds to narrative and cultural processes which span minutes to years. A meta-analysis of pertinent functional neuroimaging studies identifies brain areas of interest for future studies. Although limited by the scope of the relatively small number of studies that have examined song-related processes, the meta-analysis contributes to a framework for thinking about these processes in terms of perception–action cycles.

To date, most research on song has focused on the integration of linguistic and musical elements, whether in the binding together of pitch and syllables or, at a more temporally extended level, in the binding of melodies and lyrics. Often this has been with an eye toward the question of whether melodic information facilitates retention of linguistic information, particularly in individuals who have suffered a neurological insult. Evidence supports a view that merging novel linguistic and melodic information is an effortful process, one that is dependent on the context in which the information is associated, with a social context aiding retention. This observation provides an intriguing link between cognitive processes in the individual and the social as well as cultural functions that song and dance ultimately serve.

Functions of Song

Social and Communicative

While music and music-making often exist independently of song and dance in modern-day practice, the interrelationships of song, dance, music, and language have a long history that is tied to the communicative functions and cultural rituals which have accompanied human evolution. Both song and dance provide a basis for the coordinated action of multiple individuals and, as recent studies examining the link between synchronized and cooperative behaviors show (Kirschner and Tomasello 2010; Valdesolo et al. 2010; Wiltermuth and Heath 2009), may serve to reinforce social bonds. When embedded in the context of ritual, song and dance offer a mechanism for defining group unity with respect to external agencies, whether these are other groups, animals, or imaginary personae (see Lewis, this volume). Various other adaptive social functions can be hypothesized, such as regulation of emotion, pair bonding, management of ambiguous social situations, and coalition signaling. Song and dance thus provide a form of communication (emotional, narrative, and dramatic) that can be juxtaposed against, or combined with, the propositional communicative function of language.

Song in Relation to Instrumental Music

Although in traditional cultures, music, dance, and narrative drama are very often blended in group participation activities, in other contexts song and dance often seem to be regarded as a distinct activity from music. The tendency to distinguish song from music may be due to a tacit bias, at least in Western cultures, to make the broad term "music" synonymous with instrumental music. Although a considerable amount of music exists that has no explicit relation to song or dance, there are reasons to be skeptical about any sharp dividing lines among the domains of song, dance, and instrumental music. For example, attentive listening to music, even highly reduced forms (e.g., brief monophonic rhythms) engages, to some extent, aspects of the action systems of the brain (e.g., Brown and Martinez 2007; Grahn and Brett 2007; Janata et al. 2002a). These and other studies suggest that even the perception of music alone can engage sensorimotor processes related to actions for dancing or for performing the music. Thus while sensorimotor processes may be an obvious focus to examine song and dance, and perceptual processes for instrumental music, a broader understanding of musical experiences may be better construed in terms of a perception–action cycle (Figure 13.1).

The Perception–Action Cycle

The perception–action cycle has been proposed as a cognitive and neural framework in which to interpret interactions with the environment (Neisser 1976; Arbib 1989), and it has considerable appeal for framing our experiences with music (Overy and Molnar-Szakacs 2009; Janata 2009a; Janata and Grafton 2003; Loui et al. 2008). Broadly, the perception–action framework postulates

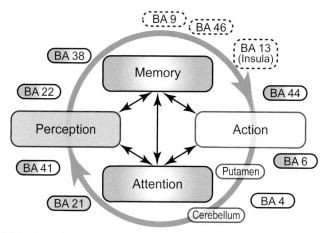

Figure 13.1 Most interactions with the environment can be thought of in terms of a perception–action cycle. This diagram emphasizes that perception and action processes need to be considered in relation not only to each other but also with respect to a variety of mnemonic and attentional processes. The Brodmann area (BA) anatomical labels arise from the meta-analysis described in this chapter and are loosely arranged according to their commonly accepted relationships to perception, action, and memory processes. Gray and white labels imply primary association with perceptual and action processes, respectively; areas denoted by dashed lines are typically associated with other processes (e.g., memory). The shading gradients highlight regions that the meta-analysis found to be involved in both perception and production tasks. Missing from this figure is a depiction of how emotional processes relate to the perception–action cycle. Presumably, the fluency with which the perception–action cycle operates in any given behavioral instance determines, to a large part, the emotional consequences with greater positive affect associated with greater fluency.

that action plans are hierarchically organized in the frontal lobe and that perception is hierarchically organized in the temporal, occipital, and parietal lobes (Fuster 2004). In this context, it is assumed that action is an integral part of musical experience, rather than associated solely with performance of a song, dance, or musical piece. This view is consistent with the notion that music is embodied but that over time it has become elaborated or abstracted into purely instrumental music.

One might ask where the action aspect of listening to a piece of instrumental music lies. The answer relies on the notion that actions can be either overt or covert, with covert action recruiting brain areas responsible for sequencing and implementing actions (e.g., Langheim et al. 2002; Shergill et al. 2001). More abstractly, the formation of expectations for specific future events (i.e., the direction of attention toward specific locations on different feature dimensions at specific moments in time) recruits cognitive control and premotor brain areas and may therefore be regarded as a form of action (Schubotz 2007). In terms of music, singing along in one's mind is a prime example of a covert action.

A variety of auditory imagery studies have found that both sensory and motor areas are recruited during this type of mental activity, both for musical material with and without lyrics (Zatorre et al. 1996; Halpern and Zatorre 1999; Zatorre and Halpern 2005; Janata 2001; Navarro Cebrian and Janata 2010a, b). Even though overt action may not be an explicit requirement in a musical situation, music that creates an urge to move and results in subtle but measurable movements, such as head bobs and toe taps (Janata et al. 2012), is indicative of pleasurable engagement of the perception–action cycle.

Whereas perception and action may form a basic level of functional organization, attention and memory play important modulatory roles in facilitating smooth functioning of the perception–action cycle (Figure 13.1). For example, one must listen attentively to others in an ensemble in order to adjust one's vocal production appropriately to either blend in or stand out from the group. Note that attention and memory are not "add-ons" to the action–perception cycle. A key idea of the cycle is that both parts are constantly interacting in service of our current goals. When listening to a familiar song, the notes and words guide the retrieval of the upcoming notes and words and allow one to sing along accurately. It is likely that the context of any particular musical interaction, along with additional factors such as expertise, will determine the complement of perception and action brain areas that is recruited. In laboratory settings and limited interactions (e.g., judgments about discrete brief stimuli), it is possible that the perception–action cycle is not engaged as strongly as under more natural circumstances in which the musical interaction engages an individual across an extended period of time and sequence of musical events. In this regard, it is important to appreciate the cyclical aspect of the perception–action cycle. One consequence of this is that successive moments in time are bound together to create a coherent ongoing experience. In addition, the primacy of either perception or action, and the associated causality of one process or the other in determining the experience, is greatly diminished. Although attention may be biased toward the perceptual or action side of the cycle depending on the context or goal of the moment, musical experiences have at their core both perception and action components.

Characterizing extended musical experiences as embedded within a perception–cycle with the attendant blurring of perception and action does not mean, however, that component processes of the experience cannot be identified and ascribed primarily perceptual or motor functions, or causality. For example, hearing a word can cause an associated lyric and melodic fragment to be remembered and produced, overtly or covertly, which may then cause retrieval of further words and notes, or associated autobiographical memories. Participating in a call and response emphasizes clear perception and action phases within a broader musical experience. Consequently, it might be expected that the specific recruitment of the different brain areas in Figure 13.1 depends on the specific context and goals and the relative coordination of perception and action processes in time.

Mechanisms of Song

The analysis of the neural structures that support song may be thought of in terms of functions and component mechanisms across a range of timescales (Figure 13.2). For example, vocal production is a specific instance of sensorimotor control. At the most elemental level, there is the requirement of binding phonemic, syllabic, and pitch information. Sensorimotor control shapes intonational and timbral aspects of the vocalization, both at the level of individual notes as well as phrase length shapes and forms (gestures; prosodic contours). Such spectrotemporal aspects of song have semantic (affective) connotations that broaden the space of semantic implications that can be achieved through the lyrics alone. Further, psychological and neural mechanisms of song need to be regarded both from the perspective of a "producer"' (who is also perceiving in the context of vocal control or audience feedback) and a "perceiver" (who may also be producing if singing along covertly).

Neural Mechanisms Supporting Song

Evidence for the way in which song is organized in the brain comes from a number of cognitive neuroscience techniques, including brain lesion methods (patient and transcranial magnetic stimulation studies) and neuroimaging methods including electroencephalography (EEG; Nunez and Srinivasan 2005) and derived event-related potentials (ERPs; Luck 2005) as well as functional magnetic resonance imaging (fMRI) and positron emission tomography (PET; Buckner and Logan 2001). Each of the different methods used to infer how any given psychological function is instantiated in the brain has advantages and disadvantages which collectively define the inferential scope afforded by each method.

A Meta-Analysis of Song-Related Perception and Production

Relatively few studies have examined the neural basis of song and given its multifaceted nature, as depicted in Figure 13.2, it would be difficult for any single study to cover the entire space of variables and interactions. As a starting point for discussing the neural basis of song, we collated most of the neuroimaging literature on song to conduct two meta-analyses emphasizing song perception and song production. A central challenge in performing meta-analyses of neuroimaging data is to summarize appropriately the data reported across the studies, given that inferences might be made about thousands of different brain regions. We used a method of activation likelihood estimation (GingerALE; Eickhoff et al. 2009; Laird et al. 2005), which generates brain maps that indicate the likelihood for any given brain region to be active under

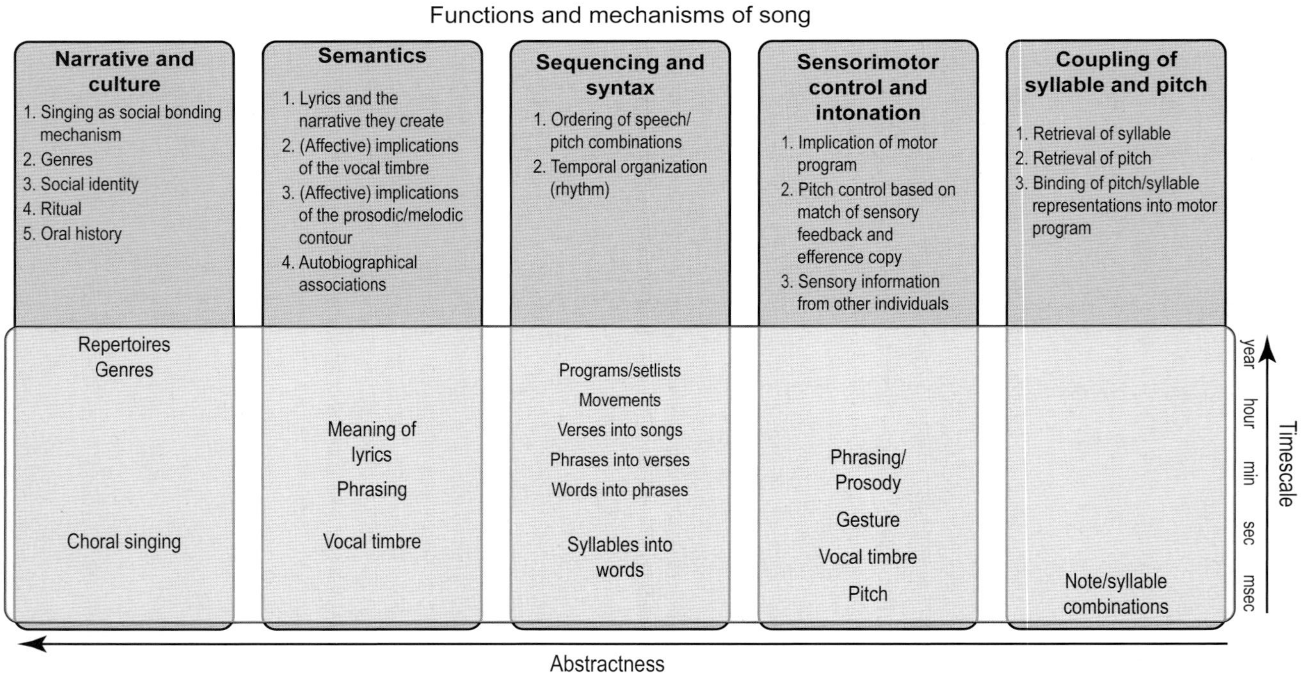

Functions and mechanisms of song

Narrative and culture	Semantics	Sequencing and syntax	Sensorimotor control and intonation	Coupling of syllable and pitch
1. Singing as social bonding mechanism 2. Genres 3. Social identity 4. Ritual 5. Oral history	1. Lyrics and the narrative they create 2. (Affective) implications of the vocal timbre 3. (Affective) implications of the prosodic/melodic contour 4. Autobiographical associations	1. Ordering of speech/pitch combinations 2. Temporal organization (rhythm)	1. Implication of motor program 2. Pitch control based on match of sensory feedback and efference copy 3. Sensory information from other individuals	1. Retrieval of syllable 2. Retrieval of pitch 3. Binding of pitch/syllable representations into motor program

Timescale: year — hour — min — sec — msec

Repertoires Genres	Meaning of lyrics	Programs/setlists Movements Verses into songs Phrases into verses Words into phrases		
	Phrasing		Phrasing/Prosody	
		Syllables into words	Gesture	
Choral singing	Vocal timbre		Vocal timbre Pitch	Note/syllable combinations

Abstractness

Figure 13.2 Schematic of many of the functions and mechanisms associated with song. Different timescales and levels of abstractness associated with the different mechanisms are emphasized. For example, the combining of pitches with syllables is a very concrete operation that occurs on a short timescale. By contrast, development of repertoires and genres occurs across much longer timescales and contributes to mechanisms supporting culture.

conceptual preconditions (e.g., listening to a musical stimulus). This provides a quantitative tool with which to identify consistent observations across the study corpus.

The view of neural circuitry provided by such meta-analyses is limited by the conceptual space delineated by the particular collection of tasks and contrasts from the studies included in the analysis. The current space of tasks and contrasts was primarily focused around the comparatively low-level processes associated with pitch–syllable coupling and the control of intonation. Criteria for selecting a study for the meta-analysis were that (a) the study be a PET or fMRI study and that (b) lists of centers of activation corresponding to specific contrasts be presented in tabular form. Particularly in regard to studies of perceptual processing of musical material, the dividing line between studies that explicitly used songs or song-relevant material (e.g., sung syllables) and those that used stimuli with which participants could conceivably sing along covertly was more difficult to draw. Further meta-analyses that explicitly incorporate the broader spectrum of music-related processes will be necessary to dissociate song-specific processes properly.

Perception

Eleven contrasts from six separate studies comprising 151 activation loci were entered into the meta-analysis of brain areas primarily involved in the perception of song elements and song, but also included melodic perception tasks:

1. Listening to sung trisyllabic words (Schön et al. 2010).
2. Listening to Chinese percussion music for which a verbal code had been learned (Tsai et al. 2010).
3. Adapting to repeated lyrics in brief sung phrases (Sammler et al. 2010).
4. Listening to familiar versus unfamiliar songs containing lyrics (Janata 2009a).
5. Listening to melodies with which subjects had been recently familiarized in the context of an auditory imagery experiment relative to unfamiliar melodies (Leaver et al. 2009).
6. Passively listening to melodies while performing melodic and harmonic discrimination tasks (Brown and Martinez 2007).

Figure 13.3 illustrates that consistent activations (blue and magenta clusters) were identified bilaterally along the superior temporal plane comprising the Brodmann areas (BA) 21, 22, and 38. Additional foci in which activations were observed in at least two different studies were located in the left Broca's area (BA 44) and right lateral prefrontal cortex (BA 9/46). Although individual studies included in this meta-analysis reported more extensive activity in premotor areas of the lateral and medial prefrontal cortices (Brown and Martinez 2007; Janata 2009b), activations of premotor areas do not invariably occur

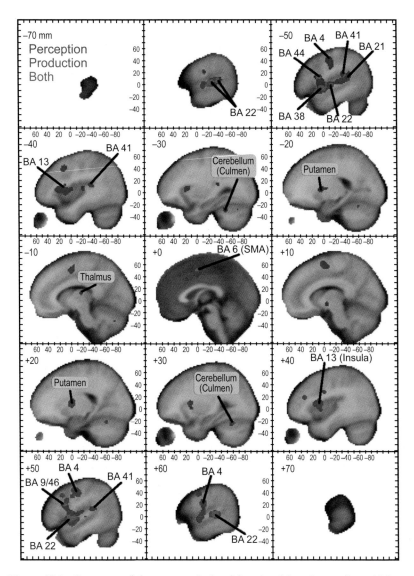

Figure 13.3 Summary of the meta-analysis of functional imaging studies which examined elements of song perception and production. Overall, these studies emphasized basic mechanisms of syllable–pitch coupling. The Brodmann area (BA) labels point to activation likelihood maxima identified by the GingerALE algorithm. BA assignments were determined using the Talairach Daemon, a database that maps each region of the brain to its associated BA. The BAs identified in this analysis were the source of the anatomical labels in Figure 13.1.

during tasks that are primarily perceptual in nature and that are unlikely to require or support covert vocalization processes.

Production

The complementary meta-analysis of production tasks that required either covert or overt singing of pitches (monosyllabic or polysyllabic) incorporated 24 contrasts from 14 studies comprising 326 activation foci. Tasks ranged from sensorimotor control over single pitch–syllable combinations (Brown et al. 2004; Perry et al. 1999; Riecker et al. 2002; Wildgruber et al. 1996; Zarate and Zatorre 2008) to the production of short musical phrases (Leaver et al. 2009; Brown et al. 2006b; Callan et al. 2006; Halpern and Zatorre 1999; Kleber et al. 2007; Riecker et al. 2000; Saito et al. 2006; Wilson et al. 2010).

Not surprisingly the production tasks recruited an extensive set of motor as well as sensory areas (Figure 13.3, red and magenta clusters). Most consistent among these (identified in more than 7 studies) were loci in the auditory cortex bilaterally (BA 41, 22), the supplementary motor area (SMA; BA 6), the anterior insula/frontal operculum (BA 13), and motor cortex (BA 4). Together, this set of regions may be regarded as a core sensorimotor network underlying simple song. Although a number of contrasts across some of the studies identified multiple foci in the cerebellum (Brown et al. 2004, 2006b; Callan et al. 2006; Perry et al. 1999; Riecker et al. 2000; Tsai et al. 2010), those foci were more distributed, perhaps reflecting task heterogeneity (Callan et al. 2007). Activations in the basal ganglia and thalamus were reported in five of the studies. Contrasts to identify brain areas, in which activity was greater during singing than during speaking, were performed in only three studies (Callan et al. 2006; Saito et al. 2006; Jeffries et al. 2003). The meta-analysis of these studies identified one locus common across all three studies in the right mid-insula ($x = 43$, $y = -6$, $z = 11$). Although the function of this area is unclear, such lateralization may be consistent with an hypothesis that speech production is predominantly lateralized to the left insula whereas song production, particularly perhaps in nonmusicians, is lateralized to the right insula (Ackermann and Riecker 2004). Note that hemispherically lateralized activity can vary considerably, depending on linguistic and musical expertise (Wong et al. 2004; Ohnishi et al. 2001; Bever and Chiarello 1974; Groussard et al. 2010).

What Can the Meta-Analysis Tell Us?

The meta-analysis identified several brain areas that were found to be active both in task contrasts that emphasize perception as well as production (magenta blobs in Figure 13.3). Results of this type blur the distinction between action and perception when trying to assign only one or the other role to these brain areas. In addition, such results highlight the fact that linking a particular area primarily to either a sensory or motor function does not rule out a role for that area in the complementary function.

Meta-analyses such as this one are limited in that they serve to identify putative brain areas (nodes) in brain networks that support a behavior; however, they do not satisfy the objective of specifying how information flows between these nodes. Information flow between the nodes is necessarily constrained by the fiber tracts that connect them. Therefore, principles that govern the underlying neuroanatomy are of great relevance to any computational models. The organization of corticostriatal circuits affords an example. Projections between cortical areas, the basal ganglia, and the substantia nigra are topographically organized in a ventral to dorsal manner mirroring a gradient of motivated behavior that spans from emotion-related processes in the ventral striatum and orbitofrontal cortex through explicit action plans and action monitoring in the dorsal anterior cingulate and dorsolateral prefrontal cortex (Haber and Knutson 2009). The basal ganglia are critically involved in motivated learning, formation of action repertoires and habits, as well as sequencing and timing of actions (Graybiel 1998; Buhusi and Meck 2005; Graybiel 2008). When examined in the context of sensorimotor integration tasks that vary in complexity, including musical tasks, the basal ganglia become engaged when sequence complexity reaches moderate levels (Janata and Grafton 2003). Aside from more obvious connections to the processing of temporal structure in music (Chapin et al. 2010; Grahn and Rowe 2009), the basal ganglia are engaged by anticipatory (dorsal striatum) and reward (ventral striatum) processes when experiencing highly pleasing self-selected music (Salimpoor et al. 2011). Together, these observations point toward the possibility of developing a model of how songs are learned and become an engrained part of individuals and, by extension, groups of individuals and cultures.

Assuming that the dynamics of information flow among the brain areas identified in the meta-analysis mediate the reception of sensory signals and production of motor signals, or at least provide a descriptor of that mediation, questions arise as to the coordinate frames in which the sensory and motor signals exist and how they are mapped to each other. For example, how are the sensory and motor representations of the sung syllable *la* represented independently and combined, perhaps within the motor-sensory interface for speech comprising lateral prefrontal and temporal lobe areas (Hickok and Poeppel 2007)? Electrophysiological techniques are perhaps best suited to answer these questions because of their excellent temporal resolution. Thus, computational models of interactions between multiple nodes in a network as well as more direct measures of neural activity from nodes within the network are needed to understand more fully the workings of the perception–action cycle in specific behavioral contexts.

Any model of the perception and production of song that is based on a specific set of loci, such as those identified in Figure 13.3, must ultimately take into account the range of functions and behaviors in which those foci are implicated. In other words, while the meta-analysis indicates involvement of cortical areas and implies engagement of cortical-striatal-thalamic loops and the cerebellum, as might be expected for moderately complex sensorimotor behaviors, the specificity of this particular constellation of cortical and subcortical areas for song-related behavior can be ascertained only when the likelihood of engaging these areas in other tasks using other materials can be assessed. The capacity for meta-analyses within large-scale databases should aid in this goal (Yarkoni et al. 2011).

As already noted, the majority of the neuroimaging studies of song have emphasized perception and production of simple song elements (combinations of pitches and syllables) or the production of (highly) familiar melodies with or without lyrics. Consequently, the neural substrates of many important functions of song listed in Figure 13.2 remain uncharacterized. Nonetheless, preliminary identification of a core song circuit provides regions of interest for further analyses and computational modeling efforts. For example, an examination of the way in which hearing and covertly singing along with a familiar song contributes to an autobiographical memory experience might involve a functional connectivity analysis using loci in the core song circuit as seed regions. A similar approach might be taken to understand how lyrics retrieved during production of familiar but unrehearsed songs are fed into the song circuit.

With the coarse meta-analysis as a backdrop, we now examine a number of the levels outlined in Figure 13.2 in slightly greater detail, referring to pertinent behavioral studies in normal and patient populations.

Lower-Level Representations

Melody

An indispensable component of song is melody. Evidence is emerging that the sizes of melodic intervals in music may be influenced by pitch patterns in speech (Patel et al. 2006) and that they share affective connotations (Curtis and Bharucha 2010). Considerable evidence has accumulated to indicate that the superior temporal lobe in the right hemisphere plays an important role in extracting and representing the sequential pitch relationships among notes— the melodic contour (Johnsrude et al. 2000; Patterson et al. 2002; Warrier and Zatorre 2004). The role of the right superior temporal gyrus (STG) in melodic processing is of particular interest in relation to the region's role in the processing of prosody (Grandjean et al. 2005), and consequently for the binding of speech and pitch information. The hemispheric lateralization of melodic processing is not strict, however, in that anterior regions of the STG in both hemispheres respond more strongly to melodic stimuli than nonmelodic stimuli (Griffiths et al. 1998; Schmithorst and Holland 2003). These observations are relevant to the discussion below of the anterior temporal lobe's role in binding lyrics and melodies together.

Timbre

A feature specific to song in particular is the timbre of the human voice. The location of voice-selective areas along the superior temporal sulcus (STS) (Belin et al. 2000), anterior to primary auditory areas, is roughly the same as those areas involved in integrated representation of syllables and pitch (Sammler et al. 2010). These areas appear to be separate from the more posterior areas, which are associated with representation of other timbres and other environmental sounds (Menon et al. 2002; Lewis et al. 2004; Halpern et al. 2004). If supported by further studies, an anatomical distinction between vocal and other timbres within superior temporal regions could be important for understanding differences in experiencing of vocal and instrumental music. The observation of increased activity more anteriorly in the temporal pole during vocal harmonization (Brown et al. 2004) suggests further functional neuroanatomical dissociations for distinct musical processes.

Binding of Representations: Coupling of Syllable and Pitch

Perhaps the most basic process associated with singing is that of producing the right syllable or word at the right pitch (note) at the right time. Several questions arise: Are the representations of words and notes separate or integrated? If separate, where are the independent codes stored, and where are they combined during perception or during production? If they are integrated, where are those integrated representations stored?

Elements of song perception have been contrasted with speech and vocal melody perception in a study by Schön et al. (2010) in which tri-syllabic words were combined with three-note melodies. All three stimulus types activate broad regions of the anterior-to-mid temporal lobe, with a tendency toward activation of the STG on the right and STS on the left. Sung words were better able to activate these areas than either the vocalized melodic fragments or spoken words alone, though it is possible that the greater activation for sung words were due to summation of separate melodic and linguistic responses rather than a true interaction that would be indicative of an integrated representation.

The binding of melodic and linguistic information has also been examined using an adaptation paradigm in which novel lyrics were paired with novel melodies, and one or the other dimension was adapted by keeping the information along that dimension constant (Sammler et al. 2010). Two regions displayed evidence of integrated melody–lyric representations: the mid-STS in the left hemisphere and a left hemisphere premotor area along the precentral gyrus. Interestingly, patterns of behavioral interference effects in the processing of bi-syllabic nonwords sung on two-tone melodic intervals show that vowels are more tightly bound with pitch information than are consonants (Kolinsky et al. 2009).

To test for automatic binding of pitches and vowels, the mismatch negativity (MMN) ERP response can be exploited. The MMN is generally regarded as a measure of automatic processing in the auditory system because it arises in response to deviant stimuli when the focus of attention is not explicitly oriented toward the stream within which the deviants are embedded. When embedded in streams of pitch–vowel combinations, the pitch–vowel combinations that deviate in both dimensions do not elicit a larger MMN than the response to a deviant on either dimension alone. If there were separate representations, a larger response would be expected for the combined pitch–vowel deviants. Thus, observing a comparable magnitude response is more consistent with integrated pitch–vowel representations (Lidji et al. 2010), at least at the level of the auditory cortex.

Taken together, the evidence suggests that pitch and syllable information are integrated within secondary auditory areas within the middle sections of the STG/STS (BA 22), though as we will see below, melodies and lyrics are not bound together obligatorily.

Sensorimotor Integration: Vocal Control, Intonation, and Amusia

Closely related to the issue of whether and how pitch and syllable representations are bound together is the issue of vocal control (i.e., achieving the proper pitch). Vocal control has been examined in several studies (Brown et al. 2004; Perry et al. 1999; Zarate and Zatorre 2008) and, as indicated in the meta-analysis, both motor and auditory areas are recruited during vocal production (for a review, see Dalla Bella and Berkowska 2009).

In addition, amusia is relevant to both the perception and production aspects of song (Ayotte et al. 2002; Loui et al. 2008, 2009). Amusia, commonly referred to as tone-deafness, has largely been considered a perceptual deficit in which the ability to follow changes in pitch (i.e., melodies) is impaired (Peretz et al. 2002). However, impaired singing ability, while not diagnostic of amusia by itself, is commonly regarded as part of the package of the melody-processing deficits observed among tone-deaf individuals, suggesting that both sensory and motor impairments may contribute to amusia.

Recent studies utilizing the technique of diffusion tensor MRI (DTI), in combination with behavioral testing, support such a characterization. DTI provides a means of estimating, *in vivo*, the anatomical connectivity between pairs of regions in the brain, and thereby facilitates anatomical comparisons between groups of individuals. Of particular relevance to issues of sensorimotor integration in speech and music is a fiber tract called the arcuate fasciculus. The arcuate fasciculus connects the pars opercularis region of the inferior frontal gyrus (BA 44 in Figure 13.3; Kaplan et al. 2010) with superior temporal areas (BA 22 in Figure 13.3). This connection is believed to provide auditory feedback control of speech (Tourville et al. 2008), and likely forms the backbone of the auditory–motor interface for speech as part of the dorsal stream for language

(Hickok and Poeppel 2007; Saur et al. 2008). (For an extended discussion of a putative parallel between speech and song production based on Guenther's model, see also Patel chapter, this volume.) The long direct segment of the arcuate fasciculus, which connects Broca's area in the lateral frontal lobe with Wernicke's area in the posterior temporal lobe, is thicker in the left hemisphere in the normal population; this observation is used to explain left hemisphere language dominance (Catani and Mesulam 2008). Interestingly, the part of the arcuate fasciculus that terminates in the STG is disproportionately thinner in the right hemisphere of amusics (Loui et al. 2009). Its smaller size in amusic individuals, in combination with both perception and production deficits, provides support for the idea that perception and production of melody and song are tightly coupled and may have a genetic and developmental basis.

Less direct evidence for the link between production and perception of song and melody comes from a number of mental imagery studies which find activation of premotor and secondary auditory areas during the imagination of notes in familiar and unfamiliar melodies (Leaver et al. 2009; Halpern and Zatorre 1999; Kleber et al. 2007; Zatorre et al. 1996; Kraemer et al. 2005; Janata 2001; Navarro Cebrian and Janata 2010a).

Song: Combining Melody and Text

Although single notes can be sung, songs arise when sung speech sounds (generally words) are sequenced together to form melodies and associated phrases. As with single pitch–syllable combinations, the question arises whether melodies and texts are kept in separate stores or whether they exist in an integrated store. Although the relationship of lyrics to melodies has not been studied extensively using brain imaging methods, a number of behavioral studies in healthy controls and (primarily aphasic) patients provide insights both into the issue of representational independence as well as likely neural substrates. Study materials have included both familiar and novel songs. On the perceptual side, tasks have typically involved recognition and recall paradigms; for action, production tasks are used from which the number of correctly produced words and notes is tallied.

There is a fair amount of evidence to suggest that text and melodies are strongly associated, though probably not fully integrated. Evidence in support of integrated representations comes from two studies in which a list of unfamiliar folk songs was played to participants, who then had to identify matches of the previously heard items (Serafine et al. 1984, 1986). If the items did not match exactly, participants had to identify whether the melody or words were old. The test items comprised several combinations of melodies and text: (a) old melody, old text; (b) old melody, new text; (c) new melody, old text; (d) old melody, old text, but in a novel combination; (e) new melody, new text. Subjects recognized the original combinations of text and melodies well and, importantly, text and melody were better recognized when they occurred in their original context. This effect held even when attention during encoding was directed explicitly toward the melodies. The surprising finding in these studies was that the binding between melodies and texts was very rapid and not related to semantic content of the text (Serafine et al. 1986).

Within familiar songs that have been heard across a significant portion of the life span, there is also evidence of tightly coupled melody–text representations. In a behavioral study using a priming paradigm, in which either a brief segment of the melody or text of a familiar song was used as a prime for another fragment of the melody or text, each modality (melody or text) effectively primed not only itself, but also the other (Peretz et al. 2004a). This was true both when the location of the prime in the song was before the target as well as when it followed the target.

In a different study (Straube et al. 2008), a nonfluent aphasic with a large left hemisphere lesion was able to produce a larger number of words to familiar songs when singing than when speaking the lyrics. This individual was unable to learn novel songs, however, even though he was able to learn the melody component quite accurately or was able to produce more words of the novel lyrics when sung to an improvised melody. Even after learning a novel melody to perfection by singing it with the syllable *la*, he was unable to sing new lyrics associated with the melody. In fact, he was able to speak 20% of the novel lyrics when the requirement to sing was removed. Such results indicate that the association between lyrics and melodies requires time to strengthen in order for the representation to appear integrated. Moreover, integration appears to depend on left hemisphere mechanisms, although the role of the right hemisphere in a comparable study remains to be explored.

Even after an integrated representation has had a chance to form, the basic representations of melody and lyrics and/or the routes to production may be maintained separately. This is seen in patient case studies in which production of lyrics from familiar melodies, whether spoken or sung, is impaired, while the ability to sing those melodies using the syllable *la* is preserved (Peretz et al. 2004a; Hebert et al. 2003). The anterior temporal lobes are likely important for representing text and melody components of familiar songs, albeit with lateralization of the text to the left hemisphere and the melodies to the right, at least in nonmusicians. Following left temporal lobe resection there is impaired recall of text, whether spoken or sung, whereas following right temporal lobe resection there is impaired melody recognition (Samson and Zatorre 1991). However, either the left or right hemisphere resections are linked to impaired recognition of the melodies when they are sung with new words. In combination, this suggests that integrated representations can develop in either hemisphere. Another case study of interest suggests that integrated representations of melody and lyrics develop partly in the left hemisphere (Steinke et al. 2001). The patient in this study, who has a lesion of the anterior temporal lobe and lateral prefrontal cortex in the right hemisphere, was able to recognize familiar melodies that are normally associated with lyrics (but are presented without lyrics) but failed to recognize familiar instrumental melodies.

The behavioral data in healthy individuals largely support a view that, at least initially, lyrics and melodies are independently stored and that their integration during recall is an effortful process. An apparent superiority of recall (after <15 minutes) of lyrics from an unfamiliar song, which were originally heard in a sung rather than spoken version, disappears when the presentation rate of the two modes is matched (Kilgour et al. 2000). There is no difference in recall of lyrics that are heard as sung during encoding and then either sung or spoken during retrieval (Racette and Peretz 2007). Repetition of a relatively simple melody appears to be an important factor for achieving accurate recall of text (Wallace 1994).

Nonetheless, the ability of melody to facilitate encoding of novel lyrics becomes apparent in compromised populations under certain circumstances. When visually presented lyrics of forty unfamiliar songs are accompanied by either spoken or sung accompaniment, healthy controls show no difference between spoken and sung forms during subsequent old/new recognition testing of visually presented lyrics, whereas individuals with probable Alzheimer's disease show a benefit of the sung presentation during encoding (Simmons-Stern et al. 2010).

The issue of melody–text integration is of particular interest from a clinical perspective. The observation that patients with expressive aphasia can accurately sing but not speak the lyrics of familiar songs can be taken as evidence of integrated melody–text representations (Yamadori et al. 1977; Wilson et al. 2006). It also suggests that word representations might be accessed more easily via song. Melodic intonation therapy for rehabilitation of speech following stroke is founded on this idea (Norton et al. 2009). For example, in a Broca's aphasic, when novel phrases were sung to the tunes of familiar melodies over a period of several weeks, the patient's ability to speak those phrases was better than his ability to speak matched phrases that were either unrehearsed or had been rehearsed by with rhythmically accented repetition (Wilson et al. 2006). The ability to couple melody and speech may depend on laterality of the damage, however, as extensive right hemisphere damage impairs production of both familiar and novel lyrics, whether sung or spoken, more severely than it does the production of the melody when sung on *la* (Hebert et al. 2003). Even in patients with left hemisphere damage, there is little benefit for word production of either familiar or novel phrases when the words are sung (Racette et al. 2006). However, if the words were sung or spoken along with an auditory model, as in a chorus, production of the words improved in the sung condition. Overall, the representation of melodies and associated lyrics in memory and the ability to access those representations appears to depend on encoding circumstances (e.g., whether the subject simply perceives or produces sung lyrics, the amount of time allowed for consolidation, and whether an auditory model is present).

Semantics and Episodic Memory

A question largely ignored in the studies mentioned above is the degree to which the words that come to be integrally associated with melodies are non-propositional as opposed to semantically laden. Lyrics in well-rehearsed songs are automatic and non-propositional, which raises the question of the degree to which songs interact with the brain's systems for acquiring and maintaining semantic information (Racette et al. 2006). In this connection, a case study documents a dissociation between (a) declarative memory systems for music and associated lyrics and (b) typical declarative memory. The study reports a densely amnesic individual who was able to learn new accordion songs following the onset of amnesia, and could readily produce, when presented with the song titles, both the music and the lyrics of songs in her repertoire (Baur et al. 2000). Nonetheless, the lyrics were semantically empty, in that her ability to perform semantic tasks (e.g., categorization) was as bad for the words in the songs as for matched words not encountered in the songs. It would seem, therefore, that the lyrics become part of a musical representation independently of their semantic connotations.

The semantic content of lyrics can nonetheless be processed while vocal music is heard. When sung melodies are terminated with semantic, melodic, or combined incongruencies, behavioral and ERP responses to those incongruencies suggest independent processing of the melody and the lyrics (Besson et al. 1998; Bonnel et al. 2001). However, when a richer tonal (harmonic) context is created by sung four-part harmonies, harmonic anomalies interact with semantic relatedness of target words in the lyrics as measured with a lexical decision task (Poulin-Charronnat et al. 2005). Even when not sung, harmonic anomalies influence the semantic processing of concurrently presented linguistic material (Steinbeis and Koelsch 2008b). Even though there is evidence that under some circumstances music engages the semantic representations systems in the brain (Koelsch et al. 2004), the exact role of lyrics in the web of associations between melody, harmony, and semantics remains unclear.

The semantic component of songs gains relevance when the construction of personal and group narratives is considered. Songs provide an excellent vehicle for joint action (singing together) which in turn provides a mechanism for reinforcing social bonds and memories of those social bonds, as evidenced by the content of music-evoked autobiographical memories (MEAMs; Janata et al. 2007). Functional neuroimaging measurements of responses to song excerpts regarded as familiar and/or memory evoking show that an extensive sensorimotor network is recruited during familiar songs, presumably reflecting covert singing along with the song of either the lyrics or the melody (Janata 2009b). Songs that were memory evoking engaged further brain areas, including the dorsal medial prefrontal cortex—areas shown to be important for autobiographical memory and, more generally, self-representations (Svoboda et al. 2006; Northoff and Bermpohl 2004; Northoff et al. 2006). Viewed in the broader context of social cognition, the medial prefrontal cortex is involved in thinking about oneself and others, and, by extension, the building of social scripts and narratives (Van Overwalle 2009). A separate analysis of the MEAM fMRI data utilized a structural description of each song, specifically the time-varying pattern of each song's movement through tonal space (Janata et al. 2002a). Here, several tonality-tracking areas were identified, notably the ventral lateral prefrontal cortex and dorsal medial prefrontal cortex. The responses within these areas to both the structural aspects of the music (transpiring on a faster timescale) and the overall autobiographical salience suggest a role of these areas for associating music with personal extramusical information. Associating music with memories, however, is unlikely to be the sole province of the medial prefrontal cortex, given the many areas that exhibit tonality tracking when listening to memory-evoking music and the many areas associated with representing various facets of autobiographical memories (Svoboda et al. 2006). When considering the ability of songs to serve as retrieval cues, it is interesting to note that extremely brief (300–400 ms) excerpts are effective for retrieval of meta-information (e.g., artist and title) related to the song (Krumhansl 2010). Thus, songs appear to exist in memory, not only in terms of their melodies and lyrics but also other representations, such as timbre.

Song in the Context of Domain-General Brain Mechanisms

As suggested at the outset, an examination of the perception and production of song can be usefully situated in a domain-general framework of the perception–action cycle and associated mnemonic, attentional, and affective processes. Such a framework suggests that there are parallels and relationships between music, song, language, and other cognitive processes beyond those examined in the studies reviewed here. Here we mention two areas that we think are particularly relevant to cross-modal associations between music (instrumental or vocal), language, movement, semantics, and affect.

Gesture and Prosody

One class of communicative movements comprises gestures, and it is useful to consider the relationship of the time-varying structures of gestures, dance movements, melodic contours, and prosody in both language and song. A generalized concept of gesture (i.e., a modality-independent concept of gesture) may have a characteristic timescale of 2–3 seconds during which a trajectory is drawn. In music, this trajectory may be characterized by a brief melodic pattern in which the articulation of individual notes and the intervals and co-articulation between notes define the gesture as well as its semantic and affective implications. There is considerable evidence for a cross-modal 2–3 s integration window in which percepts, thoughts, and actions can be arranged hierarchically and bound to support the "subjective present" (Poeppel 1997, 2009). A broad range of human actions and interactions, including music and language, exhibits organization on this timescale (Nagy 2011).

Sequences of gestures create longer phrase-length structures (spanning several seconds) that allow the semantic and affective implications of individual gestures to be elaborated into an evolving narrative. The right lateral temporal lobe, given its role in the processing of melody and prosody, is of particular interest.

Temporal Organization in Singing, Music, and Dancing

The structuring of meaning across contours, gestures, and phrases necessitates the traversal of multiple timescales (Figure 13.2). For example, vowels are short relative to syllables, which are short relative to words, which are short relative to phrases. Similarly, a single melodic interval is short relative to melodic fragments or melodic contours that span a phrase, and chords are short relative to chord progressions. To achieve a coherent experience that encompasses the shorter elements and the longer structures, it seems necessary to recruit brain areas that are capable of organizing information across multiple timescales and/or coordinate activity among brain areas specialized for representing information on different timescales. In this regard, Broca's area in the ventrolateral prefrontal cortex, spanning from BA 47 across BAs 45 and 44 to BA 6, is of interest given its role in the hierarchical organization of abstract rule or movement associations. More anterior regions (BA 45) show stronger phasic responses to superordinate action chunk boundaries, whereas simple action chunk boundaries and individual actions drive activity more posteriorly in BA 44 and BA 6, respectively (Koechlin and Jubault 2006). Hierarchical action plans maintained in the lateral prefrontal cortex presumably combine with timing information from medial premotor areas (e.g., SMA) to generate temporally and hierarchically structured behaviors (Koechlin and Jubault 2006). The involvement of Broca's area in language production, processing of musical syntax, abstract rule–action relationships in general, and connections to the lateral temporal lobe, makes it a likely substrate for controlling the integration of music, language, and meaning across multiple timescales. The homologous region (BA 44) in the right hemisphere has been implicated in dancing (Brown et al. 2006b).

Time in music and dance is a complex phenomenon composed of pattern (local timing within a few notes or time beats), phrasing (variation in time to reflect boundaries between subsequences of a passage), tempo (rate per unit time, which may increase or decrease), meter (a higher-order organization of cyclical time, with strong and weak accents), and overall duration. Each temporal feature activates a different subset of functional mechanisms linked to executive, higher cognitive, emotion, timing, auditory, and somatosensory function (Parsons et al., submitted), befitting their different computational requirements. Moreover, these subsystems are distinct from those in the perception of melody, harmony, and timbre (e.g., Peretz and Kolinsky 1993; Platel et al. 1997; Foxton et al. 2006).

Examples of such results can be seen in an fMRI study of musicians discriminating phrased and unphrased musical passages where increased activity was observed for phrase boundaries in bilateral planum temporale, left inferior frontal cortex, left middle frontal cortex, and right intraparietal sulcus (Nan et al. 2008). Likewise, in an MEG study of musicians, imposing an implicit meter (waltz or march) onto presented isochronous click sequences, localized sources in globus pallidus (lentiform nucleus), claustrum, fusiform gyrus, left dorsal premotor (BA 6), and inferior frontal gyrus (BA 47) (Fujioka et al. 2010). Studies that examined beat perception report activity in auditory areas (superior temporal cortex), premotor areas (pre-SMA/SMA and dorsal premotor cortex), and basal ganglia regions, especially in the putamen (Grahn and Brett 2007; Chen et al. 2008a, b; Grahn and McAuley 2009; Grahn and Rowe 2009; Kornysheva et al. 2010).

Dance

A variety of developmental and maturational aspects in rhythm processing have been studied (see Trehub, this volume). Studies suggest that infants can perceive beat information (Winkler et al. 2009b) and that other categories of metric information are processed in early childhood (Hannon and Johnson 2005). Moreover, studies of infants, typically those which look at time paradigms, indicate an early ability (a) to detect isochronous versus non-isochronous patterns of tones (Demany et al. 1977), (b) to generalize rhythm across auditory sequences (Trehub and Thorpe 1989), (c) to detect brief temporal gaps in short tones (Trehub et al. 1995), and (d) to detect fine variations in duration (Morrongiello and Trehub 1987) and tempo (Baruch and Drake 1997). Moreover, infants can detect categories of unique rhythms on the basis of implicit metrical structure (Hannon and Trehub 2005a).

Other studies confirm a close connection between movement rhythm and musical pattern perception (Hannon and Trehub 2005b; Phillips-Silver and Trainor 2005). In addition, it has been observed that young infants (5–24 months) make more rhythmic movements to regular sounds (such as movement) than to speech, and the degree of coordination with the music is associated with displays of positive affect (Zentner and Eerola 2010a). Other recent studies have also confirmed the tendency in young children to synchronize their drumming patterns spontaneously with others in a social setting (Kirschner and Tomasello 2009). Several behavioral studies of musical rhythm processing in adults have focused on the role of limb and body movement in integration (determination) with musical meter (Phillips-Silver and Trainor 2007, 2008;

Trainor et al. 2009; Phillips-Silver et al. 2010), making explicit the embodied aspects of engagement with music.

Genetic analyses of variation in polymorphisms of genes for the arginine vasopressin receptor 1a (*AVPR1a*) and serotonin transporter (*SLC6A4*) in a sample of dancers, competitive athletes, and nondancers and nonathletes suggest that the capacity for human dance and its relation to social communication and spiritual practice has an extended evolutionary history (Bachner-Melman et al. 2005; Ebstein et al. 2010). In traditional cultures, dance is often a kind of gesture language that depicts an instructive narrative of events and personalities of collective importance for a group (see Lewis, this volume). Dance's essential collective character is seen in rhythmic patterning, usually to a musical beat, which engenders a nearly unique interpersonal synchronization and coordination. Rhythmic synchronization (entrainment) is rare among nonhumans, but may be present in some form in certain birds (e.g., parrots) (Schachner et al. 2009; Patel et al. 2009). While dance can be an individual display, group dancing during ceremonial rituals in traditional cultures generally serves a cooperative function, reinforcing a group's unity of purpose (Lewis, this volume).

The neural, behavioral, and evolutionary basis of dance has only very recently attracted scientific attention. Brown et al. (2006b) have researched the neural basis of entrainment, meter, and coordinated limb navigation as well as the evolutionary context of dance (e.g., its adaptation as a measure of fitness in potential mates; Brown et al. 2005). Mithen (2005) has looked at dance as part of a gestural protolanguage, possibly preceding recent language evolution in humans. The neural correlates of learning to dance have been examined by Cross et al. (2006) and dance expertise has been scrutinized by Calvo-Merino et al. (2005). (Cf. the discussion of mirror neurons by Fogassi, this volume, and behavioral studies of synchronization by ensembles of dancers; Maduell and Wing 2007).

Using production of dance as an illustration, Brown et al. (2006b) examined three core aspects of dance: entrainment, metric versus nonmetric entrainment, and patterned movement. Entrainment of dance steps to music, compared to self-pacing of movement, is supported by anterior cerebellar vermis. Movement to a regular, metric rhythm, compared to movement to an irregular rhythm, implicated the right putamen in the voluntary control of metric motion. Spatial navigation of leg movement during dance, when controlling for muscle contraction, activated the medial superior parietal lobule, reflecting proprioceptive and somatosensory contributions to spatial cognition in dance. Moreover, right inferior frontal areas, along with other cortical, subcortical, and cerebellar regions, were active at the systems level. The activation of right inferior frontal areas (BA 44) suggests their involvement in both elementary motor sequencing and in dance, during both perception and production. Such findings may also bear on functional hypotheses that propose supralinguistic sequencing and syntax operations for the region broadly defined as Broca's region and its right homologue. In addition, consistent with other work on simpler, rhythmic, and motor-sensory behaviors, these data reveal an interacting network of brain areas that are active during spatially patterned, bipedal, and rhythmic movements and are integrated in dance.

As with song, we can distinguish between dance forms in which movement patterns are tightly coupled to the rhythmic structure of the music for time spans ranging from the basic metric level to phrase-length patterns (e.g., many forms of ballroom or folk dance) and those in which many of the movements appear less tightly coupled to specific musical elements (e.g., modern dance). This distinction may lead to a discussion of the meaning of movement shapes and forms, on the one hand, or the production of action patterns which do not produce sound but are nonetheless an integral part of the overall rhythmic structure of the music–dance experience.

Beyond Performances of the Individual to Pairs and Ensembles

In their natural form of paired dances, group dancing, and duet and chorus singing, song and dance involve socially coordinated joint action (Keller 2008; Sebanz et al. 2006), just as does speech in its natural form of conversation (Stivers et al. 2009; Garrod and Pickering 2004; Levinson, this volume). Some of these activities have been explored in behavioral and observational studies, for example, in duet piano performance (Keller et al. 2007), flamenco (Maduell and Wing 2007), and conversation (De Ruiter et al. 2006; Sacks et al. 1974). Recently, some of these processes have been examined with neuroscientific techniques: simultaneous dual fMRI studies of musicians performing duet singing (Parsons et al., in preparation), fMRI studies of duet percussion performances (Parsons et al., in preparation), and EEG studies of pairs of guitarists (Lindenberger et al. 2009). To achieve further progress, however, future studies will need to examine the common and distinct coordinative social, cognitive, affective, and sensorimotor processes in comparisons among paired dancing, paired conversation, and duet singing.

14

Sharing and Nonsharing of Brain Resources for Language and Music

Aniruddh D. Patel

Abstract

Several theoretical and practical issues in cognitive neuroscience motivate research into the relations of language and music in the brain. Such research faces a puzzle. Currently, evidence for striking dissociations between language and music coexists with evidence for similar processing mechanisms (e.g., Peretz 2006; Patel 2008). The intent of this chapter is to initiate a dialog about how such conflicting results can be reconciled. Clearly, a coherent picture of language–music relations in the brain requires a framework that can explain both kinds of evidence. Such a framework should also generate hypothesis to guide future research. As a step toward such a framework, three distinct ways are put forth in which language and music can be dissociated by neurological abnormalities, yet have closely related cortical processing mechanisms. It is proposed that this relationship can occur when the two domains use a related functional computation and this computation relies on (a) the same brain network, but one domain is much more robust to impairments in this network than the other, (b) the interaction of shared brain networks with distinct, domain-specific brain networks, or (c) separate but anatomically homologous brain networks in opposite cerebral hemispheres. These proposals are used to explore relations between language and music in the processing of relative pitch, syntactic structure, and word articulation in speech versus song, respectively.

Background: Why Study Language–Music Relations in the Brain?

From the standpoint of cognitive neuroscience, there are at least five distinct reasons to study language–music relations. Such studies are relevant to (a) comparative research on the neurobiological foundations of language, (b) debates over the modularity of cognitive processing, (c) evolutionary questions surrounding the origins of language and music, (d) the clinical use of music-based treatments for language disorders, and (e) educational issues concerning the impact of musical training on linguistic abilities, such as reading and second language learning.

In terms of comparative research on language, a common strategy for exploring the neurobiological foundations of human behavior is to examine animal models. For example, the brain mechanisms of decision making are increasingly being studied using animal models (Glimcher 2003). This comparative approach, however, has certain limitations when it comes to human language. Animals communicate in very diverse ways, but human language differs in important respects from all known nonhuman systems. To be sure, certain abilities relevant to language, such as vocal learning and the perception of pitch, are shared with other species and have good animal models (e.g., Bendor and Wang 2005; Jarvis 2007b; Fitch and Jarvis, this volume). Furthermore, some of the basic auditory cortical mechanisms for decoding and sequencing complex sounds may be similar in humans and other primates or birds, making these species important model systems for studying the evolutionary foundations of speech perception (Doupe and Kuhl 1999; Rauschecker and Scott 2009; Tsunada et al. 2011). Nevertheless, some of the core features of language, such as complex syntax and rich semantics, have no known counterparts in the communication systems of other species. It is notable that even with enculturation in human settings and intensive training over many years, other animals attain very modest syntactic and semantic abilities compared to humans (Fitch 2010). This sets certain limits on what we can learn about the neurobiology of high-level language processing via research on other brains.

Yet within our own brain lies a mental faculty which offers the chance for rich comparative research on high-level language processing, namely music. Like language, music involves the production and interpretation of novel, complex sequences that unfold rapidly in time. In terms of syntactic structure, these sequences have three key properties in common with language (cf. Lerdahl and Jackendoff 1983):

1. Generativity: novel sequences are built from discrete elements combined in principled ways.
2. Hierarchy: sequences have a rich internal structure, with multiple levels of organization.
3. Abstract structural relations: elements fill distinct structural roles depending on the context in which they occur. (For example, in English, the word "ball" can be the subject or object of a verb depending on context, and in tonal music the note B-flat can be the tonic or leading tone in a musical sequence, depending on context).

In terms of semantics, musical sequences, like linguistic sequences, can convey complex and nuanced meanings to listeners (Patel 2008:300–351). While the nature of musical meaning and its relationship to linguistic meaning is a topic of active discussion and debate (e.g., Antović 2009; Koelsch 2011b and responses in the same journal), the critical point is that the meanings of musical sequences are rich and varied. This stands in contrast to animal songs (e.g., birdsong, whale songs), which convey simple meanings such as "this is my territory" or "I'm seeking a mate." Thus, both in terms of syntax and semantics, music offers a rich domain for the comparative neurobiological study of language.

Turning to issues of modularity, language–music research is relevant to a persistent question in cognitive science (Fodor 1983; Elman et al. 1996; Peretz and Coltheart 2003; Patel 2003): To what extent are the functional computations underlying particular mental faculties (e.g., language or music) unique to those domains? For example, are certain aspects of linguistic syntactic processing supported by brain mechanisms shared by other types of complex sequence processing? Instrumental (nonverbal) music provides an excellent tool for studying this question, as it is a nonlinguistic system based on hierarchically organized, rule-governed sequences. Hence, if instrumental music processing and linguistic syntactic processing share neural mechanisms, this would inform modularity debates and (more importantly) provide new, comparative methods for studying the neurobiology of syntactic processing.

In terms of evolution, there is a long-standing debate over the role that music or music-like vocalizations played in the evolution of language. Darwin (1871) proposed that human ancestors sang before they spoke; that is, they had a nonsemantic "musical protolanguage" which laid the foundation for the evolution of articulate speech. His contemporary, Herbert Spencer, disagreed and argued that music was a cultural elaboration of the sounds of emotional speech (Spencer 1857). Spencer foreshadowed thinkers such as James (1884/1968) and Pinker (1997), who argue that our musicality is a byproduct of other cognitive and motor abilities, rather than an evolved trait which was selected for in evolution. This debate is very much alive today. Current proponents of the musical protolanguage theory (e.g., Mithen 2005; Fitch 2010) have updated Darwin's ideas in the light of modern research in anthropology, archeology, linguistics, and cognitive science. As noted by Fitch (2010:506), "The core hypothesis of musical protolanguage models is that (propositionally) meaningless song was once the main communication system of prelinguistic hominids." Fitch proposes that the neural mechanisms underlying song were the precursors of phonological mechanisms in spoken language, a view that predicts "considerable overlap between phonological and musical abilities (within individuals) and mechanisms (across individuals)." Fitch argues that such a prediction is not made by lexical and gestural protolanguage hypotheses. In other words, Fitch regards music–language relations in the brain as important evidence for resolving debates over the evolution of language (see Fitch and Jarvis, this volume; for a critique see Arbib and Iriki, this volume).

In terms of clinical issues, there is growing interest in the use of music as a tool for language remediation in certain developmental and acquired language disorders (e.g., dyslexia and nonfluent aphasia, Goswami 2011; Schlaug et al.

2008). To determine which types of language problems might respond to musical training, and to explore how such benefits might take place, it is important to have a basic understanding of how music and language processing are related at the neurobiological level. This same kind of basic understanding is important for educational questions surrounding the benefits of music training to linguistic abilities in normal, healthy individuals. A growing body of evidence suggests that musical training enhances certain linguistic skills in normal individuals, including reading abilities, second language learning, and hearing speech in noise (Moreno et al. 2009; Moreno et al. 2011; Slevc and Miyake 2006; Parbery-Clark et al. 2009). To optimize such effects and, ultimately, to help influence educational policy, we need a neurobiological understanding of how and why such effects occur.

Nonsharing: Neuropsychological and Formal Differences between Language and Music

As cognitive and neural systems, music and language have a number of important differences. Any systematic exploration of music–language relations in the brain should be informed by an awareness of these differences. This section discusses three types of differences between music and language: differences in acoustic structure (which are reflected in differences in hemispheric asymmetries), neuropsychological dissociations, and formal differences.

Beginning with acoustic structure, speech and music tend to exploit different aspects of sound. Speech relies heavily on rapidly changing spectral patterns in creating the cognitive categories that underlie word recognition (e.g., phonetic features, phonemes). Much music, on the other hand, relies on more slowly changing spectral patterns to create the cognitive categories that underlie music recognition (e.g., pitch classes, pitch intervals). Animal studies have revealed that different areas of auditory cortex show preferences for rapidly versus slowly changing spectral patterns (Tian and Rauschecker 1998; Rauschecker and Tian 2004), and fMRI research has pointed to distinct regions within the anterior superior temporal cortex for speech and musical instrument timbres (Leaver and Rauschecker 2010). Furthermore, a large body of evidence from patient and neuroimaging research suggests that musical pitch perception has a right-hemisphere bias in auditory cortex (e.g., Zatorre and Gandour 2007; Klein and Zatorre 2011). For example, Figure 14.1 shows the anatomical locations of lesions producing different kinds of pitch-related musical deficits, based on a review of the neuropsychological literature (Stewart et al. 2006).

Zatorre et al. (2002) have suggested that the right-hemisphere bias in musical pitch processing reflects a trade-off in specialization between the right and left auditory cortex (rooted in neuroanatomy), with right-hemisphere circuits having enhanced spectral resolution and left-hemisphere circuits having

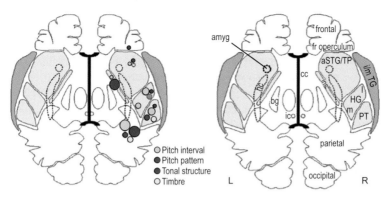

Figure 14.1 Left: A transverse view of the human brain showing the anatomical emphasis of lesions producing different types of pitch-related musical deficits as well as musical timbre deficits, based on a literature review by Stewart et al. (2006). The image has been thresholded: the presence of a colored circle corresponding to a particular function in a region indicates that at least 50% of studies of the function implicate that region (hence some studies in which left-hemisphere lesions were associated with musical pitch or timbre problems are not represented). The size of the circles indicates the relative extent to which particular brain areas contribute to each deficit. Right: Anatomical guide to the regions shown in the left image: amyg = amygdala; aSTG = anterior superior temporal gyrus; bg = basal ganglia; cc = corpus callosum; fr = frontal; hc = hippocampal; HG = Heschl's gyrus; ic = inferior colliculus; i = inferior; ins = insula; l = lateral; m = medial; PT = planum temporale; TG = temporal gyrus. Reprinted from Stewart et al. (2006) with permission from Oxford University Press.

enhanced temporal resolution.[1] Another proposal which could account for the right-hemisphere bias in musical pitch processing is Poeppel's (2003) "asymmetric sampling in time" hypothesis, which proposes that the right auditory cortex has a longer temporal integration window for acoustic analysis than the left hemisphere, which would predispose it to fine-grained pitch analysis. (These proposals are largely equivalent from a systems-theoretical standpoint, since frequency resolution and time resolution are inversely related.) The right-hemisphere bias for musical pitch processing is not just present in auditory cortex, but is also observed in frontal and temporal regions to which the auditory cortex is connected via long reciprocal fiber tracts. For example, in Tillmann et al.'s (2003) fMRI study of harmonic processing, Brodmann areas 44 and 45 in both hemispheres showed an increased response to harmonically unexpected (vs. expected) chords, with a stronger response in the right hemisphere. The right-hemisphere bias for musical pitch processing stands in sharp contrast to the well-known left-hemisphere bias for many aspects of language

---

[1] Note, however, that when pitch is used to make linguistic distinctions (e.g., between the words of a tone language), pitch perception is often associated with significant left-hemisphere cortical activations. Zatorre and Gandour (2007) propose a framework combining bottom-up and top-down processing to account for the differences in cortical activation patterns observed when pitch plays a musical versus a linguistic role in communication.

processing, such as language production and syntactic processing (Hickok and Poeppel 2007; Hagoort and Poeppel, this volume).

Apart from acoustic structure and hemispheric asymmetry, another striking difference between music and language concerns the neuropsychological dissociations between these domains. For example, individuals with "congenital amusia" have severe, lifelong problems with basic aspects of music perception, such as discriminating between simple melodies or recognizing culturally common tunes. Such individuals may, however, function normally (or even excel) in other cognitive domains, and their everyday language abilities can seem perfectly intact. Structural neuroimaging has revealed subtle abnormalities in the brains of such individuals (Hyde et al. 2007), which likely have a genetic origin (Peretz et al. 2007). These abnormalities occur in multiple regions, including right-frontal and temporal cortices and their connections via the right arcuate fasciculus (Loui et al. 2009). Thus it appears that a neurogenetic condition can selectively affect musical but not linguistic development.

Neuropsychological dissociations between music and language can also occur after brain damage (e.g., stroke), as in cases of amusia without aphasia (deficits in music perception without any obvious language problems) and aphasia without amusia (impaired linguistic abilities but spared musical abilities). Such "double dissociations" are often considered strong evidence for the nonsharing of brain resources for music and language. Since dissociations between music and language due to brain damage have been extensively discussed by Peretz and others (e.g., Peretz and Coltheart 2003; Peretz 2012), they will not be discussed further here. For the current purposes, the key point is that any purported connections between music and language processing must be able to account for such dissociations.

The final set of differences between music and language that I wish to discuss concerns the formal properties of linguistic and musical systems. Let us begin with important aspects of language not reflected in music. One recent statement of such aspects is reproduced below. This list comes from a leading researcher in the field of child language acquisition (via personal communication):

1. The defining trait of human languages is the fact that there are two (largely) parallel structures. There are (a) elementary expressions ("words," "morphemes") and rules, according to which these can be combined to form more complex expressions, and (b) these elementary expressions have a conventionally fixed meaning, and compound expressions build up their meaning according to the way in which they are constructed. (There are many violations of this "principle of compositionality" but it is always the basis). Nothing like this exists in music.
2. There are typical functions which are expressed and systematically marked in all languages (e.g., statements, questions, commands). There is nothing comparable in music, except in a very metaphorical sense.

One may perhaps interpret a musical phrase as a "question" but it is not systematically marked, as in all languages. All natural languages have negation,[2] but music does not have anything similar. Music does have many functions—and not just aesthetic—but they are quite different to those of language.
3. Linguistic expressions are systematically built on an asymmetry of various types of constructions (head-dependent element, operator–operand). Again, there is nothing comparable in music.
4. Linguistic expressions serve to express a particular cognitive content, which the interlocutor is supposed to understand. This is not true for (instrumental) music, again with minor exceptions. If you do not believe this, try to transform your last published paper into a piece of music so that someone (who has not read the paper) is able to understand its content.

While some theorists may dispute the claim that music lacks all of the above properties (e.g., in terms of Pt. 3, see Rohrmeier 2011), by and large the claims are likely to be acceptable to many experts on linguistic and musical structure (cf. Jackendoff 2009).

Shifting the focus to music, it is easy to identify properties of music that are not reflected in ordinary language. One useful compendium of common features in music is Brown and Jordania (2011). As noted by these authors, due to the vast and ever-growing diversity of world music, musical universals are "statistical universals" (i.e., widespread patterns reflecting the way music "tends to be") rather than features present in every musical utterance. Some widespread features of music not found in human language are:

- Use of discrete pitches and intervals, drawn from underlying scales.
- Rooting of songs/melodies in the tonic (ground-pitch) of whatever scale type is being used.
- Predominance of isometric (beat-based) rhythms.
- Use of repetitive rhythmic and melodic patterns.
- Frequent occurrence of coordinated group production (i.e., group singing or playing).

The above formal differences between music and language, together with the neuropsychological differences discussed earlier, clearly demonstrate that music and language are not trivial variants of each other as cognitive systems. This is precisely what makes the discovery of related processing mechanisms in the two domains interesting. Indeed, given the differences outlined above, any related processing mechanisms are likely to be fundamental to human communication.

---

[2] Negation does not mean contrast! It means to deny the truth of something.

Sharing: Three Types of Hidden Connections between Linguistic and Musical Processing

The neuropsychological and formal differences reviewed above suggest that relations between musical and linguistic cortical processing are not likely to be obvious; that is, they will be "hidden connections." In the remainder of this chapter, I illustrate three types of hidden connections that can exist between linguistic and musical processing in the brain. In each case, musical and linguistic cortical processing rely on a similar functional computation, yet musical and linguistic abilities can be dissociated by brain damage. This relationship is possible when a functional computation in language and music relies on:

1. The same brain network, but developmentally one domain (language, music) is much more robust to impairments in this network.
2. The interaction of shared brain networks with distinct, domain-specific brain networks.
3. Separate but anatomically homologous brain networks in opposite cerebral hemispheres.

The following sections illustrate these three situations, focusing on prosody, syntax, and the motor control of speech versus song, respectively.

Before proceeding, however, it is worth noting that there is one aspect of brain processing where music and language can, a priori, be expected to overlap. At subcortical auditory processing stages (e.g., in structures such as the cochlear nucleus and inferior colliculus; see Figure 14.2), the processing of music and spoken language can be expected to overlap to a large degree, since both domains rely on temporally and spectrally complex signals, often with salient harmonic structure and a discernable pitch. Support for this overlap comes from recent research, which shows that subcortical encoding of linguistic sounds is superior in individuals who have been trained on a musical instrument and that the quality of encoding is related to number of years of musical training (Kraus and Chandrasekaran 2010). In other words, learning a musical instrument appears to enhance the early auditory processing of linguistic sounds. This presumably occurs via mechanisms of neural plasticity, perhaps driven by top-down "corticofugal" pathways from the cortex to subcortical structures (cf. dashed lines in Figure 14.1). The relationship between musical training and brainstem encoding of speech is now an active area of research. Such research has important practical consequences since the quality of brainstem speech encoding has been associated with real-world language skills such as reading ability and hearing speech in noise. While there is growing empirical research in this area, there is a need for theoretical frameworks specifying *why* musical training would benefit the neural encoding of speech (for one such framework, see Patel 2011). Since the focus of this chapter is on cortical processing, language–music overlap at the subcortical level is not discussed further here.

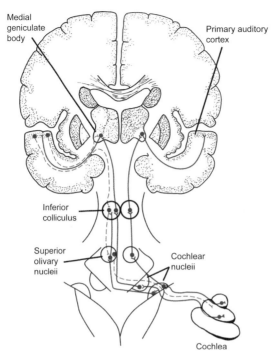

Figure 14.2 A simplified schematic of the auditory pathway between the cochlea and primary auditory cortex, showing a few of the subcortical structures involved in auditory processing, such as the cochlear nucleii in the brainstem and the inferior colliculus in the midbrain. Solid red lines show ascending auditory pathways; dashed lines show descending ("corticofugal") auditory pathways (and are shown on one side of the brain for simplicity). Reprinted from Patel and Iversen (2007) with permission from Elsevier.

Shared Networks: Relative Pitch Processing

As an example of a functional computation in language and music which may rely on the same brain network, we examine relative pitch processing. Humans effortlessly recognize the same speech intonation pattern, such as a "question" contour with a final pitch rise, when heard at different absolute frequency levels, for example, as spoken by an adult male or a small child (Ladd 2008). Humans also easily recognize the same musical melody (e.g., the "Happy Birthday" tune) when heard at different absolute frequency levels (e.g., played on a piccolo or a tuba). These skills rely on relative pitch: the ability to encode and recognize a pattern of ups and downs of pitch independent of absolute frequency level. Humans take this ability for granted: it requires no special training and is present in infancy (Trehub and Hannon 2006). Yet comparative research indicates that this ability is not universal among animals (McDermott and Hauser 2005). Starlings, for example, are animals that use acoustically complex, learned songs for communication, yet have great difficulty recognizing a tone sequence when it is transposed up or down in pitch (Bregman et al. 2012). Unlike most humans, starlings appear to favor absolute frequency cues in tone sequence recognition (e.g., Page et al. 1989). This suggests that relative pitch processing requires special brain mechanisms, which abstract the pattern of ups and downs from pitch patterns, and use these as an important cue for sound sequence recognition.

Do speech and music share brain mechanisms for relative pitch processing? Evidence from the study of congenital amusia (henceforth, amusia) is relevant to this question. Amusics have severe deficits in musical melody perception, yet their speech abilities often seem normal in everyday life. Early research on their ability to discriminate pitch contours in speech on the basis of falling versus rising intonation contours (e.g., discriminating statements from questions) suggested that they had no deficits (Ayotte et al. 2002). However, such research used sentences with very large pitch movements (e.g., statements and questions with pitch falls or rises of 5–12 semitones). More recent research with smaller, but still natural-sounding pitch movements (4–5 semitones) has demonstrated that most amusics have deficits in discriminating between statements and questions (Liu et al. 2010). Furthermore, their performance on this task correlates with their psychophysically determined thresholds for discriminating upward from downward pitch glides. These findings are consistent with the idea that speech and music share a brain network for relative pitch processing and that, in amusics, this network is impaired, leading to a higher threshold for discriminating the direction of pitch movements than in normal listeners. According to the "melodic contour deafness hypothesis" (Patel 2008:233–238), this higher threshold disrupts the development of music perception, which relies heavily on the ability to discriminate the direction of small pitch movements (since most musical intervals are 1 or 2 semitones in size). However, it does not disrupt the development of speech perception, which is quite robust to modest problems in pitch direction discrimination, as speech tends to rely on larger pitch movements and often has redundant information which can compensate for insensitivity to pitch direction. Thus the melodic contour deafness hypothesis suggests that music and language use the same brain network for relative pitch processing, but that musical and linguistic *development* place very different demands on this network, resulting in different developmental outcomes.

What brain regions are involved in computing pitch direction? Lesion studies in humans point to the importance of right-hemisphere circuits on the boundary of primary and secondary auditory cortex (Johnsrude et al. 2000). It is likely, however, that these regions are part of a larger network that extracts relative pitch patterns and (when necessary) stores them in short-term memory for the purpose of recognition or comparison. Structural neuroimaging studies of amusic individuals have revealed abnormalities in a number of frontal and temporal brain regions in both hemispheres (Hyde et al. 2007; Mandell et al. 2007), as well as in connections between right frontal and temporal regions

(Loui et al. 2009). Furthermore, functional neuroimaging shows that the right inferior frontal gyrus is underactivated in amusics during tone sequence perception (Hyde et al. 2010). Hence, right frontal and temporal regions and their connections are good candidates for being part of a network that processes relative pitch patterns. Furthermore, the intraparietal sulcus may be important in comparing pitch sequences on the basis of patterns of relative pitch (Foster and Zatorre 2010).

Identifying the different anatomical components of this network, and the role played by each, is an enterprise that could benefit from an integrated approach combining neuroimaging with computational models of relative pitch perception. Husain et al. (2004) have proposed, for example, a computational model of auditory processing that incorporates up and down frequency modulation selective units, as well as contour units, which are responsive to changes in sweep direction (Figure 14.3). The model includes auditory, superior temporal, and frontal brain regions, and builds on neurophysiological data on pitch direction processing, including several studies by Rauschecker (e.g., Rauschecker 1998). It has also been related to fMRI studies of auditory processing. This model could be used to simulate the pitch direction deficits of amusics (e.g., by disrupting the function of selected areas or by manipulating connectivity between areas) and may help guide the search for brain regions supporting relative pitch processing.

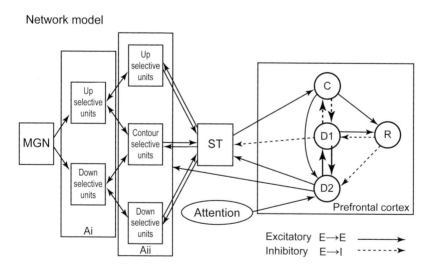

Figure 14.3 Network diagram of the computational model of auditory pattern recognition proposed by Husain et al. (2004). Regions include medial geniculate nucleus (MGN), two regions in primary auditory cortex (Ai, Aii), superior temporal gryus/sulcus (ST), and prefrontal cortex. In the prefrontal cortex region, C contains cue-sensitive units, D1 and D2 contain delay units and R contains response units. Reprinted from Husain et al. (2004) with permission from Elsevier.

Shared Networks Interacting with Domain-Specific Networks: Syntactic Processing

As an example of a functional computation in language and music that may rely on shared networks interacting with domain-specific networks, we examine structural integration in linguistic syntax and tonal harmony. In the past decade or so, a surprising amount of neuroimaging evidence has accumulated suggesting that the cortical processing of tonal harmonic relations in instrumental music shares neural resources with the cortical processing of linguistic syntax. This evidence is surprising for three reasons. First, a number of behavioral investigations of patients with musical deficits (following brain damage or due to congenital amusia) have demonstrated clear dissociations between tonal harmonic processing and linguistic syntactic processing (Peretz and Coltheart 2003). Second, tonal harmonic music (henceforth, "tonal music" or "tonality") does not have the grammatical categories that are fundamental to language, such as nouns, verbs, subjects, or objects. Third, tonality and linguistic syntax serve very different ends. The former helps articulate the nuanced ebb and flow of tension and resolution in nonpropositional sound sequences, in the service of emotional and aesthetic communication (for an example using the music of the Beatles, see Jackendoff and Lerdahl 2006). The latter, in contrast, helps articulate argument structure within referential propositions, by specifying who did what to whom, as well as where, when, and why. (For a brief introduction to the structure of tonal harmonic music, oriented toward comparison with language, see Patel et al. 2008, Section 5.2).

I begin by briefly reviewing some of the key neuroimaging evidence for overlap in the processing of tonality and linguistic syntax (for related material, see Koelsch, this volume). This evidence raises the question of why these two seemingly very different kinds of processing should overlap in this way. I offer my own perspective on this subject and outline the "shared syntactic integration resource hypothesis" or SSIRH (Patel 2003). The SSIRH makes specific, testable predictions, including predictions about interference between tonality and linguistic syntax processing. Thus far, the predictions have been supported by empirical work, but this is a young line of research and more work is needed to explore precisely how and why tonality and linguistic syntax processing are related in the human brain.

Evidence from Neuroimaging Studies

The first neurobiological study to compare tonality processing and linguistic syntactic processing directly used event-related potentials or ERPs (Patel et al. 1998). ERPs have played an important role in neurolinguistic research. Two extensively studied ERP components are the N400, which is thought to reflect the cost of integrating a word's meaning into the meaning representation of a sentence (or the ease of accessing information in semantic memory), and the

P600, which is associated with syntactic processing (e.g., syntactic violations or ambiguities, or structural complexity, Gouvea et al. 2010).[3]

Patel et al. (1998) used music-theoretic principles of harmony and key-relatedness to construct chord sequences in which a target chord in the sequence was either in key, from a nearby key, or from a distant key compared to the rest of the phrase. All target chords were well-tuned major chords which sounded consonant in isolation; in the context of a chord sequence, however, the out-of-key chords sounded contextually dissonant due to their departure from the prevailing key. Participants in the study also heard spoken sentences in which a target word was easy, difficult, or impossible to integrate into the preceding syntactic structure. The critical finding was that the out-of-key chords elicited a bilateral P600 that was statistically indistinguishable from the P600 elicited by syntactically incongruous words (i.e., in terms of latency, amplitude, and scalp distribution in the 600 ms range). This was interpreted as evidence that similar processes of structural integration were involved in tonality and linguistic syntax processing.

ERPs do not provide firm information of the physical location of underlying neural generators, and thus direct tests for shared brain regions involved in linguistic syntactic and tonal harmonic processing await studies that use localizationist techniques (such as fMRI). In recent years, a few fMRI studies have compared sentence processing to instrumental musical sequence processing (e.g., Abrams et al. 2011; Fedorenko et al. 2011; Rogalsky et al. 2011b) and have reported salient differences in activation patterns associated with the processing of structure in the two domains. However, the manipulations of musical structure in these studies did not focus specifically on tonality, and the manipulations of language structure did not specifically target syntactic complexity in meaningful sentences. Hence, the door is open for future fMRI studies which focus on comparing linguistic syntactic processing to musical tonality processing, based on established principles of structural complexity in each domain.

At the moment, the strongest neuroimaging evidence for overlap in tonality and linguistic syntax processing comes from the research done by Koelsch and colleagues on brain responses to out-of-key chords in chord sequences. Patel et al. (1998) found that such chords, in addition to eliciting a P600, also elicited a right-hemisphere antero-temporal negativity (RATN) peaking around 350 ms, reminiscent of the left anterior negativity (LAN) associated with linguistic syntactic processing. While Patel et al. (1998) studied musically trained individuals, Koelsch et al. (2000) examined ERP responses to out-of-key chords in nonmusicians (i.e., individuals with no musical training, outside of normal

---

[3] Recently, P600s and N400s have also been observed in cases where there are conflicts between semantic and syntactic information. For recent data and a theory which explains these findings while maintaining the distinction between the N400 as an index of semantic processing and the P600 as an index of syntactic processin, see Kos et al. (2010).

exposure to music in school). They discovered an early right anterior negativity (ERAN), peaking around 200 ms, in response to such chords, again reminiscent of a LAN associated with linguistic syntax processing, i.e., the early left anterior negativity or ELAN, which is associated with word category violations. (For evidence that the RATN and ERAN are two variants of a similar brain response, with different latency due to the rhythmic context in which a chord occurs, see Koelsch and Mulder 2002). In subsequent studies, Koelsch and colleagues have demonstrated that the ERAN occurs in both musicians and nonmusicians. In a series of pioneering studies on the neurobiology of tonality processing in musicians and nonmusicians (largely based on the perception of chord sequences), they have shown that the ERAN has several attributes reminiscent of syntactic processing:

1. The ERAN is elicited by *structurally* unexpected events, not by *psychoacoustically* unexpected events. Using a psychoacoustic model of sensory processing, they have shown that the ERAN is not simply a response to a sensory mismatch between the pitches of an incoming chord and a pitch distribution in short-term memory created by previous chords (Koelsch et al. 2007).
2. The ERAN is a response to departures from *learned structural norms*, not to deviations from local sound patterns in sensory memory. The ERAN is distinct from the mismatch negativity, in terms of both the nature of its eliciting events and in terms of its underlying neural generators (Koelsch 2009).
3. The amplitude of the ERAN is modulated by the degree of structural incongruity between target chord and local context (Koelsch et al. 2000).
4. Based on magnetoencephalography (MEG)[4] studies, the generators of the ERAN include the inferior part of left BA 44 in Broca's area (Maess et al. 2001), as well as its right-hemisphere homolog, which is an even stronger generator. This is of interest because left BA 44 (part of Broca's region) appears to be involved in the processing of linguistic syntax (Hagoort 2005; Meltzer et al. 2010). (Note that the right-hemispheric bias in the scalp topography of the ERAN is rather weak and is even absent in some studies; for a review, see Koelsch 2009).
5. The ERAN is abnormal in individuals with lesions in left Broca's area (Sammler et al. 2011).
6. The ERAN is absent in children with specific language impairment, a disorder which includes problems with linguistic syntax processing (Jentschke et al. 2008).
7. An ERP marker of linguistic syntactic processing (the ELAN) is enhanced in musically trained children, who also show an enhanced

---

[4] MEG measures the magnetic fields generated by bioelectric currents in cortical neurons, and unlike EEG, can be used to reconstruct the location of underlying sources of brain activity.

ERAN compared to musically untrained children (Jentschke and Koeslch 2009).
8. Intracranial EEG measurements (from epileptic patients awaiting surgery) reveal partial overlap of the sources of the ERAN and ELAN in bilateral superior temporal gyrus and, to a lesser extent, in the left inferior frontal gyrus (although electrode coverage was not extensive in frontal cortex; Sammler et al. 2009).

Resolving a Paradox

From the evidence just reviewed, it is clear that any attempt to understand the relationship between tonality processing and linguistic syntactic processing must address a paradox: Evidence from behavioral studies of patients with musical deficits (either due to brain damage or lifelong congenital amusia) points to the independence of tonality and linguistic syntax processing (Peretz 2006), while evidence from neuroimaging points to an overlap.

To resolve this paradox and to guide future empirical research on the relationship between tonality and syntax, SSIRH (Patel 2003) posits a distinction between domain-specific representations in long-term memory (e.g., stored knowledge of words and their syntactic features, and of chords and their harmonic features) and shared neural resources which act upon these representations as part of structural processing. This "dual-system" model considers syntactic processing to involve the interaction (via long-distance neural connections) of "resource networks" and "representation networks" (Figure 14.4).

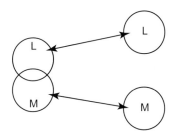

Resource networks   Representation networks

Figure 14.4 Schematic diagram of the functional relationship between shared resource networks and domain-specific representation networks, according to the SSIRH. L = Language, M = Music. The diagram represents the hypothesis that linguistic and musical long-term knowledge are stored in anatomically distinct "representation networks," which can be selectively damaged, whereas there is overlap in the "resource networks" that help activate structural information in representation networks during sequence processing. Arrows indicate functional connections between networks. Note that the circles do not necessarily imply highly focal brain areas. For example, linguistic and musical representation networks could extend across a number of brain regions, or exist as functionally segregated networks within the same brain regions. Reprinted from Patel (2008) with permission from Oxford University Press.

Such a view contrasts with an alternative view, common in artificial neural network models, that syntactic representation and processing involve a single network (e.g., Elman et al. 1996).

Why posit a dual-system model? The primary reason is that such a model can explain dissociations between tonality and linguistic syntax processing (via damage to domain-specific representation networks) as well as overlap (via activations in shared resource networks). Furthermore, the idea that complex cognitive processing involves temporally coordinated activity between spatially segregated brain regions is part of current neurobiological theory. For example, a core feature of the theory of *neural Darwinism* (Edelman 1993) is the concept of reentry, "a process of temporally ongoing parallel signaling between separate maps along ordered anatomical connections" (Edelman 1989:65). Neurobiologically realistic computational models of reentrant interactions between functionally segregated brain regions have been explored by Edelman and colleagues (e.g., Seth et al. 2004). In such models, different cortical regions have distinct functions, but higher-level processing abilities (such as object recognition) are an emergent property of a network of reentrantly connected regions. Within neurolinguistics, reentrant models involving interactions between distant cortical regions have also been proposed; for a recent example, see Baggio and Hagoort's (2011) discussion of the brain mechanisms behind the N400.

Based on evidence from neuropsychological dissociations and neuroimaging, I suggested that the domain-specific representation networks involved in language and music processing were located in temporal regions of the brain, while the shared resource networks were located in frontal regions (Patel 2003). Further, I posited that resource networks are recruited when structural integration of incoming elements in a sequence is costly; that is, when it involves the rapid and selective activation of low-activation items in representation networks. I used specific cognitive theories of syntactic processing in language (dependency locality theory; Gibson 2000) and of tonal harmonic processing in music (tonal pitch space theory; Lerdahl 2001b) to specify the notion of processing cost. In both models, incoming elements incur large processing (activation) costs when they need to be mentally connected to existing elements from which they are "distant" in a cognitive sense (e.g., in music, distant in tonal pitch space rather than in terms of physical distance in Hz). According to the SSIRH, in such circumstances, activity in frontal brain regions increases in order to rapidly activate specific low-activation representations in temporal regions via reentrant connections. Put another way, music and language share limited neural resources in frontal brain regions for the activation of stored structural information in temporal brain regions.

The SSIRH has a resemblance to another independently proposed dual-system framework for linguistic syntactic processing: the *memory, unification, and control* (MUC) model (Hagoort 2005; Hagoort and Poeppel, this volume). As noted by Baggio and Hagoort (2011:1357), "the MUC model assigns the storage of lexical items to temporal cortex and the unification of retrieved structures to frontal cortex." Hence, no permanent memory patterns are stored in frontal cortex. Instead, "unification is essentially the result of the coactivation of different tokens in inferior frontal gyrus, dynamically linked to their lexical types in middle temporal gyrus/superior temporal gyrus via persistent neuronal firing and feedback connections" (Baggio and Hagoort 2011). Figure 14.5a shows the components of the MUC model projected onto the left cerebral hemisphere, and Figure 14.5b shows some of the anatomical connections between frontal and temporal brain regions which could support interactions between these regions.

The MUC model views different parts of inferior frontal gyrus as involved in different aspects of linguistic unification (i.e., phonological, semantic, and syntactic), with semantic unification relying more heavily on BA 45/47, and syntactic on BA 44/45 (Hagoort 2005; for recent empirical data, see Snijders et al. 2009). It is thus of interest that tonal harmonic processing appears to activate the BA 44/45 region, in line with the idea that tonality processing has more in common with linguistic syntactic processing than with semantic processing.

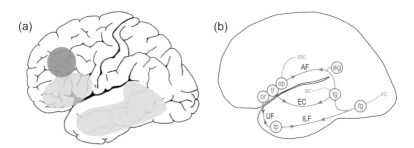

Figure 14.5 (a) Schematic of brain regions involved in Hagoort's *memory unification and control* model of language processing, projected onto the left hemisphere: Memory (yellow) in left temporal cortex, unification (blue) in left inferior frontal gyrus, and control (gray) in dorsolateral prefrontal cortex (control is involved in verbal action planning and attentional control, but is not discussed in this article). For the memory region, associated Brodmann areas include BA 21, 22, and 42. For the unification region, associated Brodmann areas include BA 44, 45, and 47. Reprinted from Hagoort (2005) with permission from Elsevier. (b) Simplified illustration of the anatomy and connectivity of the left-hemisphere language network. Cortical areas are represented as red circles: pars orbitalis (or), pars triangularis (tr), and pars opercularis (op) of the left inferior frontal gyrus; angular gyrus (ag), superior and middle temporal gyri (tg), fusiform gyrus (fg) and temporal pole (tp). White matter fibers are shown in gray, arrows emphasize bidirectional connectivity: arcuate fasciculus (AF), extreme capsule (EC), inferior longitudinal fasciculus (ILF) and uncinate fasciculus (UC). Interfaces with sensorimotor systems are shown in green: visual cortex (vc), auditory cortex (ac) and motor cortex (mc). Reprinted from Baggio and Hagoort (2011) with permission from Taylor and Francis Ltd.

One important issue for future work relating the SSIRH and MUC frameworks concerns hemispheric asymmetries. The SSIRH diagram in Figure 14.4 and the MUC diagram in Figure 14.5 represent areas and connections within one hemisphere. However, the processing of tonality likely involves frontotemporal networks in both hemispheres, though likely with a right-hemisphere bias, reflecting the fact that tonality involves structured pitch patterns. How strong is this bias, and how does it compare to the strength of left-hemisphere bias in frontotemporal interactions that support linguistic syntax? Within each domain, how is the processing of structural relations coordinated in the left and right hemispheres? Are the "resource" regions shared by language and music in the left hemisphere only (e.g., due to a strong leftward bias in linguistic syntactic processing vs. a more equal hemispheric balance for musical tonality processing)? Or do resource regions overlap in both hemispheres? Does the degree of hemispheric bias for tonality processing depend on the amount of musical training? These interesting questions await future research.

Why Link Tonality to Syntax?

The SSIRH provides a conceptual framework for understanding how tonality processing and linguistic syntactic processing might be related in the brain, yet a deeper theoretical question is: *Why* should tonality processing be related to linguistic syntactic processing? The SSIRH is framed in terms of integration cost, but in parallel models of language architecture (e.g., Jackendoff 2002; Hagoort 2005), linguistic integration (or unification) occurs at multiple levels in parallel (e.g., phonological, semantic, and syntactic). Why is tonality more cognitively akin to linguistic syntax than to phonology or semantics?

Considering phonology, it is notable that speech, like tonal music, has rule-like processes for combining sounds (Chomsky and Halle 1968). Phonological rules, however, typically involve changes to the forms of sounds as a result of the context in which they occur. For example, French has a phonological process known as voicing assimilation, in which a phoneme changes its feature "+/− voiced" depending on the local context in which it occurs. Ramus and Szenkovits (2008) provide the following illustration. In French, the voicing feature may spread backward from obstruents or fricatives to the preceding consonant: *cape grise* [kapgriz] → [kabgriz] (gray cloak). This assimilation process is both context specific (it does not occur before nasals: *cape noire* is always [kapnwar]; black cloak) and language specific (it does not occur in English, which instead shows assimilation of place of articulation: *brown bag* [brownbag] → [browmbag]). In tonal music, in contrast, principles of combination can result in an individual note changing its structural category without necessarily changing its physical characteristics in any way (e.g., the note B4 as played on a piano can have the identical physical frequency spectrum when played as the tonic of the key of B or the leading tone in the key of C, yet be perceived as sounding very different in these two contexts in terms of tension or resolution). Empirical research indicates that listeners are sensitive to changes in the structural categories of musical sounds as a function of context, even when the sounds themselves are physically identical in the different contexts (e.g., Bigand 1993).

Turning to semantics, an obvious difference between tones and words is in their referential qualities. Words are linked to concepts in a very intricate way: words have many phonological and semantic features which give them a specific position in a vast network of meaningful concepts in long-term memory. This ensures that when we hear a word, we activate the relevant, specific concept; when we wish to communicate about some concept, we use the appropriate word form. This highly complex form of cognition involves multiple brain areas in the perisylvian region (with a left-hemisphere bias) and is likely supported by a network that has been specialized over evolutionary time for this semantic processing. Consistent with this idea, recent neuroimaging research using MEG combined with structural MRI has shown that a similar frontotemporal brain network, with a left-hemisphere bias, is used for word understanding by adults and 12- to 18-month-old infants, meaning that this network is in place early in life (Travis et al. 2011).

The pitches of instrumental tonal music do not have rich semantic properties. To be sure, tonal sequences *can* sometimes convey general semantic concepts to an enculturated audience: a passage of instrumental music can sound "heroic" and may activate associated semantic concepts in the brain of musically untrained listeners (Koelsch 2011b; Koelsch et al. 2004). Even in such cases, however, one cannot pin down the semantic meaning of music in a precise fashion (Slevc and Patel 2011). Swain (1997:140) has argued that "the difference between musical and linguistic reference lies not in quality but in range"; one requires, for example, little effort to decide if Beethoven's *Appassionata* connotes "explosive fury" or "peaceful contemplation." However, deciding if it has to do with "explosive fury" or "passionate determination" is more difficult. In other words, if the distinction is binary, and concepts underlying a musical motive are diametrically opposed, musical meaning is grasped easily. If, however, there is a finer nuance, agreement among listeners is lost. Therefore, compared to linguistic meaning, the range of musical semantics is rather limited, (Swain 1997:49), as musical and linguistic structures have "varying degrees of semantic potential" (Antović 2009).

What of syntax? Here I would like to argue for significant connections between tonality and language. Notably, in both linguistic and tonal harmonic sequences, the brain interprets incoming events in terms of a small number of abstract structural categories and relations. For example, a word in a sentence is not just a semantic reference to some entity in the world, it also belongs to a structural category (e.g., noun, verb) and can enter into certain structural relations with respect to other words (e.g., subject, object). Importantly, while many languages mark structural relations by phonetically "tagging" the word in question (e.g., via distinct case markers for subjects and objects), this is not a necessary feature of language, as illustrated by English, where the same lexical form (e.g., "chapter") can be the subject or object of a sentence, depending on context.

In tonal music, pitches also become part of abstract structural categories and relations. In Western tonal harmonic music, abstract structural categories are formed at the level of chords (i.e., simultaneous, or near simultaneous, soundings of pitches). For example, particular collections of pitches define a "major chord" versus a "minor chord," but the precise physical instantiation of a chord (e.g., its component frequencies) depends on the particular key of the passage, which pitch serves as the root of the chord, the current musical tessitura (pitch range), and the "inversion" of the chord. Thus (implicitly) recognizing a certain chord type (e.g., a major chord) requires a certain kind of abstraction over different events, which can vary widely in their physical structure.

Turning from abstract structural categories to abstract structural relations, the same musical sound (e.g., a C major chord, C-E-G) can be an in-key or out-of-key chord and when it is in-key can serve different "harmonic functions." For example, it can be a "tonic chord" (i.e., the structurally most central and stable chord in a key) or when it occurs in a different key, it can be a "dominant chord" (i.e., a less stable chord built on the fifth note of scale). Tonic and dominant chords play a central role in Western tonal music, since movement from tonic to dominant back to tonic is an organizing structural progression for chord sequences. Musically untrained listeners are sensitive to the structural categories of chords (Bigand et al. 2003). This sensitivity is thought to be acquired via implicit learning (Tillmann et al. 2000) and plays an important role in feelings of tension and resolution created by chords (Lerdahl and Krumhansl 2007) and in the emotional responses to chords in musical context (Steinbeis et al. 2006). In other words, the harmonic functions of chords are abstract structural relations that influence the "meaning" of tone sequences in music. It is important to note that abstract structural relations in tonal music apply to individual pitches, not just chords. For example, a well-tuned note (e.g., A4, a pitch of 440 Hz) can be in-key or out-of-key, and when in-key can vary in the structural role it plays (e.g., a stable "tonic note" or an unstable "leading tone"), depending on the context in which it occurs (Krumhansl 1990). Just as with chords, sensitivity to abstract relations among individual pitches emerges via implicit learning in enculturated listeners and requires no formal musical training (Krumhansl and Cuddy 2010; Trainor and Corrigal 2010). Thus such sensitivity is likely to be part of many of the world's musical systems, since many such systems (e.g., the classical music of North India) have rich melodic but not harmonic structures (cf. Brown and Jordanian 2011).

Thus the reason that tonality may have something in common with linguistic syntax is that both involve the interpretation of rapidly unfolding sequences in terms of abstract structural categories and relations. Such categories and relations are *structural* because they are not bound to specific semantic meanings, nor are they necessarily signaled by the physical structure of an event, yet their existence and organization strongly influences the overall meaning of the sequence (e.g., who did what to whom in language, or patterns of tension and resolution in music).

With this perspective, the SSIRH can be framed in a more specific way. SSIRH addresses the cost of integrating an event's *structural status* with the existing structure of the sequence. In language processing, some structural integrations are more difficult to integrate than others. For example, a word may be cognitively distant from another word to which it needs to be conceptually related (e.g., long-distance noun–verb dependencies in dependency locality), or a word's category may be unexpected and result in a change in the existing structural interpretation of a sentence, as in garden path sentences (e.g., Levy 2008). Similarly, in tonal music, some structural integrations are more difficult to integrate than others. For example, a chord may be cognitively distant from the existing tonal region (e.g., an out-of-key chord) or because it is an unexpected category at that particular point in a sequence (e.g., a minor chord at a point in a sequence where a major chord is highly expected). According to the SSIRH, structural integration difficulty in language and music results in an increased activation cost, and this activation cost is "paid" by increased activity in shared frontal brain regions which are reciprocally connected to domain-specific temporal regions in which linguistic or musical representations reside.

In making this connection between linguistic syntactic processing and musical tonal harmonic processing, it is important to keep in mind that structural integration difficulty in the two domains can have quite distinct consequences. In the case of language, integration can become so difficult that it actually becomes impossible (in certain ungrammatical or highly complex sentences), and this effectively defeats a listener's attempt to assign a meaningful structure to the sentence. In tonal harmonic music, listeners generally try (implicitly) to make sense of any harmonic sequence they encounter, even when such sequences are highly complex or "ungrammatical." Perhaps this is because structurally unexpected events in music play an important role in eliciting emotional responses from listeners (Huron 2006; Steinbeis et al. 2006).

Testing the Predictions of the SSIRH

The SSIRH makes specific, testable predictions about the relationship between tonality and linguistic syntactic processing. One prediction is that since neural resources for structural integration are limited, simultaneous costly integrations in tonality and language should lead to interference. Testing this prediction requires experiments which present music and language simultaneously, and which align points to difficult structural integration in the two domains. This prediction has now been tested in five studies across three different laboratories: two studies using ERPs (Koelsch et al. 2005; Steinbeis and Koelsch 2008b) and three using behavioral methods (Fedorenko et al. 2009; Hoch et al. 2011; Slevc et al. 2009). All have supported the predictions of the SSIRH. For the sake of brevity, only one experiment is described here.

Fedorenko et al. (2009) manipulated linguistic syntactic integration difficulty via the distance between dependent words. These researchers used fully grammatical sentences of the type shown below, which differ in the structure of an embedded relative clause (italicized):

(a) The cop *that met the spy* wrote a book about the case.
(b) The cop *that the spy met* wrote a book about the case.

The sentences were sung to melodies (one note per word) which did or did not contain an out-of-key note on the last word of the relative clause: "spy" in (a), "met" in (b). According to dependency locality theory (Gibson 2000), this word is associated with a distant structural integration in (b) (between "met" and "that") but not in (a). A control condition was included for an attention-getting but nonharmonically deviant musical event: a 10 dB increase in volume on the last word of the relative clause. After each sentence, participants were asked a comprehension question, and accuracy was assumed to reflect processing difficulty.

The results revealed an interaction between musical and linguistic processing: comprehension accuracy was lower for sentences with distant versus local syntactic integrations (as expected), but crucially, this difference was larger when melodies contained an out-of-key note. The control condition (loud note) did not produce this effect: the difference between the two sentence types was of the same size as that in the conditions which did not contain an out-of-key note. These results suggest that structural integration in language and music relies on shared processing resources.

Another line of work motivated by the SSIRH concerns tonality processing in agrammatic Broca's aphasia (Patel 2003). The sensitivity of such aphasics (all with unilateral left-hemisphere lesions, though in variable areas) to tonal harmonic relations was tested using both explicit and implicit tasks (Patel et al. 2008b). The aphasics showed reduced sensitivity to such relations and, in the explicit task, the pooled performance of aphasics and controls on the tonality task predicted their performance on the linguistic syntax task (but not on a linguistic semantic task).

These initial studies call for further work to test reliability of these findings as well as to probe the specificity of the link between tonality and syntax (e.g., as opposed to tonality and semantics or tonality and phonology). There is a particular need for within-subjects fMRI research to compare brain areas involved in tonality versus linguistic syntactic processing. However, even overlapping activation regions in fMRI cannot absolutely prove the existence of shared neural circuits for tonality and linguistic syntax, due to issues of spatial resolution.

It is possible that spatially distinct networks exist at the microscopic level but occur in the same macroscopic brain region (e.g., by exhibiting different mosaics of functional organization interdigitated in the same cortical region). Thus fMRI results should be combined with evidence from other methods, including ERPs, behavioral research, patient studies, and techniques which produce transient "virtual lesions" (i.e., transcranial magnetic stimulation). That is, multiple converging methods, driven by specific hypotheses, are needed to discover the cognitive and neural operations shared by tonality and linguistic syntactic processing. These operations are worth uncovering because they are likely to be fundamental to our complex communicative abilities.

Related Computations in Homologous Areas of Opposite Hemispheres: Word Articulation

As an example of a related functional computation in language and music that may rely on homologous networks in opposite hemispheres, let us look at word articulation in speech and song. By "word articulation" I mean the brain mechanisms that convert a lexical item retrieved from memory to a sequence of sounds. Word articulation is one of the most complex motor actions produced by the brain. In both speech and song, it requires rich motor planning, the coordination of multiple articulators and the respiratory system, as well as self-monitoring and online correction based on sensorimotor information (Levelt et al. 1999). Despite this similarity between speech and song, there are salient differences between spoken and sung word articulation. For example, words in song are usually produced more slowly than spoken words (in syllables per second), and require more precise pitch control. These differences (i.e., slower rate, emphasis on pitch precision) are likely to be two reasons why song relies heavily on a right-hemisphere auditory motor network that is involved in the precise control of pitch patterns: a frontotemporal network that connects the superior temporal gyrus to the inferior frontal lobe via the arcuate fasciculus (Wan and Schlaug 2010).

What is the relationship between the brain networks involved in word articulation in speech and song? Evidence from neuropsychology is mixed. It has long been claimed that some nonfluent aphasics, who have severe difficulty with spoken word production, can produce words fluently when they sing familiar songs (Yamadori et al. 1977). However, in a study of nonfluent aphasics, Racette, Bard, and Peretz (2006) reported that word articulation abilities were no better during singing as opposed to speaking, suggesting that the same (impaired) word articulation network was involved in both domains. On the other hand, there is a long-standing observation that individuals with developmental stuttering, which is associated with left inferior frontal cortex structural anomalies (Kell et al. 2009), can often sing words with great fluency, and recent evidence indicates that transmagnetic stimulation over left Broca's region disrupts spoken but not sung word production (Stewart et al. 2001). Furthermore, neuroimaging research shows that when individuals sing songs, they activate certain brain regions that are not activated when they speak the same words. For example, Saito et al. (2006) compared overt singing to reciting song lyrics and found that the right inferior frontal gyrus, the right premotor cortex and the right anterior insula were active only during singing (cf. Peretz 2012). Hence, song and speech production show different brain activation patterns, with song production engaging certain right-hemisphere regions not activated by ordinary speech.

In sum, the neuroscientific evidence on word articulation in speech versus song is paradoxical: some evidence point to overlap (e.g., Racette et al. 2006), some to differentiation (e.g., Saito et al. 2006). A full resolution of this paradox remains to be proposed. Here I would like to suggest that some of the right-hemisphere brain circuits involved in word articulation in song are anatomically distinct from homologous left-hemisphere circuits involved in spoken word articulation, but that they carry out similar functional computations in terms of motor control. The idea of similar functional computations in homologous motor regions of opposite hemispheres is well accepted in neuroscience: after all, the left and right hands are largely controlled by homologous motor regions in opposite hemispheres. In the case of hand control, this opposite lateralization is largely driven by the decussating pattern of neuroanatomical connections between the hands and the brain. In contrast, the opposite lateralization of certain parts of the word articulation network for song versus speech may be driven by the differences in rate and pitch precision with which words are articulated in the two domains (Zatorre et al. 2002; Poeppel 2003).

To make my suggestion more concrete, I would like to place it in the context of a specific model of speech production; namely, the gradient-order DIVA (GODIVA) model (Bohland et al. 2010). GODIVA is an update of Guenther's original *directions into velocities of articulators* (DIVA) model—a neural network model of speech motor control and acquisition that offers unified explanations for a number of speech phenomena including motor equivalence, contextual variability, speaking rate effects, and coarticulation. Among computational models of speech production, GODIVA is notable for making detailed reference to known neuroanatomy and neurophysiology. A schematic of the model is provided in Figure 14.6.

A salient feature of the model, motivated by neurobiological research on speech production, is the left-hemisphere lateralization of several components. For example, the model posits that a speech sound map (SSM) exists in the left ventral premotor cortex and/or posterior inferior frontal gyrus pars opercularis ("frontal operculum" in Figure 14.6). SSM is the interface between the phonological encoding system and the phonetic/articulatory system and contains cell groups that code for well-learned speech sounds. SSM representations are functionally similar to a mental syllabary (Levelt and Wheeldon 1994; Crompton 1982), suggested by Levelt et al. (1999:5) to consist of a "repository

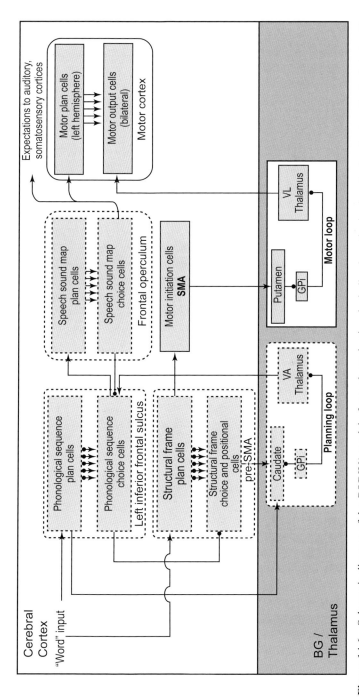

Figure 14.6 Schematic diagram of the GODIVA model of speech production, including hypothesized cortical and subcortical substrates. Boxes with dotted borders are given explicit computational treatment, whereas other boxes are treated conceptually. Lines with arrows represent excitatory pathways, and lines with filled circles represent inhibitory pathways. Lines with both arrowheads and filled circles indicate that connectivity between these modules features top-down excitatory connections as well as bottom-up inhibitory connections. For further details, see Bohland et al. (2010). Reprinted from Bohland et al. (2010) with permission from MIT Press.

of gestural scores for the frequently used syllables of the language." Using alternative terminology, SSM representations can be thought of as sensorimotor chunks or programs, learned higher-order representations of frequently specified spatiotemporal motor patterns. Recent fMRI research motivated by this model has supported the idea of syllable-level motor programs for speech which rely heavily on left ventral premotor cortex (Peeva et al. 2010).

My suggestion is that the right hemisphere has a "song sound map" (SGSM) that involves the right premotor cortex, complementary to the left-hemisphere SSM. The left-hemisphere SSM and the right-hemisphere SGSM carry out similar functional computations for word articulation in speech and song, though the SGSM part of the network normally operates at slower rates and with strong functional coupling to right-hemisphere regions involved in precise pitch control.

Such a view is relevant to recent research which has used a singing-based therapy called *melodic intonation therapy* or MIT (Albert et al. 1973) to help nonfluent aphasics recover some of their spoken language abilities. MIT embeds short phrases (e.g., "I love you") in "melodic" speech intonation patterns that rely on up-and-down movements between two discrete pitches. Patients practice such utterances intensively and regularly with a therapist, who gradually lengthens the phrases to span more syllables (Norton et al. 2009). The goal of the therapy is to improve fluency for both the trained phrases and for spontaneous, untrained utterances spoken in a normal fashion. Two features of MIT that distinguish it from nonmusical speech therapy are the use of melodic speech intonation and rhythmic tapping (i.e., while speaking the utterance, the patient also taps its syllabic rhythm using the hand unaffected by the stroke, typically the left hand).

Schlaug and colleagues have recently begun a set of studies aimed at measuring the efficacy of MIT versus a matched *speech repetition therapy* (SRT) without melodic intonation and tapping. In addition to quantifying the effects of MIT versus SRT on posttherapy measures of verbal fluency, the researchers are also measuring changes in brain physiology associated with the two therapies, by conducting fMRI and structural neuroimaging studies before and after the therapy. Of particular interest in this regard is the extent to which MIT patients shift toward using right-hemisphere "song" circuits for speech after therapy. That is, can such patients retrain right-hemisphere song networks to take over the functions of left-hemisphere speech networks? From a theoretical standpoint, this might be possible if the song network is already doing functional computations similar to the damaged speech network.

Preliminary data reported by Schlaug, Marchina, and Norton (2008) support the idea of a right-hemisphere shift in speech control with MIT; a patient who had undergone forty sessions of MIT showed substantially increased verbal fluency as well as increased speech-related activation in a right-hemisphere network involving the premotor, inferior frontal, and temporal lobes. Schlaug et al. (2009) have also found structural enhancement of the right arcuate fasciculus in patients who underwent MIT. The arcuate fasciculus is a large fiber tract that connects the frontal and superior temporal lobes and which is thought to be important for auditory sensorimotor integration.

At the moment, the relative contributions of the vocal/melodic versus rhythmic/hand movement components of MIT to these neural changes is unknown. An important issue for future research is to study how vocal melody and manual tapping might act synergistically to recruit neural plasticity in right-hemisphere word articulation circuits; that is, how vocal and manual motor circuits might be interacting in the brain (cf. Arbib 2006a). For current purposes, however, the findings of Schlaug and colleagues support the idea that right-hemisphere regions normally involved in word articulation in song can take over for damaged left-hemisphere regions normally involved in word articulation in speech. I would argue that this sort of neural plasticity is possible because the two sets of networks must have initially been carrying out similar functional computations. If this is indeed the case, it illustrates how music and language cortical processing can be related via similar functional computations in anatomically distinct circuits.

Conclusion

In this chapter, I have offered three conceptual solutions to address the paradoxical evidence on language–music relations in the brain. All three solutions illustrate how language and music may rely on related functional computations, despite neuropsychological dissociations between linguistic and musical abilities. Related functional computations used by language and music are likely to be fundamental to human cognition. The presented solutions are, however, not exhaustive. Additional solutions may emerge when other relations (e.g., in rhythmic processing) are considered. My intent in offering these solutions is to open up a dialog about the paradox of language–music relations in the brain. A full resolution of this paradox will yield significant new insights into the mechanisms underlying our species' uniquely powerful communicative abilities.

Acknowledgments

The author's work was supported by the Neurosciences Research Foundation as part of its research program on music and the brain at The Neurosciences Institute, where he was the Esther J. Burnham Senior Fellow. Helpful comments were provided by David Perlmutter, Bob Slevc, and two anonymous reviewers.

15

Action, Language, and Music

Events in Time and Models of the Brain

Michael A. Arbib, Paul F. M. J. Verschure, and Uwe Seifert

Abstract

Many accounts linking music and language to the brain represent the brain as a network of boxes, each of which has an active role in providing some resource, but once we move away from the auditory periphery there are very few models that offer finer-grain explanations of the underlying circuitry that supports these resources and their interaction. This chapter thus offers a bridge to future research by presenting a tutorial on a number of models that link brain regions to the underlying networks of neurons in the brain, paying special attention to processes which support the organization of events in time, though emphasizing more the timing or ordering of events than the organization of sequential order within a hierarchical framework. Our tour of models of the individual brain is complemented by a brief discussion of the role of brains in social interactions. The integration of cerebral activity is charted with that in other brain regions, such as cerebellum, hippocampus, and basal ganglia. The implications for future studies linking music and language to the brain are discussed which offer increased understanding of the detailed circuitry that supports these linkages. Particular emphasis is given to the fact that the brain is a learning machine continually reshaped by experience.

Introduction

Much of the important research linking music and language to the brain seeks correlations, such as what brain regions are most active when a sentence contains a semantic as compared to a syntactic anomaly, or assessing the extent to which similar brain mechanisms are active in language prosody and music. Computational modeling of the brain in other domains, however, goes beyond correlation to process. It aims to link functional properties of domains (such as action, memory, and perception) to causal interactions within and between neural systems as a basis for generating and testing hypotheses which integrate the diversity of experimental results and suggest new experiments. Given the complexity of the human brain—hundreds of brain regions, a hundred billion neurons, a million billion synapses—any model (like any field of experimental enquiry) must focus drastically, choosing particular processes and seeking a network of relevant neural structures. Moreover, given a particular model, we can ask: Which aspects of the processes are supported by the initial structure of the network, and which depend on learning processes?

Neuroscientists have long sought structural decompositions of the brain. The work of the nineteenth century neurologists labeled various regions of the brain as visual, auditory, somatosensory, or motor cortex (for a history, see Young 1970), and this localization was reinforced by the work of neuroanatomists who, around 1900 (e.g., Brodmann 1909), were able to subdivide the cerebral cortex on the basis of patterns of neurons, cytoarchitectonics. Meanwhile, the neuroanatomist Ramón y Cajal (1911) and the neurophysiologist Charles Sherrington (1906) established the neuron doctrine: the view that the brain functions in terms of discrete units, the neurons, influencing each other via synapses. The challenge for the brain theorist, then, is to map aspects of action, perception, memory, cognition, emotion, etc., not only onto the interactions of rather large entities, anatomically defined brain regions, but also on very small and numerous components, the neurons, or intermediate structures, such as columns in cerebral cortex (Mountcastle 1978). A complementary approach, the *schema theory* (Arbib et al. 1998; Shallice and Cooper 2011), seeks to decompose the overall function of a region functionally rather than structurally, with a model of the competition and cooperation of finer-scale functional entities, called schemas, as a basis for finding functional decompositions whose constituent subschema can indeed be related to the activity of brain regions or smaller structures or circuitry.

As Jeannerod et al. (1995) note, neuroscience has a well-established terminology for levels of structural analysis (e.g., brain area, layer, and column) but pays little attention to the need for a functional terminology. Schema theory (in the variant defined by Arbib 1981) provides a rigorous analysis of behavior that requires no prior commitment to hypotheses on neural localization. Schemas are units for this analysis. Perceptual schemas serve perceptual encoding, whereas motor schemas provide control units for movement. Crucially, schemas can be combined to form coordinated control programs, which control the phasing-in and phasing-out of patterns of schema coactivation, and the passing of control parameters from perceptual to motor schemas. The notion of schema is recursive: a schema might later be analyzed as a coordinated control program of finer schemas, and so on, until such time as a secure foundation of neural localization is attained. The level of activity of an instance of a perceptual schema represents a "confidence level" that the object represented by the schema is indeed present, while that of a motor schema might signal its "degree of readiness" to control a part of an action.

Mutually consistent schema instances are strengthened and reach high activity levels to constitute the overall solution of a problem, whereas instances that do not reach the evolving consensus lose activity, and thus are not part of this solution. A corollary to this view is that the instances related to a given object-oriented action are distributed. A given schema, defined functionally, might be distributed across more than one brain region; conversely, a given brain region might be involved in many schemas. Hypotheses about localization of schemas in the brain might be tested by observing the effects of lesions or functional imaging, and a given brain region can then be modeled by seeing if its known neural circuitry can indeed be shown to implement the posited schemas. An example of this approach is given here. In providing an account of the development (or evolution) of schemas, we find that new schemas often arise as modulators of existing schemas rather than as new systems with independent functional roles. Thus, schemas for control of dexterous hand movements serve to modulate less specific schemas for reaching with an undifferentiated grasp and to adapt them to the shape or the use of an object.

Modern neuroscience has probed the mechanisms of learning and memory, both in determining regional differences (e.g., the role of hippocampus in establishing episodic memories; as distinct from the roles of cerebellum and basal ganglia in procedural learning) and in establishing different forms of synaptic plasticity. Here it is common to distinguish between:

- Hebbian learning: Strengthen synapses connecting co-active cells—"what fires together wires together" (Frégnac 1995; Hebb 1949; Rauschecker 1991).

- Supervised learning: Adjust synapses to make each neuron more likely in the future to respond to input patterns in a way specified by a "teacher" (Rosenblatt 1958).

- Reinforcement learning: Adjust synapses to change the network in such a way as to increase the chance of positive reinforcement and decrease the chance of negative reinforcement in the future (Thorndike 1898; Sutton and Barto 1998). Since reinforcement may be intermittent, for the network to learn autonomously the expected future reinforcement associated with taking an action must be estimated and used in adjusting synaptic weights (Schultz 2006; Sutton 1988).

Even gene expression plays a role in neural plasticity, changing the behavior of neurons, not just the synapses that link them (see Arbib and Iriki as well as Fitch and Jarvis, this volume). *Connectionism* offers a "quasi-neural" approach to modeling psychological function. Here the focus is on *artificial* neural networks whose "neurons" are in fact highly simplified abstractions from real, biological neurons but given some power by equipping the synapses with learning rules such as those listed above. Unlike most of the models discussed in this chapter, the issue of getting a network to exhibit an observed function is little concerned with data from neuroscience.

Like behavior more generally, music and language are based on the organization of events in time. It is common to think of phonemes as the units of speech (with somewhat different units for the "phonology" of signed languages, such as hand shape and movement), though in fact production of these "units" (like syllables or moras) involve the control of multiple articulators (such as jaws, tongue, lips, and velum), and this control modifies successive "units" via co-articulation. Moreover, phonemes are complemented by other "time units" for affective and emotional expression (such as those in intonation and phrasing, etc. in prosody; see Ladd, this volume), demanding a (partial) ordering and timing within and across levels. Thus an analysis of speech production or comprehension may focus on the level of words, first asking how words are combined hierarchically and then seeking brain signatures for syntactic or semantic anomalies (see Hagoort and Poeppel, this volume) or seeking to assess changes in affective contour. Both experiments and models are, however, limited on the range of aspects to which they attend. Turning to music, "notes," "chords" (i.e., variations of pitch in melody and harmony), rhythm, and rhythmic structures constitute crucial elements in Western tonal music. These concepts, however, are not universally applicable in all musical forms; for example, in African drumming, Indian tabla music, and the noise and glides from electronic and computer music as well as sound installations. Thus, in research, there is a crucial need to broaden our understanding of these basic concepts, which may lead to wider definition of "music" itself. From the motor side, we have basic movements of the vocal articulators, hands, and body which link perception and action within hierarchies, interactions which extend dramatically as we consider instrumental as well as vocal music, and the coordination of multiple people engaged in song, music and dance.

Data from lesion studies and brain imaging focus on gross ascriptions of functional units and their processes to brain regions in humans. For data on actual neural circuitry, we must turn to other species. Of course, other species lack music and language but there are good data and corresponding models on animal systems that share elementary but important properties (e.g., perceptual processing, linking of sequential and hierarchical processes, learning, memory processes, with action generation) processing of music and language.

The remaining sections present a range of models, primarily from the Arbib or Verschure research groups. Each has a relatively narrow focus but, in their overall span, offers lessons that we believe are relevant to future models of brain processes underlying music and language. These models (Table 15.1) provide a baseline framework, with examples of "bottom-up" and "top-down" modeling approaches. We go beyond the auditory pathway and cortex to model some aspects of action, sensation, and perception, viewed as events in time, while also addressing the role of motivation and emotion for such processes.

Table 15.1 Review of the computational models of neural processing for action and perception presented in this chapter.

(Sub)Section Title	Scope of Modeling
1. Structuring Events in Time	Cortical circuits for the conversion of spatial input patterns to temporal codes
2. The Brain's Many Rhythms	Interaction of different neural rhythms may provide one mechanisms for integrated multilevel representations
3. The Construction of Events through Emotional Learning	How reward shapes behavior through classical conditioning
4. Motivation and the Development of Cognitive Maps	How reward shapes behavior through reinforcement learning
5. Representing Events for Self and Other	Modeling of the macaque canonical and mirror systems for grasping illustrates how the interaction of multiple brain regions supports learning processes that link notions of self and other
6. Integration of the Basal Ganglia and Prefrontal Cortex	Two models of the integration of the basal ganglia and prefrontal cortex in learning and recognizing sequential structure
7. From Sequences to Grammatical Constructions	Extension of the second model in Pt. 6 to the use of grammatical constructions in parsing sentences; a complementary connectionist model linking sentence processing to the "what" and "where" systems

Structuring Events in Time

The structuring of events in time plays a crucial role in both language and music, as does the interplay of varied streams of events underlying perception and production in both music and language. Our first model demonstrates how the connectivity of the cerebral cortex could serve to convert a stimulus into a temporal code, which could then provide input to further processing, whether for perception or production.

First we need some basic observations on the patterns of connections between cortical neurons. Douglas and Martin (2004) analyzed the basic laminar and tangential organization of the excitatory neurons in the visual neocortex to define a generic cortical network characterized by dense local and sparse long-range interarea connectivity (Figure 15.1a). Verschure's group showed how their *temporal population code* model could exploit the dense lateral connections in such a network to transform spatial properties of input patterns into a distinct temporal response (Wyss et al. 2003a) (Figure 15.1b, c).

Some researchers claim that local circuits perform the same operations throughout the neocortex, arguing, for instance, that the local receptive field properties of neurons can be understood in terms of specific statistical

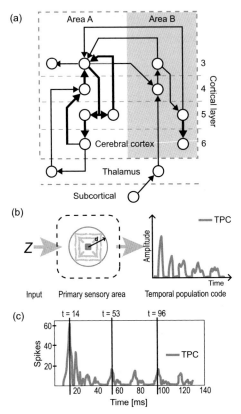

Figure 15.1 Structure and function relationships in the neocortex. (a) Cortical networks are characterized by dense local intra- and sparse long-range interarea connectivity. The thickness of the lines in this wiring diagram indicate the probability of finding connections between neurons at different layers of the cerebral cortex within an area and between cortical areas and subcortical nuclei in the cat (adapted from Douglas and Martin 2004). (b) The *temporal population code* (TPC) model proposes that the preponderance of dense lateral connectivity in the cortex gives rise to wave-like responses to provide a substrate for the rapid encoding of complex stimuli (Wyss et al. 2003a). (c) Illustration of the TPC and the spatial distribution of neuronal activity that results from the presentation of a stimulus. The input induces a dispersing wave of cortical activity shown at different time steps in the lower panels ($t = 14$, 53, and 96 ms, respectively). The trace in the upper panel shows the result of spatial averaging of the population response.

objectives, such as being optimally sparse or stable irrespective of their sensory modality (Körding et al. 2004; Olshausen and Field 1996; Wiskott and Sejnowski 2002). Indeed, the functional capabilities of the circuits of the neocortex seems highly malleable to the statistics of specific inputs. For instance, if projections to the visual and auditory cortex in the ferret are rerouted before birth to project to a different primary sensory area, the receptive field properties developing in the novel target area are identical to those found in the original target area in control animals (Sur et al. 1999; Sur and Leamy 2001). Modeling has shown that processing a continuous natural input stream generated by a mobile robot and optimizing a multilayered network of cortical-like processing elements for the statistical objective of temporal stability can account for a complete physiologically realistic visual hierarchy (Wyss et al. 2006). The same optimization principle generalizes to tactile (Hipp et al. 2005) and auditory (Lewicki 2002; Duff et al. 2007) modalities. Thus at the level of sensory preprocessing in the cerebral cortex, a theoretical case can be made that there is one generic computational infrastructure with the key variation being the detailed properties of the inputs.

It has been claimed that "that speech itself has been adapted to the optimal signal statistics the auditory system is tuned to" (Smith and Lewicki 2006). Speech has (at least) three key components: the ability to hear it, the ability to produce it, and the ability to use it as a medium for language communication. Nonhuman primates, however, have only the first of these three abilities. Thus we must ask: How did human ancestors acquire sufficient neural control over the vocal apparatus to generate "protospeech" to provide enough material for statistical shaping to condition the emergent structure of genuine speech? Even if one argues that the evolution of brain mechanisms for controlling speech production was conditioned by prior capabilities of the primate auditory system (as seems reasonable), one has to explain what is new about the human brain compared to other primates. A further problem is that human children are as adept in acquiring sign languages as they are in acquiring spoken language. Indeed, speech may be shaped as much by the top-down challenges of linguistic communication as by the bottom-up processes of auditory processing. Since it has been posited that different sensory cortices share a general architecture, it might be that adapting central mechanisms, which support language, to the peripheral mechanisms for extracting auditory events in time would equally suit those central mechanisms to exploit peripheral mechanisms for extracting visual events-in-time.

Moving beyond the sensory cortex, other models and data sets emphasize variations of the basic six-layered structure of the neocortex that are visible under the light microscope. Going further, Zilles and Amunts (2009) argue that receptor mapping of the human brain (these receptors assess transmitter molecules when they pass across synapses) reveals organizational principles of segregation of cortical and subcortical structures which extends our understanding of the brain's architecture beyond the limits of cytoarchitectonics. Thus, much more research is needed to understand how the specific local variant of cortical structure changes the way in which spatiotemporal patterns of inputs are transformed by a cortical region into temporal patterns of activity in each neuron. The actual variation (both genetically grounded and experientially based) in individual neurons of human brains lies far beyond the reach of experimentalists, but we may hope to learn enough from the neurophysiology of individual cells in animal brains to ground plausible neurobiological models of key aspects of music and language processing. For example, Rauschecker and colleagues have built on neurophysiological studies of the processing of space and motion-in-space (events in time) in the auditory system in nonhuman primates to illuminate some basic features of human speech processing, even while noting the distinct challenges of understanding the recursive structure and combinatorial power of human language (Rauschecker 2011; Rauschecker and Scott 2009). A theme to which we return below is the integration of processing in multiple cortical streams, which complements the local computations captured in the TPC model of Figure 15.1.

Is the human brain, then, just a statistical learning machine with more neurons that has been subjected to a long period of cultural evolution which has provided new external stimuli for the primate to adapt to? Or did the specific parcellation of the brain change to realize certain types of learning and integration in ways that a nonhuman brain could not achieve? Perhaps we must ask how genetics and experience (within a certain cultural milieu) impose different objective functions that might be satisfied by musical and linguistic subprocesses. In any case, future modeling must explore to what extent these subprocesses are candidates for sharing the same circuitry, or different circuits with very similar architecture. Then, with Patel (2008, this volume), we can debate whether the brain required novel paths of biological evolution to support music given those for language, or whether music simply exploits the same mechanisms as language but in a different fashion.

The Brain's Many Rhythms

The brain has many rhythms—from the basic rhythms of breathing, locomotion, and the heartbeat to various rhythms in neocortex, hippocampus, and other brain regions—which may vary with different states of awareness and attention (Buzsáki 2006). Theta oscillations are defined as activity in the 4 to 8 Hz range, the alpha rhythm operates in the 9 to 12 Hz range, while beta oscillations occur around 20 Hz. Gamma oscillations are produced when masses of neurons fire at around 40 Hz but can occur as low as 26 Hz to upwards of 70 Hz. It has been argued that transient periods of synchronized firing over the gamma waveband of neurons from different parts of the brain may integrate various cognitive processes to generate a concerted act of perception (Fries 2009). Researchers have also assessed brain rhythms when studying music and language, and some (but by no means all) view synchronization of specific rhythms as a central mechanism for neuronal information processing within and between multiple brain areas. Musical systems, like languages, vary among cultures and depend upon learning, though music rarely makes reference to the external world (see Seifert et al., this volume). Large (2010) suggests that general principles of neural dynamics might underlie music perception and musical behavior and enable children to acquire musical knowledge.

Indeed, Snyder and Large (2005) studied the relationship between short-latency gamma-band activity and the structure of rhythmic tone sequences and showed that induced (nonphase-locked) gamma-band activity predicts tone onsets and persists when expected tones are omitted. Evoked (phase-locked) gamma-band activity occurs in response to tone onsets with about 50 ms latency and is strongly diminished during tone omissions. These properties of auditory gamma-band activity correspond with perception of meter in acoustic sequences and provide evidence for the dynamic allocation of attention to temporally structured auditory sequences. This suggests that bursts of beta- and gamma-band activity that entrain to external rhythms could provide a mechanism for rhythmic communication between distinct brain areas, while attention may facilitate such integration among auditory and motor areas. Note that it is the bursts superimposed on the beta and gamma activity that carry the musical signal, rather than the underlying rhythms themselves.

With this, we turn to two modeling studies from the Verschure group: one on coupling gamma and theta rhythms in hippocampus; the other suggests how corticothalamic interactions may generate neural rhythms. We then turn from studies of rhythm generation to further studies of how variations in these rhythms correlate with aspects of music and language processing (continuing the themes of Large 2010; Snyder and Large 2005). It will be clear that these studies pose important challenges for future modeling.

Coupling Gamma and Theta in Hippocampus

We start with a model linking the theta and gamma rhythms, relating these to circuitry in the hippocampus. In rats, the theta rhythm, with a frequency range of 6–10 Hz, is easily observed in the hippocampus when a rat is engaged in active motor behavior, such as walking or exploratory sniffing, as well as during REM (rapid eye movement) sleep, but not when a rat is eating or grooming. In humans and other primates, hippocampal theta is difficult to observe but we offer a model extrapolated from the rat data. One facet of the events-in-time hypothesis focuses on the dentate gyrus (DG)-CA3 network of the hippocampus, seeing the gamma and theta oscillations as providing a basis for establishing discrete pieces of event information and the serial order of events. Lisman (2005) hypothesized that the dual gamma and theta oscillations in hippocampal neurons allows for the encoding of multiple events preserving their serial order, with the slower theta rhythm working as a global synchronization signal that initiates and binds a sequence of gamma oscillations. Each gamma cycle then carries a specific and discrete piece of information. The hypothesis suggests that the DG-CA3 network (Figure 15.2) integrates cortical representations into more elaborate event representations, sensorimotor couplets, and also orders them in time.

Entorhinal cortex has two relevant subdivisions: the grid cells of the medial division report spatial information while sensory responding neurons form the

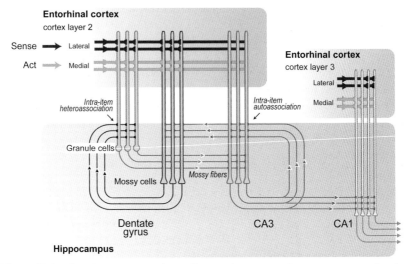

Figure 15.2 Creating representations of events by constructing sense-act couplets in the hippocampus (adapted from Lisman 2007). The hippocampus receives dual inputs from the lateral and medial entorhinal cortex that provide representations of sensory and action states respectively. A first integration of these sensory and action inputs, the latter derived from grid cells of entorhinal cortex (Fyhn et al. 2004), occurs at the level of the DG (dentate gyrus), exploiting the massively divergent projections it receives from the cortex without requiring plasticity. The DG provides inputs to the CA3 processing stage. CA3 cells are recurrently coupled to both CA3 and DG using plastic connections. This recurrent CA3 system provides a distributed declarative memory system in which high-level event encoding and acquisition can occur. The next processing stage of the hippocampus, CA1, provides a directed and local readout of this system. The activity of CA1 is projected back to the cortex. Selection among active cells in the DG-CA3 system has been coupled to the ratio of the theta-gamma cycle (de Almeida et al. 2009). Recently it was shown how the grid cell-derived action component of the basic event representations formed in the hippocampus could provide a basic metric for encoding sensory events (Rennó-Costa et al. 2010).

lateral division. Rennó-Costa et al. (2010) show how these representations are multiplexed in the responses of the neurons in the DG. Their model explains rate remapping: in an environment where the visual cues are smoothly varying, the correlation of the population response of the DG in subsequent environments smoothly degrades, or rates remap, while those in CA3 show a stronger generalization, or memory, between subsequent environments (Leutgeb et al. 2007). This suggests that the combined DG-CA3 system serves as a memory buffer that displays both instantaneous mapping capabilities in DG and plasticity-dependent classification in CA3.

Intriguingly, most animal studies emphasize the role of hippocampus in navigation (see section on Representing Events for Self and Other), while studies of humans focus on the role of the hippocampus in enabling the formation of episodic memories, with the consolidation of such memories maintained in cerebral cortex even after hippocampectomy. How does the linkage of places in physical space relate to the linkage of episodes in the space and time of an individual life? Perhaps we may need to search for the mechanisms which couple the cerebral cortex and hippocampus to integrate the sentences of a story into a coherent narrative or different fragments of a piece of music into a musical whole. However, the sequencing afforded by the theta-gamma system seems, at most, a part of what is needed to support linguistic, musical, and action sequencing.

Corticothalamic Interactions and Neural Rhythms

Corticothalamic interactions play an important role in establishing and maintaining different neural rhythms at different timescales and rhythms as, for example, in processing speech signals in the auditory cortex (see Hagoort and Poeppel, this volume). Evidence for the important role of corticothalamic interaction for brain rhythms comes from Parkinson's disease: Small changes in the excitatory drive to the thalamus induce dramatic changes in the control of brain rhythms. This is due to the fact that thalamic neurons switch from a regular low-frequency bursting mode when hyperpolarized to a more variable externally driven firing mode when depolarized. Excessive hyperpolarization (e.g., due to a lack of dopamine at the level of the striatum) will induce pathological thalamocortical dysrhythmia (Llinás et al. 1999). These pathologies show the decisive role that the thalamocortical system plays in maintaining the rhythms of the brain (Figure 15.3). A fundamental question is whether the ability of thalamic relay neurons to shift between low-frequency bursting when hyperpolarized to regular spiking when depolarized provides a substrate for processing information at varying intervals or rhythms.

Do Cortical Rhythms Play a Role in Language and Music?

Unfortunately, these models establish only a baseline for our understanding of the role of rhythms in music and language. Here we discuss some of the challenges to be met in developing models of how rhythms really enter into the brain's support of music and language, although we note that the studies reported here do not really address the way music and language enter our lives. Instead (as scientists must so often do), they seek simplified situations that can be studied in the lab but may establish data that can be assembled to paint an emerging picture of the whole. Earlier we assessed the claim that "that speech itself has been adapted to the optimal signal statistics the auditory system is tuned to" (Smith and Lewicki 2006). Now we turn to a competing claim: Hagoort and Poeppel (this volume) speculate that the organization of speech has tailored itself to the intrinsic rhythmic infrastructure that the brain provides. However, this conflates two separate ideas:

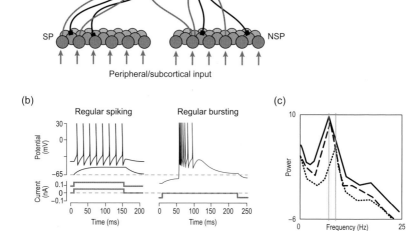

Figure 15.3 Rhythmicity in the thalamus. (a) The thalamus comprises specific (SP) and nonspecific (NSP) nuclei that receive input from the sensory periphery and subcortical areas, which are recurrently coupled to the thalamic reticular nucleus (RTN) and the cerebral cortex. The RTN provides recurrent inhibitory input to the SP and NSP. Neurons of SP are wired to the reticular nucleus in a highly specific fashion, whereas neurons of NSP show divergent connectivity. (b) Current response of a typical thalamic neuron with regular spiking resulting from depolarization (left panel) and a bursting response when hyperpolarized with an interburst frequency in the theta range (6–10 Hz). (c) This comparison of the average power spectra of the global EEG in thalamocortical dysrhythmia patients (solid) and healthy controls (dotted) with eyes closed shows that the power is enhanced in patients and the dominant peak shifted toward lower frequencies suggesting a bias due to excessive thalamic bursting (Sarnthein and Jeanmonod 2007). This power spectrum shift demonstrates the decisive role played by the thalamocortical system in the induction and maintenance of the rhythms of the brain. (Figure 15.3a and 15.3b were adapted from Henning Proske et al. 2011.)

1. Neural population rhythms provide "frames" which organize processing by individual neurons at different timescales (Figures 15.2 and 15.3).
2. Speech and music possess a rhythmic structure because of the neural rhythms intrinsic to cortex and related brain regions.

One might adopt the first yet reject the second. Note that the auditory system offers powerful mechanisms for transforming auditory signals from the time domain into frequencies and relative frequencies in the frequency domain to distinguish and predict spatiotemporal patterns. Such auditory scene analysis (Winkler et al. 2009a; Bregman 1990) is clearly a basic survival mechanism for all mammals (and others) for whom hearing *and* locating predators and prey is a crucial survival mechanism. However, noting the strong coupling of music and dance, one might link the rhythm of music to the rhythms generated by the motor pattern generators of the spinal cord and cerebellum that establish the basic patterns of rhythmic limb movements. There is no reason to expect the frequency of gamma oscillations, for example, to speed up as one sings or plays a musical instrument more quickly; however, the same mechanisms that can change our gait, speed, and direction as we move across different landscapes might well be adapted to serve variations in musical performance and perception, and accompanying body movements, as we move across different soundscapes.

Nonetheless, a number of studies have sought to relate neural rhythms to aspects of language music processing, complementing the work of Large and his colleagues reviewed earlier (Large 2010; Snyder and Large 2005). Bhattacharya et al. (2001) explored the role that long-range synchrony in the gamma band plays in music perception, viewing it as a special case of the necessary binding of spatial and temporal information in different cortical areas to build a coherent perception. They analyzed spontaneous EEG from two groups—musicians and nonmusicians—during several states: listening to music, listening to text, and at rest (eyes closed and eyes open). They found that degrees of gamma-band synchrony over distributed cortical areas were significantly higher in musicians than nonmusicians but that no differences between these two groups were found in resting conditions or while listening to a neutral text. They interpreted this as a manifestation of a more advanced musical memory of musicians in binding together several features of the intrinsic complexity of music in a dynamical way—another demonstration of the role of the brain's plasticity in extracting new skills from experience.

Carrus et al. (2011) investigated the patterns of brain oscillations during simultaneous processing of music and language using visually presented sentences and auditorily presented chord sequences. They, like Koelsch (this volume), limit their notion of music "syntax" to harmonic progressions. Irregular chord functions presented simultaneously with a syntactically correct word produced an early spectral power decrease over anterior frontal regions in the theta band and a late power increase in both the delta and the theta band over parietal regions. Syntactically incorrect words (presented simultaneously with a regular chord) elicited a similar late power increase in delta-theta band over parietal sites, but no early effect. Interestingly, the late effect was significantly diminished when the language-syntactic and "music-syntactic" irregularities occurred at the same time, suggesting that low-frequency oscillatory networks activated during "syntactic" processing may possibly be shared. Ruiz et al. (2009) investigated the neural correlates associated with the processing of irregularities in music syntax (in this limted sense) as compared with regular syntactic structures in music. They showed that an early (~200 ms) right anterior negative (ERAN) component was primarily represented by low-frequency (<8 Hz) brain oscillations. Further, they found that music-syntactical irregularities, as compared with music-syntactical regularities, were associated with

(a) an early decrease in the alpha-band (9–10 Hz) phase synchronization between right frontocentral and left temporal brain regions, and (b) a late (~500 ms) decrease in gamma-band (38–50 Hz) oscillations over frontocentral brain regions. These results indicate a weaker degree of long-range integration when the musical expectancy is violated.

Thus, the use of event-related potential (ERP) and functional magnetic resonance imaging (fMRI) for the study of language (Hagoort and Poeppel, this volume) and music (Koelsch, this volume) can be complemented by the study of oscillations and, like ERP studies, may even yield weak data on cortical localization. While intriguing, such results do little to explain what the two forms of "syntax" have in common (for a very different level of syntax for music, see Lerdahl, this volume). Perhaps the answer is that prosody rather than syntax is involved. Gordon et al. (2011) started from the idea that song composers incorporate linguistic prosody into their music when setting words to melody, tending to align the expected stress of the lyrics with strong metrical positions in the music. Gordon et al. aligned metronome clicks with all, some, or none of the strong syllables when subjects heard sung sentences. Temporal alignment between strong/weak syllables and strong/weak musical beats was associated with modulations of induced beta and evoked gamma power, which have been shown to fluctuate with rhythmic expectancies. Furthermore, targets that followed well-aligned primes elicited greater induced alpha and beta activity, and better lexical decision task performance, compared with targets that followed misaligned and varied sentences. This approach may begin to explain the mechanisms underlying the relationship between linguistic and musical rhythm in songs, and how rhythmic attending facilitates learning and recall of song lyrics. Moreover, the observations reported here coincide with a growing number of studies which report interactions between the linguistic and musical dimensions of song (for more on song, see Janata and Parsons, this volume), which likely stem from shared neural resources for processing music and speech. Still, the question remains: What additional, nonshared, resources may be necessary to support the ways in which music and language are distinct (e.g., the fact that language has a compositional semantics)?

We close this section by briefly discussing an oscillator-based model for low-level language processing by Vousden et al. (2000), a hierarchical model of the serial control of phonology in speech production. Vousden et al. (2000) analyzed phoneme movement errors (anticipations, perseverations, and exchanges) from a large naturalistic speech error corpus to derive a set of constraints that any speech production model must address. Their new computational model, a dynamic oscillator-based model of the sequencing of phonemes in speech production (OSCAR for OSCillator-Based Associative Recall) was able to account for error type proportions, movement error distance gradients, the syllable position effect, and phonological similarity effects. Although it makes contact with psycholinguistic rather than neurolinguistic data, it may point the way forward to models of processes more directly linked to language and music, yet which can make contact with the data from neurophysiology via the insights gleaned from models such as those of Figures 15.2 and 15.3. Nonetheless, a cautionary note is in order: Bohland et al. (2010) address the same problem as Vousden et al. (2000)—using a finite alphabet of learned phonemes and a relatively small number of syllable structures, speakers are able to rapidly plan and produce arbitrary syllable sequences that fall within the rules of their language. However, the approach taken by Boland et al. (2010) makes no use of oscillations, although it does make strong contact with a range of neurophysiological data, in the same style that we use to exemplify the models presented in this chapter. Moreover, their framework also generates predictions that can be tested in future neuroimaging and clinical case studies. Thus the issue of whether neural rhythms are indeed crucial to the human capability for language and music remains very much open.

The Construction of Events through Emotional Learning

Our concern with language and music has to take account not only the way in which events are ordered in time but also motivational and emotional aspects. Moreover, the brain cannot be neatly separated into separate modules for emotion and reason. Rather, event representations include emotional aspects such as the valence of stimuli. We present two related models: one (from the Verschure group) in the framework of classical conditioning; the other (from the Arbib group) a model of navigation in the framework of temporal difference learning, a form of reinforcement learning based on learning to predict an expectation of future reward and punishment subsequent on an action, rather than immediate reinforcement. The application of these models to address issues for music and language poses further challenges. (For relevant data, conceptual—as distinct from computational—models, and discussion, see chapters by Scherer, by Koelsch, and by Seifert et al., this volume.)

Classical Conditioning as a Simple Form of Emotional Learning

Many studies of emotion-linked representations employ the paradigm of Pavlovian or classical conditioning (Figure 15.4), which comprises both stimulus-stimulus and stimulus-response learning. Mowrer and Robert (1989) and Konorski (1967) hypothesized that this fundamental form of learning results from two interconnected learning systems: the nonspecific or preparatory learning system and the specific or consummatory learning systems. In the former, activation of the amygdala and the nucleus basalis induces plastic changes in the cortex so that behaviorally relevant stimuli will trigger a larger cortical response (Buonomano and Merzenich 1998; Sanchez-Montanes et al. 2000; Weinberger 2004). These enhanced cortical responses are projected to the cerebellum through the pons. In Figure 15.4, the conditioned stimulus (e.g.,

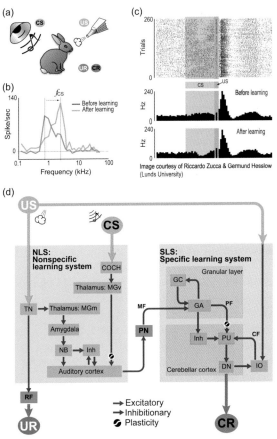

Figure 15.4 (a) The classical conditioning paradigm (Pavlov 1927). An initially neutral or conditioned stimulus (CS), here a tone, is paired with an emotionally significant or unconditioned stimulus (US), here an air puff to the eye. The US will induce a reflexive, unconditioned response (UR). During learning the amplitude-time course of the UR will be shaped to the specific properties of the CS and US to form a conditioned response (CR), here an eye blink timed to mitigate the air puff. (b) Pairing a tone CS that is within the tuning curve of a neuron in the primary auditory cortex (A1), but not at its best frequency, will induce a shift of the tuning curve after a few dozen paired CS-US presentations. These learning-induced changes in the tuning properties of A1 neurons can extend across the whole of A1 and lead to systematic changes on the organization of the tonotopic maps (Kilgard and Merzenich 1998). (c) Acquiring the timing of a CR is correlated with a pause in the activity of the Purkinje cells in the cerebellar cortex. These in turn release the neurons of the deep nucleus (DN) from inhibition and allowing them to trigger a CR in the motor nuclei of the brainstem. (d) An integrated model of the sensory and motor learning components of classical conditioning (Inderbitzin et al. 2010). CF: climbing fiber; COCH: cochlea; GA: granule cells; GC: Golgi cells; Inh: inhibitory interneurons; IO: inferior olive; NB: nucleus basalis; mCR: motor conditioned reaction; MF: mossy fiber; MGm: medial geniculate body; MGv: ventral medial geniculate body; PF: parallel fiber; PN: pontine nucleus; PU: Purkinje cell; RT: reticular formation; THAL: thalamus; TN: trigeminal nucleus.

a tone) and the unconditioned stimulus (e.g., a shock) converge at the level of the Purkinje cells, resulting in the induction of long-term depression at the parallel fiber to Purkinje cell synapse (Inderbitzin et al. 2010). The resultant reduction of Purkinje cell response induces a disinhibition of the deep nucleus leading to rebound excitation driven activity and ultimately an exactly timed response or conditioned response. This response closes a negative feedback circuit that inhibits the teaching signal triggered by the unconditioned stimulus and in this way stabilizes the learning process (Hofstötter et al. 2002); animals only learn when events violate their expectations (Rescorla and Wagner 1972).

To understand classical conditioning fully, we have to consider many cortical and subcortical components. Indeed, the involvement of the cerebellum and related circuitry in classical conditioning has long been a focus of both empirical research and modeling by many groups, with alternative theories being vigorously debated (Thompson and Steinmetz 2009; Gluck et al. 2001; Lepora et al. 2010; Llinas et al. 1997). However, it must be noted that the cerebellum has many different fine-grained subdivisions called "microcomplexes," and these are involved in a large variety of sensorimotor transformations, with the adjustment of the conditioned response to a noxious stimulus being a very special case of the more general achievement of grace in the timing and coordination of actions (chapter 9 in Arbib et al. 1998). Thus, if we wish to understand skill in the performance of speech, sign, song, instrumental music, and dance, we must understand the crucial role of the cerebellum in tuning and coordinating diverse cortical systems. Going further, Masao Ito (2012:193) has surveyed the many data on how the cerebellum supports its diverse roles and concludes with the hypothesis that the more lateral regions of the cerebellum may have evolved to support "unique roles...in our thought processes in which we manipulate ideas and concepts instead of moving body parts." These regions may thus be involved in the top-down structuring of the comprehension and production of language and music.

Motivation and the Development of Cognitive Maps

Figure 15.5 offers a different perspective on the emotional tagging of events, in this case modeling brain mechanisms supporting rat navigation. O'Keefe and Nadel (1978) distinguished the *taxon system* for navigation just by behavioral orientation (a taxis is an organism's response to a stimulus by movement in a particular direction) from the *locale system* for map-based navigation. Since activity of "place cells" in the hippocampus (O'Keefe and Dostrovsky 1971) correlates with where the rat is located rather than where the animal wants to be, Guazzelli et al. (1998) postulate that the place cells must interact with "goal cells" and a "cognitive map" located elsewhere.

A rat has a repertoire of affordances for possible actions associated with the immediate environment such as "go straight ahead" for visual sighting of a corridor, "hide" for a dark hole, and "eat" for food. Guazzelli et al.'s (1998)

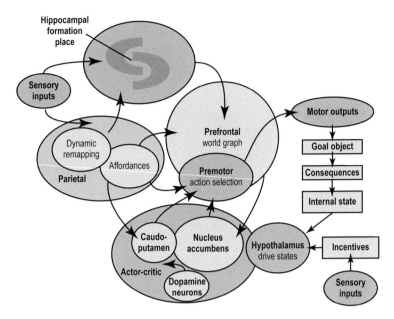

Figure 15.5 The TAM-WG model of motivation-based navigation. The TAM (taxon affordance model) explains how reinforcement learning can increase the drive-dependent success of basing premotor action selection on parietal affordances. The WG (world graph) model links a hippocampal system signaling "you are here" to a "world graph" in prefrontal cortex that provides a cognitive map which allows the animal to act on those affordances which are most consonant with longer-range plans based on world knowledge which is itself emotion/drive-dependent and updated through reinforcement learning on the basis of experience.

taxon affordance model (TAM) explains how the animal could learn to exploit affordances in the environment even when deprived of place cues from the hippocampus by a lesion of the fornix.

Temporal difference learning adjusts the mapping from the parietal representation of affordances to premotor action selection so that the consequences of action choice yields improved reinforcement across time. It is mediated by the actor-critic in Figure 15.5—the nucleus accumbens is the locus of reinforcement learning, which yields an adaptive bias signal for action selection dependent on the current internal state. Emotion (in its primordial form of motivations such as hunger, thirst, fear, etc.) came into play as the current internal state of the animal determined whether, for example, food, water, or escape would yield more positive reinforcement.

Guazzelli et al. (1998) extended TAM with a *world graph* (Arbib and Lieblich 1977), a cognitive map built as a set of nodes connected by a set of edges, where the nodes represent recognized places and the edges ways of moving from one to another. They modeled the process whereby the animal decides where to move next on the basis of its current drive state. The emphasis is on spatial maps for guiding locomotion into regions which may not be currently visible and yields exploration and latent learning without the introduction of an explicit exploratory drive. The model is inherently hierarchical: the selection of paths between places represented in the world graph frames the actual pattern of footfalls required to traverse the path. Moreover, the path the animal selects may vary greatly from occasion to occasion depending on the emotional state of the animal. The model thus exemplifies the integration of "reason" (in this case, path selection) and "emotion" as well as the integration of action and perception in behavior that raises and releases the tensions of expectation. (For a related model of insect navigation, see Mathews et al. 2010.)

Figure 15.5 is best understood by "reading" from right to left: On the right is the basic motivational system, centered on the hypothalamus. The activity level can depend on both internal state signals (e.g., low blood sugar increases the hunger signal) and sensory cues (seeing food may provide increase hunger). As the animal acts, it may or may not be successful in reaching a goal object (e.g., food); as a consequence, the animal may (in the case of food) change its internal state by becoming less or more hungry. The middle of the figure shows that sensory inputs determine parietally encoded affordances for the rat's locomotion; these provide the "menu" for premotor cortex to select appropriate commands to select actions. The "you are here" system of the hippocampus is augmented by the world graph, posited to be in prefrontal cortex. Dynamic remapping allows the animal to update its place-cell encoding on the basis of its recent movements when features of the new place are not currently visible. Temporal difference learning is again crucial, but here we might call it spatial difference learning (Arbib and Bonaiuto 2012)—choosing the next node not on the basis of immediate reward, but rather on the basis of maximizing the eventual expected reinforcement as the animal moves toward a goal that meets the needs of its current motivational state. Temporal difference learning creates predictions about future reward that may only be verified (or disappointed) after further actions are performed and uses the prediction error to update the adaptive critic's estimates, the expected future reward attendant upon taking different actions; the actor then bases its actions on this estimate. Dramatically, the notion of prediction error generated on theoretical grounds (Sutton 1988) was later found to explain the activity seen in input from dopamine neurons to the basal ganglia (Schultz 2002).

The corticobasal ganglia system is linked to other key components in regulating the reward circuit such as the amygdala, hippocampus, and the dorsal prefrontal cortex (for a review, see Haber and Knutson 2009). The ventral striatum in humans includes the nucleus accumbens of Figure 15.5. Turning to music, Menon and Levitin (2005) used functional and effective connectivity analyses of human brain imaging to show that listening to music strongly modulates activity in a network of structures involved in reward processing, including the nucleus accumbens and ventral tegmental area as well as the hypothalamus and insula. They infer that the association between dopamine release and nucleus accumbens response to music provides a physiological basis for why listening to music can be so pleasurable. The challenge, of course, is to understand why our brains evolved in such a way that auditory patterns of a particular form should have this effect.

In conclusion, note that there is a great distance between "primordial emotions" (Denton et al. 2009), such as the simple reinforcements of Figure 15.4 or the drives of Figure 15.5, and the fully nuanced richness of human emotions (see Scherer, this volume). The latter rests on the evolutionary refinement of subcortical drive-related circuits through extensive links with other structures (LeDoux 2000; Fellous and Ledoux 2005; Rolls 2005b). To what extent, then, is the simple integration of emotional learning and sensory learning, as observed in classical conditioning or motivated navigation, the substrate for the emotional labeling of language, action, and music, and what are the essential roles of supplementary mechanisms? Zald and Zatorre (2011:398) emphasize that "music may resemble biologically rewarding stimuli in its ability to engage similar neural circuitry" and sketch a conceptual model of musical reward that focuses on anticipation and prediction, as in the prediction error of temporal difference learning. Further they suggest (Zald and Zatorre 2011:404) a role for the mirror system in mediating the interpretation of sensory stimuli by simulating the intended outcome of other's motor actions, thus linking the "emotional" interpretation of music to motor feature of musical performance. (For a discussion of mirror systems, see Fogassi, this volume.)

Representing Events for Self and Other

Arbib and Iriki (this volume) outline the *mirror system hypothesis* which argues that a mirror system for recognition of hand actions supported a gestural basis for the evolution of the human language-ready brain (Arbib 2012). Here we turn to the exposition of models related to the macaque system for visual control and recognition of grasping, providing an explicit account of how the mirror system may learn to recognize the hand-object relations associated with grasps already in its repertoire. (Oztop et al. 2004 explain how such actions may be acquired by trial-and-error as distinct from imitation.)

What, Where, and Affordances in the Control of Self-Actions

Parietal area AIP (the anterior region of the intraparietal sulcus) and ventral premotor area F5 (the fifth area in an arbitrary numbering of regions of macaque frontal cortex) anchor the cortical circuit in macaque which transforms visual information on intrinsic properties of an object into hand movements for grasping it (Jeannerod et al. 1995). Discharge in most grasp-related F5 neurons correlates with an action rather than with the individual movements that form it so that one may relate these neurons in F5 to various *motor schemas* corresponding to the action associated with their discharge. Fagg and Arbib (1998) modeled the control of grasping (Figure 15.6), explaining how the size and orientation of a graspable portion of an object (its affordance for grasping) can determine the time course and spatial properties of the preshape of the grasp and its enclosure upon the affordance. Area cIPS (caudal intraparietal sulcus) provides visual input to parietal area AIP concerning the position and orientation of the object's surfaces. AIP then extracts the affordances. The *dorsal* pathway AIP → $F5_{canonical}$ → F1 (primary motor cortex) then transforms the (neural code for) affordance to the appropriate motor schema (F5) and thence to the appropriate detailed descending motor control signals (F1).

Going beyond the empirical data then available, Fagg and Arbib (1998) stressed that there may be several ways to grasp an object and thus hypothesized that object recognition (mediated by a ventral path through inferotemporal cortex, IT) can bias the computation of working memory and task constraints as well as the effect of instruction stimuli in various areas of prefrontal cortex (PFC), and that, exploiting the resultant encoding, strong connections between PFC and F5 provide the data for F5 to choose one affordance from the possibilities offered by AIP.

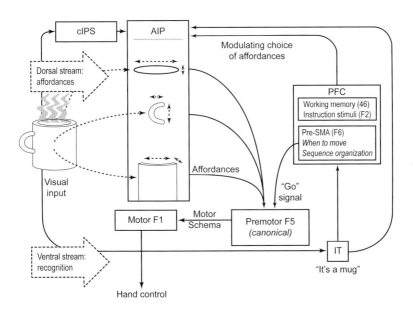

Figure 15.6 Visual inspection may reveal, via the anterior intraparietal sulcus (AIP), various possible affordances for grasping an object. Recognition of the object by the inferotemporal (IT) cortex supports the task-dependent choice of which grasp to perform in the scheduling of action, biasing the choice of motor schemas in premotor cortex (the hand area F5). Adapted from Fagg and Arbib (1998), based on anatomical suggestions by Rizzolatti and Luppino (2003).

Notice here the crucial notion of the "two visual systems": a "what" (ventral) and "where" (dorsal) system based on monkey neurophysiology (Ungerleider and Mishkin 1982), as well as a related but somewhat different notion of a "what" (ventral) and "how" (dorsal) system based on human lesion studies of object-based hand movements (Goodale and Milner 1992). The discovery of "what" and "where" pathways for the auditory system (Romanski et al. 1999) may have interesting implications for parallel studies of speech and language (Rauschecker 2005). (This will be discussed further below in the section, What and Where in the Auditory System.)

Mirror Neurons and the Recognition of Others

As explained by Fogassi (this volume), there is a subset of the F5 neurons related to grasping, the *mirror neurons*, which are active not only when the monkey executes a specific hand action but also when it observes another carrying out a similar action. These neurons constitute the "mirror system for grasping" in the monkey and we say that these neurons provide the neural code for matching execution and observation of hand movements. (By contrast, the *canonical* neurons are active for execution but not observation.) The superior temporal sulcus (STS) has neurons that discharge when the monkey observes certain biological actions (e.g., walking, turning the head, bending the torso, and moving the arms) and some discharge when the monkey observes goal-directed hand movements (Perrett et al. 1990), though not during movement execution. STS and F5 may be indirectly connected via inferior parietal area PF (BA 7b) (Seltzer and Pandya 1994) (P stands for "parietal lobe" and F indicates that it is one of a set of subdivisions labeled alphabetically.)

Just as we have embedded the F5 canonical neurons in a larger system involving both the parietal area AIP and the inferotemporal area IT, so do we now stress that the F5 mirror neurons are part of a larger system that includes (at least) parts of STS and area PF of the parietal lobe. We now discuss a model of this larger system, the *mirror neuron system* (MNS) model (Figure 15.7; Oztop and Arbib 2002). One path in Figure 15.7 corresponds to the basic pathway AIP → $F5_{canonical}$ → M1 of Figure 15.6 (but MNS does not include prefrontal influences; different models address different aspects of the whole) while intraparietal areas MIP/LIP/VIP (medial/lateral/ventral subdivisions of IP) provide object location information needed to execute a reaching movement which positions the hand for grasping. The shaded rectangle of Figure 15.7 presents the core elements for understanding the mirror system. The sight of both hand and object—with the hand moving appropriately to grasp the seen (or recently seen) object—is required for the mirror neurons attuned to the given action to fire. This requires schemas for the recognition of the shape of the hand and its motion (STS), and for analysis of the relation of the hand parameters to the location and affordance of the object (Areas 7a and 7b [≈PF]). This defines a hand-object trajectory. Oztop and Arbib (2002) showed that an

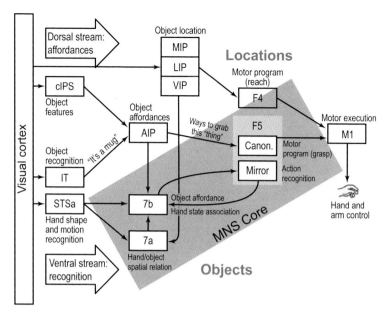

Figure 15.7 The *mirror neuron system* (MNS) model (Oztop and Arbib 2002) focuses on the circuitry highlighted by the gray diagonal rectangle and explains how, by recognizing movement of the hand relative to an object affordance for self-generated actions, mirror neurons may come to respond to similar hand-object trajectories when the movement is performed by others. AIP: anterior region the intraparietal sulcus; cIPS: caudal intraparietal sulcus: IT: inferotemporal cortex; LIP: lateral intraparietal sulcus; M1: primary motor cortex; MIP: medial intraparietal sulcus; VIP: ventral intraparietal sulcus; STSa: anterior superior temporal sulcus.

artificial neural network corresponding to PF and $F5_{mirror}$ could be trained to recognize the grasp type from this trajectory, with correct classification often being achieved well before the hand reached the object. During training, the output of the F5 canonical neurons, acting as a code for the grasp being executed by the monkey at that time, was used as the training signal for the F5 mirror neurons to enable them to learn which hand-object trajectories corresponded to the canonically encoded grasps. As a result of this training, the appropriate mirror neurons come to fire in response to the appropriate trajectories even when the trajectory is not accompanied by F5 canonical firing. As a result, F5 mirror neurons can respond to hand-object relational trajectories even when the hand is of the "other" rather than the "self" because the hand state is based on the movement of a hand relative to the object.

This modeling makes clear that mirror neurons are not restricted to recognition of an innate set of actions but can be recruited to recognize and encode an expanding repertoire of novel actions. Given the debate over innateness with regard to language acquisition, it is worth noting that in the MNS model the attention to hand-object relationships is "built in," but not the exact nature of the grasps that will be learned. Fogassi (this volume) charts the broader implications of mirror neurons, and corresponding mirror systems in the human brain, for linking action to music and language. Our concern here has been to show that portions of the puzzle are already within the domain of computational modeling of adaptive neural circuitry distributed across multiple brain regions. In the next subsection we turn our attention from vision to audition.

Our emphasis has been on mechanisms for action and perception in a single brain, and we have looked at mirror systems purely in terms of recognition of the action of others (for hypotheses on the role of mirror neurons in self-action, see Bonaiuto and Arbib 2010). However, in conversation the recognition of the other's articulatory actions is not an end in itself, but contributes to a growing structure of shared meaning which shapes how one will continue the conversation. Similarly, when the members of a chamber quartet play together, the recognition of what notes the others are playing is focused not on the appreciation of those notes in themselves but rather on adjusting the timing of one's own continuing performance. The development of an appropriate "dyadic brain theory" poses many exciting new challenges.

What and Where in the Auditory System

The two previous subsections have focused on the role of vision in the control of hand movements, yet the sensory modality most associated with music and language is that of sound. However, sign language and co-speech gestures make clear that hand movements are an integral part of language broadly considered, while playing instruments links hand movements to music. Moreover, dance involves not just the hands but all parts of our body in engaging with the sound of music. As a guide to future computational models, it will be useful to note briefly that the "what" and "where/how" systems of vision implicated in Figure 15.6 have their counterparts in the auditory system (and recall our earlier comments on auditory scene analysis).

A little background before we turn to the auditory system. Throughout most of this chapter, a "model" (conceptual or computational) provides an account of how interactions of entities within the brain (e.g., neurons, schemas, regions) mediate between inputs, internal, states, and outputs. We may refer to such models as "processing model." However, computational neuroscience and cognitive science more generally have exploited the term "model" in an additional sense: not a model *of* the brain, but a model *in* the brain. This idea goes back, at least, to Craik (1943) and relates to the general notion of perceptual schemas and motor schemas discussed earlier, the control theory concepts of feedback and feedforward, and the notion of forward and inverse models of a system (Wolpert and Kawato 1998). A forward model represents the transition from inputs to outputs in the system and can thus compute predictions and expectations about the result of applying control signals to the system. An inverse model provides a controller with the command signals which yield a desired change in output, thus (akin to mirror neurons) transforming the sensing of an action into the motor code for commanding that action.

The dual-path model of auditory cortical processing marshalls data which indicate that two processing streams originating in the lateral belt of auditory cortex subserve the two main functions of hearing: identification of auditory "objects," including speech, and localization of sounds in space. Rauschecker (2000) presented monkey calls at different spatial locations and determined the tuning of various lateral belt neurons in the monkey brain to monkey calls and spatial location. Differential selectivity supported the distinction between a spatial stream that originates in the caudal part of the superior temporal gyrus (STG) and projects to the parietal cortex, and a pattern (or object) stream originating in the more anterior portions of the lateral belt. A similar division of labor can be seen in human auditory cortex by using functional neuroimaging. This convergent evidence from work in humans and monkeys suggests that an antero-ventral pathway supports the "what" function, whereas a postero-dorsal stream supports the "where" function. While these paths are largely segregated we note that, as we demonstrated for visual control of grasping (Figure 15.6), the two streams must also be integrated to serve behavior more effectively.

The postero-dorsal stream has also been postulated to subserve some functions of speech and language in humans. Note again that we are here touching on survival-related properties of the auditory system that can be exploited for music as well as speech rather than the higher level properties of syntax and semantics. To build on this, Rauschecker and Scott (2009) have proposed the possibility that both functions of the postero-dorsal pathway can be subsumed under the same structural forward model: an efference copy sent from prefrontal and premotor cortex provides the basis for "optimal state estimation" in the inferior parietal lobe and in sensory areas of the posterior auditory cortex. More generally, they show how the study of maps and streams in the auditory cortex of nonhuman primates can illuminate studies of human speech processing. Their conceptual model connects structures in the temporal, frontal, and parietal lobes to link speech perception and production. Here we see echoes of models of neural mechanisms for action and action recognition which employs forward and inverse models, with the inverse models being in some sense akin to mirror neurons (Oztop et al. 2004, 2012)

Turning to the sounds of speech, DeWitt and Rauschecker (2012) conducted a meta-analysis of human brain imaging studies to show that preference for complex sounds emerges in the human auditory ventral stream in a hierarchical fashion, consistent with nonhuman primate electrophysiology. They showed that activation associated with the processing of short timescale patterns (i.e., phonemes) was consistently localized to left mid-STG; left mid- to anterior STG was implicated in the invariant representation of phonetic forms, responding preferentially to phonetic sounds, above artificial control sounds or environmental sounds; and activation associated with the integration of phonemes into temporally complex patterns (i.e., words) was consistently localized to left anterior STG, with encoding specificity and invariance increasing along the auditory ventral stream for temporally complex speech sounds.

Sequencing Events in the Brain

Both music and language can fundamentally be viewed as ways of ordering events in time. Although in each case, hierarchical structure is crucial, the most basic structure is a simple sequence of events. We thus turn now to two models which implicate the prefrontal cortex in memory-based responses to sensory states, actions and their combinations, and in the learning of sequences. In each case, the functional cognitive units link the circuitry of the forebrain in loops which include the basal ganglia. The second model has grounded an approach to the parsing of sentences, which gives a quasi-neural implementation of construction grammar.

Integration of the Basal Ganglia and Prefrontal Cortex

Duff et al. (2011) include prefrontal cortex and basal ganglia in a model (Figure 15.8) for the integration of sensorimotor contingencies in rules and plans. They propose that working memory, under the influence of specific reward signals, instructs the memory with respect to the validity of certain sequences. Moreover, elementary events are organized in densely coupled networks with reward signals, such as those triggered by the dopamine neurons of the ventral tegmentum (see earlier discussion: Motivation and the Development of

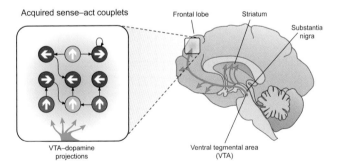

Figure 15.8 Rule and sequence learning in the prefrontal cortex. Neurons in the frontal lobe of the neocortex display memory fields to specific elements of tasks (indicated by arrows for movement directions and colors for specific cues). To assemble these acquired elementary events into a specific order and generate plans for action, the lateral interaction between these elements is modulated by specific value-dependent neuromodulators, such as the ventral tegmental area (VTA), whose neurons send dopamine (DA) reinforcement signals to varied parts of the brain. Changing the strength of these lateral connections allows the system to acquire and express specific goal-oriented behavioral sequences (see also Duff et al. 2011).

Cognitive Maps), modulating their lateral connectivity. This affects the way that event X at time t is chained with event Y at $t + 1$. In recall, a number of goal-oriented biases can facilitate correct execution of acquired sequences (Marcos et al. 2010).

Dominey et al. (1995) provided a somewhat different model (Figure 15.9), showing how different sequences of saccadic eye movements may be associated with appropriate cues by coupling a recurrent prefrontal network that encodes sequential structure with corticostriatal connections that learn to

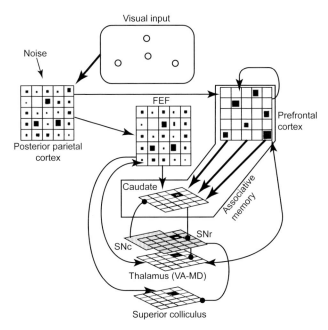

Figure 15.9 The monkey has been trained to saccade to a target on the screen (visual input) when the central fixation stimulus is removed. In the base model, the visual input is relayed to posterior parietal cortex (area LIP) and from there to the eye fields of frontal cortex (FEF) that provide input to the superior colliculus in the midbrain which, it is hypothesized, implements a winner-take-all-circuit to choose just one target to be used by brainstem circuitry to trigger a saccade. However, FEF also talks to basal ganglia, activating a pattern in the caudate which inhibits corresponding activity in the substantia nigra pars reticulata (SNr). SNr tonically inhibits superior colliculus, so the net effect is to disinhibit the same areas of superior colliculus that are excited by the direct path. This seems like useless redundancy. However, Dominey et al. (1995) show how reinforcement learning (with reinforcement provided by dopamine signals from the substantia nigra pars compacta) can allow signals from prefrontal cortex to affect differentially activity in caudate so that a given prefrontal state will cause the disinhibition to favor just one of the exhibited targets. They then build on this, plus cycling of the prefrontal cortex through different states, to show how the system can learn to associate different patterns (e.g., "red triangle" versus "blue square") of the fixation stimulus with the emission of different sequences of saccades. The loop between thalamus and FEF supports working memory.

associate different prefrontal states with appropriate actions. We show in the next section how this model has grounded one approach to the parsing of sentences, giving a quasi-neural implementation of construction grammar. Such a "transfer" raises the following questions: How does rule learning mediated by prefrontal cortex and basal ganglia contribute to the acquisition of different rule-like properties of language, music, and action? Do these different modalities require different rules and acquisition systems? If so, what are their differences?

We have focused almost exclusively on the representation and sequencing of discrete perceptual and motor events. A complementary field of motor control studies the motor pattern generators which control and coordinate muscles in the unfolding trajectories of action. Here (recall our earlier comments) the cerebellum plays a crucial role in complementing activity in cerebral cortex, brainstem, and spinal cord. The notion of forward and inverse models complements the role of control systems in predicting the effect of control signals and in finding control signals which will achieve a desired effect (Wolpert and Kawato 1998; Wolpert et al. 2003; Skipper et al. 2006; Arbib and Rizzolatti 1997). These will play an increasingly important role as we probe the mechanisms of prosody and melody.

From Sequences to Grammatical Constructions

Unfortunately, although there is a range of conceptual models of language processing in the brain, very few models are computationally implemented (for a partial review, see Arbib 2013). As a stand-in for the diversity of brain models that could illuminate the neural mechanisms underlying language processing (not to mention music), we consider a neurolinguistic model based on *construction grammar* (e.g., Fillmore and Kay 1993; Goldberg 1995). David Kemmerer has made explicit the relevance of construction grammar to neurolinguistics in presenting the major semantic properties of action verbs (Kemmerer 2006; Kemmerer et al. 2008) and argues that the linguistic representation of action is grounded in the mirror neuron system (Kemmerer and Gonzalez-Castillo 2010).

The model of Dominey et al. (2006) learns grammatical constructions as mappings from the form of some portion of a sentence to its meaning. It focuses particularly on thematic role assignment, determining who did what to whom. The grammatical structure of sentences is specified by a combination of cues including word order, grammatical function words (and or grammatical markers attached to the word roots), and prosodic structure. As shown in Figure 15.10, Dominey et al. (2006) consider constructions as templates consisting of a string of closed-class words as well as slots into which a variety of open-class elements (nouns, verbs, etc.) can be inserted to express novel meanings. (Their simplified constructions do not capture "the ball" vs. "a ball" nor do they show how the sentence could be marked for tense.) They make the simplifying assumption that the sequence of function words uniquely indexes

(a) Form: John gave the ball to Mary
Meaning: Gave(John, ball, Mary)

(b) Form: NP$_1$ V$_T$ NP$_2$ to NP$_3$
Meaning: Act$_T$(Agent, Obj, Recip)

Figure 15.10 An example (a) of applying a simple construction (b) relating parts of a sentence (form) to their semantic roles (meaning) (after Dominey et al. 2006).

the construction. However, *The boy threw Mary the bone* and *The brown dog licked the bone* both have the form <the ___ the _> but correspond to different constructions.

As shown in Figure 15.11, the cortical network at top replaces the content words by "slot indicators" and then recognizes the resultant "construction index." The system at the right recognizes the content words (currently, only nouns and verbs) and inserts them in the order received into a memory buffer called the "open class array"; each of these words triggers the neural code for its meaning, thus populating the "predicted referents array." The meaning of the sentence is then obtained by transferring these codes to the "scene event array," where the position will indicate the "action," "agent," "object," and "recipient" of the event specified by the input sentence. The "secret" of the

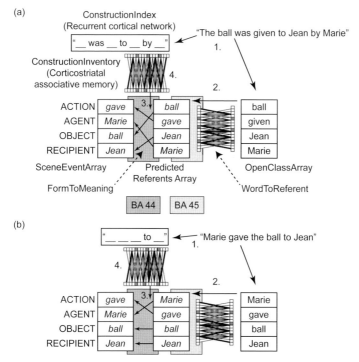

Figure 15.11 A model of recognizing constructions as a basis for assigning semantic roles: (a) a passive sentence; (b) an active sentence. Figure reprinted from Dominey et al. (2006) with permission from MIT Press.

system is that the construction index is decoded not as a sequence but as a biasing of synaptic weights for reordering the predicted referents array to the scene event array as appropriate to the current construction. Figure 15.11a shows the processing of a passive sentence whereas Figure 15.11b shows the processing of an active sentence.

Dominey et al. (2006) hypothesized that insertion of semantic content into the predicted referents array is realized in pars triangularis BA 45 and that the mapping from form to meaning corresponding to the scene event array will take place in frontal cortical regions including BA 44, 46, and 6. The model was tested by comparison of fMRI brain activation in sentence processing and nonlinguistic sequence mapping tasks (Hoen et al. 2006). Hoen et al. found that a common cortical network, including BA 44, was involved in the processing of sentences and abstract structure in nonlinguistic sequences whereas BA 45 was exclusively activated in sentence processing. This provides an interesting example of predicted brain imaging data from a processing model based on (though in this case not implemented as) detailed neural circuitry. In the current implementation, the neural network associative memory for the "construction inventory" is replaced by a procedural look-up table for computational efficiency. This renders the model too algorithmic. A more general concern is how to represent hierarchically structured sentences, binding items across different frames. Dominey's 2006 model did not really address "going hierarchical," though it did offer a simple account of the recognition of relative clauses. It should also be noted, as a challenge for future work, that the model of Figure 15.11 is not really an "events-in-time model" in that it does not take input incrementally over time, but rather requires that the whole sentence is buffered in memory before it extracts the order of the function words that characterize the current construction.

Language and the What/Where System

Having stressed the importance of the "what" and "where" systems in both visual control of hand movements and in auditory processing, we briefly consider an account of sentence processing explicitly based on "what" and "where." The model is connectionist rather than neurolinguistic, but it succeeds in addressing a wide range of psycholinguistic data on both acquisition and performance.

Figure 15.12 reflects key assumptions of Chang et al.'s (2006) model. Psycholinguistic research has shown that the influence of abstract syntactic knowledge on performance is shaped by particular sentences that have been experienced. To explore this idea, Chang et al. applied a connectionist model of sentence production to the development and use of abstract syntax. The model makes use of error-based learning to acquire and adapt sequencing mechanisms and meaning-form mappings to derive syntactic representations.

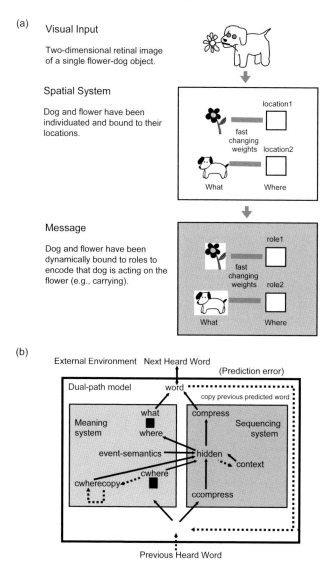

Figure 15.12 (a) Message representations consist of dynamic bindings between concepts (what) and roles (where). (b) Dual pathways: the dual-path model integrates separate meaning and sequencing systems, with restricted communication between them. Solid arrows: feedforward; dotted arrows: feedback. (Figure courtesy of Franklin Chang.)

The model is able to account for much of what is known about structural priming in adult speakers, as well as preferential looking and elicited production studies of language acquisition. As suggested by Figure 15.12a, message representations are said to consist of dynamic bindings between concepts (what) and roles (where), though it seems mistaken to say roles are a "where" function;

rather, the relations between "where's" used to establish the roles. We may see this as generalizing the action recognition model of Figure 15.7, where the relative motion of hand and object provides the key input for the mirror system to recognize the action performed by the hand. Moreover, Chang et al. associate the participants in an event with nonstandard roles rather than the more usual roles of "agent," "patient," instrument," etc. Their "XYZ roles" are designed to make the order of assignment of roles approximate the way that scenes are viewed. Chang et al. rather boldly view the above assumptions as providing part of "the universal prelinguistic basis for language acquisition." Whatever the status of this view, the search for understanding the relation between our experience of the world and our verbal description thereof seems like a fruitful approach to neurolinguistics, which locates language within a broader frame of action and perception (Lee and Barrès 2013).

The sequencing system (Figure 15.12a) is a simple recurrent network of the Elman type; that is, one in which the internal state ("context") is provided by the state of the hidden layer. Sequential prediction and production is enabled by a context or short-term memory that changes as each word is predicted or produced. The crucial point is that words are not simply sequenced on a statistical basis but are instead structured according to the constructions constrained by the XYZ roles of the meaning system. Hence, the event semantics helps the sequencing system learn language-specific frames for conveying particular sets of roles by giving the sequencing system information about the number of arguments and their relative prominence. In some sense, this is the dual of the model of Figure 15.11: going from vision to role assignment to learning how best to assemble words to express those roles (Figure 15.12) rather than extracting the sequencing of function words to activate a construction that then assigns content words to the appropriate roles. The model in Figure 15.12 learns to place words associated with more prominent roles earlier in the sentence because the environment input grammar reflects such a correlation, and the model's architecture allowed event semantics to affect the sequencing system. For subsequent refinements and extensions see, for example, Chang (2009) and Chang et al. (2012).

Discussion

Returning to music, we may see sensory consonance and dissonance in music within the frame of low-level expectations or predictions, whereas Lerdahl (2001b; this volume) built on the hierarchical event structure in Lerdahl and Jackendoff (1983) to develop a "formal" (but not computational or neural) model that generates quantitative predictions of tension and relaxation for the sequence of events in any passage of tonal music. Lerdahl and Krumhansl (2007) report psychological experiments used to test this model, focusing on the moment to moment buildup of tension in music and its resolution, and

Steinbeis and Koelsch (2008b) offer related data for neural resources in the processing of musical tension-resolution patterns. Clearly, such data challenge brain theory in a number of ways:

- What is the neurophysiological correlate of tension?
- How does this relate to the linkage of acoustic variables to musical features and these in turn to emotion?
- What processes build it up, and what processes resolve it?

These questions are very much related to "events in time." While a chord considered in isolation may have a psychological effect, the effect of the chord generally depends on the prior state of tension and resolution created by the previous music, and this depends, in turn, on the genre of the piece. Nonetheless, the challenge is posed: to understand the neural events in time that bridge between the formal model of Lerdahl, the psychological judgments assessed by Krumhansl, and the neural correlates of Steinbeis and Koelsch. Rather than address the challenge here, we turn to a related issue in language. When we converse with someone, or are about to turn a page in a book, we often anticipate the next few words to be spoken or read, respectively. Consider the following:

- "I haven't seen Jones for months. Do you know what happened to him?" "Oh, poor fellow. He kicked the …."
- "Where's Jones got to? And where did that hole in the wall come from?" "He went to the infirmary after he kicked the …."

In each case, the syntactic context of "he kicked the…" leads us to expect a noun phrase to complete the sentence, but the very different semantic contexts strongly bias which noun or noun phrase we will expect next. In the first example, the expectation may be dominated by the idiom "kick the bucket" whereas in the second, general sentence structure leads us to expect the words "wall" and "hole" with roughly equal probability. Thus in listening to an utterance, we may create syntactic as well as semantic expectations, and these will generally interact. As in the model presented in Figure 15.12, there is much work that relates grammar to expectation-based processing, and results on ERP data on N400 and P600 (see Hagoort and Poeppel, this volume) are consistent with the view that semantic and syntactic expectations, respectively, are created and have neural effects when they are violated. Gigley (1983) used a variant of *categorial grammar* (Steedman 1999; Bar-Hillel 1964) to offer a neural-like parsing mechanism which generates competing expectations: seeing a determiner, expect a noun phrase; when expecting a noun phrase, expect to see an adjective or noun. Moreover, Gigley showed how her model could be "lesioned" to yield some symptoms of agrammatic aphasia.

In relation to the theme of relating mechanisms for music and language to processes linking perception and action more generally (and recalling the earlier discussion; see section on The Construction of Events through Emotional

Learning), Schubotz and von Cramon (2002) used fMRI to explore correlates of our use of recognition of sequential patterns to predict their future course and thus to plan and execute actions based on current perceptions and previous experiences. Specifically, they showed that prediction of size in hand movements engages the superior part of the ventrolateral premotor cortex, whereas the prediction of pitch in articulation engages inferior-most ventrolateral premotor cortex. They conclude that events are mapped onto somatotopically corresponding motor schemas whenever we predict sequential perceptions. The challenge offered by the above examples, though, is to probe how the specific predictions are neurally coded and how the brain activates them: whether they correspond to the upcoming chords in a musical piece (and what, then, are the relevant motor schemas) or the upcoming words in a sentence. Bubic et al. (2010) build on such studies of general reflections on prediction, cognition, and the brain.

To continue, we distinguish between low-, mid-, and high-level approaches to language and music with an emphasis on auditory processing. Bottom-up or data-driven models mainly address low-level features associated with the acoustical signal, such as the relation between frequency and pitch perception (e.g., consonance and dissonance), timbre (e.g., sound source recognition), and binaural hearing (sound source localization). This research can exploit cell recordings in animals to gain data relevant to the early stages of perception of music and language.

Mid-level auditory scene analysis (Bregman 1990) focuses on preattentive processing mechanisms for figure-ground separation as well as simultaneous and sequential grouping of frequencies and has ties to data from single-cell recordings in animals and psychoacoustic experiments in humans. For the top-down approach of "schema-based" auditory scene analysis, only data from brain imaging or EEG studies are now available (Winkler et al. 2005; Koelsch, this volume; cf. Snyder and Alain 2007). Although there exist some connectionist models of music cognition, there is as yet no modeling approach related closely to brain areas and networks (for a review, see Purwins et al. 2008a).

In his overview of computational models of emotional processing, Fellous (2009:912) concludes: "Mainly because of the availability of experimental data, the focus of neural models of emotions has been on fear and reward processing; there have been relatively few attempts at using neural networks to model emotions in general." The situation is even worse in computational modeling of musical emotions. Experimental data on brain circuits involved in the processes linking music and emotion are only now becoming available (for an overview, see Koelsch, this volume). Future computational modeling of language, action, and music processing should take the following into account:

1. Emotions are dynamical modes of functioning and not states and are internally (memory) and externally (perception) driven.
2. There is no one predominant emotional center in the brain.

3. Emotions carry out a functional role such as redirecting or modulating of information flow in cognitive processing (cf. Fellous 2009:912–913)

Our intent in this chapter has been to provide a useful basis for future modeling of brain mechanisms for music and language processing, both catalyzing and benefiting from new experiments on the interplay of music, language, and action. We have briefly noted the importance of taking into account the biomechanics of body movements and internal models of the brain's processing of action–perception loops as well as mechanisms of social interaction. An important challenge would include modeling the integration of dance with music. We finally note (for a related discussion, see Verschure and Manzolli, this volume) the promise of using robots as tools for such research, with new experimental paradigms moving "out of the lab" to use augmented and virtual environments in connection with robots to collect data in interactive situations and test neural and schema models of the underlying processes.

Acknowledgments

Our thanks to Franklin Chang, Josef Rauschecker, and an anonymous reviewer for their thoughtful and constructive comments on an earlier draft.

16

Computational Modeling of Mind and Music

Paul F. M. J. Verschure and Jônatas Manzolli

Abstract

Music can be defined as organized sound material in time. This chapter explores the links between the development of ideas about music and those driven by the concept of the embodied mind. Music composition has evolved from symbolic notated pitches to converge onto the expression of sound filigrees driven by new techniques of instrumental practice and composition associated with the development of new interfaces for musical expression. The notion of the organization of sound material in time adds new dimensions to musical information and its symbolic representations. To illustrate our point of view, a number of music systems are presented that have been realized as exhibitions and performances. We considered these synthetic music compositions as experiments in situated aesthetics. These examples follow the philosophy that a theory of mind, including one of creativity and aesthetics, will be critically dependent on its realization as a real-world artifact because only in this way can such a theory of an open and interactive system as the mind be fully validated. Examples considered include RoBoser, a real-world composition system that was developed in the context of a theory of mind and brain called *distributed adaptive control* (DAC), and "ADA: intelligent space," where the process of music expression was transported from a robot arena to a large-scale interactive space that established communication with its visitors through multi-modal composition. Subsequently, *re(per)curso*, a mixed reality hybrid human–machine performance, is analyzed for ways of integrating the development of music and narrative. Finally, the chapter concludes with an examination of how multimodal control structures driven by brain–computer interfaces (BCI) can give rise to a brain orchestra that controls complex sound production without the use of physical interfaces. All these examples show that the production of sound material in time that is appreciated by human observers does not need to depend on the symbolically notated pitches of a single human composer but can emerge from the interaction between machines, driven by simple rules and their environment.

Introduction

One contemporary definition of music, given by the composer Edgar Varèse (1883–1965), holds that "music is organized sound." Here we expand this definition to: "music is the organization of sound material in time." Our thesis in this chapter is that in the twentieth century, sound gradually gained center stage in the musical production realm and that it is possible to compare this shift with theories of the embodiment of mind. Starting with the use of sonorities in Claude Debussy's music in the beginning of last century and concomitant with the development of more precise mechanisms for capturing and broadcasting sound via recording, microphones, speakers, and computers, passing through *musique concrète* (Schaeffer 2012), the notion of music composition diverges from symbolic notated pitches to converge onto the expression of sound filigrees. These compositional trends resulted in and were driven by the development of new techniques of instrumental practice associated with the development of new interfaces for music expression, methods for sound synthesis using digital instruments, and interactive devices used for human–machine interaction. The notion of the organization of sound material in time adds dimensions to musical information and its symbolic representations. Essentially we see a shift from the top-down specification of musical pieces in the Western musical tradition, from the single genius of the single composer to the tightly controlled production pipeline of the conductor and the orchestra to a paradigm where musical structure emerges from the interaction between music systems and their environment. Indeed, music research has evolved toward analyzing the nature of the sound phenomena and the various possibilities of extracting information from sound pressure waves via spectral analysis, and major advances have been made in the area of recognizing timbre (Peeters et al. 2010), including those of percussion instruments (Herrera et al. 2002). This emphasis on sound as such, combined with new technologies for its synthesis and combination in complex compositions, has shifted the focus from composing for a particular musical piece to designing for a potential space of musical expression, where the particular musical piece rendered is synthesized in real time and defined through the interaction between the music system and its environment. To illustrate our point of view, we present in the next sections a trajectory that starts with the so-called RoBoser,[1] a real-world composition system that generates coherent musical symbolic organization and production using trajectories produced by robots without establishing a process based on preestablished macro-level rules. Instead of using generative grammars, we tested the possibility that long-term musical forms can emerge from the iteration of simpler local rules by a system that interacts with its environment, RoBoser (Manzolli and Verschure 2005). The data streams of RoBoser were associated with the internal representations of a robot control architecture, called *distributed adaptive control* (DAC) (for a review, see Verschure 2012), especially with regard to its reactive mechanisms of attraction and repulsion

---

[1] Supplemental online information is available at http://specs.upf.edu/installations (for video and sound tracks of the RoBoser-derived systems discussed in this chapter as well as additional material references). See also http://esforum.de/sfr10/verschure.html

to light stimuli and collisions and the acquired responses to visual events. The second system we analyze is "ADA: intelligent space"; in this case the process of music expression was transported from the robot arena to a large-scale space (Eng et al. 2005a, b). We tested the interaction between the stimuli generated by the space (sound, light, computer graphics) and visitors' behaviors as a way of producing synthetic emotions and their expression in sound material (Wasserman et al. 2003). With ADA, we worked with the concept that the association of music with other forms of expression can communicate internal emotional states of the controller of the space, where emotions were defined by the ability of the space to achieve its multiple goals. In the hybrid human–machine mixed reality performance *re(per)curso* (Mura et al. 2008), we verify ways of integrating the development of music and narrative through large-scale interaction. The structural pillars of this chronicle were not a script or a storyteller's description; in this new perspective we used the concept of recursion as a possible way of constructing meaning. Precisely, in a recursive process between humans and synthetic agents, including an avatar in the virtual world, a meaningful exchange was realized through their interactions. Finally, we examine how multimodal control structures driven by brain–computer interfaces (BCI) can give rise to a "brain orchestra" that controls complex sounds production without the use of physical interfaces (Le Groux et al. 2010). The Brain Orchestra performance capitalized on the production of sound material through real-time synthetic composition systems by allowing an emotional conductor to drive the affective connotation of the composition in real-time through her physiological states, thus giving rise to an emergent narrative. All of these examples show that the production of sound material in time that is appreciated by human observers does not need to depend on the symbolically notated pitches of a single human composer but can emerge from the interaction between machines, driven by simple rules, and their environment.

In analyzing this paradign shift in the production of music, we will also assess how the broadening of concepts of music in recent years might assist us in understanding the brain, and vice versa. Much of the theoretical study of music has been based on the tonal music system developed in Europe during the eighteenth and nineteenth centuries (Christensen 2002). Such a foundation, however, excludes a large variety of music generated outside this geographic region and time frame.

A number of suggestions have been made to define the key features of tonal music: from the subjective analysis provided by Schenkerian theory (Schenker and Jonas 1979) to the work of Eugene Narmour (1990), which was inspired by gestalt psychology. Fred Lerdahl and Ray Jackendoff (1983) distinguish four hierarchical structures, or domains, in tonal music:

1. pitch material (time-span reduction),
2. metric and meter,
3. harmonic and melodic structure (prolongational reduction), and

4. motives, phrases, and sections (grouping).

In each of these domains, akin to the notion of grammar in language (see Lerdahl, this volume), two sets of rules define possible and actual "legal" output structures: well-formedness and preference rules, respectively. A practical expression of such a formal approach toward tonal music can be found in the so-called *experiments in musical intelligence* of David Cope (2004). Based on a prior analysis of harmonic relationships, hierarchical structure, and stylistic patterns of specific compositions, Cope has written algorithms to generate plausible new pieces of music in the style of the original input music. In contrast to a formal rule-based approach, others have argued from an ethnographic point of view that the notion of musical universals is questionable, and that music is largely defined as a social construct with some invariants with respect to pitch contrasts and timbre (Patel 2003). As yet, the question of what music is and whether it is based on universals of perception, cognition, emotion, and action has no clear answer (for a skeptical discussion of universals in language, see Levinson this volume). We propose that building autonomous synthetic music composition systems will be a key methodology in resolving this issue.

A further complication in elucidating the causal organization of music is the practically boundless ability of the brain to adapt to its environment at varying timescales (Bell 1999). An additional bias to this discussion stems from the grammar-oriented linguistic approach toward music, which implies that music, like language, is conceived of as being representationalist, formal, and internalist; in other words, it is "generated within the head." In sum, traditional Western music theories are formal structural theories (for a discussion on the distinction of structural and process descriptions of the computational mind, see Jackendoff 1987:37–39, 44) that apply primarily to the Western music (-score) tradition. Another distinguishing feature of the Western musical tradition is that it has adopted a very specific and limiting process model of expression: the perceiver is a passive listener who is exposed to the external genius of the composer, whose creation is mediated by a drone-like performer. The performer has mastered the interface between the musical score and the stream of sounds through a musical instrument. These instruments are designed with the goal of reliably producing pitches given physical models of sound production, often sacrificing the ecological validity of their use. This accounts for the superhuman effort it takes to fully master musical instruments. We call this the musical interface bottleneck and one could argue that it is one of the limiting factors on human creativity. Musical expression and experience is closely linked to the tools humans have developed for the production of pitches. These tools will increasingly become virtual, as opposed to physical, with deep implications for the future of music (Coessens 2011).

Symbolic artificial intelligence (AI), with its study of the disembodied mind following a computer metaphor (Newell 1990), is fully coherent with the traditional theoretical treatment of music. Both can be characterized as rationalistic and grounded in rule-based analysis. Since the mid-1980s, traditional AI has been superseded by an alternative paradigm that is embodied, situated, and action oriented (Brooks 1991; Pfeifer and Bongard 2007; Pfeifer and Scheier 1999). The associated developments in artificial life, connectionist or neural networks as well as behavior-based robotics have sparked interest in a more biologically grounded approach to the study of mind and behavior (Verschure 1993; Edelman 1987), and we expect that comparative music and language research will follow suit. The question is: What shape will process-oriented theories of mind, brain, and behavior take, and to what extent will they enable comparative research on action, language, and music?

A biologically grounded approach has advantages, since it is founded on the dynamics of real-world interaction and embodiment. While various authors have emphasized the need to complement the study of language by methods which capture the real-world aspects of mind and brain, along with their realization and expression in action, including computational and robot-based approaches (Cangelosi 2010; Steels 2010), our focus here is on an approach to music within this framework (Verschure 1996, 1998; Le Groux and Verschure 2011).

Despite the rapid development of computer music systems, some of it reviewed below, the transition still needs to be made from partial models for isolated tasks (at best evaluated with isolated empirical data) to general theories rooted in the recognition of the inherent embodiment of perception, emotion, cognition, and action (Verschure 1993; Leman 2007; Purwins et al. 2008a, b). We argue that a general framework for computational theories of music processing must be action-oriented and reflect the interaction of music-making (and language-using) systems. Implementation of compositional process prototypes using the interaction of multiple agents can give rise to a large number of generative structures producing a complex informational network. Composition is considered as an emerging organization of sound events and, more generally, as an organization of succession and overlapping elements of musical discourse (for an ethnographic perspective, see Lewis this volume). As a concrete realization of such an approach, we describe below the biologically grounded robot-based cognitive architecture, DAC, which has been proposed as an integrative and self-contained framework to understand mind, brain, and behavior (for a review, see Verschure 2012). In addition to a range of derived models in computational neuroscience (see also Arbib et al., this volume), DAC provides the foundation for a series of experiments in situated multimodal composition and performance (summarized in later sections; for a discussion of processing models grounded in neuroscience, see Arbib et al., this volume). Indeed, rule-based methods to create music have existed since Guido D'Arezzo (ca. 991–ca. 1028) proposed, in chapter XVII of his *Micrologus* (1025–1026), a formal composition process for generating melodies based on the mapping of letters extracted from Latin liturgical texts (Loy 1989). D'Arezzo used a look-up table to map vowels to one of the table's possible corresponding pitches. Other composers have also used themes derived from letter acrostics, such as the *B-A-C-H* motive (which, translated from German corresponds to B flat, A, C, B natural) used by J. S. Bach to introduce the last and unfinished fugue of the *Art of the Fugue*.

From the 1950s and into the 1990s, the programmed computer was viewed in AI and related work in cognitive science as a model of the mind; this emphasized the underlying rules and representations and overcame the peripheralism of the behaviorists "empty organism" approach while ignoring the system-oriented analysis provided by cybernetics. During this period, AI remained committed to the functionalism of the computer metaphor, but it ran into a number of stumbling blocks:

- The frame problem, which demonstrates that symbolic AI systems will succumb to the exponential growth of their world models (McCarthy and Hayes 1969).
- The symbol-grounding problem, which shows that understanding cannot be achieved on the basis of a priori symbolic specification (Searle 1980; Harnad 1990).
- The problem of situatedness, which argues that by virtue of being in the world, an agent need not represent its environment at all cost; the world also exists to represent itself (Suchman 1987).
- The frame of reference problem, which argues that when talking about internal representation, one needs to consider the point of origin; that is, whether they originated with a designer, an observer, or the agent itself (Clancey 1992; for an overview, see Verschure 1993; Pfeifer and Scheier 1999).

Essentially, these problems boil down to the issue of relying on representations that are not autonomously defined by an agent, but which are externally defined as a prior to the operation of the system and organized in an exponentially growing search space; in other words, the *problem of priors* (Verschure 1998). This raises the fundamental question of how biophysical systems, like human brains, become interpretative systems capable of using meaningful signs (Deacon 2006).

Connectionist models (i.e., networks of simplified neuron-like elements whose connections vary according to certain "learning rules") were proposed in the late 1980s to solve some of these challenges (Rumelhart and McClelland 1986). They had, however, difficulty addressing the problem of priors: a supervised learning system still needs to "know" that an error was produced, and problems of scaling up to complex processing must be faced (Fodor and Pylyshyn 1988). Unlike symbolic computational systems, connectionist models could neither rely on symbolic representation nor on recursion. Indeed, one important implication of the concept of recursion for brain-like processing is that it must decouple the operators from symbols or the process from representation. Most theories of neural coding depend on the notion of labeled lines (Kumar et al. 2010), where the connections between neurons are uniquely labeled with respect to their contribution to specific representations. This view originates in the earliest proposals on neural computation, such as the logic circuits of McCulloch and Pitts (1943) and the perceptron of Rosenblatt (1958). Ultimately every connection in the network will have a specific label that contributes to the mappings the system can perform. There are very few proposals in the literature that show how the brain could achieve such a decoupling of process and representation. One of these, the so-called *temporal population code* (Wyss et al. 2003a; Luvizotto et al. 2012), is discussed by Arbib et al. (this volume).

Another concern in declaring recursion a natural category is that it is defined in the context of a logic that is constrained and sequential. Although the brain is by no means fully understood, it is a parallel structure that relies strongly on massive input-output transformations and a fast memory capacity at varying temporal scales (Rennó-Costa et al. 2010; Strick et al. 2009). Hence, the constraints under which it operates might be rather different from those requiring recursion and symbolic processing. We must be careful not to impose standard notions of hierarchy on the brain, such as the sense–think–act cycle which dates back to the nineteenth century psychophysicists of Donders, and which still is widely adhered to in cognitive science or the aforementioned perceptron and its contemporary descendants. On anatomical grounds, there is no reason for a strict distinction between sensory and motor areas (Vanderwolf 2007). Theoretical studies have shown that the entropy in the activity of motor areas can be exploited to "sense" behavioral decision points (Wyss et al. 2003b), thus blurring the strict distinction between perception and action.

Parallel to the necessity of using massive input-output transformations and multiscale memory capacity to understand the mind and brain, Western music developed over the last century the concept of "composition of mass" or "sound clouds," as in, for example, the compositions by Iannis Xenakis (1992). Others, such as Steve Reich, used recursion to define music as a gradual process (Schwartz 1981). In his *Études pour Piano* (1985–2001), which were inspired by fractal geometry (Mandelbrot 1977), György Ligeti constructed complex sound textures based on iterations of rhythmic and melodic patterns called "pattern-mechanico compositions" (Clendinning 1993).

Even in a simple musical structure, such as in *Clapping Music* (composed by Reich in 1972 for two musicians, who perform it by clapping their hands), it is possible to observe how one temporal-scale "phase shift" produces a diversity of rhythmic perception (Colannino et. al. 2009). In Ligeti's 8th *Étude pour Piano*, a simple recursive structure (based on the reiteration, in different octaves, of the pitch "A") produces a complex stratification of pitch patterns (Roig-Francolí 1995; Palmer and Holleran 1994).

In summary, application of the mathematical-logical concept of *computation* to finite real-world systems and their models requires us to explore intrinsic problems of this approach: the symbol-grounding problem, the frame problem, the problem of biological and cultural situatedness, and the frame of reference problem. In addition, to explain mechanism and interaction, process-oriented theories of mind are needed. Moreover, computational modeling with robots is necessary to test issues of embodiment and interaction with the external world. Such an integrated control system of the agent combined with its embodiment constitutes a theory of the body–brain nexus in its own right. For instance, using the DAC architecture (see below) we have shown that a direct feedback exists between the structures that give rise to perception and the actions of an agent, or behavioral feedback (Verschure et al. 2003). Hence, without understanding the interaction history of the agent, the specifics of its perceptual structures cannot be fully understood. The interaction history, in turn, depends on details of the morphology of the agent; its action repertoire and properties of the environment. Hence, we predict that behavioral feedback plays a decisive role in how perception, cognition, action, and experience are structured in all domains of human activity including music and language.

Although it is beyond the scope of this chapter to discuss the modeling of single neurons (for a brief introduction to a variety of approaches, see Koch et al. 2003), the key point is that such models reflect the microscopic organization of the anatomy and physiology of specific areas of the nervous system. Thus, the question then becomes: How can such a level of description be scaled up to overall function (Carandini 2012)? The so-called *schema theory* offers perhaps an alternate approach (see Arbib, this volume). Schema theory proposes a hierarchical decomposition of brain *function* rather than structure, although some of the schemas in a model may well be mapped onto specific neural structures. Schema theory faces, however, the problem of any functionalist theory: in decoupling from physical realization, its models are underconstrained (Searle 1980) and thus cannot overcome the challenge of the problem of priors. A third approach is grounded in the notion of convergent validation. Here, multiple sources of constraints (e.g., behavior, anatomy, and physiology) are brought together in a computational framework (Verschure 1996): the DAC architecture, which aims at integrating across these functional and structural levels of description (discussed further below). Other examples can be found in the Darwin series of real-world artifacts proposed by Edelman (2007). Arbib (this volume) also discusses a range of other models, some explored in simulation, others using robots.

Lessons from Robotic Approaches to Music Systems

In documenting Babbage's analytical engine, Ada Lovelace Lady Byron anticipated that the mechanized computational operations of this machine could enter domains that used to be the exclusive realm of human creativity (Lovelace 1843/1973):

> Again, it [the Analytical Engine] might act upon other things besides number, were objects found whose mutual fundamental relations could be expressed by

those of the abstract science of operations...Supposing, for instance, that the fundamental relations of pitched sounds in the science of harmony and of musical composition were susceptible of such expression and adaptations, the engine might compose elaborate and scientific pieces of music of any degree of complexity or extent.

Indeed, now 170 years later, such machines do exist and computer-based music composition has reached high levels of sophistication. For instance, based on a prior analysis of harmonic relationships, hierarchical structure, and stylistic patterns of specific composition(s), Cope has written algorithms to generate plausible new pieces of music in the style of the composer behind the original pieces (Cope 2004). However, despite these advances, there is no general theory of creativity that either explains creativity or can directly drive the construction of creative machines. Definitions of creativity mostly follow a Turing test-like approach, where the quality of a process or product is declared to be creative dependent on the subjective judgment by other humans of its novelty and value (Sternberg 1999; Pope 2005; Krausz et al. 2009; for a more detailed analysis, see below). The emerging field of computational creativity defines its goals explicitly along these lines (Colton and Wiggins 2012). Although these approaches could generate interesting applications, its definition is not sufficient either as an explanation nor as a basis to build a transformative technology as the classic arguments of Searle against the use of the Turing test in AI has shown (Searle 1980): mimicry of surface features is not to be equated with emulation of the underlying generative processes. For instance, Cope's algorithms critically depend on his own prior analysis of the composer's style, automatically leading to a similar symbol grounding problem that has undercut the promise of the AI program (Harnad 1990). The question thus becomes: How can we define a creative musical system that is autonomous?

Computational models implemented in robots provide new ways of studying embodied cognition, including a role in grounding both language and music. In particular, models of perception, cognition, and action have been linked to music composition and perception in the context of *new media art* (Le Groux and Verschure 2011). In new media art, an *interactive environment* functions as a kind of "laboratory" for computational models of cognitive processing and interactive behavior. Given our earlier critique of the rationalist framework paradigm for music, where the knowledge in the head of the composer is expressed through rules and representations and projected onto the world without any feedback or interaction, it is surprising that the advent of advanced automated rule execution machines (i.e., computers) in the twentieth century has not given rise to a massive automation of music generation, based on the available formal linguistic theories and their application to music composition. (The same surprise can be extended to the field of linguistics in general, where machine-based translators are still struggling to be accepted by humans.) This raises a fundamental question: Are grammar-like theories of language more effective as post hoc descriptions of existing musical structures, or can they generate plausible and believable ones de novo? In other words, does music theory suffer from a form of the symbol-grounding problem that brought classical AI to its knees (Searle 1980)? Is a similar shift from a priori rules and representations to acquired, enactive, and embodied semantics necessary, as has befallen the stagnated cognitive revolution (Verschure and Althaus 2003)?

Human–Machine Interaction Design in New Media Art

With the advent of new technologies that have emphasized interaction and novel interfaces, alternative forms and modes of music generation have been realized (Rowe 1993; Winkler 2001). This development raises fundamental questions on the role of embodiment as well as the environment and interaction in the generation of music and musical aesthetics. In addition, it places emphasis on a more situated and externalist view. In most cases, interactive music systems depend on a human user to control a stream of musical events, as in the early example of David Rockeby's *Very Nervous System* (1982–1991). Similarly, the use of biosignals in music performance goes back to the early work on music and biofeedback control of Rosenboom (1975, 1990). This pioneering perspective on music performance was developed to integrate musical performance with the human nervous system.

The Plymouth musical brain–computer interface (BCI) project (Miranda and Wanderley 2006) was established to develop assistive technologies based on interactive music controlled by the BCI. This new technology has been proposed to support human creativity by using technology to overcome the musical interface bottleneck.

Tod Machover's *The Brain Opera* (1996) was inspired by ideas from Marvin Minsky on the agent-based nature of mind. This opera consists of an introduction, in which the audience experiments and plays with a variety of Machover's instruments (including large percussive devices that resemble neurons), followed by a 45-minute musical event orchestrated by three conductors. The music incorporates recordings made by the audience as it arrived, along with material and musical contributions made by participants via the World Wide Web. For a detailed description of the musical instruments that use sensing technology in *The Brain Opera*, see Paradiso (1999).

Recently there has been a shift in music production and new media art. Interactive computer music provides new devices for musical expression and has been developed in different contexts for different applications. Miranda and Wanderley (2006) discuss the development and musical use of *digital musical instruments* (DMIs); that is, musical instruments comprised of a gestural controller used to set the parameters of a digital synthesis algorithm in real time, through predefined mapping strategies. In addition, international forums, such as the New Interfaces for Musical Expression (NIME), have convened researchers and musicians to share knowledge on new musical interface design

(NIME 2013). In parallel, the concept of an "orchestra" has expanded to incorporate the use of computers and interfaces as musical instruments. For instance, the Stanford Laptop Orchestra (SLOrk; http://slork.stanford.edu/) is a computer-mediated ensemble made up of human-operated laptops, human performers, controllers, and a custom multi-channel speaker array which render unique computer meta-instruments.

Another ensemble that explores the use of DMIs is the McGill Digital Orchestra (Ferguson and Wanderley 2010). Here, two types of real-time information are used—digital audio information (i.e., the sonic output of an instrument) and gestural control data (i.e., the performance gestures of the instrumentalist)—and reproduced, stored, analyzed, recreated and/or transformed. This particular type of digital technology allows the musical potential of DMIs to be studied in an interdisciplinary setting and furthers the development of prototypes in close collaboration with instrument designers.

As mentioned above, much of the current musical repertoire is structured on perceptual attributes of sound that are not necessarily the ones related to pitch-oriented symbolic music. Sound attributes, such as roughness and brightness, are also considered as structural elements of current compositional trends. The new development of DMIs liberates us from having to construct instruments based on the physical properties of materials, and thus disconnects the interface from the production of pitches. As mentioned above, the body's perceptual, cognitive, motor, and kinesthetic responses have to be reconfigured to the needs and constraints concerning action and perception in this new space, and the interface can now be optimized to its user (Coessens 2011). The point could be made that the history of music has been characterized by the invention of new musical instruments or the adaptation of musical instruments from other cultures. Indeed, the use of musical instruments as such is embodied, and the shift to computer-generated sound with arbitrary interfaces will remove this form of the embodiment of musical experience. However, we emphasize that musical instruments have thus far exploited the adaptive capabilities of the human body–brain nexus as opposed to building on its specific natural confirmations. As a result, we claim that a price has been paid in terms of both the quality and the accessibility of musical expression.

The Distributed Adaptive Control Architecture

The limitations of the computer metaphor, as summarized in the problem of priors, explicitly challenge our understanding of mind and brain. The DAC architecture (Figure 16.1) was formulated in response to this challenge and provides the integrated real-time, real-world means to model perception, cognition, and action (Verschure 1992, 2003, 2012). In particular, DAC solves the problem of symbol grounding by relying on autonomous and embodied learning to provide a basis for the world model in the history of the agent itself on the basis of which cognition emerges. This is augmented with situated action

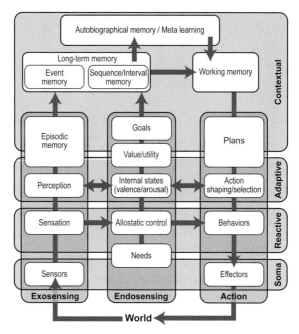

Figure 16.1 Conceptual diagram of the *distributed adaptive control* (DAC) architecture for perception, cognition, and action (for a review and examples of its many applications, see Verschure 2012), showing four tightly coupled layers: *soma, reactive, adaptive,* and *contextual*. Across these layers, three functional columns of organization can be distinguished: exosensing, defined as the sensation and perception of the *world*; endosensing, the detection and signaling of states derived from the physically instantiated *self*; and the interface to the world through *action*. At the level of the contextual layer, these three axes become tightly integrated. The arrows show the primary flow of information, mapping exo- and endosensing into action. At each level of organization, increasingly more complex and memory-dependent mappings from sensory states to actions are generated, dependent on the internal state of the agent. The somatic layer designates the body itself and defines three fundamental sources of information: sensation, needs, and actuation. The reactive layer endows a behaving system with a prewired repertoire of reflexes, enabling it to display simple adaptive behaviors, unconditioned stimuli, and responses on the basis of a number of predefined behavioral systems through a self-regulatory or allostatic model (Sanchez-Fibla et al. 2010). The adaptive layer, linked to a valence-attribution system, provides the mechanisms for the construction of the world and agent state space through the adaptive classification of sensory events and the reshaping of responses using prediction-based learning mechanisms (Duff and Verschure 2010). The contextual layer forms sequential representations of these states in short- and long-term memory in relation to the goals of the agent and its value functions. These sequential representations form plans for action which can be executed, dependent on the matching of their sensory states to states of the world further primed by memory. The contextual layer can be further described by an autobiographical memory system, thus facilitating insight and discovery. It has been demonstrated (Verschure and Althaus 2003) that the dynamics of these memory structures are equivalent to a Bayesian optimal analysis of foraging, thus providing an early example of the predictive brain hypothesis.

by mapping acquired knowledge onto policies for action through learning. DAC is one of the most elaborate real-world neuromorphic cognitive architectures available today. It is based on the assumption that a common *Bauplan* (blueprint) underlies all vertebrate brains that have ever existed (Allman and Martin 2000).

DAC proposes that for an agent to generate action and successfully survive, the agent must address the following questions:

1. *Why?* (Motivation for action in terms of needs, drives, and goals must be clarified.)
2. *What?* (Objects involved in the action must be specified.)
3. *Where?* (Location of objects must be defined from the perspective of the world and self.)
4. *When?* (Timing of an action must be specified relative to the dynamics of the world.)

These questions constitute the "H4W" problem, and each *W* represents a large set of sub-questions of varying complexity (Verschure 2012). H4W has to be answered at different spatiotemporal scales.

Further, DAC proposes that the neuraxis is conceptualized as comprising three tightly coupled layers—reactive, adaptive, and contextual—which roughly follow the division of the mammalian brain in the brainstem, diencephalon, and telencephalon:

1. At the reactive level, predefined reflexes, or motor schemas, define simple behaviors that provide basic functionality (e.g., defense, orientation, mobility, etc.). Activation of each reflex, however, is considered to be a signal for learning at subsequent levels.
2. The adaptive layer acquires representations of states of the environment, as detected by the sensors of the robot, and shapes the predefined actions to these states via classical conditioning implemented with a local prediction-based Hebbian learning rule. Essentially, the adaptive layer addresses a fundamental problem of developing a state space description of the environment (sensation) and the degrees of freedom of the body (action), conditional on the internal states of the organism (drives/emotions/goals).
3. At the level of the contextual layer, the states of the world and those of the body are integrated into sequential short- and long-term memory systems, conditional on the goal achievement of the agent. Sequential memories can be recalled through sensory matching and internal chaining among memory sequences (Figure 16.2).

At each level of organization, increasingly more complex, memory-dependent mappings from sensory states to actions are generated that are dependent on the internal state of the agent.

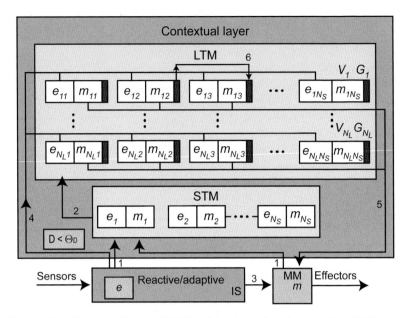

Figure 16.2 Contextual layer of DAC and the dynamics of sequencing: (1) The perceptual prototype e and the motor activity m generated by the adaptive layer are acquired and stored as a segment if the discrepancy between predicted and occurring sensory states, D, falls below its predefined threshold, Θ_D. D is the time-averaged reconstruction error of the perceptual learning system: $x - e$. (2) If a goal state is reached (e.g., reward or punishment detected), the content of short-term memory (STM) is retained in long-term memory (LTM) as a sequence conserving its order, and STM is reset. Every sequence is labeled with respect to the specific goal and internal state to which it pertains. (3) The motor population MM receives input from the IS populations according to the rules of the adaptive layer. (4) If the input of the IS population to MM is subthreshold, the values of the current CS prototypes, e, are matched against those stored in LTM. (5) The MM population receives the motor response calculated as a weighted sum over the memory segments of LTM as input. (6) The segments that contributed to the executed action will prospectively bias segments with which they are associated.

The DAC architecture is unique in the following ways:

1. It is a complete architecture integrating perception, emotion, cognition, and action (for a comparison with other cognitive architectures, see Verschure 2012).
2. It resolves the symbol-grounding problem in the context of perception, cognition, and action supporting real-world behavior.
3. Specific DAC subsystems have been generalized to identified subsystems of the brain, including the cerebellum, amygdala, cerebral cortex, and hippocampus (Duff et al. 2011; Rennó-Costa et al. 2010; Wyss et al. 2006; Hofstötter et al. 2002; Sanchez-Montanes et al. 2000; Marcos et al. 2013).

4. It can give rise to structured and complex behavior, including foraging, maze learning, and solving logical puzzles (Duff et al. 2011; Verschure and Althaus 2003).

The DAC architecture has been validated in a range of domains, including foraging, classical and operant conditioning, and adaptive social behaviors by a human accessible space (Eng et al. 2005a, b). It has also been generalized to operate with multiple sensory modalities (Mathews et al. 2009) and has been augmented with an integrated top-down bottom-up attention system (Mathews et al. 2008; Mathews and Verschure 2011).

The DAC architecture autonomously generates representations of its primary sensory inputs at the level of the adaptive layer using a fully formalized neuronal model based on the objectives of prediction and correlation (Duff and Verschure 2010; Verschure and Pfeifer 1992). Indeed, more recently, these have been proposed to be the key principles underlying the organization of the brain (Friston 2009). In the case of foraging, the dynamics of the memory structures of DAC have been demonstrated to be equivalent to a Bayesian optimal analysis (Verschure and Althaus 2003). The principle of behavioral feedback discovered with DAC (Verschure 2003), mentioned above, essentially states that because of learning, behavioral habits emerge that restrict the effective environment to which an agent is exposed. Given that perceptual structures are sensitive to the statistics of their inputs, this bias in input sampling translates into a bias in perception itself.

The DAC framework proposes that the memory systems for contextual action generation provide a substrate onto which language and music can be generated. In other words, DAC proposes that the brain has evolved to define events, states of the world, and the body, and to organize these in time—the events-in-time hypothesis. In the context of the computational mind hypothesis, the question arises as to what extent the processes implemented in DAC can account for recursion. The contextual memory of DAC does provide a substrate for compositionality, where one active sequence can be effectively embedded in others through association or can predefine heuristics for goal-oriented chaining. In addition, the interactions between short- and long-term memory are bidirectional so that active states of long-term memory can be recombined and reused through parallel interaction with short-term memory. Finally, DAC solves one important challenge to the Turing machine model: DAC is situated in the real world. Whereas the Turing machine is defined on the basis of an infinite sequential tape, DAC uses the real world as a representation of itself. Actions lead to changes in its task domain and can subsequently be perceived and remapped onto memory, leading to new actions; that is, situated recursion exploiting both the environment and memory.

An application of DAC—the RoBoser system (Manzolli and Verschure 2005)—has been used in interactive installations, performance, and music composition, to which we now turn. DAC is uniquely suited for this step because the fundamental components of an interactive performance requires the multilevel integration that DAC offers, including systems for perception of states of the world, their appraisal relative to the goals of the performance system, planning of actions, and the execution of actions. In other words, we consider the technical infrastructure required for an interactive performance as a robot turned inside-out, with the dynamics resulting from the actions of other agents in the environment: performers.

From RoBoser to ADA and XIM

An alternative to the formalistic paradigm of music composition was proposed in the RoBoser project, where the musical output of the system sonifies the dynamics of a complex autonomous real-world system (Manzolli and Verschure 2005). First exhibited in 1998 in Basel as an interactive installation, RoBoser addressed the fundamental problem of novelty in music composition. The ability to induce novelty and surprise in listeners is recognized as one of the hallmarks of human creativity expressed in music. Solutions to this problem, however, can be plainly random. Take for instance, Mozart's so-called *Würfelspiel*, where a roll of dice is used to select sections of music, which are then pieced together to form a musical composition (Mozart 1787). The *Würfelspiel* can be seen as an attempt to solve the grouping problem, as it combines musical motives into larger structures through random choices, dictated by dice. Contemporary composers have followed suit using various methods including generative grammars, cellular automata, Markov chains, genetic algorithms, and artificial neural networks (Nierhaus 2009).

In the RoBoser system, the question of whether the interaction between instantiated musical primitives and the real world can give rise to plausible musical structures is explicitly addressed. To answer this question, a synthetic multivoice composition engine, called Curvasom (Maia et al. 1999), is coupled to a mobile robot and the neuromorphic control architecture (i.e., DAC). This renders a system consistent with work on multivoiced music perception (Palmer and Holleran 1994). In addition, the RoBoser system makes the observer a participant as well as a composer through interaction, thus solving the interface bottleneck.

Curvasom is based on a number of internal heuristics that transform input states into timed events sonified using the MIDI sound protocol, or "sound functors," resulting in a set of parametric sound specifications with which the system is seeded (Manzolli and Maia 1998). The integrated composition engine sonifies a number of states of the robot and its controller. These include the peripheral states of the sensors of the robot, the default exploration mode of the robot, and its internal states, which can be driven by either predefined stimulus events or acquired ones. States of the robot may change properties of single voices (e.g., intensity or timbre) and of all voices (e.g., changes in tempo or shifts in pitch). A detailed analysis of the robot performance showed that the integrated robot-composition system was able to generate complex and coherent musical structures (Manzolli and Verschure 2005). In a subsequent generalization of this approach, the interaction was expanded to include a complete building, called ADA (Figure 16.3; see also http://specs.upf.edu/installation/547). In ADA, audiovisual expression is generated on the basis of the interaction between the visitors to the space and the neuromorphic control system of the building (Wasserman et al. 2003). Hence, a continuum of expression including sound and vision can be realized. Confirming the validity of the music generated by RoBoser, we note that the soundtrack for a movie on Ada, generated by Roboser, was awarded the prize for best soundtrack in the 2001 national Swiss movie awards, thus passing a musical Turing test (Wasserman et al. 2003).

The results obtained with the RoBoser system are significant with respect to the discussion on music and language because they show how the hierarchical structure of music can be obtained by relying on the hierarchical structure of a complex real-world system, as opposed to that of an internal grammar which mimics the complexity of overt behavior. This suggests that although the formal approach might provide for an effective description of music, it may not capture the full richness required for music generation. The plausibility of this

Figure 16.3 The ADA main space (180 m^2) engages with its visitors through interactive multimodal compositions. The hexagonal floor tiles are pressure sensitive and display colored patterns dependent on ADA's behavior modes and visitor interactions. The walls are made of semitransparent mirrors, which permit visitors in the outer corridor to view what happens inside ADA. Above the mirrors, a circular projection screen displays real-time animated graphics which, similar to the music, represent Ada's current behavior and emotional state. ADA was operational from May until October, 2002, and was visited by over 500,000 people.

assertion is strengthened by the fact that the RoBoser paradigm has been realized in real-world systems of multiple forms and functionalities, as opposed to relying solely on the descriptive coherence and internal consistency offered by grammar-based approaches. The skeptic could argue that it takes a human to make "real" music. However, we emphasize that this argument hinges on the definition of "real" which is ontologically suspect. We will skirt this issue by focusing on the pragmatics of music production and the effectivity of the RoBoser system in being accepted by human audiences and performers, together with its commitment to unraveling the generative structures behind musical aesthetics. Indeed, the RoBoser paradigm has opened the way to a view on musical composition, production, and performance that emphasizes the controlled induction of subjective states in observers in close interaction with them. It has also led to a new psychologically grounded system called SMuSe, which is based on situatedness, emotional feedback, and a biomimetic architecture (Le Groux and Verschure 2009b, 2010; Le Groux et al. 2008). With SmuSe, the listener becomes part performer and part composer. For instance, using direct neural feedback systems a so-called brain tuner was constructed to optimize specific brain states measured in the EEG through musical composition; that is, not direct sonification of the EEG waves. This highlights the differences in the design objectives between interactive–cognitive music systems and Western tonal music. Shifting from complete prespecification by a single human composer to the parameterization of musical composition, the psycho-acoustical validation of these parameters and the mapping of external states onto these musical parameters rendering novel interaction-driven compositions (Le Groux 2006; Le Groux and Verschure 2011).

The RoBoser paradigm has been generalized from the musical domain to that of multimodal composition in the ADA exhibition (Figure 16.3) and other performances, such as *re(per)curso*[2] (Mura et al. 2008)[3] and *The Brain Orchestra* (Figure 16.4) (Le Groux et al. 2010). This raises the fundamental question of whether a universal structure underlies human auditory and visual experiences based on a unified meaning and discourse system of the human brain, as opposed to being fragmented along a number of modality-specific modules (see also Seifert et al., this volume). This invariant property could be the drive of the brain to find meaning in events organized in time, or to define a narrative structure for the multimodal experiences to which it is exposed. In terms of the applications of the RoBoser paradigm, we conceptualize its aesthetic outputs as audiovisual narratives whose underlying compositional primitives and processes share many features across modalities. One can think of the synesthetic compositions of Scriabin as a classic example, but with one important difference: currently, machines can generate such interactive compositions on the

---

[2] A video of the performance is available at http://www.iua.upf.edu/repercurso/.

[3] See also Examples 1 and 2 in the online supplemental material to this volume (http://esforum.de/sfr10/verschure.html).

Figure 16.4 The *Multimodal Brain Orchestra* (MBO): four members of the MBO play virtual musical instruments solely through brain–computer interface technology. The orchestra is conducted while an emotional conductor, seated in the front right corner, is engaged to drive the affective content of a multimodal composition through means of her physiological state. Above the right-hand side of the podium, the affective visual narrative *Chekëh* is shown, while on the left-hand side, MBO-specific displays visualize the physiological states of the MBO members and the transformation of the audiovisual narrative.

basis of relatively simple principles in close interaction with different states of human observers/performers as opposed to that originating in a singular genius.

These examples illustrate that aesthetic musical experience can be, at least partially, understood as emerging from the interaction between human users and open music systems. That this insight has come so late is partially due to the fact that Western music is based on the view of the perceiver as a passive listener: someone who is exposed to the external genius of the composer that is mediated by the performer. Accordingly, the perceiver (or listener) is not an active participant. The listener, however, is very versatile and resilient with respect to the mode of perception offered (i.e., active or passive). We have demonstrated, in contrast, that appreciation of interaction correlates with the level of activity of the perceiver (Eng et al. 2005a, b). In addition, contemporary technology has reduced the instrumentalist's monopoly on producing music. Using a range of technologies, new interfaces to musical instruments have been created that are more ecologically valid (Paradiso et al. 2000). We expect these technologies to contribute to a fundamental change in the future, and envision the deployment of systems that will be able to generate individualized multimodal compositions in real time, dependent on both explicit and implicit states of active human performers and perceivers (Le Groux et al. 2008; Le Groux and Verschure 2009a, 2010). An early example of such a system is the *eXperience Induction Machine*[4], which is a direct descendent of RoBoser and ADA (Inderbitzin et al. 2009).

---

[4] See Example 3 at http://www.esforum.de/sfr10/verschure.html

Conclusion

We have juxtaposed the traditional rationalist perspective on mind and music with an emerging trend in the study of mind and brain to emphasize the interaction between an agent and its environment. The "agent" terminology of AI (Russell and Norvig 2010) may serve as a basis for reevaluating computational models of music and language processing, where this agent is embodied and instantiated in the real world.

An action-oriented approach to music subsumes many approaches: from score analysis to the modeling of music cognition. In addition, it is more congruent with research in ethnomusicology and music sociology (Clayton 2009; Clayton et al. 2005), because it does not focus on purely Western concepts of musical structure (i.e., concepts based on score-oriented reasoning and language-mediated musical structure and meaning). Music is a study of much interest for its sonic complexity and its perceptual and cognitive, affective impact.

As an example of a more situated and brain-based approach toward musical creativity and aesthetics, we have described the DAC architecture and its translation into an increasingly more complex and interactive paradigm for situated multi-modal composition and performance (Figure 16.5). We have shown that DAC has given rise to a range of novel performances and installations that redefine musical composition and expression. We have also illustrated how the unidirectional paradigm of the Western musical tradition, with its specific representationalist formalization and unidirectional production stream (from the active composer to the passive listener via the drone-like musician) can be expanded to a model where the composer becomes a designer of a space of musical expression to be explored by an active observer/perfomer. RoBoser has demonstrated that plausible musical structures can emerge that are appreciated by an audience, and that the interface bottleneck can be eliminated. This form of interaction leads to the exploitation of a situated recursion between the performer/observer and an emergent musical structure, each action operating on the musical output produced by previous actions.

In terms of radical change in music production and perception, two different aspects were highlighted: (a) new music can emerge from machines, whether or not in interaction with humans, and (b) the resulting lessons can be used to inform our understanding of the millennia-long human experience of music, as well as our attempts to probe the neural mechanisms and social interactions which support it. When Beethoven composed his symphonies, or when we sing along to a recording or with a group, or when we dance spontaneously to a piece of music, we engage in "musical narratives." Instead of caricaturing the whole of Western musical experience to mere passivity, we suggest that such experiences can be used to enrich our understanding of the brain and culture, and perhaps the whole of human experience.

The development of advanced interactive synthetic music composition also raises a fundamental issue concerning the form that a theory of music

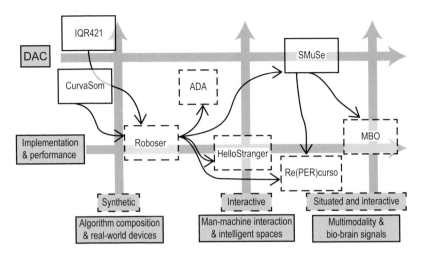

Figure 16.5 Illustration of the connections in the trajectory from the large-scale neuronal simulation environment (Bernadet and Verschure 2010) IQR421 and CurvaSom (1998) to the *Multimodal Brian Orchestra* (2009). In the top layer is the development of DAC, which integrates the systems as a multimodal narrative engine. The vertical rows indicate the integration of systems starting with a "synthetic" approach based on the integration of algorithmic composition and real-world devices (robots). Next is an "interactive" view in which the systems ADA (2002), HelloStranger (2005), and *re(per) curso* (2007) approached human–machine interaction and sentient spaces. Finally, the "situated and interactive" *Multimodal Brian Orchestra* (2009) and SMuSe system relate to multimodality and body–brain signals.

aesthetics will take (Verschure 1998; Le Groux and Verschure 2011). The synthetic approach expressed in RoBoser argues that the rules embedded in the machine encapsulate the principles of musical aesthetics. It was with this in mind that RoBoser was purposefully developed. However, by virtue of being open and interactive, these technologies also delineate the contributions of both intrinsic and external factors in the human experience of music, language, and action.

Hence, universals of music do not depend on its mode of codification, and we should be careful in not mistaking the ontological status of these codifications. We propose that interactive synthetic music systems offer promising strategies for a comparative research program on language, music, and action. Their ability to combine computational modeling, robotics, and empirical research may advance knowledge of embodied and situated perspectives on music, language, and the brain, and further our understanding of music and creativity. In particular, the rise of these new music technologies will help us to overcome the bottleneck of the instrument, which will herald a new liberalized phase of music production, and experience that emphasizes the expression of the direct creative inspiration of many as opposed to the genius of the few.

Acknowledgments

We thank Michael Arbib and Julia Lupp for their thoughtful comments to this manuscript. This work is supported by CEEDS (FP7 ICT-258749), the FAA project (TIN2010-16745), and the EFAA project (FP7 ICT-270490).

First column (top to bottom):   Jonathan Fritz, Larry Parsons, Isabelle Peretz, Gottfried Schlaug, group discussion, Laurel Trainor, John Halle
Second column: Francesca Stregapede, Josef Rauschecker, group discussion, John Halle, Jonathan Fritz, Petr Janata, David Poeppel
Third column: David Poeppel, Ani Patel, Laurel Trainor, Josef Rauschecker, Ani Patel, Isabelle Peretz, Francesca Stregapede

17

The Neurobiology of Language, Speech, and Music

Jonathan Fritz, David Poeppel, Laurel Trainor,
Gottfried Schlaug, Aniruddh D. Patel,
Isabelle Peretz, Josef P. Rauschecker, John Halle,
Francesca Stregapede, and Lawrence M. Parsons

Abstract

To clarify the domain-specific representations in language and music and the common domain-general operations or computations, it is essential to understand the neural foundations of language and music, including the neurobiological and computational "primitives" that form the basis for both, at perceptual, cognitive, and production levels. This chapter summarizes the current state of knowledge in this burgeoning cross-disciplinary field and explores results from recent studies of input codes, learning and development, brain injury, and plasticity as well as the interactions between perception, action, and prediction. These differing perspectives offer insights into language and music as auditory structures and point to underlying common and distinct mechanisms and future research challenges.

Introduction

We seek to characterize the *relations* between language and music—and their neurobiological foundations—in the hope that this will lead to real advances and a unified understanding of both. For example, such cross-domain research may elicit change in the architecture of music theory and the psychology of music; deepen our insight into the roles of vocal intonation, cadence, rhythm, rhyme, inflection in emotional expression in drama, poetry and song; shed light on the ontogeny of music and language development in children; reveal common cognitive operations in language comprehension and music perception; and clarify the specialization and the relative contributions of the right and left hemispheres to these two domains. With these possibilities in mind, our discussions summarized in this report centered on two issues: First, how can we be precise and explicit about how to relate and compare the study of language and music in a neurobiological context? Second, what are the neural foundations that form the basis for these two dimensions of human experience, and how does the neural infrastructure constrain ideas about the complex relation between language and music? For example, to what extent is neural circuitry shared between music and language, and to what extent are different circuits involved? More ambitiously, we ask at what level of analysis can we go beyond "correlational'" statements (e.g., "brain area X underpins music function Y") and strive for accounts of the underlying processes.

An important caveat before we proceed: Speech is propositional, unlike music, and speech sounds carry denotative meaning, unlike most music. However, music can evoke and elicit emotion directly without linguistic mediation. Thus, while we explore the similarities, we must also be aware of the differences between the two domains. Moreover, only a very restricted set of issues is addressed here: one that reflects the composition and expertise of our working group as well as the actual content of the discussions. By design, other chapters in this volume address the issue of structure and syntax (Thompson-Schill et al.) and meaning/semantics in music and language (Seifert et al.), with evolutionary considerations (Cross et al.). Here, we focus on the input interfaces (i.e., perception of music and speech), the relation to production or action, and questions about their neurobiological implementation. However, even within this narrow scope, there are productive areas of study that, regrettably, do not receive the attention they deserve, including the study of song and dance as possible approaches to explore the relations between music, speech, language, and the brain. We leave these important topics for further discussion and research (see, however, Janata and Parsons, this volume).

By way of overview, the chapter proceeds as follows:

1. We outline an approach to characterizing the domains of language and music (from a perception–action perspective), with the goal of identifying tentative lists of basic, fundamental elements, or "primitives," in music and language and discussing how they might be related across domains. We pursue the hypothesis that there are domain-specific representations in language and music and domain-general operations or computations, and we discuss some candidate areas, such as sequencing and attention in time.
2. We discuss some cross-cutting issues that have been investigated productively in both domains and that illustrate areas in which the brain sciences add considerable insight, including the nature of the input codes, learning and development, brain injury and plasticity, and the perception–action cycle. Anticipation-in-time or prediction is an

essential component of the sequencing process in language and music, and will be considered as part of a perception–prediction cycle.
3. Although we point to neurobiological data throughout the chapter, the section on neurobiological constraints and mechanisms focuses explicitly on some of the neural mechanisms that are either reasonably well understood or under consideration. Building on the previous sections, we distinguish between data pointing to domain-specific neural correlates versus domain-general neural correlates.
4. In the final section, we return to some of the open questions, pointing especially to the important contributions that can be made by work on dance, music, poetry, and song. However, the overall, more modest goal of the report is to highlight a set of experimental approaches that can explore the properties and neural bases of the fundamental constituents in music and language.

Structure of the Domains: Computational Primitives and Processes

To identify the relations between domains, it can prove useful to characterize the problem in two ways. One approach is to (attempt to) spell out for each domain an "elementary parts list." For example, such a list for language might include hypothesized representational primitives (e.g., phoneme, syllable, noun phrase, clause) or operational/processing or computational primitives (e.g., concatenation, linearization, dependency formation). Similarly, for music, a list might include tone (representation) and relative pitch detection (computation). The decomposition into constituent parts is necessarily incomplete and the list (Table 17.1) changes as progress is made. For example, such inventories of primitive elements will be modified as one considers the relation of music to dance or speech to song.

A second approach derives from the work of computational neuroscientists such as David Marr, who provided a way to talk about the characterization of complex systems, focusing on vision (Marr 1982). In a related perspective, Arbib (1981) employed schema theory to chart the linkages between perceptual structures and distributed motor control. The neural and cognitive systems that underlie language and music are usefully considered as complex systems that can be described at different levels. At one level, *computational goals and strategies* can be formulated, at an intermediate level the *representations and procedures* are described, while at a third level lie the *implementation* of the representations and procedures. It is assumed that these distinct levels are linked in principle, although actual close linkages may not always be practical. Commitments at one level of description (e.g., the implementation) constrain the architecture at other levels of description (structural or procedural) in principled ways.

Table 17.1 Levels of analysis: A Marr's eye view.

Implementational (domain general)	Hypothesized *implementational* (*neurobiological*) *infrastructure* • Generic forms of circuitry • General learning rules which can adapt circuits to serve one or both domains	
Algorithmic computational (domain general)	Hypothesized *computational primitives* • Constructing spatiotemporal objects (streams, gestures) • Extracting relative pitch • Extracting relative time • Discretization • Sequencing, concatenation, ordering • Grouping, constituency, hierarchy • Establishing relationships: local or long distance • Coordinate transformations • Prediction • Synchronization, entrainment, turn-taking • Concurrent processing over different levels	
Representational computational (domain specific)	Hypothesized *representational primitives*: *language* • Feature (articulatory) • Phoneme • Syllable • Morpheme • Noun phrase, verb phrase, etc. • Clause • Sentence • Discourse, narrative	Hypothesized *representational primitives*: *music* • Note (pitch and timbre) • Pitch interval (dissonance, consonance) • Octave-based pitch scale • Pitch hierarchy (tonality) • Discrete time interval • Beat • Meter • Motif/theme • Melody/satz • Piece

Pursuing the Marr-style analysis, at the top level (i.e., the level of overall goals and strategies) we find perception and active performance of music (including song and dance) as well as language comprehension and production. At the intermediate level of analysis, a set of "primitive" representations and computational processes is specified that form the basis for executing "higher-order" faculties (Table 17.1). At a lower level of analysis is wetware: the brain implementation of the computations that are described at the intermediate level. What may seem a plausible theory at one level of analysis may require drastic retooling to meet constraints provided by neural data. A special challenge is provided by development. Since the brain is highly adaptive, what is plausibly viewed as a primitive at one time in the life of the organism may be subject to change, even at relatively abstract conceptual levels (Carey 2009), and the organism may bootstrap, building perceptual and cognitive algorithms, which may form building blocks for subsequent processing in the adult.

Decomposing/Delineating Elementary Primitive Representations and Computations

The richness and complexity of music and language are well described elsewhere in this volume (see, e.g., chapters by Patel, Janata and Parsons, as well as Hagoort and Poeppel). Here, we attempt to identify domain-specific and domain-general properties of music and language, with special (but not exclusive) reference to how constituent processes may be mapped to the human brain and, in some cases, related to mechanisms also available in the brains of other species. Table 17.1 provides the overview of how these questions were discussed, and what kind of analysis and fractionation yielded (some, initial) emerging consensus.

Domain-Specific Representational Inventories: The Primitive Constituents, or "Parts List"

Though there will inevitably be disagreements about the extent to which a given concept constitutes a basic, primitive unit, we offer some candidates for a representational inventory that may underpin each domain—representations without which a successful, explanatory theory of language processing or music processing cannot get off the ground. Regardless of one's theoretical commitments, it would be problematic to develop a theory of language processing that does not contain a notion of, say, syllable or phrase; similarly, a theory of music that does not refer to, say, note or meter will most likely be irreparably incomplete.

If we focus, to begin, on language in terms of the input sound patterns of speech (and forget about written language and orthography altogether), then we seek primitives required for phonetic featural representation. These acoustic and articulatory distinctive features (Stevens 2000; Halle 2002) form the minimal units of description of spoken language from which successively higher levels of linguistic representation are derived. At the most granular level, the speech system traffics in small segment-sized (phonemic) as well as slightly larger syllable-sized units. These need to be "recovered" in the context of perception or production. Segmental and syllabic elements are combined to form morphemes or roots or, more colloquially, words; the combinations of sounds forming words are subject to phonetic and phonological constraints or rules. Although this chapter focuses on spoken language, we find that the lower-level units of signed language are somewhat different and that higher levels (say words) converge.

One of the remarkable features of language is the large number of morphemes (or words) that are stored as the "mental lexicon." Speakers of a language have tens of thousands of entries stored in long-term memory, each of which they can retrieve within ~200 ms of being uttered (given standard speaking rates). This feat of memory is impressive because the words are constructed from a relatively small set of sound elements, say in the dozens. This could plausibly lead to high confusability in online processing, but the items and their (often subtle) distinctions (e.g., /bad/ vs. /bat/ or /bad/ vs. /pad/) are stored in a format or code that permits rapid and precise retrieval. The words are sound–meaning pairings, often with rather complicated internal morphological structure (e.g., an un-assail-able pre-mise), but aspects of sound and meaning may be distributed across multiple brain regions, with strong associative links between them. Combining these elements pursuant to certain language-specific regularities (syntax) yields phrases and clauses (e.g., "the very hungry caterpillar") that yield compositional meaning. Ultimately, the information is interpreted in some pragmatic or narrative discourse context that provides common ground about knowledge of the world and licenses inferences as well as being integrable into ongoing conversation (Levinson, this volume).

The interrelations among levels are currently being investigated in linguistics and psycholinguistics (see Thompson-Schill et al. and Hagoort and Poeppel, both this volume) and fall outside the scope of this report. However, it is worth bearing in mind that when even a single sentence or phrase is uttered, all of these levels of representation are necessarily and obligatorily activated across multiple brain regions. It is, by contrast, less clear to what extent the same consequences obtain in the musical case for nonmusicians. When speech and language are used in even more multimodal contexts—say during audiovisual speech, spoken poetry or during singing—further levels are recruited, including visual representations of speech and musical representations during singing (and, presumably, motor representations during both types of actions).

Turning to music (Table 17.1, bottom right column), similar categories apply to some degree to the lower levels of speech and musical structure; namely (and minimally) grouping, meter, and pitch space. The assignment of all of these categories involves a complex interaction of primary sensory cues. Thus the constituents of music, which we will refer to, following Lerdahl and Jackendoff (1983), as the musical group, can be induced by alterations in pitch, duration, and, to a lesser degree, amplitude and timbre. Of these, duration is arguably the most determinative, with boundaries tending to be assigned between events that are relatively temporally dispersed. However, if we broaden our study of music from sound patterns to dance, then further primitives are required to link motion of the body with musical space. Moreover, music is often an ensemble activity (e.g., dancing together, singing together, an orchestral performance; see Levinson and Lewis, both this volume) which forces us to assess how the primitives within an individual's behavior are linked with those of others in the group.

Metrical structure—the periodic alternations of strong and weak temporal locations experienced as a sense of "beat"—is also largely induced by manipulation of these parameters. Length, again, is probably the most determinative, with a somewhat reduced role for pitch. (Indeed, some highly beat-oriented music, such as some forms of drumming, may lack pitch altogether.) Here, amplitude and timbre, insofar as these create a musical accent, tend to be highly determinative. Finally, with respect to harmony–pitch space, pitch is required to be primary, with listeners orienting themselves in relationship to a tonic, perceptual reference point according to which other pitch categories are defined. While Western functional harmony makes use of an extremely rich system of pitch representation, the great majority of musical styles make use of a scale, allowing for a categorical distinction between adjacent and nonadjacent motion with respect to the tonal space (steps and skips). Perhaps most prominently, a sense of closure or completion resulting from the return to a tonic is a recurrent, if not universal, property of Western musical systems.

By analogy to the language case, the music theorists and musicians in our working group guided the discussion and converged on a list of primitives in three parts: spectral and timbral elements related to pitch and pitch relations (note, pitch interval, octave-based pitch scale, pitch hierarchy, harmonic/inharmonic, timbre and texture), elements related to time and temporal relations (discrete time interval, phrase, beat, meter, polyrhythms), and elements related to larger groups (motif/theme, melody/satz, piece, cycle). Here, too, there is a hierarchy of elements. By analogy to language, exposure to a melody presumably entails the obligatory recruitment of the elements lower in the hierarchy, such as the temporal structure of the pitch-bearing elements.

Comparison of Primitives

It is, of course, tempting to draw analogies and seek parallels. Indeed, if one focuses on the lower, input-centered levels of analysis (phoneme, syllable, tone, beat, etc.), one might be seduced into seeing a range of analogies between music processing and the processing of spoken language (including, crucially, suprasegmental prosodic attributes such as stress or intonation). However, if one looks to the representational units that are more distal to the input/output signal (e.g., morphemes, lexical and compositional semantics) in language, possible analogies with music become metaphorical, loose, and sloppy. In fact, closer inspection of the parts lists suggests domain specificity of the representational inventories.

Let us briefly focus on some differences. One general issue pertains to whether (knowledge and processing of) music can be characterized in a manner similar to language. For example, sound categorization skills in language and music may be linked (Anvari et al. 2002; Slevc and Miyake 2006). Although the approach is too simple, a useful shorthand for discussion is that language consists, broadly, of words and rules (Pinker 1999); that is, meaning is created or interpreted by (a) looking up stored items and retrieving their attributes or individual meanings and (b) combining stored words according to some constraints or rules and generating—or "composing"—new meaning. Both ingredients are necessary, and much current cognitive neuroscience of language has focused on studying how and where words are stored and how and where items are combined syntactically and semantically to create new meaning (for a review, see Hickok and Poeppel 2007; Lau et al. 2008; Hinzen and Poeppel 2011; for a recent meta-analysis of word representation, see DeWitt and Rauschecker 2012).

One question that arises is whether there exists such a construct as a "stored set" of musical elements; that is, a "vocabulary" or musical lexicon that encodes simple structures underlying the construction of larger units. Presumably this would be true for musicians/experts, but even nonmusicians may be better at attending or dancing to music in a familiar genre, which suggests some sort of familiarity with building blocks and patterns of assemblage. One clear difference will be that the lexicon carries meaning, a role that tones or musical phrases do not have (at least not in the same way; see Seifert et al., this volume). It seems that this line of argumentation thus reveals another fundamental difference between the domains.

Two additional differences are worth noting. First, while duration can function as a distinctive feature of vowels and consonants in some languages, and is also one of the acoustic correlates of word and focus stress, its role tends to be minor compared to the primary role that duration plays in musical systems (for a discussion on how to distinguish prosody from the tones of the vowels of a tone language like Chinese, see Ladd, this volume). Second, whereas the levels of linguistic structure exist, roughly speaking, on successively larger temporal frames (this is somewhat less true for signed languages), the objects described within each musical category frequently exist, to a greater degree than language, on the same timescale. Thus, meter coexists with grouping, as can be seen with respect to a minimal group of four events that occur during the metrical frame of two strong beats: a "satz" unit (roughly, melodic unit) denoted by a cadence will tend to exist on a larger timescale, often four or eight beats, etc. Furthermore, pitch relationships are experienced on a variety of temporal levels, from the most local (the motion of adjacent pitches tend to be, in most styles, primarily stepwise contours) to listeners being highly attuned to the beginnings and endings of large musical groups. Similarly, harmonic syntax is also highly locally constrained by a limited repertoire of possible progressions, which, within the so-called common practice period, achieve closure by means of the cadential progression: dominant to tonic (V–I).

Thus, we conclude that an inventory of the fundamental representational elements, as sketched out in Table 17.1, reveals a domain-specific organization, especially at the highest level of analysis. As we turn to neural evidence in sections below, the claim of domain specificity is supported by dissociations between the domains observed in both imaging and lesion data. It goes without saying, however, that there must be at least some shared attributes, to which we turn next.

Domain-General, Generic Computations/Operations: Shared Attributes

In contrast to the representational inventories, we hypothesize that many of the algorithms/operations that have such primitives as their inputs are, by and large, domain general or, at least, will prove to combine generic algorithms in domain-specific ways. One way to conceptualize this is to imagine different invocations of the same neural circuitry; that is, "copies" of the same circuitry, but which operate on input representations of different types that are domain specific. For example, the task of constructing an auditory stream, of extracting relative pitch, or of sequencing or concatenating information are the types of operations that are likely to be "generic," and thus a potentially shared computational resource for processing both music and language. Some of the hypothesized shared operations are summarized in Table 17.1. (Note: we use "stream" in two senses in this chapter. We distinguish "auditory streams" as defined, for example, by the voices of different people at a cocktail party from "neuroanatomical streams" as in the two routes from primary auditory cortex to prefrontal cortex shown in Figure 17.1.) Here we offer a brief list that merits further exploration as candidate domain-general operations and then discuss two of these operations—sequencing and timing—in a bit more detail.

- *Constructing spectro-temporal auditory objects* (Griffiths and Warren 2004; Zatorre et al. 2004; Leaver and Rauschecker 2010) or *identifying auditory streams* (Shamma et al. 2011; Micheyl et al. 2005, 2007) is part of a necessary prior auditory scene analysis (Bregman 1990). The required neural circuitry is evident across species (certainly primates and vocal learners; see Fitch and Jarvis, this volume). What appears as specialization in the human brain thus arises from the interface of these more generic circuits with domain-specific input representations and/or the production and interpretation of such representations. Some of

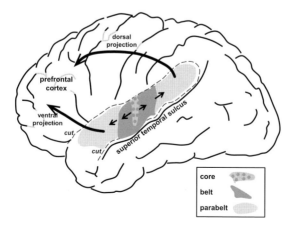

Figure 17.1 Human auditory cortex (modified from Hall and Barker 2012).

this grouping requires attention; however some scene segregation and object identification appears to occur preattentively.
- *Discretization into auditory events*: Because both speech and music typically come to the listener as a continuous stream, some form of chunking or discretization is required (e.g., for lexical access or the identification of a theme/motif).
- *Grouping*, both in terms of establishing constituency (segmentation into groups) and hierarchy (establishing relationships between components), can occur in space (as in dance, orchestras, marching bands, choruses, and cocktail parties), in time (e.g., intricate polyrhythms in African drumming), and/or feature space (e.g., timbre). In music, lower-level components emerge at the metrical level of periodicities and metrical accents (beat induction). The next level might be harmonic. A higher level arises from the grouping of constituents and may lead to musical phrases, progressions, or the satz. In language there is a similar hierarchical process, such that you go from phonology to morphology to syntax and meaning. Building on Lerdahl and Jackendoff (1983), Drake (1998) summarizes such segmentation (as well as regularity extraction) processes and distinguishes between those that appear to be universal and innate (segmentation, temporal regularity extraction) from those that are acquired or derived and culture specific. Later (see section on Learning and Development), the way in which "innate" skills emerge over the first few months or years of life is characterized.

The above three types of operations are functions of the auditory system, and probably the cortical auditory system. The nuts and bolts of these operations are the focus of much current auditory neuroscience research (Schnupp et al. 2010). We know relatively little about the precise locations and mechanisms that are involved, beyond the rather superficial insight that superior temporal areas of the auditory cortex are implicated, including the so-called core, belt, and parabelt regions. Figure 17.1 (modified from Hall and Barker 2012) illustrates the anatomy of the human auditory cortex. It provides the gyral and sulcal anatomic context, highlights the structure of the superior temporal gyrus (STG; the region above the superior temporal sulcus), and shows one of the dorsal and ventral projection schemes with the ventral route serving as a "what" pathway associated with auditory object identification (Rauschecker and Scott 2009). To date, however, there is no compelling evidence to suggest that any of these early cortical regions are selectively specialized for either speech or music processing.

- *Sequencing of constituents* must be accomplished in various contexts involving relative time, ordering, concatenation, and relative pitch contour. Placing items (elementary representations) in a sequence must be done in both domains, but clearly differs between music and language. In language, concatenation and linear order are not sufficient—certainly for syntax and semantics, where structure dictates interpretation—although it is a critical part of phonological processing. Sequencing and ordering operations have implicated auditory areas as well as inferior frontal and premotor and motor regions, including Broca's region, basal ganglia, the cerebellum, and other potential substrates. One cortical region that is consistently implicated in basic constituent building, at least in language, is the left anterior temporal lobe (see Figure 17.1). For example, recent work by Bemis and Pylkkänen (2011) shows how minimal unit building (e.g., "red boat") activates the left anterior temporal lobe across studies and imaging approaches.

- *Linking multimodal objects* in audiovisual speech (i.e., vision and hearing), song (i.e., words/speech and melody/music), and dance (i.e. music and motoric patterns). Typically, multimodal perceptuo-motor tasks have implicated three regions: posterior superior temporal sulcus (STS; Figure 17.1), the inferior parietal lobe, and inferior frontal regions, together often referred to as the dorsal auditory cortical pathway (Rauschecker and Scott 2009). Although this has been an active area of research in multisensory speech, less is known about the musical case. What is known about song and dance is reviewed by Janata and Parsons (this volume).

- *Coordinate transformations are ubiquitous* (e.g., from input spoken words in acoustic coordinates to output speech in motor coordinates; the mapping from auditory input to vocal tract output in vocal learners; the alignment between perception and action in music performance; the alignment between musical and linguistic information in song; the alignment between musical and motor information in dance). The common problem is that information in one domain, represented on coordinate system i, must be made compatible with information in another domain, represented in coordinate system j. In the case of speech, a growing literature suggests that the dorsal processing stream (Rauschecker and Scott 2009), or perhaps more specifically a temporoparietal area (Sylvian parieto-temporal, SPT) provides the cortical substrate for this computation (Hickok et al. 2003; Hickok and Poeppel 2007; Hickok 2012). It is not obvious where and how such basic and widespread operations are executed in tasks involving music perception and performance.

- *Entrainment, synchronization, alternation, interleaving, turn-taking* require that the listener form a model of a conversational (or musical) partner as well as an accurate internal model in order to synchronize all information and set up the framework for properly timed communicative alternation (Levinson 1997, this volume). The neural basis for these operations is not yet understood. While entrainment to stimulus features (e.g., the temporal structure of speech or music) is known to occur in sensory areas and has been well described and characterized (e.g., Schroeder et al. 2008), how such inter-agent alignment occurs in neural terms is not yet clear. There are occasional appeals to the mirror neuron system, but many in our working group showed little enthusiasm for mirror neurons and their promise (for a critique of mirror neuron hypothesis, see Hickok 2008; Rogalsky et al. 2011a; for a more positive view, see Fogassi as well as Arbib and Iriki, this volume, and Jeannerod 2005). However, observations of auditory responses in motor areas and motor responses in auditory areas highlight the importance of audio-motor linkages and transformations.

- *Organizing structure for use by social partners* (of particular relevance in conversation, musical ensemble playing and jazz improvisation, and in dance).

Sequencing in Music and Language: The Importance of Relative Pitch and Duration

The acoustic signals and production gestures of speech and music are physically complex and continuous. In both domains, a process of discretization in auditory cortex yields elementary units (such as tones in music or syllables in language) that serve as "elementary particles" for sequencing operations. These sequencing operations encode the order and timing of events, and also concatenate elementary events into larger chunks. Important work on the cognitive neuroscience of sequencing has been done by Janata and Grafton (2003), Dominey et al. (2009), and others; see also the "events-in-time" modeling described by Arbib et al (this volume).

While sequencing necessarily involves the encoding of "absolute features" of events (e.g., duration, frequency structure), a very important aspect of music and speech processing is the parallel encoding of these same physical features in relative terms. For example, when processing a musical melody or a spoken intonation contour, we extract not only a sequence of pitches but also a sequence of relative pitches (the sequence of ups and downs between individual pitches, independent of absolute frequency). This is what allows us to recognize the same melody or intonation contour (such as a "question" contour with a rise at the end) at different absolute frequency levels. Relative pitch seems to be "easy" for humans but not for other species. Birds, for example, show remarkable ability in absolute pitch (Weisman et al. 1998, 2010) but struggle with relative pitch (e.g., Bregman et al. 2012). Monkeys, however, almost totally lack relative pitch ability (Wright et al. 2000), but extensive training can lead to a limited form of relative pitch in ferrets (Yin et al. 2010) and monkeys (Brosch et al. 2004). The human ability to perceive relative pitch readily may mark a crucial step in the evolution of language and music.

Similarly, in sequencing we encode not only absolute duration of events but also the relative durations of successive events or onsets between events (e.g., inter-onset intervals). There are similar constraints for compression and dilation of time in speech and music. Humans are able to recognize immediately a tune, within rather stringent limits, when it is slowed down or sped up (Warren et al. 1991), because the pattern of relative durations is preserved, even though the duration of every element in the sequence has changed. Similar limits exist for the perception of speech when the speed of its delivery is decreased or accelerated (Ahissar et al. 2001). Similarly, sensitivity to relative duration in speech allows us to be sensitive to prosodic phenomena (such as phrase-final lengthening or stress contrasts between syllables) across changes in speech rate or the overall emphasis with which speech is produced. Thus when we think of acoustic sequences in music and language that need to be decoded by the listener, it is important to remember that from the brain's perspective a sequence of events is a multidimensional object or stream unfolding in time; that is, a sequence of absolute and relative attributes, with relative pitch and relative duration being the minimum set of relative attributes that are likely to be encoded.

Neural Basis of Timing and Attention in Time

Obviously, timing is critical for music and language. Both spoken language and music are, typically, extended signals with a principled structure in the time domain. A temporally evolving dynamic signal requires that the listener (or producer) can accurately analyze the information, parse it into chunks of the appropriate temporal granularity, and decode the information in the temporal windows that are generated. There are similar timescales for both language and music processing, presumably the consequence of basic neuronal time constants. Both domains require timing at the tens-of-millisecond timescale (e.g., analysis of certain phonetic features, analysis of brief notes and spaces between notes), a scale around 150–300 ms (associated with, e.g., syllabic parsing in speech), and longer scales relating to intonation contour. It is interesting to note the extent of overlap between the timescales used for elementary analytic processes in speech and music perception. Expressed in terms of modulation rate, the typical phenomena that a listener must analyze to generate an interpretable musical experience (e.g., beat, tonal induction, and melody recognition as well as phonemic and lexical analysis) range between approximately 0.5 and 10 Hz (Farbood et al. 2013). While there are faster and slower perceptual phenomena, these are roughly the temporal rates over which listeners perform optimally; that is, these rates constitute the "temporal sweet spot" both for music and for speech.

A fair amount of recent research has focused on how to represent and analyze temporal signals at a neural level. One approach emphasizes neural timekeepers in a supramodal timing network that includes cerebellum, basal ganglia, premotor and supplementary motor areas, and prefrontal cortex (Nagarajan et al. 1998; Teki et al. 2011). In contrast, another approach emphasizes the potential utility of neuronal oscillations as mechanisms to parse extended signals into units of the appropriate size. These oscillations are population effects which provide a framework for the individual activation of, and interaction between, the multitude of neurons within the population. The basic intuition is that there are intrinsic neuronal oscillations (evident in auditory areas) at the delta (1–3 Hz), theta (4–8 Hz), and low gamma (30–50 Hz) frequencies that can interact with input signals in a manner that structures signals (phase resetting) and discretizes them (sampling); this may provide (by virtue of the oscillations) a predictive context (active sensing). Figure 17.2 illustrates the neural oscillation hypothesis (for which experimental support is still controversial and provisional). For the domain of speech processing, the neuronal oscillation approach is reviewed in Giraud and Poeppel (2012), Zion Golumbic et al. (2012), and Peelle et al. (2012). How neuronal oscillations may play a role in speech perception is also briefly summarized by Hagoort and Poeppel (this volume). One challenge is to understand how neuronal oscillations may facilitate processing on these timescales, since they are also evident in typical musical signals. Electroencephalography (EEG) or magnetoencephalography (MEG) studies using musical signals with temporally manipulated structure will have to be employed while testing where and how music-related temporal structure interacts with neuronal oscillations. Recent MEG data show that intrinsic neuronal oscillations in the theta band are facilitated in left temporal cortex when the input is intelligible speech (Peelle et al. 2012). Future studies will need to explore whether these responses might be amplified in right temporal cortex when presented with musical signals. Using neurophysiological data from single neurons, EEG, and MEG as well as neuroimaging tools, we can explore mechanistic hypotheses about how neural responses might encode complex musical or linguistic signals and guide attention allocation.

Thus one overarching question is whether the same timing mechanisms are used in both domains. A different way of approaching the question is to ask: If a subject is trained in a perceptual learning paradigm on an interval using a pure-tone duration, will this timing information be equally available for timing a sound (musical note or syllable) in a musical or linguistic context? Evidence from behavioral studies (Wright et al. 1997; Wright and Zhang 2009) indicates that temporal interval discrimination generalizes to untrained markers of the interval, but not to untrained intervals. There is also evidence that training on temporal interval discrimination (two tone pips separated by an interval) generalizes (a) to duration discrimination of the same overall duration, (b) to motor tapping (for the trained duration only), and (c) from training in the somatosensory system to the auditory system (for the trained duration only). This insight may help us understand the supramodal timing representations that underlie language as well as music and dance performance and perception. However, it leaves open whether the transfer implicates a single shared brain system or coupling between domain-specific systems (see Patel, this volume, for further discussion of shared resources). Finally, native speakers of languages that use vowel duration as a phonetic cue have better naive performance on temporal

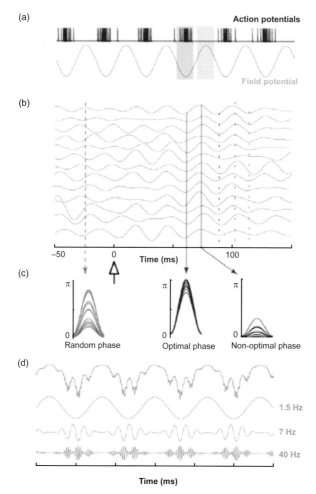

Figure 17.2 Neural oscillation hypothesis (Schroeder et al. 2008). Cortical and subcortical brain regions reveal intrinsic oscillatory neural activity on different rates/time scales, for example, between 1 Hz (delta band) and 40 Hz (gamma band). Such intrinsic oscillations are potentially in nested hierarchical relationships (d). Because the temporal structure of speech and music falls into the modulation rates of such oscillations, one hypothesis suggests that oscillatory neural mechanisms may underlie segmentation, grouping, alignment, attention allocation, and so on, and may interact with the stimulus input to generate different forms of "readout" on different timescales commensurate with the oscillations. The relation between firing patterns and excitability cycles provided by oscillations is shown in in (a); the relevance of phase is depicted in (b) and (c). Intuitively, these mechanisms allow the system to lock to the phase of (or entrain to) the temporal structure of a stimulus and generate temporal windows or units for further processing. The alignment of spikes with preferential phases of a cycle (a) illustrates the packaging of spikes by oscillations. For a detailed discussion of the relevant cellular circuitry specifically for speech processing, see Giraud and Poeppel (2012). Figure from Schroeder et al. (2008), used with permission from Elsevier.

interval discrimination than do native speakers of other languages, and musicians have better naive performance on temporal interval discrimination than nonmusicians, thus suggesting that timing is truly a domain-general capacity.

Cross-Cutting Approaches and Sources of Evidence

In this section, we touch on four areas of neurocognitive research that have provided data both on domain-specific representational and domain-general computational questions. We briefly discuss, in turn, input codes, learning and development, brain injury and plasticity, as well as the interaction between perception, action, and prediction.

The Role of Input Codes or How Input Can Determine Functional Specialization

The input code (in the afferent auditory pathway, and especially cortex) for music and spoken language is arguably the same kind of spectro-temporal representation (Griffiths et al. 1998) and is processed in parallel by distinct networks tuned to different features; for example, spectral versus temporal resolution (Zatorre et al. 2004). Trivially, at the most peripheral level, the signal that the system receives is the same: spectro-temporal variation stimulates the auditory periphery. Thus, differences between the domains must arise at more central levels. A fundamental issue concerns whether the structure of the input interacts with neuronal specializations of a certain type, such as preferences for spectral information versus temporal information or preferences for certain time constants.

Input codes may transform general-purpose auditory mechanisms into specialized ones that ultimately interact with the representations underlying music or speech. The existence of multiple specialized microsystems, even if they function in a similar way is more likely because modularization is more efficient. It is possible that domain specificity emerges from the operation of a general mechanism. However, in practice, it may be very difficult to demonstrate it because the general or "shared" mechanisms under study are likely to modularize with experience and also because dual domain-specific mechanisms may work together, as in song learning (Thiessen and Saffran 2003).

For example, the acquisition of tonal knowledge uses general principles by extracting statistical regularities in the environment (Krumhansl 1990; Tillmann et al. 2000). Although tonal encoding of pitch is specific to music, it may be built on "listeners' sensitivity to pitch distribution, [which is] an instance of general perceptual strategies to exploit regularities in the physical world" (Oram and Cuddy 1995:114). Thus, the input and output of the statistical computation may be domain specific while the underlying learning mechanism is replicated across circuitry serving both domains (Peretz 2006). Once acquired, the functioning of the system—say the tonal encoding of pitch—may be modular, by encoding musical pitch in terms of keys exclusively and automatically.

The same reasoning applies to auditory scene analysis as well as to auditory grouping. The fact that these two processing classes organize incoming sounds according to general Gestalt principles, such as pitch proximity, does not mean that their functioning is general purpose and mediated by a single processing system. They need not be. For instance, it would be very surprising if visual and auditory scene analyses were mediated by the same system; both types of analyses obey Gestalt principles. It is likely that the visual and auditory input codes adjust these mechanisms to their processing needs.

A developmental perspective (see next section) may be useful in disentangling initial states from modularized end stages, in both typical and atypical developing populations. Developmental disorders offer special insight into this debate. Advocates of a "domain-general" cognitive system may search for co-occurrence of impairments in music and language (and other spheres of cognition, such as spatial cognition). Such correlations may give clues as to the nature of the processes that are shared between music and language. It may turn out that domain specificity depends on very few processing components relative to a largely shared common cognitive background. These key components must correspond to domain- and human-specific adaptations, whereas the common background is likely to be shared with animals. Developmental disorders are particularly well placed to yield insight into both parts of the debate: that which is unique to music and language, and that which is not. It follows that much can be learned by comparing impaired and spared music, language, and cognition in individuals both within and between disorders over the course of development.

Still, somewhat separable modular components may exist for speech and music processing, both at a lower auditory-processing level and a higher cognitive level. Not surprisingly, the null hypothesis (analyzed by Patel, this volume) is that speech and music have very little in common in terms of cortical cognitive processing.

Learning and Development

Infants are born unable to understand or speak a particular language; they are also unable to understand or produce music. In both cases, language and music are acquired in an orderly sequence through everyday informal interaction with people in a cultural setting.

It is now accepted that the brain has a remarkable capacity to modify its structural and functional organization throughout the life span, in response to injuries, changes in environmental input, and new behavioral challenges. This plasticity underlies normal development and maturation, skill learning, memory, recovery from injury, as well as the consequences of sensory deprivation or environmental enrichment. Skill learning offers a useful model for studying plasticity because it can be easily manipulated in an experimental setting. In particular, music making (e.g., learning to sing or play a musical instrument) is an activity that typically begins early in life, while the brain has greatest plasticity. Often, musical learning continues throughout life (e.g., in musicians). Recent high-resolution imaging studies have demonstrated the ability of functional and structural auditory-motor networks to change and adapt in response to sensorimotor learning (Zatorre et al. 2012b).

Returning to the Marr-inspired taxonomy of levels, the representational elements of language and music are different, as shown in Figure 17.1. Those of language include phonemes, morphemes, words, and phrases whereas those of music include notes, pitch intervals, beats and meters, motifs and melodies. Despite representational differences at the higher level, music and language do rely to a large extent on shared elementary procedures that appear to be in place in prelinguistic and premusical infants (Drake 1998; Trainor and Corrigal 2010).

Indeed, it is possible that domain-specific processing develops in the brain largely through exposure to the different structures in the speech and musical input from the environment. Of course, at the same time, specialization of some brain regions for music or language likely occurs from intrinsic properties of those regions being more suited for processing structural elements of music (e.g., fine spectral structure) or speech (e.g., rapid temporal structure). Receptive language and music tend to be processed in similar regions in most people, though with some hemispheric differences, and expressive speech and musical vocalization might rely on shared auditory-motor systems (Özdemir et al. 2006).

To demonstrate the capabilities of infants (which are not, in most cases, present in neonates), consider the following examples of the early presence of a number of putatively primitive domain-general processing mechanisms (for more on this topic, see Trehub):

1. *Constructing spatiotemporal objects*: Newborns are able to discriminate their mother's voice from that of a stranger (DeCasper and Fifer 1980). At 6 months (probably younger), they can discriminate one voice from another in the context of multiple tokens from each speaker. That there is a learned component to this is evident: infants are equally good at human and monkey voice discrimination at 6 months, but better for human voices at 12 months. Infants can also discriminate timbres, and exposure to a particular timbre increases their neural response to that timbre, as measured by EEG (Friendly, Rendall, and Trainor, pers. comm.). From at least as young as two months, infants can categorize musical intervals as consonant or dissonant and prefer to listen to consonance (Trainor et al. 2002).

2. *Discretizing and sequencing* the signal: Young infants can discriminate rhythmic patterns as well as orders of pitches in a sequence (Chang and Trehub 1977; Demany et al. 1977)
3. *Relative pitch:* Infants readily recognize melodies in transposition, as evidenced by behavioral and EEG (mismatch response) studies (e.g., Tew et al. 2009; Trainor and Trehub 1992).
4. *Relative timing*: Infants recognize sequences played at somewhat faster or slower rates (Trehub and Thorpe 1989).
5. *Grouping*: Infants segregate and integrate incoming elements into perceptual streams. This has been shown both for sequential input, where higher and lower tones are grouped into separate streams (Winkler et al. 2003) as well as for simultaneous input, where one of several simultaneous tones can be "captured" into a separate stream if the simultaneous tones are preceded by several repetitions of the tone to be captured (Smith and Trainor 2011). Similarly, infants hear a harmonic that is mistuned as a separate object in a complex tone (Folland et al. 2012). The presentation of repeating patterns (e.g., short–short–long) also leads to grouping, such that group boundaries are received after the "long" elements.
6. *Hierarchical processing*: Infants perceive different meters, which require processing on at least two levels of a metrical hierarchy (Hannon and Trehub 2005a).
7. *Coordinate transformations and sensorimotor coordination*: Because young infants are not motorically mature, this is more difficult to demonstrate. The way that they are moved by their caregivers, however, affects their perception of auditory patterns, suggesting that they can transform from one reference frame to another (Phillips-Silver and Trainor 2005). When infants are presented with a repeating auditory six-beat pattern with an ambiguous meter (i.e., it had no internal accents), the pattern can be interpreted as two groups of three beats (as in a waltz) or as three groups of two beats. During a training phase, two groups of infants heard the ambiguous rhythm while they were simultaneously bounced up and down: one group on every third beat, the other on every second beat of the pattern. After this familiarization, both groups were given a preferential listening test. Infants bounced every second beat of the ambiguous pattern preferred (in the absence of bouncing) to listen to a version with accents added every second beat compared to a version with accents added every third beat. On the other hand, infants bounced on every third beat of the ambiguous pattern preferred to listen to the version with accents every third beat compared to the version with accents very second beat. The fact that infants were passively bounced and that the effect remained when they were blindfolded suggests that the vestibular system may play a role in

this. This study also suggests that the roots of common representations for music and dance may be seen in infancy.

8. *Prediction*: Auditory mismatch responses in EEG data (mismatch negativity) can be seen very early in development, even *in utero* during the last trimester (Draganova et al. 2005). In these studies, one stimulus or set of stimuli was repeated throughout; occasionally, one repetition was replaced with another, deviant stimulus. The existence of a mismatch response suggests that sensory memory is intact and that the mechanisms underlying regularity extraction and local prediction in time are available at the earliest stages of development.

9. *Entrainment/turn-taking*: Any evidence for entrainment in young infants is not widely known, although how they are moved to rhythms by their caregivers affects how they hear their metrical structure (Phillips-Silver and Trainor 2005). However, there is evidence for turn-taking in speech interactions with adults.

This short review suggests that the basic processing algorithms that enable language and musical learning in the young infant are in place as of a very early age. However, it goes without saying that the "linguistic and musical inventory" is not yet in place at this stage. That is, the representational elements are acquired, incrementally, over the course of development. In the case of speech and language, the trajectory is well known. In the first year, the learner acquires the sounds (signs) of her language, and by the end of year 1, the first single words are evident. Between the ages of two to three years, the vocabulary explosion "fills" the lexicon with items, and the first structured multiword (or multisign) utterances are generated. There is consensus that by three years of age, the neurotypical learner has the syntactic capabilities of a typical speaker (with, obviously, a more restricted vocabulary). What the steps look like for a child learning music—perhaps through song—is less clear.

The Give and Take of Language and Music

The Perception–Prediction–Action Cycle

The perception–action cycle (Neisser 1976; Arbib 1989) emphasizes that we are not bound by stimuli in our actions. In general, our perceptions are directed by our ongoing plans and intentions, though what we perceive will in turn affect our plans and actions. Within this framework, Fuster (2004) postulates that (a) action plans are hierarchically organized in the frontal lobe (Koechlin and Jubault 2006) whereas perception is hierarchically organized in the temporal, occipital, and parietal lobes, and (b) reciprocal paths link action and perception at all levels. One may recall the work of Goldman-Rakic (1991) in delineating the reciprocal connections between specific areas of frontal and parietal cortex.

Janata and Parsons (this volume) discuss this further and emphasize that attentive listening to music can engage the action systems of the brain.

A key part of the perception–action cycle is the predictive model: to prepare the next action, it is important to generate plausible expectations about the next stimulus. Activity in the auditory cortex thus represents not only the acoustic structure of a given attended sound, and other sounds in the environmental soundscape, but it also signals the predicted acoustic trajectories and their associated behavioral meaning. The auditory cortex is therefore a plastic encoder of sound properties and their behavioral significance.

Feedback and predictive (feedforward) coding is likely to function in both the dorsal and ventral auditory streams (Rauschecker and Scott 2009; Rauschecker 2011; Hickok 2012), with the direction of feedback and feedforward depending on one's vantage point within the perception–action cycle. During feedback from motor to sensory structures, an efference copy sent from prefrontal and premotor cortex (dorsal stream) could provide the basis for sensorimotor control and integration as well as "optimal state estimation" in the inferior parietal lobe and in sensory areas of the posterior auditory cortex. In contrast, forward prediction may arise from the ventral stream through an "object-based" lexical–conceptual system.

Figure 17.3 complements Figure 17.1 to provide a perspective on the interaction between auditory, premotor, and prefrontal areas using the notion of internal models. The perception–action cycle can be run either as a forward or an inverse model. The predictive, forward mapping builds on knowledge about objects processed and stored in the anterior temporal lobe via the ventral stream and continues via prefrontal and premotor cortex into parietal and posterior auditory cortex, where an error signal is generated between real and predicted input. The inverse mapping, which runs the cycle in the opposite direction, instructs the motor system and creates affordances via the dorsal stream for generating sounds that match the motor representations, including sound sequences that require concatenation in a particular order, as they are the substance of both speech and music. There is overwhelming evidence for such internal forward models in motor control (Flanagan et al. 2003; Wolpert et al. 2003; Wolpert and Kawato 1998), but the extension to both perceptual and cognitive models is more recent.

How much prediction occurs at this level of neuronal precision in the human auditory cortex as we process speech? One of the domains in which this has been addressed extensively is audiovisual speech. Both EEG and MEG research (e.g., van Wassenhove et al. 2005; Arnal et al. 2009) and fMRI-based (Skipper et al. 2007a, b, 2009) research has convincingly shown that information conveyed by facial cues provides highly predictive and specific information about upcoming auditory signals. For example, because facial dynamics slightly precede acoustic output (Chandrasekaran and Ghazanfar 2009), the content of the face signal (e.g., bilabial lip closure position) signals that a certain consonant type is coming (e.g., "b," "p," or "m"). This prediction is

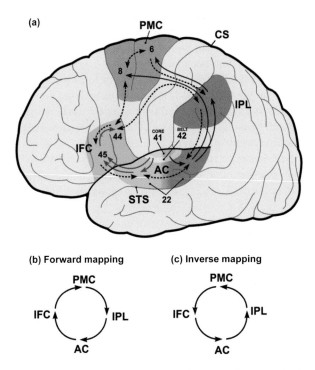

Figure 17.3 Feedforward and feedback organization (after Rauschecker and Scott 2009). (a) A schematic version of the dorsal and ventral processing streams and their basic connectivity. Dorsal projections extend from superior posterior temporal (auditory cortex, AC) through inferior parietal lobe (IPL) to inferior frontal cortex (IFC) and premotor cortex (PMC). The ventral stream projections typically extend through the extreme capsule and the uncinate fasciculus to inferior frontal areas. Superior temporal sulcus: STS; CS: central sulcus. The hypothesized forward and inverse mappings are illustrated in (b) and (c), respectively.

reflected both in response latencies (shorter for highly predictable items) and amplitudes (smaller for predictable items). Overall, whether one is considering speech or music alone as purely acoustic signals, audiovisual signals, such as speaking faces or playing musicians, or spoken sentences, or even higher-order conceptual information, it is beyond dispute that both high- and low-level information is incorporated into models (perhaps of a Bayesian flavor) that shape upcoming perception and action in a precise manner. There is as much, if not more, expectancy-driven top-down processing as there is bottom-up analysis. Of course, prediction occurs at higher levels, as when the listener predicts the next word of a sentence.

Feedback is also critical at lower levels. Speech production is known to be dependent on auditory feedback, going back to Levelt (1983) and emphasized in recent work by Houde and Nagarajan (2011) and Hickok et al. (2011).

Speech can lead to motor-induced suppression of the auditory cortex (Allu et al. 2009) and may result in noise suppression or cancellation of self-produced speech. MEG studies have long shown evidence for this putative efference copy, i.e., a predictive signal from motor to auditory cortex (Kauramäki et al. 2010; Nishitani and Hari 2002), and nonhuman primate studies have demonstrated a neurophysiological correlate (Eliades and Wang 2008). One insight that has emerged is that local, early feedforward as well as feedback processing in auditory cortical areas reflects analysis of the error signal (i.e., the mismatch between the predicted input and the actual input).

Recent research has been conducted on brain activation during total silence, based on the expectation of upcoming or anticipated sounds. For a recent example of this in musical sequences, see Leaver et al. (2009).

Timing and Turn-Taking

Elsewhere in this volume, Levinson explores "the interactive niche," which includes social interactions involved in turn-taking as well as the sequencing of actions. "Informal verbal interaction is the core matrix for human social life. A mechanism for coordinating this basic mode of interaction is a system of turn-taking that regulates who is to speak and when" (Stivers et al. 2009:10,587). The work of Levinson and his colleagues (Stivers et al. 2009) has shown that there are striking universals in the underlying pattern of response latency in conversation, providing clear evidence for a general avoidance of overlapping talk and a minimization of silence between conversational turns (an incredibly brief gap since the peak of response is within 200 ms of the end of the previous question). As Levinson observes, since it takes at least 600 ms to initiate speech production, speakers must anticipate the last words of their companion's turns and predict the content and the form of their companion's utterance in order to respond appropriately. Thus, conversation is built on detailed prediction: figuring out when others are going to speak, what they are going to say, when they are going to finish, and how to prepare your own reply. This requires encoding of the utterance they intend to make at all levels.

Another study on turn-taking (De Ruiter et al. 2006) demonstrates that the symbolic (i.e., lexical, syntactic) content of an utterance is necessary (and possibly sufficient) for projecting the moment of its completion, and thus for regulating conversational turn-taking. By contrast, and perhaps surprisingly, intonational contour is neither necessary nor sufficient for end-of-turn projection. This overlap of comprehension and production in conversation can be extremely demanding at a cognitive level.

As Hagoort and Poeppel (this volume) state:

> It might well be that the interconnectedness of the cognitive and neural architectures for language comprehension and production enables the production system to participate in generating internal predictions while in the business of

comprehending linguistic input. This prediction-is-production account, however, might not be as easy in relation to the perception of music, at least for instrumental music. With few exceptions, all of humankind are expert speakers. However, for music, there seems to be a stronger asymmetry between perception and production. Two questions result: Does prediction play an equally strong role in language comprehension and the perception of music? If so, what might generate the predictions in music perception?

Clearly, predictions can be guided if the musicians are playing a composed score of music from memory. Thus, this question is likely to be particularly important during conversational turn-taking in music, which may place even greater cognitive demands on a musician playing in an orchestra or quartet, and particularly during improvisation, as in jazz. Moreover, playing in an ensemble, singing in a choir or dancing with a partner all involve patterns of coordination that require far more delicate timing than that involved in initiating a turn in a conversation.

The true complexity of the mechanics of turn-taking is illustrated in Figure 17.4 (Menenti et al. 2012). Building on the work of Pickering and Garrod (2004), Figure 17.4 shows at how many levels of analysis two speakers have to align, ranging from sounds to highly abstract situation models. In the case of language, the nature of the necessary alignments becomes increasingly well understood. However, whether such a model is plausible (or even desirable) for musical performance or dance in a pair or group is not at all clear. Future

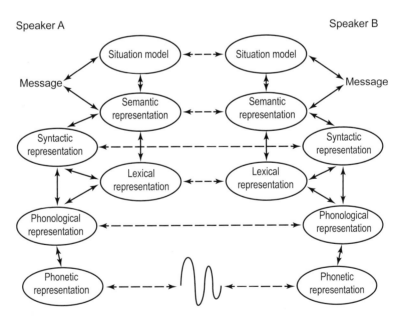

Figure 17.4 The interactive alignment model (reprinted with permission from Menenti et al. 2012).

work will have to determine if such alignments are in play at all, in ways similar to conversation models.

The neural basis for the postulated "representational parity" between production and comprehension is a topic of current research (Menenti et al. 2012). To test these predictions, future studies will record from participants engaged in online interaction.

Perception and Production Interaction in Singing

Although this topic has been discussed by Janata and Parsons (this volume), we wish to emphasize the importance of the interaction between perception and production in singing. Vocal control of song and pitch has been studied in both nonmusicians and musicians (Perry et al. 1999; Brown et al. 2004; Zarate and Zatorre 2008). Recent behavioral studies report evidence that suggests that nonmusicians with good pitch discrimination sing more accurately than those with poorer auditory skills (perceptual and vocal production skills). However, Zarate et al. (2010) gave auditory discrimination training on micromelodies to a group of nonmusicians and found that training-enhanced auditory discrimination did not lead to improved vocal accuracy although it did lead to enhanced auditory perception (Zatorre et al. 2012a). Thus, there may be a partial dissociation between auditory perceptual and vocal production abilities; that is, while it may not be possible to produce precise pitch intonation without equal perceptual abilities, the presence of perceptual ability alone does not guarantee vocal ability. (For differences in the brain systems supporting verbal and tonal working memory in nonmusicians and musicians, see Koelsch, this volume.)

Audiomotor interactions in music perception and production have been reviewed by Zatorre et al. (2007). Auditory imagery has also been described by Leaver et al. (2009), suggesting links between auditory and motor areas, premotor and supplementary motor areas, as well as prefrontal regions. Zatorre (2007) proposes that when we listen to music, we may activate the ventral premotor cortex links, associated with producing the music. However, listening also engages another neural system, in which the dorsal premotor cortex is a component, to process higher-order metrical information, which may be critical for setting up temporal and melodic expectancies at the heart of musical understanding. These topics are discussed further below in the section on the neurology of amusia.

Neurobiological Constraints and Mechanisms

The neurobiological foundations of music, speech, and language processing have been studied at virtually every level of analysis: from single unit physiology to noninvasive imaging to deficit lesion studies. Here we offer a selection of phenomena to illustrate the range of data that need to be incorporated into a synthetic understanding of shared and distinctive processes in music and language.

Domain-General Processes: Shared Neural Substrates for Language and Music

Hierarchical Organization

There is considerable evidence for hierarchical organization in the human auditory cortex (Okada et al. 2010; Obleser et al. 2010; Chevillet et al. 2011; DeWitt and Rauschecker 2012) as well as in nonhuman primates (Tian et al. 2001; Rauschecker and Scott 2009; see also Figure 17.1 and 17.3). Core regions, including primary auditory cortex, respond best to tones and contours, whereas higher areas (belt and parabelt) are selectively responsive to more complex features such as chords, band-passed noise, vocalizations, and speech. We propose that core levels encode low-level features that are combined in higher levels to yield more abstract neural codes for auditory objects, including phonemes and words for language.

Although neural representation of music is likely to be constrained by this hierarchical organization for auditory processing in the brain, one key issue is whether such hierarchical organization can be demonstrated for higher groupings in music. In the case of language, we have pretty clear ideas of what constitutes a processing hierarchy (e.g., phonology, morphology, lexical semantics, syntax, compositional semantics, discourse representation), and there is a growing body of evidence about where such functions are executed (for a review, see Hagoort and Poeppel, this volume). How each level is executed, however, is largely unknown. In the case of music, there are equally intuitive hierarchies (e.g., note, motif, rhythm, melody, piece), but the functional anatomy of the hierarchy is less clear. Trivially, auditory areas are implicated throughout; interestingly motor areas are implicated in many of the temporal, beat, and rhythm subroutines. However, a well-defined functional anatomy is still under construction, in part because many of the functions are compressed into auditory regions and the role of memorized structures is less well established (see below).

Acoustic Scene Analysis and Streaming

Auditory streaming is the perceptual parsing of acoustic sequences into "streams." This makes it possible for a listener to follow sounds from a given source despite the presence of other sounds and is critical in environments that contain multiple sound sources (Carlyon 2004). Neural mechanisms underlying streaming are common to music and language, are strongly influenced by attention, and appear to use a full range of grouping mechanisms for frequency, timbre, as well as spatial and temporal cues (Micheyl et al. 2005; Shamma et al. 2011; Wang and Brown 2006).

Real-Time, Attention-Driven Adaptive Plasticity

To understand what is going on at a neural level, it is critical to realize that the auditory cortex is not a passive detector of acoustic stimulus events. Its activity and responses nimbly change with context (reflecting task demands, attention, and learning) to provide overall functional relevance of the sound to the listener (Fritz et al. 2003). Rapid changes in auditory filter properties reconfigure the listening brain to enhance the processing of current auditory objects of interest, whether in the linguistic or musical domain, and may help segregate relevant sounds from background noise (Ahveninen et al. 2011). The attention-driven top-down capabilities are especially important in light of the top-down influences (evident in the perception–action cycle) that condition processing even at the periphery (Fritz et al. 2010; Xiao and Suga 2002; Delano et al. 2007, 2010; Leon et al. 2012). Figure 17.2 illustrates one possible hypothesis of how such attention in time can be accomplished. If either musical or speech elements arrive at predictable times (as they often do, given underlying rhythms, though more so in music), amplifying or selecting those moments can facilitate processing with attention-driven, adaptive plasticity mechanisms. A challenge for this oscillatory hypothesis arises in situations where acoustic input is less structured in time.

Pitch

There are many examples of (nearly) perfect pitch and perfect tempo in musicians. What was a bit unexpected is that there is also good evidence that mothers without musical training also demonstrate absolute pitch and tempo as they sing songs to their infants (Bergeson and Trehub 2002). It is also intriguing to note that musicians who are native speakers of a tone language are more likely to have musical absolute pitch than musicians who do *not* speak a tone language (Deutsch et al. 2006).

What is the neurobiological basis of pitch? There are well-studied neurophysiological mechanisms that can help us begin to think about how absolute pitch is encoded (Bendor and Wang 2005; Bizley et al. 2009) as well as how relative pitch is encoded, such as frequency-shift detectors or frequency-modulated, direction-sensitive neurons in the auditory cortex. Computational models have built on this work to suggest how the brain represents and remembers sequences of relative pitches (see, e.g., the model put forth by Husain et al. 2004), which incorporates multiple brain regions, including superior temporal and prefrontal regions.

Interestingly, studies of other species with complex acoustic communication (such as starlings) show that recognizing tone sequences on the basis of relative pitch is difficult for nonhuman animals (cf. Bregman et al. 2012). This raises the question of whether our system has been optimized or specialized over evolutionary time for this purpose. Animal behavioral studies are beginning to address this issue (Wright et al. 2000; Yin et al. 2010; Bregman et al. 2012) as are some initial animal neurophysiological studies (Brosch et al. 2004).

Timing and Beat Perception

A large body of work, including recent neurophysiological studies (Jaramillo and Zador 2011; Bendixen et al. 2011), shows that neurons in auditory cortex are modulated by the expected time of arrival of an incoming sound. What is the neural basis for timing and time constants?

Timing networks are widespread throughout the brain. A recent experimental and theoretical study by Bernachhia et al. (2011) suggests that there is a neuronal "reservoir" of time constants in areas of the prefrontal, cingulate, and parietal cortex that can be used to support a flexible memory system in which neural subpopulations with distinct sets of long or short memory timescales can be deployed according to task demands. Other studies (Itskov et al. 2011; Jin et al. 2009; Fritz et al. 2010) have shown similar arrays of neurons with variable time constants in cerebellum, basal ganglia and prefrontal cortex.

Interestingly, simply listening to musical rhythms activates the motor system (Chen et al. 2008a). The cerebellum, basal ganglia, dorsal premotor cortex, and prefrontal cortex have all been shown to play an important role in timing in music, and likely in language processing (Zatorre 2007). A recent imaging study (Teki et al. 2011) suggests that there are distinct neural substrates for beat-based and duration-based auditory timing encompassing a network of the inferior olive and the cerebellum that acts as a precision clock to mediate absolute, duration-based timing, and a distinct network for relative, beat-based timing incorporating a striato-thalamo-cortical network. The supplementary motor area (SMA) and pre-SMA are critical for sequencing and integration into unified sequences (Bengtsson et al. 2009; Leaver et al. 2009). However, these networks are not typically recruited during language processing, notwithstanding the quasi-rhythmic nature of spoken language (see, however, Ivry et al. 2001. However, we do not yet know exactly how the concatenation of elements into specific sequences is accomplished in musical perception and production; neither do we know how such sequences are exquisitely timed, and how sequences and their tempo are recalled. One recent study that has explicitly addressed the issue of the timing circuit, especially with regard to the role of the basal ganglia, is by Kotz et al. (2009). They develop the well-known view that the basal ganglia play a key role in sequencing motor production to argue for its role in sensory predictability in auditory language perception.

Although there is insight into neural mechanisms that may underlie timing, the neural basis for beat and meter processing is still largely unknown. The resonance hypothesis for beat and meter perception (Large 2008) proposes that beat perception emerges from the entrainment of neuronal populations oscillating at the beat frequency, giving rise to higher-order resonance at subharmonics of beat frequency, corresponding to the meter. Experimental support for the resonance hypothesis comes from a recent study showing that entrainment to beat and meter creates temporal expectancies which can be observed directly in the human EEG as a periodic response at the frequency of the beat and at subharmonics corresponding to the metrical interpretation of the beat (Nozaradan et al. 2011). Although such entrainment clearly occurs in music, it is also likely to occur in poetry and cadenced speech (see also Ladinig et al. 2009; Honing et al. 2009).

Self-Monitoring During Speech and Music

Vocal communication involves both speaking and listening, often taking place concurrently. It is important for the auditory system to simultaneously monitor feedback of one's own voice as well as external sounds coming from the acoustic environment during speaking. The self-monitoring in the audio-vocal system may play a part in distinguishing between self-generated or externally generated auditory inputs and also in detecting errors, and making compensatory corrections, in our vocal production as part of a state feedback control system (Houde and Nagarajan 2011). Neurons in the auditory cortex of marmoset monkeys are sensitive to auditory feedback during vocal production, and changes in vocal feedback alter the coding properties of these neurons and increased their sensitivity (Eliades and Wang 2003, 2005, 2008). Such self-monitoring occurs during speaking and singing as well as during instrumental performance. In addition, there is clear evidence for attenuation or suppression of neural responses to self-triggered sounds in the human auditory cortex (motor-induced suppression for nonvocal as well as vocal stimuli). This suggests the importance of internal forward-predictive models in processing sound and distinguishing between the auditory consequences of one's own actions as distinct from other externally generated acoustic events (Baess et al. 2011; Martikainen et al. 2005). Musicians have been shown to have a particularly keen ability to generate accurate forward-predictive models (Tervaniemi et al. 2009). Thus, more generally, prior expectation (based on memory or forward models) can mediate neural adaptation (Todorovic et al. 2011).

Auditory Memory

Human long-term auditory memory appears to be extraordinarily powerful when it comes to recall of poetry or music, as compared to individual acoustic stimuli, even in musicians—particularly in comparison with the striking retention observed in visual memory (Cohen et al. 2009, 2011). Visual recognition memory in monkeys is much superior to auditory recognition memory, for stimuli with no visual association (Fritz et al. 2005). These results support the hypothesis that the emergence of language and music went hand in hand with the development of improved auditory working memory and long-term memory for abstract but meaningful sounds (Aboitiz et al. 2010; Aboitiz and García 2009). Neuroimaging evidence also exists for a network of at least two distinct, and highly interconnected, neural loci for working memory, which are both part of the dorsal cortical pathway (Rauschecker and Scott 2009)—a frontal region which comprises the dorsal part of Broca's area (Brodmann area 44) and the adjacent inferior frontal sulcus—supporting syntactic working memory and a phonological working memory store in parietal cortex (Friederici 2012). Learning the pathways and dynamic interactions between these two areas will greatly aid our understanding of language processing.

There may be evidence for a dissociation of memory for melody and lyrics in song. A patient with a lesion of the right hemisphere anterior temporal lobe and right hemisphere lateral prefrontal cortex (Steinke et al. 2001) was able to recognize familiar songs when the accompanying lyrics were removed (i.e., melodies without words), but could not recognize equally familiar but purely instrumental melodies. Evidence for the integration of melody and text has been found, however, in other patients, such as expressive aphasics who can accurately sing but not speak the lyrics of familiar songs.

Exemplar-based verbal memory has been shown in the linkage of identification and memory for individual voices with word recall ability (linking "who" and "what"; Nygaard and Pisoni 1998; Perrachione et al. 2011). In terms of linking melody to instrumental timbre ("which musical instrument played that piece" or motor linkages, i.e., "how I played that piece on that instrument" and "what melody was played"), this has also been demonstrated in musical memory. Such exemplar-based memory for music has even been observed in infants (Trainor et al. 2004). The linkage of auditory memory with semantics may also be inferred from patients with semantic dementia, who have difficulty in understanding the meaning of environmental sounds and are also impaired in the recognition of melodies (Hsieh et al. 2011). Even within the context of verbal material, there is better memory for poetry than for prose (Tillmann and Dowling 2007), emphasizing the mnemonic importance of temporal organization and rhythmic structure in poetry and music, and thus linking memory for words with rhythm and rhyme.

Koelsch (this volume) reviews evidence for two auditory working memory systems: a phonological loop which supports rehearsal of phonological (verbal) information, and a tonal loop which supports rehearsal of tonal (nonverbal) information, which are differentially developed and localized in musicians and nonmusicians. Furthermore, there are different short-term, or sensory-memory storage buffers for pitch, timbre, loudness, and duration (Semal and Demany 1991; Clement et al. 1999; Jaramillo et al. 2000; Caclin et al. 2006) and different working memory networks for melodies and rhythms (Jerde et al. 2011). Studies also indicate that auditory short-term memory for complex tone patterns is enhanced in musicians (Boh et al. 2011).

Domain-Specific Processes: Neural Substrates for Music

Deficit-Lesion Characterizations: Insights from Neurological Cases

Many neuropsychological dissociations exist between language and music perception and production (Peretz 2006). One very rare but compelling neurological disorder is pure word deafness. There have been only a few dozen documented cases since the late 1800s, and very few cases in which the syndrome is "pure," but there is convergence on the general phenomenon (for reviews, see Poeppel 2001; Stefanatos 2008). Such patients have normal audiograms and largely intact peripheral auditory processing. Moreover, they are not aphasic; that is, they can read, write, and *speak*, albeit often haltingly. However, their perception of spoken language, and indeed any speech stimulus, is completely compromised. Patients are deaf to spoken words, yet, interestingly, their perception of music is relatively intact. Thus, in one single dissociation, hearing, speech, perception, language, and music are functionally fractionated. Recent cases (Stefanatos 2008; Wolmetz et al. 2011; Slevc et al. 2011) support the conjecture that the lesion pattern underlying pure word deafness is twofold. In two-thirds of the cases, the posterior aspect of the STG is affected, bilaterally. (This means patients have two lesions, sequentially.) In one-third of the cases, a deep left lateralized white matter lesion is observed; this lesion deafferents the two sides from one another, thus also implicating the integrity (or integration) of both sides for successful speech processing. Posterior STG/STS has long been thought of as the necessary tissue for speech perception, but not music perception. A recent meta-analysis of a large number of neuroimaging studies of speech perception argues, however, that the necessary site is, in fact, more anterior than previously thought (DeWitt and Rauschecker 2012). Thus, the relative contributions of each site are still under debate.

Amusia and Congenital Amusics

Parallel to pure word deafness, a well-characterized neuropsychological disorder is acquired amusia, where patients have selective difficulty with processing musical material. Brain lesions can selectively interfere with musical abilities while the rest of the cognitive system remains essentially intact (e.g. Steinke et al. 1997). Conversely, brain damage can impair musical abilities exclusively. Patients may no longer recognize melodies (presented without words) that were highly familiar to them prior to the onset of their brain damage but perform normally when recognizing spoken lyrics (and words, in general), familiar voices, and other environmental sounds (e.g., animal cries, traffic noises, and human vocal sounds). The existence of a specific problem with music alongside normal functioning of other auditory abilities, including speech comprehension, is consistent with damage to processing components that are both essential to the normal process of music recognition and specific to the musical domain (for reviews, see Peretz 2006, Peretz et al. 2009a).

Similar findings are obtained in production studies. Brain-damaged patients may lose the ability to sing familiar songs but retain the ability to recite the lyrics and speak with normal prosody. The reverse condition (i.e., impaired speech with intact vocal production) is more common or has been reported more often. Aphasic patients may remain able to sing familiar tunes and learn novel tunes but fail to produce intelligible lyrics in both singing and speaking (Racette et al. 2006). These results suggest that verbal production, whether sung or spoken, is mediated by the same (impaired) language output system, and that this speech route is distinct from both the (spared) musical and prosodic route. In sum, the autonomy of music and language processing extends to production tasks.

Similarly, individuals who suffer from lifelong musical difficulties, a condition which Peretz (2008) and Stewart (2011) refer as to congenital amusia, have normal speech comprehension and production. In contrast, they experience difficulties in recognizing instrumental melodies; they have problems hearing when someone sings out of tune or plays a "wrong" note (typically, a mistuned or out-of-key note); and the large majority sing out of tune. Amusics have difficulties recognizing hummed melodies from familiar songs, yet they can recognize the lyrics that accompany these melodies. In singing, they can recall the lyrics of familiar songs to which they can hardly produce a recognizable tune (e.g., Tremblay-Champoux et al. 2010).

Curiously, there is a paucity of research on this striking dissociation between music and speech. The only area of comparison studied so far concerns the intonation pattern of speech. In both French and English, intonation is used to convey a question or a statement. Amusics have little difficulty to distinguish these although they may show mild impairments when these pitch changes are subtle (Hutchins et al. 2010; Liu et al. 2010) or require memory (Patel et al. 2008b). Similarly, amusics may experience difficulties when comparing lexical tones taken from Mandarin or Thai (Tillmann et al. 2011). Speakers of a tonal language essentially show the same profile (e.g., Nan et al. 2010). Thus, amusics may show a deficit in processing pitch information in speech but this deficit is generally mild.

The clear-cut dissociation between music and speech seen in amusia provides a unique opportunity to address other fundamental questions related to the comparison of music and speech. For example, a current, hotly debated issue concerns the sharing (or overlap) of the processing involved in music and speech syntax. As mentioned above, a behavioral failure in the detection and discrimination of melodies by an out-of-key note is diagnostic of the presence of congenital amusia, presumably because the out-of-key note is tuned correctly but violates the tonal ("syntactic") relationships between notes in the given key of the melody. According to Patel's *shared syntactic integration resource hypothesis*" (SSIRH), discussed further below, amusics should exhibit similar difficulties with language syntax. Future research is needed to determine the analogous situation in language.

What counts as music or as nonmusical is not trivial. For example, rap music may be heard as speech, and highly dissonant music as noise. Conversely, some speech streams, such as the typical speech used by an auctioneer, may not be considered musical yet this form of chanting might be processed as music. Such ambiguous signals are not problematic for the peripheral auditory system, which does not need to decide which part of the auditory pattern is sent to music processors and which part to the language system. All information in the auditory input, including the text and the melody of an auction chant, would be sent to all music and language processors. The intervention of music- or language-specific components is determined by the aspect of the input for which the processing component is receptive. Thus, by studying the way amusics process different forms of music and speech, we may gain insight into which aspects are essential and specific to music.

Vocal control has also been studied in sensory and motor amusia (i.e., the loss or impairment of the ability to perceive or produce music or musical tones) (Ayotte et al. 2002; Loui et al. 2008, 2009). Diffusion tensor imaging has shown that in the right hemisphere of amusics, there is a thinner arcuate fasciculus: a fiber bundle connecting pars opercularis and the superior temporal areas, which is believed to provide auditory feedback control of speech. Amusics also have deficits in discriminating statements from questions (Liu et al. 2010) when there are small (4–5 semitone) pitch movements. The higher threshold for discriminating pitch movement in amusics may impair musical perception (that uses 1–2 semitone intervals) without usually affecting speech perception, which uses larger pitch movements (4–12 semitones). The areas affected are likely to include both the superior temporal and frontal areas. Evidence for domain specificity comes from patients with congenital amusia, who are impaired in short-term memory for music (pitch and timbre) but not for verbal material (Tillmann et al. 2009).

Beat Deafness

The most frequent form of congenital amusia affects the pitch dimension of music and spares, to some extent, rhythm. Recently, the reverse situation was observed in a young man in whom amusia is expressed by a marked difficulty to find and synchronize with the musical beat (Phillips-Silver et al. 2011). This case suggests a deficit of beat finding in the context of music. The subject is unable to period- and phase-lock his movement to the beat of most music pieces, and cannot detect most asynchronies of a model dancer (Phillips-Silver et al. 2011). The ability to identify or find beat has many practical uses beyond music and dance. For example, the act of rowing, marching in a group, or carrying a heavy object with others is made easier when beat is present.

Nonmusical behaviors involving temporal coordination between individuals, such as conversational turn-taking or even simply adjusting one's gait to that of a companion, require sophisticated processes of temporal prediction and movement timing (see chapters by Levinson and Fogassi, both this volume). It is of interest to know whether such processes share mechanisms with beat finding. Does temporally coordinated behavior with another person outside of the musical domain rely on the same brain network involved in tracking and predicting beats in music? Future studies aim to test whether beat-deafness impacts speech rhythm, gait adjustment, or other nonmusical rhythmic tasks.

Animal studies may also be useful in elucidating the neural pathways involved in beat perception (Patel et al. 2009). Cockatoos, parrots, and elephants have been shown to be able to synchronize their movements to a musical beat. While there is also some evidence that monkeys display drumming behavior and that brain regions preferentially activated to drumming are also activated by vocalizations (Remedios et al. 2009), it has been shown that the same types of monkeys (rhesus macaques) cannot synchronize their taps to an auditory metronome (Zarco et al. 2009).

Brain Injury and Plasticity

One empirical approach that has been valuable both in the study of music and in the study of language is the evaluation of compensatory plasticity in the nervous system. Extreme cases of plasticity can be seen following stroke or a traumatic brain injury, or in developmental disorders of deafness and blindness. Both deafness and blindness lead to compensation of sensory loss by the remaining senses (cross-modal plasticity). Visual deprivation studies in animals and neuroimaging studies in blind humans have demonstrated massive activation of normally visual areas by auditory and somatosensory input (Rauschecker 1995). While changing sensory modality, formerly visual areas in the occipital and temporal lobe retain their functional specialization in the processing of space, motion, or objects, such as faces or houses (Renier et al. 2010). Restitution of functionality impaired after an insult is paralleled by micro- and macro-structural as well as representational (functional) changes in cerebral gray and white matter. These changes can be seen in the immediate perilesional cortex as well as in homologous regions in the unimpaired healthy hemisphere. The major mechanisms of this plasticity are regeneration and reorganization. Regeneration involves axonal and dendritic sprouting and formation of new synapses, most likely induced by the production and release of various growth factors and up-regulation of genetic regulators. Reorganization involves remapping of lesional area representations onto nonlesional cortex either in the perilesional region or in the contralesional hemisphere.

One of the most typical examples of lesional plasticity is the ability of the brain, through internal or external triggers, to reorganize language functions after an injury to the language-dominant hemisphere. The general consensus is that there are two routes to recovery. In patients with small lesions in the left hemisphere, there tends to be recruitment and reorganization of the left hemispheric perilesional cortex, with variable involvement of right hemispheric homologous regions during the recovery process. In patients with large left hemispheric lesions involving language-related regions of the frontotemporal lobes, the only path to recovery may be through recruitment of homologous language and speech-motor regions in the right hemisphere, recruitment which is most effective in young children. Activation of right hemispheric regions during speech/language fMRI tasks has been reported in patients with aphasia, irrespective of their lesion size. For patients with large lesions that cover language-relevant regions in the left hemisphere, therapies that specifically engage or stimulate the homologous right hemispheric regions have the potential to facilitate the language recovery process beyond the limitations of natural recovery. It is worth remembering that the plastic reorganization is age dependent and that some damage is irreversible.

Turning to the relation of music, melodic intonation therapy (MIT) is an intonation-based treatment method for severely nonfluent or dysfluent aphasic patients who do not have sufficient perilesional cortex available anymore for local functional remapping and reorganization. MIT has been developed in response to the observation that severely aphasic patients can often produce well-articulated, linguistically accurate utterances while singing, but not during speech. MIT uses a combination of melodic and sensorimotor rhythmic components to engage the auditory-motor circuitry in the unimpaired right hemisphere and trains sound-motor mappings and articulatory functions (Schlaug et al. 2010). In expressive aphasics, song may be used for therapeutic purposes to encourage the recovery of speech. MIT therapy has been used to help nonfluent aphasics recover speech, and it appears to work by recruiting neural plasticity in right hemisphere word articulation circuitry. Similar interventions for musical dysfunctions or the use of language structures and language tools to overcome musical dysfunctions have not been developed, but this could be an interesting line of research to pursue.

Role of Temporal Frontal Neuroanatomical
Connections in Speech and Music Production

The left "perisylvian" cortex (consisting of superior temporal, inferior parietal, and inferior frontal regions) is seen as crucial for language perception and production, with various fiber pathways (see below) connecting the left superior temporal cortex ("Wernicke's area") and the left inferior frontal cortex ("Broca's area"). If someone suffers a large left hemispheric lesion, leading to aphasia, then the variability of the size of the right hemispheric language tracts might actually contribute to natural recovery of language function.

Rilling et al. (2008; see also Figure 9.6 in Hagoort and Poeppel, this volume) have presented a comparative analysis of arcuate fasciculus (AF) across three species (macaque, chimpanzee and human). In all three cases there is significant connectivity along dorsal projections. However, the extensive ventral stream projections observed in the human brain is not observed in either the chimpanzee or macaque brain.

Three different tracts connect the temporal lobe with the frontal lobe: the AF, the uncinate fasciculus, and the extreme capsule. Most is known about the AF, which connects the STG and middle temporal gyri (MTG) with the posterior inferior frontal lobe, arching around the posterior Sylvian fissure. Recent studies have suggested that the AF may be primarily involved in the mapping of sounds to articulation (in singing and spoken language) and/or to audiomotor interactions in learning and performance of instrumental music. (Earlier, we suggested it provides auditory feedback control of speech.) Some believe that the AF is direct and that there are fibers between the STG/MTG and the inferior frontal gyrus (IFG) (Loui et al. 2008, 2009; Schlaug et al. 2010), whereas Frey et al. (2008) argue that the AF is an indirect tract, as in most nonhuman primates (Petrides and Pandya 2009).

The temporal component connects to the parietal lobe, and then the superior longitudinal fasciculus connects the parietal lobe to the IFG. The AF has connections with inferior primary somatosensory, inferior primary motor, and adjacent premotor cortex. In humans, the AF is usually larger in the left than in the right hemisphere, although the right hemisphere does have a complete tract which might allow the right hemisphere to support vocal output even if the left hemisphere is lesioned. In chimpanzees, arcuate terminations are considerably more restricted than they are in humans, being focused on the STG posteriorly and on the ventral aspects of premotor cortex (BA 6) and pars opercularis (BA 44) anteriorly. In macaques, the arcuate is believed to project most strongly to dorsal prefrontal cortex rather than to Broca's area homologue.

Schlaug and colleagues have shown that the AF of musicians is larger in volume than in nonmusicians (Halwani et al. 2011); it also differs in microstructure (fractional isotropy) from nonmusicians. Moreover, in singers, the microstructural properties in the left dorsal branch of the AF are inversely correlated with the number of years of vocal training. These results suggest that musical training leads to long-term plasticity in the white matter tracts connecting auditory-motor and vocal-motor areas in the brain. To complicate matters, there may be a developmental story in which myelination and maturation of these fiber bundles in the AF influences language development (Brauer et al. 2011a).

The uncinate fasciculus is a hook-shaped fiber bundle that links the anterior portion of the temporal lobe with the orbital and inferior frontal gyri. The extreme capsule is a fiber bundle that links the temporal with more anterior portions of the IFG (Brodmann 45) and inferior prefrontal regions. Both the uncinate fasciculus and the extreme capsule are thought to be more involved in the mapping of sounds to meaning. Both fiber tracts can carry information along the ventral "what" pathway from the anterior STG to IFG (Marchina et al. 2011). We speculate that these ventral pathways are likely to be important for the processing of speech meaning whereas dorsal pathways are likely to be important in speech and musical production.

Ventral pathways are perceptual; they allow auditory object identification and association with behavioral "meaning." Dorsal pathways associate sounds with actions. In Hickok and Poeppel's model, area SPT is involved in the transformation of perceived and spoken words (Hickok and Poeppel 2004); Rauschecker and Scott (2009) emphasize the importance of the reverse transformation. Unresolved is how the architecture of the multistream models that have provided useful heuristics in speech and language research extends to music. Clearly these models are integrated and "deployed" during song, since speech/language are half the battle. However, since these models are also motor control and sensorimotor transformation models, it stands to reason that they also play a central role in performance of instrumental music, and crucially in the predictive aspects of processing. (For a complementary view of the dorsal and ventral pathways, in this case in the visual control of action, see Arbib et al., this volume. The view there is that the dorsal pathway is implicated in the parameterization of action whereas the ventral pathway can invoke object identification to support prefrontal planning of action.)

Speech and Song Production

Speech production mechanisms are intimately tied to song production (as discussed further by Janata and Parsons, this volume). Speech and language production involves a multistage process: first you must select an appropriate message, then each lexical item (a lemma) to express the desired concept, and then access the sound structure. Of course, additional stages are also necessary for the construction of hierarchically organized sentences or intonation contours. Brain activation (reviewed in detail by Indefrey 2011) includes sensory-related systems in the posterior superior temporal lobe of the left hemisphere; the interface between perceptual and motor systems is supported by a sensorimotor circuit for vocal tract actions (not dedicated to speech) that is very similar to sensorimotor circuits found in primate parietal lobe (Rauschecker and Scott 2009). The posterior-most part of the left planum temporale (SPT) has been suggested to be an interface site for the integration of sensory and vocal tract-related motor representations of complex sound sequences, such as speech and music (Hickok and Poeppel 2004, 2007; Buchsbaum et al. 2011). As such, SPT is part of a dorsal-processing stream for sensorimotor control and integration, where general sensorimotor transformations take place for eye and limb movements in the service of internal models of behavior and optimal state control

(Rauschecker and Scott 2009). The data cited above on vocalization and singing suggest that song without words builds on the same circuitry (Zarate et al. 2010; Zarate and Zatorre 2008).

A comparison of speech and singing (Özdemir et al. 2006) shows shared activation of many areas, including the inferior pre- and postcentral gyrus, STG, and STS bilaterally. This indicates the presence of a large shared network for motor preparation and execution as well as sensory feedback/control for vocal production. Hence, these results suggest a bi-hemispheric network for vocal production regardless of whether words or phrases were intoned or spoken. However, singing more than humming ("intoned speaking") showed additional right-lateralized activation of the STG, inferior central operculum, and IFG. This may explain the clinical observation that patients with nonfluent aphasia due to left hemisphere lesions are able to sing the text of a song while they are unable to speak the same words. The discussion on melodic intonation therapy above provides an important connection point here.

Potential Right Hemisphere Biases: Evidence from Neuropsychology and Neuroimaging

Based on neurological cases and neuroimaging research, evidence suggests that musical pitch perception, or at least dynamic pitch, has a right hemisphere bias in the auditory cortex. There is now evidence (Foster and Zatorre 2010) that cortical thickness in right Heschl's sulcus and bilateral anterior intraparietal sulcus can predict the ability to perform relative pitch judgments. The intraparietal sulcus is known to play a role in other transformations, since it is activated during visuospatial rotation (Gogos et al. 2010) and mental melody rotation (Zatorre et al. 2010). However, there is not universal agreement on the role of the right hemisphere in pitch; for example, there are differences in hemispheric contributions in absolute pitch and nonabsolute pitch musicians (Brancucci et al. 2009).

Left temporal areas have been shown to be important for fine intensity discrimination and fine pitch discrimination (Reiterer et al. 2005); right temporal areas are more important for other highly differential acoustic stimuli (i.e., holistic feature processing). However, in a more careful parametric study, Hyde et al. (2008) showed that the right hemisphere has higher pitch resolution than the left hemisphere. Left auditory cortex also showed greater activation during active stream segregation (Deike et al. 2010). In an interesting study that employed two discrimination tasks (tone contour vs. duration) with identical stimuli in each task condition (Brechmann and Scheich 2005), the right hemisphere auditory cortex was more strongly activated for the contour task, whereas the left hemisphere auditory cortex was more strongly activated for the duration task. It is important to emphasize, however, that the auditory cortices were bilaterally activated in both task conditions. These results indicate that there is no simplistic right or left hemisphere specialization in general auditory analysis and that, even for the same acoustic stimuli, lateralization may vary with task condition and demands.

Laterality in processing of vocalizations appears to have emerged early in evolution in primates and avian species. Neuronal responses to vocalizations in primates have been described in a network that includes STS, STG, and temporal pole, cingulate, and inferior frontal cortex. Laterality has been described in monkeys based upon imaging and lesion studies (Heffner and Heffner 1984; Harrington et al. 2001; Poremba et al. 2004; Joly et al. 2012).

Klein and Zatorre (2011) investigated categorical perception, a phenomenon that has been demonstrated to occur broadly across the auditory modality, including in the perception of speech (e.g., phonemes) and music (e.g., chords) stimuli. Several functional imaging studies have linked categorical perception of speech with activity in multiple regions of the left STS: language processing is generally left hemisphere dominant whereas, conversely, fine-grained spectral processing shows a right hemisphere bias. Klein and Zatorre found that greater right STS activity was linked to categorical processing for chords. The results suggest that the left and right STS are functionally specialized and that the right STS may take on a key role in categorical perception of spectrally complex sounds, and thus may be preferentially involved in musical processing. It is worth noting, however, that not all phonemes are categorically perceived; for instance, vowels and lexical tones of tone languages do not have categorical perception, although they are stable sound categories (Patel 2008). Conversely, there is evidence for categorical perception of tone intervals in musicians (Burns and Ward 1978).

Domain-Specific Processes: Neural Substrates for Speech

Speech Perception

Two stages can be identified in the perception of speech: phonological information (i.e., speech sounds) must be recovered and lexical-semantic information must be accessed. The recognition of speech sounds is carried out bilaterally in the superior temporal lobe (with a left hemisphere bias); the STS is bilaterally (and increasingly anteriorly) involved in phonological-level aspects (phonemes, words, and short phrases) of this process (DeWitt and Rauschecker 2012). The frontal premotor system is not involved in the perception of speech sounds per se (i.e., decoding of sounds and speech recognition in naturalistic conditions), but is important for their categorization in laboratory tasks. Currently it is unclear where conceptual access mechanisms are located in the brain, although the lateral and inferior temporal lobes (middle and inferior temporal gyri) most likely play a role.

Differences between Neural Substrates for Language and Music

Within-area differences have been found between activation for speech and music. Multivariate pattern classification analyses (Rogalsky et al. 2011b) indicate that even within the regions of blood oxygenation, level-dependent (BOLD) response overlap, speech and music elicit distinguishable patterns of activation. This raises the possibility that there are overlapping networks or even distinct and separate neural networks for speech and music that coexist in the same cortical areas. Such a view is supported by a recent fMRI study which defined language regions functionally in each subject individually and then examined the response of these regions to nonlinguistic functions, including music; little or no overlap was found (Fedorenko et al. 2011). However, as Patel (2008) observes, other studies show that musical training influences the cortical processing of language (Moreno et al. 2009) and supports the idea that there are shared networks, as seems obvious at least for "early" auditory regions.

Activation for sentences and melodies were found bilaterally in the auditory cortices on the superior temporal lobe. Another set of regions involved in processing hierarchical aspects of sentence perception were identified by contrasting sentences with scrambled sentences, revealing a bilateral temporal lobe network. Sentence perception elicited more ventrolateral activation, whereas melody perception elicited a more dorsomedial pattern, extending into the parietal lobe (Rogalsky et al. 2011b).

Patel (this volume) offers the "dual systems" SSIRH model to explain the domain-specific representations in long-term memory (i.e., stored knowledge of words and their syntactic features and stored knowledge of chords and their harmonic features) and shared neural resources that act on these representation networks (see also Patel 2003, 2011). However, although there is considerable support for the SSIRH model, there is some controversy over the degree of shared neural resources for syntactic processing in music and language. For example, Maidhof and Koelsch (2011) examined the effects of auditory selective attention on the processing of syntactic information in music and speech using event-related potentials. They suggest that their findings indicate that the neural mechanisms underlying the processing of syntactic structure of music and speech operate partially automatically and, in the case of music, are influenced by different attentional conditions. These findings, however, provide no clear support for an interaction of neural resources for syntactic processing already at these early stages. On the other hand, there is also evidence for shared mechanisms. When an acoustic (linguistic or musical) event occurs that violates the expectations of the predictive model, the brain responds with a powerful mismatch response. This can take the form of mismatch negativity for oddballs or violations of acoustic patterns, and may lead to bi-hemispheric changes in the evoked brain potentials: early left anterior negativity (ELAN) for the presentation of an unexpected syntactic word category and early right anterior negativity (ERAN)) after the presentation of a harmonically unexpected chord at the end of a sequence. The SSIRH model (Patel 2003) and several recent studies suggest that linguistic and musical syntax may indeed be co-localized and overlapping (Sammler et al. 2012).

Outlook: Challenges and Mysteries

Dance and Music

Janata and Parsons (this volume) provide a discussion of the neural mechanisms involved in music, song, and dance. To focus our efforts at the Forum, we limited our discussion primarily to a consideration of spoken rather than signed language. Still, we emphasize the importance of gesture and movement in language, both as a vibrant accompaniment to spoken language and as a signal in conversational turn-taking, in musical performance as well as in dance. Future research will need to address the dimension of body movement and integrate it with our understanding of music. In particular, it will be very important to learn how kinesthetic, proprioceptive, and visual cues are integrated with the motor and auditory systems.

Poetry and Song: Bringing Music and Language Together

"Language" and "music" actually form two poles of a continuum that includes song-like or musical speech, tonal languages, poetry, rap music, and highly syntactically structured music. Lewis (this volume) describes a fusion of language and music in the BaYaka Pygmy hunter-gatherers in the Congo, and Levinson (this volume) observes that "song in a sense is just language in a special, marked suprasegmental register or style or genre" and that "music may be an ethnocentric category" (Nettl 2000).

Parallel to the controversies over the neural representation for language and music mentioned above, there continues to be vigorous debate about the relationship between the processing of tunes and lyrics in song, and about the neural structures involved. While there is good neuroimaging and neuropsychological evidence for separate processing of lyrics and melody in song, there is also compelling evidence for integrated processing of words and music in a unified neural representation. While brain activation patterns evoked by the perception and production of song show overlap with the spoken word activation network in many studies (Janata and Parsons, this volume), other studies emphasize differences. Patel (this volume) has suggested that there is a song sound map in the right hemisphere and a speech sound map in the left hemisphere. Experimental support for such hemispheric specialization at a production level is provided by a study which found that the right IFG, right premotor cortex, and right anterior insula were active in singing only, suggesting that song production engages some right hemisphere structures not activated in normal speech (Saito et al. 2006). Right hemisphere dominance for singing has also been shown by TMS studies (Stewart et al. 2001). However, activation studies of song perception by Schön et al. (2010) and adaptation studies of Sammler et al. (2010) argue against domain specificity and show broad, bilateral activation of auditory areas in superior temporal lobe and STS for lyrical songs. In more detail, the latter results also show greater integration of lyrics in tunes in the left middle left STS, suggesting lyrics and tunes are strongly integrated at a prelexical, phonemic level. The more independent processing of lyrics in the left anterior STS may arise from analysis of meaning in the lyrics. An explanation for divergent and disparate reports in the literature may be that there are variable degrees of integration/dissociation of lyrics and melody at different stages of song perception, production, and memory (Sammler et al. 2010). Depending on the specific cognitive demands of an experimental task, text and melodies may be more or less strongly associated but not fully integrated, and the extent of integration may also vary with the degree of familiarity of the song to the listener, and the listener's attentional focus. Additional variation can also occur in other ways within the same song; for example, vowels are more tightly bound with pitch information than consonants in song perception (Kolinsky et al. 2009). There may be more variation and independent processing at the perceptual rather than the production level since lyrical and melodic features of song must be integrated in the output stage as a vocal code for singing.

Additional Problems and Challenges for Future Research

Our search for the neural and computational "primitives" underlying music and language, "domain-specific" and "domain-general" representations and computations, and our summary of current neurobiological insights into the relations between language and music have revealed a tremendous, recent surge of research and interest in this interdisciplinary field, and yielded an extraordinary treasure trove of fascinating advances, many achieved with dazzling new neuroscientific techniques. For example, we have described great advances in understanding brain development and plasticity during acquisition of language and music, insights into the neural substrates of emotional responses to music (Salimpoor et al. 2011), the relation between music and language perception and production in the perception–action–prediction cycle, the evidence for separable modular components for speech and music processing both at lower auditory levels and a higher cognitive level. Although there is compelling neuropsychological data for a neat dissociation between the neural substrates for music and language, the neuroimaging data tell a more complex story. While many neuroscientists think of music and language as distinct modular systems, another viewpoint is that they are different ends of the continuum of "musilanguage" that also includes song and poetry (Brown 2000), with music emphasizing sound as emotional meaning whereas language emphasizes sound as referential meaning. Given the range of perspectives in this field, and the fundamental questions that still remain unanswered, it is clear that are still many "gaps" in our knowledge. Thus, in an effort to spur future research, we conclude by listing areas that we feel require further study:

1. How are the memory systems for language and music both independent and interwoven? How are the lyrics and melody of familiar songs separately and conjointly stored?
2. What are the parallel and overlapping substrates for language and music acquisition during childhood development? Do structural and functional brain changes occur during the learning of speech and music?
3. What are the shared versus distinct speech and song production mechanisms?
4. What causes lateralization? Is there an overall right hemisphere lateralization for music and left hemisphere lateralization for speech?
5. What are the neural representations and multisensory mechanisms shared by dance, music, and language? Is there a common neural basis underlying the ability of dance, music, and language to evoke emotions?
6. Precisely what contributions do brain oscillations make to auditory processing in language and music? How best can these influences be explored, evaluated, and critically tested?
7. How have speech and music evolved through the prism of animal models of communication and rhythm perception?
8. What is the nature of the interaction between external acoustic inputs and anticipatory and predictive internal feedforward systems in language and music during conversation and improvisation?

Acknowledgment

We gratefully acknowledge the contributions of Petr Janata to our discussions.

Part 5: Development, Evolution, and Culture

18

Communication, Music, and Language in Infancy

Sandra E. Trehub

Abstract

Music, as considered here, is a mode of communication, one that has particular resonance for preverbal infants. Infants detect melodic, rhythmic, and expressive nuances in music as well as in the intonation patterns of speech. They have ample opportunity to use those skills because mothers shower them with melodious sounds, both sung and spoken. Infants are sensitive to distributional information in such input, proceeding from culture-general to culture-specific skills with alacrity. Mothers' arousal regulatory goals are well known, but their intuitive didactic agenda is often ignored. Regardless of the amiable and expert tutoring that most infants receive, their progress from avid consumers of music and speech to zealous producers is remarkable.

Introduction

Music is often considered a form of communication, quite different from language, of course, but one sharing a number of common properties, the most basic being the auditory–vocal channel and cultural transmission (for a review, see Fitch 2006). Although verbal utterances readily meet conventional communication criteria in having senders, messages, and receivers who share common ground with senders, the situation is rather different for music. Composers and performers may have global intentions, affective and imaginative, rather than specific referential intentions. In some social or societal contexts, common ground with receivers is more limited in musical than in linguistic contexts, which may result in mismatches between communicative intentions and interpretations. Such situations are not necessarily problematic because the indeterminacy of musical meaning may enhance the appeal of music rather than reduce it (Cross 2003b).

According to Tomasello (2008), three broad motives underlie intentional communication: requesting, informing, and sharing aimed at influencing the feelings, attitudes, or actions of others. For nonhuman primates, intentional communication is largely restricted to requesting. Human linguistic and gestural communication can express all three motives but music, as we know it, is largely restricted to sharing (for unique cross-cultural perspectives, see Lewis, this volume). Bachorowski and Owren (2003) argue that vocal affect across species influences the affect, attitudes, and behavior of the listener in ways that are often favorable to signaler and listener. They contend that such signals were shaped over evolutionary time by their consequences for both parties. Obviously, these notions were not conceived with music in mind, but they seem quite relevant to vocal music, even to nonvocal music, which has its roots in vocal music.

Scientific conceptions of music (e.g., Patel 2008; Pinker 1997) have generally ignored the communicative context or significance of music, focusing instead on origins, perception, production, memory, and neural substrates. Unfortunately, the literature in these domains is derived from a sparse sampling of musical traditions, with little consideration of cultures that differ substantially from our own (Blacking 1995; Cross et al., this volume; Grauer 2006; Lewis, this volume). In addition, music is typically studied in disembodied laboratory contexts with stimuli that have questionable ecological validity (e.g., synthesized tone patterns), thus raising questions about the generalizability of such research to the rich musical textures and contexts in our culture and others. The voice, despite its obvious biological significance and indisputable status as the original musical instrument, is largely absent from the burgeoning literature on the neuroscience of music. It is clear, however, that vocal tones produce larger and more distinctive cortical responses than instrumental tones (Gunji et al. 2003).

For much of the world today, as in our distant past, music remains a multimodal activity that unfolds in face-to-face contexts. The consequences of the multimodal experience are anything but trivial. The duration of a musician's visual gestures alter listeners' perception of the duration of notes (Schutz and Lipscomb 2007). Simultaneous video displays alter the perceived tempo and affective valence of music (Boltz et al. 2009). In addition, the act of moving while listening to music alters the metrical interpretation of music (Phillips-Silver and Trainor 2005, 2007).

Setting the grand ideas aside—what music is, where it came from, and when—there may be something to be gained from pursuing modest goals like examining music from the perspective of preverbal infants. Infants' response to music or music-like stimuli has the potential to shed light on early perceptual biases or dispositions for music and early signs of enculturation. Parental interactions with infants, when viewed through a musical lens, may reveal intuitive fine-tuning to infants' needs and dispositions.

Music Listening Skills in Infancy

Audition in early infancy, although considerably more refined than vision, is nevertheless immature. Quiet sounds that are audible to adults are often inaudible to infants—a gap that does not close until eight to ten years of age (Trehub et al. 1988). Infants' resolution of pitch and timing is also deficient relative to adult levels (Spetner and Olsho 1990; Trehub et al. 1995), but their resolution is more than adequate for the pitch and timing differences that are musically and linguistically relevant across cultures.

Relative Pitch

Despite substantial differences between infants and adults in the detection of isolated sounds, their perception of global aspects of music is surprisingly similar (Trehub 2000, 2003; Trehub and Hannon 2006). Music appreciation depends, to a considerable extent, on relational features. Like adults, infants perceive the similarity of a melody across pitch levels (Plantinga and Trainor 2005; Trehub et al. 1987), an ability that enables us to recognize familiar tunes sung by a man, woman, or child or played on a cello or piccolo. Infants also exhibit octave equivalence, or the sense of affinity between tones an octave apart (Demany and Armand 1984). Octave equivalence, which is a musical universal (or near universal), is thought to originate from the structure of the auditory system (Dowling and Harwood 1986). When men and women across cultures sing "in unison," they are generally producing pitches an octave apart (fundamental frequency ratio of 2:1). Despite the importance of octave equivalence in music, it is irrelevant in nonmusical contexts, including speech.

Infants also exhibit long-term memory for music. After brief periods of daily exposure to instrumental music for one to two weeks, infants distinguish novel music from the music to which they were familiarized (Ilari and Polka 2006; Saffran et al. 2000; Trainor et al. 2004). Because the familiar and novel music usually differ in multiple respects, it is impossible to identify the relevant cues. For simple piano melodies, however, it is clear that infants remember the relative pitch patterns or tunes but not the pitch level at which they were originally presented (Plantinga and Trainor 2005).

Relative pitch processing in infancy is demonstrable in a single, brief test session. By contrast, relational pitch processing is very difficult for nonhuman species, including songbirds, which excel at absolute pitch processing (Hulse and Cynx 1985). Nevertheless, extensive training enables European starlings to recognize transposed conspecific songs but not transposed piano melodies (Bregman et al. 2012). More intensive training (thousands of trials distributed over several months) enables rhesus monkeys to recognize octave transpositions of Western tonal melodies but not atonal or randomly generated melodies (Wright et al. 2000). The implications of the tonal advantage for rhesus monkey remain to be determined.

Absolute Pitch

Although relational features are central to music processing in human adults and infants, absolute features are also relevant. Like adults, who remember the pitch level of highly familiar pop recordings (Levitin 1994) or the theme music of familiar television programs (Schellenberg and Trehub 2003), infants remember the pitch level of familiar lullaby recordings (Volkova et al. 2006) but not familiar piano recordings (Plantinga and Trainor 2005). It is unclear whether the advantages stem from lullabies, which are universal (Trehub and Trainor 1998), or from vocal stimuli, which elicit stronger and more distinctive cortical responses than nonvocal stimuli even in the absence of linguistic or melodic content (Belin et al. 2002; Gunji et al. 2003).

Infants are also able to segment three-tone sequences from continuous sequences of tones (of equal duration and amplitude) on the basis of conditional probabilities involving absolute or relative pitch (McMullen and Saffran 2004; Saffran et al. 2005). In addition, they segment three-syllable nonsense words from continuous sequences of syllables (all of equal duration and amplitude) on the basis of conditional probabilities (Saffran et al. 1996). As one may imagine, the continuous tone and syllable sequences sound little like ordinary music or speech.

Consonance and Dissonance

Central to the music of all cultures is melody, which refers to the horizontal or linear succession of notes. Some musical cultures also make use of harmony, or the vertical dimension of music, which consists of two or more simultaneous pitches (i.e., intervals or chords). Pitch relations, whether simultaneous (i.e., harmonic) or successive (i.e., melodic), have important consequences for infant and adult listeners. From the newborn period, Western and Japanese infants listen longer to music with consonant harmonic intervals (i.e., pleasant-sounding to adults) rather than dissonant intervals (i.e., unpleasant-sounding to adults) (Masataka 2006; Trainor and Heinmiller 1998; Trainor et al. 2002; Zentner and Kagan 1998). Although infants' listening bias is usually interpreted as an innate aesthetic *preference* for consonance, that interpretation is at odds with historical and cross-cultural considerations. Over the centuries, there have been notable changes in Western conceptions of consonance, with once-dissonant intervals (e.g., major and minor sixths, major and minor thirds) becoming consonant with increasing use and familiarity (Tenney 1988). In traditional Western classical contexts, beating and roughness from simultaneous sounds close in pitch are considered objectionable; however, these qualities are highly desirable for Indonesian gamelan instruments (Vasilakis 2005).

To date, there have been no demonstrations of infant listening preferences for melodic intervals (i.e., combinations of successive tones) that Western adults consider consonant or dissonant. Nevertheless, Western infants detect subtle pitch changes more readily in tone patterns with consonant melodic (sequential) intervals, which have fundamental frequencies related by small integer ratios (e.g., 2:1, 3:2), rather than dissonant intervals, which have fundamental frequencies related by more complex ratios (e.g., 45:32) (Schellenberg and Trehub 1996; Trainor and Trehub 1993; Trehub et al. 1990). There are suggestions that the interval of a perfect fifth (7 semitones, frequency ratio of 3:2), which is cross-culturally ubiquitous (Sachs 1943), plays an anchoring role in music, thus enhancing the ability of infants and adults to encode and retain melodies (Cohen et al. 1987; Trainor and Trehub 1993).

Timbre Processing

Timbre is another critical dimension of the music listening experience. Musical timbre refers to the tone quality or color that differentiates complex sounds or sound sequences that have the same pitch, duration, and amplitude (e.g., the same musical passage played by different instruments). Oddly, timbre is defined by what it is not rather than what it is. Our memory for familiar musical performances goes well beyond absolute and relative pitch cues to include rich information about timbre. As a result, we are able to recognize which of five familiar pop recordings we are hearing from the initial 100 or 200 milliseconds (all or part of one note) (Schellenberg et al. 1999). When the responses choices are unlimited (i.e., an open-set task), we recognize familiar instrumental recordings from 500 ms excerpts (Filipic et al. 2010). When timbre is uninformative, as in piano versions of familiar songs, we need four notes, on average (approximately 2 s), to recognize a song as familiar and an additional two notes to identify it definitively (Dalla Bella et al. 2003).

Our recognition of newly familiarized melodies is reduced when the timbre is changed (e.g., piano to banjo) between exposure and test (Halpern and Müllensiefen 2008; Peretz et al. 1998); this implies that novel melodies are encoded with their respective timbres. Comparable changes of timbre between exposure and test (e.g., piano to harp) disrupt the long-term memory of six-month-olds for melodies (Trainor et al. 2004). Nevertheless, four-month-olds who receive distributed exposure to melodies played on the guitar or marimba (total of three hours) show more robust event-related potential responses to novel tones presented in the familiar timbre than in a novel timbre (Trainor et al. 2011).

What is clear, however, is that timbres are unequal in their cognitive consequences. Our memory for melodies is enhanced when the melodies are presented vocally (sung on *la la la*) rather than instrumentally (Weiss et al. 2012). Familiarity alone is unlikely to underlie the effect since we are no better at remembering melodies presented in a familiar instrumental timbre (piano) than in less familiar timbres (banjo or marimba). Biologically significant stimuli like the human voice may enhance attention or arousal, resulting in greater depth of processing (Craik and Lockhart 1972) and more durable learning. The prevalence of timbres of convenience (e.g., synthesized piano) in studies of music cognition may obscure important aspects of music processing in human listeners of all ages.

Listening in Time

Infants remember the tempo of familiar recordings (Trainor et al. 2004), as do adults (Levitin and Cook 1996). They group tones on the basis of their frequency, intensity, or harmonic structure (Thorpe and Trehub 1989), and they perceive the invariance of melodies or rhythmic patterns (e.g., *dum da da dum dum*) across variations in tempo (Trehub and Thorpe 1989). These and other perceptual grouping principles are presumed to reflect general auditory biases that have implications for the perception of speech and nonspeech patterns. Adults whose native language is English (a stress-timed language) or French (a syllable-timed language) hear alternating syllables (or square waves) of contrasting duration as iambic (weak-strong) patterns and those of contrasting intensity as trochaic (strong-weak) patterns (Hay and Diehl 2007). Language experience seems to affect these grouping biases. For example, five-month-old English-learning infants show no grouping biases for sequences of alternating short and long tones, but seven-month-olds exhibit the iambic bias of English-speaking adults (Iversen et al. 2008; Yoshida et al. 2010). By contrast, Japanese-learning infants in both age groups have inconsistent grouping preferences, just as Japanese-speaking adults do.

Temporal regularity facilitates infants' detection of pitch and rhythmic changes in musical sequences (Bergeson and Trehub 2006; Hannon et al. 2011; Trehub and Hannon 2009). Indeed, the ability to perceive a beat may be innate, as reflected in the distinctive neural responses of infants to violations of the beat structure of musical patterns (Honing et al. 2009; Winkler et al. 2009b). Synchronized (i.e., precisely coordinated) movement to sound provides unambiguous evidence of beat perception, but infants are considered incapable of such coordination, Nevertheless, glimmers of entrainment are evident during maternal singing (Longhi 2009), and rhythmic instrumental patterns generate rhythmic movement in infancy (Zentner and Eerola 2010a). When entrainment to sound begins to emerge more clearly at three or four years of age, movement is more synchronous in the context of a human model than in the context of a mechanical object that displays similar motion (Kirschner and Tomasello 2009). Adult levels of synchrony are not apparent until later childhood (McAuley et al. 2006).

Experiencing movement while listening to a musical passage affects how that passage is encoded. When infants and adults move (or are moved) on every second beat of an ambiguous rhythmic sequence (i.e., no pitch or timing accents), they encode the sequence in duple meter (i.e., accents on every second beat), as reflected in subsequent preferential listening to patterns in duple rather than triple meter (Phillips-Silver and Trainor 2005, 2007). With movement on every third beat, they encode the sequence in triple meter (i.e., accents on every third beat), as reflected in preferential listening to patterns in triple rather than duple meter. Movement in the course of music listening may affect more than its metrical interpretation. It is possible, for example, that movement generates a richer or more detailed encoding of the music, leading to greater retention of various features. Movement may also add pleasure to the listening experience, fostering greater appreciation of the music in question. One can only wonder about the consequences of movement or its absence for contemporary audiences at rock and symphonic concerts.

Musical Enculturation

Competence in music perception does not require formal lessons or systematic exposure. In fact, casual exposure to the music of one's community results in implicit knowledge comparable to that of trained musicians (Bigand 2003). Presumably, listeners become progressively attuned to regularities in the music they hear much as they become attuned to regularities in the language they hear. Such implicit knowledge enables them to detect minor performance mishaps, such as mistuned notes or timing errors, but it also has minor costs. Although adults readily detect a 1-semitone change that goes outside the key of a melody and sounds like a sour note (see Figure 18.1), they have difficulty detecting a 4-semitone change that is consistent with the key and the implied harmony (Trainor and Trehub 1992). By contrast, six-month-olds, who are unaware of culture-specific conventions involving key membership or harmony, detect both changes with equal ease. Implicit knowledge of the notes in a key is apparent by four years of age (Corrigall and Trainor 2010) and knowledge of implied harmony by seven years of age (Trainor and Trehub 1994).

Another cost of increasing exposure to the music of one's culture is the diminished ability to perceive atypical musical patterns. For example, adults detect mistuned notes in the context of the major scale, but they fail to detect comparable mistuning after brief exposure to unfamiliar (invented) scales (Trehub et al. 1999). By contrast, infants detect mistuned notes in real or invented scales, provided those scales embody unequal step sizes (e.g., the whole tones and semitones of the major scale), which are prevalent in scales across cultures. The property of unequal step size allows each note to assume a distinctive function within a scale, such as the focal or tonic note (Balzano 1980). It also facilitates the perception of tension (dissonance or instability), resolution, and corresponding affective responses (Meyer 1956; Shepard 1982). Among the factors that presumably interfered with the widespread acceptance of 12-tone music were the use of an equal-step scale (the steps of the chromatic scale) and the burden on working memory posed by 12 component tones rather than the "magical number seven, plus or minus two" (Miller 1956).

Enculturation effects are apparent considerably earlier for the metrical patterns of music than for its pitch patterns. Western adults are sensitive to simple metrical structures (i.e., accented beats at regular temporal intervals and temporal intervals related by 2:1 ratios) that are characteristic of Western music, but not to complex metrical structures (i.e., non-isochronous accents and temporal

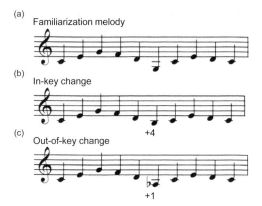

Figure 18.1 The standard or familiarization melody is shown in (a), the comparison melody with an in-key change of 4 semitones is shown in (b), and the comparison melody with an out-of-key change of 1 semitone is shown in (c). The standard and comparison melodies were always presented in transposition (i.e., different keys or pitch levels), thus requiring listeners to use relative rather than absolute pitch cues to make the discrimination. Adults had difficulty detecting the 4-semitone change (b), which maintained the musical "meaning"; however, they readily detected the 1-semitone change (c), which sounded like an error. Infants detected both changes equally well.

intervals related by 3:2 ratios) that occur in other parts of the world along with simple meters (Hannon and Trehub 2005a; Kalender et al. 2013). Interestingly, Western six-month-olds are sensitive to complex as well as simple metrical structures (Hannon and Trehub 2005a), which implies that they are not yet attuned to Western metrical structures. Sensitivity to culture-specific aspects of meter is evident, however, when the task demands are minimized. Western infants four to eight months of age listen longer to simple meters, which characterize Western music; Turkish infants listen equally to simple and complex meters, both of which occur in Turkish music, and infants from both cultures listen longer to metrically regular than to metrically irregular or nonmetric music (Soley and Hannon 2010).

By 12 months of age, infants are like their adult counterparts in detecting changes in culturally typical metrical patterns but not foreign metrical patterns (Hannon and Trehub 2005b). The implication is that twelve-month-olds have acquired implicit knowledge of conventional metrical structure on the basis of their very limited exposure to music. Despite their nascent culture-specific attunement, twelve-month-olds retain considerable perceptual flexibility. After two weeks of brief daily exposure to complex metrical patterns, twelve-month-olds detect metrical changes in music with complex meter, but adults do not (Hannon and Trehub 2005b). Adults' entrenched metrical representations seem to interfere with the perception of novel metrical structures just as culture-specific speech segmentation interferes with segmentation of a rhythmically distinct language (Cutler et al. 1986).

Music and Speech Perception: Parallels and Contrasts

The culture-general perception of music in young infants and the culture-specific perception of music in older infants have interesting parallels in speech perception. For example, six-month-old infants differentiate a number of foreign speech sounds that pose difficulty for older infants and adults, but diminished sensitivity is evident by ten or twelve months for consonants (Best et al. 1995; Werker and Tees 1984) and earlier for vowels (Kuhl et al. 1992). This progression from relative ease of differentiating sound contrasts to subsequent difficulty with the same contrasts is largely restricted to situations in which the contrasting foreign speech sounds (e.g., English *ra* and *la* for native speakers of Japanese) map onto the same phonemic category in the native language (e.g., Japanese *ra*), leading to inappropriate assimilation of the foreign sounds to native sound categories (Best et al. 1988).

To ascertain whether diminished sensitivity to foreign speech sounds could be prevented in older infants, Kuhl, Tsao, and Liu (2003) provided American nine-month-olds with five hours of Mandarin input (corresponding to approximately 26,000–42,000 syllables) over the course of four weeks. Infants who received the input from naturalistic interactions with a native speaker of Mandarin succeeded in differentiating the Mandarin speech contrasts, but those who received similar input from audio-only or audio-visual recordings did not. As noted previously, passive exposure to musical recordings (five minutes per day for two weeks) enabled twelve-month-olds to learn about foreign metrical patterns (Hannon and Trehub 2005b).

The parallels between music and speech perception in infancy do not necessarily imply common mechanisms or mechanisms that are specialized for these domains. It is possible that they arise from domain-general perceptual and learning mechanisms. Infants are sensitive to the distributional properties of speech (Maye et al. 2002; Saffran et al. 1996), tone sequences (Saffran et al. 1999), and action sequences (Baldwin et al. 2008). Furthermore, perceptual tuning or narrowing in infancy has been demonstrated for phonemes (Best et al. 1988; Kuhl et al. 1992; Werker and Tees 1984), sign language hand shapes (Palmer et al. 2012), musical meter (Hannon and Trehub 2005b), and faces (Pascalis et al. 2002; Slater et al. 2010). Such culture-specific tuning is a step toward native-like processing of language, music, and faces, all of which are biologically significant stimuli.

Rhythmic aspects of speech are apparently privileged in early processing (Nazzi and Ramus 2003). Newborns differentiate their native language-to-be from rhythmically different languages (Mehler et al. 1988). Some months later, they differentiate languages within the same rhythmic class (Nazzi et al. 2000). Newborns with prenatal exposure to English exhibit a listening preference for low-pass filtered English over Tagalog sentences that eliminate segmental cues but preserve contrasting rhythmic cues; newborns with prenatal exposure to both languages show no such preference (Byers-Heinlein et al. 2010).

Languages with contrastive auditory rhythms have corresponding differences in visual rhythms. English-learning four-month-olds distinguish English from French speech on the basis of visual cues alone; by eight months of age, only infants with exposure to both languages show comparable differentiation (Weikum et al. 2007).

Rhythmic processing seems comparably privileged in music. To date, culture-specific narrowing in infancy has been demonstrated for musical meter (Hannon et al. 2011; Hannon and Trehub 2005b) but not for pitch patterning. Although infants' perception of melodic patterns is quite remarkable (for a review, see Trehub and Hannon 2006), four years is the earliest age of differential responding to notes in or outside the key of a melody (Corrigall and Trainor 2010).

The prevailing belief is that receptive and productive aspects of language are acquired rapidly and effortlessly, without instruction, in contrast to the seemingly slow and effortful acquisition of receptive and productive aspects of music (Patel 2008; Pinker 1997). It takes years, however, before children attain full proficiency in the phonological, semantic, and syntactic aspects of their native language (Best and McRoberts 2003; Duncan 2010; Entwisle and Frasure 1974). This protracted course of development parallels the acquisition of implicit tonal and harmonic knowledge in music (Corrigall and Trainor 2010; Krumhansl 1990; Trainor and Trehub 1994).

The situation differs for prosodic aspects of speech. Culturally typical intonation patterns (i.e., speech melodies) are distinguishable in the babbling of preverbal infants from different language backgrounds (de Boysson-Bardies et al. 1984). The vocalizations of infants from English-speaking environments are dominated by falling contours, and those from French-speaking environments are dominated by rising pitch contours (Whalen et al. 1991). Remarkably, three-month-olds produce spontaneous imitations of prosodic contours during mother–infant interactions (Gratier and Devouche 2011). Even more remarkably, newborn cries are influenced by the ambient language *in utero*, with French newborns producing more rising contours and German newborns producing more falling contours (Mampe et al. 2009). These studies seem to suggest that preverbal infants accord priority to global or suprasegmental aspects of speech despite their ability to differentiate segmental aspects.

From infancy through the preschool years, a mother's verbal interactions with her child seem to have didactic as well as social-regulatory goals. Early linguistic communication in Western middle-class families is frequent, dyadic (parent–infant), child-centered, and highly simplified in form and content (Ochs and Schieffelin 1995), a strategy that is eminently more successful than conventional language instruction. (For an unusual example of didactic speech to a six-month-old infant, see Example 7 in the online the supplemental information to this chapter: http://esforum.de/sfr10/trehub.html). Parent language instruction is by no means universal. In a number of cultures (e.g., Kaluli, Kwar'ae, Samoan, Western working-class families), linguistic communication with infants and young children is situation-centered rather than child-centered, occurring frequently in multi-person contexts (Ochs and Schieffelin 1995). In fact, linguistic interactions with infants may only begin after the child starts to produce words, at which point instruction is aimed at the socialization of language use (i.e., what to say when and to whom) rather than lexical acquisition. Toddlers are guided in the production of fully formed utterances (e.g., specific formulaic expressions) well before they understand their meaning or significance. Regardless of the pedagogical approach, children acquire the relevant language and social conventions, albeit on a different timescale.

Although we know that infants hear or overhear considerable speech, we know relatively little about the amount of music they hear. In one sample of Swedish families, speech accounted for roughly 71% of maternal and sibling interactions with nine- and twelve-month-old infants; singing accounted for a mere 10% of those interactions (Eckerdal and Merker 2009). Despite infants' limited musical input, their exposure to simple, repetitive material and stable performances of such material (Bergeson and Trehub 2002) undoubtedly promotes the development of receptive skills and, ultimately, production skills.

Infants are thought to acquire their native language by virtue of their sensitivity to statistical regularities in the input. Indeed, considerable evidence supports the notion of infants as competent statistical learners (e.g., Saffran et al. 1996). However, statistical learning mechanisms may play a more modest role in everyday life than they do in laboratory settings with extremely simple artificial languages involving a tiny corpus of two- or three-syllable words of equal duration and amplitude (Johnson 2012). Interestingly, infants fail to segment "words" when the four-word artificial language under consideration consists of two disyllabic and two trisyllabic words with the usual transitional probabilities (Johnson and Tyler 2010). They also fail after prior exposure to words differing in length from the target words (Lew-Williams and Saffran 2012). Infants' success in the context of uniform syllable and word duration may be attributable to spontaneous grouping processes. When adults listen to repeating sequences of isochronous (equally timed) sounds, they perceive illusory accents at regular intervals (Fraisse 1982) and illusory pauses between the resulting perceptual groups (Thorpe and Trehub 1989), which could lead the target words to pop out of the syllable stream. Perhaps it is not surprising, then, that infants can segment variable-duration words in simple artificial languages with infant-directed prosody but not with adult-directed prosody (Thiessen et al. 2005). The attention- or arousal-enhancing properties of infant-directed speech may facilitate the extraction of statistical regularities in the input even in the absence of prosodic cues to word boundaries.

The common notion that linguistic competence is universal but musical competence is infrequent (Patel 2008; Pinker 1997) arises from comparisons of apples and oranges: everyday speech skills, on one hand, and instrumental mastery, on the other. It also results from failure to consider cultures in which music making is widespread (Blacking 1995; Lewis, this volume) or informal contexts (e.g., playgrounds, school yards) where children exhibit mastery of elaborate songs, rhymes, hand-clapping games, and dance routines (Arleo 2006; Marsh 2008). Indeed, the overwhelming majority of musically untrained adults can match pitches, sing in tune (Dalla Bella et al. 2007; Pfordresher and Brown 2007), recognize hundreds of songs, sing dozens from memory, and synchronize their rhythmic movement with music or with each other (Phillips-Silver et al. 2011; McNeill 1995). Despite obvious differences in exposure and opportunity, there is little evidence of glaring discrepancies in the relative ease of acquiring receptive and expressive music and language. As with language, informal contexts with socially valued mentors promote more enjoyable, durable, and generalizable learning than that afforded by formal instructional contexts.

Music in Everyday Life: Singing

Although infants in Western cultures experience considerably less musical compared to verbal input, the situation may differ in non-Western cultures, where child-rearing practices include co-sleeping, constant mother–infant proximity, and high levels of indulgence (Wolf et al. 1996). In general, caregivers from cultures that value interdependence exhibit more comforting interactions (e.g., touching, nursing, rocking) and fewer didactic interactions than caregivers from cultures that value independence (Fernald 1991; Morelli et al. 1992; Morikawa et al. 1988).

Lullabies—the musical counterpart of soothing speech—are universal, with play songs or action songs sometimes deferred until the toddler phase (Trehub and Trainor 1998). Naïve listeners readily distinguish foreign lullabies from nonlullabies matched on tempo and culture of origin (Trehub et al. 1993a), perhaps on the basis of their structural simplicity or repetitiveness (Unyk et al. 1992).

A separate genre of music for infants may be less important than the distinctive performing style. Among the features that mark maternal performances of play songs are higher than usual pitch, slower than usual tempo, and bright voice quality (Trainor et al. 1997; Trehub et al. 1997, 1993b). In contrast to the conversational form of maternal speech to infants—infant turns involving coos, smiles, or attentive silence—maternal songs usually continue uninterrupted from beginning to end. These performances are individually distinctive and highly stereotyped, with many mothers singing the same song at the identical pitch level and tempo on different occasions (Bergeson and Trehub 2002).

The primary functions of maternal singing are arousal regulation, engagement, and the sharing of feelings, but such singing also reveals intuitive sensitivity to infants' perceptual and informational needs. Infant-directed singing exhibits greater temporal regularity than noninfant-directed singing (Nakata and Trehub 2011), which is a reasonable accommodation to infants' enhanced processing of rhythmically or metrically regular sequences (Bergeson and Trehub 2006; Nakata and Mitani 2005; Trehub and Hannon 2009). Although infant-directed singing sacrifices expressive timing in favor of temporal regularity, it capitalizes on the potential for dynamic expressiveness. Overall, mothers sing more softly than nonmothers do, but they use a greater dynamic range, emphasizing the melodic contours of their songs (Nakata and Trehub 2011). Whether unintentionally or otherwise, mothers sing in a way that highlights the pitch and temporal structure of their songs. Mothers' intuitive didactic goals are evident in their singing to preschool children, which reveals more precise articulation of the lyrics (Bergeson and Trehub 1999).

Infant-directed singing achieves its intended goals, modulating the arousal of infant listeners (Shenfield et al. 2003) and sustaining their attention more effectively than noninfant-directed singing does (Masataka 1999; Trainor 1996). Audiovisual episodes of maternal singing are especially engaging to infants, even more engaging than comparable episodes of maternal speech (Nakata and Trehub 2004). (For an example of typical maternal singing to a six-month-old infant, see Example 1, http://esforum.de/sfr10/trehub.html.)

Music in Everyday Life: Speech

Mothers, especially those in Western cultures, speak a great deal to infants at a time when the verbal content is inaccessible. Such speech can be viewed as a vehicle for transmitting melodious sound patterns or a warm acoustical flow from mother to infant. Depending on the infant's state and the mother's intentions, maternal speech varies on a continuum from very soothing to very playful. (For an example of exuberant maternal speech to a six-month-old infant, see Example 8, http://esforum.de/sfr10/trehub.html.) Mothers' elevated pitch, expanded pitch range, and slow speaking rate give special prominence to their pitch contours, some of which are similar across cultures (Fernald et al. 1989). A finer-grained analysis, at the level of intervals (i.e., exact pitch distances between adjacent sounds), reveals pitch patterns or tunes that are individually distinctive (Bergeson and Trehub 2007).

According to Falk (2004), development of a special vocal register for infants was driven by ancestral mothers' need to keep infants content while foraging— the putting-the-baby-down hypothesis. One challenge for this hypothesis is that maternal vocalizations in such circumstances would endanger infants by attracting predators. If mothers' appraisal of infant needs underlies their vocal expressiveness, then they should vocalize more expressively when infants are obscured from view rather than in view. In fact, speech and singing are more expressive when mothers are face-to-face with infants rather than separated by an opaque curtain (Trehub et al. 2011). This finding implies that maternal affect mediates expressiveness in speech and singing, with infant feedback contributing to the potency of maternal performances.

As noted, mothers adopt a conversational or turn-taking style in their spoken interactions with preverbal infants, pausing to accept infants' contributions of coos, gurgles, yawns, and smiles. Infants are highly engaged by such speech, even when the speaker or language is unfamiliar (Cooper and Aslin 1990; Fernald 1991). Their engagement stems largely from positive vocal affect rather than features exclusive to the maternal speech register. Indeed, infants are differentially responsive to speech with high levels of positive affect whether that speech is infant- or adult-directed (Singh et al. 2002). Such happy-sounding speech attracts and maintains the attention of preverbal infants, influences their social preferences (Schachner and Hannon 2010), and facilitates their segmentation of words from the speech stream (Thiessen et al. 2005) as well as their memory for words (Singh et al. 2009).

Divergent Paths for Speech and Singing

Maternal speech and singing are generally studied as auditory phenomena, but infants are attentive to the visual gestures and movement that typically accompany speech (Munhall and Johnson 2012). For example, four- and six-month-old infants discriminate a familiar from an unfamiliar language solely on the basis of dynamic visual cues (Weikum et al. 2007). Maternal speech to preverbal infants is often more attention-getting than maternal singing by virtue of its highly variable pitch (from low pitch to falsetto), dynamics (from whisper to squeal), and timing. However, maternal singing is more visually expressive than maternal speech by virtue of considerably more smiling and rhythmic movement (Plantinga and Trainor 2005; Trehub et al. 2011). Maternal action songs (e.g., *Itsy Bitsy Spider*, *Wheels on the Bus*) also include pantomimed gestures (Eckerdal and Merker 2009).

Maternal speech and music have broadly similar functions in early infancy, revolving largely around social and emotional regulation, but their functions and forms become increasingly divergent. As noted, maternal singing is much more stereotyped than maternal speech (Bergeson and Trehub 2002). Despite mothers' endless repetition of the same songs in the same manner, there is little evidence of infant satiety. Instead, mothers' highly predictable renditions result in ever-increasing infant engagement, perhaps like adults who have enduring regard for specific musical pieces and iconic performances of those pieces (e.g., Paul McCartney's *Yesterday*, Glenn Gould's *Goldberg Variations*). Maternal speech exhibits changing functions over time, with the focus on comfort at three months, affection at six months, and attention direction and information sharing at nine months and thereafter (Kitamura and Burnham 2003).

Infants become active communication partners well before they utter their first words (Capirci and Volterra 2008; Caselli et al. 2012; Liszkowski 2008; Tomasello 2008). By eight months of age, they request a desired object by extending an open hand in its direction, and they may exhibit one or two conventional (imitated) gestures like waving "bye bye" or clapping their hands. By ten months, they repeatedly show objects but refrain from giving them to their caregiver, and a month or so later they start offering the objects. At 11–12 months, they begin pointing to objects or events, occasionally to inform but more commonly to request, and even more commonly to share interesting sights. Gestures may have priority in the first year because of their iconicity. Pointing to an object or location does not require shared conventions but rather some understanding of the communication partner as an intentional agent and potential helper (Tomasello 2008). Although referential gestures are often combined with attention-getting sounds and, later, with words, such gestures are rarely combined in the absence of exposure to a sign language (Capirci and Volterra 2008).

Some infants imitate their mother's sung pitches or tunes, with or without encouragement, well before the onset of speech (see Examples 2 and 3, http://esforum.de/sfr10/trehub.html). Toddlers often sing long stretches of song lyrics at a time when their speech is limited to rudimentary one- and two-word utterances. Mothers frequently encourage collaborative singing, pausing to give toddlers the opportunity to contribute a sound or two at critical junctures (see Example 4, http://esforum.de/sfr10/trehub.html). Presumably, infants who participate in these duets have a mental representation of the overall song, which enables them to fill in the empty slots with appropriate sounds. By two years of age, most infants produce credible, if imperfect, renditions of familiar songs (see Example 5, http://esforum.de/sfr10/trehub.html).

Prior to their efforts at song reproduction, infants commonly engage in rudimentary dancing. In the single demonstration of this phenomenon under controlled conditions (Zentner and Eerola 2010a), infants moved rhythmically to music or to repeated isochronous sounds (like a metronome) but not to infant- or adult-directed speech. Although such movements were unsynchronized with the sound, their rate of movement was correlated with the tempo of the music. Curiously, many infants smiled when their movements were temporarily aligned with the music, raising the possibility of a "sweet spot" or emotional jolt during such moments of synchrony. By 18 months of age or shortly thereafter, dancing to music is ubiquitous and immensely pleasurable (see Example 6, http://esforum.de/sfr10/trehub.html).

Ritual Culture

According to Merker (2009), human culture is unique not only because of speech and language but also because of our highly developed ritual culture, of which music is a critical component. His tripartite conception of human culture includes instrumental culture, ritual culture, and language. In instrumental culture, specific goals are primary, and the means of achieving them are secondary and variable. For cultural rituals (even those with clear goals), the means are primary, and members of a culture are expected to follow the prescribed form. From this perspective, maternal singing, with its repetition and stereotypy (Bergeson and Trehub 2002), initiates infants into ritual culture (Eckerdal and Merker 2009). Among the Kaluli of Papua New Guinea, the formulaic utterances modeled for infants and young children (Ochs and Schieffelin 1995) accomplish similar goals.

Merker (2009) argues that the richness of human ritual culture would be impossible without the capacity and motivation for expressive mimesis, which is primarily vocal but involves visual gestures as well. He also considers the motivation for learning and duplicating arbitrary patterns to be the most important behavioral adaptation of our species. This predisposition would be of limited importance unless coupled with a strong collaborative disposition (Syal and Finlay 2011; Tomasello 2008) that applied to nonritual as well as ritual domains. Syal and Finlay (2011) suggest that human language resulted from links between a powerful vocal learning system and social-motivational circuitry, aspects of which have been identified in avian vocal learners.

Speculations on Origins

Syal and Finlay's (2011) perspectives as well as those of Merker (2009) focus on the presumed vocal origins of language, in contrast to Arbib (2005a), Corballis (2010), and Tomasello (2008), who assume gestural origins. Ontogenetically, communicative gestures appear before speech, but they are soon combined with meaningless (attention-getting) vocalizations and subsequently overtaken by speech (Capirci and Volterra 2008; Caselli et al. 2012). For Tomasello (2008), ancestral changes that supported substantially increased collaboration and cooperation provided the most important prerequisites for language, but the nature of those changes remains unresolved. Perhaps meaningless, melodious vocalizations attenuated the aggressive implications of eye contact in ancestral species, paving the way for sustained physical proximity, face-to-face interaction, and multimodal communication. Dunbar and his colleagues argue that aspects of sociality created pressures across species for increased encephalization (Dunbar 2003; Shultz and Dunbar 2010) and, ultimately, for language. Because sociality is equally important for motivating the acquisition and perpetuation of cultural rituals, egalitarian societies would be expected to have more elaborate rituals than hierarchical or status-conscious societies (Lewis, this volume; Ochs and Schieffelin 1995).

In sum, viewing language and music, or speech and singing, within a communication framework has advantages for the consideration of early development in both domains. It may have comparable advantages for cross-cultural comparisons, drawing attention to context, intentions, and participants' roles—the pragmatics of language and music—in addition to linguistic and musical structure. When focusing on language and music as modes of communication, a gradual progression from simpler to more complex forms seems more reasonable than the sudden emergence that is often postulated for language (Bickerton 1995; Lieberman 1998).

Finally, there are the inevitable but insoluble chicken-and-egg questions. One intriguing proposal is that language was preceded by a musical protolanguage in which song-like strings had holistic meanings (Brown 2000; Fitch 2010; Merker 2009; Mithen 2005). The convergent evolution of song in several distantly related species (Fitch 2010; Merker 2005, 2009) is congenial with this position because it provides a path to analytic speech by furnishing relevant component processes. Others consider music a by-product of language (Patel 2008; Pinker 1997). For Patel (2008, 2010), the demonstrable influence of language on music and shared neural resources corroborate his language-first perspective. Once language became central to human culture, however, it is hardly surprising that it would influence or dominate many aspects of human activity, including music. Borrowing from Patel's (2008) characterization of music, one could argue that language rather than music was the "transformative technology," especially when aided and abetted by literacy.

Ontogenetically, the neural substrates for language and music are active in newborns but hemispheric lateralization is evident for music only (Perani et al. 2011, 2010). Interestingly, newborn neural responses to child-directed speech and hummed versions of that speech (formants removed) are robust, but flattened versions that eliminate fundamental frequency variations yield no discernible response (Perani et al. 2011), suggesting that the music in speech is critical rather than decorative.

Acknowledgments

Preparation of this article was assisted by funding from the Natural Sciences and Engineering Research Council and the Social Sciences and Humanities Research Council of Canada.

19

Evolving the Language- and Music-Ready Brain

Michael A. Arbib and Atsushi Iriki

Abstract

This chapter focuses on the evolution of the language-ready brain, offering triadic niche construction as the framework in which to see the interaction between the environmental niche, the cognitive niche, and the neural potential latent in the genome at any stage of evolution. This framework enriches the presentation of the mirror system hypothesis, which traces an evolutionary path from mirror neurons for the recognition of manual actions, via systems that support increasingly complex forms of imitation, to the emergence of pantomime, protosign, and protospeech. This hypothesis is briefly contrasted with the Darwinian musical protolanguage hypothesis, which roots the evolution of language ability in a birdsong-like ability coupled to increasing cognitive complexity. The linkage of both language and music to outward bodily expression and social interaction is stressed and, in conclusion, the evolution of the music-ready brain is discussed.

Introduction

It is easy to focus on those aspects of language captured by the words on a page or screen, but writing is so recent that an evolutionary perspective must place language in a more biological setting. Perhaps the most basic situation is that of two people using words and sentences to develop a more or less shared understanding or course of action. Of course, there are diverse speech acts (Searle 1979), but it seems reasonable to emphasize dyadic behavior related to the sharing of meaning that may relate at its most basic to states of the world and courses of action to change that state. However, the act of communication often augments the words (i.e., the part that is "really" language) through facial and bodily expressions, which can inject emotion or a sense of the relevant context into the conversation, enriching or in some cases even contradicting the meaning of the words.

What then of music? It is perhaps too easy to reduce it to the sound patterns that can be captured electronically, or an asymmetrical relation between performers and audience. However, just as language involves hands and face as well as voice, it seems more appropriate to seek the roots of music in a triple integration of voice (song), body (dance), and instruments (with, perhaps, the rhythmic beat of the drum as the most basic form). Where the turn-taking of a dyad sharing information may provide grounding for language, the grounding for music may be in a group immersed in music together, with joint activity building a social rapport (Levinson, this volume).

Various cognitive domains (e.g., vocal learning, gestures, observation learning, pedagogy) were combined and developed to permit our ancestors to adapt to their living habitat. Through evolution from a common ancestor, some domains are superior in monkeys compared to humans, whereas others (including certain combinations necessary for language and music) are superior in humans. The mechanisms underlying language and music, in some sense, form a spectrum. Consideration of *Noh*, the Japanese traditional performing art, furthers the discussion (as do opera and the tribal customs assessed by Lewis, this volume) of the relation between language, music and bodily expression. *Noh* integrates three components:

1. *Mai* (舞): dance, which is usually very slow, but occasionally quick movements of the limbs and trunk are employed, with or without masks.
2. *Utai* (謡): the plain chanting of the text.
3. *Hayashi* (囃子): musical accompaniments of drums and flutes.

Utai provides the "language" part of the whole *Noh* play, but it is often and commonly appreciated alone, independent from other components; it accompanies unique intonations which make it more than plain text, or just poetry, but somewhat less than musical song. (The word shares its pronunciation with *utai*, 歌, which means "singing.") In a *Noh* play, *Utai* appears to function as a sort of "glue" to integrate, fuse, or merge music and dance. Thus there might be some common functions or mechanisms (perhaps biological ones acquired through evolutionary processes) that are shared among language, music, and dance, with each component having its own "margin" that could be modified culturally to match the social or environmental situation. Alternatively, these three components may have evolved independently but have obtained common mechanisms due to common environmental demands (like convergent evolution), which enabled these different components to be unified at a later stage.

Niche Construction as a Bridge between Biological and Cultural Evolution

The notion of a "language-ready" brain (e.g., Arbib 2005a) argues that early *Homo sapiens* did not have language but they did have the necessary neural equipment to support its use once languages were invented—just as the human brain evolved long before there was agriculture or cities, but it equipped the species to invent these, eventually, with resultant, dramatic consequences. Whether exploring the evolution of language or of music, the challenge is thus twofold: How did *biological evolution* (natural selection) yield a "language-ready" and "music-ready" species, and how did *cultural evolution* (processes of historical and social change) yield a diversity of languages and musical genres? In addition, when we explore music and language together, we must ask: To what extent did biological evolution, which yielded the language-ready brain, also yield the music-ready brain, or vice versa?

Niche construction refers to the process whereby creatures, by altering the physical environment, change in turn the adaptive pressures that will constrain the evolution of their own and other species (Odling-Smee et al. 2003). To this we add that by creating new patterns of behavior, creatures can alter the *cultural* niche in which they evolve (Bickerton 2009; Pinker 2010; Arbib 2011). Importantly, such cultural niche construction can take place on both a biological timescale as well as on the much faster timescale on which culture evolves, with an increasing tempo in the history of *Homo sapiens*. Indeed, humans can induce such changes intentionally to create a novel environmental niche: *intentional* niche construction (Iriki and Sakura 2008). Going further, building on observations that both literacy (Petersson et al. 2000) and musical expertise can remodel the human brain as well as on experiments with teaching tool use to monkeys (Iriki and Sakura 2008), Iriki and Taoka (2012) advance the idea of *neural* niche construction; that is, new behaviors may remodel the brain in such a way that, by opening novel patterns of gene expression, creatures can take advantage of opportunities that would not otherwise be possible, thus rendering genes adaptive that can modulate the brain around this new stable state.

Iriki (2010) outlines how a novel faculty can emerge from preexisting machinery that had originally been used differently, so that "topologically similar" circuitry could be assembled based on different neural structures—as in Jarvis's comparison of vocal learning in birds and humans (discussed below)—through common genetic guidance, driven by similar interactions between brain and environment. For example, when a monkey is trained to use a rake, certain monkey intraparietal neurons, which normally code for the hand in the body schema, will come to code the hand or the tool in a bistable or polysemous way—coding for the tool when it is in use, but otherwise coding for the hand (Iriki et al. 1996). Such functional plasticity might exploit mechanisms adaptive for body growth, adapting to "sudden elongation" by the tool; however, the bistability reflects a context-dependence that is absent in the adaptation to growth. Moreover, monkeys exhibited substantial expansion (detectable in each individual monkey) of gray matter, including the intraparietal region under study, after a mere two-week tool-use training period (Quallo et al. 2009). New research using the monkey paradigm is needed to probe the concrete biological and genetic mechanisms realizing this expansion.

Once a novel bistable state has proved useful, additional resources will be invested to stabilize the system, perhaps supporting further flexibility. Humans can induce such expansion intentionally (though unaware of this effect on their brains) by consciously mastering a new skill. Triggered by (extra- or epigenetic) factors embedded in this new cognitive niche, the corresponding neural niche in the brain could be reinforced further, supporting a recursive form of intentional niche construction (Iriki and Sakura 2008; Ogawa et al. 2010).

If we are to address the biological evolution of the human capacity for language, then an understanding of how the human brain and body have evolved is indispensable. Such a study must include the various learning mechanisms which give humans a language- and/or music-ready brain; that is, a brain which endows human children to learn a language or musical genre, a capability denied to all other species. Language use cannot develop without the appropriate social context. The manifestation of language readiness rests on developing within the appropriate language-rich cultural niche as well as on development guided by the human genome. Until now, every attempt to raise nonhumans in a language-rich cultural niche has yielded only limited fragments of language-like behavior rather than language per se. Genetic differences among species, however, might not be in the form of the presence or absence of the genetic instructions for growing fundamental language structures so much as a difference in the "genetic switches" that allow component processes to turn "on" under certain conditions.

In the course of human evolution and human history, our ancestors have created new habitats: from modified hunter-gatherer environments to agricultural landscapes with villages to modern civilized technological cities. The evolution of various new cognitive capacities, including those underwriting the manufacture and use of tools as well as the production and comprehension of languages, has enabled these ecological transformations. Such new cognitive capacities, in turn, are an outcome of the dramatic expansion of the human brain and of new functional brain areas. Humans have constructed a new "niche" in each of these ecological, cognitive, and neural domains.

The Mirror System Hypothesis on the Evolution of the Language-Ready Brain

Many have contributed to the brain-based approach to the evolution of language (Deacon 1997; Corballis 2002; Lieberman 2000), and our work furthers this tradition. Since there are intriguing data on the diverse calls of monkeys (Seyfarth et al. 1980b), and since the social relations of baboons offer complexities that in some way foreshadow certain structures evident in language (Cheney and Seyfarth 2005), it certainly seems plausible that spoken language evolved directly from the vocalization system of our common ancestor with monkeys. Our work offers neuroscientific support for a gestural origins theory (Hewes 1973; Armstrong et al. 1995; Armstrong and Wilcox 2007; Corballis 2002). Further support comes from the fact that the repertoire of calls of nonhuman primates is genetically specified, whereas different groups of apes, or even individual apes, of a given species can each develop and use some group-specific gestures (Arbib 2008; Tanner and Byrne 1999). The guiding hypotheses of the mirror system hypothesis are:

1. Language did not evolve as a separate "faculty." Rather, brain mechanisms for the perceptual control of action (including detection of information embedded in the environment through observing others' action as a basis for one's own actions) provided the "evolutionary substrate" for the language-ready brain.
2. Manual dexterity provides a key to understanding human speech. Birdsong provides an example of superb vocal control but has never become the substrate for a language. By contrast, primates are precocious in their manual dexterity, and so we seek to establish that humans exploited this dexterity to become the unique primates possessing language. Indeed, speaking humans, even blind ones, make extensive use of manual co-speech gestures (McNeill 2005), and children raised appropriately can learn the sign languages of the deaf as readily as hearing children learn a spoken language.

We adhere to the view that the primary function of language is communication between members of, at least, a dyad. Language is not a property of individuals, although the capacity to participate is. We thus place *language parity* (i.e., the ability of the hearer to gather, approximately, the intended meaning of the speaker, and similarly for signed language) at stage center. Language is a shared medium, and thus parity is essential to it.

No matter how useful a word may be as a tool for cognition, it must initially be learned; thereafter, numerous conversations, in concert with thought processing, are necessary to enrich our understanding of any associated concept and our ability to make fruitful use of it. Both the external social uses of language and the internal cognitive uses of language could have provided powerful and varied adaptive pressures for further evolution of such capacities as anticipation, working memory, and autobiographic memory as language enriched both our ability to plan ahead, to consider counterfactual possibilities, and to mull over past experience to extract general lessons (Suddendorf and Corballis 2007). All this would greatly expand human capacities for intentional (cognitive) niche construction. Indeed, where we lay stress on parity in the evolution of the language-ready brain, Aboitiz and colleagues lay primary stress on the evolution of working memory systems (Aboitiz et al. 2006a, b; Aboitiz and Garcia 1997, 2009). Such alternatives complement rather than exclude each other.

A *mirror neuron*, as observed in macaque brains, is a neuron that fires vigorously, both when the animal executes an action and when it observes another execute a more or less similar action (for further details, see Fogassi, this volume). Human brain imaging (e.g., Grafton et al. 1996) shows that there is a human mirror system for grasping (i.e., a brain region activated for both grasping and observation of grasping) in or near Broca's area. But why might a mirror system *for grasping* be associated with an area traditionally viewed as involved in speech production? The fact that aphasia of signed, and not just spoken, languages may result from lesions around Broca's area (Poizner et al. 1987; Emmorey 2002) supports the view that one should associate Broca's area with *multimodal* language production rather than with speech alone.

For the argument that follows, we do not claim that grasping is unique in terms of the ability to use action recognition as part of a communication system; as counterexamples, consider the recognition of the baring of teeth or the play crouching of a dog. However, we will claim that there is a crucial difference between the limited imitation of a small species-specific repertoire (compare neonatal imitation [Meltzoff and Moore 1977; Myowa 1996] or "triggering" as in vervet alarm calls [Seyfarth et al. 1980a] or even fish schooling [Hurford 2004]) and open-ended imitation which, as we shall see below, comes to support pantomime to open up semantics.

Different brain regions—not individual neurons—may be implicated in the human brain as mirror systems for different classes of actions. Many researchers have attributed high-level cognitive functions to human mirror regions such as imitation (Buccino et al. 2004b), intention attribution (Iacoboni et al. 2005), and language (Rizzolatti and Arbib 1998). However, monkeys do not imitate to any marked degree and cannot learn language. Thus, any account of the role of human mirror systems in imitation and language must include an account of the evolution of mirror systems and their interaction with more extended systems within the human brain; that is, "beyond the mirror system."

The original mirror system hypothesis argued that the basis for language parity evolved "atop" the mirror system for grasping, rooting speech in communication based on manual gesture (Rizzolatti and Arbib 1998; Arbib and Rizzolatti 1997). In other words, a path from praxis to communication was traced. Developing this basic premise, Arbib argues (2005a, 2002, 2012) that a language-ready brain resulted from the evolution of a progression of mirror systems *and linked brain regions "beyond the mirror"* that made possible the full expression of their functionality.

Imitation

Monkeys have, at best, a very limited capacity for imitation (Visalberghi and Fragaszy 2001; Voelkl and Huber 2007) that is far overshadowed by "simple" imitation, such as exhibited by apes. Myowa-Yamakoshi and Matsuzawa (1999) observed that chimpanzees took on average 12 trials to learn to "imitate" a behavior in a laboratory setting, focusing on bringing an object into relationship with another object or the body, rather than the actual movements involved. Byrne and Byrne (1993) found that gorillas learn complex feeding strategies but may take months to do so. Consider eating nettle leaves: skilled gorillas grasp a stem firmly, strip off leaves, remove petioles bimanually, fold leaves over the thumb, pop the bundle into the mouth, and eat. The challenge for acquiring such skills is compounded because ape mothers seldom, if ever, correct and instruct their young (Tomasello 1999a). In addition, the sequence of "atomic actions" varies greatly from trial to trial. Byrne (2003) implicates *imitation by behavior parsing*, a protracted form of statistical learning whereby certain *subgoals* (e.g., nettles folded over the thumb) become evident from the repeated observation of common occurrences in most performances. In his account, a young ape may acquire the skill over many months by coming to recognize the relevant subgoals and by deriving action strategies to achieve them through trial and error.

This ability to learn the overall structure of a specific feeding behavior over numerous observations is very different from the human ability to understand any sentence of an open-ended set as it is heard and to generate another novel sentence as an appropriate reply. In many cases of praxis (i.e., skilled interaction with objects), humans need just a few trials to make sense of a relatively complex behavior if the constituent actions are familiar and the subgoals inherent in actions are readily discernible, and they can use this perception to repeat the behavior under changing circumstances. We call this ability *complex imitation*, extending the definition of Arbib (2002) to incorporate the goal-directed imitation of Wohlschläger et al. (2003).

In itself, a mirror system does not provide imitation. A monkey with an action in its repertoire may have mirror neurons active both when it executes and observes that action, yet it does not repeat the observed action nor, crucially, does it use the observation of a novel action to add that action to its repertoire. Thus, the mirror system hypothesis claims, in part, that evolution embeds a monkey-like mirror system in more powerful systems in two stages:

1. *A simple imitation system for grasping*, shared with the common ancestor of human and apes; and
2. *A complex imitation system for grasping*, which developed in the hominid line since that ancestor.

Both of these changes may represent an evolutionary advantage in supporting the transfer of novel skills between the members of a community, involving praxis rather than explicit communication.

Laboratory-raised, nonhuman primates exposed to the use of specific tools can exhibit behaviors never seen in their wild counterparts (Iriki et al. 1996; Umiltà et al. 2008). Tool-use training appears to forge novel corticocortical connections that underlie this boost in capacity, though the limited use of tools by apes does vary "culturally" from group to group (Whiten et al. 1999; for an experimental study, see Whiten et al. 2005). Although tool-use training is patently nonnaturalistic, its marked effects on brain organization and behavior could shed light on the evolution of higher intelligence in humans. Note, however, that the ability of other primates to be trained by humans in the limited use of specific tools is rather different from the human capacity to master new tools on an almost weekly basis or the human ability to invent new tools of possibly great complexity (as an interaction between the human individual and a highly evolved cultural/cognitive niche).

An important problem for the tool-using brain to solve is how to "regard" a tool (i.e., as one's own body part or merely as an external object). This could lead to two important aspects emerging in the mind of the tool user: (a) explicitly realizing that one's own body is composed of multiple parts that are spatially arranged and intentionally controllable; and (b) that those body parts could be compatible (both functionally and structurally) with external objects, namely tools. This establishment of parity (or polysemy) between body parts and tools enables tool users to incorporate (or assimilate) tools into their own body schemas. However, the knowledge that *in context A, I should first find tool B and grasp it in a certain way* is very different from having any general awareness of the notions of *context* or *tool*. What is crucial, though, is that the brain is sufficiently complex so that it can learn that, once the tool is grasped, visual attention must pass from the hand to the end-effector of the tool, and that proprioceptive feedback is now evaluated in terms of the pressure of the tool on the hand when it is used to interact with an object (Arbib et al. 2009).

Iriki (2006) argues that tool users can "objectify" their own body parts and should eventually be able to objectify themselves entirely (i.e., to observe themselves mentally from a third-person perspective). This progression should contribute greatly to the development of true imitation, by which the form of an action can be extracted (through the mirror neuron system) and treated independently of its goal, as a consciously recognized, independent object. Such "self-objectification" processes establish equivalence in the mind of the agent between other agents and the self, including understanding that tools are equally compatible for both. However, where Iriki would argue that even the macaque brain can exhibit these properties, if given extensive enough training in a novel cognitive niche, Arbib (noting, e.g., the failure, despite extensive training, to get apes to master syntax) would argue that evolutionary changes in the genome accumulated over the last 20 million years were necessary to yield a human capacity for the understanding of self and others. The mirror system hypothesis offers one such evolutionary scenario for consideration.

From Pantomime to Protosign and Protolanguage

We now explore the stages whereby our distant ancestors made the transition to *protolanguage*, in the sense of a communication system that supports the ready addition of new utterances by a group through some combination of innovation and social learning. This system is open to the addition of new "protowords," in contrast to the closed set of calls of a group of nonhuman primates, yet lacks the tools (beyond mere juxtaposition of two or three protowords) to put protowords together to create novel utterances from occasion to occasion.

Continuing, the mirror system hypothesis suggests that brain mechanisms for complex imitation evolved to support not only pantomime of manual actions but also *pantomime of actions outside the panto-mimic's own behavioral repertoire*. This, in turn, provides the basis for the evolution of systems supporting *protosign*: conventional gestures used to formalize and disambiguate pantomime.

The transition from complex imitation and the small repertoires of ape gestures (perhaps ten or so novel gestures shared by a group) to protosign involves the pantomime of grasping and manual praxic actions of nonmanual actions (e.g., flapping the arms to mime the wings of a flying bird), complemented by conventional gestures which simplify, disambiguate (e.g., to distinguish "bird" from "flying"), or extend the pantomime. Pantomime transcends the slow accretion of manual gestures by ontogenetic ritualization, providing an "open semantics" for a large set of novel meanings (Stokoe 2001). Such pantomime, however, is inefficient, both in the time taken to produce it as well as in the likelihood of misunderstanding. Conventionalized signs extend and exploit more efficiently the semantic richness opened up by pantomime. Processes like ontogenetic ritualization can convert elaborate pantomimes into a conventionalized "shorthand," just as they do for praxic actions. This capability for protosign—rather than elaborations intrinsic to the core vocalization systems—may have provided the essential scaffolding for protospeech and evolution of the human language-ready brain. Protosign and protospeech (an expanding spiral) are then conventionalized manual, facial, and vocal communicative gestures ("protowords") that are separate from pantomime.

Here we note that the use of sign language (i.e., a full human language like American Sign Language or British Sign Language, well beyond the mere stage of protosign) can be dissociated from pantomime. Sign language aphasics who can, for example, no longer recall the conventional sign for "flying" might instead pantomime the motion of a flying plane (Corina et al. 1992; Marshall et al. 2004).

Following Hewes (1973) and Bickerton (1995), we use the term *protolanguage* to indicate any of the intermediate forms that preceded true languages in our lineage. This is not to be confused with the usage in historical linguistics, where a *protolanguage* for a family of languages is a full language posited to be ancestral to the whole family, just as scholars have sought to infer the lexicon and grammar of "proto-Indo-European" as ancestral to all Indo-European languages. Here, the debate is between two views. The *holophrastic* view (Wray 1998) holds that in much of protolanguage, a complete communicative act involved a "unitary utterance" or "holophrase" whose parts had no independent meaning. Accordingly, as "protowords" were fractionated or elaborated to yield words for constituents of their original meaning, so were *constructions* developed to arrange the words to reconstitute those original meanings and many more besides. Opposing this, the *compositional* view (Bickerton 1995) hypothesizes that *Homo erectus* communicated by means of a protolanguage in which a communicative act comprises a few words in the current sense strung together without syntactic structure. Accordingly, the "protowords" (in the evolutionary sense) were so akin to the words of modern languages that languages evolved from protolanguages just by "adding syntax." Elsewhere Arbib (2008) argues that the earliest protolanguages were in great part holophrastic, and that as they developed through time, each protolanguage retained holophrastic strategies while making increasing use of compositional strategies (for a range of viewpoints, see Arbib and Bickerton 2010). We may again see an opportunity for triple niche construction: as fractionation and construction developed, so must conventions have arisen to distinguish a holophrase from a similar constructed phrase; this, in turn, would have provided an adaptive pressure to speed the production and recognition for such distinctions as a basis for acquiring a lexicon and a (proto)syntax. This may explain the fact that the human ability to discriminate complex syllable sequences and separate "words" by statistical processing is augmented by the presence of subliminal cues such as pauses (Peña et al. 2002; Mueller et al. 2008) and appears to be present even in very young infants.

From Protolanguage to Language

Arbib (2005a) argues that a brain with the above systems in place was language-ready, and that it was cultural evolution in *Homo sapiens* that yielded language incrementally. In this regard, it is worth stressing that the brain's language structures reflect more than just genetics, and that functional hemispheric specialization depends on both genetic and environmental factors (Petersson et al. 2000). Illiterate subjects are consistently more right-lateralized than literate controls, even though the two groups show a similar degree of left-right differences in early speech-related regions of superior temporal cortex. Further, the influence of literacy on brain structure related to reading and verbal working memory affects large-scale brain connectivity more than gray matter. Here a cultural factor, literacy, influences the functional hemispheric balance.

We now come to the final stage: the transition from protolanguage to language and the development of syntax and compositional semantics. This may have involved grammatically specific biological evolution. The nature of the transition to language remains hotly debated and lies beyond the scope of this chapter (cf. chapters 10 and 13 in Arbib 2012). Although the mirror system hypothesis posits that complex imitation evolved first to support the transfer of praxic skills and then came to support protolanguage, it is important to note its crucial relevance to modern-day language acquisition and adult language use. Complex imitation has two parts: (a) the ability to perceive that a novel action may be approximated by a composite of known actions associated with appropriate subgoals and (b) the ability to employ this perception to perform an approximation to the observed action, which may then be refined through practice. Both parts come into play when a child learns a language, whereas the former predominates in the adult use of language, where the emphasis shifts from mastering novel words and constructions to finding the appropriate way to continue a dialog.

Vocal Learning

Turning to the crucial role of vocal learning in spoken language (but not sign language) and sung (but not instrumental) music, it has been of great interest to explore evolutionary parallels between the evolution of human language and music and the evolution of birdsong. Jarvis (e.g., 2004, 2007a, b; see also Fitch and Jarvis, this volume) argues as follows:

1. All vocal learning species, even those with evolutionarily quite distant lineages, share neuroanatomical circuitry that is topologically similar, even when the concrete neural structures comprising each component in each species may be different.
2. A common group of genes, including the ones that guide axonal connections, are commonly, but specifically, expressed in these circuits in vocal learning species, but not in closely related vocal-nonlearning species.
3. These shared patterns of neuroanatomical circuitry and gene expressions necessary for vocal learning may be coded by common (but still not evident) sequences of genes that are language-related in humans and which start functioning upon environmental demand at different evolutionary lineages (i.e., deep homology to subserve convergent evolution).

Let us, however, consider primates in more detail. Nonhuman primates do not exhibit vocal learning. What paths may have led to the emergence of this ability in humans? For example, Jürgens (1979, 2002), working primarily with squirrel monkeys rather than macaques, found that voluntary control over the initiation and suppression of monkey vocalizations (i.e., the initiation and suppression of calls from a small repertoire, not the dynamic assemblage and coarticulation of articulatory gestures that constitutes speech) relies on the mediofrontal cortex, including the anterior cingulate gyrus. Note that these are mostly "emotional calls": they may be voluntarily turned on or off, but were not acquired through "vocal learning," and thus do not qualify as true voluntary calls for use in intentionally symbolic behavior. A major achievement of the mirror system hypothesis is to develop a plausible explanation as to why human Broca's area, which corresponds to macaque F5—rather than the vocalization area of cingulate cortex—lies at the core of language production. Ferrari et al. (2003, 2005) found that F5 mirror neurons include some for orofacial gestures involved in feeding. Moreover, some of these gestures (such as lip smack and teeth chatter) have auditory side effects which can be exploited for communication. This system has interesting implications for language evolution (Fogassi and Ferrari 2004, 2007), but is a long way from mirror neurons for speech. Intriguingly, squirrel monkey F5 does have connections to the vocal folds (Jürgens, pers. comm.), but these are solely for closing them and are not involved in vocalization (cf. Coudé et al. 2007). We thus hypothesize that the emergence of protospeech on the scaffolding of protosign involved the expansion of the F5 projection to the vocal folds to allow vocalization to be controlled in coordination with the control of the use of tongue and lips as part of the ingestive system.

In their study of spontaneous vocal differentiation of coo calls for tools and food in Japanese monkeys, Hihara et al. (2003) discuss mechanisms for the emergence of voluntary calls as the transition from cingulate cortex to lateral motor cortex to control directly the brainstem vocal center. Hihara et al. trained two Japanese monkeys to use a rake-shaped tool to retrieve distant food. They were at the same time, but independently, reinforced to make coo calls though the type of call was not restricted. After both types of conditioning were completed independently, both tasks were combined as follows:

- Condition 1: When the monkey produced a coo call (call A), the experimenter put a food reward on the table, but out of his reach. When the monkey again vocalized a coo call (call B), the experimenter presented the tool within his reach. The monkey was then able to retrieve the food using the tool.
- Condition 2: Here the tool was initially presented within the monkey's reach on the table. When the monkey vocalized a coo call (call C), the experimenter set a food reward within reach of the tool.

The intriguing fact is that the monkey spontaneously differentiated its coo calls to ask for either food or tool during the course of this training (i.e., coos A and C were similar to each other but different from call B). We stress that the coo calls sound the same to human ears: the experimenter can only differentiate them using acoustic analysis, and so there cannot be "shaping" of the calls by the experimenter. The monkeys spontaneously differentiated those calls after as little as five days. Hihara et al. speculate that this process might involve a change from emotional vocalizations into intentionally controlled ones by associating them with consciously planned tool use implicating (a) short-term vocal plasticity, (b) voluntary control of vocalization, and (c) a precursor of naming, all of which would comprise fundamentals of vocal learning, which had never been assumed to exist in macaques. Is this "naming" of a food or the tool or a holophrase akin to "give me food" or "give me the rake" (in the same sense that the vervet eagle alarm call is not at all equivalent to the noun *eagle*)? Rather than vocal learning, Roy and Arbib (2005) see this coo call modulation as an example of the unconscious linkage between limb movement and vocal articulation that was demonstrated in humans by Gentilucci et al. (2004a, b).

However, the intriguing issue of the underlying connectivity remains. Iriki notes (unpublished observation) that even in nonprimate mammals, there are a (very) few direct connections between motor cortex and brainstem vocal centers such as the periaqueductal gray. These followed from tracer studies on the transition of the corticobulbar projection during development from sucking to chewing (Iriki et al. 1988). These "rare" connections are usually regarded as an artifact or just ignored, but they are consistent in being both "few" and "always there." Thus, under some conditions, these connections could be "strengthened" by gene expression and circuit reorganizations to become major neural paths. We see here a candidate for neural niche construction; a sparse pathway strengthened by experiences that become part of the cultural niche might then be available for further strengthening by natural selection's indirect effects upon the genome. According to the mirror system hypothesis, the development of pantomime created an open-ended semantics and protosign crystallized an abstract structure of conventional gestural signs therefrom; this created the new cognitive niche in which the evolution of structures for vocal learning proved advantageous, building protospeech on the scaffolding of protosign.

The Darwinian Musical Protolanguage Hypothesis

When Darwin (1871) addressed the question of language evolution, he put the emphasis on song rather than gesture as the precursor. He laid out a three-stage theory of language evolution:

1. A greater development of protohuman cognition, driven by both social and technological factors.
2. The evolution of vocal imitation used largely "in producing true musical cadences, that is in singing" (Darwin 1871:463). Darwin suggests that this evolved as a challenge to rivals as well as in the expression of emotions. The first protolanguage would have been musical, driven by sexual selection as was birdsong, so that this capacity evolved analogously in humans and songbirds.
3. Articulate language then owed its origins to the imitation and modification, aided by signs and gestures, of various natural sounds, the voices of other animals, and man's own instinctive cries. This transition was again driven by increased intelligence. Once meaning was in place, actual words would have been coined from various sources, encompassing any of the then-current theories of word origins.

Note that the view of music here is very different from the view introduced earlier. Here the model is birdsong—of one individual singing to attract a mate or defend a territory. We are looking here at a theory of the musical origins of language evolution, not of evolution of music as a shared group activity. Continuing with this Darwin-inspired account, a key observation is that language has "musicality" or prosody as well as semantic expression. The same words may convey not only information but also emotion, while increasing social bonding and engaging attention. The *musical protolanguage hypothesis* can be simply stated as "phonology first, semantics later." It emphasizes that song and spoken language both use the vocal/auditory channel to generate complex, hierarchically structured signals that are learned and culturally shared. As for musical protolanguage, Jespersen (1922) suggested that, initially, meanings were attached to vocal phrases in a holistic, all-or-none fashion with no articulated mapping between parts of the signal and parts of the meaning; this is the notion of a *holophrase* discussed briefly above. Jespersen went beyond Darwin's vague suggestions about "increasing intelligence" to offer a specific path from irregular phrase-meaning linkages to syntactic words and sentences. Pointing to the pervasiveness of both irregularities as well as attempts (often by children) to analyze these into more regular, rule-governed processes ("over regularization"), Jespersen gave a detailed account for how such holophrases can gradually be analyzed into something more like words. The analysis of whole phrases into subcomponents occurs not just in language evolution and historical change, but in language acquisition as well. Children hear entire phrases initially as a whole and then become increasingly capable at segmenting words out of a continuous speech stream. Such ideas are part of the mirror system hypothesis which offers great attention to the role fractionation of holophrases and the attendant formation of novel constructions, while emphasizing that these processes are equally valid in the manual domain.

Mithen (2005) and Fitch (2010) have combined Darwin's model of "musical" or "prosodic" protolanguage and Jespersen's notion of holistic protolanguage to yield a multistage model that builds on Darwin's core hypothesis that protosong preceded language. The resulting model posits the following evolutionary steps and selective pressures, leading from the unlearned vocal communication system of the last common ancestor of chimpanzees and humans, to modern spoken language in all of its syntactic and semantic complexity:

1. *Phonology*: The acquisition of complex vocal learning occurred during an initial song-like stage of communication (comparable to birdsong or whale song) that lacked propositional meaning. Darwin proposed a sexually selected function while Dissanayake (2000) opts for a kin-selection model. The two suggestions need not be mutually exclusive. Accordingly, vocal imitation, which is lacking in chimpanzees, was the crucial step toward language. Fitch (2010) notes the convergent evolution with that for "song" in songbirds, parrots, hummingbirds, whales, and seals.
2. *Meaning*: The addition of meaning proceeded in two stages, perhaps driven by kin selection. Holistic mappings between whole, complex phonological signals (phrases or "songs") and whole semantic complexes (activities, repeated events, rituals) were linked by simple

association. Such a musical protolanguage was an emotionally grounded vocal communication system, not a vehicle for the unlimited expression of thought.
3. *Compositional meaning*: These linked wholes were gradually broken down into parts: individual lexical items "coalesced" from the previous wholes.
4. *Modern language*: As the language of its community grew more compositional (i.e., with wholes replaced by composites of parts), pressure for children to learn words and constructions rapidly became strong. This drove the last spurt of biological evolution to our contemporary state. Fitch (2010) suggests that this last stage was driven by kin selection for the sharing of truthful information among close relatives.

By contrast the mirror system hypothesis offers the following sequence:

1. *Complex imitation*: Brain mechanisms supporting the ability to recognize and imitate novel complex manual skills in tool use laid the basis for extending basic mirror system capabilities to the recognition of, and action on the basis of, compound behaviors.
2. *Meaning*: Rich meanings were scaffolded by pantomime and the consequent emergence of protosign in turn scaffolded the emergence of protospeech. Accordingly, it was the manually grounded need to express increasingly complex meanings that constructed the new cognitive and neural niche which favored the emergence of increased articulatory control and vocal learning.
3. *Phonology*: As complex pantomimes and vocal expressions of meaning proliferated, phonology—in the sense of duality of patterning, with "protowords," whether spoken or signed, built up from a small set of meaningless units—emerged to aid discrimination of similar protowords (for details, see Hockett 1960b; Sandler and Aronoff 2007; Arbib 2009).
4. *Compositional meaning*: The notion of a holophrastic phase from which syntax emerges is one we share with Jespersen, Mithen and Fitch. In our account, however, the transition to compositional meaning is a general property that goes back to essentially human modes of praxis (complex action recognition and complex imitation) rather than being a process evolved in response to the increasing complexity of musical protolanguage.

The *musical protolanguage hypothesis* roots the whole process in a capacity for vocal learning which we have every reason to believe was absent in our last common ancestor with chimpanzees. Note too that no nonhuman species with vocal learning has acquired the ability to use vocalization to convey propositional meaning, with the notable exception of the results obtained through very extensive training of a single African Gray parrot (Pepperberg 2002, 2006). By contrast, the mirror system hypothesis builds upon skills for imitation, tool use, and the development of novel communicative manual gestures by chimpanzees which, it seems reasonable to posit, were present in the last common ancestor (even though we stress that chimpanzee brains and bodies have evolved at least as much as those of humans over the last five to seven million years). Finally, we note that monkeys, apes, and humans exhibit a variety of social structures which provide a rich substrate of novel possibilities for communication that are absent in songbirds. These thus play an important role in the evolution of language, which complements the brain-centered approach adopted here. Whether this perspective favors an account based on gestural rather than vocal origins remains, however, a matter of intense debate (Seyfarth 2005; Seyfarth et al. 2005; Cheney and Seyfarth 2005).

What Then of Music?

As presented above, the musical protolanguage hypothesis focuses primarily on the role of musical protolanguage as a precursor for language. What then of music? We, the authors, are experts on brain mechanisms for the visual control of action and have explored its relation to language evolution and niche construction, among other topics, but have little expertise in the neuroscience of music. Thus, our intent here is to raise questions rather than to answer them. As is clear from the above, our concern for manual skill has made us especially aware of the coordination of hand and voice in normal speech, an awareness augmented by an understanding of the way in which sign language can replace speech when the auditory channel is unavailable. We are thus of the opinion that music also evolved within the context of bodily expression; in other words, the evolution of music is inseparable from the evolution of dance. Again, although vocalization is one basic form of musical expression, musical instruments have a crucial role, so that music builds on skills of manual dexterity in a manner that is even more direct than the way in which pantomime paved the way for protosign. However, the "tool use" of playing a musical instrument has a meaning very different from the practical goals of other types of tool use, or the exchange of information and the building up of shared patterns of propositional meaning that is supported by language. Music has a kind of free-floating "meaningfulness" that can attach itself to many types of group activity and can thus enrich events it accompanies with unifying, barrier-dissolving effects (Cross 2003c). Music lacks nouns, verbs, tense, negation, and embedding of meanings; it lacks "propositional meaning." Moreover, just as protolanguage is not language, so is a musical protolanguage not music. For example, music in many cultures now uses a small number of discrete frequency units, *notes*, which together make up a *scale* (Nettl 2000). A song in such a tradition allows only these units to be used. Just as for pantomime, neither protomusic nor early protolanguages need to have been constructed from a small, discrete set of elements (see remarks on phonology under the mirror system hypothesis sequence). However, a property of music which may be more fundamental in evolutionary terms is that time is typically subdivided evenly into discrete "beats" that occur at a relatively regular tempo and which are arranged according to a metrical structure of strong and weak events: the core ingredients of musical rhythm. Though this feature is not part of the musical protolanguage hypothesis we are discussing, it may well have been crucial to the evolution of music as integrated with dance.

Beat induction is the process whereby a person listening to music forms a certain metrical expectation as to when future notes will occur. As Honing (2011a) observes, without beat induction it would seem there would be no music, since it makes us capable of dancing together or playing music together. Ladinig et al. (2009) measured evoked response potentials in listeners without explicit musical training and showed that their brains encoded a strong feeling for meter: The absence of a note at the spot where a beat was expected created a "surprise signal," a *mismatch negativity*, showing that the brain had indeed induced the beat. Until recently, researchers believed that beat induction is learned, for example, by being rocked by one's parents to the beat of the music. However, a new experiment with two- to three-day-old infants found that they reacted similarly to adults (Winkler et al. 2009b; Honing et al. 2009). Turning to a comparison with primates, Zarco et al. (2009) compared the motor timing performance of human subjects and rhesus monkeys. The temporal performance of rhesus monkeys was similar to that of human subjects during both the production of single intervals and tapping in synchronization with a metronome. Overall, however, human subjects were more accurate than monkeys and showed less timing variability, especially during the self-pacing phase of a multiple interval production task. Zarco et al. conclude that the rhesus monkey remains an appropriate model for the study of the neural basis of timed motor production, but they also suggest that the temporal abilities of humans which peak in speech and music performance are not all shared with macaques.

Beat induction, however, seems more relevant to the coordination of rhythmic behavior of members of a group than to prosody of speech of members of dyad in conversation. We thus suggest that the neural mechanisms of beat induction may have evolved as part and parcel of the music-ready brain, rather than as an off-shoot of the evolution of the language-ready brain (for further discussion of the evolution of music, see Brown 2000; Cross 2003d; Fitch 2006a; Patel 2008; Cross and Morley 2008; Morley 2002, 2009).

20

Birdsong and Other Animal Models for Human Speech, Song, and Vocal Learning

W. Tecumseh Fitch and Erich D. Jarvis

Abstract

This chapter highlights the similarities and differences between learned song, in birds and other animal models, and speech and song in humans, by reviewing the comparative biology of birdsong and human speech from behavioral, biological, phylogenetic, and mechanistic perspectives. Our thesis is that song-learning birds and humans have evolved similar, although not identical, vocal communication behaviors due to shared deep homologies in nonvocal brain pathways and associated genes from which the vocal pathways are derived. The convergent behaviors include complex vocal learning, critical periods of vocal learning, dependence on auditory feedback to develop and maintain learned vocalizations, and rudimentary features for vocal syntax and phonology. The associated neural substrate is a set of specialized forebrain pathways found only in humans and other complex vocal learners, and it consists of premotor and motor forebrain systems that directly control brainstem vocal motor neurons. To develop and maintain function of the novel vocal-learning pathways, we argue that convergent molecular changes occurred on some of the same genes, including FoxP2 and axon guidance molecules. Our hypothesis is that the unique parts of the brain pathways, which control spoken language in humans and song in distantly related song-learning birds, evolved as specializations of a deeply homologous, preexisting motor system, which was inherited from their common ancestor and which controls movement and complex motor learning. The lesson learned from this analysis is that by studying the comparative behavioral neurobiology of human and nonhuman vocal-learning species, greater insight can be gained into the evolution and mechanisms of spoken language than by studying humans alone or humans only in relation to nonhuman primates.

Introduction

Many thinkers have intuited an evolutionary relationship between human language and music. Since Darwin (1871), it has been common to cite the complex songs of birds in support of some such link. For many skeptical commentators, however, the connection between music and language remains quite unclear, and the relevance of birdsong to either is even more obscure. What does seem clear is that humans evolved a capacity for music independently from birds' capacity for song, whatever similarities the two might share. Furthermore, clear and fundamental differences exist between human music and language, or between either of these and birdsong. Are there any real, relevant connections between these three domains?

We contend that there are—that multiple, fruitful links can be found between birdsong, human music, and human language via the mediating concept of what we call *complex vocal learning*. This concept is best illustrated in birds: young songbirds of many familiar species (e.g., robins, blackbirds, or finches) perceptually learn the songs of adults around them in their environment and then reproduce those songs (or variants thereof) as adults. The young of these vocal-learning species *require* auditory input to produce "normal" songs themselves. In contrast, in many other bird species (e.g., chickens, seagulls, hawks, or pigeons), adults can produce well-formed vocal communication signals without ever having heard such sounds produced by others. Thus there is a contrast, in birds, between vocal-learning species and those whose vocal repertoire develops without external input. Turning to primates, humans also have a rich capacity for vocal imitation that is lacking in other nonhuman primates: a human child exposed to certain sounds, whether words or melodies, will reliably learn to reproduce those vocalizations, whereas a chimpanzee will not. Thus an "instinct to learn" to produce sounds from the environment characterizes some species, including songbirds and people, but not others.

As humans, we recognize profound differences between language and music. Music can be played on instruments; language can be communicated via gestures. Language is often used informatively to communicate complex concepts and propositions, whereas music tends to fulfill more emotional or evocative roles, to function in mood regulation or group bonding, and to invite repetitive, ritualistic replay. Nonetheless, both music and language take the vocal output mechanism as their default in all cultures (via song or speech), and in both domains the cultural transmission and elaboration of songs or words *requires* an inborn capacity for vocal learning. Song and speech thus share a core similarity: reliance on vocal learning. From a bird's eye view, song and speech can be seen as different manifestations of the same underlying, fundamental ability: to hear a complex sound in the environment and then produce a close imitation yourself. An "ornithomorphic perspective" invites us to recognize this similarity, despite real differences.

There are at least three other reasons that we can gain insight into the evolution of music and language from studying songbirds. First, the fact that birds evolved their capacity for vocal learning independently from humans provides a statistically independent route to evaluate theories about the *selective forces* capable of driving the evolution of vocal learning. Second, from a neural viewpoint, vocal-learning circuits in humans and birds may have built upon common precursor circuits. If so, although the vocal-learning circuits evolved convergently, the precursors might nonetheless represent homologs inherited from the common ancestor of birds and mammals—a form of neural "deep homology." Third, the developmental pathways that generate these neural structures also built upon precursors, and it is increasingly clear that in some cases these precursors represent ancient, homologous traits as well. Such shared precursors, exemplified by the FoxP2 gene (discussed below), provide examples of genetic deep homology. In all three cases, birds make an excellent study species, open to experimental investigation in ways that humans can never be.

Thus despite clear and undeniable differences between spoken language and birdsong, we argue that birdsong has much to offer in the quest to understand the evolution of human music and language, and that complex vocal learning provides the centerpiece—the core mediating concept—in this endeavor. Here we explore the many ramifications of this perspective and consider a wealth of evolutionary, neural, and genetic data from birds which, in our opinion, can provide crucial insights into the evolution of our own most special abilities. To forestall possible misinterpretation, we begin by clarifying the dual issues of parallels in birdsong and human vocal behavior and their evolutionary bases before we launch into the main discussion.

Parallels in Vocal-Learning Abilities

Similarities between human speech and singing with birdsong have been recognized as relevant to the evolution of language since at least Darwin (1871). These similarities were not investigated mechanistically at the underlying cognitive and neural levels until the seminal research of Peter Marler (1955, 1970), Fernando Nottebohm (1976), and Masakazu Konishi (1965). Since then an entire field of research has advanced our understanding of the neurobiology of vocal communication in song-learning birds, including electrophysiological mechanisms (Doupe and Konishi 1991; Mooney 1992; Yu and Margoliash 1996; Hahnloser et al. 2002; Fee et al. 2004), biophysical basis of vocal production (Fee et al. 1998; Suthers et al. 1999; Goller and Larsen 2002), high throughput behavioral mechanisms (Tchernichovski et al. 2001), and genetic basis of avian vocal learning (Mello et al. 1998; Jarvis and Nottebohm 1997; Jarvis et al. 1998; Clayton 2000; Haesler et al. 2004, 2007; Warren et al. 2010). As a result, the current body of data on songbird vocal learning surpasses that of any other vocal-learning species or clade, including humans (for reviews, see Marler and Slabbekoorn 2004; Zeigler and Marler 2008). Many of these findings have been compared with findings on human vocal learning, including both spoken language (Doupe and Kuhl 1999; Jarvis 2004; Bolhuis et al. 2010) and song (Fitch 2006b).

The core similarity is that in several groups of distantly related birds (songbirds, parrots, and hummingbirds), some vocalizations are learned from the environment, as are human speech and song (Jarvis 2004). This capacity contrasts with the vast majority of other birds and mammals, including nonhuman primates (Egnor and Hauser 2004), which are thought to produce mainly innate vocalizations or small modifications to innate sounds. This fundamental similarity must, of course, be considered in the context of differences: song-learning birds are not known to use their songs to communicate combinatorial propositional meanings (i.e., semantics), although evidence indicates that semantic object communication does exist in some parrots and songbirds (Pepperberg and Shive 2001; Templeton et al. 2005). On the other hand, for propositional meaning in a broad sense, a vocalization that refers to something in the real or imaginary world is not unique for learned vocalizations: innate alarm or food calls, or sequences of calls, in nonhuman primates and vocal-nonlearning birds can have semantic content (Seyfarth et al. 1980a; Palleroni et al. 2005). On the other hand, vocal-learning birds use their learned vocalizations primarily to communicate affective meaning, such as mate attractiveness or territorial defense (Catchpole and Slater 2008). How, then, are we to take seriously the similarities between birdsong and human vocal learning?

First, to avert misunderstanding, we clarify what we mean by calls, syllables, song, and speech/spoken language in the context of vocal learning. Our definitions are drawn from the animal communication field and take an ornithomorphic perspective: Calls are isolated vocalizations that can be produced singly, whether repeated or not. Calls are termed syllables when they are part of long continuous and changing sequences of vocalizations, defined as song or speech. We do not make a sharp distinction between speech and spoken language, or song vocally produced with or without words, which for us is essentially all "vocal music." Rather, it is safer to view these behaviors as part of a continuum of potentially learned vocal behaviors. In contrast, for human vocal behavior, many scientists as well as the lay public often make a sharp distinction between song and speech, and they contrast the latter with language in a more general sense (cf. Fitch 2000, 2006a). For vocal-learning birds, scientists often do not make such a distinction: even "talking birds" are producing "song," regardless of its semantic meaning or lack thereof. Part of the reason is that the behavioral definitions of the vocalizations in animal models are more often based on the role of the brain pathways that control them, more than simple distinctions in the behavior. In songbirds, the same brain pathway is involved in production of song, learned calls, and human speech (Simpson and Vicario 1990; Jarvis 2004). We hypothesize that the difference in "singing," "calling," and "speaking" in vocal-learning birds may be in how the so-called song control brain pathways are used, not the pathways themselves. In humans it is not clear if the same brain pathways are used for producing (and perceiving) song and speech, but the latest evidence suggests that the brain

Animal Models for Human Speech, Song, and Vocal Learning   503

areas for speaking and singing show considerable overlap (Brown et al. 2004). Thus, to be conservative, in the absence of evidence showing distinct brain pathways, when we refer to "learned vocalizations" in song-learning birds as well as in humans, we are referring to all learned vocalizations, whether they are labeled song, calls, speech, or simply complex vocalizations. This may take some getting used to, and thus in the following discussion we offer more specific definitions and justifications.

Convergent Evolution and Deep Homology

Regarding the evolution of vocal learning (Figure 20.1), the overwhelming consensus is that humans and two, if not all three, of the song-learning bird lineages evolved vocal learning *convergently*, since there are so many other closely related birds and mammals without the trait, including nonhuman primates (Janik and Slater 1997; Jarvis 2004; Suh et al. 2011). This implies that the common ancestor of birds and mammals was not a vocal learner, and thus bird and human "song" is not homologous. It also means that the common ancestor of all vocal-learning birds was probably not a vocal learner (Figure 20.1a). This repeated, independent evolution of a functionally similar trait—complex vocal learning—means that convergent evolution could be a powerful route for understanding constraints on evolved systems and for testing hypotheses about evolution. Birdsong has provided rich insights on both counts.

Despite their independent evolution, the human and avian end state (learned vocalizations) appears to have entailed fundamental similarities at the behavioral and biological mechanistic levels (Doupe and Kuhl 1999; Jarvis 2004). For example, both song-learning birds and humans go through a "babbling" stage of variegated vocalizations early in life as an infant, and this appears to be necessary for adequate vocal learning (Marler and Peters 1982; Doupe and Kuhl 1999). Similarly, it has been hypothesized that vocal learning entails the evolution of direct connections from the motor cortical regions to the motor neurons in the brainstem that control the vocal organs, because humans and all three song-learning bird groups have such connections, whereas nonhuman primates and other vocal-nonlearning mammals and birds tested to date do not (Figure 20.2 and 20.3) (for recent reviews, see Jürgens 2002; Jarvis 2004; Fitch et al. 2010; Simonyan and Horwitz 2011; Arriaga and Jarvis 2013).

Similarly, both song-learning birds and humans have functionally similar forebrain areas hypothesized to be necessary for acquiring and producing learned vocalizations. Such forebrain areas cannot be found in vocal-nonlearning mammals and birds tested to date. Further, in the vocal-learning birds, these brain regions consist of seven nuclei each and are connected by similar, but not identical, networks (mostly studied in songbird and parrots; Jarvis 2004). To explain such striking convergent similarities, recent findings have revealed that the song-learning forebrain pathways in all three vocal-learning bird

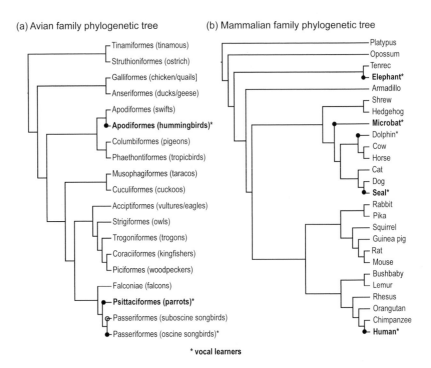

Figure 20.1 Family trees of representative living avian and mammalian species. (a) Avian tree of major orders based on DNA sequence analyses of parts of 19 genes (Hackett et al. 2008). The Latin name of each order is given along with examples of common species. The passeriforme (songbird) order is divided into their suborders to separate vocal learners (oscine songbirds) from vocal nonlearners (suboscine songbirds). Closed and open circles show the minimal ancestral nodes where vocal learning could have either evolved as independent gains or losses, respectively. Independent losses would have required at least one common vocal-learning ancestor. (b) Mammalian tree based on DNA sequence analyses of parts of 18 genes (Murphy et al. 2001), updated with additional genomic and fossil data (Murphy et al. 2007; Spaulding et al. 2009). The relationships among bats, dolphins, and carnivores (cat, dog, and seal) vary among studies. The trees are not intended to present the final dogma of mammalian and avian evolution, as there are some significant differences among studies and scientists.

groups and the speech-song brain regions in humans are all embedded within or adjacent to a motor forebrain pathway that is proposed to be homologous (at least among birds, reptiles, and mammals), indicating a possible neural deep homology in motor learning pathways for evolution of vocal learning (Brown et al. 2006b; Feenders et al. 2008). In this sense, deep homology is similar to the independent evolution of wings from the upper limbs. That is, the brain pathways for vocal learning among distantly related species are apparently not homologous in that they were not inherited from a common ancestor, but the

Animal Models for Human Speech, Song, and Vocal Learning 505

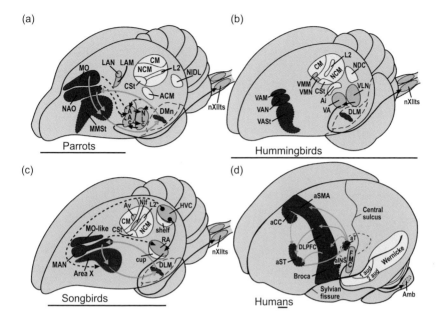

Figure 20.2 Proposed comparable vocal and auditory brain areas among vocal-learning birds (a–c) and humans (d). Left hemispheres are shown, as this is the dominant side for language in humans and for song in some songbirds. Yellow regions and black arrows indicate proposed posterior vocal pathways; red regions and gray arrows indicate proposed anterior vocal pathways; dashed lines depict connections between the two vocal pathways; blue denotes auditory regions. For simplification, not all connections are shown. The globus pallidus in the human brain, also not shown, is presumably part of the anterior pathway as in nonvocal pathways of mammals. Basal ganglia, thalamic, and midbrain (for the human brain) regions are drawn with dashed-line boundaries to indicate that they are deeper in the brain relative to the anatomical structures above them. The anatomical boundaries drawn for the proposed human brain regions involved in vocal and auditory processing should be interpreted conservatively and for heuristic purposes only. Human brain lesions and brain imaging studies do not allow functional anatomical boundaries to be determined with high resolution. Scale bar: ~7 mm. Abbreviations are listed in Appendix 20.1. Figure modified after Jarvis (2004).

motor pathway circuit from which they may have independently emerged may be a homolog, inherited from their common ancestor.

Finally, and surprisingly, it appears that some aspects of birdsong depend on similar genetic and developmental mechanisms, providing an example of genetic deep homology. A well-known example is the role of the FoxP2 gene in human speech and birdsong learning (Fitch 2009a; Scharff and Petri 2011). Natural FOXP2 mutations in humans and suppression of FoxP2 in the striatal song nucleus in songbirds prevent accurate song imitation. These add to the now widely appreciated examples of deep homology for other traits,

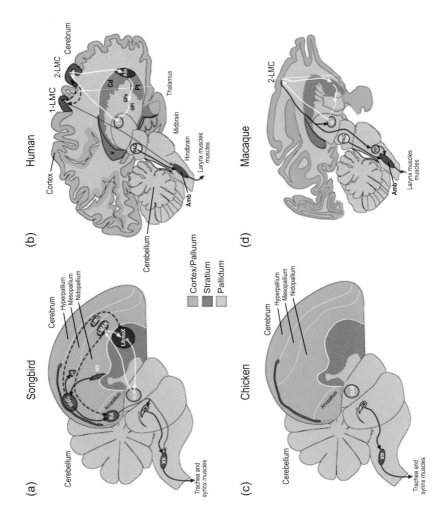

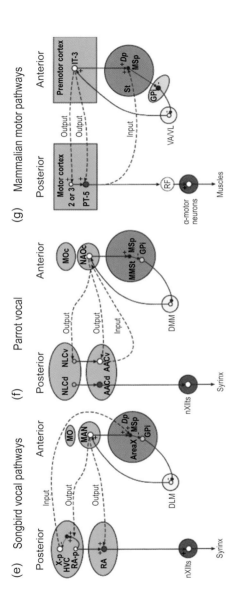

Figure 20.3 Comparative brain relationships, connectivity, and cell types among vocal learners and nonlearners. (a) Drawing of a zebra finch male brain showing song nuclei and some connectivity in 4 of the 6 song nuclei. (b) Drawing of a human brain showing proposed vocal pathway connectivity. (c) Known connectivity of a vocal-nonlearning bird showing absence of forebrain song nuclei. (d) Known connectivity of a vocal-nonlearning primates showing presence of forebrain regions that have an indirect projection to Amb, but have no known role in productive of vocalizations. (e) Songbird detailed cellular connectivity. (f) Parrot detailed cellular connectivity. (g) Mammalian motor pathway connectivity. Dashed lines indicate connections between anterior and posterior pathways; inputs and outputs are labeled relative to anterior pathways. Output from songbird MAN to HVC and RA are not from the same neurons; medial MAN neurons project to HVC, lateral MAN neurons project to RA. Excitatory neurons (°), inhibitory neurons (•), excitatory glutamate neurotransmitter release (+), inhibitory GABA release (–). GPi-like neuron in songbird AreaX and parrot MMSt. Only the direct pathway through the mammalian basal ganglia (St to GPi) is shown as this is the one most similar to AreaX connectivity (Hackett et al. 2008). Figure modified from Jarvis (2004) and Jarvis et al. (2005).

such as the HOX and PAX6 genes for independent evolution of body segments and eyes, respectively (Gehring and Ikeo 1999; Pearson et al. 2005; Fernald 2006; Shubin et al. 2009). They also indicate that deep homology may play an important role in repeated convergent evolution for complex traits, reinforcing the value of birdsong as a model for human vocal learning.

Below, we provide a detailed review of these similarities and differences in birdsong and human behavior and neurobiology. We begin with definitions and a discussion of animal models of vocal learning. We then review the history of birdsong as a model for human speech, focusing on the human and avian comparisons most relevant to students of language evolution. We briefly compare the specialized features of birdsong and speech learning, and provide an up-to-date overview of the comparative neurobiology and genetics of birdsong and human speech. We close by discussing motor theories for the convergent evolution of vocal-learning systems *within* different groups of birds and with humans. Convergent evolution provides a powerful source of ideas to test hypotheses about the evolution of spoken language at functional and mechanistic levels.

Animal Models of Vocal Learning

Defining Vocal Learning

There are multiple types and variable degrees of vocal learning in vertebrates (Janik and Slater 1997; Janik 2000). Most vertebrates are able to control (inhibit) vocal production voluntarily to some extent (Lane 1960; Sutton et al. 1973). Many species show "audience effects" in which the presence or absence of conspecifics affects vocalization (Gyger 1990; Evans and Marler 1994; Slocombe et al. 2010). Thus, in most vertebrates, production of species-typical vocalizations is not simply "reflexive" or automatic. However, the acoustic structure of such vocalizations is mostly innate.

We define innate to mean "reliably developed or canalized in a particular species" (Ariew 1999) and use instinct to denote those behavior patterns that are innate. By this very inclusive definition, "innate" is not dichotomous with "learned," and thus birdsong researchers have typically discussed song acquisition as resting upon an "instinct to learn" (Marler 1991). This is simply to say that, although a behavior may be learned, the learning mechanism itself may be, and indeed to some extent must be, innate. An innate predisposition to learn speech has also been recognized, since Darwin, to be present for spoken-language acquisition in humans (Darwin 1871; Lenneberg 1967; Locke 1995).

Employing these definitions, we can categorize some of the vocalizations produced by every known species as innate. A human baby does not need to learn to cry: the behavior is present at birth and a child requires no previous auditory input, including the hearing of others cry, to generate crying sounds.

Similarly, smiling is an innate facial display (Eibl-Eibesfeldt 1973). Again, to say that these displays are innate does not deny that learning will be involved in the deployment and use of smiling or crying later in life. It only affirms that the basic form and meaning of these signals is part of the reliably developing birthright of every normal human child. We refer to such signals as innate signals. By current information, most of the vocalizations produced by common mammals are innate: dogs bark and growl, cats meow and purr, macaques coo, and squirrel monkeys peep, trill, and twitter without a need for prior auditory input of such sounds (Newman and Symmes 1982; Romand and Ehret 1984; Owren et al. 1993; Arch-Tirado et al. 2000).

Such innate vocalizations can serve as the basis for various types of learning other than vocal learning. For instance, in a type of auditory learning, termed comprehension learning by Janik and Slater (2000), listeners can learn to associate innate vocalizations produced by others with their identity, mood, or intentions. Animals can also learn that producing certain innate vocalizations leads reliably to certain outcomes (e.g., a cat learns to direct meows to a human to elicit a food reward or a dog barks to be let out); this form of social learning is termed usage learning (Janik and Slater 2000). In addition, at least some aspects of the call morphology of an innate vocalization (e.g., its pitch, loudness, or duration) may be varied by the animal depending on context; this is called production learning (Janik and Slater 2000). The latter can be seen as a very simple example of "vocal learning," which takes as a starting point an innately given vocalization of a particular form, with minimal adjustment to the acoustic features of the sounds. This ability was recently shown in mice; under social sexually competitive conditions, males will imitate the pitch of each others' ultrasonic songs (Arriaga et al 2012). Here, we refer to this type of vocal learning, based on small modifications of innate species-typical calls, as limited vocal learning (Arriaga and Jarvis 2013).

Some vocalizations produced by some species are not in the limited vocal-learning subcategory, but actually duplicate complex novel sounds heard in the environment. The most striking examples are "talking" birds, dolphins, or seals (Janik and Slater 1997). When raised with humans, typically from a young age, these species begin to produce imitations of sounds in their environment that are not part of their innate, species-typical repertoire. We define the capacity for this type of vocal learning here as complex vocal learning: the ability to imitate novel sounds from the environment which are not part of an innate repertoire. This capacity is also often termed vocal mimicry, when animals imitate other species or environmental sounds, or simply vocal learning, when they imitate only their own species' repertoire. The latter term, however, might be misinterpreted to suggest that there are no learned modifications to innate calls, which, as the examples above show, is incorrect. The distinction between simple and complex vocal learning made above should help resolve such misinterpretations. When learned vocalizations occur in a sequence, we call the individual sound units syllables. When the learned vocalizations sometimes occur in isolation, we refer to them as calls. In this regard, both calls and syllables can be learned or innate.

Complex Vocal Learning Is Sparsely Distributed among Species

Using the definitions above, the capacity for complex vocal learning is very sparsely distributed throughout the animal kingdom (Figure 20.1). By far the most numerous in terms of species are song-learning birds (Kroodsma and Baylis 1982; Jarvis 2004). These include oscine songbirds of the suborder oscine passerines; parrots of the order Psittaciformes, also well-known vocal mimics; and hummingbirds of the suborder Trochili of order Apodiformes, which are the least studied.

A number of marine mammals are known to be capable of complex vocal learning, though in many cases the evidence is sparse enough to leave room for doubt (Schusterman 2008). These include cetaceans (whales and dolphins). Baleen whales, particularly humpback whales (*Megaptera novaeangliae*), sing long, complex songs that are shared in local regional "dialects" and which change over time; baleen whales are thus considered excellent vocal learners (Payne 2000). As shown by experiments in captivity, toothed whales such as dolphins, killer whales and belugas have the capacity to imitate novel sounds from their environment, have dialects and song sharing, and can imitate speech, respectively (Eaton 1979; Richards et al. 1984; Ford 1991; Ridgway et al. 2012). The marine group also include pinnipeds (seals, sea lions and walruses), which show abundant indications of the capacity for vocal learning (Schusterman 2008), including harbor seals (*Phoca vitulina*) which in two cases imitated human speech (Ralls et al. 1985). Based on the existence of local dialects (Van Parijs 2003), most phocids ("true" or "earless" seals) may be vocal learners, but vocal learning in this group remains little studied.

The most recent evidence for complex vocal learning in other mammalian species comes from bats, particularly the sac-winged bat *Saccopteryx bilineata* (Knörnschild et al. 2010), and elephants (Poole et al. 2005). There is good evidence for complex vocal learning in Asian elephants based on the production of whistling in human-raised individuals (Wemmer et al. 1985) and an elephant in Korea that imitates human speech (Stoeger et al. 2012).

Among primates, the capacity for complex vocal learning appears to be limited to humans (Janik and Slater 1997; Egnor and Hauser 2004; Fitch 2000, 2006b). Repeated experiments in apes raised among humans show a nearly complete lack of a capacity to learn to produce human speech (Yerkes and Yerkes 1929; Hayes and Hayes 1951). Recent suggestions of call usage learning and simple production vocal learning in chimpanzees, macaques, and marmosets (Crockford et al. 2004) do not change this conclusion, since the usage learning does not involve acoustic changes to the calls and the simple production vocal learning involves pitch modifications of species-typical calls or nonvocal but acoustic facial gestures (Reynolds Losin et al. 2008). One published suggestion of complex vocal learning in macaques (Masataka and Fujita 1989) failed to be replicated (Owren et al. 1993). Thus, according to present knowledge, humans are the only primates capable of complex vocal learning. This indicates that the capacity for complex vocal learning in our species is a recent evolutionary acquisition, which appeared sometime in the last six million years or so, since our evolutionary divergence from chimpanzees (Fitch et al. 2010).

Complex Vocal Learning Is Crucial for Speech and Song Acquisition and Production

Unlike crying or laughter, the specific vocal sound sequences produced during speech or song are learned from the child's environment. This is not to deny some innate constraints on the phonetics of speech or the range of sounds used in song. The existence of the International Phonetic Alphabet and musical notation capable of transcribing much of the world's music demonstrates that there are finite limits on the sounds used in speech and song (Jakobson et al. 1957). However, there is no truth to the widespread urban legend, often unfairly attributed to Jakobson (1968), that a babbling baby produces all the phonemes of all the world's languages: Many detailed studies show that infants produce a rather limited and stereotypic selection of phonemes during babbling (Vihman 1986; Oller and Eilers 1988; Locke 1995; MacNeilage 1998). Similarly, although the capacity to produce pitched sounds is clearly innate in humans, the selection of pitches that are considered musically appropriate is highly culture dependent and differs significantly among the world's musical traditions (Ellis 1885).

Even more important, the sequences into which individual syllables or notes are arranged into words or melodies in speech and music are extremely variable among cultures. Therefore, every normal child must learn the words and melodies typical of its local culture. This learning is necessary for both music and language to be indefinitely extensible systems, making "infinite use of finite means," and locally intelligible and shared within a culture. Indeed, the "infinitely productive" and "culturally shared" aspects of language have been offered as defining the "design features" of language by Hockett (1966) and of music by Fitch (2005). In fact, a strong distinction between learning and production of human speech and song breaks down from a neural or comparative viewpoint, because both behaviors may be different ways of expressing the same specialized evolved mechanisms for vocal learning. We hypothesize this to be the case, whether discussing neural mechanisms for production of song in humans with or without words. That is, complex vocal learning in humans is also a necessity for the culture-specific flexibility required to produce human speech and song.

Calls versus Songs

As mentioned above, calls, syllables, song, and speech or spoken language provide a continuum or spectrum of vocalization types. Most birds have a clear distinction in their vocal repertoire between *calls*, which are typically short, discrete vocalizations that are often but not always innate, and *songs*, which are more complex vocalizations that are rhythmic, often tonal, graduating from one syllable form into another, and are typically learned (Catchpole and Slater 2008). The cluck of a chicken or the quack of a duck would be classified as calls, as would the whistled alarm call of a blackbird or American robin. However, the long, complex learned vocalizations by either of the latter two species are termed, both colloquially and by scientists, song. One should not assume that if a call is produced in a vocal-learning species that it is innate. Male zebra finches produce learned calls (Simpson and Vicario 1990), and crows (a songbird) as well as budgerigars (a parrot) make many learned calls (Marler 2004; Tu et al. 2011). Whereas for songbirds this distinction is often relatively clear, there are many difficult cases in other bird clades: Should a rooster's crowing be termed "song"? What about the *unlearned* territorial or courtship vocalizations sung by males of many suboscine bird species?

We use the term song to refer only to complex learned vocalization, of the type exemplified by birdsong and human singing (cf. Fitch 2006b), and innate song to refer to limited learned or nonlearned vocalizations of the type exemplified by suboscine songbirds and some nonhuman primates (e.g., gibbon "song"). The latter also includes laughter or crying (which is part of the innate vocal repertoire of humans). According to these definitions, the main difference between human "song" and "speech" is simply that speech conveys complex propositional meanings, based on a concatenation of meaningful lexical items (words), whereas song without lyrics often does not. From the point of view of vocal control and vocal learning, these are secondary differences. Thus the musical distinctions between songs with lyrics (marrying music and language), scat singing (musically employing complex semantically meaningless syllables like "shoo be doo be"), and simple humming or chant with a fixed vowel are all variations of a broader behavior in the current discussion; that is, they are all simply "song" by our definition.

Birdsong as a Model System for Human Speech

History

The basic facts that complex vocal imitation is present in human speech and songbird song have long been recognized, as noted by Aristotle in 350 BC (Book IV, part 9 "Voice," p.111):

Of little birds, some sing a different note from the parent birds, if they have been removed from the nest and have heard other birds singing; and a mother-nightingale has been observed to give lessons in singing to a young bird, from which spectacle we might obviously infer that the song of the bird was not equally congenital with mere voice, but was something capable of modification and of improvement.

Darwin (1871:43) was also aware of these facts:

> Birds imitate the songs of their parents, and sometimes of other birds; and parrots are notorious imitators of any sound which they often hear.

He went further to suggest that birds provide an excellent model for the human instinct to learn speech (Darwin 1871:53):

> The sounds uttered by birds offer in several respects the nearest analogy to language, for all the members of the same species utter the same instinctive cries expressive of their emotions; and all the kinds which sing, exert their power instinctively; but the actual song, and even the call-notes, are learnt from their parents or foster-parents.

The modern study of birdsong learning began with the observation that birds raised in the laboratory will, in many cases, learn the songs from this environment, even if they are produced by singers of other species (Thorpe 1956, 1958). With the advent of tape recorders, this became a powerful tool to investigate the mechanisms underlying vocal learning, and indeed learning in general (Marler and Peters 1977; Nottebohm 1968, 2006)—one whose potential has not even come close to being exhausted.

Vocal-Learning Birdsong Species

The vocal-learning bird species comprise nearly half of all ~10,400 living bird species (Figure 20.1a): oscine songbirds (~4000 species), parrots (~372), and hummingbirds (~338) (Jarvis 2004; Marler and Slabbekoorn 2004; Osburn 2004). Many familiar birds, including close relatives of the vocal learners, do not appear to learn their songs, including suboscine songbirds (close relative of songbirds), falcons (possible close relative of parrots), pigeons (close relative of hummingbirds), as well as ducks, chickens, owls, seagulls, and hawks (Nottebohm and Nottebohm 1971; Kroodsma and Baylis 1982; Kroodsma and Konishi 1991). Many other clades have, however, been relatively unstudied, and it possible that there are a few more vocal learners out there to be discovered. The same can be said about mammals (Figure 20.1b). There are also sex differences. In songbird species that live in temperate climates away from the equator, learned songs are primarily sung by males during the breeding season. However, the vast majority of songbird species live nearer to the equator in tropical zones. In these species, both males and females often learn how to sing (Morton 1975).

Variation in Complexity

There are considerable differences in the complexity of songs in different bird species. Among songbirds, the zebra finch (*Taeniopygia guttata*) has become the most common laboratory model species for song learning, where the male learns how to sing and the female does not. Other species that have been intensively studied include Bengalese finches (*Lonchura striata*), nightingales (*Luscinia megarhynchos*), and canaries (*Serinus canarius*); in the latter, both males and females sing, although males sing more frequently (Marler and Slabbekoorn 2004; Zeigler and Marler 2004). Among parrots, most laboratory work has been done on budgerigars (*Melopsittacus undulatus*), and behavioral work has included African Gray parrots (Jarvis et al. 2000; Pepperberg 1999). These "model species," of course, represent only a tiny proportion of known vocal-learning species.

Zebra finches, chaffinches, and chickadees all have relatively stereotyped and repetitive songs, which made them well suited for acoustic analysis. However, there are many examples of birds with complex songs or calls (e.g., canaries, starlings, nightingales, mockingbirds, budgerigars, and many other parrots), including hierarchically structured and less predictable songs (Farabaugh et al. 1992; Catchpole and Slater 2008; Tu et al. 2011). The same is true for different species of hummingbirds (Ferreira et al. 2006). For some of these species, the complexity poses significant problems in terms of data analyses. For instance, the brown thrasher (a North American mimic thrush, closely related to the mockingbird) has so many different song types that even with extended recording, new syllables continue to appear (Kroodsma and Parker 1977). The same is true for budgerigar warble song (Farabaugh et al. 1992). To date, no one has been able to record the upper limit for different syllable types and songs of these species. This means that even the simple question of song repertoire size needs to be estimated asymptotically, exceeding 1500 different song types for the thrasher (Kroodsma and Parker 1977; Boughey and Thompson 1981). Literally tens of thousands, if not millions, of spectrograms may need to be investigated in such species to allow for a quantitative, token-type of analysis. This, of course, means that we know much more about the structure of song in simple singers with one to three song types, such as the zebra finch, than in complex singers with thousands.

"Syntax," "Semantics," and "Phonology" in Birdsong

Given the broad analogy between speech and birdsong, it is tempting to try to adopt finer-grained distinctions from linguistics to the study of birdsong (Marler and Peters 1988; Yip 2006). In particular, the term *syntax* is often used by birdsong researchers to refer to any set of rules that structures the arrangement of elements (Marler and Peters 1988; Honda and Okanoya 1999). For example, we can discuss the syntax of a programming language, where a colon or semicolon may be allowed or required in certain places, and this is independent of the question of the meaning or lack thereof of the phrase. Thus, any system that determines whether a particular string or utterance is "well formed" can be termed *syntax*, irrespective of any issues of meaning. It seems unlikely that there is an equivalent in birdsong of the way in which syntax grounds the unrestricted compositionality of semantics in human language. However, changes do exist in the "meaning" with variation of individual song phrases. For example, it was long thought that black-capped chickadee calls were innate vocalizations without meaning, until it was discovered that their alarm calls are made up of song syllables, that different sequences of syllables designate different types of predators, and the number of specific syllables designate relative predator size (Templeton et al. 2005). Similarly, it was once thought that the socially directed and undirected songs of zebra finches, which have small differences in pitch variability not easily noticed by humans, if at all, were not so important, but they are recognized as important by listening females (Woolley and Doupe 2008) and have dramatic differences in brain activation (Jarvis et al. 1998; Hessler and Doupe 1999). For the most part, the more complex the syntax, the more attractive the song to the opposite sex; this "meaning," however, is not compositional in the way that language is. Thus, the traditional term *song* seems more appropriate than *bird language* for most learned song.

Determining the distinction between syntax and phonology is more challenging. In language, the set of rules underlying the arrangement of meaningless units, such as phonemes and syllables, into larger wholes is termed *phonology*. The arrangement of morphemes (minimally meaningful units) into more complex multipart words is termed *morphology*, whereas the arrangement of meaningful words into hierarchically structured phrases and sentences is termed *syntax*. It is now clear, however, that the distinction between morphology and syntax varies considerably between languages: some languages use morphology to serve the same expressive purpose that syntax serves in other languages (Carstairs-McCarthy 2010). Thus, today, many linguists lump together, under the term morphosyntax, all arrangements of meaningful units into larger, still meaningful structures (e.g., Payne 1997). It remains unknown if birdsong has a direct equivalent of morphosyntax with meaning. At a minimum, a low level of morphosyntax seems possible in African Gray parrots, as seen in the studies of Alex (Pepperberg 1999; Pepperberg and Shive 2001): Alex was able to combine individually learned spoken words into new meanings, at a rudimentary level similar to that of a two- to three-year-old human child.

Marler and Peters (1988) used the term "phonology" to describe within-syllable structure in birdsong, essentially syllable identity, whereas they used "syntax" to denote the arrangement of syllables *into whole songs*. Because song syllables in many species are surrounded by small silences, during which the bird takes a "mini-breath" (Calder 1970), it is quite easy to demarcate objectively individual syllables in most birdsongs. Marler and Peters used learning of conspecific and heterospecific song to show that song sparrows (*Melospiza melodia*) choose which songs to attend to and imitate based both on phonology (syllable types) and syntax (trilled two-segment songs).

In evaluating the appropriateness of the term "phonology" in birdsong, phonologist Moira Yip concludes that most aspects of human phonology have never been investigated in birds (Yip 2006). Thus it seems appropriate, at present, to recognize the phonology/syntax distinction in birdsong, as used by Marler and Peters, while recognizing that fundamental differences in the interpretation of these terms may exist between humans and birds. The same might be said of human speech and song, where syntax and semantics have definitions idiosyncratic to musical structure (Koelsch et al. 2004).

The Development and Adaptive Function of Birdsong

With over 4000 oscine passerine species, it is difficult to draw valid generalizations from one or two songbird species; for virtually every rule, one might identify exceptions. By comparison, there are only ~5490 species of mammals (Osburn 2004). Below we expand on some of this diversity in song behavior for songbirds.

A Prototypical Songbird

For clarity we begin with a fictitious "typical" temperate songbird (loosely modeled on a song sparrow; Marler and Slabbekoorn 2004) and use this to explore some of the diversity of songbirds. In such species, only males sing, and there is a "critical period" for song learning: the young male must be exposed to conspecific song during a distinct and limited period if he is to sing properly as an adult. In many species, what the male learns during this critical period then "crystallizes," becoming largely fixed and invariant in the adult. Since in temperate regions, singing occurs mostly during the spring breeding season, for some species a young male first hears conspecific song while still a nestling or fledgling (typically from his own father and neighboring males) and will produce little song himself, until almost a year later, just before the next breeding season. Thus there can be a long time lag between the memorization of the auditory "template" that he will copy and the motor behavior of plastic and variable song that precedes crystallization during which his imitation is perfected.

Experiments in which birds are deafened at various ages show that hearing is necessary during both the initial exposure (obviously, since the songs are learned) and later during sensorimotor learning, when the bird practices singing (Konishi 1964). Furthermore, audition is needed to produce proper babbling-like subsong, indicating that auditory feedback is required to develop the proper motor control (Marler and Peters 1982). This fact led to the proposal that birdsong represents a cross-modal fusion of auditory and motor signals and the suggestion that song perception is in some way dependent on neural activity in motor pathways (Katz and Gurney 1981), an idea that contributed to the "motor theory of song perception" for birdsong (Williams and Nottebohm 1985). Once the bird's song has "crystallized," however, deafening has much less effect. Nonetheless, even during adulthood, auditory feedback plays a role in the long-term maintenance of proper adult song, as deafening in adults leads to long-term deterioration of song acoustic structure and syntax (Leonardo and Konishi 1999). All of these properties are similar to what happens in humans (Waldstein 1990), but not in nonhuman primates (Egnor and Hauser 2004), although humans can use somatosensory feedback to recover some speech after being deaf (Nasir and Ostry 2008).

With this mental prototype in hand, we can describe some of the important variance among bird species in their song behavior. Species can be found which differ from essentially every aspect of the "model temperate songbird" just described. These differences make possible one of the most valuable aspects of birdsong as a model system, because we can use the variation among species as a probe to understand the mechanisms, developmental basis, and evolutionary function of different aspects of song.

Open-Ended and Closed-Ended Learners

One important, and common, difference from our prototype is that many songbirds continue to learn new song material as adults (Nottebohm 1981). For example, canaries typically add seasonally new song syllables to their repertoire every year, with the greatest addition occurring in the fall, before the breeding season. This is one of the most fascinating examples where the study of birdsong has led the way in neuroscience, in this case leading to the rediscovery of adult neurogenesis (i.e., the birth of new, functional neurons in the adult brain; Nottebohm 2006). It was later discovered that in both birds and mammals, the levels of neurogenesis correlates more with use of the brain pathway than learning plasticity, and that both closed-ended and open-ended vocal learners have continued neurogenesis throughout life (van Praag et al. 1999; Li et al. 2000). Although canaries are probably the best-studied of the open-ended learning species, many other songbirds continue to learn throughout life; starlings are a prominent example (Hausberger et al. 1995; Eens 1997). Open-ended learning abilities also appear to typify parrots, who can learn new contact calls and, for some, human speech as adults (e.g., Farabaugh et al. 1994; Pepperberg and Shive 2001). In reality, although open- and closed-ended learning is discussed as a dichotomous variation in vocal-learning behavior, differences between species appear to be more continuous.

No clear explanation has yet been discovered as to what determines the differences of species on the continuum from closed- to open-ended complex vocal learning. For example, there is no difference in the presence or absence of song nuclei, no gross differences in song nuclei connectivity in zebra finches and canaries, and no clear differences in brain size (DeVoogd et al. 1993; Vates et al. 1997). There is some evidence, however, that the relative sizes of the song nuclei (particularly HVC) to brain size or the relative sizes of song nuclei to each other are correlated with repertoire size, and that increased repertoire size is correlated with opened-ended abilities (DeVoogd et al. 1993; Zeng et al. 2007). These findings have been debated, particularly within species (Gil et al. 2006; Leitner and Catchpole 2004; Garamszegi and Eens 2004). Regardless of the mechanism, larger repertoire sizes and song syntax variation appears to be more attractive to the listening female, and thus enhance chances of mating and passing genes to the next generation.

Other Variation in Song Behavior: Female Song and Duetting

It has become increasingly clear that the prototype sketched above is biased by the fact that most of the well-studied songbird species live in temperate regions, where many scientists live and where seasonality and male-only song are the general rule. Exceptions include female cardinals in North America, who defend winter territories with song (Ritchison 1986) and learn song three times faster than males (Yamaguchi 2001). In nonmigratory tropical birds, particularly those in dense forest, females often sing with males in *duets*, or closely coordinated two-voice songs produced by a mated pair (Langmore 1998; Riebel 2003). In a few species, females sing on their own during the mating season, and this singing seems to serve as courtship for males (e.g., alpine accentors; Langmore 1996). Starlings are another well-studied species where female song is typical (cf. Hausberger et al. 1995). Morton (1975) has hypothesized that vocal learning in songbirds first evolved in both sexes in equatorial zones, much like what is thought to have occurred in humans; then, as populations speciated and moved away to temperate zones, the more unstable environments selected for a division of labor and thus caused the loss of song by females. In those species where females may have lost the trait, females still select males based on the complexity of their songs.

Dialects in Birdsong

Whenever there is complex vocal learning in a species, there is a strong potential to evolve dialects, even very different repertoires. These dialects are usually geographically defined and culturally transmitted locally, whether they occur in songbirds, whales, or humans. For humans, when the dialects become so different that they are not understood by different populations, the different dialects are called different languages. An analogous distinction is made between different song repertoires for song-learning birds. This type of difference was already noted by Darwin (1871:54):

Animal Models for Human Speech, Song, and Vocal Learning

Nestlings which have learnt the song of a distinct species, as with the canary-birds educated in the Tyrol, teach and transmit their new song to their offspring. The slight natural differences of song in the same species inhabiting different districts may be appositely compared, as Barrington remarks, "to provincial dialects"; and the songs of allied, though distinct species may be compared with the languages of distinct races of man.

Since Darwin's time, birdsong dialects have become one of the best-studied examples of "animal culture" (Galef 1992; Laland and Janik 2006; Fehér et al. 2008), and the study of these dialects has become a scientific field in itself (cf. Baker and Cunningham 1985). One mechanism whereby dialects are thought to form is somewhat like gene evolution: cultural transmission of learned acoustic signals undergoes recombination and drift. Compared to gene evolution, cultural evolution can evolve much more rapidly, and thus complex vocal learners tend to have larger differences in their vocal repertoires among different populations than vocal nonlearners whose vocalizations are largely genetically determined. There is also some evidence that local dialects may be adapted to their local habitat (Nottebohm 1975), which has been proposed to allow vocal learners to potentially adapt their vocalizations more rapidly in different environments (Jarvis 2006).

Comparative Neural and Genetic Mechanisms of Birdsong and Human Speech

Avian and Mammalian Brain Organization

To compare the neural and genetic basis of birdsong with human spoken language, one must first understand the similarities and differences in their brains. The classical century-old view of bird brain evolution was based upon a theory of linear and progressive evolution, where the vertebrate cerebrum was argued to have evolved in lower to higher anatomical stages with birds at an intermediate stage. As such, the avian cerebrum was thought to consist primarily of basal ganglia territories, and these were thought to control mostly primitive behaviors. This view, although slowly challenged, was not formally changed until 2004–2005, when a forum of neuroscientists reevaluated brain terminologies and homologies for birds and other vertebrates and published a new nomenclature to reflect a modern understanding of vertebrate brain evolution (Reiner et al. 2004; Jarvis et al. 2005). This view proposes that the avian and mammalian cerebrums were inherited as a package with pallial (cortical-like), striatal, and pallidal regions from their stem amniote ancestor. The neuron populations of the avian (i.e., reptilian) pallial domain evolved to have a nuclear organization, whereas the mammalian evolved to become layered (Figure 20.3). However, the different avian nuclear groups have cell types and connectivity similar to the different cell layers of mammalian cortex which, in both vertebrate groups, perform similar functions. The striatal and pallidal domains are organized more similarly between birds and mammals, and likewise perform similar functions in both groups. Using this revised view, we can now more accurately compare the brains of avian and mammalian complex vocal learners and nonlearners.

Comparative Birdsong and Human Vocal Communication Pathways

Overview of the Songbird Vocal-Learning Pathway

Only complex vocal learners (songbirds, parrots, hummingbirds, and humans) have been found to contain brain regions in their telencephalon (forebrain) that control production of vocal behavior (Jarvis et al. 2000; Jürgens 2002). Vocal control brain regions have not yet been investigated in other mammalian complex vocal learners (cetaceans, bats, elephants, or pinnipeds). Vocal nonlearners (including chickens and cats) have only midbrain, medulla, and possibly thalamic (in cat) regions that control the acoustic structure and sequencing of innate vocalizations (Wild 1994; Farley 1997). Nonhuman primates, however, have a laryngeal premotor cortex region that is connected to other forebrain areas, which makes an indirect connection to brainstem vocal motor neurons; this region, however, is not required to produce species-specific vocalizations (Kirzinger and Jürgens 1982; Simonyan and Jürgens 2003, 2005a; Jürgens 2002). By comparing the vocal brain regions of different vocal-learning and vocal-nonlearning species, it has been possible to generate a consensus vocal-learning pathway.

Vocal-learning bird groups have seven cerebral song nuclei (Figure 20.2a–c): four posterior forebrain nuclei and three anterior forebrain nuclei (Jarvis et al. 2000). The posterior nuclei form a posterior song pathway that projects from HVC to RA, to the midbrain (DM) and medulla (12th) vocal motor neurons (Figure 20.2, black arrows) (Durand et al. 1997; Vates et al. 1997). The avian 12th motor neurons project to the muscles of the syrinx, the avian vocal organ. Vocal-nonlearning birds have DM and 12th vocal motor neurons for production of innate vocalizations, but no projections to them from the arcopallium have been found (Wild 1994; Wild et al. 1997). The anterior nuclei are part of an anterior vocal pathway loop, where a pallial song nucleus (MAN-analog) projects to the striatal song nucleus (AreaX-analog), the striatal song nucleus projects to the dorsal thalamic nucleus (aDLM), and the dorsal thalamus projects back to the pallial song nucleus (MAN-analog; Figure 20.2, gray arrows) (Durand et al. 1997; Vates et al. 1997). The pathway receives auditory input into HVC (Bauer et al. 2008).

The posterior pathway nuclei, especially HVC and RA, are required to produce learned song (Nottebohm et al. 1976). HVC is hypothesized to generate sequencing of song syllables and RA the acoustic structure of the syllables (Hahnloser et al. 2002). For these reasons the posterior pathway

Animal Models for Human Speech, Song, and Vocal Learning   521

is also called the vocal motor pathway. In contrast, the anterior song nuclei are not necessary for producing song, but are necessary for learning songs or making modifications to already learned songs (Scharff and Nottebohm 1991). Lateral MAN (LMAN) and its thalamic input, aDLM, is necessary for introducing variability into song, whereas Area X in the striatum (AreaX) is necessary for keeping song more stereotyped, when LMAN is intact (Scharff and Nottebohm 1991; Kao et al. 2005; Olveczky et al. 2005; Goldberg and Fee 2011). The anterior pathway is thought to make these changes to song by its output to the motor pathway, from LMAN to RA and medial MAN (MMAN) to HVC (Figure 20.2). In Table 20.1 we review the similarities and differences in these song pathways of vocal birds with humans and other mammals at four levels of organization: connectivity, brain function through lesions, brain activation, and genes.

Connectivity

The connectivity of avian posterior song pathways is similar to mammalian motor corticospinal pathways (Figure 20.3g). Specifically, the projection neurons of songbird HVC to RA are similar to layer 2 and 3 neurons of mammalian cortex, which send intrapallial projections to mammalian cortex layer 5 (Figure 20.3e–g) (Aroniadou and Keller 1993; Reiner et al. 2003; Jarvis 2004). The projection neurons of RA are similar to pyramidal tract (PT) neurons of lower layer 5 of mammalian motor cortex, both of which send long axonal projections out of the cortex through pyramidal tracts to synapse

Table 20.1 Comparable brain areas of vocal learners.

Modality	Vocal		Auditory	
	Songbird	Human	Birds	Human (layer)
Subdivision:				
Nidopallium	HVC	FMC – 2,3	L2	1° aud (4)
	NIf	FMC – 2,3	L1, L3,	1° aud (2, 3)
	MAN	Broca's area – 2, 3	NCM	2° aud (2, 3)
Mesopallium	Av	FMC – ?	CM	2° aud (?)
	MO	Broca – ?		
Arcopallium	RA	FMC – 5	AI	2° aud (5)
Striatum	AreaX	Cd head	CSt	CSt
Thalamus	aDLM	VL	OV	MG
	Uva			
Midbrain	DM	PAG	MLd	IC

? = uncertain relationship; layer = layered cell population on the mammalian cortex. See Appendix 20.1 for abbreviations.

onto brainstem and spinal cord premotor or motor neurons that control muscle contractions (Karten and Shimizu 1989; Keizer and Kuypers 1989; Reiner et al. 2003; Jarvis 2004). The direct projection of songbird RA to the 12th vocal motor neurons is similar to the only physical long-distance connection determined among cerebral vocal brain areas of humans, which is a projection from the human face motor cortex (FMC) directly to the brainstem vocal motor neurons, nucleus ambiguous (Amb) (Kuypers 1958). Amb projects to the muscles of the larynx, the main mammalian vocal organ (Zhang et al. 1995), and is thus the mammalian parallel of avian 12th vocal motor neurons. The monkey FMC does not project to Amb, but it does project to the hypoglossal nucleus and to other brainstem cranial motor nuclei, as found in humans. Birds also have an Amb that projects to a small larynx and which also receives a direct projection from RA in songbirds, but it is not known if this pathway controls vocalizations in birds. The hypoglossal nucleus in mammals and the non-tracheosyringeal part of the 12th nucleus in birds controls tongue muscles (Wild 1994, 1997). In this manner, the direct projection from a vocal/song motor cortex to brainstem vocal motor neurons that control the syrinx and larynx has been argued to be one of the fundamental changes that led to the evolution of learned song in birds and spoken language in humans (Fischer and Hammerschmidt 2011; Fitch et al. 2010; Jarvis 2004; Jürgens 2002; Kuypers 1958; Simonyan and Jürgens 2003; Wild 1994, 1997). However, recent experiments in mice have demonstrated that they do have a vocally active motor cortex region with layer 5 neurons that make a very sparse direction projection to Amb (Arriaga et al. 2012). This and related findings have led to a continuum hypothesis of vocal learning and associated circuits, where the presence of the direct projection for vocal-learning behavior may not be an all or none property, but a continuous property with more complex vocal learners, such as humans and the known song-learning birds, having a denser projection than the limited vocal learners, like many other species (Arriaga and Jarvis 2013).

The avian anterior vocal pathways are similar in connectivity to mammalian cortical-basal ganglia-thalamic-cortical loops (Figure 20.3) (Durand et al. 1997; Jarvis et al. 1998; Perkel and Farries 2000). Specifically, the projection neurons of the MAN-analog is similar to intratelencephalic (IT) neurons of layer 3 and upper layer 5 of mammalian premotor cortex, which send two collateral projections: one to medium spiny neurons of the striatum ventral to it, and one to other cortical regions, including motor cortex (Figure 20.3) (Jarvis 2004; Vates and Nottebohm 1995; Durand et al. 1997; Reiner et al. 2003). The avian AreaX in the striatum has a pallidal cell type which, like in the mammalian internal globus pallidus (GPi), projects to the dorsal thalamus (aDLM). The aDLM then projects back to LMAN, closing parallel loops (Figure 20.3a, e). Likewise, in mammals, the GPi projects to the ventral lateral (VL) and ventral anterior (VA) nuclei of the dorsal thalamus, which in turn projects back to layer 3 neurons of the same premotor areas that projected to the striatum, closing parallel loops (Figure 20.3g (Jacobson and Trojanowski

1975; Luo et al. 2001). Thus, vocal-learning pathways appears to follow a general design of motor learning and production pathways.

Brain Lesion Disorders of Vocal Learners

For the posterior pathway, bilateral lesions to songbird HVC and RA cause deficits that are most similar to those found after damage to the human FMC, this being muteness for learned vocalizations such as song or speech (Nottebohm et al. 1976; Valenstein 1975; Jürgens et al. 1982; Jürgens 2002; Simpson and Vicario 1990). There is also a dominant hemisphere for such an effect: the left side in canaries and humans, the right side in zebra finches (Nottebohm 1977; Williams et al. 1992). Innate sounds, such as contact and alarm calls in birds or crying, screaming, and groaning in humans, can still be produced. In contrast, lesions to the arcopallium—where song nuclei would be expected to be located if they were to exist in vocal-nonlearning birds, or to the FMC in chimpanzees and other nonhuman primates—do not affect their ability to produce vocalizations nor apparently the acoustic structure (Kirzinger and Jürgens 1982; Kuypers 1958; Jürgens et al. 1982). In noncerebral areas, lesions to avian DM of the midbrain and the presumed homologous vocal part of the mammalian PAG result in muteness in both learned and innate vocalizations; the same is true for avian 12th and mammalian Amb vocal motor neurons in the medulla (Nottebohm et al. 1976; Jürgens 1994; Seller 1981).

For the anterior song pathway, Jarvis proposed that lesions to songbird MAN cause deficits that are most similar to those found after damage to parts of the human anterior cortex, including the anterior insula, Broca's area, DLPFC, and pre-SMA: disruption of imitation and/or inducing sequencing problems, but not the ability to produce already well-learned song or speech (reviewed in Jarvis 2004). In humans, these deficits are called verbal aphasias and verbal amusias (Benson and Ardila 1996). The deficits in humans relative to songbirds, however, can be subdivided in a more complex fashion, which is unsurprising given that the original vocalizations in humans are more complex in acoustic structure and sequencing. Specifically, lesions to songbird LMAN (Bottjer et al. 1984; Kao et al. 2005; Scharff and Nottebohm 1991) or MMAN (Foster and Bottjer 2001) and lesions to the human insula and Broca's area (Benson and Ardila 1996; Dronkers 1996; Mohr 1976) lead either to a poor imitation with sparing (or even induce more stereotyped song or speech) or disruption of syntax production (as defined above for birds) or the construction of phonemes into words and words into sentences for humans (Benson and Ardila 1996). In addition, lesions to DLPFC result in uncontrolled echolalic imitation, whereas lesions to the adjacent pre-SMA and anterior cingulate result in spontaneous speech arrest, lack of spontaneous speech, and/or loss of emotional tone in speech; imitation, however, is preserved (Barris et al. 1953; Rubens 1975; Valenstein 1975; Jonas 1981). Within the basal ganglia, just as lesions to songbird AreaX cause song variability and stuttering (Kobayashi et al. 2001; Scharff and Nottebohm 1991), lesions to the human anterior striatum can lead to verbal dyspraxic aphasias and stuttering (Mohr 1976; Benson and Ardila 1996; Lieberman 2000; Cummings 1993). For the globus pallidus, well-defined lesions have been surgically generated in Parkinson's patients to alleviate motor symptoms; afterward, these patients show lexical verbal deficits (spontaneous generation of words), which suggests a problem in selecting a spoken language-specific task (Troster et al. 2003). Within the thalamus, in songbirds, lesions to aDLM lead to an immediate increase in song stereotypy similar to LMAN lesions (Goldberg and Fee 2011), whereas in humans, thalamic lesions can lead to temporary muteness followed by aphasia deficits that are sometimes greater than after lesions to the anterior striatum or premotor cortical areas (Beiser et al. 1997). Given these many parallels, we can consider such experimentally induced deficits of the anterior song pathway in songbirds to represent "song aphasias."

Brain Activation

Brain activation studies support the parallels revealed by lesion and connectivity studies among vocal-learning birds and with humans. For these studies, neural activity recordings and MRI have been used in both humans and song-learning birds. In addition, the study of activity-induced immediate early genes (IEG), such as egr1 and c-fos, has been used in birds. Because the mRNA expression of these genes is sensitive to neural activation, they are able to provide a recent history of region-specific activation. In vocal-learning birds, all seven cerebral song nuclei display singing-driven expression of IEGs (Figure 20.4a, singing) (Jarvis and Nottebohm 1997; Jarvis et al. 1998, 2000; Jarvis and Mello 2000). The singing-driven expression is motor driven, in that it does not require auditory or somatosensory input (Jarvis and Nottebohm 1997), and is associated with premotor neural firing in hearing intact and deaf animals (tested in four of the song nuclei) (Yu and Margoliash 1996; Hessler and Doupe 1999; Fee et al. 2004).

Human brain imaging work often relies on the BOLD signal measured by fMRI, which indicates regional changes in blood flow associated with neural activation, in a manner akin to immediate early gene expression. In humans, the brain areas that are most comparable to songbird HVC and RA, which are always activated with speaking and singing (using PET and fMRI), are in or near the FMC, particularly the larynx representation (Figure 20.4c, speaking) (Brown et al. 2004, 2007; Gracco et al. 2005). Similar, although not identical, to songbird anterior pathway nuclei, other human vocal brain areas appear to be activated depending upon the context in which speech or song are produced. Production of verbs and complex sentences can be accompanied by activation in all or a subregion of the strip of cortex anterior to the FMC: the anterior insula, Broca's area, DLPFC, pre-SMA, and anterior cingulate (Crosson et al. 1999; Gracco et al. 2005; Papathanassiou et al. 2000; Poeppel

Animal Models for Human Speech, Song, and Vocal Learning   525

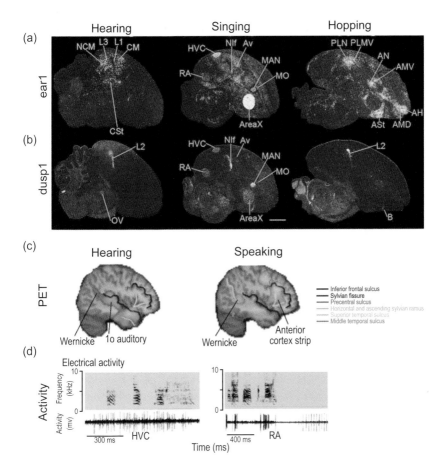

Figure 20.4 Hearing and vocalizing-driven brain activation patterns in songbirds and humans. (a) Brain expression patterns of the activity-dependent gene egr1 (white) in zebra finch males when they hear song (in the dark, standing still), sing, or hop (in a rotating wheel, when deaf, and in the dark). These are dark-field emulsion-dipped sagittal sections reacted by *in situ* hybridization. Note that the hopping-induced regions are adjacent to song nuclei. (b) The same brain sections reacted with a probe to the dusp1 activity-dependent gene. Note the lack of hopping-induced dusp1 around song nuclei, but still the presence of singing-induced dusp1 in song nuclei. [(a) and (b) modified from Horita et al. (2012).] (c) PET signals superimposed on sagittal slices showing auditory and anterior strip of activation, including FMC during speaking. The colored region (orange to yellow) is where higher activation occurs minus control conditions. Modified from Papathanassiou et al. (2000). (d) Neural activity in zebra finch HVC (interneuron) and RA (projection neuron) during singing (bottom plots), showing premotor neural firing milliseconds before song is produced (sonograph plots on top). Modified from Yu and Margoliash (1996). In panels (a)–(c), not all activated brain areas are represented in these images; anterior is to the right, dorsal is up. See Appendix 20.1 for abbreviations.

1996; Wise et al. 1999). Activation in Broca's area, DLPFC, and pre-SMA is higher when speech tasks are more complex, including learning to vocalize new words or sentences, sequencing words into complex syntax, producing non-stereotyped sentences, and thinking about speaking (Hinke et al. 1993; Poeppel 1996; Buckner et al. 1999; Bookheimer et al. 2000). Like song nuclei in birds, premotor speech-related neural activity is found in Broca's area (Fried et al. 1981). Further, similar to HVC and song arrest, low threshold electrical stimulation to the FMC, Broca's area, or the anterior SMAs cause speech arrest or generation of phonemes or words (Fried et al. 1991; Jonas 1981; Ojemann 1991, 2003). Repeated stimulation in LMAN during singing produces transient changes in amplitude and pitch (Kao et al. 2005), which leads to a testable hypothesis of whether such changes would occur in human speech when there is repeated stimulation to Broca's area and other frontal regions active during speech production.

In human noncortical areas, like the position of songbird AreaX, speech and song production are accompanied by highest activation (PET and fMRI) of the anterior striatum (Brown et al. 2004; Gracco et al. 2005; Klein et al. 1994) Further, in both songbirds and humans, singing and speech are accompanied by dopamine release, presumably from the midbrain dopamine neurons (SNC-VTA) into the anterior striatum (Sasaki et al. 2006; Simonyan et al. 2012). Low threshold electrical stimulation to ventral lateral and anterior thalamic nuclei, particularly in the left hemisphere, leads to a variety of speech responses, including word repetition, speech arrest, speech acceleration, spontaneous speech, anomia, and verbal aphasia (but also auditory aphasia) (Johnson and Ojemann 2000). The globus pallidus can also show activation during speaking (Wise et al. 1999). In nonhuman mammals and in birds, PAG and DM, and Amb and 12th, respectively, display premotor vocalizing neural firing (Dusterhoft et al. 2004; Larson et al. 1994; Zhang et al. 1995) and/or vocalizing-driven gene expression (Jarvis et al. 1998, 2000; Jarvis and Mello 2000). These findings demonstrate that it is not just cortical neurons that are involved in the production of complex learned vocalizations in humans or in songbirds, but rather an entire forebrain network, including parts of the cortex/ pallium, basal ganglia, thalamus, and brainstem.

The belief that vocal nonlearners do not have specialized forebrain areas that are active in the production of vocalizations was recently confirmed for birds in a study (Horita et al. 2012) which showed specialized singing regulation of the dusp1 gene in forebrain song nuclei of songbirds, parrots, and hummingbirds, but not in the forebrain of vocal-nonlearning birds (Figure 20.4b, for songbird only). In mammals, particularly in nonhuman primates, the anterior cingulate cortex projects indirectly to the vocal part of PAG, which in turn projects to Amb (Jürgens 2002). However, the cingulate cortex is not necessary to produce vocalization; instead, it is important in the motivation to vocalize in specific contexts (Kirzinger and Jürgens 1982; von Cramon and Jürgens 1983). Recently, two studies using IEGs in marmosets (Miller et al. 2010; Simões et al. 2010) and a PET imaging study in chimpanzees (Taglialatela et al. 2011) identified frontal cortical regions active with calling. These studies challenge the notion that nonhuman primates do not have forebrain vocal control regions outside of the cingulate cortex. However, these studies did not control for the affects of hearing oneself vocalize, and the IEG studies did not include silent controls. Further, the chimpanzee studies did not eliminate the possibility of coactivation with gesturing and only found activation with one call type: "attention getting calls." Thus, it is not yet clear if these cortical regions are motor control regions for vocalizations or for other aspects of behavior when the animals vocalize. Another study with IEG in mice found a primary layrngeal motor cortex region activated during ultrasonic song production, where lesions did not prevent production of vocalizations but resulted in more variable syllable production (Arriaga et al. 2012). One possible explanation for these potential paradoxical findings in nonhuman primates and mice is that these forebrain regions represent preadaptations for the evolution of vocal learning or that nonhuman primates may have lost part of the forebrain pathway necessary for vocal imitation (Simonyan and Horwitz 2011) and mice have a rudimentary functional vocal forebrain circuit that is part of the continuum (Arriaga and Jarvis 2013).

Taken together, the brain activation findings are consistent with the idea that songbird HVC and RA analogs are more similar in their functional properties to the laryngeal FMC in humans than to any other human brain areas; the songbird MAN, AreaX, and aDLM are respectively more similar in their properties to a strip of anterior human premotor cortex, part of the human anterior striatum, and to the human ventral lateral/anterior thalamic nucleus.

Convergent Changes in Genes

One might naturally suppose that the convergently evolved behavioral and anatomical specializations for complex vocal learning in birds and humans are associated with different molecular changes of genes expressed in those forebrain regions, since they evolved independently. These might be novel genes, novel expression patterns, or novel changes in the coding sequence or regulatory regions of existing genes. There is, however, an important distinction between IEG expression (discussed above), which is a general brain-wide indicator of activation level, and gene expression patterns during development (which build vocal-learning circuits) or in learning (which specifically support vocal learning). In other domains of biology, it has become clear that even convergently evolved characters (like eyes in mice and flies, or wings in bats and birds) often rely upon homologous developmental mechanisms and are underpinned by identical genes. This phenomenon, termed *deep homology*, appears increasingly to be common in biology (Shubin et al. 2009; Carroll 2008).

Recent genetic comparisons in birds and humans lend support to the hypothesis that deep homology also exists in the vocal domain (Feender et al. 2008; Fitch

2009a; Scharff and Petri 2011). The first involves the gene FoxP2, which codes for a transcription factor: a protein that binds to DNA and thus regulates other genes (Fisher and Scharff 2009). FoxP2 has two rare coding mutations found in humans, one of which is found in echolocating bats; such mutations are thought to be involved in the evolution of this gene's role in vocal learning and spoken language (Enard et al. 2002; Fisher and Scharff 2009; Li et al. 2007). It is important to note that FoxP2 itself is a widespread, highly conserved gene found in all birds or mammals (not just in vocal learners). Changes in this gene, or in the timing and location of expression of the gene, appear to have complex downstream effects on a cascade of neural and genetic circuits that are tied to vocal learning. Thus, FoxP2 is neither "the gene for language" nor is it "the gene for vocal learning"; it is simply one component of a genetic pathway involved in vocal learning. What is surprising is that this genetic pathway appears to be at least partially shared between humans and birds.

In support of this hypothesis, FOXP2 heterozygotic mutations in humans cause speech dyspraxias (Lai et al. 2001), the most famous case being members of the KE family, who suffer from a knock-out mutation in one copy of their FOXP2 gene (Vargha-Khadem et al. 2005). Affected family members are able to speak, but they are unable to sequence phonemes or even simple oral actions appropriately; thus they have difficulty learning and producing clear speech. Their cognitive functions are affected, but not as dramatically as speech functions. FOXP2 expression is enriched in the human striatum, and the KE family has reduced anterior striatum volume (Teramitsu et al. 2004; Vargha-Khadem et al. 2005). However, there is no evidence that mice transfected with the human FOXP2 gene show enhanced vocal skills (Enard et al. 2009).

In songbirds, FoxP2 is also enriched in the striatum and is differentially up-regulated in AreaX during the vocal-learning critical period and down-regulated in adults when they perform so-called practice undirected song (Figure 20.5a–b) (Haesler et al. 2004; Teramitsu et al. 2004; Teramitsu and White 2006). Experimental suppression of FoxP2 expression in AreaX with RNA interference leads to inaccurate vocal imitation where, like the KE family, zebra finches can still sing, but they slur the imitation of individual syllables and do not imitate syntax accurately (Figure 20.5c–f) (Haesler et al. 2007). The molecular role of FoxP2 is being heavily investigated, and current findings indicate that in the brain it regulates molecules involved in the development of neural connectivity (Vernes et al. 2011). Using genetic engineering it is possible to create neuronal precursor cells that express either the ancestral (chimpanzee-type) FoxP2 gene, or the novel human mutation. Cells expressing FoxP2 with the human mutation but not the chimpanzee version preferentially regulate the axon guidance gene Slit1, which plays a role in forming long-distance connections between neurons (Konopka et al. 2009).

Coincidentally, convergent down-regulation of Slit1 and up-regulation of its receptor Robo1 has been found in the RA analog of all three vocal-learning orders of birds (Wang et al., unpublished). Different splice variants

Animal Models for Human Speech, Song, and Vocal Learning   529

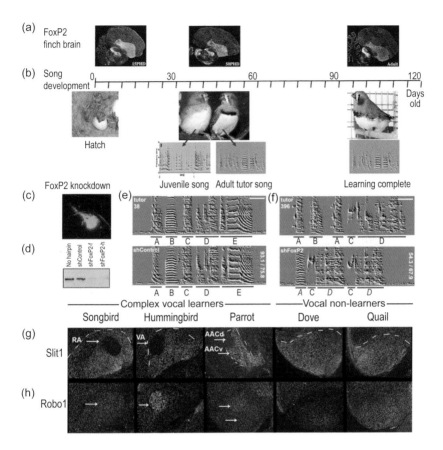

Figure 20.5 Molecular specializations in avian vocal-learning systems. (a) FoxP2 mRNA (white) increases in AreaX (red arrow) during the song-learning critical period (post hatch day 50–70). All sections are sagittal. (b) Zebra development from the egg to early adult (of a male) showing the timeline and physical appearance of the juveniles (egg and black beak stages) and of sonograms (time vs. sound frequency, in grey) showing juvenile plastic song eventually morphing to match that of the tutor. (c) FoxP2 RNAi attached to green fluorescent protein (GFP) injected into AreaX. (d) Western blot shows RNAi to FoxP2 successfully knocks down FoxP2 protein, but control samples do not. (e) Juvenile with a control RNAi sings a 93% match as an adult (lower panel) to his tutor's song (upper panel). (f) Juvenile with a FoxP2 RNAi in AreaX, only sings a 53% match as an adult (lower panel) to his tutor's song (upper panel), with inaccurate imitation of syllable structure and syllable sequencing (letters underneath sonogram). (g) Convergent down-regulation of the FoxP2 target Slit1 in the RA analog of complex vocal learners, but not of nonlearners relative to the surrounding arcopallium. (h) Convergent up-regulation of the Slit1 receptor Robo1 in the RA analog of vocal learners (highest in hummingbird). All sections are frontal. FoxP2 panels were modified after Haesler et al. (2004, 2007), with bird images courtesy of Constance Scharff. Slit1 and Robo1 panels modified after Wang et al. (unpublished).

of Robo1 mRNA, which are different combinations of parts of the RNA that make different variants of the proteins, are also enriched in human fetal frontal and auditory brain areas (Johnson et al. 2009). Mutations of Robo1 in humans are associated with dyslexia and speech sound disorders (Hannula-Jouppi et al. 2005). Enriched expression of other interacting axon guidance genes, such as neuropilin 1, have been found with specialized gene expression in songbird and parrot HVC and NLC analogs (not tested in hummingbirds) (Matsunaga et al. 2008). Specialized expression of these genes is not present in the forebrains of avian vocal-nonlearning species (ring doves and quails).

In summary, genetic studies show that there are broad overlaps in the molecular systems that underlie vocal learning in humans as well as in birds. The existence of deep homology in these systems is extremely fortunate, because it allows developmental neuroscientists to investigate genes initially uncovered in humans by studying birds in the laboratory: a much more tractable experimental system. Research on bird vocal learning will play an important role in understanding the genetic mechanisms which underlie vocal learning more generally for the simple reason that currently this is the *only* accessible model system in which complex vocal learning is present. These studies may indicate enriched gene regulation of the FoxP2 and Robo/Slit family of genes in vocal-learning species. In this regard, even though different vocal-learning birds and humans presumably evolved complex vocal learning independently, there might be constraints on which developmental circuits, and thus which genes, become co-opted for this behavior.

Ancestral Auditory System

Above we focused on the motor component of complex vocal-learning systems. This is because the motor component is specialized in vocal learners, whereas the auditory component is common among complex vocal learners and vocal nonlearners. Birds, reptiles, and mammals have relatively similar auditory pathways (Figure 20.6) (Carr and Code 2000; Vates et al. 1996; Webster et al. 1992); thus the auditory-learning pathway presumably existed before a vocal-learning pathway emerged but was not sufficient alone to drive complex vocal learning (Jarvis 2004, 2006). Carr and colleagues propose that the auditory pathway in each major tetrapod vertebrate group (amphibians, turtles, lizards, birds, and mammals) evolved independently of a common ancestor, in part because in the different vertebrate groups the cochlea nucleus in the midbrain developes from different neural rhombomeres (Christensen-Dalsgaard and Carr 2008). The weaknesses of this hypothesis are that it is possible for homologous cell types to migrate and change rhombomere locations (Jacob and Guthrie 2000), and there is no known vertebrate group that does not have a forebrain auditory pathway.

The source of auditory input into the vocal pathways of complex vocal-learning birds and humans remains somewhat unclear. For birds, proposed

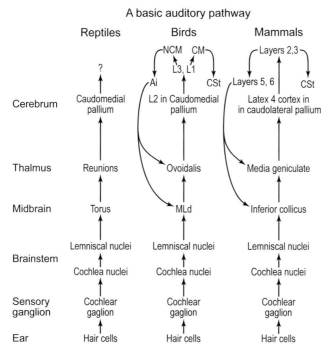

Figure 20.6 Comparative and simplified connectivity among auditory pathways in vertebrates (see Appendix 20.1 for abbreviations). Figure modified from Jarvis (2004).

routes include (Bauer et al. 2008; Fortune and Margoliash 1995; Mello et al. 1998; Vates et al. 1996; Wild 1994):

- a shelf of neurons ventral to HVC into HVC,
- anterior arcopallium cells called the "RA cup" into RA itself,
- the thalamic nucleus ovoidalis (Ov) or the caudal mesopallium (CM) into the interfacial nucleus of the nidopallium (NIf), and
- NIf dendrites in the primary forebrain auditory neurons L2.

For humans, the primary auditory cortex information is thought to be passed to secondary auditory areas (Brodmann's area 21, 22, 37) and from there to Broca's area through arcuate fasciculus fibers (Geschwind 1979), a hypothesis supported by diffusion tensor imaging experiments in humans (Friederici 2009b; Glasser and Rilling 2008). Only one of these secondary auditory projections (from Brodmann's area 22) has been found in nonhuman primates, and this projection is moderate in size in chimpanzees but very thin in macaques relative to humans (Rilling et al. 2008). These findings have been used to support an additional hypothesis to the "direct connections" hypothesis already discussed; namely, that which is different between a species that can imitate vocalizations (humans) and those that cannot (nonhuman primates) is the absence in the latter of direct input from higher cortical auditory areas into anterior speech areas (reviewed in Fitch et al. 2010).

Bilateral damage to primary auditory cortex and Wernicke's area leads to full auditory agnosia: the inability to recognize sounds consciously (e.g., speech, musical instruments, natural noises; Benson and Ardila 1996). Damage to Wernicke's area alone leads to auditory aphasias, sometimes called fluent aphasia, because the affected person produces uncontrolled nonsense-sounding sentences. One proposed reason for this symptom is that the vocal pathways may no longer receive feedback from the auditory system via the arcuate fasciculus. In songbirds, lesions to the analogous CM and caudal medial nidopallium (NCM) result in a significant decline in the ability to form auditory memories of songs heard, but do not affect the ability to sing (Gobes and Bolhuis 2007). Thus far, no one has tested whether such lesions result in fluent song aphasias (which would be best tested in a species that produces variable sequences of syllables) or in deficits for song learning. In macaques, the auditory cortex appears to be able to help form short-term auditory memories; however, unlike humans, it has been proposed that the animals have weak long-term auditory memories (Fritz et al. 2005). This proposed difference in auditory memory has been suggested as another difference between vocal-learning and vocal-nonlearning species. As yet, no one has tested whether such differences in auditory memory occur in vocal-learning and vocal-nonlearning birds.

In summary, the presence of cerebral auditory areas is not unique to vocal-learning species. Thus it is possible that the primary and secondary auditory cortex brain regions involved in speech perception in humans and song perception in birds are an ancestral homologous system found in all tetrapod vertebrates. Potential differences between vocal-learning and nonlearning species may be the absence of a direct projection from the FMC (or songbird RA) to brainstem vocal motor neurons, the weakened form or absence of a direct projection from higher secondary auditory areas to frontal motor cortical areas, and possibly weaker formation of long-term auditory memories. To support or refute these hypotheses, more comparative analyses are needed in both vocal-learning mammalian and avian complex vocal-learning and nonlearning species.

A Motor Theory for Vocal-Learning Origin

Remarkably similar systems of cerebral vocal nuclei for production of complex learned vocalizations exist in distantly related birds and humans, pathways that are not found in more closely related vocal-nonlearning relatives. This suggests that brain pathways for vocal learning in different groups evolved independently from a common ancestor under preexisting constraints. Recent experiments suggest that a possible constraint is preexisting forebrain motor pathways: using behavioral molecular mapping, Feenders et al. (2008) found that in songbirds, parrots, and hummingbirds, *all seven cerebral song-learning nuclei are embedded in discrete adjacent brain regions that are selectively activated by limb and body movements* (Figure 20.4a, singing and hopping). Similar to the relationships between song nuclei activation and singing, activation in the adjacent regions correlates with the amount of movement performed and is independent of auditory and visual input. These same movement-associated brain areas are also present in female songbirds which do not learn vocalizations and have atrophied cerebral vocal nuclei, as well as in vocal-nonlearning birds such as ring doves. Likewise, in humans, cortical areas involved in the production of spoken language are adjacent to or embedded in regions that control learned movement behavior, including dancing (Brown et al. 2006b). Based on these findings, a motor theory for the origin of vocal learning has been proposed (Feenders et al. 2008:19):

> Cerebral systems that control vocal learning in distantly related animals evolved as specializations of a preexisting motor system inherited from their common ancestor that controls movement, and perhaps motor learning.

Other evidence for this hypothesis is that the connectivity of the surrounding movement-associated areas in songbirds is similar to anterior and posterior song pathways (Figure 20.7a) (Bottjer et al. 2000; Feenders et al. 2008; Iyengar et al. 1999). The differences are that unlike HVC's projection to AreaX in songbirds, the adjacent nidopallium in zebra finches sends only a weak projection to the striatum, whereas the arcopallium adjacent to RA sends a strong collateral projection to the striatum (Bottjer et al. 2000). These differences may reflect fewer, or weaker, functional constraints on interactions between posterior and anterior motor pathways that underlie vocal learning (Jarvis 2004). Because mammalian nonvocal motor (posterior) and premotor (anterior) pathways follow a similar connectivity design, we propose that the evolution of vocal-learning brain areas for birds and humans exploited or "exapted" a more universal motor system that predated the split from the common ancestor of birds and mammals (i.e., in stem amniotes).

This preexisting forebrain motor pathway may represent a neural deep homology shared by vocal-learning systems, by analogy to the genetic deep homology discussed above. Just as the convergent evolution of wings in birds and bats co-opted the upper limbs, vocal-learning pathways among multiple bird and mammal clades would share a homologous common ancestral motor learning pathway, but are analogous in that they each could have evolved de novo from this pathway.

The proposed theory does not specify whether the forebrain vocal-learning systems formed by using a preexisting part of a motor pathway as a scaffold or duplicated a preexisting part of the pathway. However, it does not seem that a preexisting part of a motor pathway was lost. Rather, similar to gene evolution by gene duplication (Ito et al. 2007), Jarvis and colleagues propose (Feenders et al. 2008) a mechanism of motor brain pathway duplication during

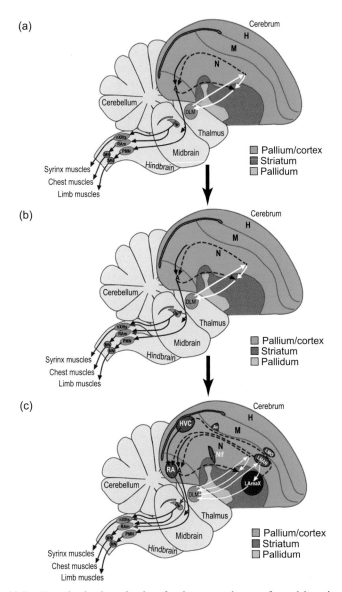

Figure 20.7 Hypothesized mechanism for the motor theory of vocal-learning origin. (a) Vocal nonlearner brain with nonvocal motor pathways, and midbrain and brainstem vocal innate pathways. White arrows depict anterior motor pathways; black arrows represent posterior motor pathway; dashed arrows depict connections between the two pathways. (b) Proposed mutational event that led to descending cortical axons from the arcopallium to synapse onto vocal motor (nXIIts) and respiratory (RAm) neurons (red arrows). (c) Vocal learner brain now with a song-learning system with parallel connectivity to its cerebral motor pathway.

embryonic development, whereby neural cells giving rise to motor pathways are replicated again, creating more cells, and then during differentiation are connected to vocal motor neurons of the brainstem (Figure 20.7).

There are alternatives to this theory in birds. For example, the song control system of seven nuclei may have evolved from an auditory pathway which then became used for vocal motor behaviors or evolved de novo without influence from an adjacent pathway (Margoliash et al. 1994; Mello et al. 1998; Bottjer and Altenau 2010). Support for these alternatives are that the descending auditory pathway has connectivity similar to the posterior song nuclei pathway and that brain areas around the anterior song nuclei show hearing song-induced IEG expression (Bottjer and Altenau 2010; Abe and Watanabe 2011). Weaknesses with this proposal, however, are that the descending auditory pathway is similar to descending motor pathways in the forebrain, and the studies above did not control for movement behavior. Feenders et al. (2008) showed that when some songbirds hear song playbacks in the dark, they hop around excitedly and show IEG expression adjacent to the song nuclei. Animals that sit still and hear the same song playbacks do not show any detectable IEG expression adjacent to the song nuclei, except part of the auditory shelf under HVC and cup of cells anterior to RA, which are only directly adjacent to song nuclei in songbirds.

The motor theory of vocal-learning origin is consistent with Robin Allott's (1992) "motor theory for language origin." Allot argued that language brain areas evolved from a preexisting motor neural system; however, he did not provide experimental evidence or flesh out the anatomical or mechanistic details of this theory. Lieberman (2002) proposed that language areas evolved from a preexisting cortical-basal-ganglia-thalamic loop, for which he deemed the basal ganglia part as the reptilian brain. We now know, however, that (a) reptilian and avian cerebrums are not made up of just basal ganglia, (b) vocal-learning birds have only part of the vocal-learning system in the basal ganglia, and (c) language brain areas involve more than just this loop (Jarvis 2004; Jarvis et al. 2005). Farries (2001) and Perkel (2004) proposed in birds, and Jarvis (2004) in birds and humans, that vocal-learning pathways in birds and humans may be similar to systems outside of the vocal pathways that intuitively could be motor pathways found in vocal-nonlearning birds and mammals, but they did not have experimental evidence to corroborate these suggestions. The findings of Feenders et al. (2008) provide supportive evidence for a motor origin theory. These findings may help answer the question of what makes vocal learning, and spoken language and singing for that matter, special (Hauser et al. 2002; Fitch et al. 2005; Okanoya 2007; Pinker and Jackendoff 2005). We suggest that at least one special feature is a cerebral vocal motor learning system that controls the vocal apparatus.

This theory has overlaps with the gestural origin of spoken-language hypothesis, where the motor learning ability of gestures in human and nonhuman primates has been argued to be a precursor behavior for motor learning of spoken language (Rizzolatti and Arbib 1998; Arbib 2005a; Pika et al. 2005; Gentilucci and Corballis 2006; Pollick and de Waal 2007). During child development, gesture production appears before speech production and is thought to enhance learning of speech and singing; adults use limb gestures automatically and often unconsciously during speech production and singing (Galantucci et al. 2006; Gentilucci and Corballis 2006; Liao and Davidson 2007).

The motor theory of vocal-learning origin hypothesis (Feenders et al. 2008) has some differences with the gestural origins hypothesis as some have proposed it (e.g., Hewes 1973; Armstrong et al. 1984; Corballis 2003). The various versions of the gestural origins hypothesis often argue that the brain pathway used to generate nonvocal gestures and generate speech are tightly linked in the left frontal cortex, and are important not only for signaling but also for syntax perception and production. In contrast, the motor theory of vocal-learning origin hypothesis argues that the speech production and gestural production pathways are anatomically parallel systems: each has similar properties and both pathways have features found in most other motor learning pathways. This proposal is more closely reminiscent of variants of the gestural hypothesis that allow a virtuous spiraling of interaction during the evolution of gestural and vocal control circuits (cf. Arbib 2005a). Further investigations into the behaviors and neural circuits for movement displays, including the recent discovery of gesturing in birds (Pika and Bugnyar 2011), may help further discriminate among these hypotheses. If the motor theory hypothesis is further supported in birds and mammals, then the evolution of complex vocal-learning and gesture-learning brain systems, via replication and differentiation of a preexisting motor learning system, might represent a general mechanism of how brain pathways for complex behavioral traits, such as speech and singing, evolve.

Conclusion: An Ornithomorphic View of Human Speech and Song

We have summarized a considerable body of behavioral, neuroscientific, and genetic work on complex vocal learning in birds and humans, and hope that this demonstrates the excellence of the songbird model as a system to investigate both the evolution and mechanistic basis for complex vocal learning. The existence of many different bird species capable of complex vocal learning provides a very powerful basis for understanding its evolution. The avian order provides species with a broad range of capabilities, from those whose call morphology is essentially innately specified (e.g., chickens), to species with relatively simple learned song (e.g., zebra finches, swamp sparrows), to species with remarkably complex learned vocalizations that rival human speech or song in complexity (e.g., thrashers or lyrebirds). This is in sharp contrast to the situation among primates—where only humans are capable of complex vocal learning, whether of speech or song—which greatly limits the scope of comparative investigations among primates.

We conclude with a brief glimpse of our own species from an ornithomorphic perspective, which we hope to have encouraged through this discussion. From the perspective of a songbird, a fundamental unity exists in human language and music: both use vocal output as their "default" modality (speech and song) and cultural transmission of both necessitates the ability for complex vocal learning. This ornithomorphic perspective invites us, in several ways, to concentrate on this ability as an important (but not sole!) explanatory factor in both domains. First, the existence of birdsong provides proof that complex vocal learning can evolve and be adaptive, independent of any linguistic use in conveying complex propositions. This clearly illustrates the plausibility of Darwin's supposition that, at some point in hominin evolution, our ancestors may have evolved a learned vocal communication system that was not employed in propositional communication, and which thus was more similar to music than to spoken language (Brown 2000; Darwin 1871; Fitch 2006a; Mithen 2005). Whether this is termed "song," a "musical protolanguage," or "musilanguage" is a matter of taste: songbirds demonstrate that this hypothesis is not implausible.

Second, birds provide a test bed for adaptive hypotheses. For instance, Darwin's familiarity with temperate-zone songbirds led him to believe that in all complex vocal-learning species, only the males sing. This sexual dimorphism led him further to suppose that sexual selection was the main driving force in the evolution of birdsong and, by analogy, in human musical protolanguage. More recent data (reviewed above) shows, however, that song in female birds is quite common (e.g., in duetting species), and in some clades (e.g., parrots) there is no obvious sexual dimorphism at all. These species show that it is necessary to disentangle Darwin's phylogenetic hypothesis (that a song-like system preceded speech) from his proposed adaptive driving force (sexual selection on males for courtship and territorial defense, cf. Fitch 2006a).

Third, completely independent of such evolutionary considerations, research on birdsong provides a rich model for understanding the proximate mechanisms which underlie vocal learning. A wide range of experiments, from song deprivation to brain lesions to manipulations of gene expression, are possible in birds that would be unthinkable in our own species, and very difficult in mammalian complex vocal learners like seals, whales, or elephants.

Fourth and most surprising, mechanistic investigations have revealed several examples of deep homology between birdsong and human vocal learning, both at the neural and genetic levels. Given the universal agreement that complex vocal learning evolved independently in birds and humans, homology in the underlying neural and genetic pathways comes as somewhat of a surprise, though, as already discussed, this now looks to be more prevalent in biology than anyone previously suspected (cf. Carroll 2008). This provides an unexpected bonus for research on human genetic underpinnings of vocal learning, because none of the other "model species" for neurogenetics (mice, zebrafish, or *Drosophila*) have yet to be shown to have clear evidence of complex vocal learning comparable to that of birds or humans (for mice, see Kikusui et al. 2011; Grimsley et al. 2011; Arriaga et al. 2012).

For all of these reasons, we confidently predict that research on birdsong will continue to play a driving role in investigations of complex vocal learning, research that has central relevance to understanding the topic of this volume: the evolution and neural basis of music and language. We are well aware that birds are not relevant to all aspects of these complex human traits. We may need to look elsewhere for model systems that can help us understand complex semantic or syntactic aspects of human language, or the evolution of our capability for instrumental music. Nonetheless, we hope to have convinced even initial skeptics that birds provide an invaluable model for understanding complex vocal learning, a capability with direct central relevance to both speech and song. For this component of spoken language and music, birds have been, and will likely remain, the model system *par excellence*.

Acknowledgments

WTF gratefully acknowledges the support of ERC Advanced Grant SOMACCA (#230604). The work of EDJ has been supported by the National Institutes of Health, the National Science Foundation, Howard Hughes Medical Institute, Whitehall Foundation, Packard Foundation, Triangle Community Foundation, Human Frontiers in Science Program, Whitehall Foundation, and the Klingenstein Foundation.

Appendix 20.1 Abbreviations used in this chapter.

AAC	central nucleus of the anterior arcopallium	CM	caudal mesopallium
aCC	anterior cingulate cortex	CSt	caudal striatum
aCd	anterior caudate	DLM	medial nucleus of dorsolateral thalamus
ACM	caudal medial arcopallium	DLPFC	dorsolateral prefrontal cortex
AI	intermediate arcopallium	DM	dorsal medial nucleus of the midbrain
aINS	anterior insula cortex	FMC	face motor cortex
Amb	nucleus ambiguous	GPi	internal globus pallidus-like neuron
aP	anterior putamen	HVC	(a letter based name)
AreaX	area X of the striatum	IC	inferior colliculus
aST	anterior striatum	IEG	immediate early genes
aT	anterior thalamus	IT-3	intratelencephalic projecting neuron of layer 3
Av	avalanche	L2	field L2
Cd	caudate	LMAN	lateral MAN

Appendix 20.1 (continued)

LMC	laryngeal motor cortex	PAG	peri aqueductal grey
LMO	lateral oval nucleus of the mesopallium	PMN	premotor neuron
MAN	magnocellular nucleus of anterior nidopallium	Pre-SMA	pre-supplementary motor area
MG	medial geniculate	Pt	putamen
MLd	mesencephalic lateral dorsal nucleus	PT-5	pyramidal tract neuron of motor cortex layer 5
MMSt	magnocellular nucleus of the anterior striatum	RA	robust nucleus of the arcopallium
MO	oval nucleus of the mesopallium	RA-p	RA-projecting neuron of HVC
MOc	oval nucleus of the mesopallium complex	SMA	supplementary motor areas
MSp	medium spiny neuron	St	striatum
NAO	oval nucleus of the anterior neostriatum	Uva	nucleus uvaeformis
NCM	caudal medial nidopallium	VA	ventral anterior nuclei of the mammalian thalamus
NDC	caudal dorsal nidopallium	VAM	vocal nucleus of the anterior mesopallium
NIDL	intermediate dorsal lateral nidopallium	VAN	vocal nucleus of the anterior nidopallium
NIf	interfacial nucleus of the nidopallium	VL	ventral lateral nuclei of the mammalian thalamus
NLC	central nucleus of the lateral nidopallium	VMM	vocal nucleus of the medial mesopallium
nXIIts	tracheosyringeal subdivision of the hypoglossal nucleus	VMN	vocal nucleus of the medial nidopallium
OV	nucleus oviodalis	X-p	X-projecting neuron of HVC

First column (top to bottom): Ian Cross, Jerome Lewis, Bjorn Merker, Katja Liebal, Dietrich Stout, Bjorn Merker, Erich Jarvis
Second column: Tecumseh Fitch, group discussion, Francisco Aboitiz, Atsushi Iriki, Sandra Trehub, group notes, Atsushi Iriki
Third column: Dietrich Stout, Erich Jarvis, Sandra Trehub, Ian Cross, Tecumseh Fitch, Francisco Aboitiz, Jerome Lewis

21

Culture and Evolution

Ian Cross, W. Tecumseh Fitch, Francisco Aboitiz,
Atsushi Iriki, Erich D. Jarvis, Jerome Lewis, Katja Liebal,
Bjorn Merker, Dietrich Stout, and Sandra E. Trehub

Abstract

This chapter captures extensive discussions between people with different forms of expertise and viewpoints. It explores the relationships between language and music in evolutionary and cultural context. Rather than trying to essentialize either, they are characterized pragmatically in terms of features that appear to distinguish them (such as language's compositional propositionality as opposed to music's foregrounding of isochronicity), and those that they evidently share. Factors are considered that constitute proximate motivations for humans to communicate through language and music, ranging from language's practical value in the organization of collective behavior to music's significant role in eliciting and managing prosocial attitudes. Possible distal motivations are reviewed for music and language, in terms of the potentially adaptive functions of human communication systems, and an assessment is made of the advantages which might accrue to flexible communicators in the light of ethological and archaeological evidence concerning the landscape of selection. Subsequently, the possible evolutionary relationships between music and language are explored within a framework supplied by six possible models of their emergence. Issues of the roles of culture and of biology in the evolution of communication systems are then addressed within the framework of triadic niche construction, and the chapter concludes by surveying available comparative and phylogenetic issues that might inform the debate.

Distinguishing Music from Language

In placing music and language within the frames of culture and evolution, one is necessarily confronted by the question: "What is intended by the terms "music" and "language?" Are we dealing with culturally shaped distinctions or biologically distinct systems? Are music and language categorically discrete human faculties, or do they constitute different manifestations of the same underlying communicative capacities? Our initial strategy is to avoid definitions in favor of identifying features that distinguish between the two domains;

in postulating distinct features of music and language, we run the risk of essentializing ethnocentric concepts or stressing between-category differences and minimizing within-category differences, in effect, reifying distinctions that may not be supported by the evidence. It must be acknowledged, however, that in all known human cultures, the available suite of behaviors includes something that appears like music, just as it includes language, though the extent to which categorical distinctions are drawn between music and language, and the factors that motivate any distinction between the two domains, differ across cultures.

A key attribute that appears to distinguish between the domains is *propositionality*. Language, unlike music, provides a way of sharing information about states of affairs by means of truth-conditional propositions and thus of coordinating action. It enables mapping between worlds, thoughts, and selves, the formulation and exchange of information, and the coordination of joint, goal-directed action. Music appears to have none of these functional benefits, but it has others that we will consider subsequently. Nevertheless, music and language share the significant feature of *generativity*. Both afford complex combinatoriality and unlimited generativity via a few simple nonblending (particulate) elements into composite, individually distinctive patterns. Such a system has been called a Humboldt system, after Wilhelm von Humboldt, who first described language in these terms (for music, see Merker 2002). Combinatoriality is also found in vocal-learning songbirds, such as the sedge warbler (*Acrocephalus schoenebaenus*) who varies the *sequencing* of his stock of some fifty different song elements to produce song patterns which essentially never repeat (Catchpole 1976). The sedge warbler's song, however, is not semanticized; the different patterns pouring out of the sedge warbler's throat are not invested with distinctive meanings. Moreover, other songbirds may only have one song, or very few variations. This is by way of contrast to the varied phoneme sequences in human speech, which may form words with learned meanings, the words in turn composing sentences, with the grammar of the language specifying how the meanings of words combine to imbue each sentence with distinctive meaning predicated on the specific assembly of phonemes/words of which they consist. This is what allows language to carry propositional meaning riding on the phonemic stream of speech, by contrast to the "note stream" of music, which has no corresponding compositionality of meaning. Music's combinatorial aspect falls closer to the sedge's warbler's use of combinatoriality to mount what we think of as an impressive aesthetic display. In any case, if the deep similarity between music and language is their hierarchical structure as yielded by Humboldt system generativity, the lack of formal semanticization in music (without lyrics) is the major contrast between music and language, fitting them for different uses in human communication. As language users, we need to share some common ground to conduct our dialogic and propositional transactions. This common ground is established largely by interaction within a shared community, being built on commonalities of knowledge and belief mediated by the propositions shared in our linguistic exchanges or our observation of such exchanges (e.g., hearsay). As an evolutionary counterpoint, we may note that there is evidence from monkeys and chimpanzees (Crockford et al. 2004; Clay and Zuberbühler 2011) of control or combination of vocalizations which result in a change of meaning, although we have here just a few such vocalizations, with neither a Humboldt system nor a compositional semantics.

This account needs, however, to be supplemented by the realization that much linguistic dialog is not concerned with the exchange of formal propositions but rather with maintaining social networks (Dunbar 1996; Wray 1998), which is to say that a significant part of linguistic interaction is relational rather than transactional. Moreover, while music cannot communicate propositional information, the idea that music has meaning is widespread across cultures. In fact, music is frequently reported as bearing meanings similar to those transmitted by linguistic means. According to Leonard Meyer (1956:265), "music presents a generic event, a "connotative complex," which then becomes particularized in the experience of the individual listener." Such experience remains individual rather than being made mutually manifest to other listeners, as would be the case for language. If music is considered an interactive or *participatory* phenomenon, in contrast to the *presentational* form that it typically takes in Western conceptions (Turino 2008), close parallels emerge between the features of music and the relational features that sustain conversational interchange. Hence the criterial distinctions between music and language as interactive media may involve the extent to which each medium requires mutually comprehensible reference and foreground features concerned with sustaining the interaction.

While music and language appear to constitute discrete categories in contemporary Western societies, for many cultures they may be best conceived of as poles of a continuum, or there are divisions into more than two categories. For example, a complex set of distinctions is provided by Seeger (1987), who notes that primary distinctions made between "communicative genres" by the Suyá people of the Amazon are between three categories:

1. *kaperní*, which more-or-less corresponds to everyday speech, where there is a priority of text over melody, text and melody being determined by speaker, with an increasing formalization in public performance;
2. *sarén*, "telling" or instructional speech, where there is a relative priority of relatively fixed texts over relatively fixed melodies; and
3. *ngére*, song, where there is a priority of melody over text, and, importantly, time, text, and melody are fixed by a nonhuman source.

As Seeger (1987:50) notes, "Melody is not a particularly good way to distinguish between Suyá speech, instruction, and song." Some manifestations of *kaperní* may appear to shade into manifestations of *sarén*; similarly, it may be difficult to distinguish between instances of *sarén* as these may, in turn, begin to shade into *ngére*. Here, modes of communication are being distinguished on the basis of their social function and their proper domain: *kaperní*, speech, is for mundane, everyday use, originating with—and being directed toward—humans; *sarén*, didactic talk, requires authority, whether deriving from present-day power structures or the invocation of a teacher from the past; whereas *ngére*, song, can constitute a special, liminally powerful medium, having nonhuman origins and being directed, in part, toward nonhuman agency. The Suyá are not alone in making such distinctions; other traditional cultures frequently embed what may appear as speech and song to Western observers in similarly complex communicative taxonomies (see Basso 1985; Feld 1982; Lewis, this volume).

We propose, therefore, that music and language constitute a continuum rather than discrete domains. This continuum can be interpreted in terms of at least two dimensions, the first running from definite to indefinite meanings and the second from greater to lesser affective potency. Music's power to form complex patterns (enabled by its generativity), its frequent repetition of elements (in comparison with language), together with its iconicity (i.e., its exploitation of biologically significant aspects of sound) endow it with an ambiguity and an immediacy that can be emotionally compelling. Language's capacity to formulate and exchange complex propositions allows it to represent an infinite variety of meanings and frees it, in principle, from the exigencies of affect. However, the discrete tones and pitch sets that supply grist for the musical mill in most cultures are rather unique to music, though a few birds (e.g., the pied butcher bird of Australia) do appear to feature them. Also, for humans the speaking voice is a highly significant biological sound whose emotional coloring draws on our repertoire of innate nonverbal emotional expressiveness. We routinely express emotion through the modality of speech rather than music; nothing compels music to convey emotion.

Given such blurrings of any strict dichotomy, it may be helpful to stress contexts of use, just as in the Suyá example above. The typical linguistic exchange is between two persons whereas, for most of its history, music has occurred in group contexts. Importantly, language and music differ in their power to coordinate human movement. There are differences in the regularity of timing between most registers of speech and most genres of music, with the latter featuring explicit use of isochrony, though it should be noted that this is a feature of both didactic talk and oratory, both oriented toward "musical" ends of capturing attention and enhancing a sense of mutual affiliation. The isochrony of music facilitates the timing of one's own movements and the prediction of others' movements, allowing for mutual co-adjustment of phase and period in simultaneous and sequential movements. Music's isochronicity and metrical structure may also underpin a greater mnemonic potential compared to language, or the musical feature of isochronicity itself may endow language with such mnemonic potential. In oral cultures, transmission of cross-generational knowledge is likely to take forms that appear musical and poetic rather than discursive (see, e.g., Rubin 1995; Tillmann and Dowling 2007).

Low-level differences in acoustic attributes may also warrant a clear distinction between music and language, or more properly, between speech and song. Music (song) and spoken language differ in their inter-event transitions—the ways in which sounds succeed one another—in terms of rhythm (the previously noted tendency toward isochronicity) and formant transition, with sharper formant transitions in speech than in song. Schlaug (see, e.g., Özdemir et al. 2006) has suggested that the same pathways are used for the perception of music and speech in contrast to parallel pathways for the production of speech and singing, the latter arising, perhaps, from rate differences between speech and music. In contrast, Jarvis (2004) has suggested that the same pathways are used in different ways to produce song and speech; the latter is true of song-learning birds, such as parrots, that can learn to sing as well as to imitate human speech.

The foregoing discussion has largely characterized music and language as an auditory-vocal phenomenon. Of course, both involve action in the form of gesture. Spoken language is typically embedded in a complex interactive matrix of gesture (see, e.g., Kendon 2004), and there are numerous signed languages. Music involves overt action, not only in its production but also as an interactive process or network of gestures among participants (Moran and Pinto 2007). Indeed, music is indissociable from dance as a cultural category in many societies (e.g., Stone 1998). As gestural media, music and language may be distinguishable in terms of timing and organization. Gestures in language tend to be sequential and timed in relation to prosody rather than an underlying rhythm, whereas those in music often involve temporal regularity and may involve simultaneity between participants. Nevertheless, there are counterexamples such as coincident gestures of participants in linguistic interaction, often at points of topical agreement in discourse (Gill et al. 2000), intermittent temporal regularity, or absence of meter in music.

Overall, no single criterial attribute, save perhaps that of propositionality, distinguishes between language and music clearly and comprehensively. As Wittgenstein (1953) noted some years ago, categories need not have criterial or defining features. Instead, instances of a category can have a "family resemblance" or one or more common attributes shared with some but not all instances of the category. As with any category (e.g., birds), there are prototypical (e.g., robins) and less prototypical (e.g., chickens) instances (Rosch 1975).

One can also ask whether the features of music and language are uniquely human. During our discussions, we listed (Table 21.1) behavioral and neural parallels that have been documented in nonhuman species. In some cases, nonhuman animals trained by humans have succeeded in recognizing many words (e.g., the dog Rico; Kaminski et al. 2004), phrases (e.g., the parrot Alex; Pepperberg 1999), as well as octave-transposed melodies (e.g., in rhesus monkeys; Wright et al. 2000). There is no indication, however, of comparable feats in the natural environment in these or other nonhuman species.

Table 21.1 Subcomponents of music and language.

1. Behavioral components:
    a. Signal
        - Perception of speech (acoustic pattern recognition system). Lexical access may be unique to humans, since speech perception (involving lexical access) involves making lexical commitments. But what about Alex, the African Gray parrot, and Rico (Fischer's dog)?
        - Production of speech and song
        - Limited vs. complex vocal learning: humans, birds
        - Opportunistic multimodality: ape gestural communication
        - Hypermeter: multilevel meter, hierarchical structure in whale song
        - Voluntary control of vocalizations
        - Instrumental, nonvocal music
    b. Structure and phonology
        - Syntax minus meaning, vocal combinatoriality (sequencing of learned syllables): any animal that has a complex song (e.g., humans and birds), but we do not know enough
        - Recursion
        - Scales
        - Relative pitch: ferrets
        - Working memory
    c. Pragmatics
        - Theory of mind (ToM), as evidenced in intentional communication
        - Extreme sociality or the motivation to share experience
        - Vocal maintenance of mother–infant bonds
        - Entrainment: frogs/insects vs. parrots (cross-modal, potentially communicative in relational terms)
        - Dyadic dialog (context of communication), face-to-face, addressed communication, deictic switch, multimodality (i.e., the extent to which contents of turns are conditional on partner's productions), agonistic versus cooperative engagement
    d. Semantics
        - Referentiality in the form of compositional semantics: unique, though precursors or minimal commonalities exist (e.g., monkey booming as signifying negation)
        - Predication (predicate/argument)
        - Cultural transmission at every level in vocal communication (extreme variability of human language)
        - Lack of signification (displacement of reference in space and time)
        - Notation

Moreover, the prevailing view is that language and music are unique products of human culture (and nature), although elements of each may be present in other species. At the same time, it is important to note that stimuli and tasks involving nonhuman participants (and even human participants) typically lack ecological validity. In general, nonhuman species have difficulty recognizing transpositions of tone sequences, so it is of particular interest that European

Table 21.1 (continued)

2. Neural components
    - Auditory forebrain pathway (Wernicke's area)
    - Forebrain vocal control path of vocal structure (including Broca's area, striatum, thalamus)
    - Direct connection from cortex to brainstem vocal-motor neurons: lateral motor area-laryngeal motor neurons
    - Between humans and nonhuman primates, there appears to be a direct connection between auditory and primary or secondary motor areas (arcuate fasciculus?)
    - Differential gene regulation and convergent mutation in genes that make direct projections in other forebrain areas that control vocalizations in both humans and songbirds (deep homology)
    - Auditory receptive field sharpness in humans
    - Lateralization: greater specialization in humans in the representation of communication sounds
    - Spindle cells in anterior cingulate cortex only in humans and great apes (also in dolphins?)
    - Heterochronicity of cortical synaptogenesis unique to humans (among primates)?
    - Does brain size matter?

starlings can be trained to recognize transpositions of conspecific songs but fail to recognize transposed piano melodies after comparable training (Bregman et al. 2012). In any case, there is no evidence of a nonhuman species, whether in the wild or trained by humans, whose members combine Humboldt system generativity with a compositional semantics.

In this chapter, we view music and language as constituting different manifestations of the human capacity to communicate—manifestations which may take very different forms in different cultural contexts. Is that partly because, outside of its cultural context, music cannot be defined unambiguously? Persons within a culture usually have no difficulty differentiating most registers of speech from most forms of music. One complicating factor is that we have a reasonable understanding of the functions of language across cultures, but we have much less understanding with respect to music. In considering the place of music and language in culture and evolution, we must address the question of what impels humans to communicate—through language or music.

Proximate Motivators for Human Communication

That humans are highly motivated to communicate is unquestionable; the issue of what may underpin that motivation is, however, less certain. Communication—at least, in the form of language—has immense value in helping groups of individuals shape their environments, individually or collectively, so as to attain goals. In the form of socially oriented or phatic talk, language can serve to build and maintain relationships in social interactions. There are, however, many other motivations to communicate that are likely to apply to a broader range of communicative systems than language alone. For vocal-learning species, there seems to be an instrinsic pleasure in vocalizing (e.g., in forms such as babbling, subsong, or imitation). For humans (and perhaps some other primate species), vocal and gestural communication serves to co-regulate affective states between the caregiver and infant, and to enhance a sense of mutual affiliation. Communication can have prosocial effects, not just for dyads but also for larger groups: we may gain pleasure from collective and synchronized performance which, in turn, reduces social uncertainty and helps bond the group, enhancing the effectiveness of group action and identity, particularly when directed against potential external threats (e.g., other groups or prospective predators). Of course, once we can behave linguistically or musically, we can be motivated to co-opt these communicative resources for other ends; "inner speech" may be deployed to reduce uncertainty in attention-based coordination (Clark 2002) or to manage communication (Allwood 2007), whereas self-directed music may be produced as a means of affect regulation, as in the *dit* songs of the Eipo (Simon 1978).

Levinson (2006; see also this volume) argues for extraordinary human sociality grounded in an innate capacity for social interaction involving unique cognitive infrastructure (see also De Ruiter et al. 2010). Others emphasize the role of culture and experience in elaborating our inherited cognitive infrastructure (e.g., Vygotsky 1978). By 12 months of age, infants engage in declarative pointing to share their interest in events with others, to make requests, or to provide helpful information (Liszkowski 2011). They also vocalize to attract parents' attention. In fact, infants vocalize well before their vocalizations are intentionally communicative, perhaps because vocalizing is intrinsically pleasurable. However, the most significant motivation for human communication is the sharing of experience; that is, wanting another to see, feel, think, or know what I see, feel, think, or know. Early pointing in infancy is of the "look at that" variety rather than the instrumental or "get me that" variety. The pleasure of vocalization as the motivator would lead to a lot more vocalization in the absence of others, but this has nothing to do with communication. More generally, young children make greater use of gesture than spoken words in their early language development (Capirci et al. 2002). Tomasello (2008) emphasizes how different this is from the instrumental form of communication in ape gestures, in which one ape tries to modify the behavior of another. The pleasure of vocalizing has a more direct utility in mother–infant interactions although such interactions proceed equally smoothly in deaf mother–infant dyads, who use gestural rather than vocal signals. Early communicative mother–infant interactions have evident functions in co-regulating the affective states of both participants. Such early experiences may underlie the ability of music to facilitate entry into states of shared intentionality or even trance-type states. These capacities may be built on a more general substrate.

Clearly, there is more that motivates humans to communicate than just vocal pleasure. We gain huge practical advantages from being able to exchange information linguistically and to coordinate our actions with others. Motivational factors may drive us not only to speak but also to sing and move together with others in dance or synchronous movement, and this may strengthen social bonds (McNeill 1995). Communication by means of language and music affords us, respectively, the capacity for information transfer as well as the formation and maintenance of group solidarity. We can use language to get what we want and to transfer information, whereas we can use music to give us pleasure and to achieve group solidarity as well as to relieve pain and suffering and to reduce stress (Knox et al. 2011). Indeed, a defining characteristic of the human species is a propensity for cooperation and prosociality (Levinson 2006; Tomasello 2008). We note, however, that much of speech does not appear to be oriented toward the transfer of information but to processes of establishing mutual affiliation with others (i.e., functions which may be hypertrophied in music). We seem motivated to order social life through language and music, but it is notable that music, rather than language, tends to be at the forefront of situations where social conflict is a potential threat to the social order (e.g., Marett 2005).

One key factor that orders the human motivation to communicate is that of culture, which plays a key role in shaping, structuring, and ordering the human motivation to communicate, although here we have an example of an expanding spiral: new means to communicate support developments in culture, and new cultural and social processes provide an ecological niche for the emergence of new communicative forms. While we may gain pleasure from communicating or synchronizing with others, different cultures sanction these behaviors in different ways, with enculturation processes shaping acceptable patterns of communication. Notable examples can be found in some traditional cultures, where silent co-presence can be privileged over relationally oriented speech (Basso 1970), as well as in a range of situations in all cultures where institutions constrain or facilitate the motivation to communicate.

While pleasure (the instrumental value of a means of information exchange) and the human benefits of interpersonal connections and group solidarity may provide proximal motivation in human communication, these forces must be situated in their broader evolutionary context, to which we now turn.

Adaptive Functions of Human Communicative Systems

The most direct evidence for the emergence of complex communicative faculties early in the hominin lineage is in the lengthy archaeological record of complex lithic technologies transmitted over multiple generations. That persistence of cultural transmission suggests that early hominin cognitions and interactions must have been characterized by intense social cooperativity and inhibition of aggression. Material technology was employed in food acquisition and preparation, including group hunting, which required the recognition of multiple levels of intention in order to second-guess prey and coordinate group hunting behavior. Also required was the capacity for planning, which involves the manipulation of nonexistent entities and the composition of structures free from the immediate constraints of the physical world. Together, all these factors create a fitness landscape within which communicative capacities—and a progressive enhancement of communicative capacities—would have been adaptive. Of course, there would have been other selection pressures for the emergence of flexible communicative capacities, perhaps arising in the context of within-group, or sexual, competition. In addition, the effects of aspects of music on arousal in nonhuman species reminds us that many of the factors that make up the modern human communicative repertoire are likely to be shared with a variety of other species. Different factors are likely to have arisen at different times under different selection pressures, and it is likely that evidence for these different evolutionary time depths is embodied in the structures and dynamics of our neural and genetic systems.

The emotional aspects of music are often conceived of as being specific to humans. However, the arousing dimension of responses to features that are evident in music may be shared by other species. For example, auditory rhythmic features arouse chickens (indexed by noradrenaline release) and affect memory consolidation (Judde and Rickard 2010; Rickard et al. 2005). The effect of subcomponents of music on other cognitive functions suggests that music can have fundamental as well as higher adaptive functions, and a comparative approach is needed to differentiate homology and analogy. Rather than taking the response to sound, in the form of music, as a starting point, perhaps learned vocal communication is being selected. In a range of species, learned vocal communication is used for mate attraction, on the basis of variability of F_0 and syntax, which raises the question of why a "supranormal stimulus" effect of vocal sounds is not more common. Given the linkage of gesture with speech or of dance with music, it is a matter of debate whether the evolution of vocal learning was the driver for the emergence of language and music or was driven in part by the evolution of other embodied systems. For example, Arbib and Iriki (this volume) discuss the hypothesis that complex imitation of manual skills underwrote the evolution of manual gesture, and that the emergence of "protosign" provided a necessary scaffolding for the emergence of vocal learning in support of semantic expressivity. Alternatively, the ability to regulate the expression of emotion, whether bodily (gestural, postural) or facial, may differentiate humans from other species. This hypothesis is rooted in our understanding of the human capacity to control the expression of emotion. At present, there is little evidence of comparable control of facial expressions and vocalization in nonhuman species, though some precursor ability has been shown in monkeys (Hihara et al. 2003). Other work (Slocombe and Zuberbühler 2007) suggests that chimpanzees have some control over the production of their vocalizations since they recruit specific group members to support them in aggressive encounters. We share with our closest relatives the capacity to produce an initial affect burst in response to situational stress (see Scherer, this volume), but we know little about their capacity to shape and redirect such affect bursts. It is certainly the case that apes can be opportunistic in exploiting different channels for communication (e.g., Leavens et al. 2004b; Liebal et al. 2004), and it may be that the multimodality which characterizes speech (and music in action) has its origins in such capacities. Humans, like all primates, mammals, and indeed most vertebrates, have a multifaceted repertoire of largely innate nonverbal emotional expressiveness, which includes a rich repertoire of specifically vocal, emotional expressivity that is neither music, nor language, but which can be drawn upon by both of these for purposes of emotional coloring (e.g., in the dynamics and prosody of emotional speech). This preexisting largely innate repertoire is the key to the biology of emotional expressiveness in humans as in other species. However, if so, it must be stressed that the differences between such capacities and human music and language are immense.

Complex behaviors—such as acts of deception, binding the exercise of capacities for adopting the perspective of others with requirements to control mutually manifest behavior (e.g., vocalization)—may have provided grounds for the emergence of signals that have reference in relation to a state to be co-opted for proto-propositional use. Here, a parallel development of speech and music may be proposed, and the relationships between the raw expression of affect and the controlled articulation of art, whether linguistic or musical, could be explored. However, reasonably stable social groups would be needed to drive this process. One way of finding evidence for these speculations is to examine the range of emotional vocalizations from "raw affect bursts" to culturally defined quasi-lexical elements. This might shed light on the way in which raw vocalizations that we share with mammals have come under increasing control, both with respect to production and desired targets for communication. We note, however, that the control of the emotional expressions we share with other primates rests on medial circuitry (anterior cingulate as modulator of brainstem circuitry), whereas much of the circuitry associated with human language and music resides more laterally in the cortex (Jürgens 2009). Moreover, it is a classic observation going back to Hughlings Jackson in the nineteenth century that an aphasic may lose the propositional use of language yet still emit imprecations. Indeed, in humans, a crucial result of evolution is that language can take over from direct, affect-induced action as a means of negotiating situations where different individuals' needs or desires are manifestly in conflict. A further factor that could have driven the emergence of something like language is an increase in the ability, and motivation, to make plans in conjunction with others. Such planning requires shared goals and manipulation of nonexistent entities, enabling the composition of structures free from the immediate constraints of the physical world. Here, the range of theories seeking to link the evolution of brain mechanisms supporting language to those supporting tool use become especially relevant, with the notion that visualization of a goal may play a crucial role in planning the means to achieve it (Stout and Chaminade 2012). Off-line planning may (but need not) render concrete phenomena less immediately relevant, affording a means to displace reference (cf. Iriki 2011). Such considerations may underlie the evolution of both language and music. Not only is language's propositionality built on reference to present and absent entities and events, but music affords an abstract domain for the construction of sound worlds that may be similarly grounded in experience yet divorced from immediate events.

The emergence of pedagogical capacities at some point in the hominin lineage may be a more specific driver for the propositional and intentional dimensions of language. Pedagogy involves the intentional alteration of one's behavior to influence the mental states (attention, knowledge, embodied skills) of other individuals. In Arbib's version of the gestural origins hypothesis (mirror system hypothesis; Arbib 2005b), the transition to intentional communication requires a pantomimic/protosign phase. It could be argued that the intentionality of non-pantomimic communication in pedagogy shows that these substages may not be needed. The counterargument is that demonstration or modeling is an important part of pedagogy in natural environments. Gesture would be critical in such circumstances and would precede verbal instruction (Zukow-Goldring 1996, 2006). The need to communicate increasingly opaque causal relations in technological pedagogy also supplies a potential selective pressure for development of propositional meaning in language, but one must not conflate later stages of language evolution with their necessary precursors. Opaque causal relations are evidenced in skill transmission in modern humans, which involves not only direct communication, but also the creation of situations conducive for learning. This requires a high level of social cohesion, including (at least in modern humans) the development of appropriate skills and motivation for caregivers. The Vygotskian zone of proximal development (e.g., Vygotsky 1978) involves adult mentoring or scaffolding, which allows the learner to go beyond what he is capable of doing on his own. It refers to the difference between what the child can do independently and what he can do with adult assistance. The former indicates his state of knowledge or skill whereas the latter indicates his potential. In essence, it concerns culturally mediated learning rather than traditional pedagogy. This need for explicit support of the child's mental development may be an additional selective pressure in the expanding spiral for language (storytelling, kinship, etc.) and music (social bonding).

Language is marked out not just by its propositionality but by its complex propositionality, which entails compositionality, hierarchical structure, and complex syntax. These constitute very general capacities that are taken to high levels in language and, in some instances, music. These features are probably important for many evolutionarily relevant behaviors, but they are visible and testable in the archaeological record of stone tools. The archeological record of tools can document the expression of a particular depth/complexity of hierarchical action organization at a particular time, which provides a *minimum* indication of past hominin capacities. Stout's work (e.g., Stout et al. 2008) provides PET and fMRI evidence of increasing activation of anterior inferior frontal gyrus (hierarchical cognition) in increasingly complex stone tool-making as well as activation of medial prefrontal cortex during observation of tool-making by experts (intention attribution). A three-year longitudinal study of tool-making skill acquisition, which involves behavioral, social, archaeological (lithic analysis), neurofunctional (fMRI), and neuroanatomical (VBM, DTI) observations, is currently in progress, one output of which will be an empirically derived action syntax of Paleolithic tool-making. This work provides a clear and testable set of hypotheses concerning the emergence of capacities for compositionality and hierarchical structure and the facilitative effects of pedagogy. If an association can be established between the presence of vocal learning and the importance of "teaching" in other animals, its implications would be substantially broadened. The mirror system hypothesis would view such skill transfer as driving gestural communication more directly, with this in turn providing scaffolding for increasingly subtle vocal communication. In any case, much of human culture, and most of animal life, proceeds without pedagogy in any explicit, formal sense. That includes the acquisition of skills in many useful arts for which observational learning with "intent participation" often suffices (Rogoff et al. 2003).

Pedagogy, in whatever form, appears to require the capacity for recognition of multiple levels of intention ("orders of intentionality") and may be tied to the emergence of that capacity. It is suggested that chimpanzees have two orders of intentionality ("I believe that you intend…"), whereas humans can manage up to five or six (Dennett 1983). In any case, there is a chicken-and-egg problem in placing language and intentionality in evolutionary perspective, as language itself promotes development of ToM abilities, as indicated by the considerable lag in deaf children's achievement of ToM milestones (Wellman et al. 2011).

One of the prime requisites for big-game hunting—a subsistence strategy of current hunter-gatherers and of several of our recent ancestor species—is the ability to second-guess prey and to coordinate group hunting behavior. A switch in the hominin lineage to social hunting, rather than scavenging, may have helped provide selection pressures for the emergence of the capacity for recognition of multiple levels of intention, though Bickerton (2009) argues that scavenging, rather than hunting, provided the ecological niche that supported the emergence of language—perhaps too mono-causal a view of human evolution. It is notable that social hunting species, such as African hunting dogs and wolves, may have higher levels of intention recognition than nonsocial hunters, most likely driven by the demands of group hunting (Nudds

1978). However, this is without a hint of leading to either music or language, so one must still seek that "something extra" in human evolution.

Social hunting necessitates close cooperation with others, and there is extensive human evidence for cooperation, collaboration, reciprocity, and shared goals. Tomasello (2008) argues that such cooperation is a precondition for the development of complex culture (i.e., involving learning in several domains) and for complex communication systems such as language and music. He also emphasizes the importance of ratcheting, so that each skill becomes the building block for others (Tomasello 1999b; Tennie et al. 2009), thus explaining why human culture is so much richer than that of chimpanzees (Whiten et al. 1999). In humans, cooperation or helping others is evident even when there is no obvious benefit to the helper. Planning becomes critical in attaining difficult goals involving two or more individuals (e.g., hunting, sharing the spoils, achieving a division of labor that increases efficiency). Moreover, effective planning is greatly assisted by effective communication. There is reported nonhuman evidence of cooperative hunting (i.e., hunting in groups or packs), but these instances of apparent cooperation may simply maximize self-interest (for evidence on the lack of reciprocity in chimpanzee food sharing, see Gilby 2006). It is therefore unclear whether group hunting involves genuine cooperation. If cooperative motives were involved, the collaborators would be unlikely to fight vigorously over the carcass, as they typically do. A major social change in our species might be revealed through the study of the social brain, or by means of social neuroscience. Indeed, the persistence over many generations of culturally transmitted behaviors, such as Acheulean technology in the *Homo* lineage, suggests that there must have been intense social cooperation and inhibition of aggression, which would predict significant frontal brain enlargement.

While these hypotheses stress the benefits conferred by linguistic and musical interaction to individuals within the group, questions remain about who accrues the advantage (individual, kin group). The aforementioned hypotheses do not necessitate group selection, but are instead concerned with standard processes of natural and sexual selection, or with standard natural selection operating within the context of cultural niche construction (see, e.g., Laland et al. 1996). We do compete within groups, and such competition is often evident in processes of sexual selection, where we find the aesthetic extravaganzas of nature such as the peacock's tail and elaborate bird song, which are intended to impress conspecifics. In line with Darwin's original suggestion that music arises as a consequence of processes of sexual selection (Darwin 1871), it is possible that aspects of music, such as pulse-based isochrony, might not have derived from general processes of cooperation but from sexual selection pressures. Our ancestral setting of male territoriality and female exogamy could have led to synchronous chorusing by analogy with what occurs in some species of crickets and cicadas. Groups of territorial males could have become more effective at attracting migrating females by extending the reach of their hooting beyond territorial boundaries during the "carnival display." The key to such an extension would be precise temporal superposition of voices, requiring predictive timing, enabled by synchrony to a common pulse (Merker et al. 2009), although such a suggestion must remain speculative in the absence of clear evidence.

It must be noted that none of the above hypotheses are mutually exclusive. Instead, different strands and factors may have been operative at different times. While behavioral, cognitive, neuroscientific, anthropological, archaeological, and ethological evidence can be used to narrow the possible problem space and make predictions concerning efficacy and general chronological ordering of various factors, these predictions may be testable by means of emerging genetic techniques. For example, the effects of sexual selection in the hominin lineage in the emergence of communicative behaviors may be tracked by exploring the prevalence of sexual dimorphism (not just behavioral, but also in terms of brain developmental control by sex steroids) by analyzing gene expression as new techniques are developed to interrogate the fossil DNA of coexisting hominin species.

Much of this discussion concerns the emergence of human communicative capacities without attempting to delineate why humans should have a plethora of communicative capacities at their disposal. While proximate, and in some instances, ultimate, adaptive functions have been sketched out for aspects of language and music, we must question why we possess at least two communicative systems that overlap so significantly in their operational characteristics.

We considered six ways of conceiving of the evolutionary relationships between language and music (Figure 21.1). While the figure appears to present language and music as discrete or unitary domains, each may best be conceived of as opportunistic conflations of a mosaic of preexisting or extant capacities which themselves have diverse origins. Nevertheless, the models have heuristic value in delineating possible evolutionary relationships between music and language, given their current status and in the light of likely precursor capacities.

Of those precursor capacities, it can be suggested that the most compelling candidate for the origin of language and music is the capacity for vocal learning. All vocal animals produce innate calls expressive of emotional states. In the case of elaborate calls (still innate) these are sometimes called song, as in nonvocal-learning songbirds (suboscines) or gibbons. In addition, a subset of these callers acquires and produces learned song (oscine birds, some cetaceans, and humans). Finally, a single species (humans) add a third something, dependent on the crux of the second (i.e., vocal production learning); namely spoken language and vocal music. All vocal-learning species produce what has been interpreted by some as "music" in the form of complex sonic patternings ("song"). If one views vocal learning as providing a general form of "music" that has value in mediating social interactions but that does not embody

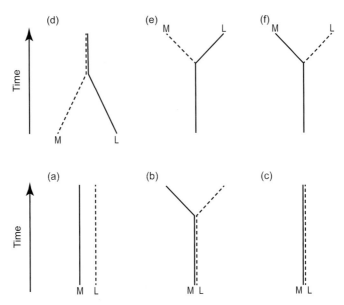

Figure 21.1 Six possible models for the evolutionary emergence of language (L) and music (M), with the timeline running from the bottom of the figure: (a) music and language have separate origins and remain distinct human faculties; (b) music and language have the same origin and diverge to become distinct faculties; (c) music and language have the same origin and remain indissociable; (d) music and language have separate and distinct origins and converge over time to share features; (e) language's origins precede those of music, which emerges as an offshoot (Herbert Spencer's view); (f) music's origins precede those of language, which emerges as an offshoot (Darwin's view).

propositionality, this may favor the last model (Figure 21.1f: language arising as a fairly late offshoot of music), with the emergence of language enabling semantic decompositionality and information transfer.

Even if we think in these terms, perhaps the distinction arises from a cultural bias, which would favor the third model (Figure 21.1c: common origins), with different cultures exploiting language and music for different ends. However, whether a culture distinguishes between language and music may not have the same perceptual consequences as cross-cultural differences in the use of color terms. It may be more relevant to aim to distinguish the ways in which music and language are bound to the evolution through natural selection of specific brain mechanisms or to processes of cultural evolution through the creation of ecological niches of cumulatively increasing social and artifactual complexity.

To understand other behaviors and capacities, broader contexts may be needed to assess the relationships between music and language. Perhaps phenomena such as language and music are different intersecting subsets of broader capacities, such as shared intentionality, or of general mimetic capacities. Moreover, to extrapolate from the kinds of enactment found in contemporary cultures to early human evolutionary history may well be unfounded. Such enactments work for contemporary humans because we are inclined to mine meaning from our physical and social environments. Perhaps that capacity, which involves a bird's eye view of the situation (in the form of a highly articulated theory of mind; Corradi-Del'Acqua et al. 2008), lies at the root of human communication. The emergence of a sense of self, a capacity to objectify ourselves and maintain a sense of self-continuity, and to relativize our experience of each other may underpin human communicative capacities.

Neither language nor music are purely vocal (or auditory); both constitute conceptual achievements that may be implemented by exploiting whatever tools are available at one's disposal (vocality, gesture, pantomime, external signing). Some of the traits that characterize both language and music, such as syntax and sequencing, are evident in other vocal-learning species. The vocal part of those traits has been inherited in the production part of the neural circuitry subserving learned vocalization in humans. The issue is to understand why humans combine compositional semantics with their vocal learning whereas other species do not. We have seen that some gestural theories favor motor learning, based on pantomime, in the development of meaningful protolanguage as a scaffolding for vocal learning, rather than postulating that our ancestors first developed meaningless "song."

Revisiting the issue of humans' exquisite control over vocalization in contrast to chimpanzees, one can ask what allowed humans to gain that control. For instance, if a chimp consistently fakes its vocalizations, it is likely to be ignored. Assuming a similar tendency in our common ancestor, how did we start to control our vocalizations? One possibility is that through "performing" to out-groups—making sounds that are out of place to deter predators (cf. Hagen and Hammerstein 2005)—early hominins derived the ability for displaced reference that is central to the linguistic faculty. For example, among the contemporary Mbendjele forest-dwelling hunter-gatherers, women sing and co-talk in the forest to deceive other animals. That cooperative behavior drives bonding within the group, and the deception is oriented outside the group. Imitation skills, including nonconscious mimicry (Lakin et al. 2003), may be especially significant in the emergence of human cooperative and communicative capacities (Lewis 2009). If individual pleasure and group bonding derive from coordinated vocalization and movement, that would create pressure for more communication, with vocalizations and gestures moving from initially holistic (Wray 1998) or social (Dunbar 1996) significance to increasing analytic status. For example, in contemporary egalitarian societies based on sharing and absence of social hierarchicality, explicit instruction is a claim to more knowledge and higher status and is thus rare. Most speech in such contemporary societies is "need-expression in the form of request," whereas much knowledge transmission is accomplished by means of pantomimetic display and mimicry (see Example 1 in the online supplemental information to this volume, http://esforum.de/sfr10/lewis.html:

Mongemba's account of an elephant hunt), which highlight expressiveness rather than efficiency of information transmission. It is notable that participants may experience a form of "transportation" as consequence of pantomimic representation, as the interaction requires displacement of the experienced world, potentially providing a trigger for the emergence of propositionality. Were such gestural, mimetic and "displacing" interactions to have part of early hominin repertoires, then a general theory linking gestural and vocal language origins with pedagogical process appears viable.

Although language and music may be functionally differentiable, that difference may be marked in such a way as to indicate its origin. For example, play interactions in canids are marked by a "play bow" to signify that the social and physical consequences of the interaction—within limits—are to be discounted. Music's "lack of consequence"—the fact that engagement with others in music sanctions types of behavior which may be socially unacceptable in other contexts—seems parallel to play as a mode of social interaction. Perhaps music constitutes an offshoot of a common communicative faculty (see Figure 21.1f), emerging through pressures imposed by increasing altriciality to co-opt juvenile, exploratory modes of thought and behavior into the adult repertoire (Cross 2003a).

Irrespective of these considerations, the major obstacle to greater clarity in our understanding of the origins of music and language is our lack of knowledge of music in cultures other than those of the contemporary West and of its relationships to other aspects of culture, including language. Most cultures have been explored as linguistic cultures, not as linguistic and musical cultures. Our knowledge of the music of those cultures is simply not commensurable with our knowledge of the languages, in part because of a lack of consensus about the key elements of music that would allow for cross-cultural comparison (despite heroic but much-criticized efforts such as those of Lomax 1968; for a sympathetic critique see Feld 1984). Until we have a sample of the rich information required to elaborate a principled theory of the relationships between what appears, from a Western, "etic" (i.e., outsider) perspective, to constitute music and language (requiring close collaboration between culture members and a range of human sciences), it will be difficult to gain any certainty about the origins of these human capacities. Our understanding of the relationship between language and music may be even more limited than we think it is. Undoubtedly, the first music was based on the voice—a biologically significant timbre—and much music across cultures continues to be based on the voice. Music is also intrinsically linked to regular and entrained collective movement—dance—in many societies. It is surprising, then, that most research on music cognition has used instrumental timbres, typically synthesized, rather than vocal timbre, and has only in recent years begun to explore music in the context of individual and collective movement. Recent work indicates, however, that adults remember melodies better when they are presented vocally (on the syllable "la") rather than instrumentally

(Weiss et al. 2012), and that joint movement, in the form of dance, can enhance memory for person attributes (Woolhouse, Tidhar, and Cross, in preparation).

Triadic Niche Construction in Relation to Music and Language Origins

A significant role in any exploration and explanation of language-music relationships is likely to be played by Iriki's theory of triadic niche construction (Iriki and Taoka 2012; Arbib and Iriki, this volume). A niche is a fragment of available environmental resources, and the process of ecological niche construction is a modification implemented by an animal to create his own niche. The interaction between the activity of the organism and its environment changes the environment, thereby changing selective pressures acting on the organism. In classic niche construction theory, there is a two-way interaction between behavior and environment. Quallo et al. (2009) have found that tool-use training in macaques led to an expansion in gray matter volume, affording extra neural machinery for the brain. This expansion in brain volume constitutes a "neural niche"—a newly available resource in the form of extra brain tissue—for future exploitation, affording the organism an increased range of responses (i.e., a "cognitive niche") introducing selective pressures which could, under some circumstances, amplify evolutionary effects.

This triad of neural niche, cognitive niche, and ecological niche are all operational for humans, allowing for an acceleration of their interaction in the course of our evolution, behavioral changes opening the door for later genetic changes. In effect, by changing the context of selection, different selection pressures come into play which may afford the possibility for new types of genetic change. If the information generated in the interaction is embedded in the structure of the environment, then it may be inherited by the next generation. In the context of human evolution, it could then be postulated that post-reproductive survival—the "grandmother" hypothesis—together with a means of transmitting knowledge critical to survival (e.g., such as language, or more particularly, mimetic and musical modes of presentation, display and participation) can allow the genetic pathway to be bypassed in the transmission of skill (Iriki 2010; Iriki and Taoka 2012). This would afford time for genetic assimilation, if it is necessary in the hominin lineage. This "Baldwinian evolution"—a mechanism that initially induced modification within the range of preprogrammed adaptation, and is then available for later mutations to optimize it—would be particularly beneficial for species with long life spans and low birth rates (e.g., in primates with humans at the extreme, who need to survive evolutionarily significant contingencies through an individual capacity to adapt). This stands in sharp contrast to species with short life spans and mass reproduction, which adapt to environmental changes through variations in their numerous offspring, expecting at least a few to survive. Both of these mechanisms, however, would aid the adaptive radiation of the species in the terrestrial ecosystem.

Comparative and Phylogenetic Issues

While triadic niche construction provides an extremely promising candidate mechanism for establishing and consolidating language and music in the human communicative repertoire, an exploration of origins requires consideration of evidence from beyond the hominin clade so as to avoid being blinkered by unacknowledged anthropocentrism (Figure 21.2a). Processes, structure, and behaviors in other species that are homologous to or convergent with those implicated in music and language are informative about their bases and manifestations in humans; after all, identification of sub-components of these complex capacities may be more directly observable in some nonhuman species. The concept of genetic or deep homology (see Fitch and Jarvis, this volume)—a genetic basis for behavioral capacities that may be common across different lineages, evidenced in the recruitment (particularly in ontogeny) of similar sets of complex genes to subserve similar functions—has significant potential to elucidate connections between types of behavioral capacity in different species: those which do not originate from a common ancestor as well as those that may be simply convergent, motivated by environmental selection pressures that operate on distantly related organisms to exploit specific types of environmental niche (Figure 21.2b).

While evolution is not progressive, there is a clear trend, at least in some lineages, toward increasing complexity, particularly in the hominin line. However, that complexity should not be considered independently of the systems that implement or enable it. With respect to song and language, when we compare, for example, a songbird with a human, we must first decide whether there is common design and then ask: How did these things emerge? Homology (i.e., the explanation that is likely the first port of call in answering the question) can be specified as either behavioral, anatomical or structural, developmental, or genetic (deep), this latter being evident in the common role played by certain genes (such as PAX6 in vision or FOXP2 in vocalization: see, e.g., White et al. 2006; Fernald 2000) in very distantly related species. We note, however, that a genetic network could have been recruited independently in two different species and may have functioned differently in different ancestor species. In the case of songbirds and humans, behavioral relationships in vocal capacity are clearly analogous rather than homologous, but may be motivated by deep homologies at the genetic level that afford the emergence of similarly functioning neural circuitry recruited for species-specific ends.

Hence it is possible to view aspects of the origins of music and language as embedded in a deep homology that is manifested at the genetic level; convergence may be occurring at the organ level (larynx in humans, syrinx in

Culture and Evolution

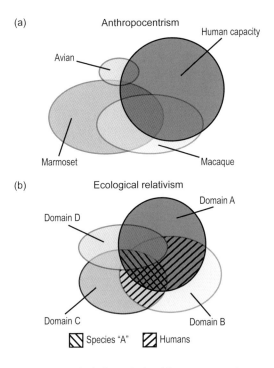

Figure 21.2 Venn diagrams depicting relationships among various cognitive capacities of different species. Sets are classified by (a) species (anthropocentrism) or (b) cognitive domains (ecological relativisim). In the anthropocentric view (a), cognitive domains are expressed as subsets within respective species set, partly overlapping with other species. In this way, humans tend to privilege only those included in the human set, making it difficult to recognize that other species may have cognitive abilities superior to humans (as shaded outside the "human set"). This perspective can lead to the misleading perception that nonhuman species are intrinsically inferior to humans. In contrast, when sets are classified by cognitive domains (b), species are depicted through a combination of subsets to illustrate inter-relationships between species' capacities. These cognitive domains and their combinations in species must be considered to have evolved through interactions with ecological conditions of habitats, thus, ecological relativism.

birds) but homology at the genetic level. The vocality that underpins speech and music may have deep homology across all vocal learners, with the motor learning circuitry being co-opted independently for vocal learning in different species. Nevertheless, vocal learning is only one of the constellations of features that can be identified as underpinning language and music. Humans' complex sociality, excessive brain (cortex) size, and capacity for cultural conservation and transformation of knowledge all seem likely to have played a significant role in shaping our communicative capacities. It would be highly desirable to track the extent to which those aspects shared with our closest nonhuman relative represent true homologies. However, we are limited by a lack of knowledge of primate evolution immediately prior to our last common ancestor, whose capacities must be extrapolated (perhaps uninformatively) from those of their descendants. Nevertheless, even in the absence of such data it might be possible to use datable divergences between existing nonhuman primate species to explore human cognitive functions such as language and music. For example, new world monkeys may provide a fertile experimental model as they have a wide range of vocal capacities as well as cooperative social structures. In the "old world," humans established their unique niche by dividing resources with other primates—apes and old world monkeys. In contrast, in the "new world," where humans did not exist, adaptive radiation should have developed differently. That is, the traits which characterize human-specific cognition, of which precursors should have derived from common ancestors and become extinct in nonhuman old world primates, might have preserved in the new world monkey lineages by deep homology and could be expressed in extant taxa through epigenetic interactions as convergent evolution. As such, new world monkeys could represent an ideal animal model to study various aspects of human-specific higher cognitive functions.

Conclusion

To return to the point made at the outset: when considering relationships between language and music from cultural and evolutionary perspectives, there is a pressing need to avoid presentist and anthropocentric biases in making inferences about cultural categories and evolutionary trajectories. Music and language may be different domains of human thought and behavior; they may be different manifestations of the same underlying capacities; or they may be the same suite of communicative capacities co-opted for different ends in different situations. They may have evolved separately or conjointly, or they may have merged or split over the course of human evolution. They or their subcomponents may be present in the repertoire of other species, or they may be unique to humans. Only by synthesizing evidence from the whole range of human sciences, in the context of investigations that are alert to cross-cultural differences in the conceptualization and implementation of communicative skills and the features shared with other species, can we achieve a degree of defensible clarity in our understanding.

Bibliography

Note: Numbers in square brackets denote the chapter in which an entry is cited.

Abbott, H. P. 2008. The Cambridge Introduction to Narrative, 2nd edition. Cambridge: Cambridge Univ. Press. [7]

Abe, K., and D. Watanabe. 2011. Songbirds possess the spontaneous ability to discriminate syntactic rules. *Nat. Neurosci.* **14**:1067–1074. [20]

Aboitiz, F. 2012. Gestures, vocalizations and memory in language origins. *Front. Evol. Neurosci.* **4**:2. [1]

Aboitiz, F., S. Aboitiz, and R. García. 2010. The phonological loop: A key innovation in human evolution. *Curr. Anthropol.* **51**:S55–S65. [17]

Aboitiz, F., and R. García. 1997. The evolutionary origin of the language areas in the human brain. A neuroanatomical perspective. *Brain Res. Rev.* **25**:381–396. [19]

———. 2009. Merging of phonological and gestural circuits in early language evolution. *Rev. Neurosci.* **20**:71–84. [17, 19]

Aboitiz, F., R. García, C. Bosman, and E. Brunetti. 2006a. Cortical memory mechanisms and language origins. *Brain Lang.* **98**:40–56. [19]

Aboitiz, F., R. García, E. Brunetti, and C. Bosman. 2006b. The origin of Broca's area and its connections from an ancestral working memory network. In: Broca's region, ed. Y. Grodzinsky and K. Amunts, pp. 3–16. Oxford: Oxford Univ. Press. [19]

Abrams, D. A., A. Bhatara, S. Ryali, et al. 2011. Decoding temporal structure in music and speech relies on shared brain resources but elicits different fine-scale spatial patterns. *Cereb. Cortex* **21**:1507–1518. [14]

Ackermann, H., and A. Riecker. 2004. The contribution of the insula to motor aspects of speech production: A review and a hypothesis. *Brain Lang.* **89**:320–328. [13]

Adolphs, R., D. Tranel, and A. R. Damasio. 2003. Dissociable neural systems for recognizing emotions. *Brain Cogn.* **52**:61–69. [4]

Aglioti, S. M., P. Cesari, M. Romani, and C. Urgesi. 2008. Action anticipation and motor resonance in elite basketball players. *Nat. Neurosci.* **11**:1109–1116. [8]

Agus, T. R., C. Suied, S. J. Thorpe, and D. Pressnitzer. 2012. Fast recognition of musical sounds based on timbre. *JASA* **131**:4124–4133. [17]

Ahissar, E., S. Nagarajan, M. Ahissar, et al. 2001. Speech comprehension is correlated with temporal response patterns recorded from auditory cortex. *PNAS* **98**:13,367–13,372. [17]

Ahveninen, J., M. Hamalainen, I. P. Jääskeläinen, et al. 2011. Attention-driven auditory cortex short-term plasticity helps segregate relevant sounds from noise. *PNAS* **108**:4182–4187. [17]

Alajouanine, T. 1948. Aphasia and artistic realization. *Brain* **71**:229–241. [4]

Albert, M. L., R. W. Sparks, and N. A. Helm. 1973. Melodic intonation therapy for aphasia. *Arch. Neurol.* **29**:130–131. [14]

Allman, J. M., and B. Martin. 2000. Evolving Brains. New York: Sci. Amer. Library. [16]

Allott, R. 1992. The motor theory of language: Origin and function. In: Language Origin: A Multidisciplinary Approach, ed. J. Wind et al., pp. 105–119. Dordrecht, The Netherlands: Kluwer. [20]

Allu, S. O., J. F. Houde, and S. S. Nagarajan. 2009. Motor induced suppression of the auditory cortex. *J. Cogn. Neurosci.* **2**:791–802. [17]

Alluri, V., P. Toiviainen, I. P. Jääskeläinen, et al. 2012. Large-scale brain networks emerge from dynamic processing of musical timbre, key, and rhythm. *NeuroImage* **59**:3677–3689. [8]

Allwood, J. 2007. Cooperation, competition, conflict and communication. Gothenburg Papers in Theoretical Linguistic 94: Göteborg Univ., Dept. of Linguistics. [21]

Andersen, R. A., and C. A. Buneo. 2002. Intentional maps in posterior parietal cortex. *Annu. Rev. Neurosci.* **25**:189–220. [4]

Andrew, R. J. 1972. The information potentially available in mammal displays. In: Nonverbal Communication, ed. R. A. Hinde, pp. 179–203. Cambridge: Cambridge Univ. Press. [5]

Antović, M. 2009. Towards the semantics of music: The twentieth century. *Lang. History* **52**:119–129. [14]

Anvari, S. H., L. J. Trainor, J. Woodside, and B. A. Levy. 2002. Relations among musical skills, phonological processing, and early reading ability in preschool children. *J. Exp. Child Psychol.* **83**:111–130. [17]

Anwander, A., M. Tittgemeyer, D. Y. von Cramon, A. D. Friederici, and T. R. Knosche. 2007. Connectivity-based parcellation of Broca's area. *Cereb. Cortex* **17**:816–825. [9]

Arbib, M. A. 1975. Artificial intelligence and brain theory: Unities and diversities. *Ann. Biomed. Eng.* **3**:238–274. [1]

———. 1981. Perceptual structures and distributed motor control. In: Handbook of Physiology: The Nervous System II. Motor Control, ed. V. B. Brooks, pp. 1449–1480. Bethesda: American Physiological Soc. [1, 8, 15, 17]

———. 1989. The Metaphorical Brain 2: Neural Networks and Beyond. New York: Wiley-Interscience. [1, 8, 13, 17]

———. 2002. The mirror system, imitation, and the evolution of language. In: Imitation in Animals and Artifacts. Complex Adaptive Systems, ed. K. Dautenhahn and C. L. Nehaniv, pp. 229–280. Cambridge, MA: MIT Press. [19]

———. 2005a. From monkey-like action recognition to human language: An evolutionary framework for neurolinguistics. *Behav. Brain Sci.* **28**:105–124; discussion 125–167. [3, 4, 7, 18, 20]

———. 2005b. The mirror system hypothesis: How did protolanguage evolve. In: Language Origins, ed. M. Tallerman. Oxford: Oxford Univ. Press. [3, 21]

———. 2006a. Aphasia, apraxia, and the evolution of the language-ready brain. *Aphasiology* **20**:1125–1155. [8, 14]

———. 2006b. A sentence is to speech as what is to action? *Cortex* **42**:507–514. [1]

———. 2008. Holophrasis and the protolanguage spectrum. *Interaction Stud.* **9**:151–165. [19]

———. 2009. Invention and community in the emergence of language: A perspective from new sign languages. In: Foundations in Evolutionary Cognitive Neuroscience: Introduction to the Discipline, ed. S. M. Platek and T. K. Shackelford, pp. 117–152. Cambridge: Cambridge Univ. Press. [19]

———. 2010. Mirror system activity for action and language is embedded in the integration of dorsal and ventral pathways. *Brain Lang.* **112**:12–24. [6, 7]

———. 2011. Niche construction and the evolution of language: Was territory scavenging the one key factor? *Interaction Stud.* **12**:162–193. [19]

———. 2012. How the Brain Got Language: The Mirror System Hypothesis. Oxford: Oxford Univ. Press. [4, 8, 15, 19]

———. 2013. Neurolinguistics. In: The Oxford Handbook of Linguistic Analysis, 2nd edition, ed. B. Heine and H. Narrog. New York: Oxford Univ. Press. [15]

Arbib, M. A., and D. Bickerton, eds. 2010. The Emergence of Protolanguage: Holophrasis vs Compositionality. Philadelphia, Amsterdam: John Benjamins. [19]
Arbib, M. A., and J. J. Bonaiuto. 2012. Multiple levels of spatial organization: World graphs and spatial difference learning. *Adap. Behav.* **20**:287–303. [15]
Arbib, M. A., A.-M. Bonnel, S. Jacobs, and S. H. Frey. 2009. Tool use and the distalization of the end-effector. *Psychol. Res.* **73**:441–462. [19]
Arbib, M. A., and M. Bota. 2003. Language evolution: Neural homologies and neuroinformatics. *Neural Netw.* **16**:1237–1260. [4]
Arbib, M. A., P. Érdi, and J. Szentágothai. 1998. Neural Organization: Structure, Function, and Dynamics. Cambridge, MA: MIT Press. [15]
Arbib, M. A., and J.-M. Fellous. 2004. Emotions: From brain to robot. *Trends Cogn. Sci.* **8**:554–561. [1, 5, 8]
Arbib, M. A., and M. B. Hesse. 1986. The Construction of Reality. Cambridge: Cambridge Univ. Press. [1, 7, 8]
Arbib, M. A., and J. Lee. 2008. Describing visual scenes: Towards a neurolinguistics based on construction grammar. *Brain Res.* **122**:146–162. [7]
Arbib, M. A., K. Liebal, and S. Pika. 2008. Primate vocalization, gesture, and the evolution of human language. *Curr. Anthropol.* **49**:1053–1076. [1]
Arbib, M. A., and I. Lieblich. 1977. Motivational learning of spatial behavior. In: Systems Neuroscience, ed. J. Metzler, pp. 221–239. New York: Academic. [15]
Arbib, M. A., A. Plangprasopchok, J. Bonaiuto, and R. E. Schuler. 2013. The neuroinformatics of brain modeling, Part I: From empirical data to brain operating principles. *Neuroinformatics*, in press [8]
Arbib, M. A., and G. Rizzolatti. 1997. Neural expectations: A possible evolutionary path from manual skills to language. *Comm. Cogn.* **29**:393–424. [15, 19]
Arch-Tirado, E., B. McCowan, J. Saltijeral-Oaxaca, I. Zarco de Coronado, and J. Licona-Bonilla. 2000. Development of isolation-induced vocal behavior in normal-hearing and deafened Guinea pig infants. *J. Speech Lang. Hearing Res.* **43**:432–440. [20]
Ariew, A. 1999. Innateness is canalization: In defense of a developmental account of innateness. In: Where Biology Meets Psychology: Philosophical Essays, ed. V. G. Hardcastle, pp. 117–138. Cambridge, MA: MIT Press. [20]
Aristotle. 350 BC. The History of Animals (translated by A. L. Peck). Loeb Classical Library, London: Heinemann. [20]
Arleo, A. 2006. Do children's rhymes reveal universal metrical patterns? In: Children's Literature: Critical Concepts in Literary and Cultural Studies, ed. P. Hunt, vol. 4, pp. 39–56. London: Routledge. [18]
Armstrong, D. F., W. C. Stokoe, and S. E. Wilcox. 1984. Signs of the origin of syntax. *Curr. Anthropol.* **35**:349–368. [20]
———. 1995. Gesture and the Nature of Language. Cambridge: Cambridge Univ. Press. [19]
Armstrong, D. F., and S. E. Wilcox. 2007. The Gestural Origin of Language. Oxford: Oxford Univ. Press. [19]
Arnal, L. H., B. Morillon, C. A. Kell, and A. L. Giraud. 2009. Dual neural routing of visual facilitation in speech processing. *Neuroscience* **29**:13,445–13,453. [17]
Arom, S. 1978. *Centrafrique: Anthologie de la Musique des Pygmées Aka*, (linear notes). Ocora. [2]
———. 1985. *Polyphonies et Polyrythmies Instrumentals d'Afrique Centrale. Structure et Méthodologie* (vols. 1 and 2). Paris: SELAF. [2]
Aroniadou, V. A., and A. Keller. 1993. The patterns and synaptic properties of horizontal intracortical connections in the rat motor cortex. *J. Neurophysiol.* **70**:1553–1569. [20]

Aronoff, M. 2007. In the beginning was the word. *Language* **83**:803–830. [11]

Aronoff, M., I. Meir, C. Padden, and W. Sandler. 2008. The roots of linguistic organization in a new language. *Interaction Stud.* **9**:133–153. [1]

Arriaga, G., Zhou, E. P., and Jarvis, E. D. 2012. Of mice, birds, and men: The mouse ultrasonic song system has some features similar to humans and song-learning birds. *PLoS ONE* **7**:e46610. [20]

Arvaniti, A., D. R. Ladd, and I. Mennen. 1998. Stability of tonal alignment: The case of Greek prenuclear accents. *J. Phonetics* **26**:3–25. [11, 12]

Ayotte, J., I. Peretz, and K. L. Hyde. 2002. Congenital amusia: A group study of adults afflicted with a music-specific disorder. *Brain* **125**:238–251. [13, 14, 17]

Bachner-Melman, R., C. Dina, and A. H. Zohar. 2005. *AVPR1a* and *SLC6A4* Gene polymorphisms are associated with creative dance performance. *PLoS Genet.* **1**: e42 [1, 13]

Bachorowski, J., and M. Owren. 2003. Sounds of emotion: The production and perception of affect-related vocal acoustics. *Ann. NY Acad. Sci.* **1000**:244–265. [18]

Baddeley, A. D. 2000. The episodic buffer: A new component of working memory? *Trends Cogn. Sci.* **11**:4. [8]

———. 2003. Working memory: Looking back and looking forward. *Nat. Rev. Neurosci.* **4**:829–839. [1]

———. 2007. Working Memory, Thought, and Action. Oxford: Oxford Univ. Press. [8]

Badre, D., R. A. Poldrack, E. J. Pare-Blagoev, R. Z. Insler, and A. D. Wagner. 2005. Dissociable controlled retrieval and generalized selection mechanisms in ventrolateral prefrontal cortex. *Neuron* **47**:907–918. [9]

Baess, P., J. Horvath, T. Jacobsen, and E. Schroger. 2011. Selective suppression of self-initiated sounds in an auditory stream: An ERP study. *Psychophysiology* **48**:1276–1283. [17]

Baggio, G., and P. Hagoort. 2011. The balance between memory and unification in semantics: A dynamic account of the N400. *Lang. Cogn. Proc.* **26**:1338–1367. [9, 14]

Bahlmann, J., R. I. Schubotz, and A. D. Friederici. 2008. Hierarchical artificial grammar processing engages Broca's area. *Neuroimage* **42**:525–534. [4]

Bahuchet, S. 1996. *Fragments pour une histoire de la Forêt Africaine et de son peuplement: Les données linguistiques et culturelles.* In: *L'Alimentation en Forêt Tropicale: Interactions Bioculturelles et Perspectives de Développement*, ed. C. M. Hladik et al., pp. 97–119. Paris: Éditions UNESCO. [2]

Baker, M. C. 2001. Atoms of Language: The Mind's Hidden Rules of Grammar. London: Basic Books. [3]

Baker, M. C., and M. A. Cunningham. 1985. The biology of bird song dialects. *Behav. Brain Sci.* **8**:85–133. [20]

Baldwin, D., A. Andersson, J. Saffran, and M. Meyer. 2008. Segmenting dynamic human action via statistical structure. *Cognition* **106**:1382–1407. [18]

Balkwill, L., and W. F. Thompson. 1999. A cross-cultural investigation of the perception of emotion in music: Psychophysical and cultural cues. *Music Percep.* **17**:43–64. [1]

Ball, P. 2010. The Music Instinct: Why Music Works and Why We Can't Do Without It. London: Bodley Head. [5]

Balzano, G. J. 1980. The group-theoretic description of 12-fold and microtonic pitch systems. *Computer Music Journal* **4**:66–84. [18]

Banse, R., and K. R. Scherer. 1996. Acoustic profiles in vocal emotion expression. *J. Pers. Soc. Psychol.* **70**:614–636. [5]

Bänziger, T., and K. R. Scherer. 2005. The role of intonation in emotional expressions. *Speech Comm.* **46**:252–267. [5]

Bar, M. 2009. The proactive brain: Memory for predictions. *Philos. Trans. R. Soc. Lond. B* **364**:1235–1243. [8]
Bar-Hillel, Y. 1964. Language and Information. Reading, MA: Addison-Wesley. [15]
Barris, R. W., M. D. Schuman, and H. R. Schuman. 1953. Bilateral anterior cingulated gyrus lesions: Syndrome of the anterior cingulate gyri. *Neurology* **3**:44–52. [20]
Baruch, C., and C. Drake. 1997. Tempo discrimination in infants. *Infant Behav. Dev.* **20**:573–577. [13]
Basso, E. 1985. A Musical View of the Universe. Philadelphia: Univ. of Pennsylvania Press. [21]
Basso, K. H. 1970. "To give up on words": Silence in Western Apache culture. *SW J. Anthropol.* **26**:213–230. [21]
Bauer, E. E., M. J. Coleman, T. F. Roberts, et al. 2008. A synaptic basis for auditory-vocal integration in the songbird. *J. Neurosci.* **28**:1509–1522. [20]
Baumeister, R., and M. Leary. 1995. The need to belong: Desire for interpersonal attachments as a fundamental human motivation. *Psychol. Bull.* **117**:497–497. [6]
Baumgartner, T., K. Lutz, C. Schmidt, and L. Jäncke. 2006. The emotional power of music: How music enhances the feeling of affective pictures. *Brain Res.* **1075**:151–164. [7]
Baur, B., I. Uttner, J. Ilmberger, G. Fesl, and N. Mai. 2000. Music memory provides access to verbal knowledge in a patient with global amnesia. *Neurocase* **6**:415–421. [13]
Beckman, M. E. 1986. Stress and Non-Stress Accent. Dordrecht: Foris. [11]
Beiser, D. G., S. E. Hua, and J. C. Houk. 1997. Network models of the basal ganglia. *Curr. Opin. Neurobiol.* **7**:185–190. [20]
Belin, P., P. E. Bestelmeyer, M. Latinus, and R. Watson. 2011. Understanding voice perception. *Br. J. Psychol.* **102**:711–725. [17]
Belin, P., R. J. Zatorre, and P. Ahad. 2002. Human temporal-lobe response to vocal sounds. *Cogn. Brain Res.* **13**:17–26. [18]
Belin, P., R. J. Zatorre, P. Lafaille, P. Ahad, and B. Pike. 2000. Voice-selective areas in human auditory cortex. *Nature* **403**:309–312. [11, 13, 17]
Bell, A. J. 1999. Levels and loops: The future of artificial intelligence and neuroscience. *Philos. Trans. R. Soc. Lond. B* **354**:2013–2020. [16]
Bemis, D. K., and L. Pylkkänen. 2011. Simple composition: A magnetoencephalography investigation into the comprehension of minimal linguistic phrases. *Neuroscience* **31**: 2801–2814. [17]
Benade, A. H. 1990. Fundamentals of Musical Acoustics, 2nd edition. New York: Courier Dover Publications. [5]
Bendixen, A., I. SanMiguel, and E. Schroger. 2011. Early electrophysiological indications for predictive processing in audition: A review. *Int. J. Psychophysiol.* **83**:120–131. [17]
Bendor, D., and X. Wang. 2005. The neuronal representation of pitch in primate auditory cortex. *Nature* **436**:1161–1165. [14, 17]
Bengtsson, S. L., F. Ullen, H. H. Ehrsson, et al. 2009. Listening to rhythms activates motor and premotor cortices. *Cortex* **45**:62–71. [17]
Ben Shalom, D., and D. Poeppel. 2008. Functional anatomic models of language: Assembling the pieces. *Neuroscientist* **14**:119–127. [9]
Benson, D. F., and A. Ardila. 1996. Aphasia: A Clinical Perspective. New York: Oxford Univ. Press. [20]
Bergeson, T. R., and S. E. Trehub. 1999. Mothers' singing to infants and preschool children. *Infant Behav. Dev.* **22**:51–64. [18]

Bergeson, T. R., and S. E. Trehub. 2002. Absolute pitch and tempo in mothers' songs to infants. *Psychol. Sci.* **13**:71–74. [18, 17]

———. 2006. Infants' perception of rhythmic patterns. *Music Percep.* **23**:345–360. [18]

———. 2007. Signature tunes in mothers' speech to infants. *Infant Behav. Dev.* **30**:648–654. [18]

Bernacchia, A., H. Seo, D. Lee, and X. J. Wang. 2011. A reservoir of time constants for memory traces in cortical neurons. *Nat. Neurosci.* **14**:366–372. [17]

Bernardet, U., and P. F. M. J. Verschure. 2009. The eXperience Induction Machine: A new paradigm for mixed reality interaction design and psychological experimentation. In: The Engineering of Mixed Reality Systems, ed. E. Dubois et al. Berlin: Springer. [16]

Bernardet, U., and P. F. M. J. Verschure. 2010. iqr: A tool for the construction of multi-level simulations of brain and behaviour. *Neuroinformatics* **8**:113–334. [16]

Bernstein, L. 1976. The Unanswered Question. Cambridge, MA: Harvard Univ. Press. [10]

Bernstein, N. A. 1996. On dexterity and its development. In: Dexterity and Its Development, ed. M. L. Latash and M. T. Turvey, pp. 3–244. Mahwah, NJ: Lawrence Erlbaum. [4]

Besson, M., F. Faita, I. Peretz, A. M. Bonnel, and J. Requin. 1998. Singing in the brain: Independence of lyrics and tunes. *Psychol. Sci.* **9**:494–498. [13]

Besson, M., A. Frey, and M. Aramaki. 2011. Is the distinction between intra- and extra-musical processing implemented in the brain? Comment on "Towards a neural basis of processing musical semantics" by Stefan Koelsch. *Phys. Life Rev.* **8**:112–113. [8]

Best, C. T., and G. W. McRoberts. 2003. Infant perception of nonnative consonant contrasts that adults assimilate in different ways. *Language and Speech* **46**:183–216. [18]

Best, C. T., G. W. McRoberts, R. LaFleur, and J. Silver-Isenstadt. 1995. Divergent developmental patterns for infants' perception of two nonnative consonant contrasts. *Infant Behav. Dev.* **18**:339–350. [18]

Best, C. T., G. W. McRoberts, and N. T. Sithole. 1988. Examination of perceptual reorganization for non-native speech contrasts: Zulu click discrimination by English speaking adults and infants. *J. Exp. Psychol. Hum. Percept. Perform.* **14**:345–360. [18]

Bestelmeyer, P., D. R. Ladd, and P. Belin. 2010. Listeners' brains respond more to native accent speakers. Poster presented at Soc. of Neuroscience 2010. San Diego [11].

Beth, E. W., and J. Piaget. 1966. Mathematical Epistemology and Psychology (translated by W. Mays). Dordrecht: Reidel. [8]

Bever, T. G., and R. J. Chiarello. 1974. Cerebral dominance in musicians and nonmusicians. *Science* **185**:537–539. [13]

Bharucha, J. J. 1994. Tonality and expectation. In: Musical Perceptions, ed. R. Aiello and J. Sloboda, pp. 213–239. Oxford: Oxford Univ. Press. [12]

Bhattacharya, J., E. Pereda, and H. Petsche. 2001. Long-range synchrony in the gamma band: Role in music perception. *J. Neurosci.* **21**:6329–6337. [15]

Bickerton, D. 1995. Language and Human Behavior. Seattle: Univ. of Washington Press. [18, 19]

———. 2009. Adam's Tongue: How Humans Made Language, How Language Made Humans. New York: Hill & Wang. [19, 21]

Bidelman, G. M., and A. Krishnan. 2011. Brainstem correlates of behavioral and compositional preferences of musical harmony. *NeuroReport* **22**:212–216. [7]

Bigand, E. 1993. Contributions of music to research on human auditory cognition. In: Thinking in Sound, ed. S. McAdams and E. Bigand, pp. 231–277. Oxford: Oxford Univ. Press. [14]

———. 2003. More about the musical expertise of musically untrained listeners. *Ann. NY Acad. Sci.* **999**:304–312. [18]

Bigand, E., R. Parncutt, and F. Lerdahl. 1996. Perception of musical tension in short chord sequences: The influence of harmonic function, sensory dissonance, horizontal motion, and musical training. *Percept. Psychophys.* **58**:125–141. [6]

Bigand, E., B. Poulin, B. Tillmann, F. Madurell, and D. A. D'Adamo. 2003. Sensory versus cognitive components in harmonic priming. *J. Exp. Psychol. Hum. Percept. Perform.* **29**:159–171. [14]

Binder, J. 2000. The new neuroanatomy of speech perception. *Brain* **123:**2371–2372. [9]

Binkofski, F., G. Buccino, K. M. Stephan, et al. 1999. A parieto-premotor network for object manipulation: Evidence from neuroimaging. *Exp. Brain Res.* **128**:210–213. [4]

Bispham, J. C. 2006. Rhythm in music: What is it? Who has it? And Why? *Music Percep.* **24**:125–134. [5]

Biswal, B., and S. S. Kannurpatti. 2009. Resting-state functional connectivity in animal models: Modulations by exsanguination. *Methods Mol. Biol.* **489**:255–274. [9]

Biswal, B., F. Z. Yetkin, V. M. Haughton, and J. S. Hyde. 1995. Functional connectivity in the motor cortex of resting human brain using echo-planar MRI. *Magn. Reson. Imaging* **34**:537–541. [9]

Bizley, J. K., K. M. M. Walker, B. W. Silverman, A. J. King, and J. W. H. Schnupp. 2009. Interdependent encoding of pitch, timbre, and spatial location in auditory cortex. *J. Neurosci.* **29**:2064–2075. [17]

Blacking, J. 1973/2000. How Musical is Man? Seattle: Univ. of Washington Press. [2]

———. 1985. Movement, dance, music and Venda girls' initiation. In: Society and the Dance: The Social Anthropology of Process and Performance, ed. P. Spencer, pp. 64–91. Cambridge: Cambridge Univ. Press. [2]

———. 1995. Music, Culture and Experience. London: Univ. of Chicago Press. [18]

Blaesing, B., M. Puttke, and T. Schack, eds. 2010. The Neurocognition of Dance. London: Psychology Press. [1]

Bloch, M. 1998. How We Think They Think: Anthropological Approaches to Cognition, Memory and Literacy. Boulder: Westview Press. [2]

Blood, A. J., and R. J. Zatorre. 2001. Intensely pleasurable responses to music correlate with activity in brain regions implicated in reward and emotion. *PNAS* **98**:11,818–11,823. [6]

Blood, A. J., R. J. Zatorre, P. Bermudez, and A. C. Evans. 1999. Emotional responses to pleasant and unpleasant music correlate with activity in paralimbic brain regions. *Nat. Neurosci.* **2**:382–387. [6, 7]

Bock, K. 2011. How much correction of syntactic errors are there, anyway? *Lang. Linguist. Compass* **5**:322–335. [9]

Boh, B., S. C. Herholz, C. Lappe, and C. Pantev. 2011. Processing of complex auditory patterns in musicians and non-musicians. *PLoS ONE* **6**:e21458. [17]

Bohland, J. W., D. Bullock, and F. H. Guenther. 2010. Neural representations and mechanisms for the performance of simple speech sequences. *J. Cogn. Neurosci.* **22**:1504–1529. [14, 15]

Bolhuis, J. J., K. Okanoya, and C. Scharff. 2010. Twitter evolution: Converging mechanisms in birdsong and human speech. *Nat. Rev. Neurosci.* **11**:747–759. [20]

Bolinger, D. 1964. Intonation: Around the edge of language. *Harvard Ed. Rev.* **34**:282–296. [5, 11]

———. 1986. Intonation and Its Parts: Melody in Spoken English. Stanford: Stanford Univ. Press. [10]

Boltz, M. G. 2001. Musical soundtracks as a schematic influence on the cognitive processing of filmed events. *Music Percep.* **18**:427–454. [8]

Boltz, M. G. 2001. 2004. The cognitive processing of film and musical soundtracks. *Mem. Cogn.* **32**:1194–1205. [7, 8]

Boltz, M. G., B. Ebendorf, and B. Field. 2009. Audiovisual interactions: The impact of visual information on music perception and memory. *Music Percep.* **27**:43–59. [7, 8, 18]

Boltz, M. G., M. Schulkind, and S. Kantra. 1991. Effects of background music on remembering of filmed events. *Mem. Cogn.* **19**:595–606. [7, 8]

Bonaiuto, J. B., and M. A. Arbib. 2010. Extending the mirror neuron system model, II: What did I just do? A new role for mirror neurons. *Biol. Cybern.* **102**:341–359. [15]

Bonini, L., S. Rozzi, F. U. Serventi, et al. 2010. Ventral premotor and inferior parietal cortices make distinct contribution to action organization and intention understanding. *Cereb. Cortex* **20**:1372–1385. [4]

Bonini, L., F. U. Serventi, L. Simone, et al. 2011. Grasping neurons of monkey parietal and premotor cortices encode action goals at distinct levels of abstraction during complex action sequences. *J. Neurosci.* **31**:5876–5886. [4]

Bonnel, A. M., F. Faita, I. Peretz, and M. Besson. 2001. Divided attention between lyrics and tunes of operatic songs: Evidence for independent processing. *Percep. Psychophys.* **63**:1201–1213. [13]

Bookheimer, S. Y., T. A. Zeffiro, T. A. Blaxton, P. W. Gaillard, and W. H. Theodore. 2000. Activation of language cortex with automatic speech tasks. *Neurology* **55**:1151–1157. [20]

Boomsliter, P., W. Creel, and S. R. Powers. 1970. Sensations of tone as perceptual forms. *Psychol. Rev.* **77**:534–545. [7]

Bordwell, D. 1996. Making Meaning: Inference and Rhetoric in the Interpretation of Cinema. Cambridge, MA: Harvard Univ. Press. [7]

———. 2006. The Way Hollywood Tells It: Story and Style in Modern Movies. Berkeley: Univ. of California Press. [7]

Bottjer, S. W., and B. Altenau. 2010. Parallel pathways for vocal learning in basal ganglia of songbirds. *Nat. Neurosci.* **13**:153–155. [20]

Bottjer, S. W., J. D. Brady, and B. Cribbs. 2000. Connections of a motor cortical region in zebra finches: Relation to pathways for vocal learning. *J. Comp. Neurol.* **420**:244–260. [20]

Bottjer, S. W., E. A. Miesner, and A. P. Arnold. 1984. Forebrain lesions disrupt development but not maintenance of song in passerine birds. *Science* **224**:901–903. [20]

Boughey, M. J., and N. S. Thompson. 1981. Song variety in the Brown Thrasher, *Toxostoma rufum*. *Z. Tierpsychol.* **56**:47–58. [20]

Bowling, D. L., K. Gill, J. D. Choi, J. Prinz, and D. Purves. 2010. Major and minor music compared to excited and subdued speech. *J. Acoust. Soc. Am.* **127**:491–503. [5]

Boyd, R., and P. J. Richerson. 2005. Solving the puzzle of human cooperation in evolution and culture. In: Evolution and Culture, ed. S. Levinson and P. Jaisson, pp. 105–132, A Fyssen Foundation Symposium. Cambridge, MA: MIT Press. [3]

Bozic, M., L. K. Tyler, D. T. Ives, B. Randall, and W. D. Marslen-Wilson. 2010. Bihemispheric foundations for human speech comprehension. *PNAS* **107**:17,439–17,444. [9]

Brancucci, A., M. di Nuzzo, and L. Tommasi. 2009. Opposite hemispheric asymmetries for pitch identification in absolute pitch and nonabsolute pitch musicians. *Neuropsychologia* **47**:2937–2941. [17]

Brass, M., R. M. Schmitt, S. Spengler, and G. Gergely. 2007. Investigating action understanding: Inferential processes versus action simulation. *Curr. Biol.* **17**:2117–2121. [4]

Brauer, J., A. Anwander, and A. D. Friederici. 2011a. Neuroanatomical prerequisites for language function in the developing brain. *Cereb. Cortex* **21**:459–466. [17]

———. 2011b. Neuroanatomical prerequisites for language functions in the maturing brain. *Cereb. Cortex* **21**:459–466. [3, 9]

Brechmann, A., and H. Scheich. 2005. Hemispheric shifts of sound representation in auditory cortex with conceptual listening. *Cereb. Cortex* **15**:578–587. [17]

Bregman, A. 1990. Auditory Scene Analysis: The Perceptual Organization of Sound. Cambridge, MA: MIT Press. [11, 12, 15, 17]

Bregman, M. R., A. D. Patel, and T. Q. Gentner. 2012. Stimulus-dependent flexibility in non-human auditory pitch processing. *Cognition* **122**:51–60. [14, 17, 18, 21]

Brodmann, K. 1909. *Vergleichende Lokalisationslehre der Großhirnrinde in ihren Prinzipien dargestellt auf Grund des Zellenbaues*. Leipzig: Barth. [1, 4, 15]

Brooks, M. (director), and M. Hertzberg (producer). 1974. Blazing Saddles. US: Warner Bros. [7]

Brooks, R. 1991. New approaches to robotics. *Science* **253**:1227–1232. [16]

Brosch, M., E. Selezneva, C. Bucks, and H. Scheich. 2004. Macaque monkeys discriminate pitch relationships. *Cognition* **91**:259–272. [17]

Brosch, T., D. Sander, and K. R. Scherer. 2007. That baby caught my eye: Attention capture by infant faces. *Emotion* **7**:685–689. [5]

Brown, S. 2000. The "musilanguage" model of music evolution. In: The Origins of Music, ed. N. L. Wallin et al., pp. 271–300. Cambridge, MA: MIT Press. [5, 10, 17–20]

Brown, S., and J. Jordania. 2011. Universals in the world's musics. *Psychol. Music* doi: 10.1177/0305735611425896 [14]

Brown, S., and M. J. Martinez. 2007. Activation of premotor vocal areas during musical discrimination. *Brain Cogn.* **63**:59–69. [2, 13]

Brown, S., M. J. Martinez, D. A. Hodges, P. T. Fox, and L. M. Parsons. 2004. The song system of the human brain. *Cogn. Brain Res.* **20**:363–375. [13, 17, 20]

Brown, S., M. J. Martinez, and L. M. Parsons. 2006a. Music and language side by side in the brain: A PET study of the generation of melodies and sentences. *Eur. J. Neurosci.* **23**:2791–2803. [13]

———. 2006b. The neural basis of human dance. *Cereb. Cortex* **16**:1157–1167. [13, 20]

Brown, S., E. Ngan, and M. Liotti. 2007. A larynx area in the human motor cortex. *Cereb. Cortex* **18**:837–845. [20]

Brown, S., and U. Volgsten. 2006. Music and Manipulation: On the Social Uses and Social Control of Music. Oxford: Berghahn Books. [6]

Brown, W. M., L. Cronk, K. Grochow, et al. 2005. Dance reveals symmetry especially in young men. *Nature* **438**:1148–1150. [13]

Bruner, J. S. 1975. The ontogenesis of speech acts. *J. Child Lang.* **2**:1–19. [3]

Brunswik, E. 1956. Perception and the representative design of psychological experiments. Berkeley: Univ. of California Press. [5]

Brunton, R. 1989. The cultural instability of egalitarian societies. *Man* **24**:637–681. [2]

Bubic, A., D. Y. von Cramon, and R. I. Schubotz. 2010. Prediction, cognition and the brain. *Front. Human Neurosci.* **4**:1–15. [8, 15]

Buccino, G., F. Binkofski, G. R. Fink, et al. 2001. Action observation activates premotor and parietal areas in a somatotopic manner: An fMRI study. *Eur. J. Neurosci.* **13**:400–404. [4]

Buccino, G., F. Lui, N. Canessa, et al. 2004a. Neural circuits involved in the recognition of actions performed by non-conspecifics: An fMRI study. *J. Cogn. Neurosci.* **16**:114–126. [4, 8]

Buccino, G., S. Vogt, A. Ritzl, et al. 2004b. Neural circuits underlying imitation learning of hand actions: An event-related fMRI study. *Neuron* **42**:323–333. [4, 19]

Buchsbaum, B. R., J. Baldo, K. Okada, et al. 2011. Conduction aphasia, sensory-motor integration, and phonological short-term memory: An aggregate analysis of lesion and fMRI data. *Brain Lang.* **119**:119–128. [17]

Buckner, R. L., W. M. Kelley, and S. E. Petersen. 1999. Frontal cortex contributes to human memory formation. *Nat. Neurosci.* **2**:311–314. [20]

Buckner, R. L., and J. M. Logan. 2001. Functional neuroimaging methods: PET and fMRI. In: Handbook of Functional Neuroimaging of Cognition, ed. R. Cabeza and A. Kingstone. Cambridge: MIT Press. [13]

Bühler, K. 1934/1988. Sprachtheorie. Jena: Fischer. [5, 8]

Buhusi, C. V., and W. H. Meck. 2005. What makes us tick? Functional and neural mechanisms of interval timing. *Nat. Rev. Neurosci.* **6**:755–765. [13]

Buonomano, D., and M. Merzenich. 1998. Cortical plasticity: From synapses to maps. *Annu. Rev. Neurosci.* **21**:149–186. [15]

Büring, D. 2007. Semantics, intonation and information structure. In: The Oxford Handbook of Linguistic Interfaces, ed. G. Ramchand and C. Reiss. Oxford: Oxford Univ. Press. [9]

Burns, E. M., and W. D. Ward. 1978. Categorical perception–phenomenon or epiphenomemon: Evidence from experiments in the perception of melodic musical intervals. *J. Acoust. Soc. Am.* **63**:456–468. [17]

Buzsáki, G. 2006. Rhythms of the Brain. Oxford: Oxford Univ. Press. [9, 15]

Byers-Heinlein, K., T. C. Burns, and J. F. Werker. 2010. The roots of bilingualism in newborns. *Psychol. Sci.* **21**:343–348. [18]

Byrne, R. W. 2003. Imitation as behavior parsing. *Philos. Trans. R. Soc. Lond. B* **358**:529–536. [19]

Byrne, R. W., and J. M. E. Byrne. 1993. Complex leaf-gathering skills of mountain gorillas, G*orilla g. beringei*: Variability and standardization. *Am. J. Primatol.* **31**:241–261. [19]

Cacioppo, J., and L. Hawkley. 2003. Social isolation and health, with an emphasis on underlying mechanisms. *Perspect. Biol. Med.* **46**:S39–S52. [6]

Caclin, A., E. Brattico, M. Tervaniemi, et al. 2006. Separate neural processing of timbre dimensions in auditory sensory memory. J. Cogn. Neurosci. **18**:1952–1972. [17]

Caggiano, V., L. Fogassi, G. Rizzolatti, P. Thier, and A. Casile. 2009. Mirror neurons differentially encode the peripersonal and extrapersonal space of monkeys. *Science* **324**:403–406. [4]

Calder, A. J., J. Keane, F. Manes, N. Antoun, and A. W. Young. 2000. Impaired recognition and experience of disgust following brain injury. *Nat. Neurosci.* **3**:1077–1088. [4]

Calder, W. A. 1970. Respiration during song in the canary, *Serinus canaria*. *Comp. Biochem. Physiol.* **32**:251–258. [20]

Call, J., and M. Tomasello. 2007. The Gestural Communication of Apes and Monkeys. Mahwah, NJ: Erlbaum. [3]

Callan, D. E., M. Kawato, L. Parsons, and R. Turner. 2007. Speech and song: The role of the cerebellum. *Cerebellum* **6**:321–327. [13]

Callan, D. E., V. Tsytsarev, T. Hanakawa, et al. 2006. Song and speech: Brain regions involved with perception and covert production. *NeuroImage* **31**:1327–1342. [6, 13]

Calvo-Merino, B., D. E. Glaser, J. Grezes, R. E. Passingham, and P. Haggard. 2005. Action observation and acquired motor skills: An FMRI study with expert dancers. *Cereb. Cortex* **15**:1243–1249. [1, 4, 8, 13]

Calvo-Merino, B., J. Grezes, D. E. Glaser, R. E. Passingham, and P. Haggard. 2006. Seeing or doing? Influence of visual and motor familiarity in action observation. *Curr. Biol.* **16**:1905–1910. [7]

Calvo-Merino, B., and P. Haggard. 2011. Neuroaesthetics of Performing Arts. In: Art and the Senses, ed. F. Bacci and D. Melcher, pp. 529–541. Oxford: Oxford Univ. Press. [8]

Calvo-Merino, B., C. Jola, D. E. Glaser, and P. Haggard. 2008. Towards a sensorimotor aesthetics of performing art. *Conscious. Cogn.* **17**:911–922. [7]

Cangelosi, A. 2010. Grounding language in action and perception: From cognitive agents to humanoid robots. *Phys. Life Rev.* **7**:139–151. [16]

Capirci, O., J. M. Iversonm, S. Montanari, and V. Volterra. 2002. Gestural, signed and spoken modalities in early language development: The role of linguistic input. *Bilingualism* **5**:25–37. [21]

Capirci, O., and V. Volterra. 2008. Gesture and speech: The emergence of a strong and changing partnership. *Gesture* **8**:22–44. [18]

Caplan, D. 2009. The Neural Basis of Syntactic Processing. In: The Cognitive Neurosciences, ed. M. B. Gazzangia, pp. 805–817. [8]

Caplan, D., N. Hildebrandt, and N. Makris. 1996. Location of lesions in stroke patients with deficits in syntactic processing in sentence comprehension. *Brain* **119**:933–949. [4]

Caplan, D., and G. S. Waters. 1996. Syntactic processing in sentence comprehension under dual-task conditions in aphasic patients. *Lang. Cogn. Proc.* **11**:525–551. [9]

Caplin, W. 1998. Classical Form. Cambridge: Cambridge Univ. Press. [10]

———. 2004. The classical cadence: Conceptions and misconceptions. *J. Am. Musicological Soc.* **57**:51–118. [6]

Carandini, M. 2012. From circuits to behavior: A bridge too far? *Nat. Neurosci.* **15**:507–509. [16]

Carey, S. 2009. The Origin of Concepts. Oxford series in cognitive development. Oxford: Oxford Univ. Press. [17]

Carlyon, R. P. 2004. How the brain separates sounds. *Trends Cogn. Sci.* **8**:465–471. [17]

Carr, C. E., and R. A. Code. 2000. The central auditory system of reptiles and birds. In: Comparative Hearing: Birds and Reptiles, ed. R. J. Dooling et al., vol. 13, pp. 197–248, R. R. Fay and A. N. Popper, series ed. New York: Springer. [20]

Carr, L., M. Iacoboni, M. C. Dubeau, J. C. Mazziotta, and G. L. Lenzi. 2003. Neural mechanisms of empathy in humans: A relay from neural systems for imitation to limbic areas. *PNAS* **100**:5497–5502. [4]

Carroll, S. B. 2008. Evo-devo and an expanding evolutionary synthesis: A genetic theory of morphological evolution. *Cell* **134**:25–36. [20]

Carrus, E., S. Koelsch, and J. Bhattacharya. 2011. Shadows of music–language interaction on low frequency brain oscillatory patterns. *Brain Lang.* **119**:50–57. [15]

Carstairs-McCarthy, A. 2010. The Evolution of Morphology. Oxford: Oxford Univ. Press. [20]

Caruana, F., A. Jezzini, B. Sbriscia-Fioretti, G. Rizzolatti, and V. Gallese. 2011. Emotional and social behaviors elicited by electrical stimulation of the insula in the macaque monkey *Curr. Biol.* **21**:195–199. [4]

Caselli, M. C., P. Rinaldi, S. Stefanini, and V. Volterra. 2012. Early action and gesture "vocabulary" and its relation with word comprehension and production. *Child Devel.* **83**:526–542. [18]

Catani, M., M. P. G. Allin, M. Husain, et al. 2007. Symmetries in human brain language pathways correlate with verbal recall. *PNAS* **104**:17,163–17,168. [9]

Catani, M., and M. Mesulam. 2008. The arcuate fasciculus and the disconnection theme in language and aphasia: History and current state. *Cortex* **44**:953–961. [13]

Catchpole, C. K. 1976. Temporal and sequential organisation of song in the sedge warbler, *Acrocephalus schoenebaenus*. *Behaviour* **59**:226–246. [21]

Catchpole, C. K., and P. J. B. Slater. 2008. Bird Song: Biological Themes and Variations, 2nd edition. Cambridge: Cambridge Univ. Press. [20]

Cattaneo, L., and G. Rizzolatti. 2009. The mirror neuron system. *Arch. Neurol.* **66**:557–560. [4]

Chandrasekaran, C., and A. A. Ghazanfar. 2009. Different neural frequency bands integrate faces and voices differently in the superior temporal sulcus. *J. Neurophysiol.* **101**:773–788. [17]

Chang, F. 2009. Learning to order words: A connectionist model of heavy NP shift and accessibility effects in Japanese and English. *J. Mem. Lang.* **61**:374–397. [15]

Chang, F., G. S. Dell, and K. Bock. 2006. Becoming syntactic. *Psychol Rev* **113**:234–272. [15]

Chang, F., M. Janciauskas, and H. Fitz. 2012. Language adaptation and learning: Getting explicit about implicit learning. *Lang. Linguist. Compass* **6**:259–278. [15]

Chang, H. W., and S. E. Trehub. 1977. Auditory processing of relational information by young infants. *J. Exp. Child Psychol.* **24**:324–331. [17]

Chapin, H., K. J. Jantzen, J. A. S. Kelso, F. Steinberg, and E. W. Large. 2010. Dynamic emotional and neural responses to music depend on performance expression and listener experience. *PLoS ONE* **5**: [8, 13]

Charry, E. S. 2000. Mande Music: Traditional and Modern Music of the Maninka and Mandinka of Western Africa. Chicago: Univ. of Chicago Press. [3]

Chen, J. L., V. B. Penhune, and R. J. Zatorre. 2008a. Listening to musical rhythms recruits motor regions of the brain. *Cereb. Cortex* **18**:2844–2854. [4, 13, 17]

———. 2008b. Moving on time: Brain network for auditory-motor synchronization is modulated by rhythm complexity and musical training. *J. Cogn. Neurosci.* **20**:226–239. [13]

———. 2009. The role of auditory and premotor cortex in sensorimotor transformations. Neurosciences and music III: Disorders and plasticity. *Ann. NY Acad. Sci.* **1169**:15–34. [4]

Chen, Y.-S., A. Olckers, T. G. Schurr, et al. 2000. Mitochondrial DNA variation in the South African Kung and Khwe and their genetic relationships to other African populations. *Am. J. Hum. Genet.* **66**:1362–1383. [2]

Cheney, D. L., and R. M. Seyfarth. 2005. Constraints and preadaptations in the earliest stages of language evolution. *Ling. Rev.* **22**:135–159. [19]

Chevillet, M., M. Riesenhuber, and J. P. Rauschecker. 2011. Functional localization of the ventral auditory "what" stream hierarchy. *J. Neurosci.* **31**:9345–9352. [17]

Chomsky, N. 1965. Aspects of the Theory of Syntax. Cambridge, MA: MIT Press. [9, 10]

———. 1981. Lectures on Government and Binding. Dordrecht: Foris. [3]

———. 1995. The Minimalist Program. Cambridge, MA: MIT Press. [10, 12]

———. 2000. Language as a natural object. In: New Horizons in the Study of Language and Mind, ed. N. Smith, pp. 106–133. Cambridge: Cambridge Univ. Press. [1]

———. 2007. Of minds and language. *Biolinguistics* **1**:1009–1037. [3]

———. 2010. Some simple evo devo theses: How true might they be for language? In: Approaches to the Evolution of Language, ed. R. K. Larson et al., pp. 45–62. Cambridge: Cambridge Univ. Press. [3]

Chomsky, N., and M. Halle. 1968. The Sound Pattern of English. New York: Harper and Row. [14]

Christensen, T. S. 2002. The Cambridge History of Western Music Theory. Cambridge: Cambridge Univ. Press. [16]

Christensen-Dalsgaard, J., and C. E. Carr. 2008. Evolution of a sensory novelty: Tympanic ears and the associated neural processing. Brain Res. Bull. **75**:365–370. [20]

Christiansen, M. H., and N. Chater. 2008 Language as shaped by the brain. Behav. Brain Sci. **31**:489–509. [3]

Clancey, W. J. 1992. The frame of reference problem in the design of intelligent machines. In: Architectures for Intelligence, Proc. 22nd Carnegie Symp. on Cognition, ed. K. v. Lehn, pp. 357–423. Hillsdale, NJ: Erlbaum. [16]

Clark, A. 2002. Minds, brains, and tools. In: Philosophy of Mental Representation, ed. H. Clapin. Oxford: Clarendon Press. [21]

Clarke, D. 2011. Music, phenomenology, time consciousness: meditations after Husserl. In: Music and Consciousness: Philosophical, Psychological, and Cultural Perspectives, ed. D. Clarke and E. Clarke, pp. 1–28. Oxford: Oxford Univ. Press. [8]

Clay, Z., and K. Zuberbühler. 2011. Bonobos extract meaning from call sequences. PLoS ONE **6**:e18786. [21]

Clayton, D. F. 2000. The genomic action potential. Neurobiol. Learn. Mem. **74**:185–216. [20]

Clayton, M. 2009. The social and personal functions of music in cross-cultural perspective. In: Handbook of Music and Emotion, ed. S. Hallam et al., pp. 879–908. Oxford: Oxford Univ. Press. [8, 16]

Clayton, M., R. Sager, and U. Will. 2005. In time with the music: The concept of entrainment and its significance for ethnomusicology. Eur. Meetings in Ethnomusicology **11**:3–75. [16]

Clément, S., L. Demany, and C. Semal. 1999. Memory for pitch versus memory for loudness. J. Acoust. Soc. Am. **106**:2805–2811. [17]

Clendinning, J. P. 1993. The Pattern-meccanico compositions of György Ligeti. Perspectives of New Music **31**:192–234. [16]

Coessens, K. 2011. Where am I? Body and mind reviewed in the context of situatedness and virtuality. Intl. J. Interdisc. Social Sci. **5**:65–74. [16]

Cohen, A. J. 1990. Understanding musical soundtracks. Empirical Studies of the Arts **8**:111–124. [7]

———. 1993. Associationism and musical soundtrack phenomena. Contemp. Music Rev. **9**:163–178. [7]

———. 1999. The functions of music in multimedia: A cognitive approach. In: Music, Mind, and Science, ed. S. W. Yi, pp. 53–69. Seoul, Korea: Seoul National Univ. Press. [7]

———. 2005. How music influences the interpretation of film and video: Approaches from experimental psychology. Selected Rep. Ethnomusicol. **12**:15–36. [7]

———. 2009. Music in performance arts: Film, theatre and dance. In: The Oxford Handbook of Music Psychology, ed. S. Hallam et al., pp. 441–451. New York: Oxford Univ. Press. [7]

———. 2010. Music as a source of emotion in film. In: Handbook of Music and Emotion: Theory, Research, Applications, ed. P. Juslin and J. Sloboda, pp. 879–908. New York: Oxford Univ. Press. [7, 8]

Cohen, A. J. 2013. Film music from the perspective of cognitive science. In: Oxford Handbook of Music in Film and Visual Media, ed. D. Neumeyer. Oxford: Oxford Univ. Press. [7]

Cohen, A. J., L. A. Thorpe, and S. E. Trehub. 1987. Infants' perception of musical relations in short transposed tone sequences. *Can. J. Psychol.* **41**:33–47. [18]

Cohen, M. A., K. K. Evans, T. S. Horowitz, and J. M. Wolfe. 2011. Auditory and visual memory in musicians and nonmusicians. *Psychon. Bull. Rev.* **18**:586–591. [17]

Cohen, M. A., T. S. Horowitz, and J. M. Wolfe. 2009. Auditory recognition memory is inferior to visual recognition memory. *PNAS* **106**:6008–6010. [17]

Cohn, N., M. Paczynski, R. Jackendoff, P. J. Holcomb, and G. R. Kuperberg. 2012. (Pea)nuts and bolts of visual narrative: Structure and meaning in sequential image comprehension. *Cogn. Psychol.* **65**:1–38. [8]

Colannino, J., F. Gómez, and G. T. Toussaint. 2009. Analysis of emergent beat-class sets in Steve Reich's Clapping Music and the Yorbua bell timeline. *Persp. New Music* **47**:111–134. [16]

Colton, S., and G. A. Wiggins. 2012. Computational creativity: The final frontier? In: ECAI 2012: 20th European Conference on Artificial Intelligence, ed. L. De Raedt et al., pp. 21–26. Amsterdam: IOS Press. [16]

Cook, N. 1987a. A Guide to Music Analysis. London: Dent. [6]

———. 1987b. The perception of large-scale tonal closure. *Music Percep.* **5**:197–205. [6, 10]

———. 1998. Analysing Musical Multimedia. Oxford: Oxford Univ. Press. [7]

Cooke, M. 2008. A History of Film Music. Cambridge: Cambridge Univ. Press. [7]

Cooper, R. P., and R. N. Aslin. 1990. Preference for infant-directed speech in the first month after birth. *Child Devel.* **61**:1584–1595. [18]

Cope, D. 2004. Virtual Music: Computer Synthesis of Musical Style. Cambridge, MA: MIT Press. [16]

Copland, A. 1939. What to Listen for in Music. New York, NY: McGraw-Hill. [7]

———. 1940. The aims of music for films. *New York Times on the web*, March 10. [7]

Corballis, M. C. 2002. From Hand to Mouth, the Origins of Language. Princeton, NJ: Princeton Univ. Press. [19]

———. 2003. From mouth to hand: Gesture, speech and the evolution of right-handedness. *Behav. Brain Sci.* **26**:199–260. [20]

———. 2007. Recursion, language, and starlings. *Cognitive Science* **31**:697–704. [1, 19]

———. 2010. Mirror neurons and the evolution of language. *Brain Lang.* **112**:25–35. [18]

Corina, D. P., H. Poizner, U. Bellugi, et al. 1992. Dissociation between linguistic and nonlinguistic gestural systems: A case for compositionality. *Brain Lang.* **43** 414–447. [4, 8, 19]

Corradi-Del'Acqua, C., K. Ueno, A. Ogawa, et al. 2008. Effects of shifting perspective of the self: An fMRI study. *NeuroImage* **40**:1902–1911. [21]

Corrigall, K. A., and L. J. Trainor. 2010. Musical enculturation in preschool children: Acquisition of key and harmonic knowledge. *Music Percep.* **28**:195–200. [18]

Costa, M., P. Fine, and P. Ricci Bitti. 2004. Interval distributions, mode, and tonal strength of melodies as predictors of perceived emotion. *Music Percep.* **22**:1–14. [5]

Coudé, G., P. F. Ferrari, F. Rodà, et al. 2007. Neuronal responses during vocalization in the ventral premotor cortex of macaque monkeys. Society for Neuroscience Annual Meeting (San Diego California), Abstract 636.3. [19]

Coudé, G., P. F. Ferrari, F. Rodà, et al. 2011. Neurons controlling voluntary vocalization in the macaque ventral premotor cortex. *PLoS ONE* **6**:e26822. [4]

Craig, A. D. 2002. How do you feel? Interoception: The sense of the physiological condition of the body. *Nat. Rev. Neurosci.* **3**:655–666. [4]

Craik, F. I. M., and R. S. Lockhart. 1972. Levels of processing: A framework for memory research. *J. Verb. Learn. Verb. Behav.* **11**:671–684. [18]

Craik, K. J. W. 1943. The Nature of Explanation. Cambridge: Cambridge Univ. Press. [15]

Crick, F., and C. Koch. 1990. Towards a neurobiological theory of consciousness. *Semin. Neurosci.* **2**:263–275. [7]

Crockford, C., I. Herbinger, L. Vigilant, and C. Boesch. 2004. Wild chimpanzees produce group-specific calls: A case for vocal learning? *Ethology* **110**:221–243. [20, 21]

Croft, W. 2001. Radical Construction Grammar: Syntactic Theory in Typological Perspective. Oxford: Oxford Univ. Press. [1]

Crompton, A. 1982. Syllables and segments in speech production. In: Slips of the Tongue and Language Production, ed. A. Cutler, pp. 109–162. Berlin: Mouton. [14]

Cross, E. S., A. F. de Hamilton, and S. T. Grafton. 2006. Building a motor simulation de novo: Observation of dance by dancers. *Neuroimage* **31**:1257–1267. [4, 13]

Cross, I. 2001a. Music, cognition, culture and evolution. *Ann. NY Acad. Sci.* **930**:28–42. [1]

———. 2001b. Music, mind and evolution. *Psychol. Music* **29**:95–102. [1]

———. 2003a. Music and biocultural evolution. In: The Cultural Study of Music: A Critical Introduction, ed. M. Clayton et al. London: Routledge. [1, 21]

———. 2003b. Music and evolution: Causes and consequences. *Contemp. Music Rev.* **22**:79–89. [18, 21]

———. 2003c. Music as a biocultural phenomenon. *Ann. NY Acad. Sci.* **999**:106–111. [19]

———. 2003d. Music, cognition, culture and evolution. In: The Cognitive Neuroscience of Music, ed. I. Peretz and R. J. Zatorre, pp. 42–56. Oxford: Oxford Univ. Press. [19]

———. 2005. Music and meaning, ambiguity and evolution. In: Musical Communication, ed. D. Miell et al. Oxford: Oxford Univ. Press. [1]

———. 2006. Music and social being. *Musicology Australia* **28**:114–126. [1]

———. 2008. Musicality and the human capacity for culture. *Musicae Scientiae* **12**:147. [6, 7]

———. 2010. The evolutionary basis of meaning in music: Some neurological and neuroscientific implications. In: Neurology of Music, ed. F. C. Rose, pp. 1–15. London: Imperial College Press. [8]

———. 2011. The meanings of musical meanings (Comment). *Phys. Life Rev.* **8**:116–119. [6]

———. 2012. Music as social and cognitive process. In: Language and Music as Cognitive Systems, ed. P. Rebuschat et al., pp. 315–328. Oxford: Oxford Univ. Press. [8]

Cross, I., and I. Morley. 2008. The evolution of music: Theories, definitions and the nature of the evidence. In: Communicative Musicality, ed. S. Malloch and C. Trevarthen, pp. 61–82. Oxford: Oxford Univ. Press. [6, 19]

Cross, I., and G. E. Woodruff. 2009. Music as a communicative medium. In: The Prehistory of Language, ed. R. Botha and C. Knight, pp. 77–98, Studies in the Evolution of Language, K. R. Gibson and J. R. Hurford, series ed. Oxford: Oxford Univ. Press. [7, 8]

Crosson, B., J. R. Sadek, J. A. Bobholz, et al. 1999. Activity in the paracingulate and cingulate sulci during word generation: An fMRI study of functional anatomy. *Cereb. Cortex* **9**:307–316. [20]

Crowe, S. F., K. T. Ng, and M. E. Gibbs. 1990. Memory consolidation of weak training experiences by hormonal treatments. *Pharmacol. Biochem. Behav.* **37**:728–734. [8]

Crystal, D. 1974. Paralinguistics. In: Current Trends in Linguistics, ed. T. A. Sebeok, pp. 265–295. The Hague: Mouton. [5]

Crystal, D., and R. Quirk. 1964. Systems of Prosodic and Paralinguistic Features in English. The Hague: Mouton. [5]

Cummings, J. L. 1993. Frontal-subcortical circuits and human behavior. *Arch. Neurol.* **50**:873–880. [20]

Curtis, M. E., and J. J. Bharucha. 2010. The minor third communicates sadness in speech, mirroring its use in music. *Emotion* **10**:335–348. [13]

Cutler, A., D. Dahan, and W. van Donselaar. 1997. Prosody in the comprehension of spoken language: A literature review. *Lang. Speech* **40**:141–201. [11]

Cutler, A., J. Mehler, D. Morris, and J. Segui. 1986. The syllable's differing role in the segmentation of French and English. *J. Mem. Lang.* **25**:385–400. [18]

Dael, N., M. Mortillaro, and K. R. Scherer. 2012. Emotion expression in body action and posture. *Emotion* **12**:1085–1101. [5]

Dalla Bella, S., and M. Berkowska. 2009. Singing and its neuronal substrates: Evidence from the general population. *Contemp. Music Rev.* **28**:279–291. [13]

Dalla Bella, S., J.-F. Giguère, and I. Peretz. 2007. Singing proficiency in the general population. *J. Acoust. Soc. Am.* **121**:1182–1189. [18]

Dalla Bella, S., I. Peretz, and N. Aronoff. 2003. Time course of melody recognition: A gating paradigm study. *Percep. Psychophys.* **65**:1019–1028. [18]

Daltrozzo, J., and D. Schön. 2009. Conceptual processing in music as revealed by N400 effects on words and musical targets. *J. Cogn. Neurosci.* **21**:1882–1892. [6]

Damasio, A. R. 1999. The Feeling of What Happens: Body and Emotion in the Making of Consciousness. New York: Harcourt Brace & Co. [1]

Dantzer, R., J. C. O'Connor, G. G. Freund, R. W. Johnson, and K. W. Kelley. 2008. From inflammation to sickness and depression: When the immune system subjugates the brain. *Nat. Rev. Neurosci.* **9**:46–56. [6]

Darwin, C. 1871. The Descent of Man and Selection in Relation to Sex, 1st edition. London: John Murray. [3, 14, 19–21]

———. 1872/1998. The Expression of the Emotions in Man and Animals, 2nd edition. London: Murray. [1, 4, 5, 8]

———. 1876. The Descent of Man and Selection in Relation to Sex, 2nd edition. London: John Murray. [10]

Davidson, R. J., K. R. Scherer, and H. Goldsmith, eds. 2003. Handbook of the Affective Sciences. Oxford: Oxford Univ. Press. [5]

Davis, M. H., M. R. Coleman, A. R. Absalom, et al. 2007. Dissociating speech perception and comprehension at reduced levels of awareness. *PNAS* **104**:16,032–16,037. [9]

Dawkins, R. 1976. The Selfish Gene. Oxford: Oxford Univ. Press. [8]

Deacon, T. W. 1997. The Symbolic Species: The Co-evolution of Language and the Brain. New York: W. W. Norton. [8, 19]

———. 1998. Language evolution and neuromechanisms. In: A Companion to Cognitive Science, ed. W. Bechtel and G. Graham, pp. 212–225, Blackwell Companions to Philosophy. Malden, MA: Blackwell. [8]

———. 2003. Universal grammar and semiotic constraints. In: Language Evolution, ed. M. H. Christiansen and S. Kirby, pp. 111–139. Oxford: Oxford Univ. Press. [1]

———. 2006. The aesthetic faculty. In: The Artful Mind: Cognitive Science and the Riddle of Human Creativity, ed. M. Turner, pp. 21–53. Oxford: Oxford Univ. Press. [8, 16]

de Almeida, L., M. Idiart, and J. E. Lisman. 2009. A second function of gamma frequency oscillations: An E%-max winner-take-all mechanism selects which cells fire. *J. Neurosci.* **29**:7497–7503. [15]

de Boysson-Bardies, B., L. Sagart, and C. Durand. 1984. Discernible differences in the babbling of infants according to target language. *J. Child Lang.* **11**:1–5. [18]

DeCasper, A. J., and W. P. Fifer. 1980. Of human bonding: Newborns prefer their mothers' voices. *Science* **208**:1174–1176. [17]

Decety, J., and T. Chaminade. 2003. When the self represents the other: A new cognitive neuroscience view on psychological identification. *Conscious. Cogn.* **12**:577–596. [4]

Decety, J., and J. Grèzes. 2006. The power of simulation: Imagining one's own and other's behavior. *Brain Res.* **1079**:4–14. [4]

Dediu, D., and S. C. Levinson. 2013. On the antiquity of language: The reinterpretation of Neandertal linguistic capacities and its consequences, in press. [3]

Deese, J. 1984. Thought into Speech: The Psychology of a Language. Englewood Cliffs: Prentice-Hall. [9]

Dehaene, S., and L. Cohen. 2007. Cultural recycling of cortical maps. *Neuron* **56**:384–398. [3]

Dehaene-Lambertz, G., A. Motavont, A. Jobert, et al. 2010. Language or music, mother or Mozart? Structural and einvironmental influences on infants' language networks. *Brain Lang.* **114**:53–65. [8]

Deike, S., H. Scheich, and A. Brechmann. 2010. Active stream segregation specifically involves the left human auditory cortex. *Hearing Res.* **265**:30–37. [17]

Delano, P. H., D. Elgueda, C. M. Harname, and L. Robles. 2007. Selective attention to visual stimuli reduces cochlear sensitivity in chinchillas. *J. Neurosci.* **27**:46–53. [17]

Delano, P. H., D. Elgueda, F. Ramirez, L. Robles, and P. E. Maldonado. 2010. A visual cue modulates the firing rate and latency of auditory-cortex neurons in the chinchilla. *J. Physiology (Paris)* **104**:190–196. [17]

Deliège, I., M. Mélen, D. Stammers, and I. Cross. 1996. Musical schemata in real time listening to a piece of music. *Music Percep.* **14**:117–160. [8]

DeLong, K. A., T. P. Urbach, and M. Kutas. 2005. Probabilistic word pre-activation during language comprehension inferred from electrical brain activity. *Nat. Neurosci.* **8**:1117–1121. [9]

Demany, L., and F. Armand. 1984. The perceptual reality of tone chroma in early infancy. *J. Acoust. Soc. Am.* **76**:57–66. [18]

Demany, L., B. McKenzie, and E. Vurpillot. 1977. Rhythm perception in early infancy. *Nature* **266**:718–719. [13, 17]

Dennett, D. 1983. Intentional systems in cognitive ethology: The "Panglossian Paradigm" defended. *Behav. Brain Sci.* **6**:343–390. [21]

Denton, D. A., M. J. McKinley, M. Farrell, and G. F. Egan. 2009. The role of primordial emotions in the evolutionary origin of consciousness. *Conscious. Cogn.* **18**:500–514. [15]

De Renzi, E. 1989. Apraxia. In: Handbook of Neuropsychology, vol. 2, ed. F. Boller and J. Grafman, pp. 245–263. Amsterdam: Elsevier. [4]

De Ruiter, J. P., H. Mitterer, and N. J. Enfield. 2006. Projecting the end of a speaker's turn: A cognitive cornerstone of conversation. *Language* **82**:515–535. [13, 17]

De Ruiter, J. P., M. L. Noordzij, S. Newman-Norlund, et al. 2010. Exploring the cognitive infrastructure of communication. *Interaction Stud.* **11**:51–77. [21]

Desain, P., and H. Honing. 2003. The formation of rhythmic categories and metric priming. *Perception* **32**:341–365. [12]

Deutsch, D. 1995. Musical illusions and paradoxes. http://philomel.com/musical_illusions/. (accessed Nov. 25, 2012). [1, 12]

Deutsch, D., T. Henthorn, E. Marvin, and H. Xu. 2006. Absolute pitch among American and Chinese conservatory students: Prevalent differences and evidence for a speech-related critical period. *J. Acoust. Soc. Am.* **119**:719–722. [17]

de Villiers, J. G., and J. E. Pyers. 2002. Complements to cognition: A longitudinal study of the relationship between complex syntax and false-belief-understanding. *Cognitive Development* **17**:1037–1060. [1]

DeVoogd, T. J., J. R. Krebs, S. D. Healy, and A. Purvis. 1993. Relations between song repertoire size and the volume of brain nuclei related to song: Comparative evolutionary analyses amongst oscine birds. *Proc. R. Soc. Lond. B* **254**:75–82. [20]

DeWitt, I., and J. P. Rauschecker. 2012. Phoneme and word recognition in the auditory ventral stream. *PNAS* **109**:E505–514. [15, 17]

Dibben, N. 1994. The cognitive reality of hierarchic structure in tonal and atonal music. *Music Percep.* **12**:1–25. [10]

Dissanayake, E. 2000. Antecedents of the temporal arts in early mother-infant interaction. In: The Origins of Music, ed. N. L. Wallin et al., pp. 389–410. Cambridge, MA: MIT Press. [19]

Dominey, P. F., M. A. Arbib, and J.-P. Joseph. 1995. A model of corticostriatal plasticity for learning oculomotor associations and sequences. *J. Cogn. Neurosci.* **7**:311–336. [15]

Dominey, P. F., M. Hoen, and T. Inui. 2006. A neurolinguistic model of grammatical construction processing. *J. Cogn. Neurosci.* **18**:2088–2107. [15]

Dominey, P. F., T. Inui, and M. Hoen. 2009. Neural network processing of natural language: II. Towards a unified model of corticostriatal function in learning sentence comprehension and non-linguistic sequencing. *Brain Lang.* **109**:80–92. [17]

Donald, M. 2001. A Mind So Rare: The Evolution of Human Consciousness. New York: W. W. Norton. [8]

Douglas, R. J., and K. A. C. Martin. 2004. Neuronal circuits of the neocortex. *Annu. Rev. Neurosci.* **27**:419–451. [15]

Doupe, A. J., and M. Konishi. 1991. Song-selective auditory circuits in the vocal control system of the zebra finch. *PNAS* **88**:11,339–11,343. [20]

Doupe, A. J., and P. K. Kuhl. 1999. Birdsong and human speech: Common themes and mechanisms. *Annu. Rev. Neurosci.* **22**:567–631. [14, 20]

Dowling, W. J., and D. L. Harwood. 1986. Music Cognition. Orlando: Academic. [18]

Draganova, R., H. Eswaran, P. Murphy, et al. 2005. Sound frequency change detection in fetuses and newborns, a magnetoencephalographic study. *Neuroimage* **28**:354–361. [17]

Drake, C. 1998. Psychological processes involved in the temporal organization of complex auditory sequences: Universal and acquired processes. *Music Percep.* **16**:11–26. [17]

Drevets, W., J. Price, and M. Bardgett. 2002. Glucose metabolism in the amygdala in depression: Relationship to diagnostic subtype and plasma cortisol levels. *Pharmacol. Biochem. Behav.* **71**:431–447. [6]

Dronkers, N. F. 1996. A new brain region for coordinating speech articulation. *Nature* **384**:159–161. [20]

Dryer, M. S. S. 2008. Relationship between the order of object and verb and the order of adposition and noun phrase. In: The World Atlas of Language Structures, ed. M. Haspelmath et al. Munich: Max Planck Digital Library, http://wals.info/feature/95. [3]

Duff, A., M. Sanchez-Fibla, and P. F. M. J. Verschure. 2011. A biologically based model for the integration of sensory-motor contingencies in rules and plans: A prefrontal cortex based extension of the distributed adaptive control architecture. *Brain Res. Bull.* **85**:289–304. [15, 16]

Duff, A., and P. F. M. J. Verschure. 2010. Unifying perceptual and behavioral learning with a correlative subspace learning rule. *Neurocomputing* **73**:1818–1830. [16]

Duff, A., R. Wyss, and P. F. M. J. Verschure. 2007. Learning temporally stable representations from natural sounds: Temporal stability as a general objective underlying sensory processing. *Lecture Notes in Computer Science: Artificial Neural Networks ICANN 2007* **4669**:129–138. [15]

Dunbar, R. 1996. Grooming, Gossip and the Evolution of Language. Cambridge, MA: Harvard Univ. Press. [21]

———. 2003. The social brain: Mind, language and society in evolutionary perspective. *Annu. Rev. Anthropol.* **32**:163–181. [18]

Duncan, L. G. 2010. Phonological development from a cross-linguistic perspective. In: Reading and Dyslexia in Different Orthographies, ed. N. Brunswick et al., pp. 43–68. New York: Psychology Press. [18]

Dunn, M., S. Greenhill, S. C. Levinson, and R. Gray. 2011. Evolved structure of language shows lineage-specific trends in word-order "universals." *Nature* **473**:79–82. [3]

Dunn, M., A. Terrill, G. Reesink, R. Foley, and S. C. Levinson. 2005. Structural phylogenetics and the reconstruction of ancient language history. *Science* **23**:2072–2075. [3]

Durand, S. E., J. T. Heaton, S. K. Amateau, and S. E. Brauth. 1997. Vocal control pathways through the anterior forebrain of a parrot, *Melopsittacus undulatus*. *J. Comp. Neurol.* **377**:179–206. [20]

Durkheim, E. 1938. The Rules of Sociological Method (translated by S. A. Solovay and J. H. Mueller London: Collier-MacMillan. [8]

Dusterhoft, F., U. Hausler, and U. Jurgens. 2004. Neuronal activity in the periaqueductal gray and bordering structures during vocal communication in the squirrel monkey. *Neuroscience* **123**:53–60. [20]

Eaton, R. L. 1979. A beluga whale imitates human speech. *Carnivore* **2**:22–23. [20]

Ebstein, R. P., S. Israel, S. H. Chew, S. Zhong, and A. Knafo. 2010. Genetics of human social behavior. *Neuron* **65**:831–844. [13]

Eckerdal, P., and B. Merker. 2009. Music and the "action song" in infant development: An interpretation. In: Communicative Musicality: Exploring the Basis of Human Companionship, ed. S. Malloch and C. Trevarthen, pp. 241–262. New York: Oxford Univ. Press. [18]

Eckersley, P., G. F. Egan, S. Amari, et al. 2003. Neuroscience data and tool sharing: A legal and policy framework for neuroinformatics. *J. Integr. Neurosci.* **1**:149–166. [8]

Edelman, G. M. 1987. Neural Darwinism: The Theory of Neuronal Group Selection. New York: Basic Books. [16]

———. 1989. The Remembered Present. New York: Basic Books. [14]

———. 1993. Neural Darwinism: Selection and reentrant signaling in higher brain function. *Neuron* **10**: [14]

———. 2007. Learning in and from brain-based devices. *Science* **318**:1103–1105. [16]

Edelman, G. M., and G. Tononi. 2000. Consciousness: How Matter Becomes Imagination. London: Penguin. [7]

Eens, M. 1997. Understanding the complex song of the European starling: An integrated approach. *Adv. Study Behav.* **26**:355–434. [20]

Egnor, S. E., and M. D. Hauser. 2004. A paradox in the evolution of primate vocal learning. *Trends Neurosci.* **27**:649–654. [20]

Eibl-Eibesfeldt, I. 1973. The expressive behaviour of the deaf-and blind-born. In: Social Communication and Movement, ed. M. Von Cranach and J. Vine, pp. 163–194. London: Academic. [20]

Eickhoff, S. B., A. R. Laird, C. Grefkes, et al. 2009. Coordinate-based activation likelihood estimation meta-analysis of neuroimaging data: A random-effects approach based on empirical estimates of spatial uncertainty. *Hum. Brain Mapp.* **30**:2907–2926. [13]

Ekman, P. 1972. Universals and cultural differences in facial expression of emotion. In: Nebraska Symposium on Motivation, ed. J. R. Cole, pp. 207–283. Lincoln: Univ. of Nebraska Press. [5]

———. 1992. Facial expressions of emotion: New findings, new questions. *Psychol. Sci.* **3**:34–38. [1]

———. 1999. Basic emotions. In: Handbook of Cognition and Emotion, ed. T. Dalgleish and M. Power, pp. 45–60. Wiley. [1, 6]

Ekman, P., and W. Friesen. 1969. The repertoire of nonverbal behavior: Categories, origins, usage, and coding. *Semiotica* **1**:49–98. [5, 11]

Eldar, E., O. Ganor, R. Admon, A. Bleich, and T. Hendler. 2007. Feeling the real world: Limbic response to music depends on related content. *Cereb. Cortex* **17**:2828–2840. [7, 8]

Eliades, S. J., and X. Wang. 2003. Sensory-motor interaction in the primate auditory cortex during self-initiated vocalizations. *J. Neurophysiol.* **89**:2194–2207. [17]

———. 2005. Dynamics of auditory-vocal interaction in monkey auditory cortex. *Cereb. Cortex* **15(10)**:1510–1523 [17]

———. 2008. Neural substrates of vocalization feedback monitoring in primate auditory cortex. *Nature* **453**:1102–1106. [17]

Elliott, T. M., and F. E. Theunissen. 2009. The modulation transfer function for speech intelligibility. *PLoS Comput. Biol.* **5**:e1000302. [9]

Ellis, A. J. 1885. On the musical scales of various nations. *J. Soc. Arts* **33**:485. [20]

Ellsworth, P. C., and K. Scherer. 2003. Appraisal processes in emotion. *Handbook Affec. Sci.* **572**:V595. [8]

Elman, J. L., E. A. Bates, M. H. Johnson, et al. 1996. Rethinking Innateness: A Connectionist Perspective on Development. Cambridge, MA: MIT Press. [14]

Emmorey, K. 2002. Language, Cognition, and the Brain: Insights from Sign Language Research. Mahwah, NJ: Lawrence Erlbaum. [19]

Enard, W., S. Gehre, K. Hammerschmidt, et al. 2009. A humanized version of Foxp2 affects cortico-basal ganglia circuits in mice. *Cell* **137**:961–971. [20]

Enard, W., M. Przeworski, S. E. Fisher, et al. 2002. Molecular evolution of FOXP2, a gene involved in speech and language. *Nature* **418**:869–872. [20]

Eng, K., R. J. Douglas, and P. F. M. J. Verschure. 2005a. An interactive space that learns to influence human behaviour. *IEEE Transactions on Systems, Man and Cybernetics Part A* **35**:66–77. [16]

Eng, K., M. Mintz, and P. F. M. J. Verschure. 2005b. Collective human behavior in interactive spaces. In: Intl. Conference on Robotics and Automation (ICRA 2005), pp. 2057–2062. Barcelona: IEEE. [16]

Entwisle, D. R., and N. E. Frasure. 1974. A contradiction resolved: Children's processing of syntactic cues. *Devel. Psychol.* **10**:852–857. [18]

Erickson, T. D., and M. E. Mattson. 1981. From words to meaning: A semantic illusion. *Journal of Verbal Learning and Verbal Behavior* **20**:540–551. [9]

Ericsson, K. A., and W. Kintsch. 1995. Long-term working memory. *Psychol. Rev.* **102**:211–245. [8]

Euler, L. 1739. *Tentamen novae theoriae musicae ex certissismis harmoniae principiis dilucide expositae*. St. Petersberg: Saint Petersberg Academy. [12]

Evans, C. S., and P. Marler. 1994. Food calling and audience effects in male chickens, *Gallus gallus*: Their relationships to food availability, courtship and social facilitation. *Anim. Behav.* **47**:1159–1170. [20]

Evans, N., and S. C. Levinson. 2009. The myth of language universals: Language diversity and its importance for cognitive science. *Behav. Brain Sci.* **32**:429–448. [3]

Fadiga, L., L. Craighero, G. Buccino, and G. Rizzolatti. 2002. Speech listening specifically modulates the excitability of tongue muscles: A TMS study. *Eur. J. Neurosci.* **15**:399–402. [4]

Fadiga, L., L. Craighero, and A. D'Ausilio. 2009. Broca's area in language, action, and music. *Ann. NY Acad. Sci.* **1169**:448–458. [4, 8]

Fadiga, L., L. Fogassi, G. Pavesi, and G. Rizzolatti. 1995. Motor facilitation during action observation: A magnetic stimulation study. *J. Neurophysiol.* **73** 2608–2611. [4]

Fagg, A. H., and M. A. Arbib. 1998. Modeling parietal-premotor interactions in primate control of grasping. *Neural Netw.* **11**:1277–1303. [1, 15]

Fahlenbrach, K. 2008. Emotions in sound: Audiovisual metaphors in the sound design of narrative films. *Projections* **2**:85–103. [7]

Falk, D. 2004. Prelinguistic evolution in early hominins: Whence motherese? *Behav. Brain Sci.* **27**:491–503. [18]

———. 2009. Finding our tongues: Mothers, infants, and the origins of language. New York: Basic Book. [8]

Farabaugh, S. M., E. D. Brown, and R. J. Dooling. 1992. Analysis of warble song in the budgerigar, *Melopsittacus undulatus*. *Bioacoustics* **4**:111–130. [20]

Farabaugh, S. M., A. Linzenbold, and R. J. Dooling. 1994. Vocal plasticity in budgerigars, *Melopsittacus undulatus*: Evidence for social factors in the learning of contact calls. *J. Comp. Psychol.* **108**:81–92. [20]

Farbood, M. M., G. Marcus, and D. Poeppel. 2013. Temporal dynamics and the identification of musical key. *J. Exp. Psychol. Hum. Percept. Perform.*, Jan 14. [Epub ahead of print] [17]

Farley, G. R. 1997. Neural firing in ventrolateral thalamic nucleus during conditioned vocal behavior in cats. *Exp. Brain Res.* **115**:493–506. [20]

Farries, M. A. 2001. The oscine song system considered in the context of the avian brain: Lessons learned from comparative neurobiology. *Brain Behav. Evol.* **58**:80–100. [20]

Fazio, P., A. Cantagallo, L. Craighero, et al. 2009. Encoding of human action in Broca's area. *Brain* **132**:1980–1988. [4]

Fedorenko, E., M. K. Behr, and N. Kanwisher. 2011. Functional specificity for high-level linguistic processing in the human brain. *PNAS* **108**:16,428–16,433. [14, 17]

Fedorenko, E., A. D. Patel, D. Casasanto, J. Winawer, and E. Gibson. 2009. Structural integration in language and music: Evidence for a shared system. *Mem. Cogn.* **37**:1–9. [6, 14]

Fee, M. S., A. A. Kozhevnikov, and R. H. Hahnloser. 2004. Neural mechanisms of vocal sequence generation in the songbird. *Ann. NY Acad. Sci.* **1016**:153–170. [20]

Fee, M. S., B. Shraiman, B. Pesaran, and P. P. Mitra. 1998. The role of nonlinear dynamics of the syrinx in the vocalizations of a songbird. *Nature* **395**:67–71. [20]

Feenders, G., M. Liedvogel, M. Rivas, et al. 2008. Molecular mapping of movement-associated areas in the avian brain: A motor theory for vocal learning origin. *PLoS ONE* **3**:e1768. [20]

Fehér, O., P. P. Mitra, K. Sasahara, and O. Tchernichovski. 2008. Evolution of song culture in the zebra finch. In: The Evolution of Language, ed. A. Smith et al., pp. 423–424. Singapore: World Scientific Publ. [20]

Feld, S. 1982. Sound and Sentiment: Birds, Weeping, Poetics and Song in Kaluli Expression. Philadelphia: Univ. of Pennsylvania Press. [2, 21]

———. 1984. Sound structure as social structure. *Ethnomus. Forum* **28**:383–409. [21]

Feld, S., and A. F. Fox. 1994. Music and language. *Annu. Rev. Anthropol.* **23**:25–53. [2]

Fellous, J.-M. 2009. Emotion: Computational modeling. In: Encyclopedia of Neuroscience, ed. L. R. Squire, vol. 3, pp. 909–913. Oxford Academic. [15]

Fellous, J.-M., and J. E. Ledoux. 2005. Toward basic principles for emotional processing: What the fearful brain tells the robot. In: Who Needs Emotions: The Brain Meets the Robot, ed. J.-M. Fellous and M. A. Arbib, pp. 79–115. New York: Oxford Univ. Press. [15]

Ferguson, S., and M. M. Wanderley. 2010. The McGill Digital Orchestra: An interdisciplinary project on digital musical instruments. *J. Interdisc. Music Stud.* **4**:17–35. [16]

Fernald, A. 1991. Prosody in speech to children: Prelinguistic and linguistic functions. *Ann. Child Devel.* **8**:43–80. [18]

Fernald, A., T. Taeschner, J. Dunn, et al. 1989. A cross-language study of prosodic modifications in mothers' and fathers' speech to preverbal infants. *J. Child Lang.* **16**:477–501. [18]

Fernald, R. D. 2000. Evolution of eyes. *Curr. Opin. Neurobiol.* **10**:444–450. [21]

———. 2006. Casting a genetic light on the evolution of eyes. *Science* **313**:1914–1918. [20]

Ferrari, P. F., V. Gallese, G. Rizzolatti, and L. Fogassi. 2003. Mirror neurons responding to the observation of ingestive and communicative mouth actions in the monkey ventral premotor cortex. *Eur. J. Neurosci.* **17**:1703–1714. [4, 19]

Ferrari, P. F., E. Kohler, L. Fogassi, and V. Gallese. 2000. The ability to follow eye gaze and its emergence during development in macaque monkeys. *PNAS* **97**:13,997–14,002. [4]

Ferrari, P. F., C. Maiolini, E. Addessi, L. Fogassi, and E. Visalberghi. 2005. The observation and hearing of eating actions activates motor programs related to eating in macaque monkeys. *Behav. Brain Res.* **161**:95–101. [19]

Ferreira, A. R. J., T. V. Smulders, K. Sameshima, C. V. Mello, and E. D. Jarvis. 2006. Vocalizations and associated behaviors of the sombre hummingbird (*Trochilinae*) and the rufous-breasted hermit (*Phaethornithinae*). *Auk* **123**:1129–1148. [20]

Ferreira, F. 2007. The "good enough" approach to language comprehension. *Lang. Ling. Compass* **1**:71–83. [12]

Ferreira, F., G. D. K. Bailey, and V. Ferraro. 2002. Good-enough representations in language comprehension. *Curr. Dir. Psychol. Sci.* **11**:11–15. [9]

Filipic, S., B. Tillmann, and E. Bigand. 2010. Judging familiarity and emotion from very brief musical excerpts. *Pschol. Bull. Rev.* **17**:335–341. [18]

Fillmore, C. J., and P. Kay. 1993. Construction Grammar Coursebook (reading material for Ling. X20). Berkeley: Univ. of California Press. [15]

Fincher, D. (director), and K. Spacey (producer). 2010. The Social Network. US: Sony Pictures. [7]

Finnegan, M. 2009. Political bodies: Some thoughts on women's power among Central African hunter-gatherers. *Radical Anthropol. Group J.* **3**:31–37. [2]

Fiorentino, R., and D. Poeppel. 2007. Compound words and structure in the lexicon. *Lang. Cogn. Proc.* **22**:953–1000. [9]

Fischer, J., and K. Hammerschmidt. 2011. Ultrasonic vocalizations in mouse models for speech and socio-cognitive disorders: Insights into the evolution of vocal communication. *Genes Brain Behav.* **10**:17–27. [20]

Fisher, S. E., and C. Scharff. 2009. FOXP2 as a molecular window into speech and language. *Trends Genet.* **25**:166–177. [20]

Fitch, W. T. 2000. The evolution of speech: A comparative review. *Trends Cogn. Sci.* **4**:258–267. [20]

———. 2005. The evolution of music in comparative perspective. In: The Neurosciences and Music II: From Perception to Performance, ed. G. Avanzini et al., pp. 29–49. New York: NY Acad. Sci. [6, 20]

———. 2006a. The biology and evolution of music: A comparative perspective. *Cognition* **100**:173–215. [1, 6, 8, 10, 18, 20]

———. 2006b. Production of vocalizations in mammals. In: Encyclopedia of Language and Linguistics, ed. K. Brown, pp. 115–121. Oxford: Elsevier. [20]

———. 2009a. The biology and evolution of language: "Deep homology" and the evolution of innovation. In: The Cognitive Neurosciences IV, ed. M. Gazzaniga, pp. 873–883. Cambridge, MA: MIT Press. [20]

———. 2009b. Fossil cues to the evolution of speech. In: The Cradle of Language, ed. R. Botha and C. Knight, pp. 112–134. Oxford: Oxford Univ. Press. [3]

———. 2010. The Evolution of Language. Cambridge: Cambridge Univ. Press. [10, 11, 14, 18–20]

———. 2011. The biology and evolution of rhythm: Unraveling a paradox. In: Language and Music as Cognitive Systems, ed. P. Rebuschat et al. Oxford: Oxford Univ. Press. [5]

Fitch, W. T., and B. Gringas. 2011. Multiple varieties of musical meaning: Comment on "Towards a neural basis of processing musical semantics" by Stefan Koelsch. *Phys. Life Rev.* **8**:108–109. [6]

Fitch, W. T., M. D. Hauser, and N. Chomsky. 2005. The evolution of the language faculty: Clarifications and implications. *Cognition* **97**:179–210. [20]

Fitch, W. T., L. Huber, and T. Bugnyar. 2010. Social cognition and the evolution of language: Constructing cognitive phylogenies. *Neuron* **65**:795–814. [20]

Flanagan, J. R., P. Vetter, R. S. Johansson, and D. M. Wolpert. 2003. Prediction precedes control in motor learning. *Curr. Biol.* **13**:146–150. [17]

Fodor, J. A. 1983. Modularity of Mind. Cambridge, MA: MIT Press. [14]

Fodor, J. A., and E. Lepore. 2002. The Compositionality Papers. Oxford: Oxford Univ. Press. [9]

Fodor, J. A., and Z. W. Pylyshyn. 1988. Connectionism and cognitive architecture: A critical analysis. *Cognition* **28**:3–71. [16]

Fogassi, L., and P. F. Ferrari. 2004. Mirror neurons, gestures and language evolution. *Interaction Stud.* **5**:345–363. [4, 19]

———. 2007. Mirror neurons and the evolution of embodied language. *Curr. Dir. Psychol. Sci.* **16**:136–141. [4, 19]

Fogassi, L., P. F. Ferrari, B. Gesierich, et al. 2005. Parietal lobe: From action organization to intention understanding. *Science* **308**:662–667. [4]

Fogassi, L., V. Gallese, L. Fadiga, et al. 1996. Coding of peripersonal space in inferior premotor cortex (area F4). *J. Neurophysiol.* **76**:141–157. [4]

Folland, N. A., B. E. Butler, N. A. Smith, and L. J. Trainor. 2012. Processing simultaneous auditory objects: Infants' ability to detect mistunings in harmonic complexes. *J. Acoust. Soc. Am.* **131**:993–997. [17]

Fonagy, I. 1983. La Vive Voix. Paris: Payot. [5]

Ford, J. K. B. 1991. Vocal traditions among resident killer whales, *Orcinus orca*, in coastal waters of British Columbia. *Can. J. Zool.* **69**:1454–1483. [20]

Fortune, E. S., and D. Margoliash. 1995. Parallel pathways converge onto HVc and adjacent neostriatum of adult male zebra finches, *Taeniopygia guttata*. *J. Comp. Neurol.* **360**:413–441. [20]

Foster, E. F., and S. W. Bottjer. 2001. Lesions of a telencephalic nucleus in male zebra finches: Influences on vocal behavior in juveniles and adults. *J. Neurobiol* **46**:142–165. [20]

Foster, N. E. V., and R. J. Zatorre. 2010. A role for the intraparietal sulcus in transforming musical pitch information. *Cereb. Cortex* **20**:1350–1359. [14, 17]

Foster, S. 2010. Choreographing Empathy: Kinesthesia in Performance: Routledge. [8]

Foxton, J. M., R. K. Nandy, and T. D. Griffiths. 2006. Rhythm deficits in 'tone deafness'. *Brain Cogn.* **62**:24–29. [13]

Fraisse, P. 1982. Rhythm and tempo. In: The Psychology of Music, ed. D. Deutsch, pp. 149–180. New York: Academic. [18]

Frazier, L. 1987. Sentence processing: A tutorial review. In: Attention and Performance XII, ed. M. Coltheart, pp. 559–585. London: Erlbaum. [9]

Frégnac, Y. 1995. Hebbian synaptic plasticity: Comparative and developmental aspects. In: The Handbook of Brain Theory and Neural Networks, ed. M. A. Arbib, pp. 459–464, A Bradford Book. Cambridge, MA: MIT Press. [15]

Frey, A., C. Marie, L. Prod'Homme, et al. 2009. Temporal Semiotic Units as Minimal Meaningful Units in Music? An Electrophysiological Approach. *Music Percep.* **26**:247–256. [8]

Frey, S., J. S. Campbell, G. B. Pike, and M. Petrides. 2008. Dissociating the human language pathways with angular resolution diffusion fiber tractography. *J. Neurosci.* **28**:11,435–11,444. [17]

Frick, R. W. 1985. Communicating emotion: The role of prosodic features. *Psychol. Bull.* **97**:412–429. [5]

Fried, I., A. Katz, G. McCarthy, et al. 1991. Functional organization of human supplementary motor cortex studied by electrical stimulation. *J. Neurosci.* **11**:3656–3666. [20]

Fried, I., G. A. Ojemann, and E. E. Fetz. 1981. Language-related potentials specific to human language cortex. *Science* **212**:353–356. [20]

Friederici, A. 2012. The cortical language circuit: from auditory perception to sentence comprehension. *Trends Cogn. Neurosci.* **16**:262–268. [17]

Friederici, A. D. 2002. Towards a neural basis of auditory sentence processing. *Trends Cogn. Sci.* **6**:78–84. [6]

———. 2004. Event-related brain potential studies in language. *Curr. Neurol. Neurosci. Rep.* **4**:466–470. [6]

———. 2009a. Allocating functions to fiber tracts: Facing its indirectness. *Trends Cogn. Sci.* **13**:370–371. [9]

———. 2009b. Pathways to language: Fiber tracts in the human brain. *Trends Cogn. Sci.* **13**:175–181. [20]

Friederici, A. D., J. Bahlmann, S. Heim, R. I. Schubotz, and A. Anwander. 2006. The brain differentiates human and non-human grammars: Functional localization and structural connectivity. *PNAS* **103**:2458–2463. [6]

Friederici, A. D., S. A. Ruschemeyer, A. Hahne, and C. J. Fiebach. 2003. The role of left inferior frontal and superior temporal cortex in sentence comprehension: Localizing syntactic and semantic processes. *Cereb. Cortex* **13**:170–177. [9]

Friedrich, R., and A. D. Friederici. 2009. Mathematical logic in the human brain: Syntax. *PLoS ONE* **4**:e5599. [6]

Fries, P. 2005. A mechanism for cognitive dynamics: Neuronal communication through neuronal coherence. *Trends Cogn. Sci.* **9**:474–480. [9]

———. 2009. Neuronal gamma-band synchronization as a fundamental process in cortical computation. *Annu. Rev. Neurosci.* **32**:209–224. [7, 15]

Frijda, N. H. 1986. The Emotions. Cambridge: Cambridge Univ. Press. [5, 8]

———. 2007. Klaus Scherer's article on "What are emotions?" [Comments]. *Soc. Sci. Inf.* **46**:381–443. [5]

Frijda, N. H., and K. R. Scherer. 2009. Emotion definition (psychological perspectives). In: Oxford Companion to Emotion and the Affective Sciences, ed. D. Sander and K. R. Scherer, pp. 142–143. Oxford: Oxford Univ. Press. [5]

Frijda, N. H., and L. Sundararajan. 2007. Emotion refinement. *Perspect. Psychol. Sci.* **2**:227–241. [5]

Friston, K. 2009. The free-energy principle: A rough guide to the brain? *Trends Cogn. Sci.* **13**:293–301. [16]

———. 2010. The free-energy principle: A unified brain theory? *Nat. Rev. Neurosci.* **11**:127–138. [8]

Frith, C. D. 2007. The Social Brain? *Philos. Trans. R. Soc. Lond. B* **362**:671–678. [8]

———. 2008. Social Cognition. *Philos. Trans. R. Soc. Lond. B* **363**:2033–2039. [8]

Fritz, J., S. V. David, S. Radtke-Schuller, P. Yin, and S. A. Shamma. 2010. Adaptive, behaviorally-gated, persistent encoding of task-relevant information in ferret frontal cortex. *Nat. Neurosci.* **13**:1011–1019. [17]

Fritz, J., M. Mishkin, and R. C. Saunders. 2005. In search of an auditory engram. *PNAS* **102**:9359–9364. [17, 20]

Fritz, J., S. A. Shamma, M. Elhilali, and D. Klein. 2003. Rapid task-related plasticity of spectrotemporal receptive fields in primary auditory cortex. *Nat. Neurosci.* **6**:1216–1223. [17]

Fritz, T., S. Jentschke, and N. Gosselin. 2009. Universal recognition of three basic emotions in music. *Curr. Biol.* **19**:573–576. [6, 8]

Frye, N. 1982. The Great Code: The Bible and Literature. New York: Harcourt Brace Jovanovich. [1]

Fujii, N., S. Hihara, and A. Iriki. 2008. Social cognition in premotor and parietal cortex. *Soc. Neurosci.* **3**:250–260. [4]

Fujioka, T., B. R. Zendel, and B. Ross. 2010. Endogenous neuromagnetic activity for mental hierarchy of timing. *J. Neurosci.* **30**:3458–3466. [13]

Fujiyama, S., K. Ema, and S. Iwamiya. 2012. Effect of the technique of conflict between music and moving picture using Akira Kurosawa's movies (in Japanese). In: Proc. of the Spring Meeting of Japanese Society of Music Perception and Cognition, pp. 85–70. Kyoto: Japanese Soc. of Music Perception and Cognition. [7]

Fuller, T. 1640. The History of the Worthies of England. London: Nuttall and Hodgson. [5]

Fuster, J. M. 2004. Upper processing stages of the perception-action cycle. *Trends Cogn. Sci.* **8**:143–145. [1, 8, 13, 17]

———. 2009. Cortex and memory: Emergence of a new paradigm. *J. Cogn. Neurosci.* **21**:2047–2072. [1]

Fyhn, M., S. Molden, M. P. Witter, E. I. Moser, and M. B. Moser. 2004. Spatial representation in the entorhinal cortex. *Science* **305**:1258–1264. [15]

Gabrielson, A., and P. Juslin. 2003. Emotional expression in music. In: Handbook of Affective Sciences, ed. R. J. Davidson, pp. 503–534. New York: Oxford Univ. Press. [6]

Galantucci, B., C. A. Fowler, and M. T. Turvey. 2006. The motor theory of speech perception reviewed. *Psychon. Bull. Rev.* **13**:361–377. [20]

Galef, B. G., Jr. 1992. The question of animal culture. *Human Nature* **3**:157–178. [20]
Gallese, V. 2011. Mirror Neurons and Art. In: Art and the Senses, ed. F. Bacci and D. Melcher, pp. 455–463. Oxford: Oxford Univ. Press. [8]
Gallese, V., L. Fadiga, L. Fogassi, and G. Rizzolatti. 1996. Action recognition in the premotor cortex. *Brain* **119**:593–609. [4]
———. 2002. Action representation and the inferior parietel lobule. In: Common Mechanisms in Perception and Action: Attention and Performance, vol. 19, ed. W. Prinz and B. Hommel, pp. 334–355. Oxford: Oxford Univ. Press. [4]
Gallese, V., and A. Goldman. 1998. Mirror neurons and the simulation theory of mind-reading. *Trends Cogn. Sci.* **2**:493–501. [4]
Garamszegi, L. Z., and M. Eens. 2004. Brain space for a learned task: Strong intraspecific evidence for neural correlates of singing behavior in songbirds. *Brain Res. Rev.* **44**:187–193. [20]
Garrido, M. I., J. M. Kilner, S. J. Kiebel, and K. J. Friston. 2007. Evoked brain responses are generated by feedback loops. *PNAS* **104**:20,961–20,966. [9]
Garrod, S., and M. J. Pickering. 2004. Why is conversation so easy? *Trends Cogn. Sci.* **8**:8–11. [13]
Gazzaley, A., and A. C. Nobre. 2012. Top-down modulation: Bridging selective attention and working memory. *Trends Cogn. Sci.* **16**:129–135. [7]
Gazzola, V., L. Aziz-Zadeh, and C. Keysers. 2006. Empathy and the somatotopic auditory mirror system in humans. *Curr. Biol.* **16**:1824–1829. [4, 8]
Gebauer, G. 2013. *Wie können wir über emotionen sprechen?* In: *Emotion und Sprache*, ed. G. Gebauer et al. Weilerswist-Metternich: Velbrück, in press. [6]
Gehring, W. J., and K. Ikeo. 1999. Pax 6: Mastering eye morphogenesis and eye evolution. *Trends Genet.* **15**:371–377. [20]
Gentilucci, M., and M. C. Corballis. 2006. From manual gesture to speech: A gradual transition. *Neurosci. Biobehav. Rev.* **30**:949–960. [4, 20]
Gentilucci, M., P. Santunione, A. C. Roy, and S. Stefanini. 2004a. Execution and observation of bringing a fruit to the mouth affect syllable pronunciation. *Eur. J. Neurosci.* **19**:190–202. [19]
Gentilucci, M., S. Stefanini, A. C. Roy, and P. Santunione. 2004b. Action observation and speech production: Study on children and adults. *Neuropsychologia* **42**:1554–1567. [4, 19]
Gentner, T. Q., K. M. Fenn, D. Margoliash, and H. C. Nusbaum. 2006. Recursive syntactic pattern learning by songbirds. *Nature* **440**:1204–1207. [1]
Gernsbacher, M. A. 1990. Language Comprehension as Structure Building. Mahwah, NJ: Erlbaum. [8]
Geschwind, N. 1979. Specializations of the human brain. *Sci. Am.* **241**:180–199. [20]
Gibbs, M. E., and K. T. Ng. 1979. Behavioural stages in memory formation. *Neurosci. Lett.* **13**:279–283. [8]
Gibson, E. 2000. The dependency locality theory: A distance-based theory of linguistic complexity. In: Image, Language, Brain, ed. A. Marantz et al., pp. 95–126. Cambridge, MA: MIT Press. [14]
Gigley, H. M. 1983. HOPE-AI and the dynamic process of language behavior. *Cogn. Brain Theory* **6**:39–88. [15]
Gil, D., M. Naguib, K. Riebel, A. Rutstein, and M. Gahr. 2006. Early condition, song learning, and the volume of song brain nuclei in the zebra finch, *Taeniopygia guttata. J. Neurobiol.* **66**:1602–1612. [20]
Gilby, I. C. 2006. Meat sharing among the Gombe chimpanzees: Harassment and reciprocal exchange. *Anim. Behav.* **71**:953–963. [21]

Gill, K. Z., and D. Purves. 2009. A biological rationale for musical scales. *PLoS ONE* 4:e8144. [5]
Gill, S. P., M. Kawamori, Y. Katagiri, and A. Shimojima. 2000. The role of body moves in dialogue. *Intl. J. Lang. Comm.* 12:89–114. [21]
Giraud, A. L., A. Kleinschmidt, D. Poeppel, et al. 2007. Endogeneous cortical rhythms determine hemispheric dominance for speech. *Neuron* 56:1127–1134. [9]
Giraud, A. L., and D. Poeppel. 2012. Cortical oscillations and speech processing: Emerging computational principles and operations. *Nat. Neurosci.* 15:511–517. [9, 17]
Gjerdingen, R. O. 1996. Courtly behaviors. *Music Percep.* 11:335–370. [10]
Glasser, M. F., and J. K. Rilling. 2008. DTI tractography of the human brain's language pathways. *Cereb. Cortex* 18:2471–2482. [20]
Glennie, E. 1993. Hearing essay. http://www.evelyn.co.uk/Resources/Essays/Hearing%20Essay.pdf. (accessed 6 Jan 2013). [1]
Glimcher, P. W. 2003. Decisions, Uncertainty, and the Brain: The Science of Neuroeconomics. Cambridge, MA: MIT Press. [14]
Gluck, M. A., M. T. Allen, C. E. Myers, and R. F. Thompson. 2001. Cerebellar substrates for error correction in motor conditioning. *Neurobiol. Learn. Mem.* 76:314–341. [15]
Gobes, S. M., and J. J. Bolhuis. 2007. Birdsong memory: A neural dissociation between song recognition and production. *Curr. Biol.* 17:789–793. [20]
Goffman, E. 1981. Forms of Talk. Philadelphia: Univ. of Pennsylvania Press. [3]
Gogos, A., M. Gavrilescu, S. Davison, et al. 2010. Greater superior than inferior parietal lobule activation with increasing rotation angle during mental rotation: An fMRI study. *Neuropsychologia* 48:529–535. [17]
Goldberg, A. E. 1995. Constructions: A Construction Grammar Approach to Argument Structure. Chicago: Univ. of Chicago Press. [15]
———. 2003. Constructions: A new theoretical approach to language. *Trends Cogn. Sci.* 7:219–224. [1]
Goldberg, J. H., and M. S. Fee. 2011. Vocal babbling in songbirds requires the basal ganglia-recipient motor thalamus but not the basal ganglia. *J. Neurophysiol.* 105:2729–2739 [20]
Goldman-Rakic, P. S. 1991. Parallel systems in the cerebral cortex: The topography of cognition. In: Natural and Artificial Computation, ed. M. A. Arbib and J. A. Robinson, pp. 155–176. Cambridge, MA: MIT Press. [17]
Goldsmith, J. 1990. Autosegmental and Metrical Phonology. Oxford: Blackwell. [11]
Goldstein, L., D. Byrd, and E. Saltzman. 2006. The role of vocal tract gestural action units in understanding the evolution of phonology. In: From Action to Language via the Mirror System, ed. M. A. Arbib, pp. 215–49. Cambridge: Cambridge Univ. Press. [1]
Goller, F., and O. N. Larsen. 2002. New perspectives on mechanisms of sound generation in songbirds. *J. Comp. Physiol. A* 188:841–850. [20]
Goodale, M. A., and A. D. Milner. 1992. Separate visual pathways for perception and action. *Trends Neurosci.* 15:20–25. [15]
Gorbman, C. 1987. Unheard Melodies: Narrative Film Music. Bloomington, IN: Indiana Univ. Press. [7]
Gordon, R. L., C. L. Magne, and E. W. Large. 2011. EEG correlates of song prosody: A new look at the relationship between linguistic and musical rhythm. *Front. Psychol.* 2:352. [15]
Gosselin, N., I. Peretz, E. Johnsen, and R. Adolphs. 2007. Amygdala damage impairs emotion recognition from music. *Neuropsychologia* 45:236–244. [6]

Goswami, U. 2011. A temporal sampling framework for developmental dyslexia. *Trends Cogn. Sci.* **15**:3–10. [14]

Goudbeek, M., and K. R. Scherer. 2010. Beyond arousal: Valence and potency/control in the vocal expression of emotion. *J. Acoust. Soc. Am.* **128**:1322–1336. [5]

Gouvea, A. C., C. Phillips, N. Kazanina, and P. D. 2010. The linguistic processes underlying the P600. *Lang. Cogn. Proc.* **25**:149–188. [14]

Gracco, V. L., P. Tremblay, and B. Pike. 2005. Imaging speech production using fMRI. *Neuroimage* **26**:294–301. [20]

Grafton, S. T., M. A. Arbib, L. Fadiga, and G. Rizzolatti. 1996. Localization of grasp representations in humans by PET: 2. Observation compared with imagination. *Exp. Brain Res.* **112**:103–111. [19]

Grahn, J. A., and M. Brett. 2007. Rhythm and beat perception in motor areas of the brain. *J. Cogn. Neurosci.* **19**:893–906. [2, 12, 13]

Grahn, J. A., and J. D. McAuley. 2009. Neural bases of individual differences in beat perception. *Neuroimage* **47**:1894–1903. [13]

Grahn, J. A., and J. B. Rowe. 2009. Feeling the beat: Premotor and striatal interactions in musicians and nonmusicians during beat perception. *J. Neurosci.* **29**:7540–7548. [13]

Grainger, J., A. Rey, and S. Dufau. 2008. Letter perception: From pixels to pandemonium. *Trends Cogn. Sci.* **12**:381–387. [9]

Grandjean, D., D. Sander, G. Pourtois, et al. 2005. The voices of wrath: Brain responses to angry prosody in meaningless speech. *Nat. Neurosci.* **8**:145–146. [13]

Grandjean, D., and K. R. Scherer. 2008. Unpacking the cognitive architecture of emotion processes. *Emotion* **8**:341–351. [5]

Gratier, M., and E. Devouche. 2011. Imitation and repetition of prosodic contour in vocal interaction at 3 months. *Devel. Psychol.* **47**:67–76. [18]

Grauer, V. 2006. Echoes of our forgotten ancestors. *World of Music* **48**:5–59. [18]

———. 2007. New perspectives on the Kalahari debate: A tale of two genomes. *Before Farming* **2**:4. [2]

Graybiel, A. M. 1998. The basal ganglia and chunking of action repertoires. *Neurobiol. Learn. Mem.* **70**:119–136. [13]

———. 2008. Habits, rituals, and the evaluative brain. *Annu. Rev. Neurosci.* **31**:359–387. [13]

Green, J. 2010. Understanding the score: Film music communicating to and influencing the audience. *J. Aesthetic Ed.* **44**:81–94. [7]

Greenberg, J. 1966. Some universals of grammar with particular reference to the order of meaningful elements. In: Universals of Language, ed. J. Greenberg, pp. 73–113. Cambridge, MA: MIT Press. [3]

Greenberg, S. 2005. From here to utility. In: The Integration of Phonetic Knowledge in Speech Technology, ed. W. J. Barry and W. A. V. Dommelen, pp. 107–133. Dordrecht: Springer. [9]

Gregoriou, G. G., E. Borra, M. Matelli, and G. Luppino. 2006. Architectonic organization of the inferior parietal convexity of the macaque monkey. *J. Comp. Neurol.* **496**:422–451. [4]

Gregory, R. L. 2004. Empathy. In: The Oxford Companion to the Mind, edited by R. L. Gregory. Oxford: Oxford Univ. Press. [8]

Grèzes, J., N. Costes, and J. Decety. 1998. Top-down effect of strategy on the perception of human biological motion: A PET investigation. *Cogn. Neuropsychol.* **15**:553–582. [4]

Grice, P. 1975. Logic and conversation. In: Syntax and Semantics, vol. 3, ed. P. Cole and J. L. Morgan, pp. 41–58. New York: Academic Press. [7]

———. 1989. Studies in the Way of Words. Cambridge, MA: Harvard Univ. Press. [9]
Grieser Painter, J., and S. Koelsch. 2011. Can out-of-context musical sounds convey meaning? An ERP study on the processing of meaning in music. *Psychophysiology* **48**:645–655. [6]
Griffiths, T. D., C. Buchel, R. S. J. Frackowiak, and R. D. Patterson. 1998. Analysis of temporal structure in sound by the human brain. *Nat. Neurosci.* **1**:422–427. [13, 17]
Griffiths, T. D., and J. D. Warren. 2004. What is an auditory object? *Nat. Rev. Neurosci.* **5**:887–892. [17]
Grimsley, J. M., J. J. Monaghann, and J. J. Wenstrup. 2011. Development of social vocalizations in mice. *PLoS ONE* **6**:e17460. [20]
Grodzinsky, Y., and K. Amunts. 2006. Broca's region: Mysteries, facts, ideas, and history. New York: Oxford Univ. Press. [1]
Grossberg, S. 1995. The attentive brain. *Am. Sci.* **83**:438–449. [7]
———. 2007. Consciousness CLEARS the mind. *Neural Netw.* **20**:1040–1053. [7]
Groussard, M., R. La Joie, G. Rauchs, et al. 2010. When music and long-term memory interact: Effects of musical expertise on functional and structural plasticity in the hippocampus. *PLoS ONE* **5**:e13225. [13]
Grube, M., F. E. Cooper, P. F. Chinnery, and T. D. Griffiths. 2010. Dissociation of duration-based and beat-based auditory timing in cerebellar degeneration. *PNAS* **107**:11,597–11,601. [12]
Guazzelli, A., F. J. Corbacho, M. Bota, and M. A. Arbib. 1998. Affordances, motivation, and the world graph theory. *Adap. Behav.* **6**:435–471. [15]
Guilford, T., and R. Dawkins. 1991. Receiver psychology and the evolution of animal signals. *Anim. Behav.* **42**:114. [5]
Gunji, A., S. Koyama, R. Ishii, et al. 2003. Magnetoencephalographic study of the cortical activity elicited by human voice. *Neurosci. Lett.* **348**:13–16. [18]
Gussenhoven, C. 2004. The Phonology of Tone and Intonation. Cambridge: Cambridge Univ. Press. [5]
Gyger, M. 1990. Audience effects on alarm calling. *Ethol. Ecol. Evol.* **2**:227–232. [20]
Haber, S. N., and B. Knutson. 2009. The reward circuit: Linking primate anatomy and human imaging. *Neuropsychopharm.* **35**:4–26. [13, 15]
Hackett, S. J., R. T. Kimball, S. Reddy, et al. 2008. A phylogenomic study of birds reveals their evolutionary history. *Science* **320**:1763–1768. [20]
Haesler, S., C. Rochefort, B. Georgi, et al. 2007. Incomplete and inaccurate vocal imitation after knockdown of FoxP2 in songbird basal ganglia nucleus AreaX. *PLoS Biol.* **5**:e321. [20]
Haesler, S., K. Wada, A. Nshdejan, et al. 2004. FoxP2 expression in avian vocal learners and non-learners. *J. Neurosci.* **24**:3164–3175. [20]
Hagen, E. H., and P. Hammerstein. 2005. Evolutionary biology and the strategic view of ontogeny: Genetic strategies provide robustness and flexibility in the life course. *Res. Human Devel.* **1**:87–101. [21]
Hagoort, P. 2003. How the brain solves the binding problem for language: A neurocomputational model of syntactic processing. *NeuroImage* **20**:S18–S29. [9]
———. 2005. On Broca, brain, and binding: A new framework. *Trends Cogn. Sci.* **9**:416–423. [3, 9, 14]
Hagoort, P., L. Hald, M. Bastiaansen, and K. M. Petersson. 2004. Integration of word meaning and world knowledge in language comprehension. *Science* **304**:438–441. [9]
Hahnloser, R. H. R., A. A. Kozhevnikov, and M. S. Fee. 2002. An ultra-sparse code underlies the generation of neural sequences in a songbird. *Nature* **419**:65–70. [20]

Hall, D., and D. Barker. 2012. Coding of basic acoustical and perceptual components of sound in human auditory cortex. In: The Human Auditory Cortex, ed. D. Poeppel et al., pp. 165–197. New York: Springer. [17]

Halle, M. 2002. From Memory to Speech and Back: Papers on Phonetics and Phonology, 1954–2002 (illustrated ed.). Berlin: Walter de Gruyter. [17]

Halliday, M. A. K. 1967. Notes on transitivity and theme in English. Part 2. *J. Ling.* **3**:177–274. [9]

Halpern, A. R., and D. Müllensiefen. 2008. Effects of timbre and tempo change on memory for music. *Q. J. Exp. Psychol.* **61**:1371–1384. [18]

Halpern, A. R., and R. J. Zatorre. 1999. When that tune runs through your head: A PET investigation of auditory imagery for familiar melodies. *Cereb. Cortex* **9**:697–704. [13]

Halpern, A. R., R. J. Zatorre, M. Bouffard, and J. A. Johnson. 2004. Behavioral and neural correlates of perceived and imagined musical timbre. *Neuropsychologia* **42**: [13]

Halwani, G. F., P. Loui, T. Ruber, and G. Schlaug. 2011. Effects of practice and experience on the arcuate fasciculus: Comparing singers, instrumentalists and nonmusicians. *Front. Psychol.* **2**:156. [17]

Hannon, E. E., and S. P. Johnson. 2005. Infants use meter to categorize rhythms and melodies: Implications for musical structure learning. *Cogn. Psychol.* **50**:354–377. [13]

Hannon, E. E., G. Soley, and R. S. Levine. 2011. Constraints on infants' musical rhythm perception: Effects of interval ratio complexity and enculturation. *Devel. Sci.* **14**:865–872. [18]

Hannon, E. E., and S. E. Trehub. 2005a. Metrical categories in infancy and adulthood. *Psychol. Sci.* **16**:48–55. [13, 18, 17]

———. 2005b. Tuning in to musical rhythms: Infants learn more readily than adults. *PNAS* **102**:12,639–12,643. [13,18]

Hannula-Jouppi, K., N. Kaminen-Ahola, M. Taipale, et al. 2005. The axon guidance receptor gene ROBO1 is a candidate gene for developmental dyslexia. *PLoS Genet.* **1**:e50. [20]

Hanson, A. R., and E. M. Riseman. 1978. VISIONS: A computer system for interpreting scenes. In: Computer Vision Systems, ed. A. R. Hanson and E. M. Riseman, pp. 129–163. New York: Academic Press. [1]

Hargreaves, D. J., and A. C. North. 2010. Experimental aesthetics and liking for music. In: Handbook of Music and Emotion: Theory, Research, Applications, ed. P. Juslin and J. A. Sloboda, pp. 515–546. New York: Oxford Univ. Press. [5]

Harlow, H. 1958. The nature of love. *Am. Psychol.* **13**:673–685. [6]

Harnad, S. 1990. The symbol grounding problem. *Physica* **42**:335–346. [16]

Harrington, I. A., R. S. Heffner, and H. E. Heffner. 2001. An investigation of sensory deficits underlying the aphasia-like behavior of macaques with auditory cortex lesions. *NeuroReport* **12**:1217–1221. [17]

Hasson, U., O. Furman, D. Clark, Y. Dudai, and L. Davachi. 2008a. Enhanced intersubject correlations during movie viewing correlate with successful episodic encoding. *Neuron* **57**:452–462. [8]

Hasson, U., O. Landesman, B. Knappmeyer, et al. 2008b. Neurocinematics: The neuroscience of film. *Projections* **2**:1–26. [7]

Hasson, U., Y. Nir, I. Levy, G. Fuhrmann, and R. Malach. 2004. Intersubject synchronization of cortical activity during natural vision. *Science* **303**:1634–1640. [8]

Hast, M. H., J. M. Fischer, A. B. Wetzel, and V. E. Thompson. 1974. Cortical motor representation of the laryngeal muscles in Macaca mulatta. *Brain Res.* **73**:229–240. [4]

Hauptmann, M. 1853. *Die Natur der Harmonik und der Metrik: Zur Theorie der Musik.* Leipzig: Breitkopf & Härtel. [10]

Hausberger, M., M.-A. Richard-Yris, L. Henry, L. Lepage, and I. Schmidt. 1995. Song sharing reflects the social organization in a captive group of European starlings, *Sturnus vulgaris. J. Comp. Psychol.* **109**:222–241. [20]
Hauser, M. D. 1996. The Evolution of Communication. Cambridge, MA: MIT Press. [5]
Hauser, M. D., N. Chomsky, and W. T. Fitch. 2002. The faculty of language: What is it, who has it, and how did it evolve? *Science* **298**:1569–1579. [1, 10, 20]
Hawk, S., G. Van Kleef, A. Fischer, and J. Van der Schalk. 2009. Worth a thousand words: Absolute and relative decoding of nonlinguistic affect vocalizations. *Emotion* **9**:293–305. [5]
Hay, J. S. F., and R. L. Diehl. 2007. Perception of rhythmic grouping: Testing the iambic/trochaic law. *Percep. Psychophys.* **69**:113–122. [18]
Hayes, B. 1989. The prosodic hierarchy in poetry. In: Phonetics and Phonology: Rhythm and Meter, ed. P. Kiparsky and G. Youmans. New York: Academic. [10]
Hayes, K. J., and C. Hayes. 1951. The intellectual development of a home-raised chimpanzee. *Proc. Am. Philos. Soc.* **95**:105–109. [20]
Hazanavicius, M. (director), and T. Langmann (producer). 2011. The Artist. US: The Weinstein Co. [7]
Hebb, D. O. 1949. The Organization of Behaviour. New York: John Wiley and Sons. [5, 15]
Hebert, S., A. Racette, L. Gagnon, and I. Peretz. 2003. Revisiting the dissociation between singing and speaking in expressive aphasia. *Brain* **126**:1838–1850. [13]
Heffner, H. E., and R. S. Heffner. 1984. Temporal lobe lesions and perception of species-specific vocalizations by macaques. *Science* **226**:75–76. [17]
Heim, I., and A. Kratzer. 1998. Semantics in Generative Grammar. New York: Blackwell. [9]
Helmholtz, H. L. F. 1863/1954. On the Sensations of Tone as a Physiological Basis for the Theory of Music. New York: Dover. [5]
Hennenlotter, A., C. Dresel, F. Castrop, et al. 2009. The link between facial feedback and neural activity within central circuitries of emotion: New insights from botulinum toxin-induced denervation of frown muscles. *Cereb. Cortex* **19**:537–542. [1]
Henning Proske, J., D. Jeanmonod, and P. F. M. J. Verschure. 2011. A computational model of thalamocortical dysrhythmia. *Eur. J. Neurosci.* **33**:1281–1290. [15]
Herrera, P., A. Yeterian, and F. Gouyon. 2002. Automatic classification of drum sounds: A comparison of feature selection methods and classification techniques. In: Music and Artificial Intelligence, ed. C. Anagnostopoulou et al., vol. 2445, pp. 69–80. Heidelberg: Springer. [16]
Herrington, R. (director), and J. Silver (producer). 1989. Road House. US: United Artists. [7]
Herrmann, C. S., M. H. J. Munk, and A. K. Engel. 2004. Cognitive functions of gamma-band activity: Memory match and utilization. *Trends Cogn. Sci.* **8**:347–355. [7]
Hessler, N. A., and A. J. Doupe. 1999. Singing-related neural activity in a dorsal forebrain-basal ganglia circuit of adult zebra finches. *J. Neurosci.* **19**:10,461–10,481. [20]
Hewes, G. W. 1973. Primate communication and the gestural origin of language. *Curr. Anthropol.* **12**:5–24. [19, 20]
Hickok, G. 2008. Eight problems of the mirror neuron theory of action understanding in monkeys and humans. *J. Cogn. Neurosci.* **21**:1229–1243. [17]
———. 2012. Computational neuroanatomy of speech production. *Nat. Rev. Neurosci.* **13**:135–145. [17]
Hickok, G., B. Buchsbaum, C. Humphries, and T. Muftuler. 2003. Auditory-motor interaction revealed by fMRI: Speech, music, and working memory in area SPT. *Spatial J. Cogn. Neurosci.* **15**:673–682. [17]

Hickok, G., J. Houde, and F. Rong. 2011. Sensorimotor integration in speech processing: Computational basis and neural organization. *Neuron* **69**:407–422. [17]
Hickok, G., and D. Poeppel. 2000. Towards a functional neuroanatomy of speech perception. *Trends Cogn. Sci.* **4**:131–137. [9]
———. 2004. Dorsal and ventral streams: A framework for understanding aspects of the functional anatomy of language. *Cognition* **92S**:67–99. [9, 17]
———. 2007. The cortical organization of speech processing. *Nat. Rev. Neurosci.* **8**:393–402. [9, 13, 14, 17]
Hihara, S., H. Yamada, A. Iriki, and K. Okanoya. 2003. Spontaneous vocal differentiation of coo-calls for tools and food in Japanese monkeys. *Neurosci. Res.* **45**:383–389. [19, 21]
Hillecke, T., A. Nickel, and H. V. Bolay. 2005. Scientific perspectives on music therapy. *Ann. NY Acad. Sci.* **1060**:271–282. [6]
Himmelmann, N., and D. R. Ladd. 2008. Prosodic Description: An Introduction for Fieldworkers. *LD&C* **2**:244–274. [11]
Hinke, R. M., X. Hu, A. E. Stillman, et al. 1993. Functional magnetic resonance imaging of Broca's area during internal speech. *NeuroReport* **4**:675–678. [20]
Hinzen, W., and D. Poeppel. 2011. Semantics between cognitive neuroscience and linguistic theory: Guest editors' introduction. *Lang. Cogn. Proc.* **26**:1297–1316. [9, 17]
Hipp, J., W. Einhäuser, J. Conradt, and P. König. 2005. Learning of somatosensory representations for texture discrimination using a temporal coherence principle. *Network* **16**:223–238. [15]
Hitchcock, A. (producer and directory). 1960. Psycho. US: Shamley Productions. [7]
Hjemlslev, L. 1953. Prolegomena to a Theory of Language, supplement to the *Intl. J. Am. Ling.* **19(1)**. Baltimore: Waverly Press. [11]
Hoch, H., B. Poulin-Charronnat, and B. Tillmann. 2011. The influence of task-irrelevant music on language processing: Syntactic and semantic structures. *Front. Psychol.* **2**:112. [14]
Hockett, C. F. 1958. A Course in Modern Linguistics. New York: Macmillan. [11]
———. 1960a. Logical considerations in the study of animal communication. In: Animal Sounds and Communication, ed. W. E. Lanyon and W. N. Tavolga, pp. 392–430. Washington, D.C.: American Institute of Biological Sciences. [6]
———. 1960b. The origin of speech. *Sci. Am.* **203**:88–96. [1, 19]
———. 1966. The problem of universals in language. In: Universals of Language, ed. J. Greenberg, pp. 1–29. Cambridge, MA: MIT Press. [20]
———. 1973. Man's Place in Nature. New York: McGraw-Hill. [11]
Hockett, C. F., and R. Ascher. 1964. The human revolution. *Curr. Anthro.* **5**:135–147. [11]
Hodges, D. A. 2010. Psychophysiological measures. In: Handbook of Music and Emotion: Theory, Research, Applications, ed. P. Juslin and J. A. Sloboda, pp. 279–312. New York: Oxford Univ. Press. [5]
Hodgkin, A. L., and A. F. Huxley. 1952. A quantitative description of membrane current and its application to conduction and excitation in nerve. *J. Physiol.* **117**:500–544. [1]
Hoen, M., M. Pachot-Clouard, C. Segebarth, and P. F. Dominey. 2006. When Broca experiences the Janus syndrome: An ER-fMRI study comparing sentence comprehension and cognitive sequence processing. *Cortex* **42**:605–623. [15]
Hoenig, K., and L. Scheef. 2005. Mediotemporal contributions to semantic processing: fMRI evidence from ambiguity processing during semantic context verification. *Hippocampus* **15**:597–609. [9]
Hofstötter, C., M. Mintz, and P. F. M. J. Verschure. 2002. The cerebellum in action: A simulation and robotics study. *Eur. J. Neurosci.* **16**:1361–1376. [8, 15, 16]

Honda, E., and K. Okanoya. 1999. Acoustical and syntactical comparisons between songs of the white-backed munia (*Lonchura striata*) and its domesticated strain, the Bengalese finch (*Lonchura striata var.domestica*). *Zoolog. Sci.* **16**:319–326. [20]

Honing, H. 2010. *De ongeletterde luisteraar. Over muziekcognitie, muzikaliteit en methodologie*. Amsterdam: Amsterdam Univ. Press. [1]

———. 2011a. *The Illiterate Listener: On Music Cognition, Musicality and Methodology*. Amsterdam: Amsterdam Univ. Press. [19]

———. 2011b. *Musical Cognition. A Science of Listening*. New Brunswick, NJ: Transaction Publishers. [1]

Honing, H., O. Ladinig, G. P. Haden, and I. Winkler. 2009. Is beat induction innate or learned? Probing emergent meter perception in adults and newborns using event-related brain potentials. *Ann. NY Acad. Sci.* **1169**:93–96. [17–19]

Horita, H., M. Kobayashi, W.-C. Liu, et al. 2013. Repeated evolution of differential regulation of an activity-dependent gene dusp1 for a complex behavioral trait. *PLoS ONE*, in press. [20]

Houde, J. F., and S. S. Nagarajan. 2011. Speech production as state feedback control. *Front. Human Neurosci.* **5**:82. [17]

Hsieh, S., M. Hornberger, O. Piguet, and J. R. Hodges. 2011. Neural basis of music knowledge: Evidence from the dementias. *Brain* **134**:2523–2534. [17]

Hudson, H. (director), and D. Putnam (producer). 1981. Chariots of Fire. UK: Allied Stars Ltd., Enigma Productions. [7]

Hulse, S. H., and J. Cynx. 1985. Relative pitch perception is constrained by absolute pitch in songbirds, *Mimus*, *Molothrus*, and *Sturnus*. *J. Comp. Psychol.* **99**:176–196. [18]

Hurford, J. R. 2004. Language beyond our grasp: What mirror neurons can, and cannot, do for language evolution. In: Evolution of Communication Systems: A Comparative Approach, ed. D. K. Oller and U. Griebel, pp. 297–313. Cambridge, MA: MIT Press. [19]

Huron, D. 2001. Is music an evolutionary adaptation? *Ann. NY Acad. Sci.* **930**:43–61. [6]

———. 2006. *Sweet Anticipation: Music and the Psychology of Expectation*. Cambridge, MA: MIT Press. [5, 7, 8, 10, 12, 14]

Husain, F. T., M. A. Tagamets, S. J. Fromm, A. R. Braun, and B. Horwitz. 2004. Relating neuronal dynamics for auditory object processing to neuroimaging activity: A computational modeling and an fMRI study. *NeuroImage* **21**:1701–1720. [14, 17]

Hutchins, S., and I. Peretz. 2011. Perception and action in singing. *Prog. Brain Res.* **191**:103–118. [17]

Hutchins, S., J. M. Zarate, R. Zatorre, and I. Peretz. 2010. An acoustical study of vocal pitch matching in congenital amusia. *J. Acoust. Soc. Am.* **127**:504–512. [17]

Hutchison, W. D., K. D. Davis, A. M. Lozano, R. R. Tasker, and J. O. Dostrovsky. 1999. Pain-related neurons in the human cingulate cortex. *Nature Neurosci.* **2**:403–405. [4]

Hyde, K. L., J. P. Lerch, R. J. Zatorre, et al. 2007. Cortical thickness in congenital amusia: When less is better than more. *J. Neurosci.* **27**:13,028–13,032. [14]

Hyde, K. L., I. Peretz, and R. J. Zatorre. 2008. Evidence for the role of the right auditory cortex in fine pitch resolution. *Neuropsychologia* **46**:632–639. [17]

Hyde, K. L., R. J. Zatorre, and I. Peretz. 2010. Functional MRI evidence of an abnormal neural network for pitch processing in congenital amusia. *Cereb. Cortex* **21**:292–299. [14]

Hyman, L. 2006. Word-prosodic typology. *Phonology* **23**:225–257. [11]

Iacoboni, M., I. Molnar-Szakacs, V. Gallese, et al. 2005. Grasping the intentions of others with one's own mirror neuron system. *PLoS Biol.* **3**:529–535. [4, 19]

Iacoboni, M., R. P. Woods, M. Brass, et al. 1999. Cortical mechanisms of human imitation. *Science* **286**:2526–2528. [4]

Ilari, B., and L. Polka. 2006. Music cognition in early infancy: Infants' preferences and long-term memory for Ravel. *Intl. J. Music Ed.* **24**:7–19. [18]

Ilie, G., and W. F. Thompson. 2006. A comparison of acoustic cues in music and speech for three dimensions of affect. *Music Percep.* **23**:319–329. [5]

———. 2011. Experiential and cognitive changes following seven minutes exposure to music and speech. *Music Percep.* **28**:247–264. [5]

Indefrey, P. 2004. *Hirnaktivierungen bei syntaktischer Sprachverarbeitung: Eine Meta-Analyse.* In: Neurokognition der Sprache, ed. H. M. Müller and G. Rickheit, pp. 31–50. Tübingen: Stauffenburg Verlag. [9]

———. 2011. The spatial and temporal signatures of word production components: A critical update. *Front. Psychol.* **2**:255. [17]

Indefrey, P., and A. Cutler. 2004. Prelexical and lexical processing in listening. In: The Cognitive Neurosciences III, ed. M. S. Gazzaniga, pp. 759–774. Cambridge, MA: MIT Press. [9]

Indefrey, P., and W. J. Levelt. 2004. The spatial and temporal signatures of word production components. *Cognition* **92**:101–144. [9]

Inderbitzin, M., I. Herreros-Alonso, and P. F. M. J. Verschure. 2010. An integrated computational model of the two phase theory of classical conditioning. In: 2010 Intl. Joint Conf. on Neural Networks, pp. 1–8. Barcelona: IEEE. [8, 15]

Inderbitzin, M., S. Wierenga, A. Väljamäe, U. Bernardet, and P. F. M. J. Verschure. 2009. Social cooperation and competition in the mixed reality space eXperience Induction Machine (XIM). *Virtual Reality* **13**:153–158. [16]

Iriki, A. 2006. The neural origins and implications of imitation, mirror neurons and tool use. *Curr. Opin. Neurobiol.* **16**:660–667. [19]

———. 2010. Neural re-use, a polysemous and redundant biological system subserving niche-construction. *Behav. Brain Sci.* **33**:276–277. [19, 21]

———. 2011. And yet it thinks.... In: Brain Science and Kokoro: Asian Perspectives on Science and Religion, ed. P. L. Swanson, pp. 21–38. Nagoya: Nanzan Institute for Religion and Culture Press. [21]

Iriki, A., S. Nozaki, and Y. Nakamura. 1988. Feeding behavior in mammals: Corticobulbar projection is reorganized during conversion from sucking to chewing. *Devel. Brain Res.* **44**:189–196. [19]

Iriki, A., and O. Sakura. 2008. The neuroscience of primate intellectual evolution: Natural selection and passive and intentional niche construction. *Philos. Trans. R. Soc. Lond. B* **363**:2229–2241. [19]

Iriki, A., M. Tanaka, and Y. Iwamura. 1996. Coding of modified body schema during tool use by macaque postcentral neurones. *NeuroReport* **7**:2325–30. [19]

Iriki, A., and M. Taoka. 2012. Triadic niche construction: A scenario of human brain evolution extrapolating tool-use and language from control of the reaching actions. *Philos. Trans. R. Soc. Lond. B* **367**:10–23. [19, 21]

Ito, H., Y. Ishikawa, M. Yoshimoto, and N. Yamamoto. 2007. Diversity of brain morphology in teleosts: Brain and ecological niche. *Brain Behav. Evol.* **69**:76–86. [20]

Ito, M. 2012. The Cerebellum: Brain for an Implicit Self. Upper Saddle River, NJ: FT Press. [15]

Itskov, V., C. Curto, E. Pastalkova, and G. Buzsaki. 2011. Cell assembly sequences arising from spike threshold adaptation keep track of time in the hippocampus. *J. Neurosci.* **31**:2828–2834. [17]

Iversen, J. R., A. D. Patel, and K. Ohgushi. 2008. Perception of rhythmic grouping depends on auditory experience. *J. Acoust. Soc. Am.* **124**:2263–2271. [18]
Ivry, R. B., T. C. Justus, and C. Middleton. 2001. The cerebellum, timing, and language: Implications for the study of dyslexia. In: Dyslexia, Fluency, and the Brain, ed. M. Wolf, pp. 198–211. Timonium, MD: York Press. [17]
Iyengar, S., S. S. Viswanathan, and S. W. Bottjer. 1999. Development of topography within song control circuitry of zebra finches during the sensitive period for song learning. *J. Neurosci.* **19**:6037–6057. [20]
Izard, C. E. 1977. Human Emotions. New York: Plenum. [5]
Jackendoff, R. 1987. Consciousness and the Computational Mind. Cambridge, MA: MIT Press. [16]
———. 1997. The Architecture of the Language Faculty. Cambridge, MA: MIT Press. [9]
———. 1999. The representational structures of the language faculty and their interactions. In: The Neurocognition of Language, ed. C. M. Brown and P. Hagoort, pp. 37–79. Oxford: Oxford Univ. Press. [9]
———. 2002. Foundations of Language. New York: Oxford Univ. Press. [9, 10, 14]
———. 2009. Parallels and nonparallels between language and music. *Music Percep.* **26**:195–204. [14]
Jackendoff, R., and F. Lerdahl. 2006. The capacity for music: What is it, and what's special about it? *Cognition* **100**:33–72. [1, 10, 14]
Jacob, J., and S. Guthrie. 2000. Facial visceral motor neurons display specific rhombomere origin and axon pathfinding behavior in the chick. *J. Neurosci.* **20**:7664–7671. [20]
Jacob, P., and M. Jeannerod. 2005. The motor theory of social cognition: A critique. *Trends Cogn. Sci.* **9**:21–25. [4]
Jacobson, S., and J. Q. Trojanowski. 1975. Corticothalamic neurons and thalamocortical terminal fields: An investigation in rat using horseradish peroxidase and autoradiography. *Brain Res.* **85**:385–401. [20]
Jakobson, R. 1968. Child Language, Aphasia, and Phonological Universals. The Hague: Mouton. [20]
Jakobson, R., G. Fant, and M. Halle. 1952. Preliminaries to Speech Analysis: The Distinctive Features and Their Correlates. Cambridge, MA: MIT Press. [10]
———. 1957. Preliminaries to Speech Analysis: The Distinctive Features and Their Acoustic Correlates. Cambridge, MA: MIT Press. [20]
James, W. 1884/1968. What is an emotion? *Mind* **9**:189–190. [1, 5, 14]
Janata, P. 2001. Brain electrical activity evoked by mental formation of auditory expectations and images. *Brain Topography* **13**:169–193. [13]
———. 2009a. Music and the self. In: Music That Works, ed. R. Haas and V. Brandes, pp. 131–141. Wien: Springer. [13]
———. 2009b. The neural architecture of music-evoked autobiographical memories. *Cereb. Cortex* **19**:2579–2594. [13]
Janata, P., J. L. Birk, J. D. V. Horn, et al. 2002a. The cortical topography of tonal structures underlying Western music. *Science* **298**:2167–2170. [6, 10, 13]
Janata, P., and S. T. Grafton. 2003. Swinging in the brain: Shared neural substrates for behaviors related to sequencing and music. *Nat. Neurosci.* **6**:682–687. [6, 13, 17]
Janata, P., B. Tillmann, and J. J. Bharucha. 2002b. Listening to polyphonic music recruits domain general attention and working memory circuits. *Cogn. Affect. Behav. Neurosci.* **2**:121–140. [2, 6, 13]
Janata, P., S. T. Tomic, and J. M. Haberman. 2012. Sensorimotor coupling in music and the psychology of the groove. *J. Exp. Psychol. Gen.* **141**:54–75.[13]

Janata, P., S. T. Tomic, and S. K. Rakowski. 2007. Characterisation of music-evoked autobiographical memories. *Memory* **15**:845–860. [13]

Janik, V. M. 2000. Whistle matching in wild bottlenose dolphins, *Tursiops truncatus*. *Science* **289**:1355–1357. [20]

Janik, V. M., and P. J. B. Slater. 1997. Vocal learning in mammals. *Adv. Study Behav.* **26**:59–99. [20]

———. 2000. The different roles of social learning in vocal communication. *Anim. Behav.* **60**:1–11. [20]

Jaramillo, M., P. Paavilainen, and R. Näätänen. 2000. Mismatch negativity and behavioural discrimination in humans as a function of the magnitude of change in sound duration. *Neurosci. Lett.* **290**:101–104. [17]

Jaramillo, S., and A. M. Zador. 2011. The auditory cortex mediates the perceptual effects of acoustic temporal expectation. *Nat. Neurosci.* **14**:246–251. [17]

Jarvis, E. D. 2004. Learned birdsong and the neurobiology of human language. *Ann. NY Acad. Sci.* **1016**:749–777. [19–21]

———. 2006. Selection for and against vocal learning in birds and mammals. *Orinthol. Sci.* **5**:5–14. [20]

———. 2007a. The evolution of vocal learning systems in birds and humans. In: Evolution of Nervous Systems, ed. J. H. Kaas, pp. 213–227. Oxford: Academic. [19]

———. 2007b. Neural systems for vocal learning in birds and humans: A synopsis. *J. Ornithol.* **148**:S35–S44. [14, 19]

Jarvis, E. D., O. Güntürkün, L. Bruce, et al. 2005. Avian brains and a new understanding of vertebrate brain evolution. *Nat. Rev. Neurosci.* **6**:151–159. [20]

Jarvis, E. D., and C. V. Mello. 2000. Molecular mapping of brain areas involved in parrot vocal communication. *J. Comp. Neurol.* **419**:1–31. [20]

Jarvis, E. D., and F. Nottebohm. 1997. Motor-driven gene expression. *PNAS* **94**:4097–4102. [20]

Jarvis, E. D., S. Ribeiro, M. L. da Silva, et al. 2000. Behaviourally driven gene expression reveals song nuclei in hummingbird brain. *Nature* **406**:628–632. [20]

Jarvis, E. D., C. Scharff, M. R. Grossman, J. A. Ramos, and F. Nottebohm. 1998. For whom the bird sings: Context-dependent gene expression. *Neuron* **21**:775–788. [20]

Jeannerod, M. 1988. The Neural and Behavioural Organization of Goal-Directed Movements. Oxford: Oxford Univ. Press. [4]

———. 2005. How do we decipher other's minds? In: Who Needs Emotions? The Brain Meets the Robot, ed. J.-M. Fellous and M. A. Arbib, pp. 147–169. Oxford: Oxford Univ. Press. [8, 17]

Jeannerod, M., M. A. Arbib, G. Rizzolatti, and H. Sakata. 1995. Grasping objects: The cortical mechanisms of visuomotor transformation. *Trends Neurosci.* **18**:314–320. [15]

Jeannerod, M., and B. Biguer. 1982. Visuomotor mechanisms in reaching within extrapersonal space. In: Advances in the Analysis of Visual Behavior, ed. D. J. Ingle et al., pp. 387–409. Cambridge, MA: MIT Press. [1]

Jeffries, K. J., J. B. Fritz, and A. R. Braun. 2003. Words in melody: An H-2 O-15 PET study of brain activation during singing and speaking. *NeuroReport* **14**:749–754. [13]

Jentschke, S., and S. Koelsch. 2009. Musical training modulates the development of syntax processing in children. *NeuroImage* **47**:735–744. [6, 14]

Jentschke, S., S. Koelsch, S. Sallat, and A. Friederici. 2008. Children with specific language impairment also show impairment of music-syntactic processing. *J. Cogn. Neurosci.* **20**:1940–1951. [14]

Jerde, T. A., S. K. Childs, S. T. Handy, J. C. Nagode, and J. V. Pardo. 2011. Dissociable systems of working memory for rhythm and melody. *Neuroimage* **57**:1572–1579. [17]

Jespersen, O. 1922. Language: Its Nature, Development and Origin. London: George Allen and Unwin. [5, 19]

Jin, D. Z., N. Fujii, and A. M. Graybiel. 2009. Neural representation of time in cortico-basal ganglia circuits. *PNAS* **106**:19,156–19,161. [17]

Johansson, S. 2005. Origins of Language: Constraints on hypotheses. Amsterdam: John Benjamins. [5]

John, E. R. 2002. The neurophysics of consciousness. *Brain Res. Rev.* **39**:1–28. [7]

———. 2003. A theory of consciousness. *Curr. Dir. Psychol. Sci.* **12**:244–250. [7]

Johnson, E. K. 2012. Bootstrapping language: Are infant statisticians up to the job? In: Statistical Learning and Language Acquisition, ed. P. Rebuschat and J. Williams, pp. 55–90. Boston: Mouton de Gruyter. [18]

Johnson, E. K., and M. D. Tyler. 2010. Testing the limits of statistical learning for word segmentation. *Devel. Sci.* **13**:339–345. [18]

Johnson, M. B., Y. I. Kawasawa, C. E. Mason, et al. 2009. Functional and evolutionary insights into human brain development through global transcriptome analysis. *Neuron* **62**:494–509. [20]

Johnson, M. D., and G. A. Ojemann. 2000. The role of the human thalamus in language and memory: Evidence from electrophysiological studies. *Brain Cogn.* **42**:218–230. [20]

Johnsrude, I. S., V. B. Penhune, and R. J. Zatorre. 2000. Functional specificity in the right human auditory cortex for perceiving pitch direction. *Brain* **123**:155–163. [13, 14]

Joly, O., F. Ramus, D. Pressnitzer, W. Vanduffel, and G. A. Orban. 2012. Interhemispheric differences in auditory processing revealed by fMRI in awake rhesus monkeys. *Cereb. Cortex* **22**:838–853. [17]

Jonaitis, E., and J. Saffran. 2009. Learning harmony: The role of serial statistics. *Cogn. Sci.* **33**:951–968. [6]

Jonas, S. 1981. The supplementary motor region and speech emission. *J. Comm. Disorders* **14**:349–373. [20]

Judde, S., and N. S. Rickard. 2010. The effect of post-learning presentation of music on long-term word list retention. *Neurobiol. Learn. Mem.* **94**:13–20. [8, 21]

Jürgens, U. 1979. Neural control of vocalizations in nonhuman primates. In: Neurobiology of Social Communication in Primates, ed. H. D. Steklis and M. J. Raleigh, pp. 11–44. New York: Academic. [19]

———. 1994. The role of the periaqueductal grey in vocal behaviour. *Behav. Brain Res.* **62**:107–117. [20]

———. 1998. Neuronal control of mammalian vocalization, with special reference to the squirrel monkey. *Naturwiss.* **85**:376–388. [21]

———. 2002. Neural pathways underlying vocal control. *Neurosci. Biobehav. Rev.* **26**:235–258. [6, 19, 20]

———. 2009. The neural control of vocalization in mammals: A review. *J. Voice* **23**:1–10. [5, 21]

Jürgens, U., A. Kirzinger, and D. von Cramon. 1982. The effects of deep-reaching lesions in the cortical face area on phonation: A combined case report and experimental monkey study. *Cortex* **18**:125–139. [20]

Juslin, P. N. 2001. Communicating emotion in music performance: A review and a theoretical framework. In: Music and Emotion: Theory and Research, ed. P. N. Juslin and J. A. Sloboda, pp. 309–337. Oxford: Oxford Univ. Press. [4]

———. 2009. Emotional responses to music. In: The Oxford Handbook of Music Psychology ed. S. Hallam et al., pp. 131–140. Oxford, UK: Oxford Univ. Press. [8]

Juslin, P. N., and P. Laukka. 2003. Communication of emotions in vocal expression and music performance: Different channels, same code? *Psychol. Bull.* **129**:770–814. [5, 6]
Juslin, P. N., S. Liljeström, D. Västfjäll, and L.-O. Lundqvist. 2010. How does music evoke emotion? Exploring the underlying mechanisms. In: Handbook of Music and Emotion: Theory, Research, Applications ed. P. N. Juslin and J. Sloboda, pp. 605–642. Oxford: Oxford Univ. Press. [8]
Juslin, P. N., and K. R. Scherer. 2005. Vocal expression of affect. In: The New Handbook of Methods in Nonverbal Behavior Research, ed. J. A. Harrigan et al., pp. 65–135. New York: Oxford Univ. Press. [5]
Juslin, P. N., and J. Sloboda, eds. 2010. Handbook of Music and Emotion: Theory, Research, Applications. Oxford: Oxford Univ. Press. [8]
Juslin, P. N., and D. Västfjäll. 2008a. All emotions are not created equal: Reaching beyond the traditional disputes. *Behav. Brain Sci.* **31**:600–612. [5]
———. 2008b. Emotional responses to music: The need to consider underlying mechanisms. *Behav. Brain Sci.* **31**:559–621. [4, 8]
Kalender, B., S. E. Trehub, and E. G. Schellenberg. 2013. Cross-cultural differences in meter perception. *Psychol. Res.* **77**:196–203. [18]
Kaminski, J., J. Call, and J. Fischer. 2004. Word learning in a domestic dog: Evidence for "fast mapping." *Science* **304**:1682–1683. [21]
Kanedera, N., T. Arai, H. Hermansky, and M. Pavel. 1999. On the relative importance of various components of the modulation spectrum for automatic speech recognition. *Speech Comm.* **28**:43–55. [9]
Kant, I. 1790/2001. Kritik der Urteilskraft. Hamburg: Meiner. [5]
Kao, M. H., A. J. Doupe, and M. S. Brainard. 2005. Contributions of an avian basal ganglia-forebrain circuit to real-time modulation of song. *Nature* **433**:638–643. [20]
Kaplan, E., M. A. Naeser, P. I. Martin, et al. 2010. Horizontal portion of arcuate fasciculus fibers track to pars opercularis, not pars triangularis, in right and left hemispheres: A DTI study. *Neuroimage* **52**:436–444. [13]
Kappas, A., U. Hess, and K. R. Scherer. 1991. Voice and emotion. In: Fundamentals of Nonverbal Behavior, ed. R. S. Feldman and B. Rimé, pp. 200–238. Cambridge: Cambridge Univ. Press. [5]
Karbusicky, V. 1986. *Grundriss der musikalischen Semantik*. Darmstadt: Wissenschaftliche Buchgesellschaft. [8]
Karlin, F., and R. Wright. 2004. On the Track: A Guide to Contemporary Film Scoring, 2nd edition. New York: Routledge. [7]
Karlsson, F. 2007. Constraints on multiple center-embedding of clauses. *J. Ling.* **43**:365–392. [3]
Karten, H. J., and T. Shimizu. 1989. The origins of neocortex: Connections and lamination as distinct events in evolution. *J. Cogn. Neurosci.* **1**:291–301. [20]
Katz, J., and D. Pesetsky. 2011. The Identity Thesis for Language and Music, unpublished manuscript. http://ling.auf.net/lingBuzz/000959. (accessed 23 July 2011). [10]
Katz, L. C., and M. E. Gurney. 1981. Auditory responses in the zebra finch's motor system for song. *Brain Res.* **221**:192–197. [20]
Kauramäki, J., I. P. Jääskeläinen, R. Hari, et al. 2010. Transient adaptation of auditory cortex organization by lipreading and own speech production. *J. Neurosci.* **30**:1314–1321. [17]
Kay, R., M. Cartmill, and M. Balow. 1998. The hypoglossal canal and the origin of human vocal behavior. *PNAS* **95**:5417–5419. [5]
Keenan, E. L. 1979. On surface form and logical form. *Stud. Ling. Sci.* **8**:163–203. [9]

Keiler, A. 1977. The syntax of prolongation, I. *In Theory Only* **3**:3–27. [10]

Keizer, K., and H. G. Kuypers. 1989. Distribution of corticospinal neurons with collaterals to the lower brain stem reticular formation in monkey, Macaca fascicularis. *Exp. Brain Res.* **74**:311–318. [20]

Kell, C. A., K. Neumann, K. von Kriegstein, et al. 2009. How the brain repairs stuttering. *Brain* **132**:2747–2760. [14]

Keller, P. E. 2008. Joint action in music performance. In: Enacting Intersubjectivity: A Cognitive and Social Perspective on the Study of Interactions, ed. F. Morganti et al., pp. 205–221. Amsterdam: IOS Press. [13]

Keller, P. E., G. Knoblich, and B. H. Repp. 2007. Pianists duet better when they play with themselves: On the possible role of action simulation in synchronization. *Conscious. Cogn.* **16**:102–111. [13]

Kelley, A. E. 2005. Neurochemical networks encoding emotion and motivation: An evolutionary perspective. In: Who Needs Emotions: The Brain Meets the Robot, ed. J.-M. Fellous and M. A. Arbib, pp. 29–77. Oxford, New York: Oxford Univ. Press. [1]

Kelly, C., L. Q. Uddin, Z. Shehzad, et al. 2010. Broca's region: Linking human brain functional connectivity data and non-human primate tracing anatomy studies. *Eur. J. Neurosci.* **32**:383–398. [9]

Kemmerer, D. 2006. Action verbs, argument structure constructions, and the mirror neuron system. In: Action to Language Via the Mirror Neuron System, ed. M. A. Arbib, pp. 347–373. Cambridge: Cambridge Univ. Press. [15]

Kemmerer, D., J. G. Castillo, T. Talavage, S. Patterson, and C. Wiley. 2008. Neuroanatomical distribution of five semantic components of verbs: Evidence from fMRI. *Brain Lang.* **107**:16–43. [15]

Kemmerer, D., and J. Gonzalez-Castillo. 2010. The two-level theory of verb meaning: An approach to integrating the semantics of action with the mirror neuron system. *Brain Lang.* **112**:54–76. [15]

Kendall, R. A. 2008. Stratification of musical and visual structures II: Visual and pitch contours. In: Tenth International Conference on Music Perception and Cognition. Hokkaido, Japan. [7]

Kendon, A. 1972. Some relationships between body motion and speech: An analysis of an example. In: Studies in Dyadic Communication, ed. A. Siegman and B. Pope, pp. 177–210. Elmsford, NY: Pergamon. [2]

———. 1980. Gesticulation and speech: Two aspects of the process of utterance. In: The Relationship of Verbal and Nonverbal Communication, ed. M. Key, pp. 207–227. The Hague: Mouton. [2]

———. 1997. Gesture. *Annu. Rev. Anthropol.* **26**:109–128. [2]

———. 2004. Gesture: Visible Action as Utterance. Cambridge: Cambridge Univ. Press. [21]

Kiehl, K. A., K. R. Laurens, and P. F. Liddle. 2002. Reading anomalous sentences: An event-related fMRI study of semantic processing. *NeuroImage* **17**:842–850. [9]

Kikusui, T., K. Nakanishi, R. Nakagawa, et al. 2011. Cross fostering experiments suggest that mice songs are innate. *PLoS ONE* **6**:e17721. [20]

Kilgard, M. P., and M. Merzenich. 1998. Cortical map reorganization enabled by nucleus basalis activity. *Science* **279**:1714–1718. [8, 15]

Kilgour, A. R., L. S. Jakobson, and L. L. Cuddy. 2000. Music training and rate of presentation as mediators of text and song recall. *Mem. Cogn.* **28**:700–710. [13]

Kim, K.-H., and S.-I. Iwaymiya. 2008. Formal congruency between Telop patterns and sounds effects. *Music Percep.* **25**:429–448. [7]

Kintsch, W. 1998a. Comprehension: A Paradigm for Cognition. Cambridge: Cambridge Univ. Press. [7]

Kintsch, W. 1998b. The role of knowledge in discourse comprehension: A construction-integration model. *Psychol. Rev.* **95**:163–182. [8]

Kirschner, S., and M. Tomasello. 2009. Joint drumming: Social context facilitates synchronization in preschool children. *J. Exp. Child Psychol.* **102**:299–314. [6, 13, 18]

———. 2010. Joint music making promotes prosocial behavior in 4-year-old children. *Evol. Hum. Behav.* **31**:354–364. [13]

Kirzinger, A., and U. Jürgens. 1982. Cortical lesion effects and vocalization in the squirrel monkey. *Brain Res.* **233**:299–315. [20]

Kisliuk, M. 2001. Seize the Dance: BaAka Musical Life and the Ethnography of Performance. Oxford: Oxford Univ. Press. [2]

Kitamura, C., and D. Burnham. 2003. Pitch and communicative intent in mother's speech: Adjustments for age and sex in the first year. *Infancy* **4**:85–110. [18]

Kleber, B., N. Birbaumer, R. Veit, T. Trevorrow, and M. Lotze. 2007. Overt and imagined singing of an Italian aria. *Neuroimage* **36**:889–900. [13]

Klein, D., R. J. Zatorre, B. Milner, E. Meyer, and A. C. Evans. 1994. Left putaminal activation when speaking a second language: Evidence from PET. *NeuroReport* **5**:2295–2297. [20]

Klein, M. E., and R. J. Zatorre. 2011. A role for the right superior temporal sulcus in categorical perception of musical chords. *Neuropsychologia* **49**:878–887. [14, 17]

Klein, R. G. 1999. The Human Career: Human Biological and Cultural Origins. Chicago: Chicago Univ. Press. [3]

Knörnschild, M., M. Nagy, M. Metz, F. Mayer, and O. von Helversen. 2010. Complex vocal imitation during ontogeny in a bat. *Biology Lett.* **6**:156–159. [20]

Knox, D., S. Beveridge, L. A. Mitchell, and R. A. R. MacDonald. 2011. Acoustic analysis and mood classification of pain-relieving music. *J. Acoust. Soc. Am.* **130**:1673–1682. [21]

Kobayashi, K., H. Uno, and K. Okanoya. 2001. Partial lesions in the anterior forebrain pathway affect song production in adult Bengalese finches. *NeuroReport* **12**:353–358. [20]

Koch, C., C.-H. Mo, and W. Softky. 2003. Single-cell models. In: The Handbook of Brain Theory and Neural Networks, 2nd edition, ed. M. A. Arbib, pp. 1044–1049. Cambridge, MA: MIT Press. [16]

Koch, H. C. 1793. *Versuch einer Anleitung zur Composition.* Leipzig: A. F. Böhme. [10]

Koechlin, E., and T. Jubault. 2006. Broca's area and the hierarchical organization of human behavior. *Neuron* **50**:963–974. [6, 13, 17]

Koelsch, S. 2000. Brain and Music: A Contribution to the Investigation of Central Auditory Processing with a New Electro-physiological Approach. Leipzig: Risse. [6]

———. 2005. Neural substrates of processing syntax and semantics in music. *Curr. Opin. Neurobiol.* **15**:1–6. [6]

———. 2006. Significance of Broca's area and ventral premotor cortex for music-syntactic processing. *Cortex* **42**:518–520. [6, 8]

———. 2009. Music-syntactic processing and auditory memory: Similarities and differences between ERAN and MMN. *Psychophysiology* **46**:179–190. [14]

———. 2010. Towards a neural basis of music-evoked emotions. *Trends Cogn. Sci.* **14**:131–137. [6, 7]

———. 2011a. Towards a neural basis of music perception: A review and updated model. *Front. Psychol.* **2**:110. [6, 8]

———. 2011b. Towards a neural basis of processing musical semantics. *Phys. Life Rev.* **8**:89–105. [6, 14]
———. 2011c. Transitional zones of meaning and semantics in music and language. Reply to comments on "Towards a neural basis of processing musical semantics." *Phys. Life Rev.* **11**:125–128. [8]
———. 2012. Brain and Music. Chichester: Wiley. [6]
Koelsch, S., T. Fritz, D. Y. von Cramon, K. Müller, and A. D. Friederici. 2006. Investigating emotion with music: An fMRI study. *Hum. Brain Mapp.* **27**:239–250. [4, 6]
Koelsch, S., T. Grossmann, T. C. Gunter, A. Hahne, and A. Friederici. 2003. Children processing music: Electric brain responses reveal musical competence and gender differences. *J. Cogn. Neurosci.* **15**:683–693. [1]
Koelsch, S., T. C. Gunter, A. D. Friederici, and E. Schröger. 2000. Brain indices of music processing: "Non-musicians" are musical. *J. Cogn. Neurosci.* **12**:520–541. [6, 14]
Koelsch, S., T. C. Gunter, D. Von Cramon, et al. 2002a. Bach speaks: A cortical "language-network" serves the processing of music. *NeuroImage* **17**:956–966. [6]
Koelsch, S., T. C. Gunter, M. Wittforth, and D. Sammler. 2005. Interaction between syntax processing in language and music: An ERP study. *J. Cogn. Neurosci.* **17**:1565–1577. [6, 14]
Koelsch, S., and S. Jentschke. 2010. Differences in electric brain responses to melodies and chords. *J. Cogn. Neurosci.* **22**:2251–2262. [6]
Koelsch, S., S. Jentschke, D. Sammler, and D. Mietchen. 2007. Untangling syntactic and sensory processing: An ERP study of music perception. *Psychophysiology* **44(3)**:476–490. [6, 14]
Koelsch, S., E. Kasper, D. Sammler, et al. 2004. Music, language and meaning: Brain signatures of semantic processing. *Nat. Neurosci.* **7**:302–307. [6, 13, 14, 20]
Koelsch, S., S. Kilches, N. Steinbeis, and S. Schelinski. 2008. Effects of unexpected chords and of performer's expression on brain responses and electrodermal activity. *PLoS ONE* **3**:e2631. [6]
Koelsch, S., and J. Mulder. 2002. Electric brain responses to inappropriate harmonies during listening to expressive music. *Clin. Neurophysiol.* **113**:862–869. [14]
Koelsch, S., B. H. Schmidt, and J. Kansok. 2002b. Effects of musical expertise on the early right anterior negativity: An event-related brain potential study. *Psychophysiology* **39**:657–663. [6]
Koelsch, S., and W. A. Siebel. 2005. Towards a neural basis of music perception. *Trends Cogn. Sci.* **9**:578–584. [6]
Koelsch, S., W. A. Siebel, and T. Fritz. 2010. Functional neuroimaging. In: Handbook of Music and Emotion: Theory, Research, Applications, 2nd edition, ed. P. Juslin and J. Sloboda, pp. 313–346. Oxford: Oxford Univ. Press. [6]
Kohler, E., C. Keysers, M. A. Umiltà, et al. 2002. Hearing sounds, understanding actions: Action representation in mirror neurons. *Science* **297**:846–848. [4]
Kolinsky, R., P. Lidji, I. Peretz, M. Besson, and J. Morais. 2009. Processing interactions between phonology and melody: Vowels sing but consonants speak. *Cognition* **112**:1–20. [13, 17]
Konishi, M. 1964. Effects of deafening on song development in two species of juncos. *Condor* **66**:85–102. [20]
———. 1965. The role of auditory feedback in the control of vocalization in the white-crowned sparrow. *Z. Tierpsychol.* **22**:770–783. [20]
Konopka, G., J. M. Bomar, K. Winden, et al. 2009. Human-specific transcriptional regulation of CNS development genes by FOXP2. *Nature* **462**:213–217. [20]

Konorski, J. 1967. Integrative Activity of the Brain: An Interdisciplinary Approach. Chicago: Univ. of Chicago Press. [15]

Körding, K., C. Kayser, W. Einhäuser, and P. König. 2004. How are complex cell properties adapted to the statistics of natural stimuli? *Neurophysiol.* **91**:206–212. [15]

Kornysheva, K., D. Y. von Cramon, T. Jacobsen, and R. I. Schubotz. 2010. Tuning-in to the beat: Aesthetic appreciation of musical rhythms correlates with a premotor activity boost. *Hum. Brain Mapp.* **31**:48–64. [13]

Kos, M., T. Vosse, D. van den Brink, and P. Hagoort. 2010. About edible restaurants: Conflicts between syntax and semantics as revealed by ERPs. *Front. Psychol.* **1**:222. [14]

Kotschi, T. 2006. Information structure in spoken discourse. In: Encyclopedia of Language and Linguistics, ed. K. Brown, pp. 677–683. Oxford: Elsevier. [9]

Kotz, S. A., and M. Schwartze. 2010. Cortical speech processing unplugged: A timely subcortico-cortical framework. *Trends Cogn. Sci.* **14**:392–399. [9]

Kotz, S. A., M. Schwartze, and M. Schmidt-Kassow. 2009. Non-motor basal ganglia functions: A review and proposal for a model of sensory predictability in auditory language perception. *Cortex* **45**:982–990. [17]

Kraemer, D. J. M., C. N. Macrae, A. E. Green, and W. M. Kelley. 2005. Musical imagery: Sound of silence activates auditory cortex. *Nature* **434**:158–158. [13]

Kraskov, A., N. Dancause, M. M. Quallo, S. Shepherd, and R. N. Lemon. 2009. Corticospinal neurons in macaque ventral premotor cortex with mirror properties: A potential mechanism for action suppression? *Neuron* **64**:922–930. [4]

Kraus, N., and B. Chandrasekaran. 2010. Music training for the development of auditory skills. *Nat. Rev. Neurosci.* **11**:599–605. [14]

Krausz, M., D. Dutton, and K. Bardsley, eds. 2009. The Idea of Creativity. Leiden Brill Academic Publ. [16]

Kravitz, D. J., K. S. Saleem, C. I. Baker, and M. Mishkin. 2011. A new neural framework for visuospatial processing. *Nat. Rev. Neurosci.* **12**:217–230. [9]

Krolak-Salmon, P., M. A. Henaff, J. Isnard, et al. 2003. An attention modulated response to disgust in human ventral anterior insula. *Ann. Neurol.* **53**:446–453. [4]

Kroodsma, D. E., and J. R. Baylis. 1982. Appendix: A world survey of evidence for vocal learning in birds. In: Acoustic Communication in Birds, vol. 2: Song Learning and Its Consequences, pp. 311–337. New York: Academic. [20]

Kroodsma, D. E., and M. Konishi. 1991. A suboscine bird (eastern phoebe, *Sayornis phoebe*) develops normal song without auditory feedback. *Anim. Behav.* **42**:477–487. [20]

Kroodsma, D. E., and L. D. Parker. 1977. Vocal virtuosity in the brown thrasher. *Auk* **94**:783–785. [20]

Krumhansl, C. L. 1990. Cognitive Foundations of Musical Pitch. New York: Oxford Univ. Press. [10, 12, 14, 18, 17]

———. 1997. An exploratory study of musical emotions and psychophysiology. *Can. J. Exp. Psychol.* **51**:336–353. [6]

———. 2010. Plink: "Thin slices" of music. *Music Percep.* **27**:337–354. [1, 13]

Krumhansl, C. L., and L. L. Cuddy. 2010. A theory of tonal hierarchies in music. In: Music Perception: Current Research and Future Directions, ed. M. R. Jones et al., pp. 51–86. New York: Springer. [14]

Kubik, G. 1979. Pattern perception and recognition in African music. In: The Performing Arts: Music and Dance, ed. J. Blacking and J. W. Kealiinohomoku. The Hague: Mouton. [2]

Kubovy, M., and M. Yu. 2012. Multistability, cross-modal binding and the additivity of conjoined grouping principles. *Philos. Trans. R. Soc. Lond. B* **367**:954–964. [7]
Kubrick, S. (producer and director). 1968. A Space Odyssey. US: MGM. [7]
———. 1989. The Shining. US: Producers Circle. [7]
Kuhl, P., F. M. Tsao, and H. M. Liu. 2003. Foreign-language experience in infancy: Effects of short-term exposure and social interaction on phonetic learning. *PNAS* **100**:9096–9101. [18]
Kuhl, P., K. Williams, F. Lacerda, K. Stevens, and B. Lindblom. 1992. Linguistic experience alters phonetic perception in infants by 6 months of age. *Science* **255**:606–608. [18]
Kumar, A., S. Rotter, and A. Aertsen. 2010. Spiking activity propagation in neuronal networks: Reconciling different perspectives on neural coding. *Nat. Rev. Neurosci.* **11**:615–627. [16]
Kuperberg, G. R. 2007. Neural mechanisms of language comprehension: Challenge to syntax. *Brain Res.* **1146**:23–49. [8]
Kuperberg, G. R., D. A. Kreher, A. Swain, D. C. Goff, and D. J. Holt. 2011. Selective emotional processing deficits to social vignettes in schizophrenia: An ERP study. *Schizophrenia Bulletin* **37**:148–163. [1]
Kutas, M., and K. D. Federmeier. 2011. Thirty years and counting: Finding meaning in the N400 component of the event-related brain potential (ERP). *Annu. Rev. Psychol.* **62**:621–647. [9]
Kuypers, H. G. J. M. 1958. Some projections from the peri-central cortex to the pons and lower brain stem in monkey and chimpanzee. *J. Comp. Neurol.* **100**:221–255. [20]
Labov, W. 1963. The social motivation of a sound change. *Word* **19**:273–309. [11]
Ladd, D. R. 2008. Intonational Phonology, 2nd edition. Cambridge: Cambridge Univ. Press. [5, 10, 11, 14]
———. 2011. Phonetics in phonology. In: The Handbook of Phonological Theory, 2nd edition, ed. J. A. Goldsmith et al., pp. 348–373. Blackwell. [11]
———. 2012. What is duality of patterning, anyway? *Lang. Cogn.* **4**:261–273. [11]
Ladd, D. R., K. Silverman, F. Tolkmitt, G. Bergmann, and K. R. R. Scherer. 1985. Evidence for the independent function of intonation contour type, voice quality, and F0 range in signalling speaker affect. *J. Acoust. Soc. Am.* **78**:435–444. [5]
Ladinig, O., H. Honing, G. Háden, and I. Winkler. 2009. Probing attentive and preattentive emergent meter in adult listeners without extensive music training. *Music Percep.* **26**:377–386. [17, 19]
Lai, C. S., S. E. Fisher, J. A. Hurst, F. Vargha-Khadem, and A. P. Monaco. 2001. A forkhead-domain gene is mutated in a severe speech and language disorder. *Nature* **413**:519–523. [20]
Laird, A. R., P. M. Fox, C. J. Price, et al. 2005. ALE meta-analysis: Controlling the false discovery rate and performing statistical contrasts. *Hum. Brain Mapp.* **25**:155–164. [13]
Lakin, J. L., V. E. Jefferis, C. M. Cheng, and T. L. Chartrand. 2003. The chameleon effect as social glue: Evidence for the evolutionary significance of nonconscious mimicry. *J. Nonverbal Behav.* **27**:145–162. [21]
Lal, H. 1967. Operant control of vocal responding in rats. *Psychon. Sci.* **8**:35–36. [21]
Laland, K. N., and V. M. Janik. 2006. The animal cultures debate. *Trends Ecol. Evol.* **21**:542–547. [20]
Laland, K. N., F. J. Odling-Smee, and M. W. Feldman. 1996. The evolutionary consequences of niche construction: A theoretical investigation using two-locus theory. *J. Evol. Biol* **9**:293–316. [21]

Lane, H. 1960. Control of vocal responding in chickens. *Science* **132**:37–38. [20]

Langheim, F. J. P., J. H. Callicott, V. S. Mattay, J. H. Duyn, and D. R. Weinberger. 2002. Cortical systems associated with covert music rehearsal. *Neuroimage* **16**:901–908. [13]

Langmore, N. E. 1996. Female song attracts males in the alpine accentor *Prunella collaris*. *Proc. R. Soc. Lond. B* **263**:141–146. [20]

———. 1998. Functions of duet and solo songs of female birds. *Trends Ecol. Evol.* **13**:136–140. [20]

Large, E. W. 2008. Resonating to musical rhythm: Theory and experiment. In: The Psychology of Time, ed. S. Grondin, pp. 189–232. West Yorkshire: Emerald. [17]

———. 2010. Neurodynamics of Music: Music Perception, ed. M. Riess Jones et al., vol. 36, Springer Handbook of Auditory Research, pp. 201–231. New York: Springer. [15]

Larson, C. R., Y. Yajima, and P. Ko. 1994. Modification in activity of medullary respiratory-related neurons for vocalization and swallowing. *J Neurophysiol* **71**:2294–2304. [20]

Lashley, K. 1951. The problem of serial order in behavior. In: Cerebral Mechanisms in Behavior, ed. L. A. Jeffress, pp. 112–136. New York: John Wiley. [1]

Lau, E. F., C. Phillips, and D. Poeppel. 2008. A cortical network for semantics:(De) constructing the N400. *Nat. Rev. Neurosci.* **9**:920–933. [17]

Lau, E. F., C. Stroud, S. Plesch, and C. Phillips. 2006. The role of structural prediction in rapid syntactic analysis. *Brain Lang.* **98**:74–88. [9]

Leavens, D. A., W. D. Hopkins, and R. K. Thomas. 2004a. Referential communication by chimpanzees, *Pan troglodytes*. *J. Comp. Psychol.* **118**:48–57. [4]

Leavens, D. A., A. Hostetter, M. J. Weasley, and W. D. Hopkins. 2004b. Tactical use of unimodal and bimodal communication by chimpanzees, *Pan troglodytes*. *Anim. Behav.* **67**:467–476. [21]

Leaver, A. M., and J. P. Rauschecker. 2010. Cortical representation of natural complex sounds: Effects of acoustic features and auditory object category. *J. Neurosci.* **30**:7604–7612. [14, 17]

Leaver, A. M., J. Van Lare, B. Zielinski, A. R. Halpern, and J. P. Rauschecker. 2009. Brain activation during anticipation of sound sequences. *J. Neurosci.* **29**:2477–2485. [13, 17]

Lebel, C., and C. Beaulieu. 2009. Lateralization of the arcuate fasciculus from childhood to adulthood and its relation to cognitive abilities in children. *Hum. Brain Mapp.* **30**:3563–3573. [9]

LeDoux, J. E. 1996. The Emotional Brain. New York: Simon & Schuster. [8]

———. 2000. Emotion circuits in the brain. *Annu. Rev. Neurosci.* **23**:155–184. [1, 4, 5, 8, 15]

———. 2004. Emotional brain. In: The Oxford Companion to the Mind, edited by R. L. Gregory. Oxford: Oxford Univ. Press. [8]

———. 2012. Rethinking the emotional brain. *Neuron* **73**:653–676. [8]

Lee, J., and V. Barrès. 2013. From visual scenes to language and back via Template Construction Grammar. *Neuroinformatics*, in press. [15]

Le Groux, S. 2006. From Sense to Sound: Mapping High-Level Sonic Percepts to Sound Generation, Masters Thesis, Universitat Pompeu Fabra. [16]

Le Groux, S., J. Manzolli, and P. F. M. J. Verschure. 2010. Disembodied and collaborative musical interaction in the multimodal brain orchestra. *NIME2010* http://specs.upf.edu/files/free_download_bibliofiles/LeGroux__2010__brain_orchestra__NIME.pdf. (accessed 8 Dec 2012). [16]

Le Groux, S., A. Valjamae, J. Manzolli, and P. F. M. J. Verschure. 2008. Implicit Physiological Interaction for the Generation of Affective Music. *ICMC* http://classes.berklee.edu/mbierylo/ICMC08/defevent/papers/cr1718.pdf. (accessed 8 Dec 2012). [16]

Le Groux, S., and P. F. M. J. Verschure. 2009a. Neuromuse: Training your brain through musical interaction. Proc. Intl. Conf. on Auditory Display. http://www.icad.org/Proceedings/2009/LeGrouxVerschure2009.pdf. (accessed 8 Dec 2012). [16]

———. 2009b. Situated interactive music system: Connecting mind and body through musical interaction. In: Proc. Intl. Computer Music Conference. Montreal [16]

———. 2010. Adaptive music generation by reinforcement learning of musical tension. *Sound and Music Computing* http://smcnetwork.org/files/proceedings/2010/24.pdf. (accessed 8 Dec 2012). [16]

———. 2011. Music is everywhere: A situated approach to music composition. In: Situated Aesthetics: Art Beyond the Skin, ed. R. Manzotti, pp. 189–210. Exeter: Imprint Academic. [16]

Leitner, S., and C. K. Catchpole. 2004. Syllable repertoire and the size of the song control system in captive canaries, *Serinus canaria. J. Neurobiol.* **60**:21–27. [20]

Leman, M. 2007. Embodied Music Cognition and Mediation Technology. Cambridge, MA: MIT Press. [16]

Lenneberg, E. H. 1967. Biological Foundations of Language. New York: Wiley. [20]

Leon, A., D. Elgueda, M. A. Silva, C. M. Harname, and P. H. Delano. 2012. Auditory cortex basal activity modulates cochlear responses in chinchillas. *PLoS ONE* **7**:e36203. [17]

Leonardo, A., and M. Konishi. 1999. Decrystallization of adult birdsong by perturbation of auditory feedback. *Nature* **399**:466–470. [20]

Lepora, N. F., J. Porrill, C. Yeo, and P. Dean. 2010. Sensory prediction or motor control? Application of Marr-Albus type models of cerebellar function to classical conditioning. *Front. Comp. Neurosci.* **4**:140. [15]

Lerdahl, F. 2001a. The sounds of poetry viewed as music. *Ann. NY Acad. Sci.* **930**:337–354. [6, 10]

———. 2001b. Tonal Pitch Space. New York: Oxford Univ. Press. [10, 12, 14, 15]

Lerdahl, F., and R. Jackendoff. 1983. A Generative Theory of Tonal Music. Cambridge, MA: MIT Press. [7, 10, 14, 15–17]

Lerdahl, F., and C. L. Krumhansl. 2007. Modeling tonal tension. *Music Percep.* **24**:329–366. [6, 10, 12, 14, 15]

Lesser, V. R., R. D. Fennel, L. D. Erman, and D. R. Reddy. 1975. Organization of the HEARSAY-II speech understanding system. *IEEE Trans. Acoustics, Speech, Sign. Proc.* **23**:11–24. [1]

Leutgeb, J. K., S. Leutgeb, M.-B. Moser, and E. I. Moser. 2007. Pattern separation in the dentate gyrus and CA3 of the hippocampus. *Science* **315**:961–966. [15]

Levelt, W. J. M. 1983. Monitoring and self-repair in speech. *Cognition* **14**:41–104. [17]

———. 1989. Speaking: From Intention to Articulation. Cambridge, MA: MIT Press. [9]

Levelt, W. J. M. 1999. Producing spoken language: A blueprint of the speaker. In: The Neurocognition of Language, ed. C. M. Brown and P. Hagoort, pp. 83–122. Oxford: Oxford Univ. Press. [9]

Levelt, W. J. M., A. Roelofs, and A. S. Meyer. 1999. A theory of lexical access in speech production. *Behav. Brain Sci.* **22**:1–38. [14]

Levelt, W. J. M., and L. Wheeldon. 1994. Do speakers have access to a mental syllabary? *Cognition* **50**:239–269. [14]

Levenson, R. W., P. Ekman, and W. V. Friesen. 1990. Voluntary facial action generates emotion-specific autonomic nervous system activity. *Psychophysiology* **27**:363–384. [1]

Leventhal, H., and K. R. Scherer. 1987. The relationship of emotion to cognition: A functional approach to a semantic controversy. *Cogn. Emot.* **1**:3–28. [5]

Levinson, J. 1996. Film music and narrative agency. In: Post-Theory: Reconstructing Film Studies, ed. D. Bordwell and N. Carroll, pp. 248–282. Madison: Univ. of Wisconsin Press. [7]

———. 1997. Music in the Moment. Ithaca: Cornell Univ. Press. [7, 17]

Levinson, S. C. 1981. Some pre-observations on the modelling of dialogue. *Discourse Proc.* **4**:93–116. [3]

———. 1983. Pragmatics. Cambridge: Cambridge Univ. Press. [3]

———. 1988. Putting linguistics on a proper footing: Explorations in Goffman's participation framework. In: Goffman: Exploring the Interaction Order, ed. P. Drew and A. Wootton, pp. 161–227. Oxford: Polity Press. [3]

———. 2006. On the human interaction engine. In: Roots of Human Sociality: Culture, Cognition and Interaction, ed. N. J. Enfield and S. C. Levinson, pp. 39–69. Oxford: Berg. [21]

Levitin, D. J. 1994. Absolute memory for musical pitch: Evidence from the production of learned melodies. *Percep. Psychophys.* **56**:414–423. [18]

Levitin, D. J., and P. R. Cook. 1996. Memory for musical tempo: Additional evidence that auditory memory is absolute. *Percep. Psychophys.* **58**:927–935. [17, 18]

Levy, R. 2008. Expectation-based syntactic comprehension. *Cognition* **106**:1126–1177. [14]

Lewicki, M. S. 2002. Efficient coding of natural sounds. *Nature Neurosci.* **5**:356–363. [15]

Lewis, J. 2002. Forest hunter-gatherers and their world: A study of the Mbendjele Yaka Pygmies and their secular and religious activities and representations. PhD thesis, Univ. of London, London. [2]

———. 2008. Ekila: Blood, bodies and egalitarian societies. *J. R. Anthropol. Inst.* **14**:297–315. [2]

———. 2009. As well as words: Congo Pygmy hunting, mimicry and play. In: The Cradle of Language, ed. R. Botha and C. Knight, vol. 2, pp. 232–252. Oxford: Oxford Univ. Press. [21]

Lewis, J. W., F. L. Wightman, J. A. Brefczynski, et al. 2004. Human brain regions involved in recognizing environmental sounds. *Cereb. Cortex* **14**:1008–1021. [13]

Lew-Williams, C., and J. Saffran. 2012. All words are not created equal: Expectations about word length guide statistical learning. *Cognition* **122**:241–246. [18]

Leyhausen, P. 1967. *Biologie von Ausdruck und Eindruck. Teil 1.* [Biology of expression and impression]. *Psychologische Forschung* **31**:113–176. [5]

Li, G., J. Wang, S. J. Rossiter, G. Jones, and S. Zhang. 2007. Accelerated FoxP2 evolution in echolocating bats. *PLoS ONE* **2**:e900. [20]

Li, X. C., E. D. Jarvis, B. Alvarez-Borda, D. A. Lim, and F. Nottebohm. 2000. A relationship between behavior, neurotrophin expression, and new neuron survival. *PNAS* **97**:8584–8589. [20]

Liao, M.-Y., and J. Davidson. 2007. The use of gesture techniques in children's singing. *Intl. J. Music* **25**:82–94. [20]

Liberman, A. M., and I. G. Mattingly. 1985. The motor theory of speech perception revised. *Cognition* **21**:1–36. [4]

Liberman, M. 1979. The Intonational System of English (publication of 1975 PhD thesis, MIT), Garland Press, New York. [11]

Liberman, M., and A. Prince. 1977. On stress and linguistic rhythm. *Ling. Inq.* **8**:249–336. [10, 11]

Lidji, P., P. Jolicoeur, R. Kolinsky, et al. 2010. Early integration of vowel and pitch processing: A mismatch negativity study. *Clin. Neurophysiol.* **121**:533–541. [13]

Liebal, K., J. Call, and M. Tomasello. 2004. Use of gesture sequences in chimpanzees, *Pan troglodytes*. *Amer. J. Primatol.* **64**:377–396. [21]

Lieberman, P. 1998. Eve Spoke: Human Language and Human Evolution. New York: W. W. Norton. [18]

———. 2000. Human Language and Our Reptilian Brain, The Subcortical Bases of Speech Syntax and Thought. Cambridge, MA: Harvard Univ. Press. [19, 20]

———. 2002. On the nature and evolution of the neural bases of human language. *Am. J. Phys. Anthropol.* **119**:36–62. [20]

Lindenberger, U., S. C. Li, W. Gruber, and V. Muller. 2009. Brains swinging in concert: Cortical phase synchronization while playing guitar. *BMC Neuroscience* **10**:22. [13]

Lipscomb, S. D. 2005. The perception of audio-visual composites: Accent structure alignment of simple stimuli. *Selected Rep. Ethnomusicol.* **12**:37–67. [7]

Lisman, J. E. 2005. The theta/gamma discrete phase code occuring during the hippocampal phase precession may be a more general brain coding scheme. *Hippocampus* **15**:913–922. [15]

———. 2007. Role of the dual entorhinal inputs to hippocampus: A hypothesis based on cue/action (non-self/self) couplets. *Prog. Brain Res.* **163**:615–625. [15]

Liszkowski, U. 2008. Before L1: A differentiated perspective on infant gestures. *Gesture* **8**:180–196. [18]

———. 2011. Three lines in the emergence of prelinguistic communication and social cognition. *J. Cogn. Education and Psychology* **10**:32–43. [21]

Liu, F., A. D. Patel, A. Fourcin, and L. Stewart. 2010. Intonation processing in congenital amusia: Discrimination, identification, and imitation. *Brain* **133**:1682–1693. [14, 17]

Llinás, R., E. J. Lang, and J. P. Welsh. 1997. The cerebellum, LTD, and memory: Alternative views. *Learn. Mem.* **3**:445–455. [15]

Llinás, R., U. Ribary, D. Jeanmonod, E. Kronberg, and P. Mitra. 1999. Thalamocortical dysrhythmia: A neurological and neuropsychiatric syndrome characterized by magnetoencephalography. *PNAS* **96**:15,222–15,227. [15]

Locke, J. L. 1995. Development of the capacity for spoken language. In: The Handbook of Child Language, ed. P. Fletcher and B. MacWhinney, p. 786. Oxford: Blackwell. [20]

Lomax, A. 1968. Folk Song Style and Culture. Washington, D.C.: AAAS. [21]

London, J. 2004. Hearing in Time: Psychological Aspects of Musical Meter. New York: Oxford Univ. Press. [10]

Longhi, E. 2009. "Songese": Maternal structuring of musical interaction with infants. *Psychol. Music* **37**:195–213. [18]

Longuet-Higgins, C., and M. Steedman. 1971. On interpreting Bach. *Mach. Intell.* **6**:221–241. [12]

Lotto, A. J., G. S. Hickok, and L. L. Holt. 2009. Reflections on mirror neurons and speech perception. *Trends Cogn. Sci.* **13**:110–114. [4]

Loui, P., D. Alsop, and G. Schlaug. 2009. Tone-deafness: A disconnection syndrome? *J. Neurosci.* **29**:10,215–10,220. [13, 14, 17]

Loui, P., F. H. Guenther, C. Mathys, and G. Schlaug. 2008. Action-perception mismatch in tone-deafness. *Curr. Biol.* **18**:R331–R332. [13, 17]

Lovelace, A. A. 1843/1973. Sketch of the Analytical Engine invented by Charles Babbage, by L. F. Menabrea, Officer of the Military Engineers, with notes upon the memoir by the Translator. Taylor's Scientific Memoirs, vol. 3, article 39, pp. 666–731. Reprinted in Origins of Digital Computers: Selected Papers, ed. B. Randell. Heidelberg: Springer. [16]

Luck, S. J. 2005. An Introduction to the Event-Related Potential Technique. Cambridge: MIT Press. [13]

Lui, F., G. Buccino, D. Duzzi, et al. 2008. Neural substrates for observing and imagining non-object-directed actions. In: The Mirror Neuron System (special issue of *Social Neurosci.*), ed. C. Keysers and L. Fadiga, pp. 261–275. New York: Psychology Press. [4]

Luo, H., Z. Liu, and D. Poeppel. 2010. Auditory cortex tracks both auditory and visual dynamics using low-frequency neuronal phase modulation. *PLoS Biol.* 1–13. [7]

Luo, H., and D. Poeppel. 2007. Phase patterns of neuronal responses reliably discriminate speech in human auditory cortex. *Neuron* **54**:1001–1010. [9]

Luo, M., L. Ding, and D. J. Perkel. 2001. An avian basal ganglia pathway essential for vocal learning forms a closed topographic loop. *J. Neurosci* **21**:6836–6845. [20]

Luvizotto, A., C. Rennó-Costa, and P. F. M. J. Verschure. 2012. A wavelet based neural model to optimize and read out a temporal population code. *Front. Comp. Neurosci.* **6**:doi 10.3389/fncom.2012.00021. [16]

MacKay, D. M. 1972. Formal analysis of communicative processes. In: Non-verbal communication, ed. R. A. Hinde, pp. 3–26. Cambridge: Cambridge Univ. Press. [1]

MacNeilage, P. F. 1998. The frame/content theory of evolution of speech production. *Behav. Brain Sci.* **21**:499–546. [20]

MacSweeney, M., C. M. Capek, R. Campbell, and B. Woll. 2008. The signing brain: The neurobiology of sign language. *Trends Cogn. Sci.* **12**:432–440. [4]

Maduell, M., and A. M. Wing. 2007. The dynamics of ensemble: The case for flamenco. *Psychol. Music* **35**:591–627. [13]

Maess, B., S. Koelsch, T. C. Gunter, and A. D. Friederici. 2001. Musical syntax is processed in Broca's area: An MEG study. *Nat. Neurosci.* **4**:540–545. [4, 6, 8, 14]

Maia, A., P. D. Valle, J. Manzolli, and L. N. S. Pereira. 1999. A computer environment for polymodal music. *Org. Sound* **4**:111–114. [16]

Maidhof, C., and S. Koelsch. 2011. Effects of selective attention on syntax processing in music and language. *J. Cogn. Neurosci.* **23**:2252–2267. [17]

Makuuchi, M., J. Bahlmann, A. Anwander, and A. D. Friederici. 2009. Segregating the core computational faculty of human language from working memory. *PNAS* **106**:8362–8367. [6]

Malloch, S., and C. Trevarthen, eds. 2009. Communicative Musicality: Exploring the Basis of Human Companionship. Oxford: Oxford Univ. Press. [3]

Mampe, B., A. D. Friederici, A. Cristophe, and K. Wermke. 2009. Newborn's cry melody is shaped by their native language. *Curr. Biol.* **19**:1–4. [18]

Mandelbrot, B. B. 1977. Fractals: Form, Chance, and Dimension. San Francisco: W. H. Freeman. [16]

Mandell, J., K. Schulze, and G. Schlaug. 2007. Congenital amusia: An auditory-motor feedback disorder? *Restor. Neurol. Neurosci.* **25**:323–334. [14]

Manzolli, J., and A. Maia, Jr. 1998. Sound functor applications. In: Proc. of 5th Brazilian Symposium on Computer Music. Belo Horizonte, Brazil [16]

Manzolli, J., and P. F. M. J. Verschure. 2005. Roboser: A real-world composition system. *Computer Music Journal* **29**:55–74. [16]

Maratos, A., C. Gold, X. Wang, and M. Crawford. 2008. Music therapy for depression. *Cochrane Database Syst. Rev.* **1**:1–16. [6]

Marchina, S., L. L. Zhu, A. Norton, et al. 2011. Impairment of speech production predicted by lesion load of the left arcuate fasciculus. *Stroke* **42**:2251–2256. [17]

Marcos, E., A. Duff, M. S. Fibla, and P. F. M. J. Verschure. 2010. The neuronal substrate underlying order and interval representations in sequential tasks: A biologically based robot study. Intl. Joint Conf. on Neural Networks. [15]

Marcos, E., P. Pani, and et al. 2013. Neural variability in premotor cortex is modulated by trial history and predicts behavioral performance. *Neuron*, in press. [16]

Marett, A. 2005. Songs, Dreamings, and Ghosts: The Wangga of North Australia. Hanover, CT: Wesleyan Univ. Press. [21]

Margoliash, D., E. S. Fortune, M. L. Sutter, et al. 1994. Distributed representation in the song system of oscines: Evolutionary implications and functional consequences. *Brain Behav. Evol.* **44**:247–264. [20]

Marler, P. 1955. Characteristics of some animals calls. *Nature* **176**:6–8. [20]

———. 1970. Birdsong and speech development: Could there be parallels? *Am. Sci.* **58**:669–673. [20]

———. 1991. The instinct to learn. In: The Epigenesis of Mind: Essays on Biology and Cognition, ed. S. Carey and R. Gelman, pp. 37–66. Hillsdale, NJ: Lawrence Erlbaum. [20]

———. 2004. Bird calls: Their potential for behavioral neurobiology. *Ann. NY Acad. Sci.* **1016**:31–44. [20]

Marler, P., and S. Peters. 1977. Selective vocal learning in a sparrow. *Science* **198**:519–521. [20]

———. 1982. Subsong and plastic song: Their role in the vocal learning process. In: Acoustic Communication in Birds, vol. 2: Song Learning and its Consequences, ed. D. E. Kroodsma et al., pp. 25–50. New York: Academic. [20]

———. 1988. The role of song phonology and syntax in vocal learning preferences in the song sparrow, *Melospiza melodia*. *Ethology* **77**:125–149. [20]

Marler, P., and H. Slabbekoorn, eds. 2004. Nature's Music: The Science of Birdsong. San Diego: Elsevier Academic. [20]

Marler, P., and R. Tenaza. 1977. Signaling behavior of apes with special reference to vocalization. In: How Animals Communicate, ed. T. A. Sebeok, pp. 965–1033. Bloomington, IN: Indiana Univ. Press. [5]

Marr, D. 1982. Vision: A Computational Approach. San Francisco: Freeman. [17]

Marsh, K. 2008. The Musical Playground: Global Tradition and Change in Children's Songs and Games. Oxford: Oxford Univ. Press. [18]

Marshall, J., J. Atkinson, E. Smulovitch, A. Thacker, and B. Woll. 2004. Aphasia in a user of British Sign Language: Dissociation between sign and gesture. *Cogn. Neuropsychol.* **21**:537–554. [4, 8, 19, 20]

Marshall, S. K., and A. J. Cohen. 1988. Effects of musical soundtracks on attitudes toward animated geometric figures. *Music Percep.* **6**:95–112. [7]

Marslen-Wilson, W. D. 1984. Function and process in spoken word-recognition. In: Attention and Performance X: Control of Language Processes, ed. H. Bouma and D. G. Bouwhuis. Hillsdale, NJ: Erlbaum. [9]

———. 2007. Morphological processes in language comprehension. In: The Oxford Handbook of Psycholinguistics, ed. M. Gareth, pp. 175–194. Oxford: Oxford Univ. Press. [9]

Martikainen, M. H., K. Kaneko, and R. Hari. 2005. Suppressed responses to self-triggered sounds in the human auditory cortex. *Cereb. Cortex* **15**:299–302. [17]

Masataka, N. 1999. Preference for infant-directed singing in 2-day-old hearing infants of deaf parents. *Devel. Psychol.* **35**:1001–1005. [18]

———. 2006. Preference for consonance over dissonance by hearing newborns of deaf parents and of hearing parents. *Devel. Sci.* **9**:46–50. [18]

Masataka, N., and K. Fujita. 1989. Vocal learning of Japanese and rhesus monkeys. *Behaviour* **109**:191–199. [3, 20]

Matelli, M., G. Luppino, and G. Rizzolatti. 1985. Patterns of cytochrome oxidase activity in the frontal agranular cortex of the macaque monkey. *Behav. Brain Res.* **18**:125–136. [4]

———. 1991. Architecture of superior and mesial area 6 and the adjacent cingulate cortex in the macaque monkey. *J. Comp. Neurol.* **311**:445–462. [4]

Mathews, Z., S. Bermudez i Badia, and P. F. M. J. Verschure. 2008. Intelligent motor decision: From selective attention to a Bayesian world model. In: Proc. of Intelligent Systems Conference. Varna, Bulgaria: IEEE. [16]

Mathews, Z., M. Lechon, M. J. Blanco Calvo, et al. 2009. Insect-like mapless navigation based on head direction cells and contextual learning using chemo-visual sensors. In: IROS 2009, pp. 2243–2250. St. Louis: IEEE. [16]

Mathews, Z., and P. F. M. J. Verschure. 2011. PASAR-DAC7: An integrated model of prediction, anticipation, sensation, attention and response for artificial sensorimotor systems. *Information Sciences* **186**:1–19. [16]

Mathews, Z., P. F. M. J. Verschure, and S. Bermúdez i Badia. 2010. An insect-based method for learning landmark reliability using expectation reinforcement in dynamic environments. 2010 IEEE Intl. Conference on Robotics and Automation, pp. 3805–3812. Anchorage: IEEE. [15]

Matsunaga, E., M. Kato, and K. Okanoya. 2008. Comparative analysis of gene expressions among avian brains: A molecular approach to the evolution of vocal learning. *Brain Res. Bull.* **75**:474–479. [20]

Mattheson, J. 1739. *Der volkommene Capellmeister*. Hamburg: C. Herold. [10]

Maye, J., J. F. Werker, and L. Gerken. 2002. Infant sensitivity to distributional information can affect phonetic discrimination. *Cognition* **82**:B101–B111. [18]

McAuley, J. D., M. R. Jones, S. Holub, H. Johnston, and N. S. Miller. 2006. The time of our lives: Lifespan development of timing and event tracking. *J. Exp. Psychol. Gen.* **135**:348–367. [18]

McCarthy, J., and P. J. Hayes. 1969. Some philosophical problems from the standpoint of artificial intelligence. *Mach. Intell.* **4**:463–502. [16]

McClelland, J., and J. L. Elman. 1986. The TRACE model of speech perception. *Cogn. Psychol.* **18**:1–86. [9]

McCulloch, W. S., and W. Pitts. 1943. A logical calculus of the ideas immanent in nervous activity. *Bull. Math. Biophys.* **5**:115–137. [16]

McDermott, J., and M. D. Hauser. 2005. The origins of music: Innateness, development, and evolution. *Music Percep.* **23**:29–59. [14]

McGaugh, J. L. 1966. Time dependent processes in memory storage. *Science* **153**:1351–1358. [8]

———. 2000. Memory: A century of consolidation. *Science* **287**:248–251. [8]

McGuiness, A., and K. Overy. 2011. Music, consciousness, and the brain: Music as shared experience of an embodied present. In: Music and Consciousness: Philosphical, Psychological, and Cultural Perspectives, ed. D. Clarke and E. Clarke, pp. 245–262. Oxford: Oxford Univ. Press. [8]

McMullen, E., and J. R. Saffran. 2004. Music and language: A developmental comparison. *Music Percep.* **21**:289–311. [18]

McNeill, D. 1985. So you think gestures are nonverbal? *Psychol. Rev.* **92**:350–371. [2]
———. 1992. Hand and Mind. Chicago: Univ. of Chicago Press. [2, 4]
———. 2005. Gesture and Thought. Chicago: Univ. of Chicago Press. [19]
McNeill, W. H. 1995. Keeping Together in Time: Dance and Drill in Human History. Cambridge, MA: Harvard Univ. Press. [18, 21]
Mega, M. S., J. L. Cummings, S. Salloway, and P. Malloy. 1997. The limbic system: An anatomic, phylogenetic, and clinical perspective. In: The Neuropsychiatry of Limbic and Subcortical Disorders, ed. S. Salloway et al., pp. 3–18. Washington, D.C.: American Psychiatric Press. [6]
Mehler, J., P. Jusczyk, G. Lambertz, et al. 1988. A precursor of language acquisition in young infants. *Cognition* **29**:143–178. [18]
Mello, C. V., G. E. Vates, S. Okuhata, and F. Nottebohm. 1998. Descending auditory pathways in the adult male zebra finch, *Taeniopygia guttata*. *J. Comp. Neurol.* **395**:137–160. [20]
Meltzer, J. A., J. J. McArdle, R. J. Schafer, and A. R. Braun. 2010. Neural aspects of sentence comprehension: Syntactic complexity, reversibility, and reanalysis. *Cereb. Cortex* **20**:1853–1864. [14]
Meltzoff, A. N., and M. K. Moore. 1977. Imitation of facial and manual gestures by human neonates. *Science* **198**:75–78. [1, 19]
Menenti, L., S. M. E. Gierhan, K. Segaert, and P. Hagoort. 2011. Shared language: Overlap and segregation of the neuronal infrastructure for speaking and listening revealed by functional MRI. *Psychol. Sci.* **22**:1173–1182. [9]
Menenti, L., M. Pickering, and S. Garrod. 2012. Toward a neural basis of interactive alignment in conversation. *Front. Neurosci.* **6**:PMCID: PMC3384290. [17]
Menninghaus, W. 2011. *Wozu Kunst: Ästhetik nach Darwin* [Why art? Aesthetics in the wake of Darwin]. Berlin: Suhrkamp. [5]
Menon, V., and D. J. Levitin. 2005. The rewards of music listening: Response and physiological connectivity of the mesolimbic system. *NeuroImage* **28**:175–184. [15]
Menon, V., D. J. Levitin, B. K. Smith, et al. 2002. Neural correlates of timbre change in harmonic sounds. *Neuroimage* **17**:1742–1754. [13]
Menuhin, Y., and C. W. Davis. 1979. The Music of Man. Toronto: Methuen. [1]
Merker, B. 2002. Music, the missing Humboldt system. *Musicae Scientiae* **6**:3–21. [1, 10, 21]
———. 2005. The conformal motive in birdsong, music and language. *Ann. NY Acad. Sci.* **1060**:17–28. [18]
———. 2009. Ritual foundations of human uniqueness. In: Communicative Musicality: Exploring the Basis of Human Companionship, ed. S. Malloch and C. Trevarthen, pp. 45–59. New York: Oxford Univ. Press. [18]
Merker, B., G. S. Madison, and P. Eckerdal. 2009. On the role and origin of isochrony in human rhythmic entrainment. *Cortex* **45**:4–17. [21]
Merritt, M. 1976. On questions following questions (on service encounters). *Lang. Soc.* **5**:315–357. [3]
Mertens, P. 2004. The prosogram: Semi-automatic transcription of prosody based on a tonal perception model. Proc. of Speech Prosody, March 23–26, 2004, ed. B. Bel and I. Marlien: Nara, Japan. [10]
Metz, C. 1974. Film Language: A Semiotics of the Cinema. New York: Oxford Univ. Press. [7]
Meyer, L. B. 1956. Emotion and Meaning in Music. Chicago: Univ. of Chicago Press. [5–8, 10, 18, 21]

Micheyl, C., R. P. Carlyon, A. Gutschalk, et al. 2007. The role of auditory cortex in the formation of auditory streams. *Hearing Res.* **229**:116–131. [17]

Micheyl, C., B. Tian, R. P. Carlyon, and J. P. Rauschecker. 2005. Perceptual organization of tone sequences in the auditory cortex of awake macaques. *Neuron* **48**:139–148. [17]

Milestone, L. (director and producer). 1939. Of Mice and Men. US: Hal Roach Studios. [7]

Miller, C. T., A. Dimauro, A. Pistorio, S. Hendry, and X. Wang. 2010. Vocalization-induced CFos expression in marmoset cortex. *Front. Integr. Neurosci.* **14**:128. [20]

Miller, G. A. 1956. The magical number seven, plus or minus two: Some limits on our capacity for processing information. *Psychol. Rev.* **63**:81–97. [18]

Miller, J. 2006. Focus. In: Encyclopedia of Language and Linguistics, ed. K. Brown. Oxford: Elsevier. [9]

Miranda, E. R., and M. M. Wanderley. 2006. New Digital Musical Instruments: Control and Interaction beyond the Keyboard. Middleton: A-R Editions. [16]

Miranda, R., and M. Ullman. 2007. Double dissociation between rules and memory in music: An event-related potential study. *NeuroImage* **38**:331–345. [6]

Mithen, S. 2005. The Singing Neanderthals: The Origins of Music, Language, Mind and Body. London: Weidenfeld and Nicholson. [3, 5, 8, 13, 14, 18–20]

Mohr, J. P. 1976. Broca's area and Broca's aphasia. In: Studies in Neurolinguistics, ed. H. Whitaker and H. A. Whitaker, vol. 1, pp. 201–235. New York: Academic. [20]

Molnar-Szakacs, I., and K. Overy. 2006. Music and mirror neurons: From motion to "e"motion. *Soc. Cogn. Affect. Neurosci.* **1**:235–241. [4, 8]

Montague, E. 2011. Phenomenology and the "hard problem" of consciousness and music. In: Music and Consciousness: Philosophical, Psychological, and Cultural Perspectives, ed. D. Clarke and E. Clarke, pp. 29–46. Oxford: Oxford Univ. Press. [8]

Montermurro, R. 1996. Singing lullabies to unborn children: Experience in village Vilamarxant, Spain. *Pre-Peri-Nat. Psychol. J.* **11**:9–16. [2]

Mooney, R. 1992. Synaptic basis for developmental plasticity in a birdsong nucleus. *J. Neurosci.* **12**:2462–2477. [20]

Moors, A., P. Ellsworth, N. Frijda, and K. R. Scherer. 2013. Appraisal theories of emotion: State of the art and future development. *Emotion Rev.* **5**, in press. [5]

Moran, J., and A. Gode. 1986. On the Origin of Language. Chicago: Univ. of Chicago Press. [5]

Moran, L. E. L., and R. E. Pinto. 2007. Automatic extraction of the lips shape via statistical lips modeling and chromatic feature. In: Proc. of Electronics, Robotics and Automotive Mechanics Conf., CERMA 07, pp. 241–246. IEEE Computer Soc. [21]

Morelli, G. A., B. Rogoff, D. Oppenheim, and D. Goldsmith. 1992. Cultural variation in infants' sleeping arrangements: Questions of independence. *Devel. Psychol.* **28**:604–613. [18]

Moreno, S., E. Bialystok, R. Barac, et al. 2011. Short-term music training enhances verbal intelligence and executive function. *Psychol. Sci.* **22**:1425–1433. [14]

Moreno, S., C. Marques, A. Santos, et al. 2009. Musical training influences linguistic abilities in 8-year-old children: More evidence for brain plasticity. *Cereb. Cortex* **19**:712–723. [14, 17]

Morikawa, H., N. Shand, and Y. Kosawa. 1988. Maternal speech to prelingual infants in Japan and the United States: Relationships among functions, forms, and referents. *J. Child Lang.* **15**:237–256. [18]

Morillon, B., K. Lehongre, R. S. Frackowiak, et al. 2010. Neurophysiological origin of human brain asymmetry for speech and language. *PNAS* **107**:18,688–18,693. [9]

Morley, I. 2002. Evolution of the physiological and neurological capacities for music. *Cambridge Arch. J.* **12**:195–216. [19]

———. 2009. A grand gesture: vocal and corporeal control in melody, rhythm and emotion. In: Language and Music as Cognitive Systems, ed. P. Rebuschat et al. Oxford: Oxford Univ. Press. [19]

———. 2011. The Prehistory of Music: The Evolutionary Origins and Archaeology of Human Musical Behaviours. Oxford: Oxford Univ. Press. [3]

Morris, C. W. 1938. Foundations of the Theory of Signs. Chicago: Univ.of Chicago Press. [8]

———. 1964. Signification and Significance: A Study of the Relation of Signs and Values. Cambridge, MA: MIT Press. [8]

———. 1971. Writings on the General Theory of Signs. The Hague: Mouton. [5]

Morris, R. D. 1993. New directions in the theory and analysis of musical contour. *Music Theory Spec.* **15**:205–228. [10]

Morrongiello, B. A., and S. E. Trehub. 1987. Age-related-changes in auditory temporal perception. *J. Exp. Child Psychol.* **44**:413–426. [13]

Morton, E. 1975. Ecological sources of selection on avian sounds. *Am. Natural.* **109**:17–34. [20]

———. 1977. On the occurrence and significance of motivation-structural rules in some bird and mammal sounds. *Am. Natural.* **111**:855–869. [5]

Moss, H. E., S. Abdallah, P. Fletcher, et al. 2005. Selecting among competing alternatives: Selection and retrieval in the left inferior frontal gyrus. *Cereb. Cortex* **15**:1723–1735. [9]

Mountcastle, V. B. 1978. An organizing principle for cerebral function: The unit module and the distributed system. In: The Mindful Brain, ed. G. M. Edelman and V. B. Mountcastle, pp. 7–50. Cambridge, MA: MIT Press. [15]

Mowrer, S. B., and R. Robert. 1989. Contemporary Learning Theories: Pavlovian Conditioning and the Status of Traditional Learning Theory. Hillsdale, NJ: Lawrence Erlbaum. [15]

Mozart, W. A. 1787. *Musikalisches Würfelspiel: Anleitung so viel Walzer oder Schleifer mit zwei Würfeln zu componieren ohne musikalisch zu seyn noch von der Composition etwas zu verstehen.* In: Köchel Catalog of Mozart's Work, p. KV1 Appendix 294d or KV6 516f. [16]

Mueller, J. L., J. Bahlmann, and A. D. Friederici. 2008. The role of pause cues in language learning: The emergence of event-related potentials related to sequence processing. *J. Cogn. Neurosci.* **20**:892–905. [19]

Mukamel, R., A. D. Ekstrom, J. Kaplan, M. Iacoboni, and I. Fried. 2010. Single-neuron responses in humans during observation and execution of actions. *Curr. Biol.* **20**:1–7. [4]

Munhall, K. G., and E. K. Johnson. 2012. Speech perception: When to put your money where the mouth is. *Curr. Biol.* **22**:R190–R192. [18]

Münsterberg, H. 1916/1970. The Photoplay: A Psychological Study. New York: Arno. [7]

Mura, A., B. Rezazadeh, A. Duff, et al. 2008. re(PER)curso: An interactive mixed reality chronicle. In: SIGGRAPH, pp. 1–1. Los Angeles: ACM. [16]

Murata, A., L. Fadiga, L. Fogassi, et al. 1997. Object representation in the ventral premotor cortex (area F5) of the monkey. *J. Neurophysiol.* **78**:2226–2230. [4]

Murphy, W. J., E. Eizirik, W. E. Johnson, et al. 2001. Molecular phylogenetics and the origins of placental mammals. *Nature* **409**:614–618. [20]

Murphy, W. J., T. H. Pringle, T. A. Crider, M. S. Springer, and W. Miller. 2007. Using genomic data to unravel the root of the placental mammal phylogeny. *Genome Res.* **17**:413–421. [20]

Myowa, M. 1996. Imitation of facial gestures by an infant chimpanzee. *Primates* **37**:207–213. [19]

Myowa-Yamakoshi, M., and T. Matsuzawa. 1999. Factors influencing imitation of manipulatory actions in chimpanzees, *Pan troglodytes. J. Comp. Psychol.* **113**:128–136. [19]

Nagarajan, S. S., D. T. Blake, B. A. Wright, N. Byl, and M. M. Merzenich. 1998. Practice-related improvements in somatosensory interval discrimination are temporally specific but generalize across skin location, hemisphere and modality. *Neuroscience* **18**:1559–1570. [17]

Nagy, E. 2011. Sharing the moment: The duration of embraces in humans. *J. Ethol.* **29**:389–393. [13]

Nakata, T., and C. Mitani. 2005. Influences of temporal fluctuation on infant attention. *Music Percep.* **22**:401–409. [18]

Nakata, T., and S. E. Trehub. 2004. Infants' responsiveness to maternal speech and singing. *Infant Behav. Dev.* **27**:455–464. [18]

———. 2011. Timing and dynamics in infant-directed singing. *Psychomusicology* **21**:130–138. [18]

Nan, Y., T. R. Knosche, S. Zysset, and A. D. Friederici. 2008. Cross-cultural music phrase processing: An fMRI study. *Hum. Brain Mapp.* **29**:312–328. [13]

Nan, Y., Y. Sun, and I. Peretz. 2010. Congenital amusia in speakers of a tone language: Association with lexical tone agnosia. *Brain* **133**:2635–2642. [17]

Napoli, D. J., and R. Sutton-Spence. 2010. Limitations on simultaneity in sign language. *Language* **86**:647–662. [11]

Narmour, E. 1990. The Analysis and Cognition of Basic Melodic Structures: The Implication-Realization Model. Chicago: Univ. of Chicago Press. [8, 16]

———. 1991. The top-down and bottom-up systems of musical implication: Building on Meyer's theory of emotional syntax. *Music Percep.* **9**:1–26. [7]

Nasir, S. M., and D. J. Ostry. 2008. Speech motor learning in profoundly deaf adults. *Nat. Neurosci.* **11**:1217–1222. [20]

Nauck-Börner, C. 1988. *Strukturen des musikalischen Gedächtnisses: Anmerkungen zu formalen Modellen der Repräsentation. Musikpsychologie* **5**:55–66.

Navarro Cebrian, A., and P. Janata. 2010a. Electrophysiological correlates of accurate mental image formation in auditory perception and imagery tasks. *Brain Res.* **1342**:39–54. [13]

———. 2010b. Influences of multiple memory systems on auditory mental image acuity. *J. Acoust. Soc. Am.* **127**:3189–3202. [13]

Nazzi, T., P. W. Jusczyk, and E. K. Johnson. 2000. Language discrimination by English-learning 5-month-olds: Effects of rhythm and familiarity. *J. Mem. Lang.* **43**:1–19. [18]

Nazzi, T., and F. Ramus. 2003. Perception and acquisition of linguistic rhythm by infants. *Speech Comm.* **41**:233–243. [18]

Neisser, U. 1976. Cognition and Reality: Principles and Implications of Cognitive Psychology. San Francisco: W. H. Freeman. [1, 8, 13, 17]

Nelson, K. 2005. Emerging levels of consciousness in early human development. In: The Missing Link in Cognition: Origins of Self-Reflective Consciousness, ed. H. S. Terrace and J. Metcalfe. New York: Oxford Univ. Press. [7]

Nespor, M., and I. Vogel. 1986. Prosodic Phonology. Dordrecht: Foris. [11]

Nesse, R. M., and P. C. Ellsworth. 2009. Evolution, emotions, and emotional disorders. *Am. Psychol.* **64**:129–139. [5]

Nettl, B. 2000. An ethnomusicologist contemplates universals in musical sound and musical culture. In: The Origin of Music, ed. N. Wallin et al., pp. 463–472. Cambridge, MA: MIT Press. [2, 3, 17, 19]

Newell, A. 1990. Unified Theories of Cognition. Cambridge, MA: Harvard Univ. Press. [16]

Newman, J. D., and D. Symmes. 1982. Inheritance and experience in the acquisition of primate acoustic behavior. In: Primate Communication, ed. C. T. Snowdon et al., pp. 259–278. Cambridge: Cambridge Univ. Press. [20]

NIME. 2013. Intl. Conference on New Interfaces for Musical Expression. http://www.nime.org/. (accessed 6 Dec 2012). [16]

Nishitani, N., and R. Hari. 2002. Viewing lip forms: Cortical dynamics. *Neuron* **36**:1211–1220. [17]

Nobre, A., J. Coull, C. Frith, and M. Mesulam. 1999. Orbitofrontal cortex is activated during breaches of expectation in tasks of visual attention. *Nat. Neurosci.* **2**:11–12. [6]

Noordzij, M. L., S. Newman-Norlund, R. Newman-Norlund, et al. 2010. Neural correlates of intentional communication. *Front. Neurosci.* **4**:10.3389/fnins.2010.00188. [3]

Norris, D. 1994. Shortlist: A connectionist model of continuous speech recognition. *Cognition* **52**:189–234. [9]

Northoff, G., and F. Bermpohl. 2004. Cortical midline structures and the self. *Trends Cogn. Sci.* **8**:102–107. [13]

Northoff, G., A. Heinzel, M. Greck, et al. 2006. Self-referential processing in our brain: A meta-analysis of imaging studies on the self. *NeuroImage* **31**:440–457. [13]

Norton, A., L. Zipse, S. Marchina, and G. Schlaug. 2009. Melodic intonation therapy: Shared insights on how it is done and why it might help. *Ann. NY Acad. Sci.* **1169**:431–436. [13, 14]

Nottebohm, F. 1968. Auditory experience and song development in the chaffinch, *Fringila coelebs. Ibis* **110**:549–568. [20]

———. 1972. The origins of vocal learning. *Am. Natural.* **106**:116–140. [21]

———. 1975. Continental patterns of song variability in *Zonotrichia capensis*: Some possible ecological correlates. *Am. Natural.* **109**:605–624. [20]

———. 1976. Vocal tract and brain: A search for evolutionary bottlenecks. *Ann. NY Acad. Sci.* **280**:643–9. [20]

———. 1977. Asymmetries in neural control of vocalizations in the canary. In: Lateralization in the Nervous System, ed. S. Harnad et al., pp. 23–44. New York: Academic. [20]

———. 1981. A brain for all seasons: Cyclical anatomical changes in song control nuclei of the canary brain. *Science* **214**:1368–1370. [20]

———. 2006. The road we travelled: Discovery, choreography, and significance of brain replaceable neurons. *Ann. NY Acad. Sci.* **1016**:628–658. [20]

Nottebohm, F., and M. E. Nottebohm. 1971. Vocalizations and breeding behaviour of surgically deafened ring doves, *Streptopelia risoria. Anim. Behav.* **19**:313–327. [20]

Nottebohm, F., T. M. Stokes, and C. M. Leonard. 1976. Central control of song in the canary, *Serinus canarius. J. Comp. Neurol.* **165**:457–486. [20]

Nozaradan, S., I. Peretz, M. Missal, and A. Mouraux. 2011. Tagging the neuronal entrainment to beat and meter. *J. Neurosci.* **31**:10,234–10,240. [17]

Nudds, T. D. 1978. Convergence of group size strategies by mammalian social carnivores. *Am. Natural.* **112**:957–960. [21]

Nunez, P. L., and R. Srinivasan. 2005. Electric Fields of the Brain: The Neurophysics of EEG, 2nd edition. New York: Oxford Univ. Press. [13]

Nygaard, L. C., and D. B. Pisoni. 1998. Talker-specific learning in speech perception. *Percept. Psychophys.* **60**:355–376. [17]

Obleser, J., A. M. Leaver, J. VanMeter, and J. P. Rauschecker. 2010. Segregation of vowels and consonants in human auditory cortex: Evidence for distributed hierarchical organization. *Front. Psychol.* **1**:232. [17]

Ochs, E., and B. B. Schieffelin. 1995. Language acquisition and socialization: Three developmental stories and their implications. In: Language, Culture, and Society, 2nd edition, ed. B. G. Blount, pp. 470–512. Prospect Heights, IL: Waveland Press. [18]

Odling-Smee, F. J., K. N. Laland, and M. W. Feldman. 2003. Niche Construction: The Neglected Process in Evolution. Princeton, NJ: Princeton Univ. Press. [19]

Ogawa, A., Y. Yamazaki, K. Ueno, K. Cheng, and A. Iriki. 2010. Inferential reasoning by exclusion recruits parietal and prefrontal cortices. *NeuroImage* **52**:1603–1610. [19]

Ohala, J. J. 1984. An ethological perspective on common cross-language utilization of F0 of voice. *Phonetica* **41**:1–16. [11]

Öhman, A., and S. Mineka. 2001. Fears, phobias, and preparedness: Toward an evolved module of fear and fear learning. *Psychol. Rev.* **108**:483–522. [5]

Ohnishi, T., H. Matsuda, T. Asada, et al. 2001. Functional anatomy of musical perception in musicians. *Cereb. Cortex* **11**:754–760. [13]

Ohtani, K. 1991. Japanese approaches to the study of dance. *Yearbook of Trad. Music* **23**:23–33. [2]

Ojemann, G. A. 1991. Cortical organization of language. *J. Neurosci.* **11**:2281–2287. [20]

———. 2003. The neurobiology of language and verbal memory: Observations from awake neurosurgery. *Intl. J. Psychophysiol.* **48**:141–146. [20]

Okada, K., F. Rong, J. Venezia, et al. 2010. Hierarchical organization of human auditory cortex: Evidence from acoustic invariance in the response to intelligible speech. *Cereb. Cortex* **20**:2486–2495. [17]

Okanoya, K. 2007. Language, evolution and an emergent property. *Curr. Opin. Neurobiol.* **17**:1–6. [20]

O'Keefe, J., and J. O. Dostrovsky. 1971. The hippocampus as a spatial map: Preliminary evidence from unit activity in the freely moving rat. *Brain Res.* **34**:171–175. [15]

O'Keefe, J., and L. Nadel. 1978. The Hippocampus as a Cognitive Map. Oxford: Oxford Univ. Press. [15]

Olivier, E., and S. Furniss. 1999. Pygmy and bushman music: A new comparative study. In: Central African Hunter-Gatherers in a Multidisciplinary Perspective: Challenging Elusiveness, ed. K. Biesbrouck et al., pp. 117–132. Leiden: Research School for Asian, African and Amerindian Studies. [2]

Oller, D. K., and R. Eilers. 1988. The role of audition in infant babbling. *Child Devel.* **59**:441–449. [20]

Olshausen, B., and D. Field. 1996. Emergence of simple-cell receptive field properties by learning a sparse code for natural images. *Nature* **381**:607–609. [15]

Olveczky, B. P., A. S. Andalman, and M. S. Fee. 2005. Vocal experimentation in the juvenile songbird requires a basal ganglia circuit. *PLoS Biol.* **3**:e153. [20]

Opitz, B., and S. Kotz. 2011. Ventral premotor cortex lesions disrupt learning of sequential grammatical structures. *Cortex* **48**:664–673. [6]

Oram, N., and L. L. Cuddy. 1995. Responsiveness of Western adults to pitch distributional information in melodic sequences. *Psychol. Res.* **57**:103–118. [17]

O'Regan, J. K., and A. Noe. 2001. A sensorimotor account of vision and visual consciousness. *Behav. Brain Res.* **24**:939–973; discussion 973–1031. [8]

Ortony, A., D. Norman, and W. Revelle. 2005. Affect and proto-affect in effective functioning. In: Who Needs Emotions, ed. J.-M. Fellous and M. A. Arbib, pp. 173–202. Oxford: Oxford Univ. Press. [8]

Osburn, L. 2004. Number of Species Identified on Earth. *Curr. Res.* http://www.currentresults.com/Environment-Facts/Plants-Animals/number-species.php. (accessed 24 August 2011). [20]

Ouattara, K., A. Lemasson, and K. Zuberbühler. 2009. Campbell's monkeys concatenate vocalizations into context-specific call sequences. *PNAS* **106**:22,026–22,031. [5]

Overy, K., and I. Molnar-Szakacs. 2009. Being together in time: Musical experience and the mirror neuron system. *Music Percep.* **26**:489–504. [6, 7, 8, 13]

Owren, M. J., J. A. Dieter, R. M. Seyfarth, and D. L. Cheney. 1993. Vocalizations of rhesus (*Macaca mulatta*) and Japanese (*M. fuscata*) macaques cross-fostered between species show evidence of only limited modification. *Devel. Psychobiol.* **26**:389–406. [20]

Özdemir, E., A. Norton, and G. Schlaug. 2006. Shared and distinct neural correlates of singing and speaking. *NeuroImage* **33**:628–635. [17, 21]

Oztop, E., and M. A. Arbib. 2002. Schema design and implementation of the grasp-related mirror neuron system. *Biol. Cybern.* **87**:116–140. [15]

Oztop, E., N. S. Bradley, and M. A. Arbib. 2004. Infant grasp learning: A computational model. *Exp. Brain Res.* **158**:480–503. [1, 15]

Oztop, E., M. Kawato, and M. A. Arbib. 2012. Mirror neurons: Functions, mechanisms and models. *Neuroscience Letters* doi: 10.1016/j.neulet.2012.10.005. [15]

Page, S. C., S. H. Hulse, and J. Cynx. 1989. Relative pitch perception in the European starling, *Sturnus vulgaris*: Further evidence for an elusive phenomenon. *J. Exp. Psychol. Anim. B.* **15**:137–146. [14]

Pagel, M. 2009. Human language as a culturally transmitted replicator. *Nat. Rev. Genet.* **10**:405–415. [3]

Pakosz, M. 1983. Attitudinal judgments in intonation: Some evidence for a theory. *J. Psycholinguist. Res.* **2**:311–326. [5]

Palleroni, A., M. Hauser, and P. Marler. 2005. Do responses of galliform birds vary adaptively with predator size? *Anim. Cogn.* **8**:200–210. [20]

Palmer, C. 1997. Music performance. *Annu. Rev. Psychol.* **48**:115–138. [1]

Palmer, C., and S. Holleran. 1994. Pitch, harmonic, Melodic, and frequency height influences in the perception of multivoiced music. *Perception and Psychophysics* **6**:301–312. [16]

Palmer, S. B., L. Fais, R. M. Golinkoff, and J. F. Werker. 2012. Perceptual narrowing of linguistic sign occurs in the first year of life. *Child Devel.* doi: 10.1111/j.1467-8624.2011.01715.x. [18]

Pandya, D. N., and B. Seltzer. 1982. Intrinsic connections and architectonics of posterior parietal cortex in the rhesus monkey. *J. Comp. Neurol.* **204**:204–210. [4]

Panksepp, J., and C. Trevarthen. 2009. The neuroscience of emotion in music. In: Communicative Musicality: Exploring the Basis of Human Companionship, ed. S. Malloch and C. Trevarthen, pp. 105–146. Oxford: Oxford Univ. Press. [8]

Papathanassiou, D., O. Etard, E. Mellet, et al. 2000. A common language network for comprehension and production: A contribution to the definition of language epicenters with PET. *Neuroimage* **11**:347–357. [20]

Paradiso, J. A. 1999. The Brain Opera technology: New instruments and gestural sensors for musical interaction and performance. *J. New Music Res.* **28**:130–149. [16]

Paradiso, J. A., K. Hsiao, J. Strickon, J. Lifton, and A. Adler. 2000. Sensor systems for interactive surfaces. *IBM Systems Journal* **39**:892–914. [16]

Parbery-Clark, A., E. Skoe, and N. Kraus. 2009. Musical experience limits the degradative effects of background noise on the neural processing of sound. *J. Neurosci.* **29**:14,100–14,107. [14]

Parr, L. A., B. M. Waller, A. M. Burrows, K. M. Gothard, and S. J. Vick. 2010. Brief communication: MaqFACS: A muscle-based facial movement coding system for the rhesus macaque. *Am. J. Phys. Anthropol.* **143**:625–630. [1]

Parr, L. A., B. M. Waller, and M. Heintz. 2008. Facial expression categorization by chimpanzees using standardized stimuli. *Emotion* **8**:216–231. [1]

Partee, B. H. 1984. Compositionality. In: Varieties of Formal Semantics, ed. F. Veltman and F. Landmand. Dordrecht: Foris. [9]

Pascalis, O., N. de Haan, and C. A. Nelson. 2002. Is face processing species-specific during the first year of life? *Science* **296**:1321–1323. [18]

Patel, A. D. 2003. Language, music, syntax and the brain. *Nat. Neurosci.* **6**:674–681. [4, 6, 9, 14, 16, 17]

———. 2005. The relation of music to the melody of speech and to syntactic processing disorders in aphasia. *Ann. NY Acad. Sci.* **1060**:59–70. [4]

———. 2006. Musical rhythm, linguistic rhythm, and human evolution. *Music Percep.* **24**:99–103. [6]

———. 2008. Music, Language, and the Brain. Oxford: Oxford Univ. Press. [3, 5, 6, 8, 10, 14, 15, 17–19]

———. 2010. Music, biological evolution, and the brain. In: Emerging Disciplines, ed. M. Bailar, pp. 91–144. Houston, TX: Rice Univ. Press. [18]

———. 2011. Why would musical training benefit the neural encoding of speech? The OPERA hypothesis. *Front. Psychol.* **2**:142. [14, 17]

Patel, A. D., E. Gibson, J. Ratner, M. Besson, and P. Holcomb. 1998. Processing syntactic relations in language and music: An event-related potential study. *J. Cogn. Neurosci.* **10**:717–733. [8, 14]

Patel, A. D., and J. R. Iversen. 2007. The linguistic benefits of musical abilities. *Trends Cogn. Sci.* **11**:369–372. [14]

Patel, A. D., J. R. Iversen, M. R. Bregman, and I. Schuiz. 2009. Experimental evidence for synchronization to a musical beat in a nonhuman animal. *Curr. Biol.* **19**:827–830. [1, 6, 8, 13, 17]

Patel, A. D., J. R. Iversen, and J. C. Rosenberg. 2006. Comparing the rhythm and melody of speech and music: The case of British English and French. *J. Acoust. Soc. Am.* **119**:3034–3047. [13]

Patel, A. D., J. R. Iversen, M. Wassenaar, and P. Hagoort. 2008a. Musical syntactic processing in agrammatic Broca's aphasia. *Aphasiology* **22**:776–789. [9, 14]

Patel, A. D., M. Wong, J. Foxton, A. Lochy, and I. Peretz. 2008b. Speech intonation perception deficits in musical tone deafness (congenital amusia). *Music Percep.* **25**:357–368. [17]

Patel, S., K. R. Scherer, E. Bjorkner, and J. Sundberg. 2011. Mapping emotions into acoustic space: The role of voice production. *Biol. Psychol.* **87**:93–98. [5]

Patterson, R. D., S. Uppenkamp, I. S. Johnsrude, and T. D. Griffiths. 2002. The processing of temporal pitch and melody information in auditory cortex. *Neuron* **36**:767–776. [13]

Pavlov, I. P. 1927. Conditioned Reflexes: An Investigation of the Physiological Activity of the Cerebral Cortex. Oxford: Oxford Univ. Press. [15]

Payne, K. B. 2000. The progressively changing songs of humpback whales: A window on the creative process in a wild animal. In: The Origins of Music, ed. N. L. Wallin et al., pp. 135–150. Cambridge, MA: MIT Press. [20]

Payne, T. E. 1997. Describing Morphosyntax: A Guide for Field Linguists. New York: Cambridge Univ. Press. [20]

Pazzaglia, M., N. Smania, E. Corato, and S. M. Aglioti. 2008. Neural underpinnings of gesture discrimination in patients with limb apraxia. *J. Neurosci.* **28**:3030–3041. [4]

Pearson, J. C., D. Lemons, and W. McGinnis. 2005. Modulating Hox gene functions during animal body patterning. *Nat. Rev. Genet.* **6**:893–904. [20]

Peelle, J. E., J. Gross, and M. H. Davis. 2012. Phase-locked responses to speech in human auditory cortex are enhanced during comprehension. *Cereb. Cortex,* May 17. [Epub ahead of print]. [17]

Peeters, G., B. L. Giordano, P. Susini, N. Misdariis, and S. McAdams. 2010. The timbre toolbox: Extracting audio descriptors from musical signals *J. Acoust. Soc. Am.* **130**:2902–2916. [16]

Peeva, M. G., F. H. Guenther, J. A. Tourville, et al. 2010. Distinct representations of phonemes, syllables, and supra-syllabic sequences in the speech production network. *NeuroImage* **50**:626–638. [14]

Peirce, C. 1931/1958. The Collected Papers of Charles Sanders Peirce. Cambridge, MA: Harvard Univ. Press. [6, 8]

Peña, M., L. L. Bonatti, M. Nespor, and J. Mehler. 2002. Signal-Driven Computations in Speech Processing. *Science* **298**:604–607. [19]

Penfield, W., and M. E. Faulk. 1955. The insula: Further observations on its function. *Brain* **78**:445–470. [4]

Pepperberg, I. M. 1999. The Alex Studies: Cognitive and Communicative Abilities of Grey Parrots. Cambridge, MA: Harvard Univ. Press. [20, 21]

———. 2002. Cognitive and communicative abilities of grey parrots. *Curr. Dir. Psychol. Sci.* **11**:83–87. [19]

———. 2006. Ordinality and inferential abilities of a grey parrot, *Psittacus erithacus. J. Comp. Psychol.* **120**:205–216. [19]

Pepperberg, I. M., and H. R. Shive. 2001. Simultaneous development of vocal and physical object combinations by a grey parrot, *Psittacus erithacus*: Bottle caps, lids, and labels. *J. Comp. Psychol.* **115**:376–384. [20]

Perani, D., M. C. Saccuman, P. Sciffo, et al. 2010. Functional specializations for music processing in the human newborn brain. *PNAS* **107**:4758–4763. [18]

———. 2011. Neural language networks at birth. *PNAS* **108**:16,056–16,061. [18]

Peretz, I. 2006. The nature of music from a biological perspective. *Cognition* **100**:1–32. [14, 17]

———. 2008. Musical disorders: From behavior to genes. *Curr. Dir. Psychol. Sci.* **17**:329–333. [17]

———. 2010. Towards a neurobiology of musical emotions. In: Handbook of Music and Emotion: Theory, Research, Applications ed. P. N. Juslin and J. Sloboda, pp. 99–126. Oxford: Oxford Univ. Press. [8]

———. 2012. Music, language, and modularity in action. In: Language and Music as Cognitive Systems, ed. P. Rebuschat et al., pp. 254–268. Oxford: Oxford Univ. Press. [14]

Peretz, I., J. Ayotte, R. J. Zatorre, et al. 2002. Congenital amusia: A disorder of fine-grained pitch discrimination. *Neuron* **33**:185–191. [13]

Peretz, I., E. Bratico, M. Järvenpaa, and M. Tervaniemi. 2009a. The amusic brain: In tune, out of key, and unaware. *Brain* **132**:1277–1286. [17]

Peretz, I., and M. Coltheart. 2003. Modularity of music processing. *Nat. Neurosci.* **6**:688–691. [10, 14]

Peretz, I., S. Cummings, and M. P. Dubé. 2007. The genetics of congenital amusia (or tone-deafness): A family-aggregation study. *Am. J. Hum. Genet.* **81**:582–588. [14]

Peretz, I., L. Gagnon, S. Hebert, and J. Macoir. 2004a. Singing in the brain: Insights from cognitive neuropsychology. *Music Percep.* **21**:373–390. [13]

Peretz, I., D. Gaudreau, and A.-M. Bonnel. 1998. Exposure effects on music preference and recognition. *Mem. Cogn.* **26**:884–902. [18]

Peretz, I., N. Gosselin, P. Belin, et al. 2009b. Music lexical networks: The cortical organization of music recognition. *Ann. NY Acad. Sci.* **1169**:256–265. [4]

Peretz, I., and R. Kolinsky. 1993. Boundaries of separability between melody and rhythm in music discrimination: A neuropsychological perspective. *Q. J. Exp. Psychol. A* **46**:301–325. [13]

Peretz, I., M. Radeau, and M. Arguin. 2004b. Two-way interactions between music and language: Evidence from priming recognition of tune and lyrics in familiar songs. *Mem. Cogn.* **32**:142–152. [13]

Perfetti, C. A., and G. A. Frishkoff. 2008. The neural bases of text and discourse processing. In: Handbook of the Neuroscience of Language, ed. B. Stemmer and H. A. Whitaker, pp. 165–174. New York: Academic. [8]

Perkel, D. J. 2004. Origin of the anterior forebrain pathway. *Ann. NY Acad. Sci.* **1016**:736–748. [20]

Perkel, D. J., and M. A. Farries. 2000. Complementary "bottom-up" and "top-down" approaches to basal ganglia function. *Curr. Opin. Neurobiol.* **10**:725–731. [20]

Perrachione, T. K., S. N. Del Tufo, and J. D. Gabrieli. 2011. Human voice recognition depends on language ability. *Science* **333**:595. [17]

Perrett, D. I., M. H. Harries, R. Bevan, et al. 1989. Frameworks of analysis for the neural representation of animate objects and actions. *J. Exp. Biol.* **146**:87–113. [4]

Perrett, D. I., J. K. Hietanen, M. W. Oram, and P. J. Benson. 1992. Organization and functions of cells responsive to faces in the temporal cortex. *Philos. Trans. R. Soc. Lond. B* **335**:23–30. [4]

Perrett, D. I., A. J. Mistlin, M. H. Harries, and A. J. Chitty. 1990. Understanding the visual appearance and consequence of hand actions. In: Vision and Action: The Control of Grasping, ed. M. A. Goodale, pp. 163–180. Norwood, N.J: Ablex. [15]

Perry, D. W., R. J. Zatorre, M. Petrides, et al. 1999. Localization of cerebral activity during simple singing. *NeuroReport* **10**:3979–3984. [13, 17]

Petersson, K. M., A. Reis, S. Askelof, A. Castro-Caldas, and M. Ingvar. 2000. Language processing modulated by literacy: A network analysis of verbal repetition in literate and illiterate subjects. *J. Cogn. Neurosci.* **12**:364–82. [19]

Petrides, M., G. Cadoret, and S. Mackey. 2005. Orofacial somatomotor responses in the macaque monkey homologue of Broca's area. *Nature* **435**:1235–1238. [4]

Petrides, M., and D. N. Pandya. 2009. Distinct parietal and temporal pathways to the homologues of Broca's area in the monkey. *PLoS Biol.* **7**:e1000170. [17]

Pfeifer, R., and J. Bongard. 2007. How the Body Shapes the Way We Think: A New View of Intelligence. Cambridge, MA: MIT Press. [16]

Pfeifer, R., and C. Scheier. 1999. Understanding Intelligence. Cambridge, MA: MIT Press. [16]

Pfordresher, P. Q., and S. Brown. 2007. Poor pitch singing in the absence of "tone deafness." *Music Percep.* **25**:95–115. [18]

Phillips-Silver, J., C. A. Aktipis, and G. A. Bryant. 2010. The ecology of entrainment: Foundations of coordinated rhythmic movement. *Music Percep.* **28**:3–14. [13]

Phillips-Silver, J., P. Toiviainen, N. Gosselin, et al. 2011. Born to dance but beat deaf: A new form of congenital amusia. *Neuropsychologia* **49**:961–969. [18, 17]

Phillips-Silver, J., and L. J. Trainor. 2005. Feeling the beat: Movement influences infants' rhythm perception. *Science* **308**:1430. [13, 17, 18]
———. 2007. Hearing what the body feels: Auditory encoding of rhythmic movement. *Cognition* **105**:533–546. [13, 18]
———. 2008. Vestibular influence on auditory metrical interpretation. *Brain Cogn.* **67**:94–102. [13]
Piaget, J. 1954. The Construction of Reality in the Child. New York: Norton [8]
———. 1971. Biology and knowledge: An essay on the relations between organic regulations and cognitive processes (Translation of 1967 essay: *Biologie et connaissance*: *Essai sur les relations entre les régulations organiques et les processus cognitifs*, Paris: Gallimard.). Edinburgh: Edinburgh Univ. Press. [1, 8]
Pickering, M. J., and S. Garrod. 2004. Towards a mechanistic psychology of dialogue. *Behav. Brain Res.* **27**:169–226. [17]
Pierrehumbert, J. 1980. The Phonetics and Phonology of English Intonation. Ph.D. thesis, MIT, Cambridge, MA. [10]
Pika, S., and T. Bugnyar. 2011. The use of referential gestures in ravens (*Corvus corax*) in the wild. *Nature Comm.* **2**:560. [20]
Pika, S., K. Liebal, and M. Tomasello. 2005. Gestural communication in subadult bonobos, *Pan paniscus*: Repertoire and use. *Am. J. Primatol.* **65**:39–61. [20]
Pike, K. L. 1943. Phonetics. Ann Arbor: Univ. of Michigan Press. [11]
Pinel, J. P. J. 2006. Psicobiologia. Bologna: Società editrice Il Mulino. [4]
Pinker, S. 1997. How the Mind Works. London: Allen Lane. [14, 18]
———. 1999. Words and Rules. New York: Basic Books. [17]
———. 2010. The cognitive niche: Coevolution of intelligence, sociality, and language. *PNAS* **107**:S8993–S8999. [19]
Pinker, S., and P. Bloom. 1990. Natural language and natural selection. *Behav. Brain Sci.* **13**:707–784. [3]
Pinker, S., and R. Jackendoff. 2005. The faculty of language: What's special about it? *Cognition* **95**:201–236. [20]
Plantinga, J., and L. J. Trainor. 2005. Memory for melody: Infants use a relative pitch code. *Cognition* **98**:1–11. [18]
Platel, H., C. Price, J. C. Baron, et al. 1997. The structural components of music perception: A functional anatomical study. *Brain* **120**:229–243. [13]
Poeppel, D. 1996. A critical review of PET studies of phonological processing. *Brain Lang.* **55**:317–385. [20]
———. 2001. Pure word deafness and the bilateral processing of the speech code. *Cogn. Sci.* **25**:679–693. [17]
———. 2003. The analysis of speech in different temporal integration windows: Cerebral lateralization as asymmetric sampling in time. *Speech Comm.* **41**:245–255. [14]
Poeppel, D., W. J. Idsardi, and V. van Wassenhove. 2008. Speech perception at the interface of neurobiology and linguistics. *Philos. Trans. R. Soc. Lond. B* **363**:1071–1086. [9]
Poeppel, E. 1997. A hierarchical model of temporal perception. *Trends Cogn. Sci.* **1**:56–61. [13]
———. 2009. Pre-semantically defined temporal windows for cognitive processing. *Philos. Trans. R. Soc. Lond. B* **364**:1887–1896. [13]
Poizner, H., E. Klima, and U. Bellugi. 1987. What the Hands Reveal about the Brain. Cambridge, MA: MIT Press. [19]
Pollick, A. S., and F. B. de Waal. 2007. Ape gestures and language evolution. *PNAS* **104**:8184–8189. [20]

Poole, J. H., P. L. Tyack, A. S. Stoeger-Horwath, and S. Watwood. 2005. Animal behaviour: Elephants are capable of vocal learning. *Nature* **434**:455–456. [20]
Pope, R. 2005. Creativity: Theory, History, Practice. New York: Routledge. [16]
Poremba, A., M. Malloy, R. C. Saunders, et al. 2004. Species-specific calls evoke asymmetric activity in the monkey's temporal poles. *Nature* **427**:448–451. [17]
Porges, S. W. 1997. Emotion: An evolutionary by-product of the neural regulation of the autonomic nervous system. In: The Integrative Neurobiology of Affiliation, ed. C. S. Carter et al., vol. 807, pp. 62–77. *Annals NY Acad. Sci.* [5]
———. 2001. The Polyvagal Theory: Phylogenetic substrates of a social nervous system. *Intl. J. Psychophysiol.* **42**:123–146. [5]
Poulin-Charronnat, B., E. Bigand, F. Madurell, and R. Peereman. 2005. Musical structure modulates semantic priming in vocal music. *Cognition* **94**:B67–B78. [13]
Prather, J. F., S. Peters, S. Nowicki, and R. Mooney. 2008. Precise auditory-vocal mirroring in neurons for learned vocal communication. *Nature* **451**:305–310. [4]
Prendergast, R. M. 1992. Film Music: A Neglected Art, 2nd edition. New York: Norton. [7]
Pribram, K. H. 1960. A review of theory in physiological psychology. *Annu. Rev. Psychol.* **11**:1–40. [1]
Price, C. J., R. J. Wise, E. A. Warburton, et al. 1996. Hearing and saying: The functional neuro-anatomy of auditory word processing. *Brain* **119**:919–931. [4]
Pulvermüller, F., and L. Fadiga. 2009. Active perception: Sensorimotor circuits as a cortical basis for language. *Nat. Rev. Neurosci.* **11**:351–360. [4]
Purwins, H., M. Grachten, P. Herrera, et al. 2008a. Computational models of music perception and cognition II: Domain-specific music processing. *Phys. Life Rev.* **5**:169–182. [15, 16]
Purwins, H., P. Herrera, M. Grachten, et al. 2008b. Computational models of music perception and cognition I: The perceptual and cognitive processing chain. *Phys. Life Rev.* **5**:151–168. [16]
Quallo, M. M., C. J. Price, K. Ueno, et al. 2009. Gray and white matter changes associated with tool-use learning in macaque monkeys. *PNAS* **106**:18,379–18,384. [19, 21]
Quiroga Murcia, C., G. Kreutz, and S. Bongard. 2011. Endokrine und immunologische Wirkungen von Musik. In: Psychoneuroimmunologie und Psychotherapie, ed. C. Schubert, pp. 248–262. Stuttgart: Schattauer. [6]
Racette, A., C. Bard, and I. Peretz. 2006. Making non-fluent aphasics speak: Sing along! *Brain* **129**:2571–2584. [13, 14, 17]
Racette, A., and I. Peretz. 2007. Learning lyrics: To sing or not to sing? *Mem. Cogn.* **35**:242–253. [13]
Radcliffe-Brown, A. R. 1922. The Andaman Islanders. Cambridge: Cambridge Univ. Press. [2]
Ralls, K., P. Fiorelli, and S. Gish. 1985. Vocalizations and vocal mimicry in captive harbor seals, *Phoca vitulina*. *Can. J. Zool.* **63**:1050–1056. [20]
Rameau, J.-P. 1726. Nouveau Système de Musique Théorique. Paris: Ballard. [10]
Ramón y Cajal, S. 1911. Histologie du Systeme Nerveux de L'homme et des Vertebres. Paris: A. Maloine (English translation by N. and L. Swanson, Oxford Univ. Press, 1995). [15]
Ramus, F., and G. Szenkovits. 2008. What phonological deficit? *Q. J. Exp. Psychol.* **61**:129–141. [14]
Randel, D. M. 2003. Form. In: The Harvard Dictionary of Music, 4th edition, ed. D. M. Randel, pp. 329–330. Cambridge, MA: Belknap Press. [8]

Raos, V., M. A. Umiltà, A. Murata, L. Fogassi, and V. Gallese. 2006. Functional properties of grasping-related neurons in the ventral premotor area F5 of the macaque monkey. *J. Neurophysiol.* **95**:709–729. *J. Neurophysiol.* **95**:709–729. [4]

Rasch, R. 2002. Tuning and temperament. In: The Cambridge History of Western Music Theory, ed. T. Christensen, pp. 193–222. Cambridge: Cambridge Univ. Press. [11]

Rauschecker, J. P. 1991. Mechanisms of visual plasticity: Hebb synapses, NMDA receptors, and beyond. *Physiol. Rev.* **71**:587–615. [15]

———. 1995. Compensatory plasticity and sensory substitution in the cerebral cortex. *Trends Neurosci.* **18**:36–43. [17]

———. 1998. Parallel processing in the auditory cortex of primates. *Audiol. Neurootol.* **3**:86–103. [14]

———. 2005. Vocal gestures and auditory objects. *Behav. Brain Sci.* **28**:143–144. [15]

———. 2011. An expanded role for the dorsal auditory pathway in sensorimotor integration and control. *Hearing Res.* **271**:16–25. [15, 17]

Rauschecker, J. P., and S. K. Scott. 2009. Maps and streams in the auditory cortex: Nonhuman primates illuminate human speech processing. *Nat. Neurosci.* **12**:718–724. [14, 15, 17]

Rauschecker, J. P., and B. Tian. 2004. Processing of band-passed noise in the lateral auditory belt cortex of the rhesus monkey. *J. Neurophysiol.* **91**:2578–2589. [14]

Raynauld, I. 2001. Dialogues in early silent sound screenplays: What actors really said. In: The Sounds of Early Cinema, ed. R. Abel and R. Altman, pp. 61–78. Indianapolis, IN: Indiana Univ. Press. [7]

Reason, M., and D. Reynolds. 2010. Kinesthesia, empathy, and related pleasures: An inquiry into audience experiences of watching dance. *Dance Res. J.* **42**:49–75. [8]

Reich, U. 2011. The meanings of semantics: Comment. *Phys. Life Rev.* **8**:120–121. [6, 8]

Reiner, A., Y. Jiao, N. Del Mar, A. V. Laverghetta, and W. L. Lei. 2003. Differential morphology of pyramidal tract-type and intratelencephalically projecting-type corticostriatal neurons and their intrastriatal terminals in rats. *J. Comp. Neurol.* **457**:420–440. [20]

Reiner, A., A. V. Laverghetta, C. A. Meade, S. L. Cuthbertson, and S. W. Bottjer. 2004. An immunohistochemical and pathway tracing study of the striatopallidal organization of Area X in the male zebra finch. *J. Comp. Neurol.* **469**:239–261. [20]

Reiterer, S. M., M. Erb, C. D. Droll, et al. 2005. Impact of task difficulty on lateralization of pitch and duration discrimination. *NeuroReport* **16**:239–242. [17]

Remedios, R., N. K. Logothetis, and C. Kayser. 2009. Monkey drumming reveals common networks for perceiving vocal and nonvocal communication sounds. *PNAS* **106**:18,010–18,015. [17]

Remijsen, B. Nilotic prosody: Research on the rich prosodic systems of Western Nilotic languages 2012. Available from http://www.ling.ed.ac.uk/nilotic/. [11]

Renier, L. A., I. Anurova, A. G. De Volder, et al. 2010. Preserved functional specialization for spatial processing in the middle occipital gyrus of the early blind. *Neuron* **68**:138–148. [17]

Rennó-Costa, C., J. E. Lisman, and P. F. M. J. Verschure. 2010. The mechanism of rate remapping in the dentate gyrus. *Neuron* **68**:1051–1058. [15, 16]

Repp, B. H. 1992. Diversity and commonality in music performance: Analysis of timing microstructure in Schumann's *Träumerei*. Haskins Laboratories Status Report on Speech Research **SR-111/1112**:227–260. [1]

Rescorla, R. A., and A. R. Wagner. 1972. A theory of Pavlovian conditioning: Variations in the effectiveness of reinforcement and nonreinforcement. In: Classical Conditioning II: Current Research and Theory, ed. A. A. Black and W. F. Prokasy, pp. 64–99. New York: Appleton-Century-Crofts. [15]

Revonsuo, A., and J. Newman. 1998. Binding and consciousness. *Conscious. Cogn.* **8**:123–127. [7]

Reynolds, D., and M. Reason. 2012. Knowing me, knowing you: Autism, kinesthetic empathy and applied performance. In: Kinesthetic Empathy in Creative and Cultural Practices, ed. D. Reynolds and M. Reason, pp. 35–50. Bristol: Intellect Ltd. [8]

Reynolds Losin, E. A., J. L. Russell, H. Freeman, A. Meguerditchian, and W. D. Hopkins. 2008. Left hemisphere specialization for oro-facial movements of learned vocal signals by captive chimpanzees. *PLoS ONE* **3**:e2529. [20]

Richards, D. G., J. P. Wolz, and L. M. Herman. 1984. Vocal mimicry of computer-generated sounds and vocal labeling of objects by a bottlenosed dolphin, *Tursiops truncatus*. *J. Comp. Psychol.* **98**:10–28. [20]

Richman, B. 2000. How music fixed "nonsense" into significant formulas: On rhythm, repetition and meaning. In: The Origins of Music, ed. N. Wallin et al., pp. 271–300. Cambridge MA: MIT Press. [2]

Rickard, N. S., S. R. Toukhsati, and S. E. Field. 2005. The effect of music on cognitive performance: Insight from neurobiological and animal studies. *Behav. Cogn. Neurosci. Rev.* **4**:235–261. [8, 21]

Rickard, N. S., W. W. Wong, and L. Velik. 2012. Relaxing music counters heightened consolidation of emotional memory. *Neurobiol. Learn. Mem.* **97**:220–228. [8]

Ridgway, S., D. Carder, M. Jeffries, and M. Todd. 2012. Spontaneous human speech mimicry by a cetacean. *Curr. Biol.* **22**:R860–R861. [20]

Riebel, K. 2003. The "mute" sex revisited: Vocal production and perception learning in female songbirds. *Adv. Study Behav.* **33**:49–86. [20]

Riecker, A., H. Ackermann, D. Wildgruber, G. Dogil, and W. Grodd. 2000. Opposite hemispheric lateralization effects during speaking and singing at motor cortex, insula and cerebellum. *NeuroReport* **11**:1997–2000. [13]

Riecker, A., D. Wildgruber, G. Dogil, W. Grodd, and H. Ackermann. 2002. Hemispheric lateralization effects of rhythm implementation during syllable repetitions: An fMRI study. *Neuroimage* **16**:169–176. [13]

Riemann, H. 1877. *Musikalische Syntaxis*. Leipzig: Breitkopf & Hartel. [6]

——. 1893. *Vereinfachte Harmonielehre: Oder, die Lehre von den tonalen Funktionen der Akkorde*. London: Augener. [10]

Rilling, J., M. F. Glasser, T. M. Preuss, et al. 2008. The evolution of the arcuate fasciculus revealed with comparative DTI. *Nat. Neurosci.* **11**:426–428. [9, 17, 20]

Rilling, J., D. Gutman, and T. Zeh. 2002. A neural basis for social cooperation. *Neuron* **35**:395–405. [6]

Ritchison, G. 1986. The singing behavior of female northern cardinals. *Condor* **88**:156–159. [20]

Rizzolatti, G., and M. A. Arbib. 1998. Language within our grasp. *Trends Neurosci.* **21**:188–194. [4, 7, 19, 20]

Rizzolatti, G., R. Camarda, L. Fogassi, et al. 1988. Functional organization of inferior area 6 in the macaque monkey. II. Area F5 and the control of distal movements. *Exp. Brain Res.* **71**:491–507. [4]

Rizzolatti, G., and L. Craighero. 2004. The mirror neuron system. *Ann. Rev. Neurosci.* **27**:169–192. [4]

Rizzolatti, G., L. Fadiga, L. Fogassi, and V. Gallese. 2002. From mirror neurons to imitation: Facts and speculations. In: The Imitative Mind: Development, Evolution and Brain Bases, ed. A. Meltzoff and W. Prinz, pp. 247–266. Cambridge: Cambridge Univ. Press. [4]

Rizzolatti, G., L. Fadiga, V. Gallese, and L. Fogassi. 1996. Premotor cortex and the recognition of motor actions. *Brain Res.* **3**:131–141. [4]

Rizzolatti, G., and G. Luppino. 2003. Grasping movements: Visuomotor transformations. In: The Handbook of Brain Theory and Neural Networks, 2nd edition, ed. M. A. Arbib, pp. 501–504. Cambridge, MA: MIT Press. [15]

Rizzolatti, G., and C. Sinigaglia. 2010. The functional role of the parietal-frontal mirror circuit: Interpretations and misinterpretations. *Nat. Rev. Neurosci.* **11**:264–274. [8]

Roach, J. R. 1985. The Player's Passion: Studies in the Science of Acting. Ann Arbor: Univ. of Michigan Press. [5]

Rodd, J. M., M. H. Davis, and I. S. Johnsrude. 2005. The neural mechanisms of speech comprehension: fMRI studies of semantic ambiguity. *Cereb. Cortex* **15**:1261–1269. [9]

Rodd, J. M., O. A. Longe, B. Randall, and L. K. Tyler. 2010. The functional organisation of the fronto-temporal language system: Evidence from syntactic and semantic ambiguity. *Neuropsychologia* **48**:1324–1335. [9]

Rogalsky, C., T. Love, D. Driscoll, S. W. Anderson, and G. Hickok. 2011a. Are mirror neurons the basis of speech perception? Evidence from five cases with damage to the purported human mirror system. *Neurocase* **17**:178–187. [17]

Rogalsky, C., F. Rong, K. Saberi, and G. Hickok. 2011b. Functional anatomy of language and music perception: Temporal and structural factors investigated using functional magnetic resonance imaging. *J. Neurosci.* **31**:3843–3852. [14, 17]

Rogoff, B., R. Paradise, R. M. Arauz, M. Correa-Chavez, and C. Angelillo. 2003. Firsthand learning through intent participation. *Annu. Rev. Psychol.* **54**:175–203. [21]

Rohrmeier, M. 2005. Towards modelling movement in music: Analysing properties and dynamic aspects of pc set sequences in Bach's chorales, Univ. of Cambridge, Cambridge. [6]

———. 2011. Toward a generative syntax of tonal harmony. *J. Math. Music* **18**:1–26. [6, 10, 14]

Rohrmeier, M., and I. Cross. 2008. Statistical properties of tonal harmony in Bach's chorales. In: Proc. 10th Intl. Conf. on Music Perception and Cognition. Sapporo: Hokkaido Univ. [6]

Rohrmeier, M., and S. Koelsch. 2012. Predictive information processing in music cognition: A critical review. *Intl. J. Psychophys.* **83**:164–175. [6]

Roig-Francolí, M. A. 1995. Harmonic and formal processes in Ligeti's net-structure compositions. *Music Theory Spec.* **17**:242–267. [16]

Rolls, E. T. 2005a. Emotion Explained. Oxford: Oxford Univ. Press. [8]

———. 2005b. What are emotions, why do we have emotions, and what is their computational role in the brain? In: Who Needs Emotions: The Brain Meets the Robot, ed. J.-M. Fellous and M. A. Arbib, pp. 117–146. New York: Oxford Univ. Press. [4, 15]

———. 2011. Functions of human emotional memory: The brain and emotion. In: The Memory Process: Neuroscientific and Humanistic Perspectives, ed. S. Nalbantian et al., pp. 173–191. Cambridge, MA: MIT Press. [8]

Rolls, E. T., and F. Grabenhorst. 2008. The orbitofrontal cortex and beyond: From affect to decision-making. *Prog. Neurobiol.* **86**:216–244. [6]

Romand, R., and G. Ehret. 1984. Development of sound production in normal, isolated, deafened kittens during the first postnatal months. *Devel. Psychobiol.* **17**:629–649. [20]

Romanski, L. M., B. Tian, J. Fritz, et al. 1999. Dual streams of auditory afferents target multiple domains in the primate prefrontal cortex. *Nature Neurosci.* **2**:1131–1136. [15]

Rosch, E. 1975. Cognitive representations of semantic categories. *J. Exp. Psychol. Gen.* **104**:192–233. [21]

Rosenbaum, D. A., R. G. Cohen, S. A. Jax, D. J. Weiss, and R. van der Wel. 2007. The problem of serial order in behavior: Lashley's legacy. *Hum. Mov. Sci.* **26**:525–554. [4]

Rosenbaum, P. 1967. The Grammar of English Predicate Complement Constructions. Cambridge, MA: MIT Press. [12]

Rosenblatt, F. 1958. The perceptron: A probabilistic model for information storage and organization in the brain. *Psychol. Rev.* **65**:386–408. [15, 16]

Rosenboom, D., ed. 1975. Biofeedback and the Arts, Results of Early Experiments. Vancouver: Aesthetic Research Center of Canada Publications. [16]

———. 1990. Extended Musical Interface with the Human Nervous System, Leonardo Monograph Series No. 1. Berkeley: Intl. Soc. for the Arts, Science and Technology. [16]

Ross, D., J. Choi, and D. Purves. 2007. Musical intervals in speech. *PNAS* **104**:9852. [6]

Rossano, F., P. Brown, and S. C. Levinson. 2009. Gaze, questioning and culture. In: Conversation Analysis: Comparative Perspectives, ed. J. Sidnell, pp. 187–249. Cambridge: Cambridge Univ. Press. [3]

Rousseau, J. J. 1760/1852. *Essai sur l'origine des langues*. In: Oeuvres Completes de J. J. Rousseau, vol. 3. Paris: Furne. [10]

Rowe, C. 1999. Receiver psychology and the evolution of multicomponent signals. *Anim. Behav.* **58**:921–931. [5]

Rowe, R. 1993. Interactive Music Systems: Machine Listening and Composing. Cambridge, MA: MIT Press. [16]

Roy, A. C., and M. A. Arbib. 2005. The Syntactic Motor System. *Gesture* **5**:7–37. [19]

Rozzi, S., R. Calzavara, A. Belmalih, et al. 2006. Cortical connections of the inferior parietal cortical convexity of the macaque monkey. *Cereb. Cortex* **16**:1389–1417. [4]

Rozzi, S., P. F. Ferrari, L. Bonini, G. Rizzolatti, and L. Fogassi. 2008. Functional organization of inferior parietal lobule convexity in the macaque monkey: Electrophysiological characterization of motor, sensory and mirror responses and their correlation with cytoarchitectonic areas. *Eur. J. Neurosci.* **28**:1569–1588. [4]

Rubens, A. B. 1975. Aphasia with infarction in the territory of the anterior cerebral artery. *Cortex* **11**:239–250. [20]

Rubin, D. C. 1995. Memory in Oral Traditions: The Cognitive Psychology of Epic, Ballads, and Counting-out Rhymes. Oxford: Oxford Univ. Press. [21]

Ruiz, M. H., S. Koelsch, and J. Bhattacharya. 2009. Decrease in early right alpha band phase synchronization and late gamma band oscillations in processing syntax in music. *Hum. Brain Mapp.* **30**:1207–1225. [15]

Rumelhart, D. E., and J. L. McClelland. 1986. Parallel Distributed Processing: Explorations in the Microstructures of Computing. Cambridge, MA: MIT Press. [16]

Ruschemeyer, S. A., S. Zysset, and A. D. Friederici. 2006. Native and non-native reading of sentences: An fMRI experiment. *NeuroImage* **31**:354–365. [9]

Russell, J. A. 2003. Core affect and the psychological construction of emotion. *Psychol. Rev.* **110**:145–172. [5]

Russell, S. J., and P. Norvig. 2010. Artificial Intelligence: A Modern Approach, 3rd edition. Upper Saddle River, NJ: Prentice Hall. [16]

Sachs, C. 1943. The Rise of Music In The Ancient World. New York: Norton. [18]
Sacks, H., E. A. Schegloff, and G. Jefferson. 1974. Simplest systematics for organization of turn-taking for conversation. *Lang. Soc.* **50**:696–735. [3, 13]
Saffran, J. R., R. N. Aslin, and E. L. Newport. 1996. Statistical learning by 8-month-old infants. *Science* **274**:1926–1928. [18]
Saffran, J. R., E. K. Johnson, R. N. Aslin, and E. L. Newport. 1999. Statistical learning of tone sequences by human infants and adults. *Cognition* **70**:27–52. [18]
Saffran, J. R., M. M. Loman, and R. R. W. Robertson. 2000. Infant memory for musical experiences. *Cognition* **77**:B15–B23. [18]
Saffran, J. R., K. Reeck, A. Neibuhr, and D. Wilson. 2005. Changing the tune: The structure of the input affects infants' use of absolute and relative pitch. *Devel. Sci.* **8**:1–7. [18]
Sahin, N. T., S. Pinker, S. S. Cash, D. Schomer, and E. Halgren. 2009. Sequential processing of lexical, grammatical, and phonological information within Broca's area. *Science* **326**:445–449. [9]
Saito, Y., K. Ishii, K. Yagi, I. Tatsumi, and H. Mizusawa. 2006. Cerebral networks for spontaneous and synchronized singing and speaking. *NeuroReport* **17**:1893–1897. [13, 14, 17]
Salimpoor, V. N., M. Benovoy, K. Larcher, A. Dagher, and R. J. Zatorre. 2011. Anatomically distinct dopamine release during anticipation and experience of peak emotion to music. *Nat. Neurosci.* **14**:257–262. [3, 8, 13, 17]
Sammler, D. 2008. The Neuroanatomical Overlap of Syntax Processing in Music and Language Evidence from Lesion and Intracranial ERP Studies. PhD thesis, Univ. of Leipzig. [6]
Sammler, D., A. Baird, R. Valabregue, et al. 2010. The relationship of lyrics and tunes in the processing of unfamiliar songs: A functional magnetic resonance adaptation study. *J. Neurosci.* **30**:3572–3578. [13, 17]
Sammler, D., S. Koelsch, T. Ball, et al. 2009. Overlap of musical and linguistic syntax processing: Intracranial ERP evidence. *Ann. NY Acad. Sci.* **1169**:494–498. [14]
———. 2012. Co-localizing linguistic and musical syntax with intracranial EEG. *NeuroImage* **64**:134–146. [17]
Sammler, D., S. Koelsch, and A. D. Friederici. 2011. Are left fronto-temporal brain areas a prerequisite for normal music-syntactic processing? *Cortex* **47**:659–673. [14]
Samson, S., and R. J. Zatorre. 1991. Recognition memory for text and melody of songs after unilateral temporal lobe lesion: Evidence for dual encoding. *J. Exp. Psychol. Learn. Mem. Cogn.* **17**:793–804. [13]
Sanchez-Fibla, M., U. Bernardet, E. Wasserman, et al. 2010. Allostatic control for robot behavior regulation: A comparative rodent-robot study. *Advances in Complex Systems* **13**:377–403. [16]
Sanchez-Montanes, M. A., P. F. M. J. Verschure, and P. König. 2000. Local and global gating of synaptic plasticity. *Neural Comp.* **12**:519–529. [15, 16]
Sander, D., D. Grandjean, and K. R. Scherer. 2005. A systems approach to appraisal mechanisms in emotion. *Neural Netw.* **18**:317–352. [5]
Sandler, W., and M. Aronoff. 2007. Is phonology necessary for language? Emergence of Language Structures Workshop. Unpublished talk, UCSD (February 6, 2007). [19]
Sandler, W., I. Meir, C. Padden, and M. Aronoff. 2005. The emergence of grammar: Systematic structure in a new language. *PNAS* **102**:2661–2665. [1]
Sänger, J., V. Müller, and U. Lindenberger. 2012. Intra- and interbrain synchronization and network properties when playing guitar in duets. *Front. Human Neurosci.* **6**:312. [8]

Sarnthein, J., and D. Jeanmonod. 2007. High thalamocortical theta coherence in patients with Parkinson's disease. *J. Neurosci.* **27**:124. [15]

Sasaki, A., T. D. Sotnikova, R. R. Gainetdinov, and E. D. Jarvis. 2006. Social context-dependent singing-regulated dopamine. *J. Neurosci.* **26**:9010–9014. [20]

Saur, D., B. W. Kreher, and S. Schnell. 2008. Ventral and dorsal pathways for language. *PNAS* **105**:18,035–18,040. [13]

Sauter, D., F. Eisner, P. Ekman, and S. Scott. 2010. Cross-cultural recognition of basic emotions through nonverbal emotional vocalizations. *PNAS* **107**:2408–2412. [5]

Saxe, R., and A. Wexler. 2005. Making sense of another mind: The role of the right temporo-parietal junction. *Neuropsychologia* **43**:1391–1399. [4]

Schachner, A., T. F. Brady, I. M. Pepperberg, and M. D. Hauser. 2009. Spontaneous motor entrainment to music in multiple vocal mimicking species. *Curr. Biol.* **19**:831–836. [13]

Schachner, A., and E. E. Hannon. 2010. Infant-directed speech drives social preferences in 5-month-old infants. *Devel. Psychol.* **47**:19–25. [18]

Schacter, D. L., D. R. Addis, and R. L. Buckner. 2007. Remembering the past to imagine the future: The prospective brain. *Nat. Rev. Neurosci.* **8**:657–661. [8]

Schaeffer, P. 2012. In Search of a Concrete Music (translated from the 1952 original, À la *recherche d'une musique concrète*, by J. Dack). California Studies in 20th-Century Music. Los Angeles: Ahmanson Foundation. [16]

Scharff, C., and F. Nottebohm. 1991. A comparative study of the behavioral deficits following lesions of various parts of the zebra finch song system: Implications for vocal learning. *J. Neurosci.* **11**:2896–2913. [20]

Scharff, C., and J. Petri. 2011. Evo-devo, deep homology and FoxP2: Implications for the evolution of speech and language. *Philos. Trans. R. Soc. Lond. B* **366**:2124–2140. [20]

Schegloff, E. A. 1984. On some gestures' relation to talk. In: Structures of Social Action: Studies in Conversation Analysis, ed. J. M. Atkinson and J. Heritage, pp. 266–296. Cambridge: Cambridge Univ. Press. [2]

———. 2007. A Primer for Conversation Analysis: Sequence Organization. Cambridge: Cambridge Univ. Press. [3]

Schellenberg, E. G., P. Iverson, and M. C. McKinnon. 1999. Name that tune: Identifying popular recordings from brief excerpts. *Psychon. Bull. Rev.* **6**:641–646. [18]

Schellenberg, E. G., and S. E. Trehub. 1996. Natural musical intervals: Evidence from infant listeners. *Psychol. Sci.* **7**:272–277. [18]

———. 2003. Good pitch memory is widespread. *Psychol. Sci.* **14**:262–266. [18]

Schellenberg, M. 2009. Singing in a tone language: Shona. In: Selected Proc. of the 39th Annual Conf. on African Linguistics, ed. A. Ojo and L. Moshi, pp. 137–144. Somerville, MA: Cascadilla Proceedings Project. [11]

Schenker, H. 1935. *Der freie Satz: Neue musikalische Theorien und Phantasien* III. Vienna: Universal. [10]

Schenker, H., and O. Jonas. 1979. Free Composition (*Der freie Satz*). London: Longman Publishing Group. [16]

Scherer, K. R. 1984. On the nature and function of emotion: A component process approach. In: Approaches to Emotion, ed. K. R. Scherer and P. E. Ekman, pp. 293–317. Hillsdale, NJ: Erlbaum. [5]

———. 1985. Vocal affect signalling: A comparative approach. In: Advances in the Study of Behavior, ed. J. Rosenblatt et al., vol. 15, pp. 189–244. New York: Academic. [5]

———. 1986. Vocal affect expression: A review and a model for future research. *Psychol. Bull.* **99**:143–165. [5]
———. 1988. On the symbolic functions of vocal affect expression. *J. Lang. Soc. Psychol.* **7**:79–100. [5]
———. 1991. Emotion expression in speech and music. In: Music, Language, Speech, and Brain. Wenner-Gren Center Intl. Symp. Series, ed. J. Sundberg et al. London: Macmillan. [5]
———. 1994a. Affect bursts. In: Emotions: Essays on Emotion Theory, ed. S. van Goozen et al., pp. 161–196. Hillsdale, NJ: Erlbaum. [5]
———. 1994b. Toward a concept of "modal emotions." In: The Nature of Emotion: Fundamental Questions, ed. P. Ekman and R. J. Davidson, pp. 25–31. New York/Oxford: Oxford Univ. Press. [5]
———. 1995. Expression of emotion in voice and music. *J. Voice* **9**:235–248. [5, 6, 11]
———. 2000. Emotional expression: A royal road for the study of behavior control. In: Control of Human Behavior, Mental Processes, and Awareness, ed. A. Grob and W. Perrig, pp. 227–244. Hillsdale, NJ: Lawrence Erlbaum. [5]
———. 2001. Appraisal considered as a process of multilevel sequential checking. In: Appraisal Processes in Emotion: Theory, Methods, Research, ed. K. R. Scherer et al., pp. 92–120. New York: Oxford Univ. Press. [5]
———. 2003. Vocal communication of emotion: A review of research paradigms. *Speech Comm.* **40**:227–256. [5]
———. 2004a. Feelings integrate the central representation of appraisal-driven response organization in emotion. In: Feelings and Emotions, ed. A. S. R. Manstead et al., pp. 136–157. Cambridge: Cambridge Univ. Press. [5]
———. 2004b. Which emotions can be induced by music? What are the underlying mechanisms? And how can we measure them? *J. New Music Res.* **33**:239–251. [5]
———. 2005. What are emotions? And how can they be measured? *Soc. Sci. Inf.* **44**:693–727. [5]
———. 2009. The dynamic architecture of emotion: Evidence for the component process model. *Cogn. Emot.* **23**:1307–1351. [5, 8]
Scherer, K. R., E. Clark-Polner, and M. Mortillaro. 2011. In the eye of the beholder? Universality and cultural specificity in the expression and perception of emotion. *Intl. J. Psychol.* **46**:401–435. [5]
Scherer, K. R., and E. Coutinho. 2013. How music creates emotion: A multifactorial process approach. In: The Emotional Power of Music, ed. R. Cochrane et al. Oxford: Oxford Univ. Press, in press. [5, 8]
Scherer, K. R., and H. Ellgring. 2007. Are facial expressions of emotion produced by categorical affect programs or dynamically driven by appraisal? *Emotion* **7**:113–130. [5]
Scherer, K. R., T. Johnstone, and G. Klasmeyer. 2003. Vocal expression of emotion. In: Handbook of the Affective Sciences, ed. R. J. Davidson et al., pp. 433–456. New York and Oxford: Oxford Univ. Press. [5]
Scherer, K. R., and A. Kappas. 1988. Primate vocal expression of affective states. In: Primate Vocal communication, ed. D. Todt et al., pp. 171–194. Heidelberg: Springer. [5]
Scherer, K. R., D. R. Ladd, and K. Silverman. 1984. Vocal cues to speaker affect: Testing two models. *J. Acoust. Soc. Am.* **76**:1346–1356. [5, 11]
Scherer, K. R., and J. Oshinsky. 1977. Cue utilization in emotion attribution from auditory stimuli. *Motiv. Emot.* **1**:331–346. [5]

Scherer, K. R., and K. Zentner. 2001. Emotional effects of music: Production rules. In: Music and Emotion: Theory and Research, ed. P. N. Juslin and J. A. Sloboda, pp. 361–392. Oxford: Oxford Univ. Press. [5, 8]

Scherer, K. R., M. R. Zentner, and D. Stern. 2004. Beyond surprise: The puzzle of infants' expressive reactions to expectancy violation. *Emotion* **4**:389–402. [5]

Schlaug, G., S. Marchina, and A. Norton. 2008. From singing to speaking: Why singing may lead to recovery of expressive language function in patients with Broca's aphasia. *Music Percep.* **25**:315–323. [14]

———. 2009. Evidence for plasticity in white matter tracts of patients with chronic Broca's aphasia undergoing intense intonation-based speech therapy. *Ann. NY Acad. Sci.* **1169**:385–394. [4, 14]

Schlaug, G., A. Norton, S. Marchina, L. Zipse, and C. Y. Wan. 2010. From singing to speaking: Facilitating recovery from nonfluent aphasia. *Future Neurology* **5**:657–665. [17]

Schmithorst, V. J., and S. K. Holland. 2003. The effect of musical training on music processing: A functional magnetic resonance imaging study in humans. *Neurosci. Lett.* **348**:65–68. [13]

Schnupp, J., I. Nelken, and A. King. 2010. Auditory Neuroscience : Making Sense of Sound. Cambridge, MA: MIT Press. [17]

Schön, D., R. Gordon, A. Campagne, et al. 2010. Similar cerebral networks in language, music and song perception. *Neuroimage* **51**:450–461. [13, 17]

Schorr, A. 2001. Appraisal: The evolution of an idea. In: Appraisal Processes in Emotion: Theory, Methods, Research, ed. K. R. Scherer et al., pp. 20–34. New York: Oxford Univ. Press. [5]

Schröder, M. 2003. Experimental study of affect bursts. *Speech Comm.* **40**:99–116. [5]

Schroeder, C. E., P. Lakatos, Y. Kajikawa, S. Partan, and A. Puce. 2008. Neuronal oscillations and visual amplification of speech. *Trends Cogn. Sci.* **12**:106–113. [9, 12, 17]

Schubert, E. 2001. Continuous measurement of self–report emotional response to music. In: Music and Emotion: Theory and Research, ed. P. N. Juslin and J. A. Sloboda, pp. 393–414. Oxford: Oxford Univ. Press. [8]

———. 2010. Continuous self-report methods. In: Handbook of Music and Emotion: Theory, Research, Applications, ed. P. N. Juslin and J. Sloboda, pp. 223–253. Oxford: Oxford Univ. Press. [8]

Schubotz, R. I. 2007. Prediction of external events with our motor system: Towards a new framework. *Trends Cogn. Sci.* **11**:211–218. [6, 13]

Schubotz, R. I., and D. Y. von Cramon. 2002. Predicting perceptual events activates corresponding motor schemes in lateral premotor cortex: An fMRI study. *NeuroImage* **15**:787–796. [15]

Schultz, W. 2002. Getting formal with dopamine and reward. *Neuron* **36**:241–263. [15]

———. 2006. Behavioral theories and the neurophysiology of reward. *Annu. Rev. Psychol.* **57**:87–115. [15]

Schusterman, R. J. 2008. Vocal learning in mammals with special emphasis on pinnipeds. In: Evolution of Communicative Flexibility: Complexity, Creativity, and Adaptability in Human and Animal Communication, ed. D. K. Oller and U. Griebel, pp. 41–70. Cambridge, MA: MIT Press. [20]

Schutz, M., and S. Lipscomb. 2007. Hearing gestures, seeing music: Vision influences perceived tone duration. *Perception* **36**:888–897. [18]

Schwartz, S. 1981. Steve Reich: Music as a gradual process, Part II. *Perspectives of New Music* **20**:225–286. [16]

Scientific American. 1996. Interview with Tod Machover: The composer from MIT's Media Lab discusses his Brain Opera. http://www.scientificamerican.com/article.cfm?id=interview-with-tod-machov. (accessed 6 Dec 2012). [16]

Scott, S. K., and I. S. Johnsrude. 2003. The neuroanatomical and functional organization of speech perception. *Trends Neurosci.* **26**:100–107. [9]

Scruton, R. 2009. The Aesthetics of Music. Oxford: Oxford Univ. Press. [5]

Searle, J. R. 1979. Expression and Meaning: Studies in the Theory of Speech Acts. Cambridge: Cambridge Univ. Press. [19]

———. 1980. Minds, brains and programs. *Behav. Brain Sci.* **3**:417–457. [16]

Seashore, C. 1938/1967. The Psychology of Music. New York: Dover. [5]

Sebanz, N., H. Bekkering, and G. Knoblich. 2006. Joint action: Bodies and minds moving together. *Trends Cogn. Sci.* **10**:70–76. [8, 13]

Seeger, A. 1987. Why Suyá Sing: A Musical Anthropology of an Amazonian People. Cambridge: Cambridge Univ. Press. [21]

———. 1994. Music and dance. In: Companion Encyclopedia of Anthropology: Humanity, Culture and Social Life, ed. T. Ingold, pp. 686–705. London: Routledge. [2]

Seifert, U. 2011. Signification and significance: Music, brain, and culture. Comment on "Towards a neural basis of processing musical semantics" by S. Koelsch. *Phys. Life Rev.* **8**:122–124. [8]

Seifert, U., and J. H. Kim. 2006. Musical meaning: Imitation and empathy. Proc. of the 9th Intl. Conf. on Music Perception and Cognition (ICMPC9), ed. M. Baroni et al., pp. 1061–1070. Bologna: ICMPC. [8]

Selkirk, E. O. 1984. Phonology and Syntax: The Relation between Sound and Structure. Cambridge, MA: MIT Press. [10, 11]

Seller, T. 1981. Midbrain vocalization centers in birds. *Trends Neurosci.* **12**:301–303. [20]

Seltzer, B., and D. N. Pandya. 1994. Parietal, temporal, and occipital projections to cortex of the superior temporal sulcus in the rhesus monkey: A retrograde tracer study. *J. Comp. Neurol.* **343**:445–463. [15]

Semal, C., and L. Demany. 1991. Dissociation of pitch from timbre in auditory short-term memory. *J. Acoust. Soc. Am.* **89**:2404–2410. [17]

Serafine, M. L., R. G. Crowder, and B. H. Repp. 1984. Integration of melody and text in memory for songs. *Cognition* **16**:285–303. [13]

Serafine, M. L., J. Davidson, R. G. Crowder, and B. H. Repp. 1986. On the nature of melody text integration in memory for songs. *J. Mem. Lang.* **25**:123–135. [13]

Seth, A. K., J. L. McKinstry, G. M. Edelman, and J. L. Krichmar. 2004. Visual binding through reentrant connectivity and dynamic synchronization in a brain-based device. *Cereb. Cortex* **14**:1185–1199. [14]

Seyfarth, R. M. 2005. Continuities in vocal communication argue against a gestural origin of language. *Behav. Brain Sci.* **28**:144–145. [19]

Seyfarth, R. M., D. L. Cheney, and T. J. Bergman. 2005. Primate social cognition and the origins of language. *Trends Cogn. Sci.* **9**:264–266. [19]

Seyfarth, R. M., D. L. Cheney, and P. Marler. 1980a. Monkey responses to three different alarm calls: Evidence of predator classification and semantic communication. *Science* **210**:801–803. [5, 19, 20]

———. 1980b. Vervet monkey alarm calls: Semantic communication in a free-ranging primate. *Anim. Behav.* **28**:1070–1094. [19]

Shallice, T., and R. Cooper. 2011. The Organisation of Mind. New York: Oxford Univ. Press. [15]

Shamma, S. A., M. Elhilali, and C. Micheyl. 2011. Temporal coherence and attention in auditory scene analysis. *Trends Neurosci.* **32**:114–123. [17]

Shenfield, T., S. E. Trehub, and T. Nakata. 2003. Maternal singing modulates infant arousal. *Psychol. Music* **31**:365–375. [18]

Shepard, R. N. 1982. Structural representations of musical pitch. In: Psychology of Music ed. D. Deutsch, pp. 343–390. New York: Academic. [18]

Shepherd, S. V., J. T. Klein, R. O. Deaner, and M. L. Platt. 2009. Mirroring of attention by neurons in macaque parietal cortex. *PNAS* **9**:9489–9494. [4]

Shergill, S. S., E. T. Bullmore, M. J. Brammer, et al. 2001. A functional study of auditory verbal imagery. *Psychol. Med.* **31**:241–253. [13]

Sherrington, C. S. 1906. The Integrative Action of the Nervous System. New Haven and London: Yale Univ. Press. [15]

Shevy, M. 2007. The mood of rock music affects evaluation of video elements differing in valence and dominance. *Psychomusicology* **19**:57–78. [7]

Shore, B. 1996. Culture in Mind: Cognition, Culture and the Problem of Meaning. New York: Oxford Univ. Press. [2]

Showers, M. J. C., and E. W. Lauer. 1961. Somatovisceral motor patterns in the insula. *J. Comp. Neurol.* **117**:107–115. [4]

Shubin, N., C. Tabin, and S. Carroll. 1997. Fossils, genes and the evolution of animal limbs. *Nature* **388**:639–648. [20]

———. 2009. Deep homology and the origins of evolutionary novelty. *Nature* **457**:818–823. [20]

Shultz, S., and R. Dunbar. 2010. Encephalization is not a universal macroevolutionary phenomenon in mammals but is associated with sociality. *PNAS* **107**:21,582–21,586. [18]

Silva, D. 2006. Acoustic evidence for the emergence of tonal contrast in contemporary Korean. *Phonology* **23**:287–308. [11]

Simmons-Stern, N. R., A. E. Budson, and B. A. Ally. 2010. Music as a memory enhancer in patients with Alzheimer's disease. *Neuropsychologia* **48**:3164–3167. [13]

Simões, C. S., P. V. Vianney, M. M. de Moura, et al. 2010. Activation of frontal neocortical areas by vocal production in marmosets. *Front. Integr. Neurosci.* **4**:123. [20]

Simon, A. 1978. Types and functions of music in the eastern highlands of West-Irian (New Guinea). *Ethnomus. Forum* **2**:441–455. [21]

Simons, D. J., and D. T. Levin. 1997. Change blindness. *Trends Cogn. Sci.* **1**:261–267. [8]

Simons, D. J., and R. A. Rensink. 2005. Change blindness: Past, present, and future. *Trends Cogn. Sci.* **9**:16–20. [8]

Simon-Thomas, E. R., D. J. Keltner, D. Sauter, L. Sinicropi-Yao, and A. Abramson. 2009. The voice conveys specific emotions: Evidence from vocal burst displays. *Emotion* **9**:838–846. [5]

Simonyan, K., and B. Horwitz. 2011. Laryngeal motor cortex and control of speech in humans. *Neuroscientist* **17**:197–208. [20]

Simonyan, K., B. Horwitz, and E. D. Jarvis. 2012. Dopamine regulation of human speech and bird song: A critical review. *Brain Lang.*PMID: 22284300. [20]

Simonyan, K., and U. Jürgens. 2003. Efferent subcortical projections of the laryngeal motorcortex in the rhesus monkey. *Brain Res.* **974**:43–59. [4, 20]

———. 2005a. Afferent cortical connections of the motor cortical larynx area in the rhesus monkey. *Neuroscience* **130**:133–149. [20]

———. 2005b. Afferent subcortical connections into the motor cortical larynx area in the rhesus monkey. *Neuroscience* **130**:119–131. [20]

Simpson, H. B., and D. S. Vicario. 1990. Brain pathways for learned and unlearned vocalizations differ in zebra finches. *J. Neurosci* **10**:1541–1556. [20]

Singer, T., and C. Lamm. 2009. The social neuroscience of empathy. *Ann. NY Acad. Sci.* **1156**:81–96. [6]

Singer, T., B. Seymour, J. O'Doherty, et al. 2004. Empathy for pain involves the affective but not the sensory components of pain. *Science* **303**:1157–1162. [4]

Singer, T., B. Seymour, J. P. O'Doherty, et al. 2006. Empathic neural responses are modulated by the perceived fairness of others. *Nature* **439**:466–469. [4]

Singh, L., J. L. Morgan, and C. Best. 2002. Infants' listening preferences: Baby talk or happy talk? *Infancy* **3**:365–394. [18]

Singh, L., S. Nestor, C. Parkih, and A. Yull. 2009. Influences of infant-directed speech on infant word recognition. *Infancy* **14**:654–666. [18]

Skipper, J. I., S. Goldin-Meadow, H. C. Nusbaum, and S. L. Small. 2007a. Speech-associated gestures, Broca's area, and the human mirror system. *Brain Lang.* **101**:260–277. [17]

———. 2009. Gestures orchestrate brain networks for language understanding. *Curr. Biol.* **19**:661–667. [17]

Skipper, J. I., H. C. Nusbaum, and S. L. Small. 2006. Lending a helping hand to hearing: Another motor theory of speech perception. In: Action to Language Via the Mirror Neuron System, ed. M. A. Arbib, pp. 250–285. Cambridge: Cambridge Univ. Press. [15]

Skipper, J. I., V. van Wassenhove, H. C. Nusbaum, and S. L. Small. 2007b. Hearing lips and seeing voices: How cortical areas supporting speech production mediate audiovisual speech perception. *Cereb. Cortex* **17**:2387–2399. [17]

Slater, A., P. C. Quinn, D. J. Kelly, et al. 2010. The shaping of the face space in early infancy: Becoming a native face processor. *Child Devel. Persp.* **4**:205–211. [18]

Slevc, L. R., R. C. Martin, A. C. Hamilton, and M. F. Joanisse. 2011. Speech perception, rapid temporal processing, and the left hemisphere: A case study of unilateral pure word deafness. *Neuropsychologia* **49**:216–230. [17]

Slevc, L. R., and A. Miyake. 2006. Individual differences in second language proficiency: Does musical ability matter? *Psychol. Sci.* **17**:675–681. [14, 17]

Slevc, L. R., and A. D. Patel. 2011. Meaning in music and language: Three key differences. *Phys. Life Rev.* **8**:110–111. [6, 8, 14]

Slevc, L. R., J. C. Rosenberg, and A. D. Patel. 2009. Making psycholinguistics musical: Self-paced reading time evidence for shared processing of linguistic and musical syntax. *Psychon. Bull. Rev.* **16**:374–381. [6, 14]

Sloboda, J. A. 2000. Individual differences in music performance. *Trends Cogn. Sci.* **4**:397–403. [1]

Sloboda, J. A. A. 1992. Empirical studies of emotional response to music. In: Cognitive Bases of Musical Communication, ed. M. R. Jones and S. Holleran, pp. 33–46. Washington, DC: American Psychological Association. [5]

———. 2010. Music in everyday life: The role of emotions. In: Handbook of Music and Emotion: Theory, Research, Applications, ed. P. N. Juslin and J. A. Sloboda, pp. 493–514. Oxford/New York: Oxford Univ. Press. [5]

Slocombe, K. E., T. Kaller, L. Turman, et al. 2010. Production of food-associated calls in wild male chimpanzees is dependent on the composition of the audience. *Behav. Ecol. Sociobiol.* **64**:1959–1966. [20]

Slocombe, K. E., and K. Zuberbühler. 2007. Chimpanzees modify recruitment screams as a function of audience composition. *PNAS* **104**:17,228–17,233. [21]

Sloman, A., R. Chrisley, and M. Scheutz. 2005. The architectural basis of affective states and processes. In: Who Needs Emotions, ed. J.-M. Fellous and M. A. Arbib, pp. 203–444. Oxford: Oxford Univ. Press. [8]

Smith, A. D. M., M. Schouwstra, B. de Boer, and K. Smith, eds. 2010. The Evolution of Language: EVOLANG8. Singapore: World Scientific Publ. [5]
Smith, E. C., and M. S. Lewicki. 2006. Efficient auditory coding. *Nature* **439**:978–982. [15]
Smith, N. A., and L. J. Trainor. 2011. Auditory stream segregation improves infants' selective attention to target tones amid distrators. *Infancy* **16**:655–668. [17]
Snijders, T. M., T. Vosse, G. Kempen, et al. 2009. Retrieval and unification of syntactic structure in sentence comprehension: An fMRI study using word-category ambiguity. *Cereb. Cortex* **19**:1493–1503. [9, 14]
Snyder, B. 2000. Music and Memory. Cambridge, MA: MIT Press. [8]
———. 2009. Memory for music. In: The Oxford Handbook of Music Psychology, ed. S. Hallam et al., pp. 107–117. Oxford Univ. Press. [8]
Snyder, J. S., and C. Alain. 2007. Toward a neurophysiological theory of auditory stream segregation. *Psychol. Bull.* **133**:780–799. [15]
Snyder, J. S., and E. W. Large. 2005. Gamma-band activity reflects the metric structure of rhythmic tone sequences. *Cogn. Brain Res.* **24**:117–126. [15]
Soley, G., and E. E. Hannon. 2010. Infants prefer the musical meter of their own culture: A cross-cultural comparison. *Devel. Psychol.* **46**:286–292. [18]
Solomyak, O., and A. Marantz. 2010. Evidence for early morphological decomposition in visual word recognition. *J. Cogn. Neurosci.* **22**:2042–2057. [9]
Sony Pictures Digital. The Social Network: Original Score (Trent Reznor and Atticus Ross) 2010. Available from http://www.thesocialnetwork-movie.com/awards/#/music. [7]
Southgate, V., M. H. Johnson, T. Osborne, and G. Csibra. 2009. Predictive motor activation during action observation in human infants. *Biology Lett.* **5**:769–772. [1]
Spaulding, M., M. A. O'Leary, and J. Gatesy. 2009. Relationships of Cetacea (*Artiodactyla*) among mammals: Increased taxon sampling alters interpretations of key fossils and character evolution. *PLoS ONE* **4**:e7062. [20]
Spencer, H. 1857. The Origin and Function of Music (reprinted in 1868 Essays: Scientific, Political, and Speculative). London: Williams and Norgate. [2, 14]
Spetner, N. B., and L. W. Olsho. 1990. Auditory frequency resolution in human infancy. *Child Devel.* **61**:632–652. [18]
Spielberg, S. (director), R. D. Zanuck, and D. Brown (producers). 1988. Jaws. US: Universal Pictures. [7]
Spiers, H. J., and E. A. Maguire. 2006. Spontaneous mentalizing during an interactive real world task. *Neuropsychologia* **44**:1674–1682. [8]
———. 2007. Decoding human brain activity during real-world experiences. *Trends Cogn. Sci.* **11**:356–365. [8]
Spreckelmeyer, K. N., M. Kutas, T. P. Urbach, E. Altenmüller, and T. F. Münte. 2006. Combined perception of emotion in pictures and musical sounds. *Brain Res.* **1070**:160–170. [7]
Sridharan, D., D. Levitin, C. Chafe, J. Berger, and V. Menon. 2007. Neural dynamics of event segmentation in music: Converging evidence for dissociable ventral and dorsal networks. *Neuron* **55**:521–532. [8]
Stam, R. 2000. Film Theory: An Introduction. Malden, MA: Blackwell. [7]
Steedman, M. 1999. Categorial grammar. In: The Encyclopedia of Cognitive Sciences, ed. R. Wilson and F. Keil. Cambridge, MA: MIT Press. [15]
———. 2000. Information structure and the syntax-phonology interface. *Ling. Inq.* **31**:649–689. [11]

Steels, L. 2010. Adaptive language games with robots. In: Computing Anticipatory Systems, ed. D. M. Dubois. Melville, NY: American Institute of Physics. [16]

Stefanatos, G. A. 2008. Speech perceived through a damaged temporal window: Lessons from word deafness and aphasia. *Seminars in Speech and Language* **29**:239–252. [17]

Stein, M., A. Simmons, J. Feinstein, and M. Paulus. 2007. Increased amygdala and insula activation during emotion processing in anxiety-prone subjects. *Am. J. Psychiatry* **164**:2. [6]

Steinbeis, N., and S. Koelsch. 2008a. Comparing the processing of music and language meaning using EEG and fMRI provides evidence for similar and distinct neural representations. *PLoS ONE* **3**:e2226. [6]

———. 2008b. Shared neural resources between music and language indicate semantic processing of musical tension-resolution patterns. *Cereb. Cortex* **18**:1169–1178. [6, 13, 14, 15]

Steinbeis, N., S. Koelsch, and J. A. A. Sloboda. 2006. The role of harmonic expectancy violations in musical emotions: Evidence from subjective, physiological, and neural responses. *J. Cogn. Neurosci.* **18**:1380–1393. [6, 14]

Steinke, W. R., L. L. Cuddy, and R. R. Holden. 1997. Dissociation of musical tonality and pitch memory from nonmusical cognitive abilities. *Can. J. Exp. Psychol.* **51**:316–334. [17]

Steinke, W. R., L. L. Cuddy, and L. S. Jakobson. 2001. Dissociations among functional subsystems governing melody recognition after right-hemisphere damage. *Cogn. Neuropsychol.* **18**:411–437. [13, 17]

Stern, T. 1957. Drum and whistle "languages": An analysis of speech surrogates. *Am. Anthropol.* **59**:487–506. [6]

Sternberg, R. S., ed. 1999. Handbook of Creativity. Cambridge: Cambridge Univ. Press. [16]

Stevens, K. N. 2000. Acoustic Phonetics (illustrated ed.). Cambridge, MA: MIT Press. [17]

Stewart, L. 2011. Characterizing congenital amusia. *Q. J. Exp. Psychol.* **64**:625–638. [17]

Stewart, L., K. von Kriegstein, J. D. Warren, and T. D. Griffiths. 2006. Music and the brain: Disorders of musical listening. *Brain* **129**:2533–2553. [14]

Stewart, L., V. Walsh, U. Frith, and J. Rothwell. 2001. Transcranial magnetic stimulation produces speech arrest but not song arrest. *Ann. NY Acad. Sci.* **930**:433–435. [14, 17]

Stilwell, R. J. 2007. The fantastical gap between diegetic and nondiegetic. In: Beyond the Soundtrack: Representing Music in Cinema ed. D. Goldmark et al., pp. 184–202. Berkeley: Univ. of California Press. [7]

Stivers, T., N. J. Enfield, P. Brown, et al. 2009. Universals and cultural variation in turn-taking in conversation. *PNAS* **106**:10,587–10,592. [3, 13, 17]

Stoeger, A. S., D. Mietchen, S. Oh, et al. 2012. An Asian elephant imitates human speech. *Curr. Biol.* **22**:2144–2148. [20]

Stokoe, W. C. 2001. Language in Hand: Why Sign Came Before Speech. Washington, DC: Gallaudet Univ. Press. [19]

Stone, R. M. 1998. Garland Encyclopedia of World Music: Africa. vol. 1. New York: Garland Publishing. [21]

Stout, D. 2011. Stone toolmaking and the evolution of human culture and cognition. *Philos. Trans. R. Soc. Lond. B* **366**:1050–1059. [1, 8]

Stout, D., and T. Chaminade. 2009. Making tools and making sense: Complex, intentional behaviour in human evolution. *Cambridge Arch. J.* **19**:85–96. [8]

Stout, D., and T. Chaminade. 2012. Stone tools, language and the brain in human evolution. *Philos. Trans. R. Soc. Lond. B* **367**:75–87. [21]

Stout, D., N. Toth, K. Schick, and T. Chaminade. 2008. Neural correlates of Early Stone Age toolmaking: Technology, language and cognition in human evolution. *Philos. Trans. R. Soc. Lond. B* **363**:1939–1949. [21]

Straube, T., A. Schulz, K. Geipel, H. J. Mentzel, and W. H. R. Miltner. 2008. Dissociation between singing and speaking in expressive aphasia: The role of song familiarity. *Neuropsychologia* **46**:1505–1512. [13]

Strick, P. L., R. P. Dum, and J. A. Fiez. 2009. Cerebellum and nonmotor function. *Annu. Rev. Neurosci.* **32**:413–434. [16]

Sturt, P., M. Pickering, and M. Crocker. 1999. Structural change and reanalysis difficulty in language comprehension. *J. Mem. Lang.* **40**:136–150. [6]

Suchman, L. A. 1987. Plans and Situated Actions. Cambridge: Cambridge Univ. Press. [16]

Suddendorf, T., and M. C. Corballis. 2007. The evolution of foresight: What is mental time travel, and is it unique to humans? *Behav. Brain Sci.* **30**:299–351. [19]

Suh, A., M. Paus, M. Kiefmann, et al. 2011. Mesozoic retroposons reveal parrots as the closest living relatives of passerine birds. *Nature Comm.* **2**:443. [20]

Sundberg, J., S. Patel, E. Björkner, and K. R. Scherer. 2011. Interdependencies among voice source parameters in emotional speech. *IEEE TAC* **99**:2423–2426. [5]

Sur, M., A. Angelucci, and J. Sharma. 1999. Rewiring cortex: The role of patterned activity in development and plasticity of neocortical circuits. *J. Neurobiol.* **41**:33–43. [15]

Sur, M., and C. A. Leamy. 2001. Development and plasticity of cortical areas and networks. *Nat. Rev. Neurosci.* **2**:251–261. [15]

Suthers, R. A., F. Goller, and C. Pytte. 1999. The neuromuscular control of birdsong. *Philos. Trans. R. Soc. Lond. B* **354**:927–939. [20]

Sutton, D., C. Larson, E. M. Taylor, and R. C. Lindeman. 1973. Vocalization in rhesus monkeys: Conditionability. *Brain Res.* **52**:225–231. [20]

Sutton, R. S. 1988. Learning to predict by the methods of temporal differences. *Mach. Learn.* **3**:9–44. [15]

Sutton, R. S., and A. G. Barto. 1998. Reinforcement Learning: An Introduction. Cambridge, MA: MIT Press. [15]

Svoboda, E., M. C. McKinnon, and B. Levine. 2006. The functional neuroanatomy of autobiographical memory: A meta-analysis. *Neuropsychologia* **44**:2189–2208. [13]

Swain, J. 1997. Musical Languages. New York: W. W. Norton. [14]

Syal, S., and B. L. Finlay. 2011. Thinking outside the cortex: Social motivation in the evolution and development of language. *Devel. Sci.* **14**:417–430. [18]

Taglialatela, J. P., J. L. Russell, J. A. Schaeffer, and W. D. Hopkins. 2011. Chimpanzee vocal signaling points to a multimodal origin of human language. *PLoS ONE* **6**:e18852. [20]

Tan, S. L., and M. P. Spackman. 2005. Listeners' judgments of the musical unity of structurally altered and intact musical compositions. *Psychol. Music* **33**:133–153. [8]

Tan, S. L., M. P. Spackman, and M. A. Bezdek. 2007. Viewers' interpretation of film characters' emotions: Effects of presenting film music before or after a character is shown. *Music Percep.* **25**:135–152. [7, 8]

Tan, S. L., M. P. Spackman, and E. M. Wakefield. 2008. Source of film music (diegetic or nondiegetic) affects viewers' interpretation of film. In: Tenth Intl. Conf. on Music Perception and Cognition. Hokkaido, Japan. [7]

Tanenhaus, M. K., M. J. Spivey-Knowlton, K. M. Eberhard, and J. C. Sedivy. 1995. Integration of visual and linguistic information in spoken language comprehension. *Science* **268**:1632–1634. [9]

Tanji, J. 2001. Sequential organization of multiple movements: Involvement of cortical motor areas. *Ann. Rev. Neurosci.* **24**:631–651. [4]

Tanji, J., and E. Hoshi. 2008. Role of the lateral prefrontal cortex in executive behavioral control. *Physiol. Rev.* **88**:37–57. [4]

Tanner, J. E., and R. Byrne. 1999. The development of spontaneous gestural communication in a group of zoo-living lowland gorillas. In: The Mentalities of Gorillas and Orangutans, Comparative Perspectives, ed. S. T. Parker et al., pp. 211–239. Cambridge: Cambridge Univ. Press. [19]

Tchernichovski, O., P. P. Mitra, T. Lints, and F. Nottebohm. 2001. Dynamics of the vocal imitation process: How a zebra finch learns its song. *Science* **291**:2564–2569. [20]

Teki, S., M. Grube, S. Kumar, and T. D. Griffiths. 2011. Distinct neural substrates of duration-based and beat-based auditory timing. *Neuroscience* **31**:3805–3812. [17]

Tembrock, G. 1975. Die Erforschung des tierlichen Stimmausdrucks (Bioakustik). In: *Biophonetik*, ed. F. Trojan. Mannheim: Bibliographisches Institut. [5]

Templeton, C. N., E. Greene, and K. Davis. 2005. Allometry of alarm calls: Black-capped chickadees encode information about predator size. *Science* **308**:1934–1937. [20]

Tenney, J. 1988. A History of Consonance and Dissonance. New York: Excelsior Music Publishing. [18]

Tennie, C., J. Call, and M. Tomasello. 2009. Ratcheting up the ratchet: On the evolution of cumulative culture. *Phil. Trans. R. Soc. Lond. B* **364**:2405–2415. [21]

Teramitsu, I., L. C. Kudo, S. E. London, D. H. Geschwind, and S. A. White. 2004. Parallel FoxP1 and FoxP2 expression in songbird and human brain predicts functional interaction. *J. Neurosci.* **24**:3152–3163. [20]

Teramitsu, I., and S. A. White. 2006. FoxP2 regulation during undirected singing in adult songbirds. *J. Neurosci.* **26**:7390–7394. [20]

Tervaniemi, M., S. Kruck, W. De Baene, et al. 2009. Top-down modulation of auditory processing: effects of sound context, musical expertise and attentional focus. *Eur. J. Neurosci.* **30**:1636–1642. [17]

Tettamanti, M., G. Buccino, M. C. Saccuman, et al. 2005. Listening to action-related sentences activates fronto-parietal motor circuits. *J. Cogn. Neurosci.* **17**:273–281. [4]

Tew, S., T. Fujioka, C. He, and L. Trainor. 2009. Neural representation of transposed melody in infants at 6 months of age. *Ann. NY Acad. Sci.* **1169**:287–290. [17]

Thiessen, E. D., E. Hill, and J. R. Saffran. 2005. Infant-directed speech facilitates word segmentation. *Infancy* **7**:53–71. [18]

Thiessen, E. D., and J. R. Saffran. 2003. When cues collide: Use of stress and statistical cues to word boundaries by 7- to 9-month-old infants. *Devel. Psychol.* **29**:706–716. [17]

Thompson, R. F., and J. E. Steinmetz. 2009. The role of the cerebellum in classical conditioning of discrete behavioral responses. *Neuroscience* **162**:732–755. [15]

Thompson, W. F., F. A. Russo, and D. Sinclair. 1994. Effects of underscoring on the perception of closure in filmed events. *Psychomusicology* **13**:9–27. [7]

Thompson-Schill, S. L., M. Bedny, and R. F. Goldberg. 2005. The frontal lobes and the regulation of mental activity. *Curr. Opin. Neurobiol.* **15**:219–224. [9]

Thompson-Schill, S. L., M. D'Esposito, G. K. Aguirre, and M. J. Farah. 1997. Role of left inferior prefrontal cortex in retrieval of semantic knowledge: A reevaluation. *PNAS* **94**:14,792–14,797. [9]

Thorndike, E. L. 1898. Animal intelligence: An experimental study of the associative process in animals. *Psychol. Rev.* Monograph Supplements, No. 8. [15]
Thorpe, L. A., and S. E. Trehub. 1989. Duration illusion and auditory grouping in infancy. *Devel. Psychol.* **25**:122–127. [18]
Thorpe, S., D. Fize, and C. Marlot. 1996. Speed of processing in the human visual system. *Nature* **381**:520–522. [9]
Thorpe, W. H. 1956. Learning and Instinct in Animals. London: Methuen. [20]
———. 1958. The learning of song patterns by birds, with especial reference to the song of the chaffinch *Fringilla coelebs*. *Ibis* **100**:535–570. [20]
Tian, B., and J. P. Rauschecker. 1998. Processing of frequency-modulated sounds in the cat's posterior auditory field. *J. Neurophysiol.* **79**:2629–2642. [14]
Tian, B., D. Reser, A. Durham, A. Kustov, and J. P. Rauschecker. 2001. Functional specialization in rhesus monkey auditory cortex. *Science* **292**:290–293. [17]
Tikka, P. 2008. Enactive Cinema: Simulatorium Eisensteinense. PhD Diss., Univ. of Art and Design, Helsinki. [8]
Tillmann, B. 2005. Implicit investigations of tonal knowledge in nonmusician listeners. *Ann. NY Acad. Sci.* **1060**:100–110. [6]
Tillmann, B., J. J. Bharucha, and E. Bigand. 2000. Implicit learning of tonality: A self-organizing approach. *Psychol. Rev.* **107**:885–913. [14, 17]
Tillmann, B., E. Bigand, N. Escoffier, and P. Lalitte. 2006. The influence of musical relatedness on timbre discrimination. *Eur. J. Cogn. Psychol.* **18**:343–358. [6]
Tillmann, B., and W. J. Dowling. 2007. Memory decreases for prose, but not for poetry. *Mem. Cogn.* **35**:628–639. [17, 21]
Tillmann, B., P. Janata, and J. J. Bharucha. 2003. Activation of the inferior frontal cortex in musical priming. *Cogn. Brain Res.* **16**:145–161. [14]
Tillmann, B., I. Peretz, and S. Samson. 2011. Neurocognitive approaches to memory in music: Music is memory. In: The Memory Process: Neuroscientific and Humanistic Perspectives, ed. S. Nalbantian et al., pp. 377–394. Cambridge, MA: MIT Press. [8, 17]
Tillmann, B., K. Schulze, and J. M. Foxton. 2009. Congenital amusia: A short-term memory deficit for non-verbal but not verbal sounds. *Brain Cogn.* **71**:259–264. [17]
Todorovic, A., F. van Ede, E. Maris, and F. P. De Lange. 2011. Prior expectation mediates neural adaptation to repeated sounds in the auditory the cortex: An MEG study. *J. Neurosci.* **31**:9118–9123. [17]
Tomasello, M. 1999a. Emulation learning and cultural learning. *Behav. Brain Sci.* **21**:703–704. [19]
———. 1999b. The human adaptation for culture. *Annu. Rev. Anthropol.* **28**:509–529. [1, 21]
———. 2008. Origins of Human Communications. Cambridge, MA: MIT Press. [18, 21]
———. 2009. Why We Cooperate (with responses by Carol Dweck, Joan Silk, Brian Skyrms, and Elizabeth Spelke). Cambridge, MA: MIT Press. [1]
Tomasello, M., J. Call, and B. Hare. 1998. Five primate species follow the visual gaze of conspecifics. *Anim. Behav.* **55**:1063–1069. [4]
Tomasello, M., J. Call, J. Warren, et al. 1997. The ontogeny of chimpanzee gestural signals. In: Evolution of Communication, ed. S. Wilcox et al., pp. 224–259. Amsterdam: John Benjamins. [1]
Tomasello, M., M. Carpenter, J. Call, T. Behne, and H. Moll. 2005. Understanding and sharing intentions: The origins of cultural cognition. *Behav. Brain Sci.* **28**:675–691. [6]
Tomkins, S. S. 1962. Affect, Imagery, Consciousness: The Positive Affects, vol. 1. New York: Springer. [5]

Tooby, J., and L. Cosmides. 1990. The past explains the present: Emotional adaptations and the structure of ancestral environments. *Ethol. Sociobiol.* **11**:375–424. [5]

Toukhsati, S. R., N. S. Rickard, E. Perini, K. T. Ng, and M. E. Gibbs. 2004. Noradrenaline involvement in the memory-enhancing effects of exposure to a complex rhythm stimulus following discriminated passive avoidance training in the young chick. *Behav. Brain Res.* **159**:105–111. [8]

Tourville, J. A., K. J. Reilly, and F. H. Guenther. 2008. Neural mechanisms underlying auditory feedback control of speech. *Neuroimage* **39**:1429–1443. [13]

Townsend, D., and T. Bever. 2001. Sentence Comprehension: The Integration of Habits and Rules. Cambridge, MA: MIT Press. [6]

Trainor, L. J. 1996. Infant preferences for infant-directed versus noninfant-directed play songs and lullabies. *Infant Behav. Dev.* **19**:83–92. [18]

Trainor, L. J., E. D. Clark, A. Huntley, and B. Adams. 1997. The acoustic basis of preferences for infant-directed singing. *Infant Behav. Dev.* **20**:383–396. [18]

Trainor, L. J., and K. Corrigal. 2010. Music acquisition and effects of musical experience. In: Music Perception: Current Research and Future Directions, ed. M. R. Jones et al., pp. 89–127. New York: Springer. [14, 17]

Trainor, L. J., X. Gao, J. Lei, K. Lehotovaara, and L. R. Harris. 2009. The primal role of the vestibular system in determining music rhythm. *Cortex* **45**:35–43. [13]

Trainor, L. J., and B. M. Heinmiller. 1998. The development of evaluative responses to music: Infants prefer to listen to consonance over dissonance. *Infant Behav. Dev.* **21**:77–88. [18]

Trainor, L. J., K. Lee, and D. J. Bosnyak. 2011. Cortical plasticity in 4-month-old infants: Specific effects of experience with musical timbres. *Brain Topography* **24**:192–203. [18]

Trainor, L. J., and S. E. Trehub. 1992. A comparison of infants' and adults' sensitivity to Western musical structure. *J. Exp. Psychol. Hum. Percept. Perform.* **18**:394–402. [17, 18]

———. 1993. Musical context effects in infants and adults: Key distance. *J. Exp. Psychol. Hum. Percept. Perform.* **19**:615–626. [18]

———. 1994. Key membership and implied harmony in Western tonal music: Developmental perspectives. *Percep. Psychophys.* **56**:125–132. [18]

Trainor, L. J., C. D. Tsang, and V. H. W. Cheung. 2002. Preference for sensory consonance in 2- and 4-month-old infants. *Music Percep.* **20**:187–194. [17, 18]

Trainor, L. J., L. Wu, and C. D. Tsang. 2004. Long-term memory for music: Infants remember tempo and timbre. *Devel. Sci.* **7**:289–296. [17, 18]

Travis, K. E., M. K. Leonard, T. T. Brown, et al. 2011. Spatiotemporal neural dynamics of word understanding in 12- to 18-month-old-infants. *Cereb. Cortex* **8**:1832–1839. [14]

Trehub, S. E. 2000. Human processing predispositions and musical universals. In: The Origins of Music, ed. N. L. Wallin et al., pp. 427–448. Cambridge, MA: MIT Press. [1, 18]

———. 2003. The developmental origins of musicality. *Nat. Neurosci.* **6**:669–673. [1, 6, 18]

Trehub, S. E., and E. E. Hannon. 2006. Infant music perception: Domain-general or domain-specific mechanisms? *Cognition* **100**:73–99. [14, 18]

———. 2009. Conventional rhythms enhance infants' and adults' perception of music. *Cortex* **45**:110–118. [18]

Trehub, S. E., J. Plantinga, and F. A. Russo. 2011. Maternal singing to infants in view or out of view. In: Society for Research in Child Development. Montreal. [18]

Trehub, S. E., E. G. Schellenberg, and S. B. Kamenetsky. 1999. Infants' and adults' perception of scale structure. *J. Exp. Psychol. Hum. Percept. Perform.* **25**:965–975. [18]

Trehub, S. E., B. A. Schneider, and J. L. Henderson. 1995. Gap detection in infants, children, and adults. *J. Acoust. Soc. Am.* **98**:2532–2541. [13, 18]

Trehub, S. E., B. A. Schneider, B. A. Morrongiello, and L. A. Thorpe. 1988. Auditory sensitivity in school-age children. *J. Exp. Child Psychol.* **46**:273–285. [18]

Trehub, S. E., and L. A. Thorpe. 1989. Infants' perception of rhythm: Categorization of auditory sequences by temporal structure. *Can. J. Psychol.* **43**:217–229. [13, 18, 17]

Trehub, S. E., L. A. Thorpe, and B. A. Morrongiello. 1987. Organizational processes in infants' perception of auditory patterns. *Child Devel.* **58**:741–749. [18]

Trehub, S. E., L. A. Thorpe, and L. J. Trainor. 1990. Infants' perception of good and bad melodies. *Psychomusicology* **9**:5–19. [18]

Trehub, S. E., and L. J. Trainor. 1998. Singing to infants: Lullabies and play songs. *Adv. Infancy Res.* **12**:43–77. [18]

Trehub, S. E., A. M. Unyk, S. B. Kamenetsky, et al. 1997. Mothers' and fathers' singing to infants. *Devel. Psychol.* **33**:500–507. [18]

Trehub, S. E., A. M. Unyk, and L. J. Trainor. 1993a. Adults identify infant-directed music across cultures. *Infant Behav. Dev.* **16**:193–211. [18]

———. 1993b. Maternal singing in cross-cultural perspective. *Infant Behav. Dev.* **16**:285–295. [18]

Treisman, A., and H. Schmidt. 1982. Illusory conjunctions in the perception of objects. *Cogn. Psychol.* **14**:107–141. [7]

Tremblay-Champoux, A., S. Dalla Bella, J. Phillips-Silver, M.-A. Lebrun, and I. Peretz. 2010. Singing proficiency in congenital amusia: Imitation helps. *Cogn. Neuropsychol.* **27**:463–476. [17]

Troster, A. I., S. P. Woods, and J. A. Fields. 2003. Verbal fluency declines after pallidotomy: An interaction between task and lesion laterality. *Appl. Neuropsychol.* **10**:69–75. [20]

Tsai, C. G., C. C. Chen, T. L. Chou, and J. H. Chen. 2010. Neural mechanisms involved in the oral representation of percussion music: An fMRI study. *Brain Cogn.* **74**:123–131. [13]

Tsunada, J., J. H. Lee, and Y. E. Cohen. 2011. Representation of speech categories in the primate auditory cortex. *J. Neurophysiol.* **105**:2631–2633. [14]

Tsuru, D. 1998. Diversity of spirit ritual performances among the Baka pygmies in south-eastern Cameroon. *Afr. Study Monogr.* **25**:47–84. [2]

Tu, H. W., M. S. Osmanski, and R. J. Dooling. 2011. Learned vocalizations in budgerigars, *Melopsittacus undulatus*: The relationship between contact calls and warble song. *J. Acoust. Soc. Am.* **129**:2289–2297. [20]

Turino, T. 2008. Musia as Social Life: The Politics of Participation. Chicago: Univ. of Chicago Press. [21]

Tyler, L. K., W. D. Marslen-Wilson, B. Randall, et al. 2011. Left inferior frontal cortex and syntax: Function, structure and behaviour in left-hemisphere damaged patients. *Brain* **134**:415–431. [9]

Tymoczko, D. 2006. The geometry of musical chords. *Science* **313**:72–74. [10]

———. 2011. A Geometry of Music. New York: Oxford Univ. Press. [10]

Umiltà, C. A. 1995. Manuale di Neuroscienze. Bologna: Società editrice Il Mulino. [4]

Umiltà, M. A., L. Escola, I. Intskirveli, et al. 2008. When pliers become fingers in the monkey motor system. *PNAS* **105**:2209–2213. [4, 19]

Umiltà, M. A., E. Kohler, V. Gallese, et al. 2001. I know what you are doing: A neurophysiological study. *Neuron* **31**:155–165. [4]

Ungerleider, L. G., and M. Mishkin. 1982. Two cortical visual systems. In: Analysis of Visual Behavior, ed. D. J. Ingle et al. Cambridge, MA: MIT Press. [15]

Unyk, A. M., S. E. Trehub, L. J. Trainor, and E. G. Schellenberg. 1992. Lullabies and simplicity: A cross-cultural perspective. *Psychol. Music* **20**:15–28. [18]

Valdesolo, P., J. Ouyang, and D. DeSteno. 2010. The rhythm of joint action: Synchrony promotes cooperative ability. *J. Exp. Soc. Psychol.* **46**:693–695. [13]

Valenstein, E. 1975. Nonlanguage disorders of speech reflect complex neurologic apparatus. *Geriatrics* **30**:117–121. [20]

Van Berkum, J. J. A., C. M. Brown, P. Zwitserlood, V. Kooijman, and P. Hagoort. 2005. Anticipating upcoming words in discourse: Evidence from ERPs and reading times. *J. Exp. Psychol. Learn. Mem. Cogn.* **31**:443–467. [9]

Van Berkum, J. J. A., Van den Brink, D., Tesink, C., Kos, M., and Hagoort, P. 2008. The neural integration of speaker and message. *J. Cogn. Neurosci.* **20**:580–591. [9]

Vanderwolf, C. H. 2007. The Evolving Brain: The Mind and the Neural Control of Behavior. Heidelberg: Springer Verlag. [16]

Van Overwalle, F. 2009. Social cognition and the brain: A meta-analysis. *Hum. Brain Mapp.* **30**:829–858. [13]

Van Parijs, S. M. 2003. Aquatic mating in pinnipeds: A review. *Aquatic Mammals* **29**:214–226. [21]

Van Petten, C., and M. Kutas. 1990. Interactions between sentence context and word frequency in event-related brain potentials. *Mem. Cogn.* **18**:380–393. [6]

van Praag, H., G. Kempermann, and F. H. Gage. 1999. Running increases cell proliferation and neurogenesis in the adult mouse dentate gyrus. *Nat. Neurosci.* **2**:266–270. [20]

van Wassenhove, V., K. W. Grant, and D. Poeppel. 2005. Visual speech speeds up the neural processing of auditory speech. *PNAS* **102**:1181–1186. [17]

Vargha-Khadem, F., D. G. Gadian, A. Copp, and M. Mishkin. 2005. FOXP2 and the neuroanatomy of speech and language. *Nat. Rev. Neurosci.* **6**:131–138. [20]

Vasilakis, P. N. 2005. Auditory roughness as means of musical expression. *Selected Rep. Ethnomusicol.* **12**:119–144. [18]

Västfjäll, D. 2010. Indirect perceptual, cognitive, and behavioural measures. In: Handbook of Music and Emotion: Theory, Research, Applications, ed. P. N. Juslin and J. A. Sloboda, pp. 255–278. Oxford: Oxford Univ. Press. [5]

Vates, G. E., B. M. Broome, C. V. Mello, and F. Nottebohm. 1996. Auditory pathways of caudal telencephalon and their relation to the song system of adult male zebra finches. *J. Comp. Neurol.* **366**:613–642. [20]

Vates, G. E., and F. Nottebohm. 1995. Feedback circuitry within a song-learning pathway. *PNAS* **92**:5139–5143. [20]

Vates, G. E., D. S. Vicario, and F. Nottebohm. 1997. Reafferent thalamo-"cortical" loops in the song system of oscine songbirds. *J. Comp. Neurol.* **380**:275–290. [20]

Vernes, S. C., P. L. Oliver, E. Spiteri, et al. 2011. Foxp2 regulates gene networks implicated in neurite outgrowth in the developing brain. *PLoS Genet.* **7**:e1002145. [20]

Verney, T., and P. Weintraub. 2002. Pre-Parenting: Nurturing Your Child from Conception. New York: Simon & Schuster. [2]

Verschure, P. F. M. J. 1992. Taking connectionism seriously: The vague promise of sub-symbolism and an alternative. In: Proc. of 14th Ann. Conf. of the Cognitive Science Society, pp. 653–658. Hillsdale, NJ: Erlbaum. [16]

———. 1993. Formal minds and biological brains. *IEEE Expert* **8**:66–75. [16]

Verschure, P. F. M. J. 1996. Connectionist explanation: Taking positions in the mind-brain dilemma. In: Neural Networks and a New Artificial Intelligence, ed. G. Dorffner, pp. 133–188. London: Thompson. [16]
———. 1998. Synthetic epistemology: The acquisition, retention, and expression of knowledge in natural and synthetic systems. In: Proc. of the World Congress on Computational Intelligence, pp. 147–153. Anchorage: IEEE. [16]
———. 2003. Environmentally mediated synergy between perception and behaviour in mobile robots. *Nature* **425**:620–624. [16]
———. 2012. The distributed adaptive control theory of the mind, brain, body nexus. *BICA* **1**:55–72. [8, 16]
Verschure, P. F. M. J., and P. Althaus. 2003. A real-world rational agent: Unifying old and new AI. *Cognitive Science* **27**:561–590. [16]
Verschure, P. F. M. J., and R. Pfeifer. 1992. Categorization, representations, and the dynamics of system-environment interaction: A case study in autonomous systems. In: From Animals to Animats: Proc. of the Intl. Conf. on Simulation of Adaptive Behavior, ed. J. A. Meyer et al., pp. 210–217. Cambridge, MA: MIT Press. [16]
Verschure, P. F. M. J., T. Voegtlin, and R. J. Douglas. 2003. Environmentally mediated synergy between perception and behaviour in mobile robots. *Nature* **425**:620–624. [16]
Vick, S.-J., B. M. Waller, L. A. Parr, M. Smith Pasqualini, and K. A. Bard. 2007. A cross-species comparison of facial morphology and movement in humans and chimpanzees using the facial action coding system (FACS). *J. Nonverb. Behav.* **31**:1–20. [1]
Vigneau, M., V. Beaucousin, P. Y. Herve, et al. 2010. What is right-hemisphere contribution to phonological, lexico-semantic, and sentence processing? Insights from a meta-analysis. *NeuroImage* **54**:577–593. [9]
Vihman, M. M. 1986. Individual differences in babbling and early speech: Predicting to age three. In: Precursors of Early Speech, ed. B. Lindblom and R. Zetterström, pp. 95–112. New York: Stockton Press. [20]
Visalberghi, E., and D. Fragaszy. 2001. Do monkeys ape? Ten years after. In: Imitation in Animals and Artifacts, ed. K. Dautenhahn and C. Nehaniv, pp. 471–500. Cambridge, MA: MIT Press. [4, 19]
Voelkl, B., and L. Huber. 2007. Imitation as faithful copying of a novel technique in marmoset monkeys. *PLoS ONE* **2**:e611. [19]
Voeltz, F. K. E., and C. Kilian-Hatz. 2001. Ideophones. Amsterdam: John Benjamins. [11]
Volkova, A., S. E. Trehub, and E. G. Schellenberg. 2006. Infants' memory for musical performances. *Devel. Sci.* **9**:584–590. [18]
von Cramon, D., and U. Jürgens. 1983. The anterior cingulate cortex and the phonatory control in monkey and man. *Neurosci. Biobehav. Rev.* **7**:423–425. [20]
Vosse, T., and G. Kempen. 2000. Syntactic structure assembly in human parsing: A computational model based on competitive inhibition and lexicalist grammar. *Cognition* **75**:105–143. [9]
Vousden, J. I., G. D. A. Brown, and T. A. Harley. 2000. Serial control of phonology in speech production: A hierarchical model. *Cogn. Psychol.* **41**:101–175. [15]
Vygotsky, L. 1978. Mind in Society: The Development of Higher Psychological Processes. Cambridge, MA: Harvard Univ. Press. [21]
Wagner, A. D., E. J. Pare-Blagoev, J. Clark, and R. A. Poldrack. 2001. Recovering meaning: Left prefrontal cortex guides controlled semantic retrieval. *Neuron* **31**:329–338. [9]
Waldstein, R. S. 1990. Effects of postlingual deafness on speech production: Implications for the role of auditory feedback. *J. Acoust. Soc. Am.* **88**:2099–2114. [20]

Wallace, W. T. 1994. Memory for music: Effect of melody on recall of text. *J. Exp. Psychol. Learn. Mem. Cogn.* **20**:1471–1485. [13]
Waller, B. M., and R. I. M. Dunbar. 2005. Differential behavioural effects of silent bared teeth display and relaxed open mouth display in chimpanzees, *Pan troglodytes*. *Ethology* **111**:129–142. [1]
Waller, B. M., M. Lembeck, P. Kuchenbuch, A. M. Burrows, and K. Liebal. 2012. GibbonFACS: A muscle-based facial movement coding system for Hylobatids. *Intl. J. Primatol.* **33**:809–821. [1]
Wallin, N., B. Merker, and S. Brown, eds. 2000. The Origins of Music. Cambridge, MA: MIT Press. [3, 5]
Wan, C. Y., and G. Schlaug. 2010. Music making as a tool for promoting brain plasticity across the life span. *Neuroscientist* **16**:566–577. [14]
Wang, D. L., and G. J. Brown. 2006. Computational Auditory Scene Analysis: Principles, Algorithms, and Applications. Hoboken: Wiley-IEEE Press. [17]
Wang, L., M. C. M. Bastiaansen, Y. Yang, and P. Hagoort. 2012. Information structure influences depth of semantic processing: Event-related potential evidence for the Chomsky illusion. *PLoS ONE* **7**:e47917. [9]
Warner-Schmidt, J., and R. Duman. 2006. Hippocampal neurogenesis: Opposing effects of stress and antidepressant treatment. *Hippocampus* **16**:239–249. [6]
Warren, R. M., D. A. Gardner, B. S. Brubaker, and J. A. Bashford, Jr. 1991. Melodic and nonmelodic sequences of tone: Effects of duration on perception. *Music Percep.* **8**:277–289. [17]
Warren, W. C., D. F. Clayton, H. Ellegren, et al. 2010. The genome of a songbird. *Nature* **464**:757–762. [20]
Warrier, C. M., and R. J. Zatorre. 2004. Right temporal cortex is critical for utilization of melodic contextual cues in a pitch constancy task. *Brain* **127**:1616–1625. [13]
Wasserman, K., J. Manzolli, K. Eng, and P. F. M. J. Verschure. 2003. Live soundscape composition based on synthetic emotions: Using music to communicate between an interactive exhibition and its visitors. *IEEE MultiMedia* **10**:82–90. [16]
Webster, D. B., A. N. Popper, and R. R. Fay. 1992. The Mammalian Auditory Pathway: Neuroanatomy. Heidelberg: Springer. [20]
Weikum, W. M., A. Vouloumanos, J. Navarra, et al. 2007. Visual language discrimination in infancy. *Science* **316**:1159. [18]
Weinberger, N. M. 1999. Music and the auditory system. In: The Psychology of Music, 2nd edition, ed. D. Deutsch. New York: Academic. [10]
———. 2004. Specific long-term memory traces in primary auditory cortex. *Nat. Rev. Neurosci.* **5**:279–290. [15]
Weisman, R. G., L. Balkwill, M. Hoeschele, et al. 2010. Absolute pitch in boreal chickadees and humans: Exceptions that test a phylogenetic rule. *Learning and Motivation* **41**:156–173. [17]
Weisman, R. G., M. Njegovan, C. Sturdy, et al. 1998. Frequency-range discriminations: Special and general abilities in zebra finches (*Taeniopygia guttata*) and humans (*Homo sapiens*). *J. Comp. Psychol.* **112**:244–258. [17]
Weiss, M. W., S. E. Trehub, and E. G. Schellenberg. 2012. Something in the way she sings: Enhance memory for vocal melodies. *Psychol. Sci.* **20**:1076–1078. [18, 21]
Wellman, H. M., F. Fang, and C. C. Peterson. 2011. Sequential progressions in a theory-of-mind scale: Longitudinal perspectives. *Child Devel.* **82**:780–792. [21]
Wemmer, C., H. Mishra, and E. Dinerstein. 1985. Unusual use of the trunk for sound production in a captive Asian elephant: A second case. *J. Bombay Nat. History Soc.* **82**:187. [20]

Wennerstrom, A. 2001. The Music of Everyday Speech: Prosody and Discourse Analysis. Oxford: Oxford Univ. Press. [5]

Werker, J. F., and R. C. Tees. 1984. Cross-language speech perception: Evidence for perceptual reorganization during the first year of life. *Infant Behav. Dev.* 7:49–63. [18]

Whalen, D. H., A. G. Levitt, and Q. Wang. 1991. Intonational differences between the reduplicative babbling of French- and English-learning infants. *J. Child Lang.* 18:501–516. [18]

White, S. A., S. E. Fisher, D. H. Geschwind, C. Scharff, and T. E. Holy. 2006. Singing mice, songbirds, and more: Models for FOXP2 function and dysfunction in human Speech and language. *J. Neurosci.* 26:10,376 –10,379. [21]

Whiten, A. 1998. Imitation of the sequential structure of actions by chimpanzees, *Pan troglodytes*. *J. Comp. Psychol.* 112:270–281. [4]

Whiten, A., J. Goodall, C. McGrew, et al. 1999. Cultures in chimpanzees. *Nature* 399:682–685. [19, 21]

Whiten, A., V. Horner, and F. B. de Waal. 2005. Conformity to cultural norms of tool use in chimpanzees. *Nature* 437:737–40. [19]

Whiten, A., N. McGuigan, S. Marshall-Pescini, and L. M. Hopper. 2009. Emulation, imitation, over-imitation and the scope of culture for child and chimpanzee. *Philos. Trans. R. Soc. Lond. B* 364:2417–2428. [4]

Wicker, B., C. Keysers, J. Plailly, et al. 2003. Both of us disgusted in my insula: The common neural basis of seeing and feeling disgust. *Neuron* 40:655–664. [4]

Widdess, R. 2012. Music, meaning and culture. *Empir. Musicol. Rev.* 7:88–94. [2]

Wild, J. M. 1994. The auditory-vocal-respiratory axis in birds. *Brain Behav. Evol.* 44:192–209. [20]

———. 1997. Neural pathways for the control of birdsong production. *J. Neurobiol.* 33:653–670. [20]

Wild, J. M., D. Li, and C. Eagleton. 1997. Projections of the dorsomedial nucleus of the intercollicular complex (DM) in relation to respiratory-vocal nuclei in the brainstem of pigeon (*Columba livia*) and zebra finch (*Taeniopygia guttata*). *J. Comp. Neurol.* 377:392–413. [20]

Wildgruber, D., H. Ackermann, U. Klose, B. Kardatzki, and W. Grodd. 1996. Functional lateralization of speech production at primary motor cortex: A fMRI study. *NeuroReport* 7:2791–2795. [13]

Willems, R. M., K. Clevis, and P. Hagoort. 2011. Add a picture for suspense: Neural correlates of the interaction between language and visual information in the perception of fear. *Social Cogn. Affec. Neurosci.* 6:404–416. [8]

Willems, R. M., A. Özyürek, and P. Hagoort. 2007. When language meets action: The neural integration of gesture and speech. *Cereb. Cortex* 17:2322–2333. [9, 12]

———. 2008. Seeing and hearing meaning: Event-related potential and functional magnetic resonance imaging evidence of word versus picture integration into a sentence context. *J. Cogn. Neurosci.* 20:1235–1249. [9]

Williams, H., L. A. Crane, T. K. Hale, M. A. Esposito, and F. Nottebohm. 1992. Right-side dominance for song control in the zebra finch. *J. Neurobiol.* 23:1006–1020. [20]

Williams, H., and F. Nottebohm. 1985. Auditory repsonses in avian vocal motor neurons: A motor theory for song perception in birds. *Science* 229:279–282. [20]

Wilson, S. J., D. F. Abbott, D. Lusher, E. C. Gentle, and G. D. Jackson. 2010. Finding your voice: A singing lesson from functional imaging. *Hum. Brain Mapp.* 2115–2130. [13]

Wilson, S. J., K. Parsons, and D. C. Reutens. 2006. Preserved singing in aphasia: A case study of the efficacy of melodic intonation therapy. *Music Percep.* 24:23–35. [13]

Wiltermuth, S. S., and C. Heath. 2009. Synchrony and cooperation. *Psychol. Sci.* **20**:1–5. [6, 13]
Winkler, I., S. L. Denham, and I. Nelken. 2009a. Modeling the auditory scene: Predictive regularity representations and perceptual objects. *Trends Cogn. Sci.* **13**:532–540. [15]
Winkler, I., G. P. Háden, O. Ladinig, I. Sziller, and H. Honing. 2009b. Newborn infants detect the beat in music. *PNAS* **106**:2468–2471. [13, 18, 19]
Winkler, I., E. Kushnerenko, J. Horvath, et al. 2003. Newborns can organize the auditory world. *PNAS* **100**:11,812–11,815. [17]
Winkler, I., R. Takegata, and E. Sussman. 2005. Event-related brain potentials reveal multiple stages in the perceptual organization of sound. *Cogn. Brain Res.* **25**:291–299. [15]
Winkler, T. 2001. Composing Interactive Music: Techniques and Ideas Using Max. Cambridge, MA: MIT Press. [16]
Wise, R. J., J. Greene, C. Buchel, and S. K. Scott. 1999. Brain regions involved in articulation. *Lancet* **353**:1057–1061. [20]
Wiskott, L., and T. Sejnowski. 2002. Slow feature analysis: Unsupervised learning of invariances. *Neural Comp.* **14**:715–770. [15]
Wittgenstein, L. 1953. Philosophical Investigations (translated by G. E. M. Anscombe). Oxford: Basil Blackwell. [21]
———. 1984. *Philosophische Untersuchungen.* Frankfurt: Suhrkamp. [6]
Wohlschläger, A., M. Gattis, and H. Bekkering. 2003. Action generation and action perception in imitation: An instance of the ideomotor principle. *Philos. Trans. R. Soc. Lond. B* **358**:501–515. [19]
Wolf, A. W., B. Lozoff, S. Latz, and R. Paludetto. 1996. Parental theories in the management of young children's sleep in Japan, Italy, and the United States. In: Parents' Cultural Belief Systems: Their Origins, Expressions, and Consequences, ed. S. Harkness and C. M. Super, pp. 364–384. New York: Guilford Press. [18]
Wolmetz, M., D. Poeppel, and B. Rapp. 2011. What does the right hemisphere know about phoneme categories? *J. Cogn. Neurosci.* **23**:552–569. [17]
Wolpert, D. M., K. Doya, and M. Kawato. 2003. A unifying computational framework for motor control and social interaction. *Philos. Trans. R. Soc. Lond. B* **358**:593–602. [15, 17]
Wolpert, D. M., and M. Kawato. 1998. Multiple paired forward and inverse models for motor control. *Neural Netw.* **11**:1317–1329. [15, 17]
Womelsdorf, T., J. M. Schoffelen, R. Oostenveld, et al. 2007. Modulation of neuronal interactions through neuronal synchronization. *Science* **316**:1609–1612. [9]
Wong, P. C. M., and R. L. Diehl. 2002. How can the lyrics of a song in a tone language be understood? *Psychol. Music* **30**:202–209. [11]
Wong, P. C. M., L. M. Parsons, M. Martinez, and R. L. Diehl. 2004. The role of the insular cortex in pitch pattern perception: The effect of linguistic contexts. *J. Neurosci.* **24**:9153–9160. [13]
Woodburn, J. 1982. Egalitarian societies. *Man* **17**:431–451. [2]
Woolley, S. C., and A. J. Doupe. 2008. Social context-induced song variation affects female behavior and gene expression. *PLoS Biol.* **6**:e62. [20]
Wray, A. 1998. Protolanguage as a holistic system for social interaction. *Lang. Comm.* **18**:47–67. [19, 21]
Wright, A. A., J. J. Rivera, S. H. Hulse, M. Shyan, and J. J. Neiworth. 2000. Music perception and octave generalization in rhesus monkeys. *J. Exp. Psychol. Gen.* **129**:291–307. [8, 17, 18, 21]

Wright, B. A., D. V. Buonomano, H. W. Mahncke, and M. M. Merzenich. 1997. Learning and generalization of auditory temporal-interval discrimination in humans. *J. Neurosci.* **17**:3956–3963. [17]

Wright, B. A., and Y. Zhang. 2009. A review of the generalization of auditory learning. *Philos. Trans. R. Soc. Lond. B* **364**:301–311. [17]

Wundt, W. 1900. *Völkerpsychologie. Eine Untersuchung der Entwicklungsgesetze von Sprache, Mythos und Sitte*, vol. 1: *Die Sprache*. Leipzig: Kröner. [5]

Wyss, R., P. Konig, and P. F. M. J. Verschure. 2003a. Invariant representations of visual patterns in a temporal population code. *PNAS* **100**:324–329. [15, 16]

———. 2003b. Involving the motor system in decision making. *Proc. Roy. Soc. Lond. B* **10**:1098. [16]

———. 2006. A model of the ventral visual system based on temporal stability and local memory. *PLoS Biol.* **4**:120. [15, 16]

Xenakis, I. 1992. Formalized Music: Thought and Mathematics in Music. New York: Pendragon Press. [16]

Xiang, H., H. M. Fonteijn, D. G. Norris, and P. Hagoort. 2010. Topographical functional connectivity pattern in the Perisylvian language networks. *Cereb. Cortex* **20**:549–560. [9]

Xiao, Z., and N. Suga. 2002. Modulation of cochlear hair cells by the auditory cortex in the mustached bat. *Nat. Neurosci.* **5**. [17]

Yamadori, A., Y. Osumi, S. Masuhara, and M. Okubo. 1977. Preservation of singing in Broca's aphasia. *J. Neurol. Neurosurg. Psychiatr.* **40**:221–224. [13, 14]

Yamaguchi, A. 2001. Sex differences in vocal learning in birds. *Nature* **411**:257–258. [20]

Yarkoni, T., R. A. Poldrack, T. E. Nichols, D. C. Van Essen, and T. D. Wager. 2011. Large-scale automated synthesis of human functional neuroimaging data. *Nat. Methods* **8**:665–670. [13]

Yerkes, R. M., and A. W. Yerkes. 1929. The Great Apes. New Haven: Yale Univ. Press. [20]

Yin, P., J. B. Fritz, and S. A. Shamma. 2010. Do ferrets perceive relative pitch? *J. Acoust. Soc. Am.* **127**:673–680. [17]

Yip, M. J. 2006. The search for phonology in other species. *Trends Cogn. Sci.* **10**:442–446. [4, 11, 20]

———. 2010. Structure in human phonology and in birdsong: A phonologists's perspective. In: Birdsong, Speech and Language. Converging Mechanisms, ed. J. J. Bolhuis and M. Everaert. Cambridge, MA: MIT Press. [11]

Yoshida, K. A., J. R. Iversen, A. D. Patel, et al. 2010. The development of perceptual grouping biases in infancy: A Japanese–English cross-linguistic study. *Cognition* **115**:356–361. [18]

Young, R. M. 1970. Mind, Brain and Adaptation in the Nineteenth Century: Cerebral Localization and its Biological Context from Gall to Ferrier. Oxford: Oxford Univ. Press. [15]

Yu, A. C., and D. Margoliash. 1996. Temporal hierarchical control of singing in birds. *Science* **273**:1871–1875. [20]

Zacks, J. M., and J. P. Magliano. 2011. Film, narrative, and cognitive neuroscience. In: Art and the Senses, ed. F. Bacci and D. Melcher, pp. 435–454. Oxford: Oxford Univ. Press. [8]

Zacks, J. M., N. K. Speer, K. M. Swallow, T. S. Braver, and J. R. Reynolds. 2007. Event perception: A mind-brain perspective. *Psychol. Rev.* **133**:273–293. [8]

Zacks, J. M., N. K. Speer, K. N. Swallow, and C. J. Maley. 2010. The brain's cutting room floor: Segmentation of narrative cinema. *Front. Human Neurosci.* **4**:1–15. [8]

Zacks, J. M., and K. M. Swallow. 2007. Event segmentation. *Curr. Dir. Psychol. Sci.* **16**:80–84. [8]

Zald, D. H., and R. J. J. Zatorre. 2011. Music. In: Neurobiology of Sensation and Reward, ed. J. A. Gottfried. CRC Press. [8, 15]

Zangwill, N. 2004. Against emotion: Hanslick was right about MUSIC. *Br. J. Aesthetics* **44**:29–43.

Zarate, J. M., K. Delhommeau, S. Wood, and R. J. Zatorre. 2010. Vocal accuracy and neural plasticity following micromelody-discrimination training. *PLoS ONE* **5**:e11181. [17]

Zarate, J. M., and R. J. Zatorre. 2008. Experience-dependent neural substrates involved in vocal pitch regulation during singing. *Neuroimage* **40**:1871–1887. [13, 17]

Zarco, W., H. Merchant, L. Prado, and J. C. Mendez. 2009. Subsecond timing in primates: Comparison of interval production between human subjects and rhesus monkeys. *J. Neurophysiol.* **102**:3191–3202. [17, 19]

Zarlino, G. 1558. *Le Istituioni Harmoniche*. Venice: Franceschi. [10]

Zatorre, R. J. 2007. There's more to auditory cortex than meets the ear. *Hearing Res.* **229**:24–30. [17]

Zatorre, R. J., P. Belin, and V. B. Penhune. 2002. Structure and function of auditory cortex: Music and speech. *Trends Cogn. Sci.* **6**:37–46. [14]

Zatorre, R. J., M. Bouffard, and P. Belin. 2004. Sensitivity to auditory object features in human temporal neocortex. *Neuroscience* **24**:3637–3642. [17]

Zatorre, R. J., J. L. Chen, and V. B. Penhune. 2007. When the brain plays music: Auditory-motor interactions in music perception and production. *Nat. Rev. Neurosci.* **8**:547–558. [8, 17]

Zatorre, R. J., K. Delhommeau, and J. M. Zarate. 2012a. Modulation of auditory cortex response to pitch variation following training with microtonal melodies. *Front. Psychol.* **3**:544. [17]

Zatorre, R. J., R. D. Fields, and H. Johansen-Berg. 2012b. Plasticity in gray and white: Neuroimaging changes in brain structure during learning. *Nat. Neurosci.* **15**:528–536. [17]

Zatorre, R. J., and J. T. Gandour. 2007. Neural specializations for speech and pitch: Moving beyond the dichotomies. *Philos. Trans. R. Soc. Lond. B* **363**:1087–1104. [14]

Zatorre, R. J., and A. R. Halpern. 2005. Mental concerts: Musical imagery and auditory cortex. *Neuron* **47**:9–12. [13]

Zatorre, R. J., A. R. Halpern, and M. Bouffard. 2010. Mental reversal of imagined melodies: A role for the posterior parietal cortex. *J. Cogn. Neurosci.* **22**:775–789. [17]

Zatorre, R. J., A. R. Halpern, D. W. Perry, E. Meyer, and A. C. Evans. 1996. Hearing in the mind's ear: A PET investigation of musical imagery and perception. *J. Cogn. Neurosci.* **8**:29–46. [13]

Zeigler, H. P., and P. Marler, eds. 2004. Behavioral Neurobiology of Birdsong, vol. 1016. New York: NY Acad. Sci. [20]

———, eds. 2008. Neuroscience of Birdsong. Cambridge: Cambridge Univ. Press. [20]

Zeng, S. J., T. Szekely, X. W. Zhang, et al. 2007. Comparative analyses of song complexity and song-control nuclei in fourteen oscine species. *Zoolog. Sci.* **24**:1–9. [20]

Zentner, M., and T. Eerola. 2010a. Rhythmic engagement with music in infancy. *PNAS* **107**:5768–5773. [13, 18]

———. 2010b. Self-report measures and models of musical emotions. In: Handbook of Music and Emotion: Theory, Research, Applications, ed. P. N. Juslin and J. A. Sloboda. Oxford: Oxford Univ. Press. [5]

Zentner, M., D. Grandjean, and K. R. Scherer. 2008. Emotions evoked by the sound of music: Characterization, classification, and measurement. *Emotion* **8**:494–521. [5, 6]

Zentner, M., and J. Kagan. 1998. Infants' perception of consonance and dissonance in music. *Infact Behav. Devel.* **21**:483–492. [18]

Zhang, S. P., R. Bandler, and P. J. Davis. 1995. Brain stem integration of vocalization: Role of the nucleus retroambigualis. *J. Neurophysiol.* **74**:2500–2512. [20]

Zilles, K., and K. Amunts. 2009. Receptor mapping: Architecture of the human cerebral cortex. *Curr. Opin. Neurol.* **22**:331–339. [15]

Zion Golumbic, E. M., D. Poeppel, and C. E. Schroeder. 2012. Temporal context in speech processing and attentional stream selection: A behavioral and neural perspective. *Brain Lang.* **122**:151–261. [17]

Zukow-Goldring, P. 1996. Sensitive caregivers foster the comprehension of speech: When gestures speak louder than words. *Early Dev. Parent.* **5**:195–211. [21]

———. 2006. Assisted imitation: Affordances, effectivities, and the mirror system in early language development. In: Action to Language via the Mirror Neuron System, ed. M. A. Arbib, pp. 469–500. Cambridge: Cambridge Univ. Press. [21]

———. 2012. Assisted imitation: First steps in the seed model of language development. *Lang. Sci.* **34**:569–582. [1]

Zwaan, R. A., and G. A. Radvansky. 1998. Situation models in language comprehension and memory. *Psychol. Bull.* **123**:162. [8]

Zwitserlood, P. 1989. The locus of the effects of sentential-semantic context in spoken-word processing. *Cognition* **32**:25–64. [9]

Subject Index

absolute pitch 26, 428, 443, 454, 470
  in infancy 465, 466
action 3, 4, 204, 207, 208, 229, 303, 309, 310, 315, 375, 384, 397, 399, 400, 408, 413, 418, 436, 485
  attribution 74–77
  coordination 19, 77, 103, 159, 167, 542
  recognition model 379, 388
  sequencing 72, 73, 92, 101, 105, 148
action–perception cycle 4–12, 75, 207–210, 218
ADA: intelligent space 393, 395, 409–411, 413
aesthetic emotions 12, 32, 107, 128–131, 138, 211, 215, 216
affect bursts 121–124, 132–134, 137–139
affective meaning 22, 51, 214, 502
affective prosody 153, 160, 172
African Gray parrots 495, 514, 515, 545, 546
agrammatic aphasia 389
alliteration 23, 271, 278
allophony 277, 279
amusia 106, 319, 320, 334, 343, 447–449
  congenital 334, 338–340, 343, 447–449
  verbal 523
amygdala 26, 96, 104, 161–163, 198, 206, 212, 333, 371, 375, 406
anthropocentrism 560–562
anthropology 50, 51, 77, 331
anticipation. *See* expectation
anticipatory schemas 208
anxiety 14, 109, 163, 166, 189
aphasia 106, 334, 451, 486
  agrammatic 389
  auditory 526, 532
  Broca's 152, 245, 322, 350
  nonfluent 331, 354, 454
  song 524, 532
  verbal 523, 524, 526
appraisal 114–117, 131, 210, 211, 215, 217. *See also* component process model (CPM)
apraxia 106
arcuate fasciculus 29, 70, 106, 241, 248–250, 319, 320, 334, 345, 351, 355, 449, 452, 531, 532, 547
AreaX 523, 526–528
artificial intelligence 396–398, 402, 412
Artist, The 179
associations 111, 154, 173–176, 183–185, 188–190. *See also* congruence association model with working narrative (CAM-WN)
attention 6, 25, 114, 128, 173, 186, 195, 224, 309, 310, 426
  neural basis of 429–432
  shared 88, 89
audiovisual speech 422, 427, 437, 438
auditory agnosia 532
auditory brain areas
  cortex 362, 367, 372, 381, 425–428, 437, 439, 442–444
  forebrain pathway 530, 531, 547
  interspecies comparison 505–507
auditory memory 445–447
auditory processing 29, 251–253, 299, 336–339, 363, 386, 390, 425, 426, 442, 447, 459
  dual-path model 381, 387
automatization 9–11
autosegmental phonology 282, 283
avian brain organization 507, 519–522
avian family phylogenetic tree 504
avian vocal learning 501, 520, 529, 530

babbling 472, 503, 511, 516, 548
basal ganglia 38, 299, 315, 316, 326, 333, 357, 359, 361, 375, 382–384, 427, 429, 444, 519, 522, 523, 526, 535
bats 504, 510, 520, 528

BaYaka Pygmies 45–66, 457
  musical socialization 54–56
  spirit plays 53–59
beat 23, 142, 167, 269, 270, 292, 422, 429
  deafness 449, 450
  induction 299, 426, 497
  perception 326, 444, 445, 468, 470
Bhaktapur Stick Dance 50–52
birdsong 20, 104, 105, 137, 274, 286, 331, 493, 494, 499–540
  adaptive function of 516–519
  compared to human vocal learning 501–503, 520–527
  dialects 518–520
  history 512–516
  neural mechanisms of 519–532
Blazing Saddles 177, 178, 185, 197
bonding 96, 134, 137, 204, 219, 308, 494, 500, 546, 548, 549, 552, 557
brain–computer interface (BCI) 393, 395, 402, 411
Brain Opera, The 402
Brain Orchestra, The 410–412
brainstem 167, 212, 336, 337, 372, 383, 384, 404, 492, 503, 520, 522, 526, 532, 534, 547
Broca's aphasia 152, 245, 322, 350
Broca's area 28, 37, 72, 99, 100, 106, 216, 219, 241, 242, 245, 250, 313, 325, 342, 446, 451, 486, 521–524, 531, 547
Brodmann areas 28, 84–86, 147, 148, 206, 242, 309, 313, 314, 333, 345, 446, 452, 531

cadence 143, 145, 156, 157, 259, 266, 267, 291, 302, 424
caesura 285, 286
calls 484, 488, 491, 492, 510, 512, 514, 515, 527, 555
  defined 502
center embedding 76
cerebellum 38, 103, 212, 299, 315, 317, 357, 359, 369, 373, 384, 406, 427, 429, 444

cerebral cortex 206, 243, 253, 361–363, 367, 368, 384, 406
cetaceans 510, 520, 555
Chariots of Fire 181–183, 186, 188, 199
chords 257, 258, 261, 266, 267, 292–294, 348, 389, 442
  progression 276, 300, 325, 341
classical conditioning 212, 361, 371–373, 376, 404
cognitive maps 373–376
combinatoriality 213, 236, 542, 546
complex vocal learning 500, 501, 510–512, 517, 527, 536–538, 546
  defined 509
component process model (CPM) 113–115, 117, 119, 131, 228
compositionality 47, 217, 235–238, 248, 290, 334, 407, 515, 542, 553
compositional semantics 14, 15, 20, 23, 218, 287, 490, 547, 557
computer music 360, 397, 402, 403. *See also* brain–computer interface (BCI)
  Brain Orchestra 410
  eXperience Induction Machine 411
congenital amusia 334, 338–340, 343, 447–449
congruence 173, 176, 183–190, 196
congruence association model with working narrative (CAM-WN) 173, 194–201, 227
consciousness 25, 117, 118, 187, 191, 196, 198, 204, 211, 226, 229
consonance 21, 24, 184, 187, 268, 388, 420, 434, 466
construction grammar 16, 382, 384
conversation 19, 31, 47, 67, 76–78, 121, 137, 204, 221, 242, 301, 302, 328, 439, 543, 546. *See also* face-to-face communication; turn-taking
  musical 428, 440
cooperation 13, 25, 55, 71–74, 159, 168, 308, 478, 549, 550, 554
co-pathy 166, 167
cortical oscillations 253, 299
creativity 57, 60, 61, 63, 393, 396, 400–402, 408, 412, 413
cultural evolution 12, 15, 69, 71, 78–80, 364, 482–484, 490, 519, 556

cultural transmission 64, 69, 169, 463, 500, 519, 537, 546, 550

dance 18, 45–49, 59, 63, 137, 204, 307, 308, 326–328, 380, 418, 419, 422, 427, 482, 496, 533, 545, 550, 558, 559
  in BaYaka 55–58
  in infants 474, 477
  mirroring in 101–105
  Stick Dance of Bhaktapur 50–52
  temporal organization 325, 326
Darwinian musical protolanguage hypothesis 40, 481, 493–496
deep homology 41, 491, 501, 503–508, 527, 530, 533, 537, 547, 560–562
Denisovans 69, 71
dependency locality theory 344, 350
depression 163–166
dialects 510, 518, 519
diegesis 177, 179, 189, 190, 197
diffusion tensor imaging (DTI) 248, 250, 319, 449, 531
Dinka 283
disambiguation 36, 295, 301
disgust 12, 97, 98, 124, 126, 211
dissonance 21, 24, 162, 164, 184, 187–189, 257, 266, 388, 420, 466, 469
distributed adaptive control (DAC) 39, 211, 228, 393, 394, 397
  architecture 400, 403–408
dorsal thalamic nucleus (aDLM) 520–522, 524, 527
duality of patterning 14, 15, 35, 273, 274, 285, 286
duration 271, 272, 299, 325, 326, 422, 424, 428–430, 446, 464, 468
dusp1 gene 525, 526
dyslexia 331, 530

early left anterior negativity (ELAN) 251, 252, 342, 343, 457
early right anterior negativity (ERAN) 146–149, 151, 152, 156–158, 342, 343, 369, 457

electroencephalography (EEG) 29, 114, 149, 219, 311, 328, 342, 343, 410, 430, 434–437
elephant hunt 49, 53, 558
elephants 450, 510, 520, 537
embodiment 7, 30, 394, 397, 400, 402, 403
emotion 10–14, 26, 107–140, 159, 160, 204, 205, 210, 375, 376, 390, 391, 395, 397, 418, 494, 544, 550
  acoustic cues for 125, 126
  aesthetic 12, 32, 107, 128–131, 138, 211, 215, 216
  evolution of 108–111
  music-evoked 21, 23, 109, 118, 128–130, 141, 153, 160–168, 198, 215, 390
  neural correlates 162–164
  regulation of 111, 118, 128, 135, 167, 476, 548
  system, architecture of 111–118
  tripartite emotion expression and perception (TEEP) model 119–121, 138
  utilitarian 12, 32, 107, 128, 130, 138, 211, 215, 216
empathy 79, 83, 96, 97, 122, 131, 137, 166, 175, 215, 216, 221. *See also* co-pathy
  aesthetic 222, 228
enculturation 60–63, 330, 464, 469, 470, 549
enjambment 285, 286
entrainment 215, 216, 219, 229, 327, 420, 427, 436, 445, 468, 546
episodic memory 323, 359, 366
ethnomusicology 50, 51, 60, 69, 412
event-related potential (ERP) 213, 234, 235, 251, 252, 311, 319, 323, 340, 341, 370
events-in-time hypothesis 365, 407, 428
evolution
  of emotion 108–111
  of language 71, 72, 218, 249, 286, 331
  of language and music 555–557
  of speech and music 131–138

expectation 6, 24, 34, 49, 116, 120, 191, 203, 208, 226, 252, 309, 376, 389, 445
  in music 142, 143, 148, 161, 217, 258, 267, 294, 497
eXperience Induction Machine 411
expressive aphasia 322
external semantics in film 33, 173, 183–186, 189
extramusical meaning 22, 153–155

F5 neurons 84, 90, 99, 100, 150, 376–379, 491, 492
face-to-face communication 31, 47, 67, 72, 137, 221, 546
facial expression 12–14, 47, 96, 111, 122, 124, 132, 171, 210, 281, 437, 438, 481, 509, 550
feelings 11, 13, 117, 118, 160, 205, 210, 219
film 138, 220–226
  external semantics in 33, 173, 183–186, 189
  music 173–202, 222, 226
  structure of 191–194
foraging 405, 407, 475, 553
forebrain 382, 499, 503, 507, 520, 526, 527
  auditory pathway 530, 531, 547
  motor pathway 504, 532, 533, 535
foundational cultural schema 30, 45, 52, 57, 63, 64
FoxP2 41, 499, 501, 505, 528–530
FOXP2 71, 505, 528, 560
frame of reference problem 398–400
functional magnetic resonance imaging (fMRI) 219, 225, 311, 313, 324, 326, 328, 332, 339, 341, 350, 370, 390, 437, 456, 524, 526, 553

gamma oscillations 199, 252, 253, 364–366, 369, 431
gaze following 10, 72, 88, 89
generative grammar 68, 259, 394, 408
generative theory of tonal music (GTTM) 144, 226, 258, 259, 262, 269, 270, 272

generativity 169, 330, 542, 544, 547
Geneva Emotional Music Scale (GEMS) 129, 130
gestural origins hypothesis 484, 535, 536, 552
gesture 18, 23, 47, 49, 79, 106, 137, 246, 247, 297, 300, 324, 325, 327, 380, 428, 457, 485, 489, 545, 550, 552, 557
  ape 72, 546, 548
  infant 476, 477, 548
  orofacial 28, 88, 90, 99, 105, 491
  role of mirroring 91, 99, 102
GingerALE algorithm 311, 314
goal-directed motor acts 5, 32, 83, 87–91, 378–380, 487
GODIVA model 352, 353
grasping 376–378, 486–489
Gricean signaling 70, 74, 75, 79
grids 19, 35, 257, 260, 271
  metrical 261–264, 269–272
group identity 19, 45, 46, 49, 168, 229, 308, 316, 327, 548, 549
grouping 18, 35, 260–262, 265, 271, 272, 278, 290–292, 396, 408, 422, 424, 426, 431, 433, 435, 468

habit 77, 111, 209, 220, 229, 316, 407
harmony 19, 131, 144, 257–259, 269, 272, 294, 323, 326, 340, 341, 348, 423
  infant perception 466, 469
  unexpected 161–163
Hayashi 482
headed hierarchies 260, 262, 271
hearing. *See* listening
hemisphere bias 243, 332, 333, 346, 347, 454, 455
hierarchical structure 36, 290–294, 296, 303, 401, 553
  in language 35, 273, 274, 286
  in music 395, 396, 409
  in phonology 283–285
hippocampus 38, 163, 164, 198, 206, 253, 357, 365–367, 373–375, 406
hocket 53, 58, 60, 61, 64
holophrase 489, 490, 494

Homo erectus 71
Homo heidelbergensis 71, 72
hummingbirds 494, 510, 513, 514, 520, 533
hypothalamus 26, 96, 168, 375

imitation 83, 84, 100, 102, 221, 486, 493, 496, 523, 548
  complex 487–489, 495
  in infants 10–12, 55, 477
  vocal 493, 494, 500, 512, 527, 528
indexical meaning 22, 153, 154, 212, 214, 281, 300
infancy 24, 40, 121, 169, 347, 434–436, 463–480, 548
  beat perception in 18, 326
  imitation in 10–12, 477
  maternal singing 21, 167, 468, 474–478
  socialization 54–56
information
  set of streams 297–301
  structure 225, 239, 278, 289, 290
  transfer 549, 556, 558
innate vocalizations 502, 508, 509, 515, 520, 523
input codes 39, 417, 418, 432, 433
instrumental music 6, 7, 43, 240, 259, 308, 309, 331, 340, 341, 347, 440, 453, 465, 482, 496, 546, 558
  digital 402, 403, 410–412
intention
  attribution 486, 550, 553
  coding 92–94
  recognition 67, 70, 75, 78, 111, 122, 141, 154, 159
intermerdiate-term memory 226–228
internal semantics of film 183–190
intonation 73, 123, 131, 132, 136, 170, 268, 269, 278, 281, 289, 298, 300, 311, 313, 319, 354, 360
  contour 252, 253, 338, 428, 429, 439
intramusical meaning 22, 36, 153, 156–158
isochrony 170, 544, 554

Jaws 182, 183, 188, 189

joint action 31, 48, 67, 77–80, 83, 88, 95, 103, 159, 323, 328, 482, 542, 559

kapernί 543, 544
kin selection 494, 495
knowledge-free structuring 142, 143

laments 49, 78
language
  ambiguity in 294–297
  components of 546, 547
  comprehension 144, 233, 238–240, 247, 420, 439, 440
  defined 14–17, 20, 67
  differences to music 332–335
  diversity 68–71
  evolution of 71, 72, 218, 249, 286, 331
  gestural origins theory 484, 535, 536, 552
  in infancy 463–480
  mirroring in 99–101
  origins of 72, 478, 479
  processing 14, 34, 70, 148–152, 233, 235, 240–248, 251, 336, 421
language–music continuum. *See* music–language continuum
language-ready brain 34, 40, 68, 248–254, 376, 481–498
lateralization 248, 249, 298, 315, 317, 321, 352, 455, 459, 479, 547
leading tone 294, 330, 346, 348
learning 27, 359, 386, 403–414, 433–436. *See also* vocal learning
  emotional 371–376
  motor 41, 499, 504, 523, 535, 557, 561
  reinforcement 212, 371, 374, 383
  usage 509, 510
left anterior negativity (LAN) 149–152, 157, 341, 342
left inferior frontal cortex (LIFC) 241, 244–248, 251
leitmotif 154, 175, 176, 185, 190, 226
lexical tone 35, 273, 298, 448, 455
limbic system 11, 96, 98, 104, 135, 162–168, 225
linguistic meaning 330, 347
linguistic syntax 257–272, 340, 348

listening 18, 143, 153, 159, 161,
  164–166, 308, 309, 313, 324, 435,
  437, 441, 444
 neural activity during 524–526
 skills in infancy 464–469
 temporal regularity 468, 475
literacy 479, 483, 490
long-term memory 8, 143, 191, 194–
  196, 198–200, 227, 234, 247, 343,
  347, 404–407, 422, 446, 456, 532
 in infants 465, 467
lullabies 21, 466, 474
lyrics 320–323, 370, 446, 457–459

magnetoencephalography (MEG) 29,
  147, 198, 326, 342, 347, 430, 437,
  439
Mai 482
mammalian brain organization 519–522
 motor pathway 507
mammalian family phylogenetic tree
  504
manual dexterity 485, 496
Maori 49
Mbendjele 557
McGuill Digital Orchestra 403
meaning 7, 14, 16, 48–50, 184, 203–207,
  215, 290, 385, 387, 418, 424, 494,
  495
 affective 22, 51, 214, 502
 affective–propositional axis 212–217
 associative 173, 175, 176, 188–190,
   192–194
 extramusical 22, 153–155
 indexical 22, 153, 154, 212, 214, 281,
   300
 intramusical 22, 36, 153, 156–158
 linguistic 330, 347
 musical 153–160, 188–190, 330, 331,
   347
 propositional 203, 214, 292, 494–496,
   502, 512, 542, 552
 sensorimotor-symbolic axis 217–220
 shared 49, 83, 228, 380, 481
 structure axis 220–228
melodic contour deafness hypothesis
  338

melodic intonation therapy (MIT) 322,
  354, 355, 451
melody 144, 268, 283, 317, 320, 326,
  338, 446, 457–459, 466, 543
 perception 227, 318, 456
 role in flim music 174, 177, 181, 182,
   189, 193
memory 206, 215, 221, 229, 309,
  310, 324, 345, 359, 366, 446. *See
  also* long-term memory
 auditory 445–447
 autobiographic 485
 declarative 323
 episodic 323, 359, 366
 intermediate-term 226–228
 music-evoked autobiographical 323,
   324
 retrieval 235–240
 short-term 338, 342, 388, 406, 407
 working 8, 16, 27, 33, 144, 195, 224,
   227, 247, 377, 382, 446, 469, 485,
   490, 546
memory, unification, and control (MUC)
 model 236, 241, 242, 251, 254, 344,
   345
mental lexicon 235–237, 239, 240, 421
Merina of Madagascar 51
meter 18, 19, 23, 51, 142–145, 325, 422,
  424, 444, 468, 470–472, 545, 546
metrical grid 261–264, 269–272
middle temporal gyrus (MTG) 244, 245,
  249, 345, 452
mirroring 32, 67, 79, 83, 88–92, 216,
  218, 220–222, 228
 in dance and music 101–105
 in language 99–101
mirror neuron 378–380
 defined 485
mirror system 10–12, 70, 75, 94, 100,
  195, 199, 218–220, 384, 428, 486
 in humans 89–92
 hypothesis 481, 484–491, 493–496,
   553
 model 378–380
 in monkeys 84–88
 role in emotion 95–98
mismatch negativity (MMN) 142, 151,
  152, 319, 342, 436, 456, 497

mokondi massana 53, 56–59
mood regulation 26, 500, 509
morphophonemics 276
morphosyntax 70, 286, 515
mother–infant interactions 40, 73, 472–477, 546, 548
motor cortex 84, 90, 100, 167, 216, 315, 345, 379, 521, 522
motor learning 41, 499, 504, 523, 535, 557, 561
motor schemas 7, 8, 20, 209, 358, 376, 377, 380, 390, 404
movement 469, 477, 482, 492, 499, 533, 544
  in music 46–48, 199
  synchronized 468, 474, 549, 557, 559
music. *See also* computer music
  ambiguity in 293, 294
  components of 546, 547
  defined 20, 393, 394, 463
  differences to language 332–335
  expectancy in 24, 142, 143, 148, 161
  function 22, 48–50, 165–168
  in infancy 463–480
  meaning in 153–160, 188–190, 330, 331, 347
  mirroring in 101–105
  neural substrates 307–328, 447–455
  perception 141–172, 240, 245, 334, 338, 364, 369, 440, 441, 447, 471–474, 545
  sequential ordering in 266, 267
  social functions of 165–168
  structure 22, 51, 52, 60–63, 130, 131, 148, 153, 161, 176, 186–188, 189, 190, 422
musicality 17–20, 46, 169, 185, 189, 331
musical protolanguage 331, 493–497, 537
  hypothesis 481, 493–496
musical semantics 34, 153, 213, 347, 546
musical syntax 34, 35, 141–153, 167, 257–272, 330, 347, 369
  components 260–262
  defined 257

processing 32, 142, 145–152
musical training 147, 169, 330, 332, 336, 341, 346, 348, 452
music composition 38, 393, 394, 397, 398, 401, 407, 412
music-evoked autobiographical memories (MEAM) 323, 324
music–language continuum 20–24, 56, 64, 141, 169–172, 544
music making 20, 30, 168, 169, 410, 412, 413, 434
  co-pathy 166, 167
music therapy 29, 163–165, 167, 331, 332

N5 146, 147, 156–158
N400 150–152, 154–157, 213, 251, 340, 344, 389
narrative 33, 34, 43, 73, 171, 180, 221–226, 302, 308, 323–325, 412
  in film music 173–201
  working 33, 173, 191, 192, 194–201
natural selection 25, 483, 493, 554, 556
navigation 327, 366, 371, 373, 376
  taxon affordance model (TAM) 374
Neanderthals 69, 71, 134, 135
neocortex 362, 363, 382
neural Darwinism 344
neural oscillation hypothesis 430, 431
neural plasticity 26, 336, 355, 359, 369, 433–436, 443, 450, 483, 517
neural rhythms 29, 252–254, 364–371, 367–369
neuraxis 404–414
neurogenesis 164, 517
neutralization 277, 279
ngére 544
niche construction
  cultural 483, 554
  intentional 483–485
  neural 483, 493
  triadic 40, 41, 481, 490, 541, 559, 560
Noh play 482
nondiegetic music 177–180, 189, 197
nonfluent aphasia 331, 354, 454
nucleus accumbens 97, 162, 163, 206, 374–376

ontogenetic ritualization 10, 489
oratory 49, 544
orofacial gestures 28, 88, 90, 99, 105, 491
OSCillator-Based Associative Recall 370
oscine songbirds 504, 510, 513, 555

P600 213, 251, 341, 389
pantomime 40, 47, 90, 91, 213, 476, 481, 486, 489, 493, 496, 552, 557
paralanguage 49, 77
paralimbic system 162–172
Parkinson's disease 367, 524
parrots 327, 450, 494, 502, 503, 507, 510, 513, 514, 517, 520, 533, 545
 African Gray 495, 514, 515, 545, 546
pedagogy 482, 552, 553
perception 3–7, 24, 70, 99, 103, 196, 197, 208, 309, 310, 313–315, 361, 375, 397–401, 406, 412, 420, 436, 465. *See also* tripartite emotion expression and perception model
 beat 468, 470
 bias 407, 464, 468
 music 141–172, 240, 245, 334, 338, 364, 369, 440, 441, 447, 471–474, 545
 song 311–314, 316, 318, 458, 517, 532
 speech 100, 101, 418, 430, 447, 471–474, 532, 545, 546
perception–action cycle 307–310, 316, 436, 437
perception–prediction–action cycle 436–439
perceptual schema 7–10, 208–210, 358, 380
performance 19, 46–48, 52–55, 71, 77, 78, 122, 130, 369, 373, 376, 386, 395, 396, 402, 408, 427, 430, 453, 557. *See also* spirit plays
 role of mirroring in 104
 synchronized 29, 168, 220, 548
perisylvian cortex 101, 240, 241, 347, 451
 left 244, 248, 250, 251

phoneme 225, 259, 260, 273, 276, 286, 292, 299, 360
 suprasegmental 279, 280
phonemic principle 275, 276
phonetics 35, 260, 273–288, 352, 511
phonology 15, 34, 35, 41, 70, 100, 105, 273–288, 297, 298, 346, 494, 495, 546
 autosegmental 282, 283
 in birdsong 514–516
 hierarchical structures in 283–285
 model for processing 243
phonotactics 276, 279
phrasing 10, 239, 278, 298, 325, 360
pinnipeds 510, 520
pitch 170, 176, 186, 187, 279, 282, 283, 292, 348, 396, 423, 446, 454. *See also* tonal pitch space
 absolute 26, 428, 443, 454, 465, 466
 contour 22, 126, 268, 271, 272, 338
 perception 332, 390, 441, 469
 processing 332, 333
 relations 258, 267, 269, 271, 317
 relative 38, 329, 337–339, 419, 420, 425, 426, 428, 429, 435, 443, 454, 465, 466, 546
 role in emotion 21, 23, 123, 124, 134
planning 7, 209, 210, 212, 218, 303, 345, 408, 550–554
poetry 23, 78, 130, 170, 259, 261, 269, 271, 272, 278, 284–286, 422, 445, 446, 457, 459, 482
polyphony 53–55, 57, 60–62, 64, 142
positron emission tomography (PET) 27, 29, 162, 311, 313, 524–527, 553
prefrontal cortex (PFC) 89, 90, 101, 103, 106, 147, 162, 206, 211, 212, 220, 223–225, 242, 339, 375, 377, 382–384, 429
 dorsolateral 316, 345
 dorsomedial 223, 323, 324
 medial 224, 324, 553
 ventral 84, 85, 89, 91, 94, 99, 148, 216, 222, 352, 354
 ventrolateral 324, 325
problem of priors 398, 400, 403–414
production learning 509, 555
propositionality 542, 545, 552, 556, 558

propositional meaning 203, 214, 292, 494–496, 502, 512, 542, 552
propositional semantics 153, 170, 171
prosody 21, 23, 35, 49, 77, 123, 124, 130, 139, 215, 239, 273–288, 298, 324, 336, 360, 370, 494, 545
  affective 153, 160, 172
  infant-directed 473
  processing of 317
protolanguage 40, 72, 73, 479, 488–490, 557
  musical 331, 481, 493–496, 493–497, 537
protosign 40, 481, 489, 492, 493, 495, 496, 550, 552
protospeech 40, 481, 489, 492, 493, 495
protosyntax 135
pure word deafness 447

recurrent sound patterns 271, 272
recursion 14, 16, 76, 144, 217, 268, 285, 290, 395, 398, 399, 407, 412, 546
reinforcement learning 212, 371, 374, 383
relative pitch 38, 419, 420, 425, 426, 428, 429, 435, 443, 454, 546
  in infants 465, 466
  processing of 329, 337–339
relaxation 258, 266, 267, 271, 284, 388
re(per)curso 393, 410
reward processing 375, 382, 390
rhyme 23, 234, 271, 278, 474
rhythm 18, 21–23, 47, 79, 103, 131, 134, 216, 271, 283, 294, 299, 327, 368, 399, 497, 545
  in infancy 468, 471–477
  role in film 177, 183, 186, 188
  role in socialization 54–56
ritual 12, 30, 45, 50–58, 63, 134, 137, 204, 219, 308, 327, 477, 478
ritualization 10, 90, 102, 120, 135, 489
Road House 177, 189, 193
Robo1 528–530
RoBoser 393, 394, 407, 409, 410, 412
robotics 363, 397, 400–411, 413
Rossel Islanders, Papua New Guinea 77, 78

sarén 543, 544
schema 209, 267, 358, 359, 378
  anticipatory 208
  foundational cultural 31, 45, 52, 57, 63, 64
  motor 7, 8, 20, 209, 358, 376, 377, 380, 390, 404
  perceptual 7–10, 208–210, 358, 380
  social 9, 27, 42, 220
  theory 8–10, 195, 197, 358, 400
semantics 203, 297, 298, 323, 324, 330, 331, 546
  in birdsong 514–516
  compositional 14, 15, 20, 23, 218, 287, 490, 547, 557
  internal 183–190
  musical 34, 153, 213, 347, 546
  propositional 153, 170, 171
semantic unification 246–248, 345
semiotics 212–215
set of streams 297–301
sexual selection 493, 494, 518, 537, 554, 555
shared attention 88, 89
shared goals 159, 167, 168, 551, 554
shared intentionality 549, 556
shared meaning 49, 83, 228, 380, 481
shared syntactic integration resource hypothesis (SSIRH) 245, 340, 343–345, 349–351, 448, 456, 457
Shining, The 187–189, 192, 199
short-term memory 338, 342, 388, 406, 407
sign language 6, 15, 47, 67, 72, 91, 213, 238, 274, 276, 279, 300, 363, 380, 471, 477, 489, 496
  Bedouin 15, 286
sign quality 22, 153, 154, 159, 171
silent film 174, 178, 179, 193
  The Artist 179
simulation theory of mind 93, 221, 222
singing 283, 422, 441, 482, 536. See also birdsong
  in birds 513, 516, 518, 524–526
  collaborative 335, 477
  impact on infants 473–475
  maternal 21, 167, 468, 474–478
  right hemisphere dominance 458

singing
  similarities with birdsong 501–503
  temporal organization in 325, 326
Slit1 528–530
SMuSe system 410, 413
social bonds 204, 308, 323. *See also* bonding
social cognition 154, 165, 166, 219, 221, 324
social interaction 4, 8, 12, 19, 40, 72, 88, 107–110, 119, 127, 133–139, 205, 208, 218, 220, 228, 439, 478, 546, 555, 561
  capacity for 548
  role of emotions 210
socialization 49, 50, 121, 473
  musical 54–56
social learning 70, 71, 488, 509
Social Network, The 174, 181–183, 186–189, 199
social organization 49, 53, 54, 56, 111
social schemas 9, 27, 42, 220
song 20–22, 31, 49, 67, 167, 170, 285, 299, 300, 336, 427, 482, 493, 501, 543
  aphasia 524, 532
  BaYaka 56–61
  brain areas for 521–527
  defined 512
  Dinka 283
  functions of 307–309
  in infancy 474–477
  lyrics and melody 320–323
  neural mechanisms of 311–324, 331, 352
  nuclei 518, 520, 524, 526, 533
  perception 311–314, 316, 318, 458, 517, 532
  production 311, 315, 316, 453, 457–459, 511, 526
  Rossel Islander 77, 78
  sound map 354, 457
  word articulation in 351, 352
songbird 8, 23, 71, 72, 500, 533, 537, 542, 555, 560
  HVC 520, 521, 523–527, 531
  learning pathways 502–505, 520–522
  oscine 504, 510, 513

prototypical 516, 517
speech 15, 22, 47, 49, 59, 77, 78, 108, 167, 170, 271, 332, 346, 502, 512, 543, 550
  audiovisual 422, 427, 437, 438
  components 363
  comprehension 233–239, 360
  contour 267–269
  emotion in 23, 104, 122–139
  evolution of 71, 72, 121, 135
  maternal 474–476
  neural basis of 455–457, 519–532
  perception 100, 101, 418, 430, 447, 532, 545, 546
  in infancy 471–474
  processing 242–244, 252–254, 336, 364, 367
  production 315, 352, 353, 360, 363, 370, 438, 439, 453, 511–513, 526, 536
  rate 125, 234, 251, 278, 429
  rhythmic structure 368
  sound map (SSM) 352, 354, 457
  Suyá 543, 544
  therapy 354, 454
  word articulation 351–355
spirit plays 53–59
Stanford Laptop Orchestra 403
Stick Dance of Bhaktapur 50–52
storytelling 49, 191, 552
stress 269–271, 284, 296, 299
  grid 261–263
structure 220, 289–304, 349, 362, 400, 419–432, 546. *See also* hierarchical structure; music structure
  axes of 279–282
  events in time 361–364
  knowledge-free 142, 143
  large-scale 205, 206, 220, 223–228
  set of streams 297–299
  surface 290, 291, 295
suboscines 504, 512, 513, 555
superior temporal gyrus (STG) 147, 216, 225, 244, 317–319, 333, 343, 345, 351, 381, 426
superior temporal sulcus (STS) 166, 216, 222, 224, 243, 244, 247, 318, 378, 379, 426, 427, 438

supplementary motor area (SMA) 216, 315, 325, 326, 429, 441, 444, 539
Suyá 46, 543, 544
syllable 225, 234, 253, 260, 273, 276–280, 284, 299, 302
  coupling with pitch 313–315, 318, 320
  stress 261, 269–272
symbol grounding 398, 399, 403
synchronization 216, 228, 364, 365, 420, 427
  in dance 326, 327
  group 204
synchrony 198, 468, 477, 555
  in film 185, 189
syntactic processing 106, 156, 157, 169, 213, 233–235, 244, 245, 301, 331, 340–351, 369, 456
  interactions, language and music 148–153
  music 32, 142, 145–152
syntax 16, 21, 34, 73, 80, 101, 289–293, 295–298, 336, 347, 370, 422, 495, 552, 557
  in birdsong 514–518, 536
  development 386, 490
  language 76, 77, 105, 167, 257–272, 340, 348
  morpho- 70, 286, 515
  musical 34, 35, 141–153, 167, 257–272, 330, 347, 369
  proto- 135

taxon affordance model (TAM) 374
temporal population code (TPC) 399
  model 361, 362, 364
tension 124, 129, 161, 180, 185, 220, 226, 258, 266, 267, 271, 272, 284, 340, 348, 388, 389, 469
thalamus 315, 367, 368, 383, 526
theory of mind 70, 71, 74, 165, 220, 393, 400, 546, 553, 557
theory theory 92, 93
theta oscillations 252, 299, 302, 364, 365, 368, 430
timbre 228, 271, 299, 311, 326, 390, 394, 396, 408, 422, 426, 434, 446, 558

processing 332, 333, 467
timing 104, 170, 176, 269, 299, 357, 373, 404, 439–441, 544, 545
  neural basis for 251, 252, 325, 372, 429–432, 444–445
  relative 435
tonality 126, 132, 269, 272, 348, 420
  processing 324, 340–351
tonal music 17, 42, 73, 170, 204, 226, 257, 266, 267, 270, 289, 348, 360, 394–396, 399, 410, 412, 423, 465, 466, 469, 470, 543
tonal pitch space (TPS) 144, 258, 259, 267, 268, 293, 344
tone 278, 279, 281, 282, 283
  lexical 35, 273, 298, 448, 455
  sequence recognition 282, 338, 339
tonic 259, 261, 265, 266, 271, 280, 294, 335, 348
tool use 16, 71, 85, 303, 483, 496, 552, 553
training 487, 488, 559
triadic niche construction 40, 41, 481, 490, 541, 559, 560
tripartite emotion expression and perception (TEEP) model 119–121, 138
turn-taking 31, 47, 67, 72–74, 79, 204, 242, 295, 302, 420, 427, 436, 439–441, 457, 476
  interactive alignment model 440

uncinate fasciculus 250, 345, 438, 452, 453
unification 235–238, 240, 242, 244–248, 344–346
  semantic 246–248, 345
universal grammar 17, 31, 67–70, 269
usage learning 509, 510
Utai 482
utilitarian emotions 12, 32, 107, 128, 130, 138, 211, 215, 216

ventral premotor cortex 84, 85, 89, 91, 94, 99, 148, 216, 222, 352, 354
verbal amusia 523
verbal aphasia 523, 524, 526
vocal capacity 560, 562

vocal control 99, 319, 320, 448, 449, 485, 512, 557
  brain activation patterns 525
  brain regions 520–522, 547
vocal expression 79, 99, 118, 122–125, 127, 132, 495
  of emotion 131–138
  maternal 475
vocal imitation 493, 494, 500, 512, 527, 528
vocal learning 8, 22, 31, 41, 67–70, 75, 105, 167, 330, 483, 491–493, 499–540, 557, 561. *See also* complex vocal learning
  brain areas 504–507
  capacity 495, 555
  defined 508–510
  evolution of 500, 503–508, 550
  human vs. birdsong 501–503
  motor theory for 532–536
voice
  discrimination 425, 434, 449
  onset time (VOT) 276–279, 298
  pitch 268, 281
  quality 47, 130, 171, 188, 189, 318, 474

Water Passion after St. Matthew 42–44
Wernicke's area 100, 101, 241, 320, 451, 532, 547
Western tonal music. *See* tonal music
whale song 272, 331, 494, 510, 546
word
  accent 278, 280, 298, 299
  articulation 38, 329, 351–355, 451
  order 68, 69, 266, 289, 384
  recognition 234, 251, 332
  stress 278, 280
working memory 8, 16, 27, 33, 144, 195, 224, 227, 247, 377, 382, 446, 469, 485, 490, 546
working narrative 33, 173, 191, 192, 194–201, 227

Yesterday 262–266, 476
yodeling 53, 55, 60

zebra finch 507, 512, 514, 515, 518, 523, 525, 528, 533